DATE DUE

Transition Metal Reagents and Catalysts

Transition Metal Reagents and Catalysts

Innovations in Organic Synthesis

Jiro Tsuji

Emeritus Professor,
Tokyo Institute of Technology
Tokyo, Japan

JOHN WILEY & SONS, LTD
Chichester • New York • Brisbane • Toronto • Singapore

Copyright © 2000 John Wiley & Sons, Ltd,
Baffins Lane, Chichester,
West Sussex PO19 1UD, England

National 01243 779777
International (+ 44) 1243 779777
e-mail (for orders and customer service enquiries): cs-books@wiley.co.uk
Visit our Home Page on http://www.wiley.co.uk
or http://www.wiley.com

Reprinted November 2000

Other Wiley Editorial Offices

John Wiley & Sons, Inc., 605 Third Avenue,
New York, NY 10158-0012, USA

WILEY-VCH Verlag GmbH, Pappelallee 3,
D-69469 Weinheim, Germany

Jacaranda Wiley Ltd, 33 Park Road, Milton,
Queensland 4064, Australia

John Wiley & Sons (Asia) Pte Ltd, 2 Clementi Loop #02-01,
Jin Xing Distripark, Singapore 129809

John Wiley & Sons (Canada) Ltd, 22 Worcester Road,
Rexdale, Ontario M9W 1L1, Canada

British Library Cataloguing in Publication Data

A catalogue record for this book is available from the British Library

ISBN 0 471 63498 0

Typeset in $10\frac{1}{2}/12\frac{1}{2}$pt Times by Techset, Salisbury, Wiltshire.
Printed and bound in Great Britain by Bookcraft (Bath) Limited, Midsomer Norton, Avon
This book is printed on acid-free paper responsibly manufactured from sustainable forestry, in which at
least two trees are planted for each one used for paper production.

CONTENTS

PREFACE

Use of transition metal compounds or complexes as catalysts or reagents in organic synthesis is an exciting field of research, and numerous novel reactions which are impossible to achieve by conventional synthetic methods have already been discovered. They are extensively employed in a wide range of areas of preparative organic chemistry. Total syntheses of many complex molecules have been achieved efficiently in much shorter steps, which was unbelievable ten years ago. Applications of transition metal catalysts and reagents to organic synthesis are still being actively investigated, and these days we can hardly open an organic chemistry journal that does not contain examples of these reactions. Now the research on the application of transition metal complexes to organic synthesis is in its golden age. Without doubt, in the last decade, the introduction of transition metal catalysts and reagents has caused revolutionary change in organic synthesis.

Today, the knowledge of organotransition metal chemistry is indispensible for synthetic organic chemists. However, the organotransition metal chemistry is clearly different mechanistically from traditional organic chemistry. I undertook to write this book in order to give a birds-eye view of the broad field of organotransition metal chemistry applied to organic synthesis. I intended to give a better understanding of the present arts of this chemistry to many synthetic organic chemists, who are not very familiar with organotransition metal chemistry, but eagerly wish to apply transition metal-catalyzed reactions to their synthetic works. I have tried to accomplish this task first by giving a simple mechanistic explanation in chapter 2. Then a number of important types of reactions classified mainly by representative substrates such as organic halides and allylic derivatives are surveyed with pertinent examples. For this purpose, I cited many references; these were selected from a much larger number which I have collected over the years. I wanted to make the book as comprehensive as possible by selecting those references which reported original ideas and new reactions, or evident synthetic utility. Synthetic utility is clearly biased towards catalytic rather than stoichiometric reactions. The overall task of selecting good examples to include was very difficult. It was done based on my own knowledge and understanding of the chemistry, and hence there must be many significant omissions. I apologize for the errors and incorrect citations which must inevitably be present in a book written by a single author.

In 1997, I wrote a book with a similar title in Japanese, and the present book is an expanded English edition. However, I replaced many old examples in the Japanese edition with new ones and added many more in order to make the book up-to-date.

I wish to acknowledge valuable suggestions and corrections given by Professor H. Nozaki who read the whole manuscript. I also thank my wife Yoshiko for her help during the preparation of the manuscript.

Jiro Tsuji
Kamakura, Japan

ABBREVIATIONS

acac	acetylacetonate
Ar	aryl
atm	atmospheric pressure
BBEDA	N, N'-bis(benzylidene)ethylenediamine
9-BBN	9-borabicyclo[3.3.1]nonyl
9-BBN-H	9-hydroborabicyclo[3.3.1]nonane
BINAP	2,2'-bis(diphenylphosphino)-1,1'-binaphthyl
Bn	benzyl
Boc	t-butoxycarbonyl
BPPFA	1-[1,2-bis(diphenylphosphino)ferrocenyl]ethyldimethylamine
BPPM	(2S,4S)-N-t-butoxycarbonyl-4-(diphenylphosphino)-α-(diphenylphosphino-methyl)pyrrolidine
bpy	2,2'-bipyridyl; 2,2'-bipyridine
BQ	1,4-benzoquinone
BSA	N,O-bis(trimethylsilyl)acetamide
BTMSA	bis(trimethylsilyl)acetylene
Bz	benzoyl
CAN	ceric ammonium nitrate
cat.	catalyst
CDT	1,5,9-cyclododecatriene
COD	1,5-cyclooctadiene
COT	cyclooctatetraene
CTAB	cetyltrimethylammonium bromide
Cp	cyclopentadienyl
Cy	cyclohexyl
DBA	dibenzylideneacetone
DABCO	1,4-diazabicylo[2.2.2]octane
DBU	1,8-diazabicyclo[5.4.0]undec-7-ene
DCHPE	bis(dicyclohexylphosphino)ethane
de	disatereomeric excess
DEAD	diethyl azodicarboxylate
DIPPP	bis(diisopropylphosphino)propane
DIOP	2,3-O-isopropylidene-2,3-dihydroxy-1,4-bis(diphenylphosphino)butane

DMAD	dimethyl acetylenedicarboxylate
DME	1,2-dimethoxyethane
DMF	dimethylformamine
DMSO	dimethylsulfoxide
DPEN	1,2-diphenylethylenediamine
DPMSPP (TPPMS)	diphenyl(*m*-sulfophenyl)phosphine
DPPB	bis(diphenylphosphino)butane
DPPE	bis(diphenylphosphino)ethane
DPPF	1,1'-bis(diphenylphosphino)ferrocene
DPPP	bis(diphenylphosphino)propane
EBTHI	ethylenebis(tetrahydroindene)
ee	enantiomeric excess
EWG	electron-withdrawing group
HMPA	hexamethylphosphoric triamide
L	any unidentate ligand, often Ph_3P
LDA	lithium diisopropylamide
MA	maleic anhydride
Mes	mesityl
NMP	*N*-methylpyrrolidone
MOM	methoxymethyl
MOP	monodentate optically active phosphine
Ms	methanesulfonyl (mesyl)
MTO	methyltrioxorhenium
NBD	norbornadiene
Nf	nonaflate
Nu	nucleophile
OAc	acetate anion
phen	1,10-phenanthroline
PhH	benzene
PhMe	toluene
PHMS	poly(hydromethylsiloxane)
PMB	*p*-methoxybenzyl
PMHS	polymethylhydrosiloxane
PPFA	*N,N*-dimethyl-1,2-(diphenylphosphino)ferrocenylethylamine
i-Pr-BPE	1,2-bis(*trans*-2,5-diisopropylphospholano)ethane
py	pyridine
R	alkyl group
RCM	ring-closing metathesis
ROMP	ring-opening metathesis polymerization
Sia	siamyl; *sec*-isoamyl, or 1,2-dimethylpropyl
TASF	tris(diethylamino)sulfonium difluoro(trimethyl)silicate
TBAC	tetrabutylammonium chloride
TBAF	tetrabutylammonium fluoride
TBDMS (TBS)	*tert*-butyldimethylsilyl
TCPC	2,3,4,5-tetrakis(methoxycarbonyl)palladacyclopentadiene
TDMPP	tri(2,6-dimethoxyphenyl)phosphine
Tf	trifluoromethanesulfonyl (triflyl)

TFP	tri(2-furyl)phosphine
THP	tetrahydropyran
TMPP	trimethylolpropane phosphite, or 4-ethyl-2,6,7-trioxa-1-phosphobicyclo-[2,2,2]octane,
TMSPP (TPPTS)	tri(*m*-sulfophenyl)phosphine
TMEDA	*N,N,N,N*-tetramethyl-1,2-ethylenediamine
TMM	trimethylenemethane
TMS	trimethylsilyl
o-Tol	*o*-tolyl
Ts	tosyl, *p*-toluenesulfonyl
TsOH	*p*-toluenesulfonic acid
TTMPP	tri(2,4,6-trimethoxyphenyl)phosphine
tu	thiourea

1

PIONEERING INDUSTRIAL PROCESSES USING HOMOGENEOUS TRANSITION METAL CATALYSTS

Some main group metals have long been used for organic synthesis. Most importantly Grignard reagents, introduced in the beginning of the twentieth century, have widespread application in organic synthesis. Compared with the extensive use of Grigrard reagents and other main group metal reagents, transition metal compounds have received little attention from synthetic organic chemists. Their application as catalysts to organic syntheses started much later. The application of transition metal compounds, particularly as homogeneous catalysts, was initiated by the following three industrial processes and related reactions, developed from the 1930s to 1950s without an understanding of their mechanisms:

1. Carbonylation of alkenes and alkynes catalysed by metal carbonyls, typically $Co_2(CO)_8$, $Ni(CO)_4$ and $Fe(CO)_5$ to produce aldehydes, carboxylic acids, esters, and alcohols.
2. Production of polyethylene and polypropylene by the Ziegler–Natta catalysts, prepared from Ti chlorides and organoalanes.
3. Production of acetaldehyde from ethylene by the Wacker process using $PdCl_2$ and $CuCl_2$ as catalysts.

These commercial processes led to the development of other synthetic reactions catalyzed by transition metal complexes. The impact and effect of these processes on organic synthesis are surveyed briefly.

1.1 Carbonylation of Alkenes and Alkynes Catalysed by Metal Carbonyls to Produce Aldehydes, Carboxylic Acids, Esters and Alcohols

In 1925, Fischer and Tropsch developed a process for producing a mixture of saturated and unsaturated hydrocarbons, and oxygenated products such as alcohols and esters by the reaction of synthesis gas (a mixture of CO and H_2) using heterogeneous catalysts of Fe and Co (eq. 1.1) [1].

$$CO + H_2 \xrightarrow{\text{Fe or Co}} \text{hydrocarbon} \tag{1.1}$$

This process is called the Fischer–Tropsch process and attracted attention as an important method of commercial production of synthetic oil. In 1938 Rölen, one of Fischer's co-workers, tried the reaction of alkene with synthesis gas using a Co catalyst and found the formation of aldehydes. This reaction is now called the oxo reaction or hydroformylation, because hydrogen and a formyl group add to an alkene bond [2,3]. At present, butyraldehyde is produced by the hydroformylation of propylene on a large scale. 2-Ethyl-1-hexanol is produced by aldol condensation of butyraldehyde and subsequent hydrogenation of the resulting enal. Bis(2-ethylhexyl) phthalate is utilized as a plasticizer for poly(vinyl chloride). At first, $Co_2(CO)_8$ was used as the catalyst under homogeneous conditions. Then an Rh complex was found to be a more efficient catalyst. Rh is much more active, and hence its high cost is easily offset. $HRh(CO)(Ph_3P)_3$ or $RhCl(Ph_3P)_3$ is the catalyst precursor (eq. 1.2).

$$\diagup\!\!\!\!\diagdown + CO + H_2 \xrightarrow{\text{Co}_2(\text{CO})_8} \diagup\!\!\!\!\diagdown\!\!\diagdown CHO + \diagdown\!\!\!\diagup\!\!\!\diagdown_{CHO} \tag{1.2}$$

In the 1930s, the Reppe group developed commercial processes for the production of carboxylic acids and esters by the carbonylation of alkynes and alkenes using metal carbonyls [4]. In particular, an industrial process for producing acrylic acid or ester by the carbonylation of highly explosive acetylene, catalysed by extremely toxic $Ni(CO)_4$, was established (eq. 1.3).

$$CH\!\equiv\!CH + CO + MeOH \xrightarrow{\text{Ni(CO)}_4} \diagup\!\!\!\!\diagdown CO_2Me \tag{1.3}$$

Another commercial process, for 1-butanol production by reductive carbonylation of propylene with water, catalysed by $Fe(CO)_5$, was developed by the Reppe group (eq. 1.4).

$$\diagup\!\!\!\!\diagdown + 3\,CO + 2\,H_2O \xrightarrow{\text{Fe(CO)}_5} \diagup\!\!\!\!\diagdown\!\!\diagup\!\!\diagdown_{OH} + 2\,CO_2 \tag{1.4}$$

The discovery of these carbonylation processes enabled the industrial production of aldehydes, carboxylic acids or esters, and alcohols from alkenes and alkynes using Fe,

Co and Ni carbonyl catalysts. These processes have stimulated development of carbonylation as an important unit reaction, catalysed by transition metal compounds.

Later, in the 1970s, a new commerial process for AcOH by the Rh-catalyzed carbonylation of MeOH in the presence of HI, the Monsanto process, was developed (eq. 1.5) [5]. It is significant that MeOH, which is a saturated compound, was shown to be carbonylated via MeI in this process, rather than unsaturated substrates such as alkenes and alkynes.

$$CH_3OH + CO \xrightarrow{\text{Rh–HI}} CH_3CO_2H \qquad (1.5)$$

1.2 Production of Polyethylene and Polypropylene by Ziegler–Natta Catalysts

Ziegler started his research on organolithium compounds before the second world war and synthesized n-BuLi for the first time. After the war, his research extended to organoaluminium compounds and he synthesized Et_3Al. Then a commercial process for producing higher alcohols by the oligomerization of ethylene using Et_3Al was developed. During this research the famous Ni effect on the reaction of Et_3Al with ethylene was discovered [6], namely that 1-butene is formed selectively and no oligomers are produced by the reaction of ethylene with Et_3Al, when a small amount of a Ni compound is present in the reaction system (eq. 1.6).

$$2\ CH_2{=}CH_2 \xrightarrow{\text{Ni(acac)}_2 - \text{Et}_3\text{Al}} \wedge\!\!\!\wedge \qquad (1.6)$$

The discovery of the Ni effect led to the invention of polyethylene production catalysed by $TiCl_4$ combined with Et_3Al, the so-called the Ziegler catalyst, in 1953. Soon after, the process for isotactic polypropylene was invented by Natta using a slightly modified catalyst prepared from $TiCl_3$ and Et_3Al, which is called the Natta catalyst (eq. 1.7) [7].

$$n\ CH_2{=}CH_2 \xrightarrow{\text{TiCl}_4 - \text{Et}_3\text{Al}} \text{polyethylene}$$

$$n\ \wedge\!\!\!\wedge \xrightarrow{\text{TiCl}_3 - \text{Et}_3\text{Al}} \text{polypropylene} \qquad (1.7)$$

The impact of Ziegler–Natta catalysis was enormous. The combination of $TiCl_4$ as a transition metal compound with Et_3Al as a main group metal compound opened the possibility for transmetallation. This is one of the most important unit reactions in transition metal-catalysed reactions. Also, production of isotactic polypropylene is a harbinger of stereocontrolled reactions catalysed by transition metal complexes, leading finally to asymmetric catalysis.

The Ziegler–Natta chemistry was extended to the polymerization of butadiene to produce polybutadiene using similar catalysts. However, Wilke found that cyclic oligomers, such as COD and CDT, rather than linear polybutadiene, are formed by changing ratios of TiCl$_4$ and Et$_3$Al [8]. Several years before Wilke's discovery, Reed gave the first report on the formation of COD from butadiene using a catalyst derived from Ni(CO)$_4$ [9]. However, its catalytic activity was lower due to the strong coordination of CO. Wilke subsequently found that naked Ni(0) or Ni(0) complexes of phosphines are active catalysts for cyclodimerization and trimerization. Based on these discoveries, the chemistry of π–allylnickel was developed.

The Reppe group reported in 1948 the formation of benzene and cyclooctatetraene by Ni-catalysed cyclotrimerization and cyclotetramerization of acetylene [10].

The Ni-catalyzed cyclizations of butadiene and acetylene opened a fruitful field of cycloaddition of various unsaturated compounds to afford various cyclic compounds. These cyclizations are now understood by the formation of metallacycles as intermediates (eq. 1.8).

$$(1.8)$$

Alkene metathesis, a remarkable reaction catalyzed by transition metal catalysts, can be traced back to Ziegler–Natta chemistry as its origin [11]. In 1964, Natta *et al.* reported a new type polymerization of cyclopentene using Mo- or W-based catalyst, without knowing the mechanism. This was the first example of ring-opening metathesis polymerization (ROMP; eq. 1.9) [12].

$$n \quad \xrightarrow{\text{WCl}_5 - \text{Et}_2\text{AlCl}} \quad \text{metathesis polymerization} \qquad (1.9)$$

1.3 Production of Acetaldehyde from Ethylene by the Wacker Process

For long time, Pd has been used mainly as a heterogeneous catalyst for the hydrogenation of unsaturated bonds. A revolution in Pd chemistry occurred with the development of homogeneous Pd catalysts. The first example was the invention of the Wacker process in 1959, by which ethylene is oxidized to acetaldehyde using PdCl$_2$ and CuCl$_2$ as catalysts in aqueous solution (eq. 1.10) [13].

$$\text{CH}_2{=}\text{CH}_2 + 1/2\,\text{O}_2 \quad \xrightarrow{\text{PdCl}_2\,/\,\text{CuCl}_2} \quad \text{CH}_3\text{CHO} \qquad (1.10)$$

In 1894 Philips found that, when ethylene is passed into an aqueous solution of PdCl$_2$, Pd(II) is reduced to Pd(0), which precipitates as black powder [14] and this reaction was used for quantitative analysis of Pd(II) [15]. Ethylene is oxidized to acetaldehyde at the same time. This reaction is the basis of the Wacker process.

Soon after the invention of the Wacker process the formation of vinyl acetate by the oxidative acetoxylation of ethylene using $Pd(OAc)_2$ was discovered by Moiseev [16], and the industrial production of vinyl acetate based on this reation was developed. At present, vinyl acetate is produced commercially by a gas-phase reaction of ethylene, acetic acid and O_2 using Pd catalyst supported on alumina or silica (eq. 1.11).

$$CH_2{=}CH_2 \ + \ AcOH \ + \ 1/2\,O_2 \ \xrightarrow{\ Pd-Alumina\ } \ CH_2{=}CH{\underset{OAc}{\diagdown}} \ + \ H_2O \quad (1.11)$$

1.4 Preparation of Organotransition Metal Complexes

In 1951, ferrocene was synthesized by Pauson [17] and Miller [18]. Soon after this synthesis, two groups led by Wilkinson and Fischer, independently reported that ferrocene has a stable carbon–iron π-bond [19]. This was the first example of a true organotransition metal complex containing a carbon–metal bond. Since then, numerous organotransition metal complexes have been prepared. The importance of these complexes as intermediates of many synthetic reactions has been discovered. More importantly, some transition metal complexes were found to behave as precursors of active catalysts.

The industrial processes and related reactions described above, combined with the progress of organometallic chemistry, have stimulated further remarkable development in applying transition metal complexes to organic synthesis. Various novel synthetic methods, which are impossible by conventional means, have been discovered, bringing revolution in organic synthesis.

References

1. C. Masters, *Adv. Organometal. Chem.*, **17**, 61 (1979).
2. R. L. Pruett, *Adv. Organometal. Chem.*, **17**, 1 (1979).
3. B. Cornils, W. A. Hermann and M. Basch, *Angew. Chem., Int. Ed. Engl.*, **33**, 2144 (1994); J. T. Morris, *Chem. in Britain*, 38 (1993).
4. W. Reppe, *Liebigs Ann. Chem.*, **582**, 1 (1953); W. Reppe and H. Kröper, *Liebigs Ann. Chem.*, **582**, 38 (1953); Review of a historical account of chemistry of metal carbonyls, L. Mond, C. Langer and F. Quincke, *J. Organometal. Chem.*, **383**, 1 (1990).
5. J. F. Roth, J. H. Caddock, A. Hershman and F. E. Paulik, *Chem Tech.*, 347 (1971).
6. K. Ziegler, H. G. Gellert, E. Holzkamp, G. Wilke, E. W. Duck and W. R. Kroll, *Liebigs Ann. Chem.*, **629**, 172 (1960); K. Fischer, K. Jonas, P. Misbach, R. Stabba and G. Wilke, *Angew. Chem., Int. Ed. Engl.*, **12**, 943 (1973).
7. K. Ziegler and G. Natta, *Angew. Chem. Int. Ed. Engl.*, **76**, 545 (1964).
8. G. Wilke, *J. Organometal. Chem.*, **200**, 349 (1980). *Angew. Chem., Int. Ed. Engl.*, **27**, 186 (1988).
9. H. W. B. Reed, *J. Chem. Soc.*, 1931 (1954).
10. W. Reppe, O. Schlichting, K. Klager and T. Toepel, *Liebigs Ann. Chem.*, **560**, 1 (1948).

11. N. Calderon, J. P. Lawrence and E. A. Ofstead, *Adv. Organometal. Chem.*, **17**, 449 (1979); N. Calderon, H. Y. Chen and K. W. Scott, *Tetrahedron Lett.*, 3327 (1967).
12. G. Natta, G. Dall'Asta and G. Mazzanti, *Angew, Chem., Int. Ed. Engl.*, **3**, 723 (1964).
13. J. Smidt, W. Hafner, R. Jira, R. Sieber, J. Sedlmeier and A. Sabel, *Angew. Chem., Int. Ed. Engl.*, **1**, 80 (1962).
14. F. C. Philips, *Am. Chem. J.*, **16**, 255 (1894).
15. S. C. Ogburn and W. C. Brastow, *J. Am. Chem. Soc.*, **55**, 1307 (1933).
16. I. I. Moiseev, M. N. Vargaftik and Ya. K. Syrkin, *Dokl. Akad. Nauk SSSR*, **133**, 377 (1960).
17. T. J. Kealy and P. L. Pauson, *Nature*, **168**, 1039 (1951).
18. S. A. Miller, J. A. Tebboth and J. F. Tremaine, *J. Chem. Soc.*, 632 (1952).
19. G. Wilkinson, *J. Organometal. Chem.*, **100**, 273 (1975).

2
BASIC CHEMISTRY OF TRANSITION METAL COMPLEXES AND THEIR REACTION PATTERNS

Organic reactions involving transition metal compounds proceed via complex formation; that is, coordination of a reactant molecule to a low-valent transition metal is essential for the reaction to occur. In order to explain how synthetic reactions involving transition metal complexes proceed, it is important to understand the fundamental behaviour of complexes and their reaction patterns.

2.1 Formation of Transition Metal Complexes

Metallic Mg is used for the preparation of Grignard reagents. However, transition metal complexes formed by the coordination of ligands (L) to metals are used for synthetic reactions. The transition metal itself is used rarely. The change of properties of transition metals, brought about by complex formation, is considerable. For example, metallic Ni has a very high melting point and is insoluble in organic solvents, whereas $Ni(CO)_4$ is a volatile, extremely toxic liquid (b.p. 43 °C) and is soluble in organic solvents [1]. Although Pd and Pt are stable noble metals, their complexes $Pd(Ph_3P)_4$ and $Pt(Ph_3P)_4$ are greenish-yellow crystals and soluble in organic solvents.

Four molecules of CO coordinate to Ni to form $Ni(CO)_4$, but $Ni(CO)_5$ is never formed. The stoichiometry of complex formation can be understood by the 18-electron rule. According to this rule, a stable complex with an electron configuration of the next highest noble gas is obtained when the sum of d electrons of metals and electrons donated from ligands equals 18. Complexes that obey the 18-electron rule are said to be coordinatively saturated. Ni(0) has the following ground state electronic structure, and forms complexes using $3d^8$ and $4s^2$ electrons equally (10 d electrons, or d^{10}):

$$1s^2 2s^2 2p^6 3s^2 3p^6 3d^8 4s^2$$

Similarly, Pd(0) and Pt(0) form complexes using their d^{10} electrons. The numbers of d electrons of major transition metals used for the complex formation are shown in Table 2.1. Coordinatively saturated complexes are formed by the donation of electrons from the ligands until total numbers of the electrons reach 18.

Well-known complexes that obey the 18-electron rule are shown below. Typical ligands, such as CO, phosphine and alkenes, donate two electrons each. The total number of d electrons of $Ni(CO)_4$ can be calculated as $10 + (2 \times 4) = 18$, and hence $Ni(CO)_5$ cannot be formed. In $Co_2(CO)_8$, the number of d electrons from Co(0) is nine and four CO molecules donate eight electrons. Furthermore, a Co–Co bond is formed by donating one electron each. Therefore, the total is $9 + 8 + 1 = 18$ electrons, to satisfy the 18-electron rule. The relationship between the coordination numbers and numbers of d electrons of metal carbonyls is shown in Tables 2.2 and 2.3.

Number of electrons in ferrocene **1** can be counted in the following way. In ferrocene, Fe is Fe(II) and has six d-electrons. The cyclopentadienyl anion donates six electrons (2×2 from two double bonds and two electrons from the anion), and $6 + (4 \times 2) + (2 \times 2) = 18$ electrons satisfy the rule. In another calculation Fe, regarded as Fe(0), offers eight electrons and the cycloptendienyl radical supplies one electron. Therefore, total electron count is $8 + (4 \times 2) + (1 \times 2) = 18$.

Cp₂Fe
ferrocene

1

Fe(II)	6e
Cp anion × 2	12e
	18e

In bis-π-allylnickel **2**, Ni(II) has 8e and the two allyl anions supply four electons each: $8 + (4 \times 2) = 16$. The following calculation is also possible: if Ni(0) supplies 10e and the two allyl radicals supply three electrons each, the total number is $10 + (3 \times 2) = 16$. Therefore, this complex is coordinatively unsaturated.

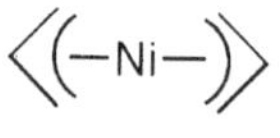

$\langle\!\langle -Ni- \rangle\!\rangle$

2

Ni(II)	8e
allyl anion × 2	8e
	16e

Table 2.1 Numbers of d electrons of transition metals

Valency	Cr Mo W	Mn Tc Re	Fe Ru Os	Co Rh Ir	Ni Pd Pt
0	6	7	8	9	10
1	5	6	7	8	9
2	4	5	6	7	8
3	3	4	5	6	7

Table 2.2 Complexes that obey the 18-electron rule

Complex	Coordination number	Number of electrons
$Pd(PPh_3)_4$	Pd(0)	10e
	4PPh_3	8e
		18e
$Fe(CO)_5$	Fe(0)	10e
	4 CO	8e
		18e
$Mo(CO)_6$	Mo(0)	6e
	6 CO	10e
		18e
$Ni(CO)_4$	Ni(0)	10e
	4 CO	8e
		18e
$Co_2(CO)_8$	Co(0)	9e
	4 CO	8e
	Co–Co	1e
		18e

Table 2.3 Numbers of d electrons and cordination numbers of metal carbonyls

Number of d electrons	Coordination number	Examples
6	6	$Cr(CO)_6$
8	5	$Fe(CO)_5$
10	4	$Ni(CO)_4$

When a reaction of an organic compound (either promoted or catalysed by a transition metal complex) occurs, the reactant must coordinate to the metal. For coordination of the reactant to the metal prior to its reaction, the complex must be coordinatively unsaturated so as to offer a vacant site in order to make the coordination of the reactants possible. Therefore, $Ni(CO)_4$ or $Fe(CO)_5$ should be made coordinatively unsaturated by liberating some of the coordinated CO by heating or irradiation. When $Pd(Ph_3P)_4$ is used as a catalyst, it becomes an unsaturated complex by liberating two molecules of Ph_3P in solution. Furthermore, in this complex, Ph_3P is a kind of 'innocent' or 'spectator' ligand, which does not take part in synthetic reactions directly, but which modifies the reactivity of the metal. Various phosphines, namely arylphosphines, alkylphosphines, and bidentate phosphines, acting as innocent ligands, have different steric effects and electron-donating abilities, and electron density of the central metal changes depending on the kind of the ligand involved. Thus in many cases different catalytic activity is observed by the same metal depending on the innocent ligands. From this consideration, clearly subtle design of the best complexes suitable for a desired reaction becomes important, although it is not always easy to do this.

2.2 Fundamental Reactions of Transition Metal Complexes; Comparison of Transition Metal-catalysed Reactions with Grignard Reactions

Six fundamental reactions of transition metal complexes are briefly explained in order to demonstrate how reactions either promoted or catalysed by transition metal complexes proceed. In the reaction schemes throughout this book, some of the spectator ligands that do not participate in the reactions are omitted for simplicity.

2.2.1 Oxidative Addition

The term 'oxidative' may sound strange to organic chemists who are not familiar with organometallic chemistry. The use of this term in organometallic chemistry has a meaning different from the 'oxidation' used in organic chemistry, such as the oxidation of secondary alcohols to ketones. Thus, oxidative addition means the reaction of a molecule X–Y with a low-valent coordinatively unsaturated metal complex $M(n)L_m$ under bond cleavage, forming two new bonds **3**. As two previously nonbonding electrons of the metal are involved in the new bonding, the metal increases its formal oxidation state by two units, namely, $M(n)$ is oxidized to $M(n+2)$, and increases the coordination number of the metal centre by two. In oxidative addition, it is defined that the electrons in the two new bonds belong to the two ligands, and not to the metal.

This process is similar to the formation of Grignard reagents **4** from alkyl halides and Mg(0). In the preparation of Grignard reagents, Mg(0) is oxidized to Mg(II) by the oxidative addition of alkyl halides to form two covalent bonds.

$$M(n) \ + \ X{-}Y \xrightarrow{\text{oxidative addition}} X{-}M(n{+}2){-}Y$$
$$\textbf{3}$$

$$CH_3{-}I + Mg(0) \xrightarrow{\text{oxidative addition}} CH_3{-}Mg(II){-}I$$
$$\textbf{4}$$

Another example, which shows clear difference between oxidation in organic chemistry and oxidative addition in organometallic chemistry, is the oxidative addition of a hydrogen molecule to $M(n)$ to form $M(n+2)$ dihydride **5**. In other words, $M(n)$ is oxidized to $M(n+2)$ with hydrogen. This sounds strange to organic chemists, because hydrogen is a reducing agent in organic chemistry.

$$M(n) \ + \ H{-}H \xrightarrow{\text{oxidative addition}} H{-}M(n{+}2){-}H$$
$$\textbf{5}$$

Oxidative addition is facilitated by higher electron density of the metals and, in general, σ-donor ligands such as R_3P and H^- attached to M facilitate oxidative addition. On the other hand, π-accepter ligands such as CO and alkenes tend to suppress oxidative addition.

A number of different polar and nonpolar covalent bonds are capable of undergoing the oxidative addition to $M(n)$. The widely known substrates are C−X (X = halogen and pseudohalogen). Most frequently observed is the oxidative addition of organic halides of sp^2 carbons, and the rate of addition decreases in the order C−I > C−Br >> C−Cl >>> C−F. Alkenyl halides, aryl halides, pseudohalides, acyl halides and sulfonyl halides undergo oxidative addition (eq. 2.1).

Substrates with halogen bonds

$$\text{(2.1)}$$

X = halogen, pseudohalogen

The following compounds with H−C and H−M′ bonds undergo oxidative addition to form metal hydrides. This is examplified by the reaction of **6**, which is often called *ortho*-metallation, and occurs on the aromatic C−H bond at the *ortho* position of such donar atoms as N, S, O and P. Reactions of terminal alkynes and aldehydes are known to start by the oxidative addition of their C−H bonds. Some reactions of carboxylic acids and active methylene compounds are explained as starting with oxidative addition of their O−H and C−H bonds.

Substrates with hydride (hydrogen) bonds

H−H R_3Si−H R_3Sn−H R_2B−H $\underset{R}{\overset{O}{\parallel}}{}\!\!-\!\!H$ R≡≡−H

$$R\!-\!\underset{\underset{E}{|}}{\overset{\overset{E}{|}}{C}}\!-\!H \xrightarrow{\ ML_n\ } R\!-\!\underset{\underset{E}{|}}{\overset{\overset{E}{|}}{C}}\!-\!M\!-\!H \qquad RCO_2\!-\!H \xrightarrow{\ ML_n\ } RCO_2\!-\!M\!-\!H$$

E = EWG

6

Metal–metal bonds M′−M′ such as R_2B−BR_2 and R_3Si−SiR_3 undergo oxidative addition, (where M′ represents main group metals; eq. 2.2).

Substrates with metal–metal bonds

R_3Si−SiR_3 R_3Sn−SnR_3 R_2B−BR_2 R_3Sn−SiR_3 etc   (2.2)

RM′−−M′R + M $\xrightarrow{\text{oxidative addition}}$ RM′−−M−−M′R

Oxidative addition involves cleavage of the covalent bonds as described above. In addition, oxidative addition of a wider sense occurs without bond cleavage. For example, π-complexes of alkenes **7** and alkynes **9** are considered to form η^2 complexes **8** and **10** by oxidative addition. Note that to specify the numbers of carbon atoms that interact with the metal center, the prefix η^n is used before the ligand formula to imply bonding to n carbons. Two distinct metal–carbon bonds are formed, and the resulting alkene and alkyne complexes are more appropriately described as the metallacyclopropane **8**, and the alkyne complex **9** may be regarded as the matallacyclopropene **10**. Thus the coordination of the alkene and alkyne results in the oxidation of the metal. The metallacyclobutane **11** is formed by the oxidative addition of cyclopropane with bond cleavage.

Oxidative cyclization is another type of oxidative addition without bond cleavage. Two molecules of ethylene undergo transition metal-catalysed addition. The intermolecular reaction is initiated by π-complexation of the two double bonds, followed by cyclization to form the metallacyclopentane **12**. This is called oxidative cyclization. The oxidative cyclization of the α,ω-diene **13** affords the metallacyclopentane **14**, which undergoes further transformations. Similarly, the oxidative cyclization of the α,ω-enyne **15** affords the metallacyclopentene **16**. Formation of the five-membered ring **18** occurs stepwise (**12, 14** and **16** likewise) and can be understood by the formation of the metallacyclopropene or metallacyclopropane **17**. Then the insertion of alkyne or alkene to the three-membered ring **17** produces the metallacyclopentadiene or metallacyclopentane **18**.

The term oxidative cyclization is based on the fact that two-electron oxidation of the central metal occurs by the cyclization. The same reaction is sometimes called 'reductive cyclization'. This term is based on alkene or alkyne bonds, because the alkene double bond in **13** is reduced to the alkane bond **14**, and the alkyne **15** bond is reduced to the alkene bond **16** by the cyclization. Cyclizations of alkynes and alkenes catalyzed by transition metal complexes proceed by oxidative cyclization. In particular, low-valent complexes of early transition metals have a high tendency to obtain the highest possible oxidation state, and hence they react with alkynes and alkenes forming rather stable metallacycles by oxidative addition or oxidative cyclization.

Oxidative cyclization

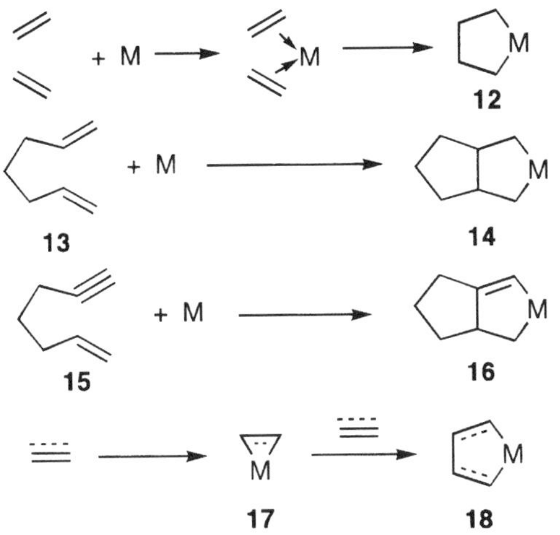

In the cyclization of conjugated dienes, typically butadiene, coordination of two molecules of butadiene gives rise to the bis-π-allyl complex **20**. The distance between terminals of two molecules of butadiene becomes closer by π-coordination **19** to metals such as Ni(0) and Pd(0), and the oxidative cyclization generates 1-metalla-2,5-divinylcyclopentane **21** and 1-metalla-3,7-cyclononadiene **22**. The bis-π-allyl complex **20** may be represented by the resonance of **21** and **22**. These complexes are intermediates of the formation of COD and CDT catalyzed by Ni(0) complexes, typically Ni(cod)$_2$. At this point it is worth explaining the rule for the uses of capital and small letters for the abbreviation of ligands. Capital letters are used for abbreviation of ligands themselves, for example COD, DPPE, BINAP, Rh-BINAP, and DBA. According to the IUPAC rules of nomenclature, the abbreviation of ligands are written with small letters when they are components of transition metal complexes, for example, Ni(cod)$_2$, Pd(dppe)Cl$_2$ and Pd$_2$(dba)$_3$. When THF and DMF are parts of the complexes of definite structure, they are written as thf and dmf, as in RuCl$_2$[(S)-binap]$_2$(dmf)$_n$.

Oxidative addition occurs with coordinatively unsaturated complexes. As a typical example, the saturated Pd(0) complex Pd(Ph$_3$P)$_4$ (four-coordinate, 18 electrons) undergoes reversible dissociation *in situ* in solution to give the unsaturated 14-electron species Pd(Ph$_3$P)$_2$ (**23**), which is capable of undergoing the oxidative addition. Various σ-bonded palladium complexes **24** are formed by the oxidative addition. In many cases, dissociation of ligands to supply a vacant coordination site is the first step of catalytic reactions.

Pd(PPh₃)₄ →(2 PPh₃) Pd(PPh₃)₂ **23** →(R–X) **24**

18 electrons   14 electrons   16 electrons

Similar to the formation of allylmagnesium chloride (**25**), the oxidative addition of allyl halides to transition metal complexes generates allylmetal complexes **26**. However, in the latter case, a π-bond is formed by the donation of π-electrons of the double bond, and resonance of the σ-allyl and π-allyl bonds in **26** generates the π-allyl complex **27** or η^3-allyl complex. The carbon–carbon bond in the π-allyl complexes has the same distance as that in benzene. Allyl Grignard reagent **25** is prepared by the reaction of allyl halide with Mg metal. However, the π-allyl complexes of transition metals are prepared by the oxidative addition of not only allylic halides, but also esters of allylic alcohols (carboxylates, carbonates, phosphates), allyl aryl ethers and allyl nitro compounds. Typically, the π-allylpalladium complex **28** is formed by the oxidative addition of allyl acetate to Pd(0) complex.

25

26   **27**

X = Cl, Br, -OCOR, OPh, NO₂

28

2.2.2 Insertion

The reaction of Grignard reagents with a carbonyl group can be understood as an insertion reaction of an unsaturated C=O bond of carbonyl group into an Mg–carbon bond as shown by **29** to form Mg alkoxide **30**. Similarly, various unsaturated ligands, such as alkenes, alkynes and CO, formally insert into an adjacent metal–ligand bond of transition metal complexes as shown by **31** to give **32**. When the adjacent ligand is σ-alkyl or -aryl (X = carbon in **32**), the process is called 'carbometallation', forming a carbon–carbon bond (A = carbon). If the ligand is a hydride (X = H), a C–H bond is formed, and the process is called 'hydrometallation'. The term 'insertion' is somewhat misleading. The insertion should be understood as the migration of the adjacent ligand from the metal to the metal-bound unsaturated ligand, generating a vacant site as shown by **33**.

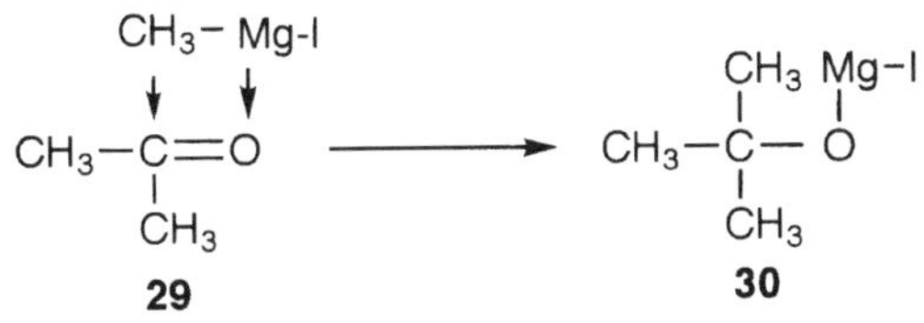

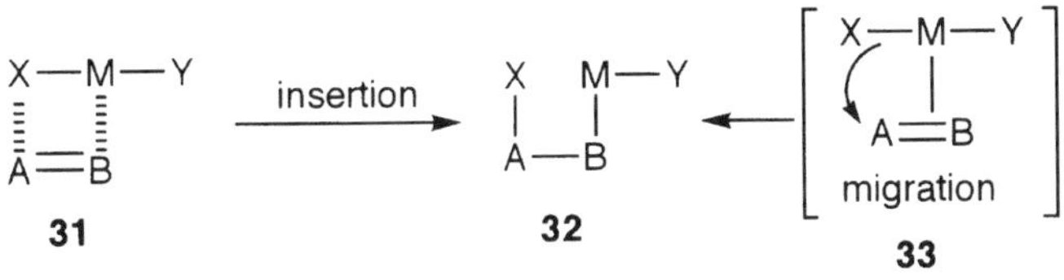

The insertion is reversible. Two types of the insertion are known. They are α,β- (or 1,2-) and α,α- (or 1,1-) insertions. Most widely observed is the α,β-insertion of unsaturated bonds, such as alkenes and alkynes. The unsaturated bonds shown in eq. 2.3 undergo α,β-insertion.

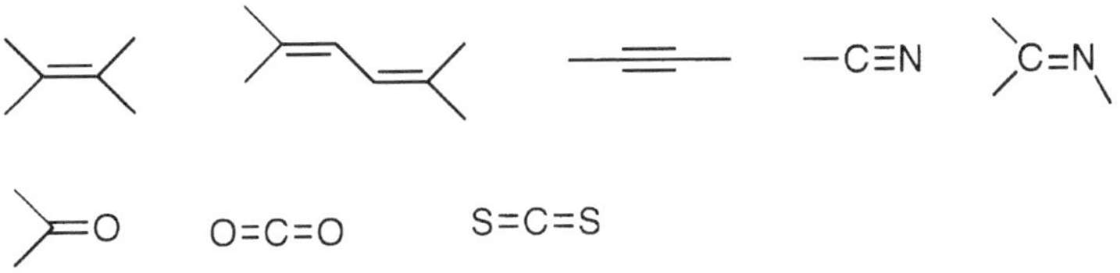

$$(2.3)$$

The insertion of alkene to metal hydride (hydrometallation of alkene) affords the alkylmetal complex **34**, and insertion of alkyne to an $M-R$ (R = alkyl) bond forms the vinyl metal complex **35**. The reaction can be understood as the *cis* carbometallation of alkenes and alkynes.

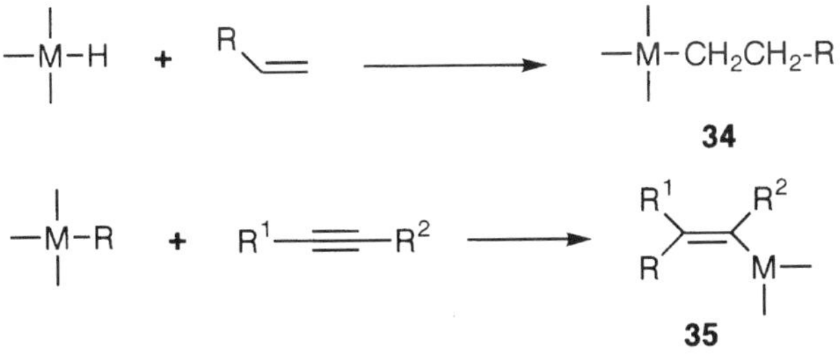

π-Allyl complexes are formed by the reaction of conjugated dienes with complexes. The insertion of one of the double bonds of butadiene to a $Co-H$ bond leads to the π-allylpalladium complex **36**. This is an intermediate of the formation of linear oligomers involving hydride shift. Also the π-allyl complex **37** is formed by the insertion of butadiene to the $Ph-Pd$ bond. The insertion is usually highly stereospecific.

Rates of the insertion are controlled by several factors. For example, the insertion of an alkene to Pd complex is faster when a cationic complex is formed. The addition of an Ag salt to a chloro complex generates a cationic complex and hence the insertion is

$$H\text{-}Co(CO)_4 \;+\; \diagup\!\!\!\diagdown \;\longrightarrow\; \left[\begin{array}{c} H\text{-}CH_2 \\ Co(CO)_n \end{array} \right] \longrightarrow \begin{array}{c} CH_3 \\ \text{-}Co(CO)_n \end{array}$$

36

$$Ph\text{-}Pd\text{-}X \;+\; \diagup\!\!\!\diagdown \;\longrightarrow\; \left[\begin{array}{c} Ph \\ Pd\text{-}X \end{array} \right] \longrightarrow \begin{array}{c} Ph \\ \text{-}Pd \diagdown X \end{array}$$

37

accelerated. For the insertion, *cis* coordination is necessary. Thus the *trans* acyl-alkene complex **38** must be isomerized to the rather unstable *cis* complex **39** to give the insertion product **40**. Coordination of a bidentate ligand forms the *cis* complex **41** by chelation, and the insertion is possible without the *trans* to *cis* isomerization. This effect explains partly an accelerating effect of the bidentate ligands, which force the *cis* coordination of reacting molecules.

38   **39**   **40**

41

CO is a representative species which undergoes α,α-insertion. Its insertion to a metal–carbon bond affords the acylmetal complexes **42**. The CO insertion is understood to occur by migration of the alkyl ligand to a coordinated CO. Mechanistically the CO insertion is regarded as 1,2-alkyl migration to the *cis*-bound CO (migratory insertion). The migration is reversible, and an important step in carbonylation. SO_2, isonitriles and carbenes are other species that undergo the α,α-insertion.

Both Mg and transition metal complexes similarly undergo oxidative addition and insertion. Whereas the main reaction path of Grignard reagents is the insertion of a

$$M\text{-}R \;+\; \overset{\alpha}{C}\!\equiv\!\overset{\alpha}{O} \;\xrightarrow{\;\alpha,\alpha\text{-insertion}\;}\; M\text{-}\underset{O}{\overset{\|}{C}}\text{-}R$$

42

$$\underset{L_nM\text{-}CO}{\overset{R}{|}} \;\xrightarrow{\;(migration)\;}\; L_nM\text{-}C\diagdown\!\!\!\overset{R}{\underset{O}{}}$$

carbonyl group, transition metal complexes can undergo both oxidative addition and insertion with a variety of π-bonds. It should be emphasized that the insertion can occur sequentially several times. For example, insertion of an alkene to a C−M or H−M bond, to afford alkyl complex **43**, is followed by CO insertion to generate the acyl complex **44**. Sometimes, further insertions of another alkene and CO take place. Particularly useful is the formation of polycyclic compounds by sequential intramolecular insertions of several alkenyl and alkynyl bonds. Several bonds are formed without adding other reagents and changing reactions conditions. These reactions are called either domino, cascade or tandem reactions. Among these terms, 'domino' is the most appropriate [2]. The polymerization of ethylene, catalysed by the Ziegler catalyst, is understood to occur by very rapid sequential insertion of nearly 20 000 molecules of ethylene to the alkyltitanium bonds **45** per second to form the polymeric product **46**.

2.2.3 Transmetallation

Ziegler discovered the selective formation of 1-butene from ethylene promoted by AlR_3 without undergoing the oligomerization of ethylene in the presence of a small amount of an Ni salt. This was called the Ni effect [3] which lead to the great discovery of the Ziegler catalyst, prepared by the combination of $TiCl_4$ and Et_3Al.

The change of reactivity of Grignard reagents by the addition of a catalytic amount of transition metal compounds has been known for many years. Coupling of Grignard reagent with alkyl halides, induced by the addition of a cobalt salt, has been known as the Kharasch reaction. Also, selective 1,4-addition of Grignard reagents to enones by addition of a small amount of CuI is an established synthetic method. In these reactions, the transfer of an alkyl or hydride group from Mg to a transiton metal (Co or Cu) takes place, and this process is called the transmetallation. Organometallic compounds M′−R and hydrides M′−H of main group metals (M′ = Mg, Zn, B, Al, Sn, Si, Hg) react with transition metal complexes A−M−X formed by the oxidative addition. The organic group or hydride is transferred to the transition metals by exchanging X with R or H. In other words, alkylation of the metals or hydride formation takes place. The driving force of the transmetallation is ascribed to the difference in electronegativity of the two metals, and the main group metal M′ must be more electropositive than the transition metal M. The oxidative addition–transmetallation sequence is widely known. Reaction of benzoyl chloride with Pd(0) gives benzoylpalladium chloride (**47**), and subsequent transmetallation with methyltributyltin generates benzoylmethylpalladium (**48**). Formation of acetophenone (**49**) by the reductive elimination of **48** is a typical example.

M—X + M'—R ⇌ M⟨ᴿ⟩M' ⇌ M—R + M'—X
            (with X bridging)

M = transition metal
M' = main group metal

$$Ph\text{-}C(=O)\text{-}Cl + Pd(0) \xrightarrow[\text{addition}]{\text{oxidative}} Ph\text{-}C(=O)\text{-}Pd\text{-}Cl \quad (47)$$

$$\xrightarrow[\text{transmetallation}]{MeSnBu_3 \quad Bu_3SnCl} Ph\text{-}C(=O)\text{-}Pd\text{-}Me \quad (48)$$

$$\xrightarrow[\text{elimination}]{\text{reductive}} Ph\text{-}C(=O)\text{-}Me \ (49) \ + \ Pd(0)$$

2.2.4 *Reductive Elimination*

Similar to 'oxidative', the term 'reductive' used in organometallic chemistry has a meaning different from reduction in organic chemistry. Reductive elimination is a unimolecular decomposition pathway, and the reverse of oxidative addition. Reductive elimination involves loss of two ligands of *cis* configuration from the metal center of **50**, and their combination to form a single elimination product **51**. In other words, the coupling of the two groups coordinated to the metal liberates the product **51** in the last step of a catalytic cycle. By reductive elimination, both the coordination number and the formal oxidation state of the metal $M(n + 2)$ are reduced by two units to generate $M(n)$. Thus, $M(n + 2)$ is reduced to $M(n)$ and the reaction is named 'reductive' elimination. The regenerated $M(n)$ species can undergo oxidative addition again. Thus a catalytic cycle becomes possible by the reductive elimination step. No reductuve elimination occurs in Grignard reactions. Without reductive elimination, the reaction ends as a stoichiometric one. This is a big difference between the reactions of transition metal complexes and main group metal compounds.

For example, in the carbonylation reaction the acylmetal complex **52**, formed by the insertion of CO, undergoes reductive elimination to give the carbonyl compound **53** as a product, and the catalytic species $M(n)$ is regenerated

$$X\text{-}A\text{-}B\text{-}M(n+2)\underset{Y}{|} \xrightarrow{\text{reductive elimination}} X\text{-}A\text{-}B\text{-}Y + M(n)$$

50   **51**

$$RCH_2CH_2\text{-}C(=O)\text{-}M(n+2)L_m\underset{X}{|} \longrightarrow RCH_2CH_2\text{-}C(=O)\text{-}X + M(n)\text{-}L_m$$

52   **53**

The reductive elimination of A—B proceeds if A and B are mutually *cis*. In other words, reductive elimination is possible from *cis* complexes. If the groups to be eliminated are *trans,* they must first rearrange to *cis.* The *cis*-diethyl complex **54** gives butane, whereas ethylene and ethane are formed from the *trans*-diethyl complex **55** via elimination of a β-hydrogen to generate the metal hydride **56**, and its reductive elimination. Also, the reductive elimination of the Pd$-$C(sp^2) bond is faster than that of the Pd$-$C(sp^3) bond. Thus the reductive elimination of *cis*-PdMe(Ph) (PEt$_2$Ph)$_2$ (**57**) proceeds rapidly at room temperature, whereas heating is necessary for the generation of ethane from *cis*-PdMe$_2$(PEt$_2$Ph)$_2$ (**58**). Reduced electron density of the central metals promotes reductive elimination, and addition of strong π-accepter ligands, such as CO and electron-deficient alkenes, promotes reductive elimination.

$$
\begin{array}{c}
\text{C}_2\text{H}_5 \\
| \\
\text{L}-\text{Pd}-\text{C}_2\text{H}_5 \\
| \\
\text{L} \\
\textbf{54}
\end{array}
\longrightarrow \quad \text{C}_2\text{H}_5\text{-C}_2\text{H}_5
\qquad\qquad
\begin{array}{c}
\text{C}_2\text{H}_5 \\
| \\
\text{L}-\text{Pd}-\text{L} \\
| \\
\text{C}_2\text{H}_5 \\
\textbf{55}
\end{array}
\longrightarrow
\left[
\begin{array}{c}
== \\
| \\
\text{L}-\text{Pd}-\text{H} \\
| \\
\text{C}_2\text{H}_5 \\
\textbf{56}
\end{array}
\right]
\longrightarrow \quad \text{C}_2\text{H}_4 + \text{C}_2\text{H}_6
$$

$$
\begin{array}{c}
\text{Et}_2\text{PhP} \quad \text{CH}_3 \\
\diagdown \quad \diagup \\
\text{Pd} \\
\diagup \quad \diagdown \\
\text{Et}_2\text{PhP} \quad \text{Ph} \\
\textbf{57}
\end{array}
\xrightarrow{\text{rt}} \quad \text{Ph}-\text{CH}_3
\qquad\qquad
\begin{array}{c}
\text{Et}_2\text{PhP} \quad \text{CH}_3 \\
\diagdown \quad \diagup \\
\text{Pd} \\
\diagup \quad \diagdown \\
\text{Et}_2\text{PhP} \quad \text{CH}_3 \\
\textbf{58}
\end{array}
\xrightarrow{\text{heat}} \quad \text{H}_3\text{C}-\text{CH}_3
$$

2.2.5 Elimination of β-Hydrogen and α-Hydrogen (Dehydrometallation)

Another reaction of the last step in catalytic cycle is *syn* elimination of hydrogen from carbon in β-position to the metal in alkyl metal complexes to give rise to the metal hydride H$-$M$-$X and an alkene. This process is termed elimination of β-hydrogen or simply β-elimination. Insertion of alkene to a metal hydride to form alkyl metal and the elimination of β-hydrogen from the alkyl metal are reversible steps. The elimination of β-hydrogen generates an alkene. Both the hydrogen and the alkene coordinate to the metal as shown by **59**, increasing the coordination number of the metal by one. Therefore, the β-elimination requires coordinative unsaturation of metal complexes. The β-hydrogen eliminated should be *syn* to the metal.

$$
\begin{array}{c}
\text{H} \\
| \\
\text{R}-\text{C}\text{---}\text{CH}_2 \\
| \quad\quad | \\
\text{H} \quad\quad \text{M}-\text{X}
\end{array}
\longrightarrow
\begin{array}{c}
\text{R}-\text{CH}{=}\text{CH}_2 \\
| \\
\text{H}-\text{M}-\text{X} \\
\textbf{59}
\end{array}
$$

The reductive elimination and the elimination of β-hydrogen are competitive. The elimination of β-hydrogen takes place with *trans* dialkylmetal complexes such as **55**. The reductive elimination is favoured by coordination of bidentate phosphine ligands which have larger bite angles to force the *cis* coordination. Thus bidentate ligands with large bite angles, such as dppf and dppb **60**, accelerate the reductive elimination more than do the bidentate dppe **61** and monodentate Ph$_3$P.

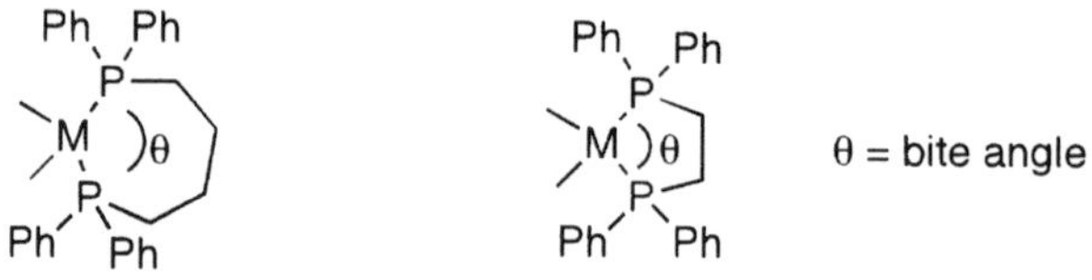

60   **61**   θ = bite angle

The catalytic cycle of the Ni-catalysed dimerization of ethylene to give 1-butene (**65**) is explained by the insertion of ethylene to the nickel hydride **62** twice to form the ethyl complex **63** and the butyl complex **64**. The elimination of β-hydrogen gives 1-butene (**65**), and regenerates the Ni—H species **62**. The reaction is chemoselective. Curiously, no further insertion of ethylene to **64** occurs.

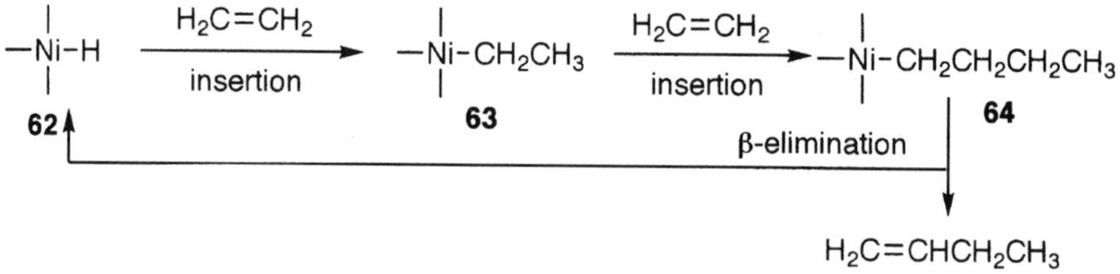

$$H_2C=CHCH_2CH_3$$
65

The carbonyl compounds **67** are formed by the elimination of β-hydrogen from the metal alkoxides **66**.

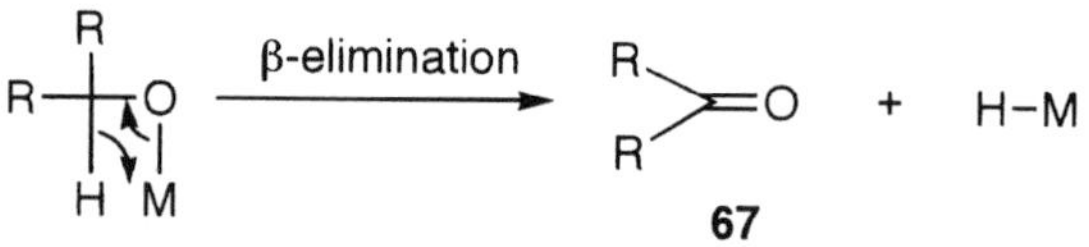

The elimination of α-hydrogen is not general and observed only with limited numbers of metal complexes. The elimination of α-hydrogen from the methyl group in the dimethylmetal complex **68** generates the metal hydride **69** and a carbene that coordinates to the metal. Liberation of methane by the reductive elimination generates the carbene complex **70**. Formation of carbene complexes of Mo and W is a key step in alkene metathesis. The α-elimination is similar to the 1,2-hydride shift observed in organic reactions.

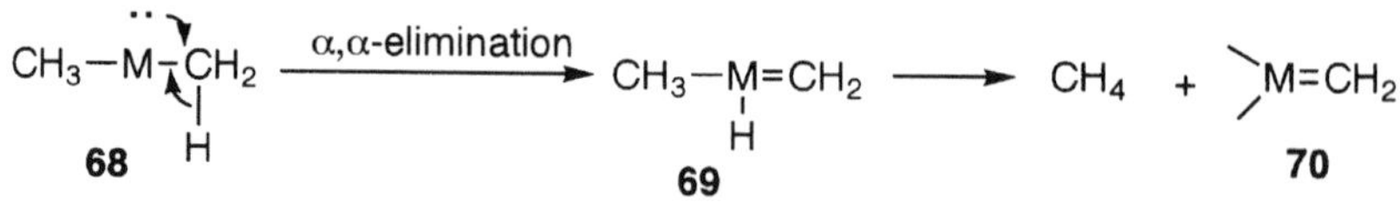

2.2.6 Nucleophilic Attack on Ligands Coordinated to Transition Metals

Many useful reactions that are entirely different from ordinary organic reactions can be achieved by using transition metal complexes. The effect of the coordination is noteworthy. Unsaturated organic compounds such as CO, alkenes, alkynes and arenes are rather unreactive towards nucleophiles because they are electron rich. However, their reactivity is inverted when these unsaturated molecules coordinate to electron-deficient metals. The coordination decreases the electron density of the unsaturated molecules, which become reactive toward nucleophilic attack and, as a result, the coordination provides novel synthetic methods. The reaction of nucleophiles with the coordinated unsaturated compounds is one of the most characteristic and useful reactions of transition metal complexes. In particular, complexes having strong π-accepter ligands, typically CO, or cationic complexes are highly electrophilic and accept nucleophiles. Mechanistically, some nucleophiles attack the ligand after coordination to the metal, and the process is understood as the insertion of the ligand. Direct attack of nucleophiles on the ligand is also possible.

Various nucleophiles can attack coordinated alkenes. Typically the attack of OH anion on ethylene coordinated to Pd(II) as shown by **71** takes place in the Wacker process to afford acetaldehyde (**72**) [4]. Also COD (**73**), coordinated to PdCl$_2$, was shown to be attacked by carbon nucleophiles such as malonate to give **74**. This reaction is the first example of carbopalladation of alkenes [5].

The reaction of carbon nucleophiles such as malonate with π-allylpalladium **75** is well-known [6]. The nucleophilic attack of an alkoxide on the dienepalladium complex **76** produces the substituted π-allyl complex **77** which is attacked again by a nucleophile to give **78**.

Arenes are inert to nucleophilic attack and normally undergo electrophilic substitution. However, arenes coordinate to Cr(CO)$_6$ to form the η^6-arenechromium tricarbonyl complex **79**, and facile nucleophilic attack on the arene generates the anionic η^5-cyclohexadienyl complex **80**, from which substituted arene **81**, or cyclohexadiene is obtained by oxidative decomplexation. In this reaction, strongly electron-withdrawing CO accelerates the attack of the nucleophiles on aromatic rings.

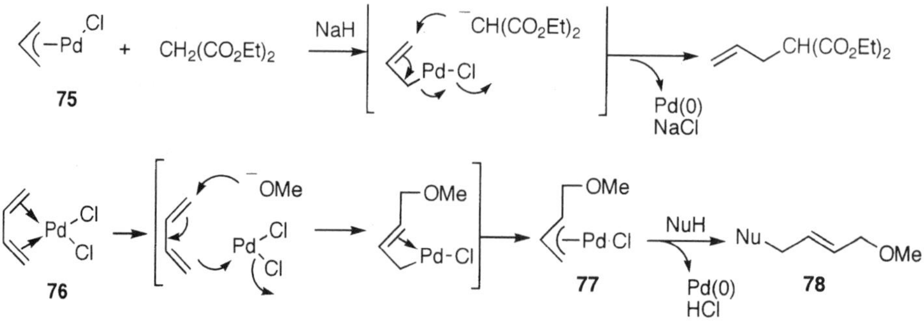

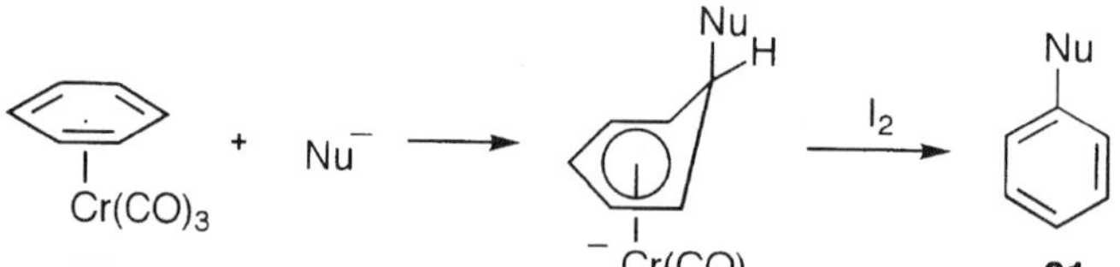

Coordinated CO in metal carbonyls is reactive towards nucleophiles, offering a good method to introduce carbonyl groups into organic substrates. Neutral metal carbonyls are attacked by strong nucleophiles such as alkyllithium to give acyl ate complex **82**. The attack of electrophiles at the metal of the ate complexes **82** or C-alkylation of metal enolates gives the neutral acyl complexes **83**, which undergo reductive elimination to afford ketones, aldehydes and carboxylic acid derivatives **84**. However, the attack of electrophiles at oxygen, or O-alkylation of the enolate **85**, is a useful preparative method for the carbene complex **86**.

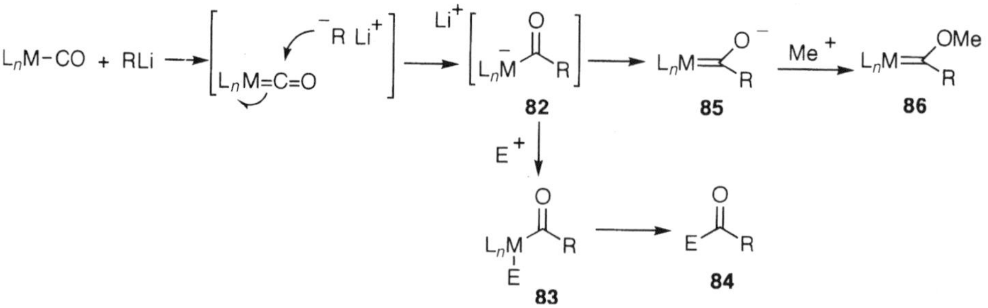

Carbamoyl or alkoxycarbonyl complexes **87** are obtained by the attack of amines or alkoxides to metal carbonyls. They are important intermediates of carbonylation reactions and undergo insertion of alkene and alkyne.

Direct cleavage of the acyl–metal bonds **88** with alcohols and amines gives esters **89** and amides. This corresponds to the last and key step of the carbonylation process.

No change of the formal oxidation states of the metals occurs in most of these nucleophilic attacks. However, an exception is palladium in the π-allyl complex **90**, which accepts two electrons by the nucleophilic attack and is reduced to Pd(0) state, directly or indirectly. This process offers the chance of undergoing the oxidative addition again, as shown below. The reduction of the metals is an essential factor for catalytic cycles. Pd is a noble metal and Pd(0) is more stable than Pd(II). In this respect, Pd is unique. In contrast, however, attack on the allylmetal complex by an electrophile such as aldehyde may proceed as shown by the reaction of **91** to generate an oxidized metal ion, which cannot enter into the catalytic cycle. Thus the reaction ends as a simple stoichiometric one. For example, π-allylnickel complex **91** is attacked by electrophiles, giving Ni(II), and the reaction is stoichiometric.

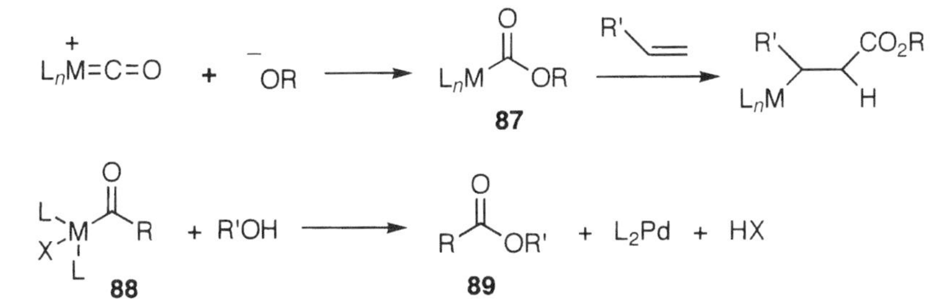

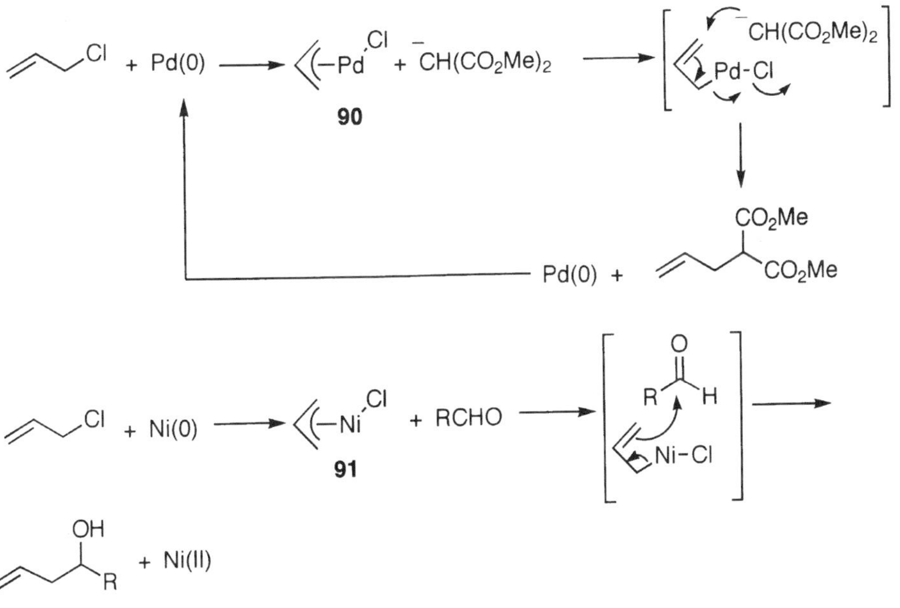

2.2.7　Termination of the Metal-promoted or -catalysed Reactions and a Catalytic Cycle

The Grignard reaction proceeds via oxidative addition and insertion. The reaction product **93** is isolated after hydrolysis of the insertion product **92** with dilute aqueous

HCl, giving $MgCl_2$, and it is practically impossible to reduce the generated Mg(II) to Mg(0) *in situ,* and hence the Grignard reaction is stoichiometric. In other words, Mg(0) is oxidized to Mg(II) by the Grignard reaction. However, reactions involving transition metal complexes proceed with a catalytic amount of the metal compounds in many cases whenever they are attacked by nucleophiles.

$$Me\overset{\displaystyle Me}{\underset{\displaystyle Me}{\vert}}\!\!-\!O\text{-}Mg\text{-}I \xrightarrow{\ \ HCl,\ H_2O\ \ } Me\overset{\displaystyle Me}{\underset{\displaystyle Me}{\vert}}\!\!-\!OH\ +\ MgCl_2$$

$$\textbf{92} \qquad\qquad\qquad\qquad \textbf{93}$$

The catalytic reaction that can be carried out with a small amount of expensive metal complex is the most useful feature of synthetic reactions involving transition metal complexes. In catalytic reactions, the active catalytic species must be regenerated in the last step of the reaction. Reductive elimination and elimination of β-hydrogen are two key reactions that can regenerate the catalytic species, making the whole reaction catalytic. Not all transition metal complexes undergo the catalytic reactions.

As shown in Scheme 2.1, the catalytic cycle of the metal(0) catalyst **94** is understood by combination of the aforementioned unit reactions. The reductive elimination of **97** regenerates M(0) **94**, which undergoes oxidative addition to afford **96** and starts the new catalytic cycle, then subsequent insertion gives **98** or transmetallation affords **97**. The catalytic species M(0) **94** can be reproduced from X−M−H **95**, which is a β-elimination product of **98**. The metal hydride **95** itself can also serve as a catalytic species through the insertion of alkenes. The ability of transition metals to undergo facile shuttling between two or more oxidation states contributes to making these reactions catalytic.

As a typical example of the catalytic cycle, the Pd-catalysed reaction of benzoyl chloride (**99**) with $MeSnBu_3$ to afford acetophenone (**100**) is explained by the

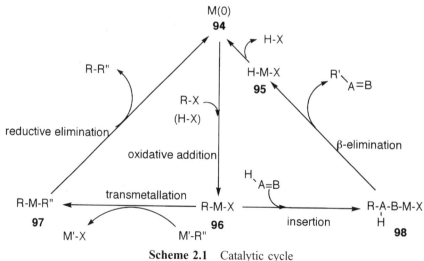

Scheme 2.1 Catalytic cycle

sequence of oxidative addition to generate the acyl complex **101**, transmetallation to give **102**, and its reductive elimination.

2.3 Effects of Ligands

As described in the preceding section, the effects of coordinated ligands is remarkable. Another example of the effect of coordination is added here. The oxidative addition of covalent molecules to a metal complex, as described before, affords a σ-complex in which the metal is connected to hydrogen, carbon, silicon and a halogen by forming covalent bonds. The chemical properties of these σ-bonds is different, depending on the metal species and the ligands. For example, the acidity of hydrogens bonded to the metals is changed considerably by ligands. The acidity decreases when CO, acting as a strongly electron-withdrawing π-accepter ligand, is displaced by electron-donating phosphines. For example, $H-Co(CO)_4$ is a strong acid, similar to a mineral acid, whereas $H-Co(CO)_3Ph_3P$ has an acidity similar to acetic acid. $H-Co(Ph_3P)_4$ does not show acidity and is considered as a source of a hydride. Also, $H-Rh(CO)_4$ is more acidic than $H-Co(CO)_4$.

Changes of pK_a for benzoic acid and phenol by the π-coordination of $Cr(CO)_3$ are shown in eq. 2.4. The $Cr(CO)_3$ group, due to the electron-withdrawing effect of CO, attracts electrons from aromatic rings as strongly as a nitro group, and its coordination to benzoic acid and phenol increases their acidities. These examples show the strong electronic effects of ligands.

pK_a 5.68   4.77   4.48   11.02   7.09

When chiral ligands are used for complex formation, there is a possibility of asymmetric synthesis. Remarkable advances have been made in asymmetric catalysis using various optically active phosphine ligands. So far, good results have been obtained mainly using bidentate phosphine ligands which have C_2 symmetry. Some chelating nitrogen compounds have also been used for asymmetric catalysis. Some chiral ligands without C_2 symmetry that have been used for asymmetric catalysis with considerable success are listed on the inside of the back cover of this book.

References

1. L. Mond, C. Langer and F. Quincke, *J. Organometal. Chem.*, **383**, 1 (1990).
2. L. F. Tietze, *Chem. Rev.*, **96**, 115 (1996).
3. K. Fisher, K. Jonas, P. Misbach, R. Stabba and G. Wilke, *Angew. Chem., Int. Ed. Engl.*, **12**, 943 (1973).
4. J. Smidt, W. Hafner, R. Jira, R. Sieber and J. Sedlmeier, *Angew. Chem., Int. Ed. Engl.*, **2**, 80 (1962).
5. J. Tsuji and H. Takahasjhi, *J. Am. Chem. Soc.*, **87**, 3275 (1965).
6. J. Tsuji, H. Takahashi and M. Morikawa, *Tetrahedron Lett.*, 4387 (1965).

Bibliography

1. P. M. Maitlis, *The Organic Chemistry of Palladium*, Vols 1 and 2, Academic Press, 1971.
2. J. Tsuji, *Organic Synthesis with Palladium Compounds*, Springer Verlag, 1980.
3. P. M. Henry, *Palladium Catalyzed Oxidation of Hydrocarbons*, Reidel, 1980.
4. S. G. Davies, *Organotransition Metal Chemistry: Applications to Organic Synthesis*, Pergamon Press, 1982.
5. R. F. Heck, *Palladium Reagents in Organic Syntheses*, Academic Press, 1985.
6. A. J. Pearson, *Metallo-organic Chemistry*, John Wiley & Sons, 1985.
7. A, Yamamoto, *Organotransition Metal Chemistry*, Academic Press, 1986.
8. J. P. Collman, L. S. Hegedus, J. R. Norton and R. G. Finke, *Principles and Applications of Organotransition Metal Chemistry*, University Science Books, 1987.
9. P. J. Harrington, *Transition Metals in Total Synthesis*, John Wiley & Sons, 1990.
10. F. J. McQuillin, *Transition Metals Organometallics for Organic Synthesis*, Cambridge University Press 1991.
11. H. M. Colquhoun, D. J. Thompson and M. V. Twigg, *Carbonylation*, Plenum Press, 1991.
12. L. S. Hegedus, *Transition Metals in the Synthesis of Complex Organic Molecules*, University Science Books, 1994.
13. A. J. Pearson, *Iron Compounds in Organic Synthesis*, Academic Press, 1994.
14. J. Tsuji, *Palladium Reagents and Catalysts; Innovations in Organic Synthesis*, John Wiley, 1995.
15. F. Diederich and P. J. Stang (Eds), *Metal-catalyzed Cross-Coupling Reactions*, Wiley-VCH, 1998.
16. J. Tsuji (Ed), *Perspectives in Organopalladium Chemistry for the XXI Century*, Elsevier, 1999.

3

REACTIONS OF ORGANIC HALIDES AND PSEUDOHALIDES

3.1 Reaction Patterns of Aryl, Alkenyl and Benzyl Halides and Pseudohalides

Various organic halides, particularly aryl and alkenyl halides, are the most important building blocks of organic synthesis involving transition metal complexes, particularly Pd and Ni catalysts. Numerous new reactions of aryl and alkenyl halides catalysed by transition metal complexes have been discovered, which are impossible to achieve by other means. Grignard reagents are prepared by treating alkyl halides with Mg metal in ether. Alkyl halides are more reactive toward Mg metal than aryl and alkenyl halides. Organic halides of sp^2 carbons are less reactive and react only with activated Mg in THF. However, transition metal complexes undergo oxidative addition more easily to halides of sp^2 carbons than those of sp^3 carbons. Based on this high reactivity, numerous useful reactions of alkenyl and aryl halides have been discovered. Whereas mainly iodides and bromides are used for Pd-catalysed reactions, chlorides can be used with Ni catalysts. Reaction of fluorides is rare [1]. Alkyl halides are less reactive toward transition metal complexes. In addition, alkylmetal compounds, even when they are formed, undergo facile elimination of β-hydrogen, and no further reaction occurs.

Innovations in the chemistry of aromatic compounds have occurred by recent development of many novel reactions of aryl halides or pseudohalides catalysed or promoted by transition metal complexes. Pd-catalysed reactions are the most important [2,29]. The first reaction step is generation of the arylpalladium halide by oxidative addition of halide to Pd(0). Formation of phenylpalladium complex **1** as an intermediate from various benzene derivatives is summarized in Scheme 3.1.

In addition to halides, some pseudohalides undergo facile oxidative addition to Pd and Ni complexes. Trifluoromethanesulfonates (triflates), namely aryl triflates **3** derived from phenols and enol triflates of carbonyl compounds, are most useful.

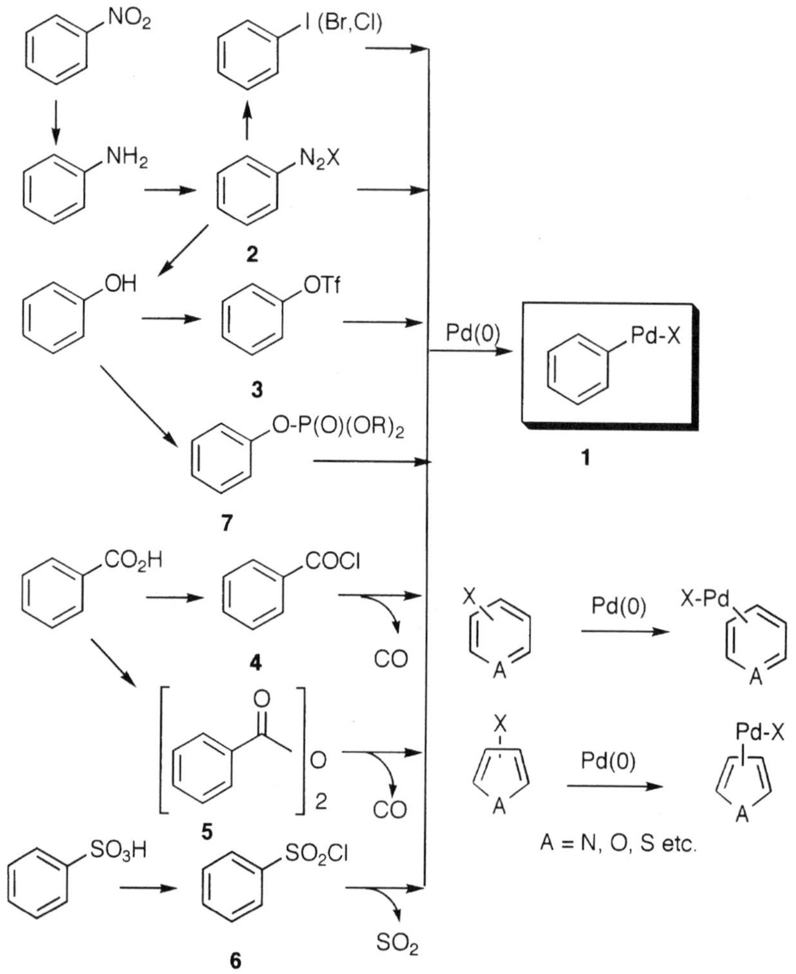

Scheme 3.1   Formation of arylpalladium intermediate by oxidative addition

Mesylates are used for Ni-catalysed reactions. Arenediazodium salts **2** are very reactive pseudohalides undergoing facile oxidative addition to Pd(0). They are more easily available than aryl iodides or triflates. Also, acyl (aroyl) halides **4** and aroyl anhydrides **5** behave as pseudohalides after decarbonylation under certain conditions. Sulfonyl chlorides **6** react with evolution of SO_2. Even aryl phosphates **7** can be used. Allylic halides are reactive, but their reactions via π-allyl complexes are treated in Chapter 4. Based on the reactions of those pseudohalides, several benzene derivatives such as aniline, phenol, benzoic acid and benzenesulfonic acid can be used for the reaction, in addition to phenyl halides. In Scheme 3.1, reactions of benzene as a parent ring compound are summarized. Needless to say, the reactions can be extended to various aromatic compounds including heteroaromatic compounds whenever their halides and pseudohalides are available.

The phenylpalladium intermediate **1** undergoes further transformations. Either insertion of unsaturated bonds or transmetallation with organometallic compounds of main group metals takes place. Further transformations of the intermediates **8–10**, formed by the insertion of conjugate diene, allene and internal alkyne, occur. Finally the reactions terminate by reductive elimination, elimination of β-hydrogen, or trapping with various nucleophiles. In Scheme 3.2, the transformations of the phenylmetal intermediate **1** via insertion are summarized. Transmetallation and nucleophilic substitution by carbon, oxygen, and nitrogen nucleophiles are shown in Scheme 3.3. Most of these transformations are not possible without the presence of catalysts, and these schemes clearly show the innovative developments in aromatic substitution reactions of organic halides catalysed by Pd catalysts.

Alkenyl halides and their pseudohalides also react with Pd(0) to form the alkenylpalladium intermediates **11**, and their transformations are summarized in Scheme 3.4. In addition to alkenyl halides, the enol triflates **12** undergo oxidative addition, showing that carbonyl compounds are useful starting compounds for Pd-catalysed reactions.

Alkynyl halides **13** undergo the insertion and transmetallation after oxidative addition, as summarized in Scheme 3.5.

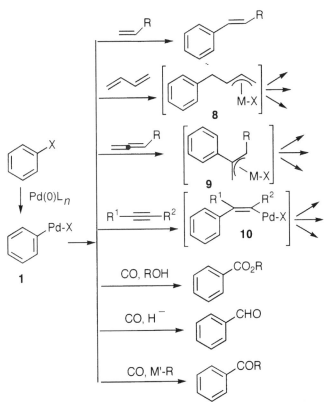

Scheme 3.2 Insertion to phenylpalladium intermediate **1**

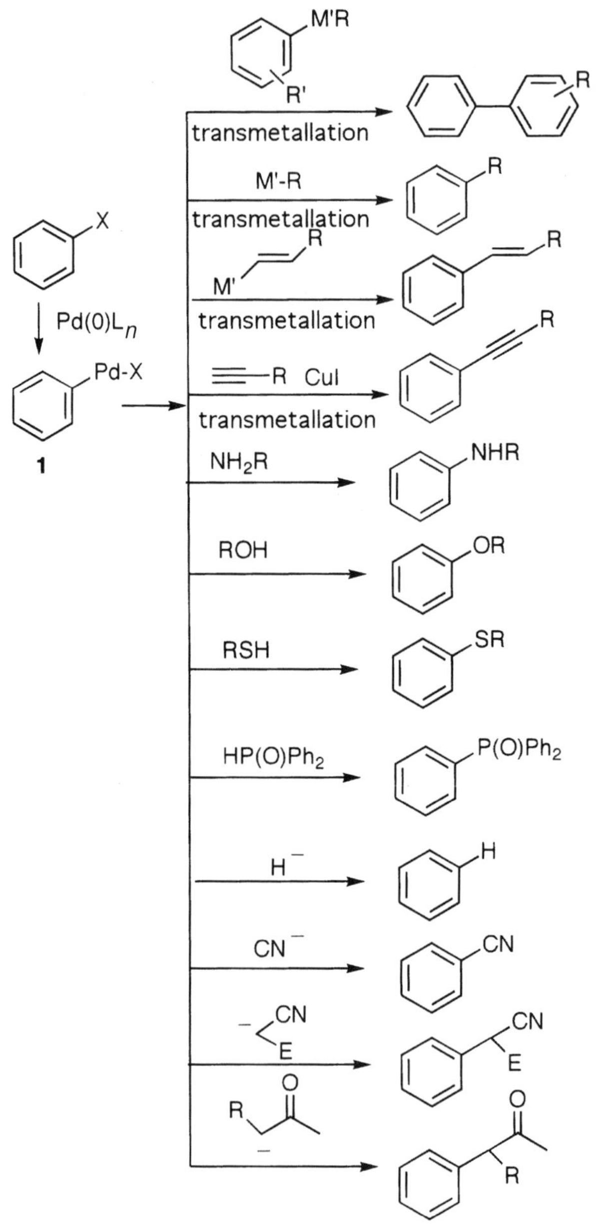

M'R

transmetallation

M'-R

transmetallation

M'

transmetallation

≡—R CuI

transmetallation

NH₂R

ROH

RSH

HP(O)Ph₂

H⁻

CN⁻

—⟨CN / E

R‿C(O)CH₃

M' = Li, Mg, Zn, B, Al, Sn, Si, Hg

Scheme 3.3 Transformation of phenylpalladium intermediate **1**

Ni(0) complexes react with halides and pseudohalides. Their reactions are somewhat different from those of Pd(0). Chlorides add to Ni(0) much more easily than to Pd(0). Even C—O bonds such as aryl alkyl ether bonds are cleaved with Ni(0) under certain conditions. Not only triflates, but also mesylates react with Ni(0). Oxidative addition to Ni(0) and subsequent transformations are summarized in Scheme 3.6.

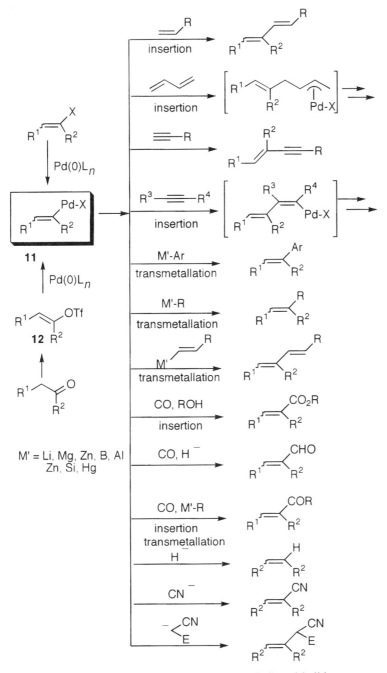

Scheme 3.4 Pd-catalysed reactions of alkenyl halides

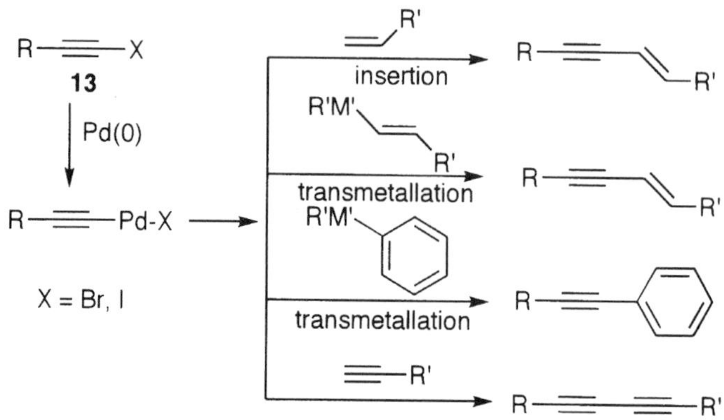

Scheme 3.5   Pd-catalysed reaction of alkynyl halides

Attention should be paid to the reactivity of **1** and **11**, which react with nucleophiles. It is well known that Grignard reagents react with electrophiles forming Mg(II), whereas Pd complexes generate Pd(0) after reacting with nucleophiles. Grignard reactions cannot be carried out catalytically, because it is difficult to reduce Mg(II) to Mg(0) *in situ*. However, formation of Pd(0) by the reaction of nucleophiles shows the possibility of catalytic reactions. This is the most important characteristic of Pd chemistry. Ni compounds react with both electrophiles (stoichiometric) and nucleophiles (catalytic), depending on the substrates.

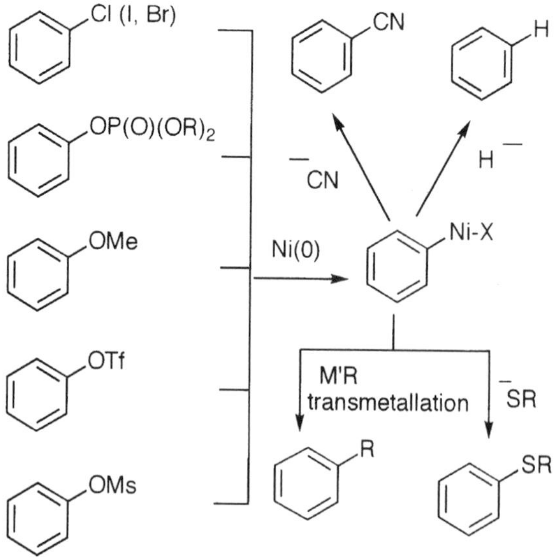

Scheme 3.6   Ni-catalysed reactions

3.2 Cross-coupling with Alkenes (Carbometallation of Alkenes)

3.2.1 Intermolecular Reactions

All reactions described in this section are explained by: (i) the oxidative addition of a halide to generate the arylpalladium halide **14**; (ii) insertion of an alkene to form **15**, which is regarded as carbopalladation of alkenes; and (iii) formation of the new alkene **16** by elimination of β-hydrogen (dehydropalladation). The reaction was reported independently by Mizoroki [3] and by Heck [4], and is called the Mizoroki–Heck or Heck reaction [5].

Aryl and alkenyl iodides and bromides undergo facile oxidative addition to Pd(0) or Ni(0) complexes. Reactivity of halides is in the following order; I > Br >>> Cl. The reaction is carried out in the presence of bases such as tertiary amines and sodium acetate. $Pd_2(dba)_3$, $Pd(Ph_3P)_4$, or even Pd on carbon can be used as Pd(0) catalysts. $Pd(OAc)_2$ is reduced easily to Pd(0) *in situ* with various reducing agents such as phosphine, CO, ROH, and alkenes, and used as a convenient source of Pd(0) catalyst. The Pd(0)-catalyzed reactions of aryl iodides can be carried out even in the absence of a phosphine ligand. The reaction of bromides generally requires phosphine ligands. Phosphonium salt is formed in some extent by the reaction of Ph_3P with iodides and bromides. In this case, use of tri-*o*-tolyphosphine (**19**) as a hindered ligand is recommended [6]. *p*-Iodobromobenzene reacts with methyl acrylate selectively in the absence of ligand to give methyl *p*-bromocinnamate (**17**), which then reacts with styrene by addition of (*o*-Tol)$_3$P to give **18** [7]. Sometimes, higher yields are obtained in the Heck reaction by the addition of tetrabutylammonium chloride [8].

It is worth noting at this point that, for stable DBA (dibenzylideneacetone) Pd(0) complexes, three types, namely $Pd(dba)_2$, $Pd_2(dba)_3$, and $Pd_2(dba)_3$–$CHCl_3$ are known. They are essentially the same complex. DBA coordinates to Pd(0) as a monodentate ligand, but not a bidentate ligand, and $Pd_2(dba)_4 = Pd(dba)_2$ is formed at first in its preparation. However, $Pd_2(dba)_4$ should be written more correctly as $Pd_2(dba)_3(dba)$. This means that three molecules of DBA coordinate to two atoms of Pd, and one molecule of DBA does not coordinate directly to Pd. When $Pd_2(dba)_4$ is recrystallized from $CHCl_3$, one DBA molecule is replaced with $CHCl_3$ to give deep violet crystals of $Pd_2(dba)_3–CHCl_3$, which is commercially available [8a].

The long life of the Pd catalyst (high turnover number, TON) is crucial to useful catalytic reactions. The Pd complex **20**, formed by the orthopalladation of $(o\text{-Tol})_3P(\mathbf{19})$, is proposed to be a very reactive catalyst, and a turnover number of 200 000 has been achieved in the Heck reation of bromoarenes without precipitation of metallic Pd [9]. The complex **20** is active even for chloroarenes. However, it was found that Pd salts such as $PdCl_2$ or $Pd(OAc)_2$ in the presence of Ph_4PCl, but not Ph_3P, show much higher catalytic activity than the complex **20** in Heck reactions, which are carried out in DMF or NMP using AcONa as a suitable base in the presence of a small amount of N,N-dimethylglycine (DMG) [10]. In the coupling of bromobenzene with acrylate to give cinnamate **21** using 0.01 mol % of the catalyst, a TON of 9800 was achieved. Also, an excess of trialkyl phosphites was found to show efficient catalysis for the Heck reaction [11].

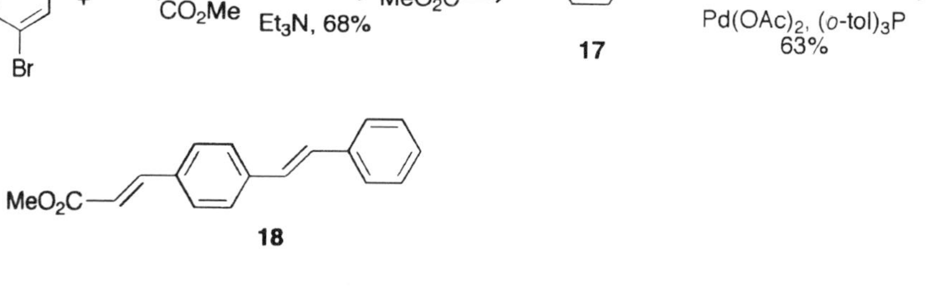

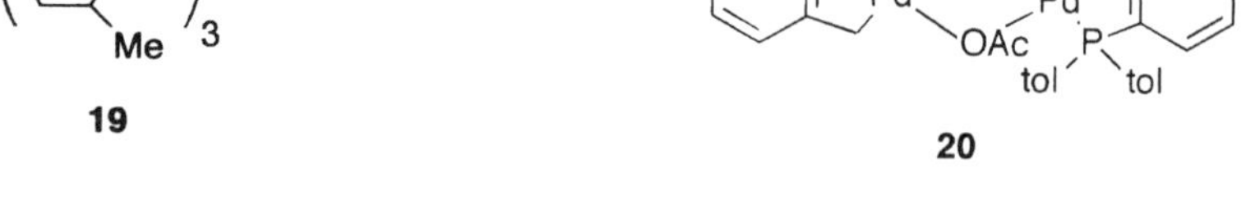

A dramatic increase of the TON is observed in the Heck reaction under high pressure. A value of 770 000 was achieved in the coupling reaction of iodobenzene with 2,3-dihydropyrrole under 8 kbar at 100 °C. High pressure seems to stabilize the catalyst [12].

Chlorides are inert to Pd(0) under normal reaction conditions [13]. Chloroarenes react by the use of more electron-donating bidentate ligands, such as dippp, under somewhat severe conditions [14]. Chlorides react easily with Ni(0) complexes. Also, *t*-Bu₃P is a good ligand for the Heck reaction of chloroarenes [14a].

Alkenes with electron-withdrawing groups react most satisfactorily. Typically methyl cinnamate is prepared in good yield by the reaction of acrylate with iodobenzene [3,4]. Even Pd on carbon is used with or without phosphine ligand in some cases. Water seems to be a good solvent. The substrates such as **22** which are soluble in a basic solution react smoothly in water, even in the absence of a ligand [15]. It is also observed that the Heck reaction is accelerated in an aqueous solution using water-soluble TMSPP (**XLVII**) as a ligand (see the inside of the back cover) [16].

Several pseudohalides are used for the reaction. Aryl triflates prepared from phenols, and enol triflates prepared from carbonyl compounds are widely used [16a]. Coupling of the dienol triflate **23**, prepared from the enone, with acrylate affords the conjugated trienyl ester **24** [17]. Aryldiazonium salts **27** are good sources of arylpalladiums **28**. They are very reactive and their reaction with alkenes proceeds without phosphine ligand [18]. The diazonium salts **27** are usually prepared from nitro compounds **25** via anilines **26**. Therefore, this is the indirect substitution of a nitro group with an alkene. It should also be mentioned that some aryl iodides, bromides and phenols, as a source of triflates, are prepared via the diazonium salts **27**, and hence the direct Pd-catalysed reactions of the diazonium salts are more convenient in such cases. The styrene derivative **30** is prepared from *p*-chloroaniline (**29**) after the diazotization with butyl nitrite in acetic acid under an ethylene atmosphere [19].

Acyl halides undergo the oxidative addition to Pd(0) with or without decarbonylation. Aroyl chlorides undergo Pd-catalysed decarbonylation at high temperature [20]. In the presence of a base, the decarbonylation of aroyl chlorides proceeds under milder conditions, and arylpalladium chlorides are formed as intermediates which undergo the alkene insertion. Based on the fact that aroyl chlorides undergo oxidative addition to Pd(0) more easily than aryl halides, the reaction of *p*-bromobenzoyl chloride (**31**) with methyl acrylate, catalysed by Pd(OAc)₂ in the absence of phosphine ligand, gives methyl *p*-bromocinnamate (**33**) chemoselectively after the decarbonylation via **32**.

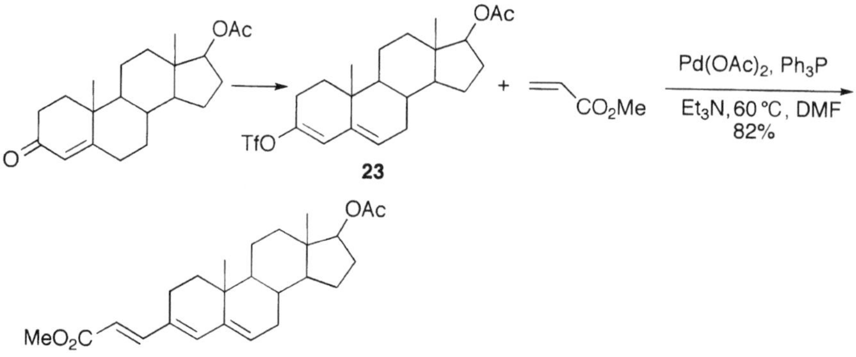

23

24

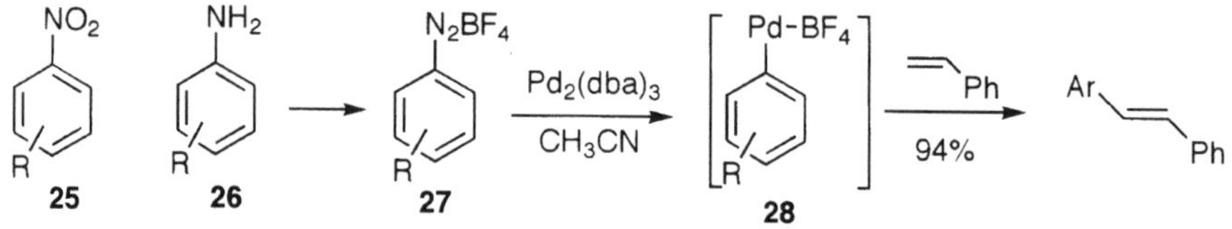

25   **26**   **27**   **28**

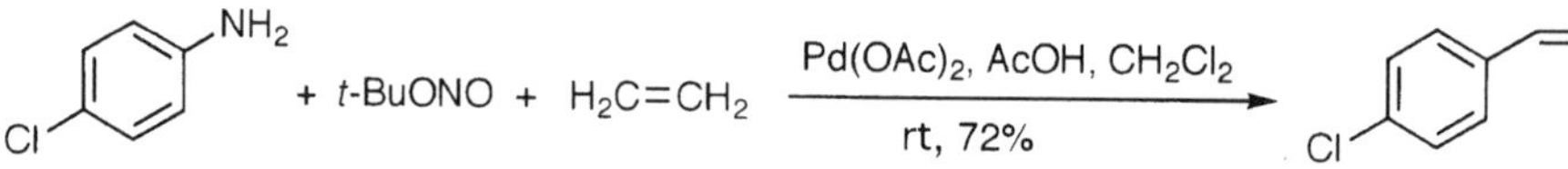

29   **30**

Further reaction of the bromide **33** with acrylonitrile in the presence of Ph$_3$P affords the disubstituted product **34** [21]. Selection of amines used in the reaction is critical. Fumaroyl dichloride (**35**) undergoes oxidative addition, decarbonylation and insertion of acrylate to produce octatrienedioate (**36**) [22].

Aromatic carboxylic anhydrides can be used for the Heck reaction without adding a base. The reaction of benzoic anhydride (**37**) with acrylate proceeds in the presence of 1.0 mol % of NaBr without phosphine ligand in NMP at 160 °C [23]. The reaction of anhydrides in the absence of bases, and hence without forming halide salts after the reaction, is attractive from a practical standpoint.

Asymmetric Heck reactions have been carried out with considerable success. The arylation of 2,3-dihydrofuran (**38**) with phenyl triflate using BINAP (**XXXI**) as a chiral ligand gave 2-phenyl-2,3-dihydrofuran (**42**) with 96% ee. Addition of H−Pd−X to the primary product **40** gives the intermediate **41**, and β-elimination affords the dihydrofuran **42** with 96% ee in 71% yield as the major product in the presence of 1,8-bis(dimethylamino)naphthalene (**39**) as a base. Another dihydrofuran **40** with 67% ee was obtained in 7%, showing that one enantiomer of **40** is converted to **42** with high selectivity [24].

In the reaction of allylic alcohols, elimination of β-hydrogen from an OH-bearing carbon takes place to give saturated carbonyl compounds, rather than arylated allylic alcohols [25,26]. The reaction of methallyl alcohol (**43**) with bromobenzene affords αmethyldihydrocinnamaldehyde (**45**) via **44**. The reaction of the prochiral *meso* form of 2-cyclopenten-1,4-diol (**46**) with the (*Z*)-alkenyl iodide **47** affords the cyclopentanone **49** with complete diastereoselectivity, which is a useful prostaglandin intermediate [27]. The fact that the reaction of the corresponding (*E*)-alkenyl iodide is not selective shows that the formation of the five-membered intermediate complex **48** is crucial for the selectivity [28].

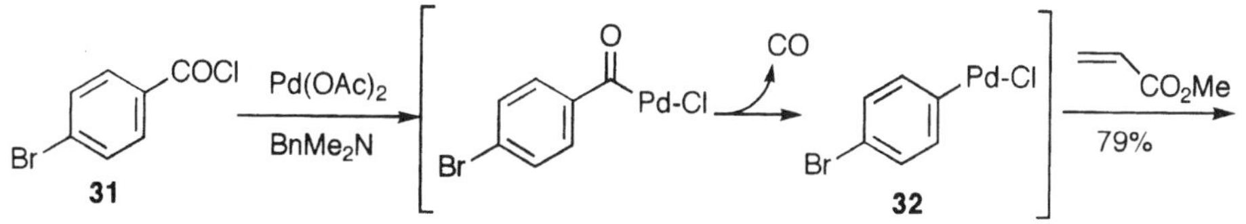

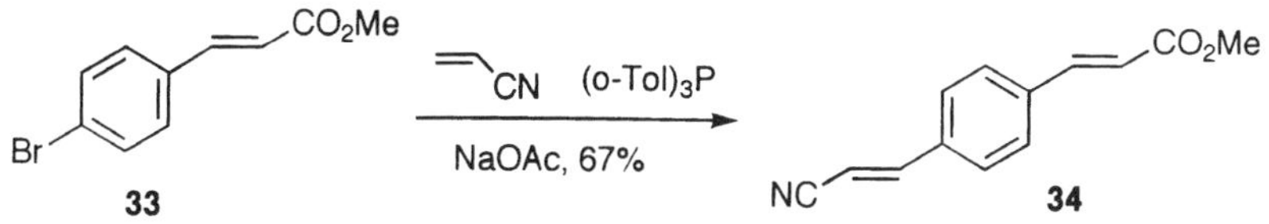

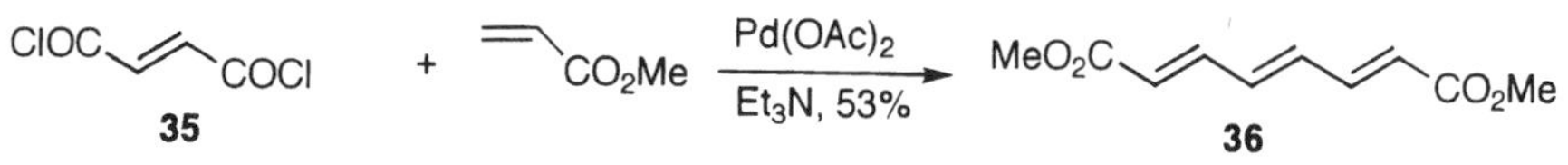

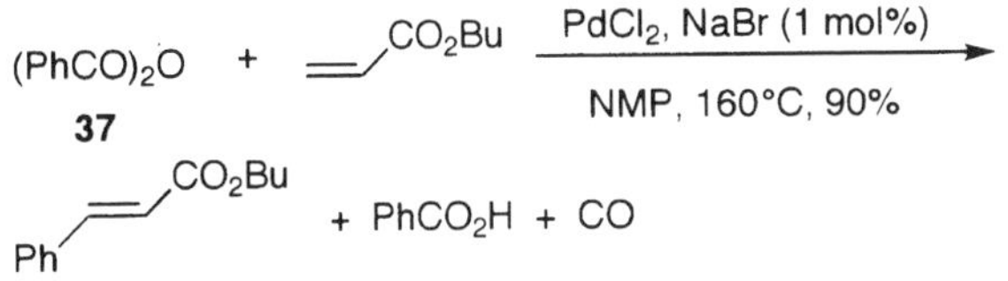

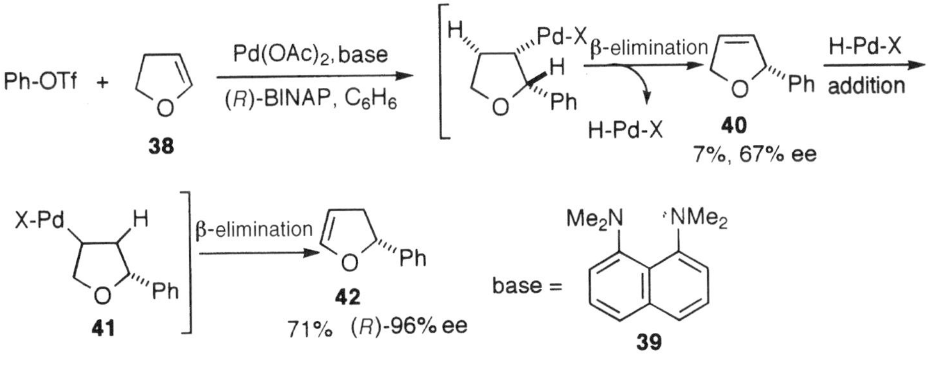

In addition to allylic alcohols, other unsaturated alcohols react with halides to give carbonyl compounds. Although the reaction was slow (three days), the reaction of 10-undecen-1-ol (**50**) with iodobenzene afforded the aldehyde **52** in a high yield. In this reaction, reversible elimination of H–Pd–I and its readdition (reinsertion) in reverse regiochemistry are repeated several times until irreversible elimination of hydrogen from the oxygen-bearing carbon in **51** occurs to afford **52** as a main product and **53** as a minor product [29].

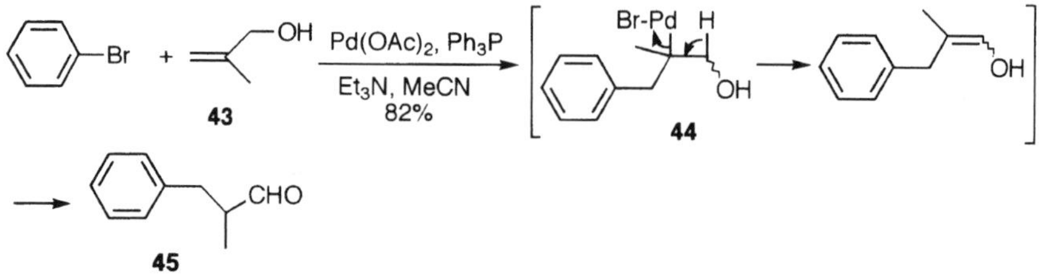

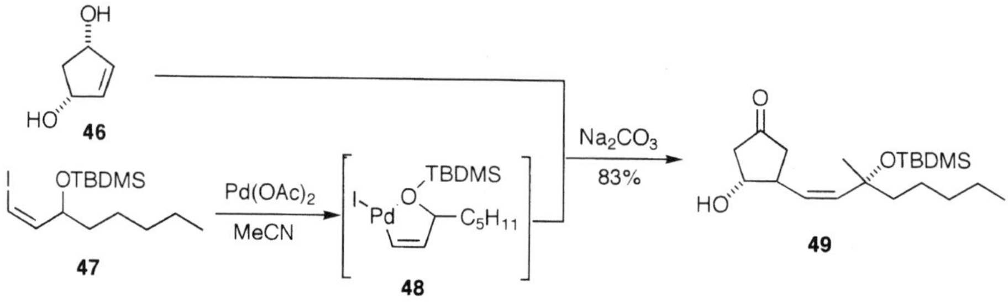

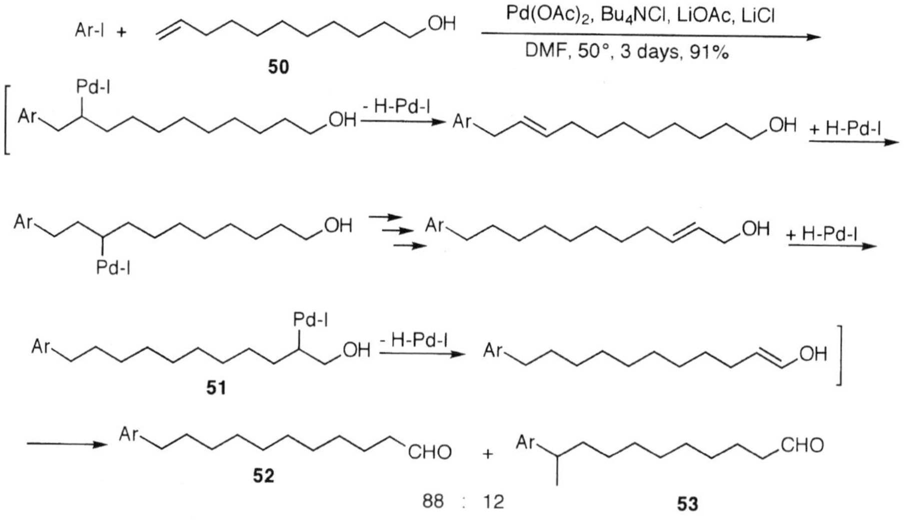

Reactions of halides with 1,2-, 1,3- and 1,4-dienes generate π-allylpalladium intermediates, which react further with nucleophiles. The reaction of 1,3-dienes with aryl and alkenyl halides is explained by the following mechanism. The insertion of 1,3-diene to the aryl or alkenylpalladium bond generates π-allylpalladium complexes **54**, which react further in several ways. As expected, nucleophiles such as carbon nucleophiles, amines, and alcohols attack the π-allylpalladium **54** to form the 1,4-

addition product **55**. Transmetallation with main group metal reagents affords **56**. In the absence of nucleophiles, elimination of β-hydrogen takes place to give the substituted 1,3-dienes **57**, which react again with aryl halides to form the π-allylpalladium **58** in some cases. Subsequent β-elimination affords the 1,4-diaryl-1,3-dienes **59**.

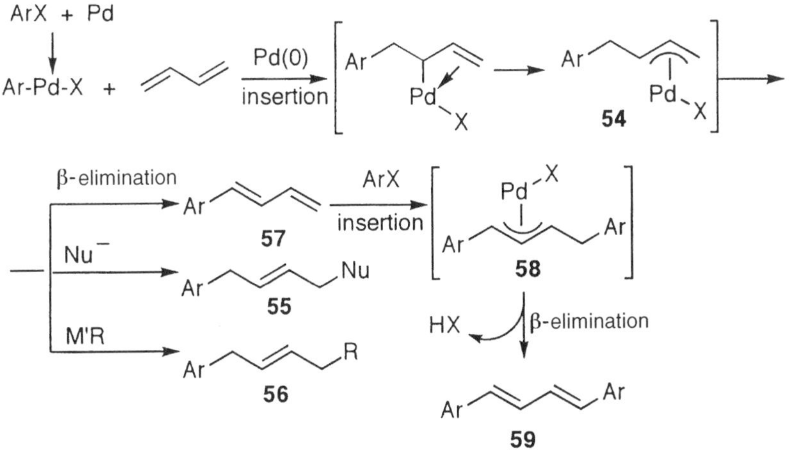

When 1,2-diene (allene) derivative **60** is treated with aryl halide in the presence of Pd(0), the aryl group is introduced at the central carbon by the insertion of one of the allenic double bonds to form the π-allylpalladium intermediate **61**, which is attacked by an amine to give the allylic amine **62**. A good ligand for the reaction is dppe [30].

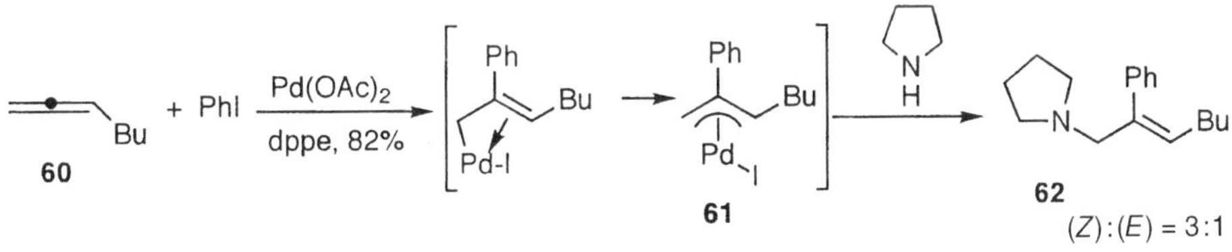

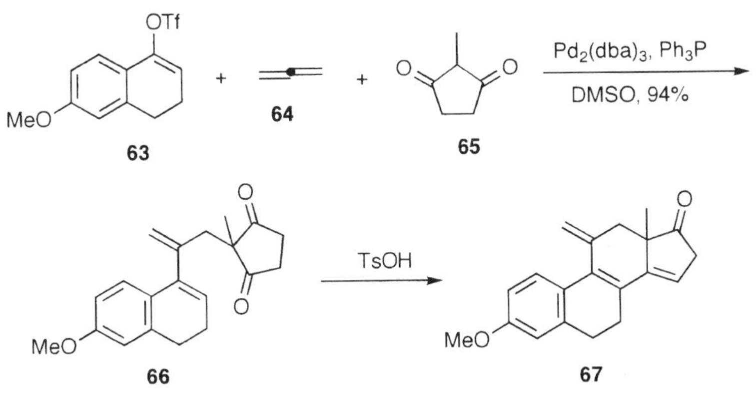

The Pd-catalyzed three-component reaction of the enol triflate **63**, allene (**64**,) and 2-methyl-1,3-cyclopentanedione (**65**) gives **66** in high yield, which is converted to the steroid skeleton **67** [31].

A series Pd-catalyzed annulation reactions have been developed by the reaction of *o*-heterosubstituted iodoarenes **68** and **69** with 1,2-, 1,3- and 1,4-dienes. In the reaction of *o*-iodoaniline with 1,3-cyclohexadiene (**70**), the π-allyl intermediate **71** reacts with the aniline moiety to give **72** [32, 33]. Carboannulation of the malonate derivative **73** with the 1,3-diene **74** gives **75** in high yield [33a].

1,2-Dienes (allenes) are also used for heteroannulation with **68** and **69**. The eight-membered nitrogen heterocycle **78** is constructed by the reaction of 1,2-undecadiene (**77**) with *o*-(3-aminopropyl)iodobenzene (**76**) [34]. The lactones are prepared by trapping the π-allyl intermediates with carboxylic acids as an oxygen nucleophile. The unsaturted lactone **81** is prepared by the reaction of β-bromo-α,β-unsaturated carboxylic acid **79** with the allene **80** [35]. In the carboannulation of **82** with 1,4-cyclohexadiene (**83**), the 1,3-diene **85** is generated by β-elimination of **84**, and the addition of H-PdX forms the π-allylpalladium **86**, which attacks the malonate to give **87** [36].

3.2.2   Intramolecular Reactions

Whereas the intermolecular Heck reaction is limited to unhindered alkenes, the intramolecular version permits the participation of even hindered substituted alkenes, and many cyclic compounds can be prepared by the intramolecular Heck reaction [37]. The stereospecific synthesis of an A ring synthon of 1α-hydroxyvitamin D has been carried out. Cyclization of the (*E*)-alkene **88** gives the (*E*)-*exo*-diene **90**, and the (*Z*)-alkene **91** affords the (*Z*)-*exo*-diene **92** [38]. These reactions are stereospecific, and can be understood by *cis* carbopalladation to form **89** and the *syn*-elimination mechanism.

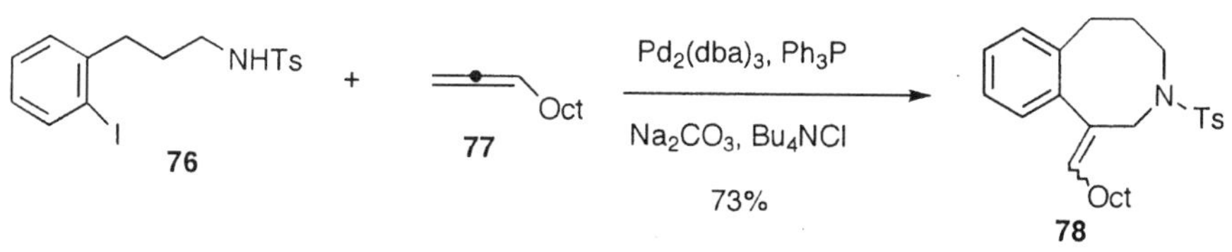
NHTs
I
76
+
Oct
77
Pd2(dba)3, Ph3P
Na2CO3, Bu4NCl
73%
N-Ts
Oct
78

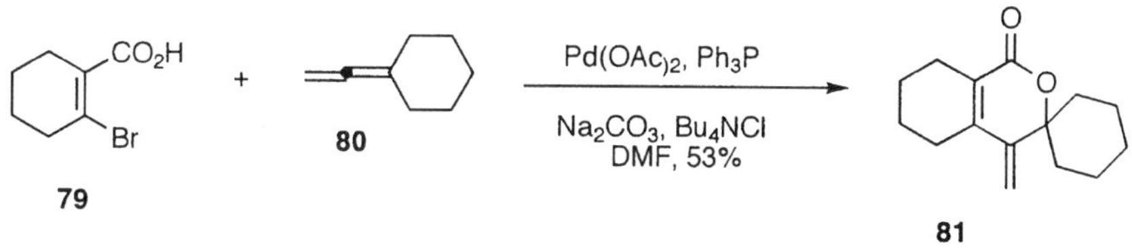
CO2H
Br
79
+
80
Pd(OAc)2, Ph3P
Na2CO3, Bu4NCl
DMF, 53%
O
O
81

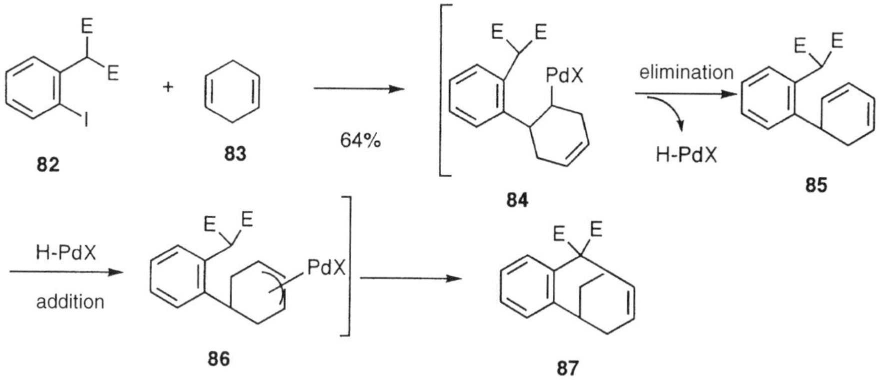
E
E
I
82
+
83
64%
E E
PdX
84
elimination
H-PdX
E E
85
H-PdX
addition
E E
PdX
86
E E
87

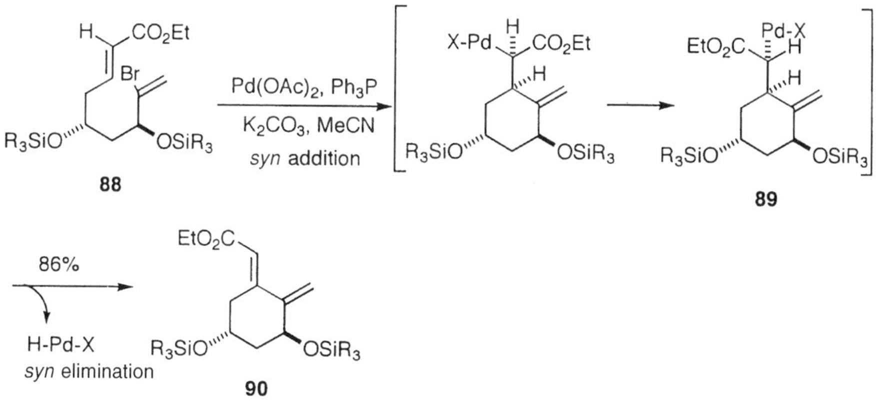
H
CO2Et
Br
R3SiO
OSiR3
88
Pd(OAc)2, Ph3P
K2CO3, MeCN
syn addition
X-Pd
H
CO2Et
H
R3SiO
OSiR3
EtO2C
Pd-X
H
H
R3SiO
OSiR3
89
86%
H-Pd-X
syn elimination
EtO2C
R3SiO
OSiR3
90

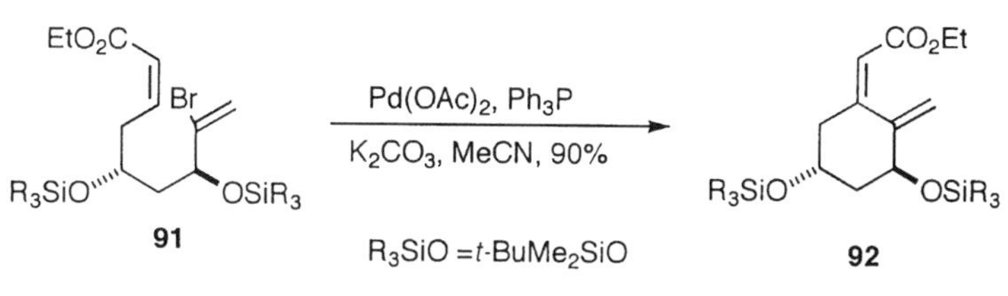
EtO2C
Br
R3SiO
OSiR3
91
Pd(OAc)2, Ph3P
K2CO3, MeCN, 90%
R3SiO =t-BuMe2SiO
CO2Et
R3SiO
OSiR3
92

Asymmetric cyclization using chiral ligands offers powerful synthetic methods for the preparation of optically active compounds [39]. After early attempts [40,41], satisfactory optical yields have been obtained in a number of cases. Synthesis of the optically active *cis*-decalin system [42] was carried out with high enantioselectivity based on the differentiation of enantiotopic C=C double bonds [43]. The cyclization of the triflate **93** gave the *cis*-decalin **94** with 95% ee in 78% yield using (*R*)-BINAP. A mixture of 1,2-dichloroethane and *t*-BuOH is the best solvent, and the asymmetric synthesis of vernolepin (**96**) via Danishefsky's key intermediate **95** has been achieved [44].

The highly efficient asymmetric cyclization of **97** using (*R*)-BINAP as a chiral ligand based on the differentiation of enantiotopic faces gave the tetralin system **98** with 93% ee and has been applied to the synthesis of (−)-eptazocine (**99**) [45].

Construction of the B ring of the congested taxol molecule **101** has been carried out by the intramolecular Heck reaction of the enol triflate **100** [46].

An elegant and efficient stereocontrolled total synthesis of strychnine has been achieved by applying intramolecular Diels–Alder and Heck reations as key reactions [47]. An unusual *exo*-Diels–Alder reaction of **102** afforded **103**, which was converted to the vinyl iodide **104**. The Heck reaction of **104** using Pd(OAc)₂ gave the hexacyclic strychnan system **105** smoothly in 74% yield. Hydrolysis of **105** afforded isostrychnine, which was isomerized to strychnine (**106**) under basic conditions.

The Ni-catalysed Heck reaction is rather rare. Although the attempted Ni(0)-catalysed cyclization of iodide **107** gave a mixture of many products, the pentacyclic nitrone **108** was isolated in 40% yield by the domino Heck-type reaction, reductive

100 **101**

102 **103** **104** **105** **106**

cyclization of the α-(*o*-nitrophenyl) ketone moiety, and isomerization of alkene as one-pot reactions using Ni(cod)$_2$ (6.6 equivalents). The iodide **107** was converted to dehydrotubifoline (**109**) by the treatment with Ni(cod)$_2$ in the presence of LiCN [48].

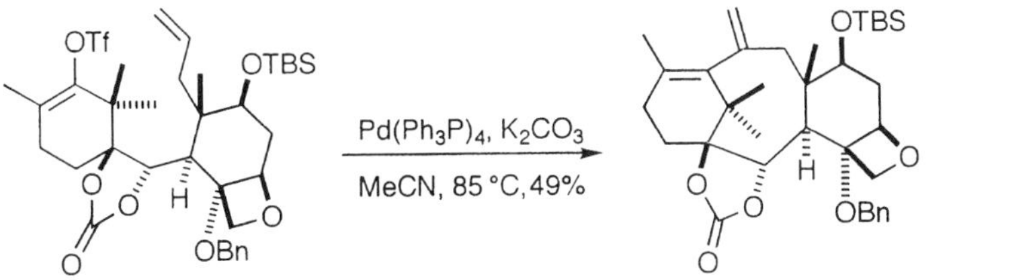

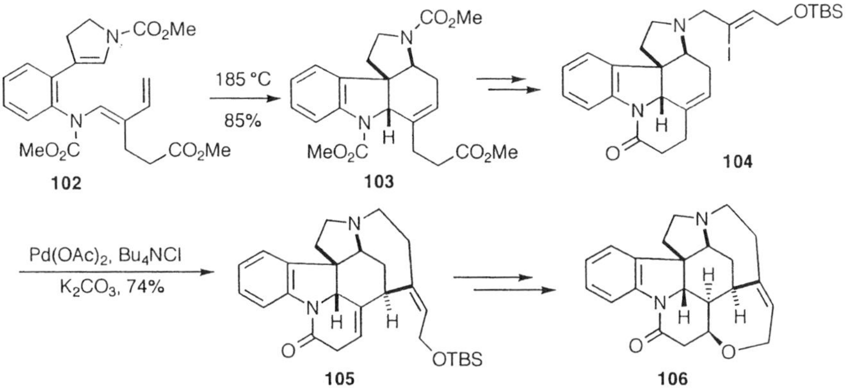

It was reported that the stable Ni(0) complexes Ni[(PhO)$_3$P]$_4$ and Ni[(EtO)$_3$P]$_4$ are active catalysts for the Heck reactions of **110** and **111** [49]

107 **108** **109**

110

111

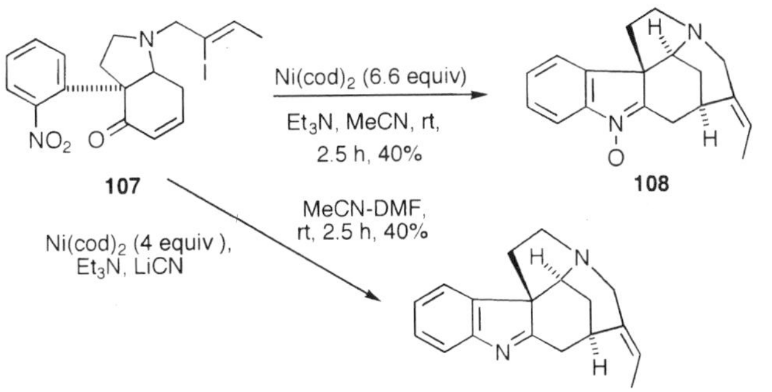

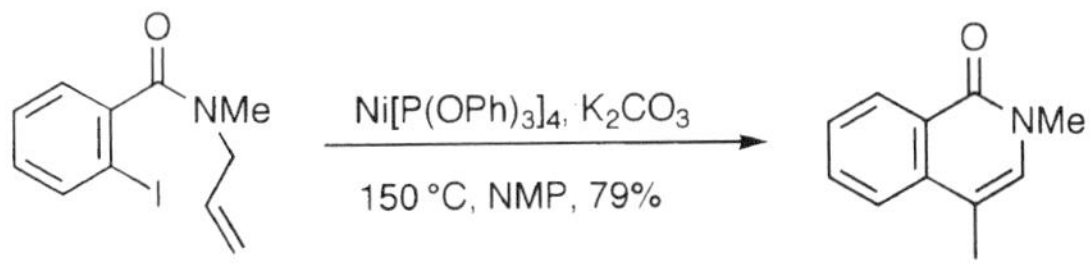

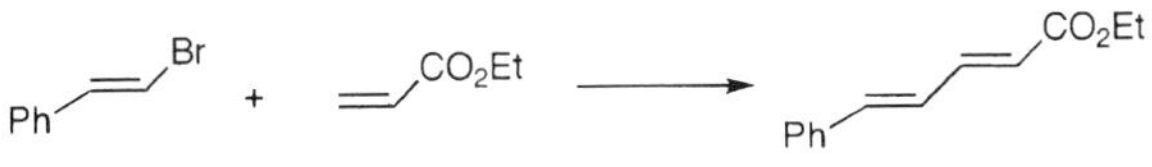

The Heck reaction of 1,3-diene systems via π-allylpalladium is also useful. This cyclization, which forms very congested quaternary carbon centers involving the intramolecular insertion of di-, tri- and tetrasubstituted alkenes, is particularly useful for natural products synthesis. In the synthesis of morphine, bis-cyclization of the octahydroisoquninoline precursor **112** by the intramolecular Heck reaction proceeded using palladium trifluoroacetate and 1,2,2,6,6-pentamethylpiperidine (PMP). The insertion of the diene system forms the π-allylpalladium intermediate **113**, which attacks the phenol intramolecularly to form the benzofuran ring **114**. Based on this method, the elegant total syntheses of (−)- and (+)-dihydrocodeinone, and (−)- and (+)-morphine (**115**) have been achieved [50].

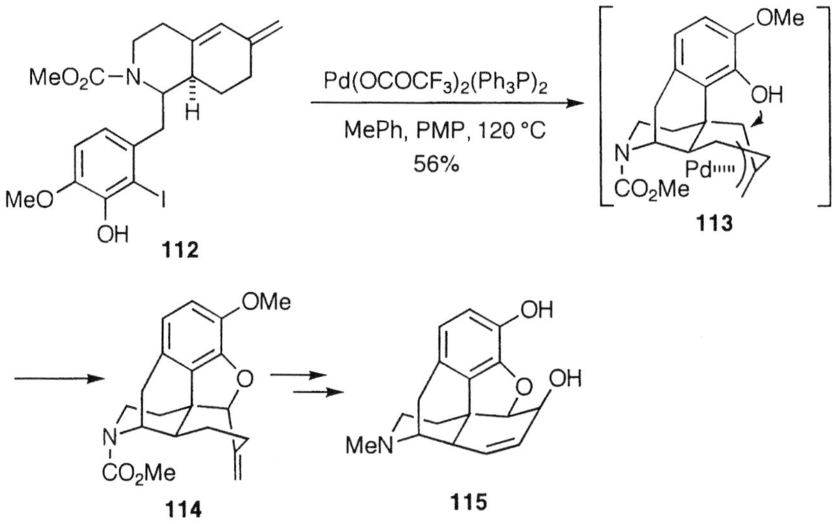

The differentiation of enantiotopic C=C double bonds in the intramolecular Heck reaction of 1,3-diene **116** using (*S*)-BINAP, and subsequent regioselective carbanion capture of the π-allylpalladium intermediate **117** gave **118** with 87% ee with complete diastereo- and regioselectivity. $\Delta^{9,(12)}$-Capnellene (**119**) was synthesized from **118**

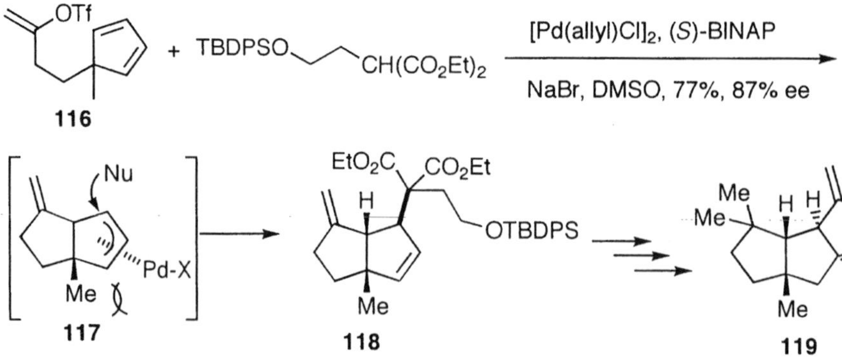

[51]. The choice of solvents is crucial in the asymmetric cyclizations. In this case, DMSO gives the best results.

The consecutive domino insertion of double bonds of halo polyenes lacking β-hydrogens bonded to sp^3 carbon produces polycyclic compounds. This offers a powerful synthetic method for preparing naturally occurring macrocyclic and polycyclic compounds. Novel total syntheses of many naturally occurring complex molecules have been achieved by synthetic designs based on this methodology. An interesting and useful application is the intramolecular polycyclization of polyalkenes by domino insertions of alkenes to give polycyclic compounds. In the domino cyclization of **120**, quarternary carbons and the neopentyl-type palladium complex **121** are formed by insertion of the 1,1-disubstituted alkene. The β-elimination reaction from **121** is not possible. That is, **121** is a living species and undergoes further insertion to give **122**. The key step in the total synthesis of scopadulcic acid B (**124**) is the Pd-catalysed construction of the tricyclic system **123** containing the bicyclo[3.2.1]octane substructure. The tricyclic product **123** was obtained in 83% yield from **120** and converted to **124** [52].

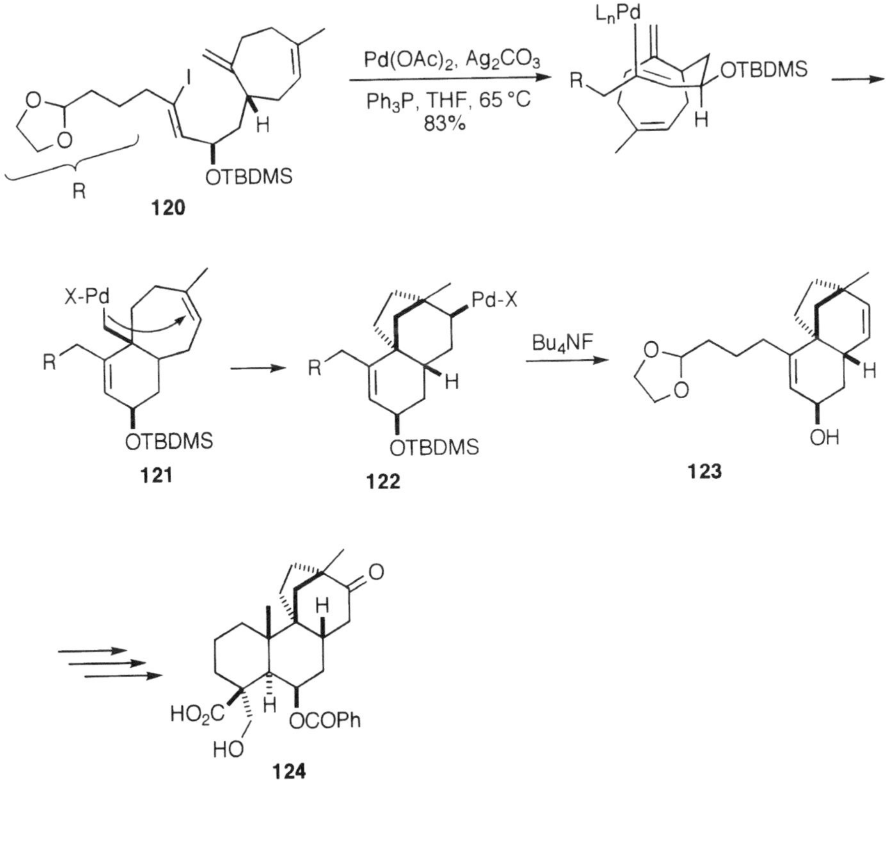

3.3 Reactions with Alkynes

3.3.1 *Cross-coupling with Terminal Alkynes to Form Alkenyl- and Arylalkynes*

Coupling of copper acetylides with halides is known as the Castro reaction [53]. Pd(0) is active for the coupling of terminal alkynes **126** with the aryl or alkenyl halides **125** or **128** to give arylalkynes **127** or conjugated enynes **129** [54]. The Pd(0)-catalysed coupling, which is called Sonogashira reaction, proceeds most efficiently by the addition of CuI as a cocatalyst [55,56]. CuI activates the alkynes **126** by forming the copper acetylide **130**, which undergoes transmetallation with arylpalladium halide **131** to form the alkynyl-arylpalladium species **132**. Its reductive elimination gives the coupling product **133** as the final step, with regeneration of Pd(0) and CuI. The Pd(0)/CuI-catalysed reaction proceeds in the presence of amines. However, the coupling proceeds smoothly without CuI when water-soluble TMSPP (**XLVII**) is used as a ligand in an aqueous solution [57]. Also, coupling without CuI gives arylalkynes **127** or enynes **129** in high yields when the reaction is carried out in piperidine or pyrrolidine. It was claimed that the use of these amines is crucial, and poor results are obtained without CuI when Et_3N, Pr_2NH, Et_2NH on morpholine are used [58]. The coupling without CuI is also possible by the addition of tetrabutylammonium salts [59]. Interestingly, the Pd-catalysed reaction of terminal alkynes with alkenyl chlorides, which are inert in many Pd(0)-catalysed reactions, proceeds smoothly without special activation of the chlorides.

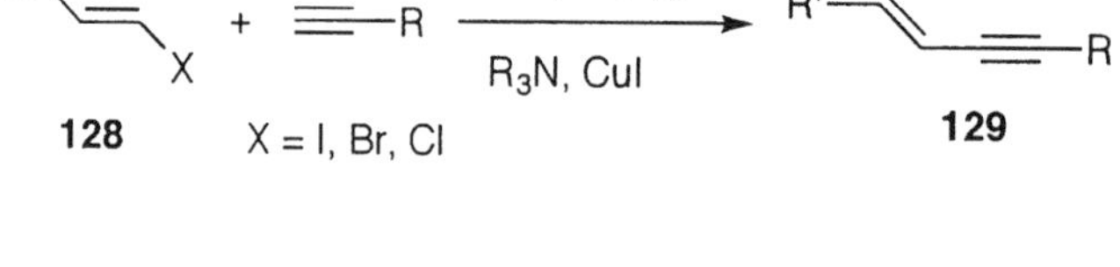

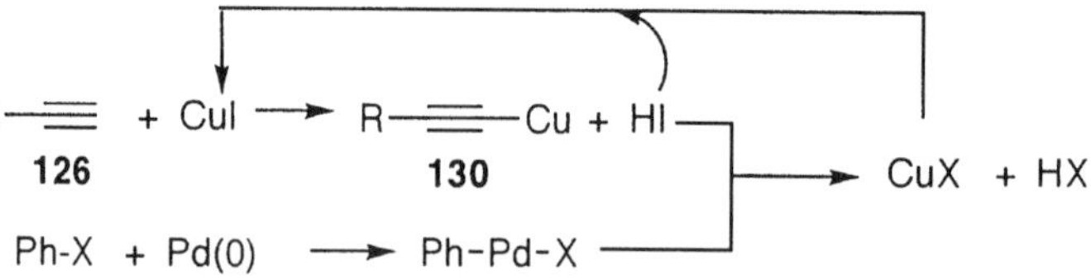

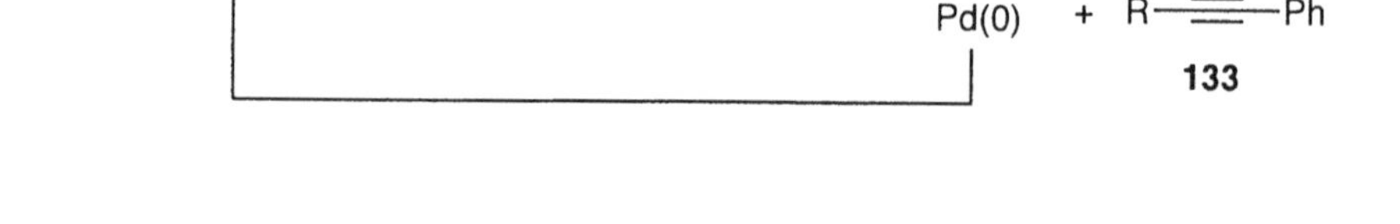

Monosubstitution of acetylene itself to prepare terminal alkynes is not easy. Therefore, trimethylsilylacetylene (**134**) is used as a protected acetylene. After the coupling, the silyl group is removed by the treatment with fluoride anion. The hexasubstitution of hexabromobenzene (**135**) with **134** afforded hexaethynylbenzene (**136**) after desilylation in total yield of 28%. The product was converted to tris(benzocyclobutadieno)benzene (**137**) using a Co catalyst (see Section 7.2.1). Hexabutadiynylbenzene was prepared similarly [60]. As another method, terminal alkynes **139** are prepared in excellent yields by the coupling of commercially available ethynyl Grignard (**138**) or ethynylzinc bromide with halides, without protection and deprotection [61].

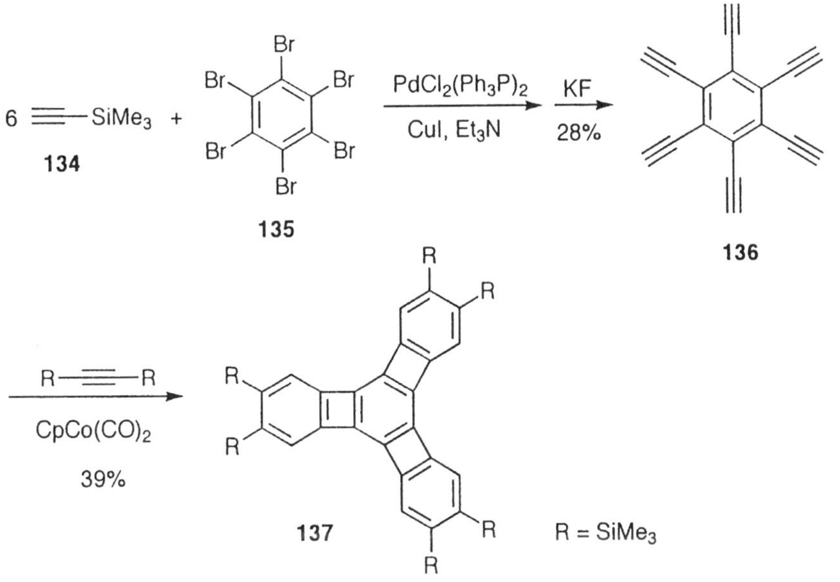

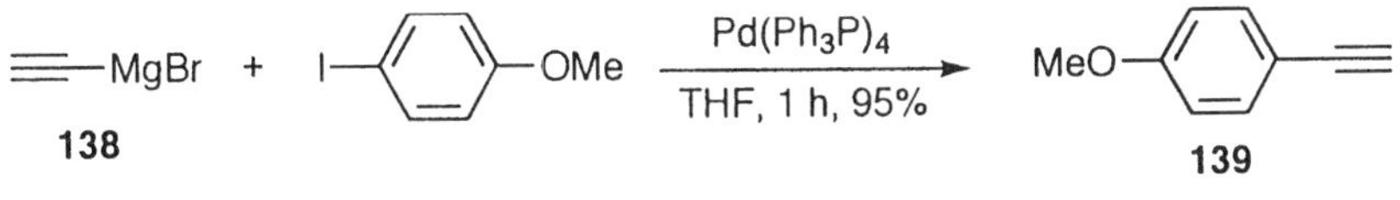

This coupling has widespread use in the construction of enediyne systems present in naturally occurring anticancer antibiotics [62]. Although (*Z*)-1,2-dichloroethylene (**141**) is inert in most of the Pd-catalysed reactions, the Pd–CuI-catalysed reaction of (*Z*)-1,2-dichloroethylene (**141**) with terminal alkynes proceeds smoothly. No clear explanation is given for the high reactivity of dichloroethylene. The coupling has wide synthetic applications, particularly for the synthesis of enediyne structures [63]. Typically, this reaction is successfully applied to the construction of the highly strained enediyne structure **146** present in naturally occurring antitumor antibiotics, such as espermicin or calichemicin. (*Z*)-1,2-Dichloroethylene (**141**) reacts stepwise with two different terminal alkynes, **140** and **143**, to afford **142** and finally **144**. The Nicholas reaction (see Section 9.3) of **145** gives **146** [64].

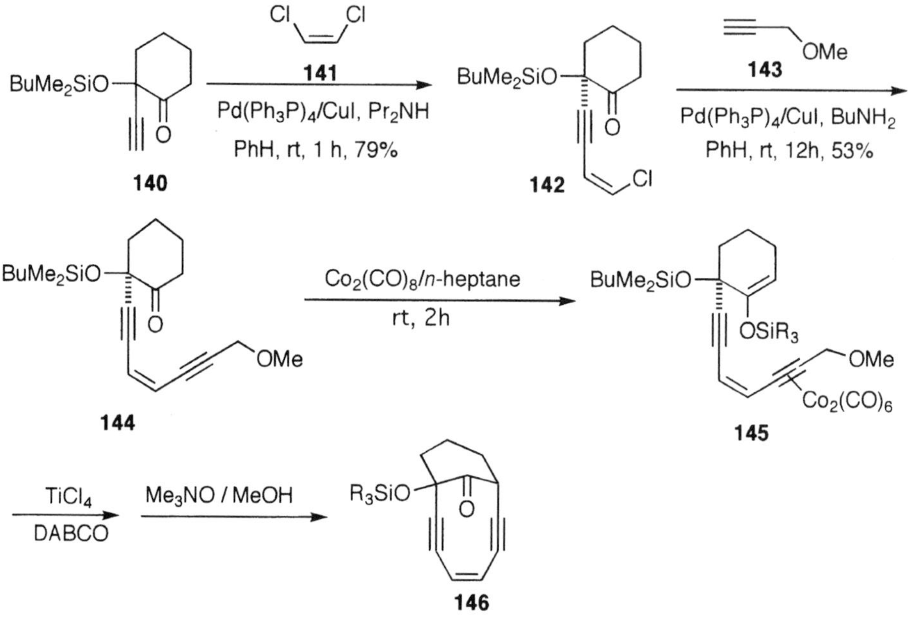

The intramolecular domino reaction of alkenyl bromide with the terminal alkyne in **147** generates **148**, which undergoes the intramolecular Diels–Alder reaction to afford the highly strained dynemicin A structure **149** in one step, although yield is somewhat low [65].

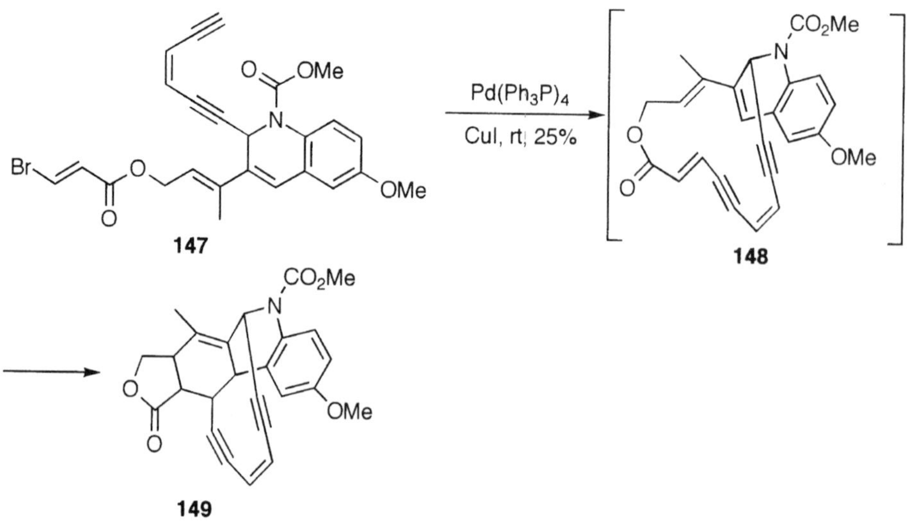

In addition to coupling via Cu acetylides generated *in situ* as mentioned above, the coupling of terminal alkynes has been carried out smoothly using actylides of Zn and other metals as an alternative method of arylation and alkenylation of alkynes [66]. Sn[67], Zn[68] and Mg[69] acetylides are used frequently as activated alkynes, rather than the alkynes themselves, and their reaction with halides proceeds without using CuI.

In the total synthesis of harveynone (**152**), reaction of the iodide **150** bearing labile functionality took place with the tin acetylide **151**, and it was claimed that no reaction occurs with the free amine [70]. However, this claim is not always true, and the coupling of the similar iodide **153** was carried out with the corresponding free terminal alkyne in the presence of diisopropylamine [71]. The coupling of the Mg acetylide **155** with vinyl carbamate **154** to give the enyne **156** is catalysed by a Ni complex [72]. It is true that free alkynes give better results than the corresponding metal acetylides in some cases [73].

TMS group is used for protection of terminal alkynes. However, alkynylsilanes themselves can be used for the coupling with aryl and alkenyl triflates using Pd–CuCl as a catalyst [74]. Thus the internal alkyne **160** is prepared by stepwise reactions of two different triflates **157** and **159** with trimethylsilylacetylene (**134**) via **158**.

As an alternative method, the highly strained enediyne system **163** was constructed by the coupling of the alkynyl bis-iodides **161** with (*Z*)-bis(trimethylstannyl)ethylene (**162**) [75]. No cyclization occurs when there is the double bond, instead of the epoxide.

The alkynyl iodide **164** undergoes the cross coupling with a terminal alkyne to give the 1,3-diyne **165** [76]. No homocoupling product is formed. This reaction offers a good synthetic route to unsymmetric 1,3-diynes.

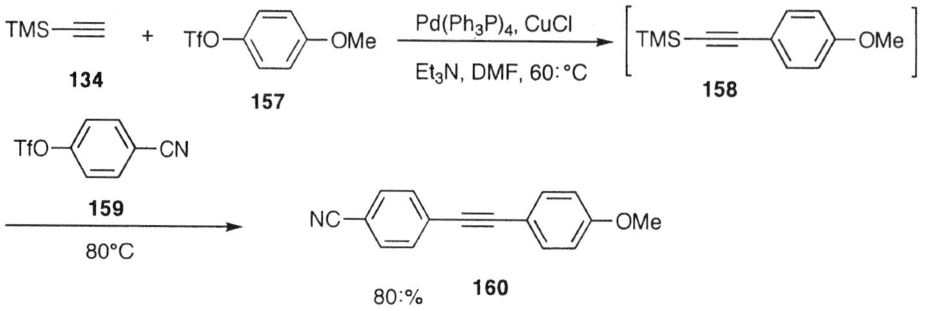

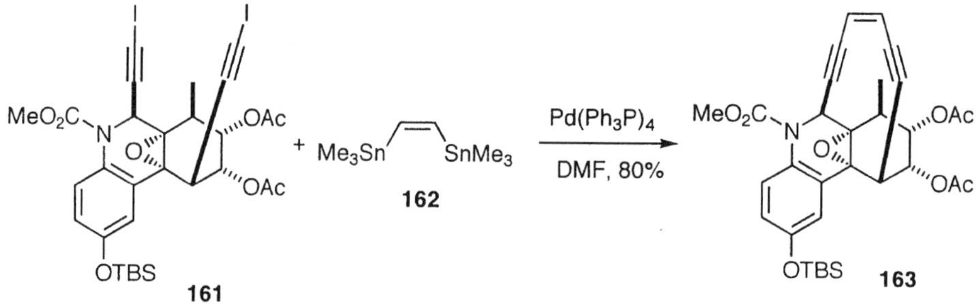

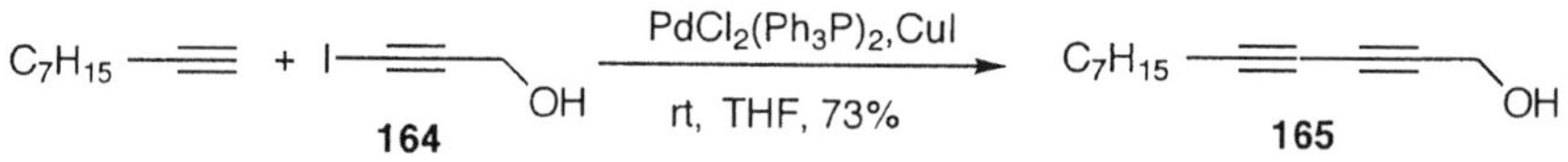

3.3.2   Reactions of Internal and Terminal Alkynes via Insertion

Internal alkynes **166** insert to some Pd–carbon bonds to generate the alkenyl–Pd bonds **167**. This process can be regarded as the *syn* addition of organopalladium species to alkynes, or the carbopalladation of alkynes. Whereas the alkene insertion is followed by facile dehydropalladation (elimination of a β-hydrogen, whenever there is a β-hydrogen) and generation of Pd(0) species, the alkyne insertion produces the thermally stable alkenylpalladium species **167**, and further transformations are required before termination of the catalytic reaction. In other words, the generated alkenylpalladium species **167** undergo no β-elimination (i.e. alkynes **168** or allenes **169** are not formed) even in the presence of β-hydrogen. The alkyne insertion is a 'living' process. Thus the alkenylpalladium species **167** are capable of undergoing further insertion or anion capture before termination, as summarized in Schemes 3.7 and 3.8.

Terminal alkynes **170** undergo the substitution with aryl and alkenyl halides to form arylalkynes and enynes in the presence of CuI, as described in Section 3.3.1. However, the insertion of terminal alkynes **170** occurs in the absence of CuI, and the alkenylpalladium complex **171** is formed as an intermediate, which cannot be terminated by itself and must undergo further reactions, such as alkene insertion or anion capture. These reactions of terminal and internal alkynes with intermediates **172** and **173** are summarized in Schemes 3.7 and 3.8.

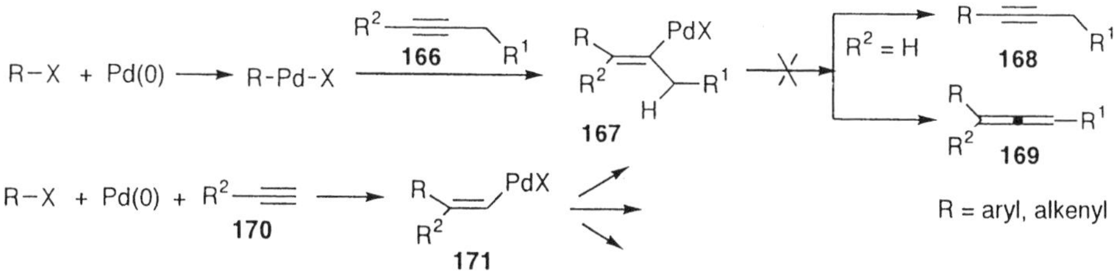

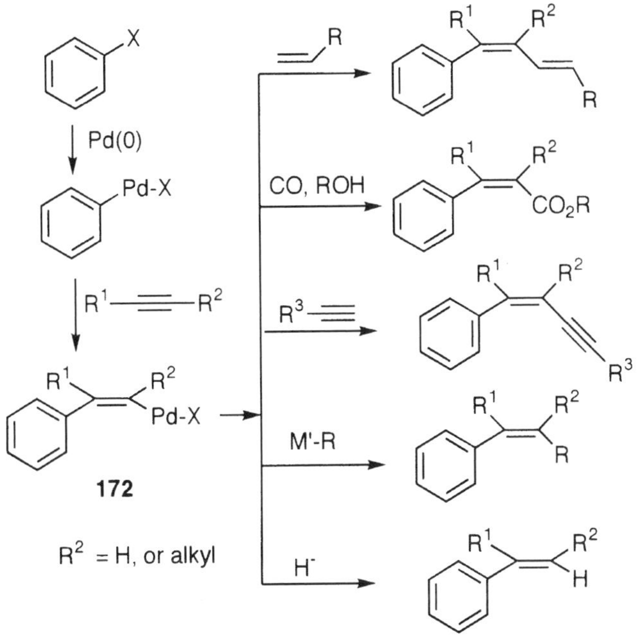

Scheme 3.7 Insertion of inner alkynes to aryl halides and their transformation

In the construction of the conjugated triene system **177** in vitamin D, the intermolecular insertion of the terminal triple bond of the 1,7-enyne **175** to the alkenylpalladium, formed from **174**, occurs at first to form the alkenylpalladium **176**. Further intramolecular insertion of the terminal double bond in **176**, followed by β-elimination yielded the triene system **177** in 76% yield [77].

Intramolecular version can be extended to polycyclization as a one-pot reaction. In the so-called Pd-catalysed domino carbopalladation of trienediyne **178**, the first step is the oxidative addition to alkenyl iodide. Then the intramolecular alkyne insertion generates **179**. One alkyne and two alkene insertions are followed. The last step is the elimination of β-hydrogen. In this way, the steroid skeleton **180** is constructed from the linear trienediyne **178** [78].

The dienyne **181** undergoes domino 6-*exo*-dig, 5-*exo*-trig and 3-*exo*-trig cyclizations to give the tetracycle **185** exclusively [79]. As the neopentylpalladium **183**, which has no β-hydrogen, is formed after the insertion of the disubstituted terminal alkene in **182**, cyclopropanation occurs to give **184**.

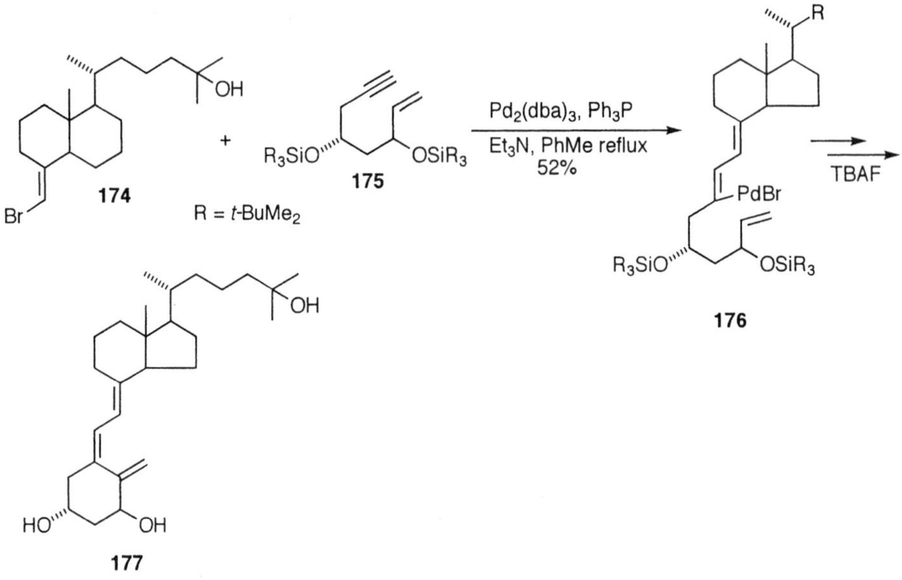
OH
Br
174
+
R3SiO
OSiR3
175
R = t-BuMe2
Pd2(dba)3, Ph3P
Et3N, PhMe reflux
52%
R
PdBr
R3SiO
OSiR3
176
TBAF
OH
HO
OH
177

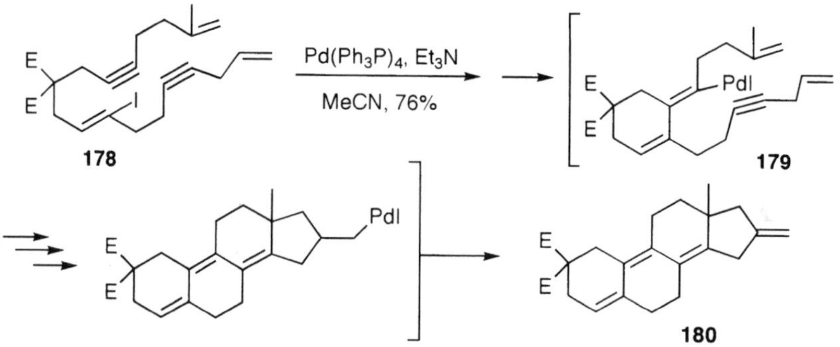
E
E
I
178
Pd(Ph3P)4, Et3N
MeCN, 76%
E
E
PdI
179
E
E
PdI
E
E
180

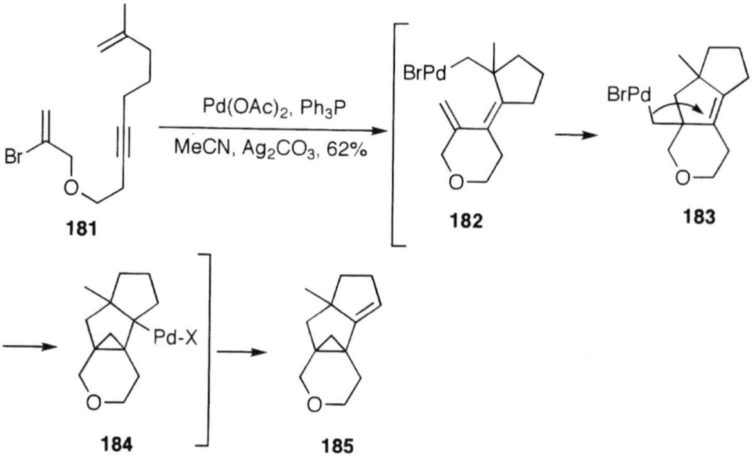
Br
O
181
Pd(OAc)2, Ph3P
MeCN, Ag2CO3, 62%
BrPd
O
182
BrPd
O
183
Pd-X
O
184
O
185

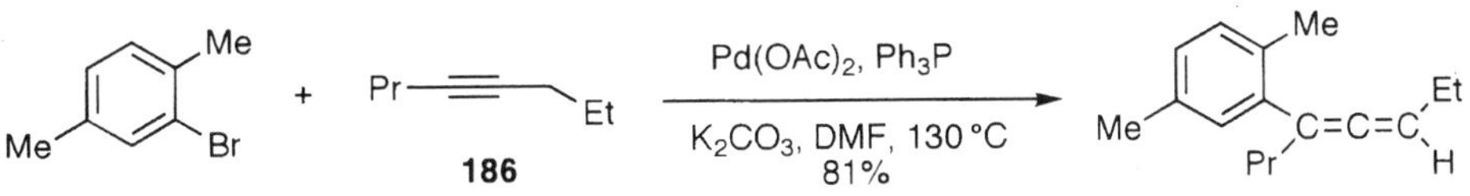

Scheme 3.8 Insertion of inner alkynes to alkenyl halides and their transformation

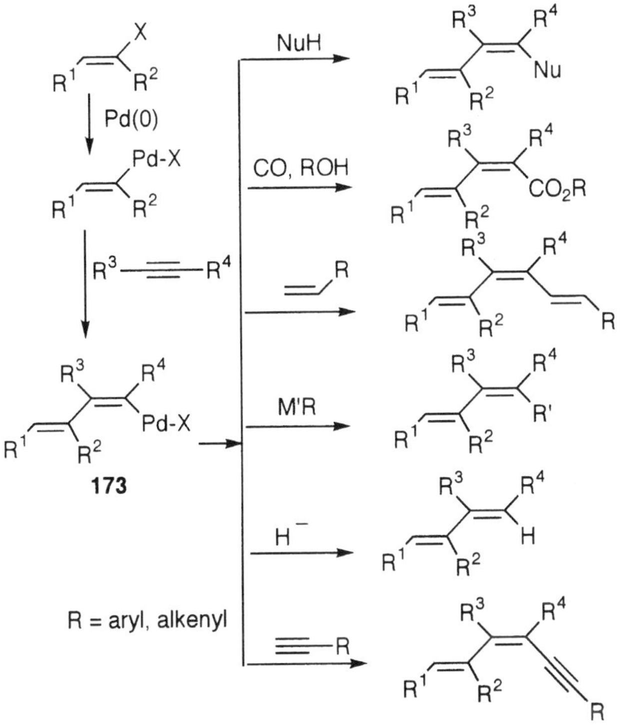

As mentioned in Section 3.3.1, allenes **169** are not formed from alkenylpalladium **167**. However, aryl-substituted allenes **187** are obtained predominantly by the coupling of aryl bromides with dialkylacetylenes **186** [80].

The alkyne insertion reaction is terminated by anion capture. As examples of the termination by the anion capture, the alkenylpalladium intermediate **189**, formed by the intramolecular insertion of **188**, is terminated by hydrogenolysis with formic acid to give the terminal alkene **192**. Palladium formate **190** is formed, and decarboxylated to give the hydridopalladium **191**, reductive elimination of which gives the alkene **192** [81]. Similarly the intramolecular insertion of **193** is terminated by transmetallation of **194** with the tin acetylide **195** (or alkynyl anion capture) to give the dienyne **196** [82].

Various heterocyclic compounds are prepared by heteroannulation using aryl iodides **68** and **69**, and internal alkynes. Although the mechanism is not clear, alkenylpalladiums, formed by insertion of alkynes, are trapped by nucleophiles intramolecularly. The reaction of *o*-iodoaniline (**68**) with alkynes in the presence of phosphine-free Pd offers a useful synthetic route to 2,3-disubstituted indoles **198** via **197** [83]. Furans are prepared from alcohol, phenol and enol derivatives of iodobenzenes. The halenaquinone framework **200** was constructed by the intramolecular version applied to the enol in **199** [84].

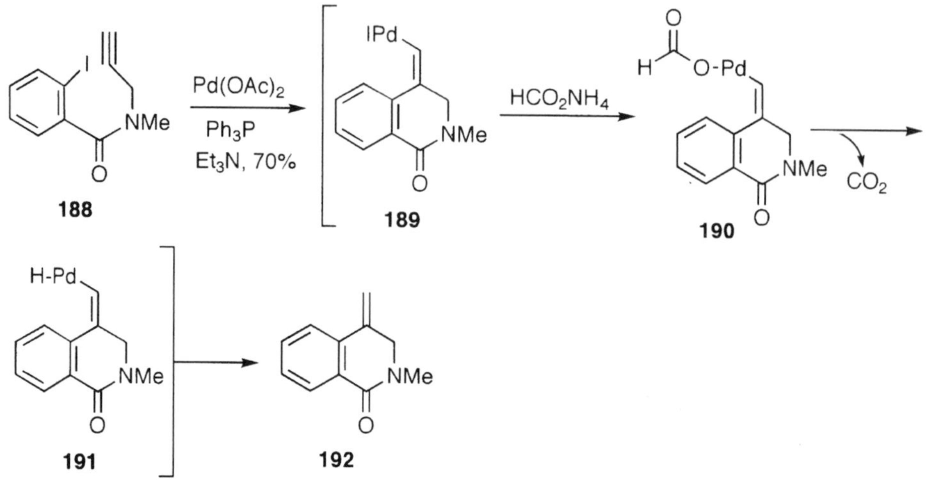

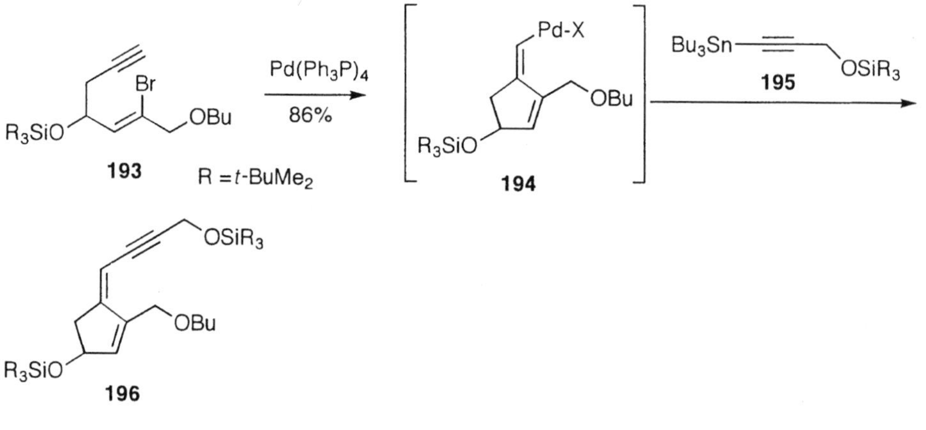

Isoquinolines **202** are prepared easily in high yield by the iminoannulation of *t*-butylimine (**201**) of *o*-iodobenzaldehyde with internal alkynes [85].

An interesting synthetic route to 2,3-substituted indenones **205** was discovered by the annulation of alkynes with *o*-iodobenzaldehyde (**203**) [86]. In this reaction, the CH bond of the aldehyde may oxidatively add to Pd(0) to generate the hydridopalladium **204**, which produces indenones **205** by reductive elimination.

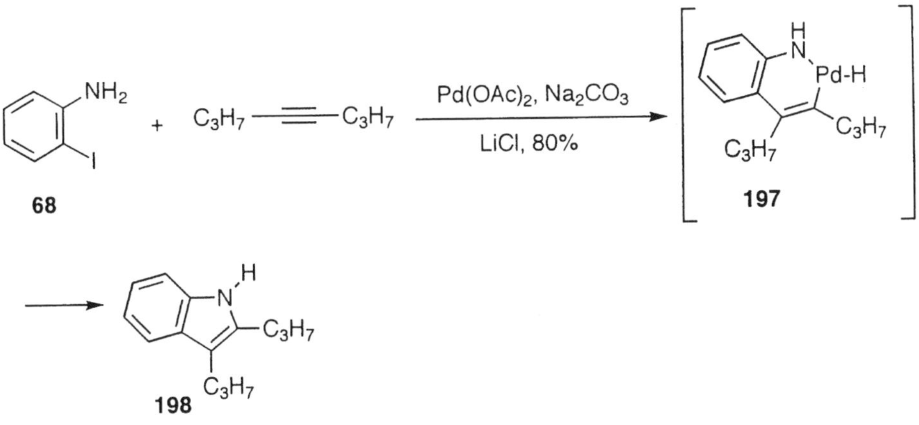

NH2
I
68
C3H7
C3H7
Pd(OAc)2, Na2CO3
LiCl, 80%
H
N
Pd-H
C3H7
C3H7
197
H
N
C3H7
C3H7
198

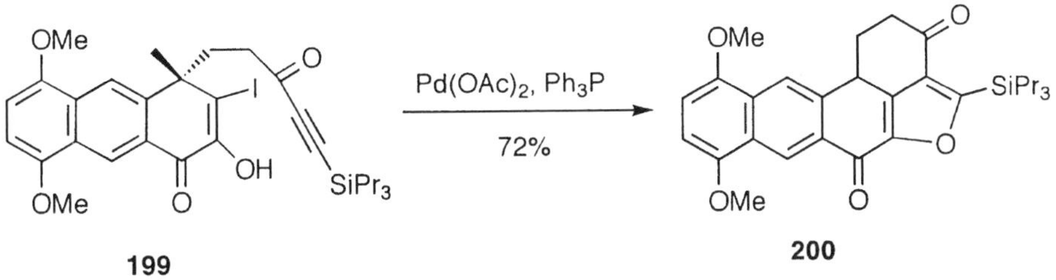

OMe
OMe
I
O
OH
O
SiPr3
199
Pd(OAc)2, Ph3P
72%
OMe
OMe
O
SiPr3
O
O
200

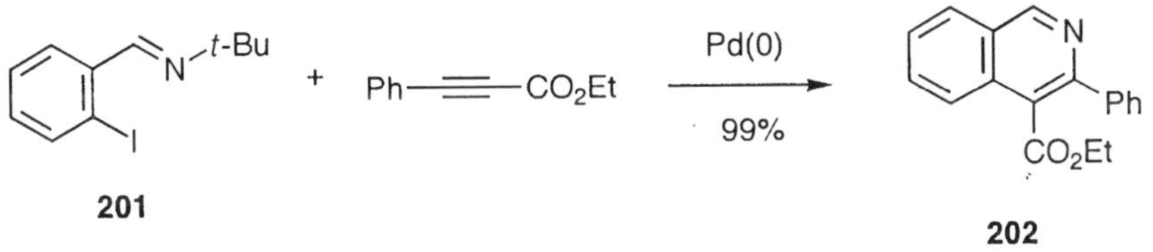

N
t-Bu
I
201
Ph
CO2Et
Pd(0)
99%
N
Ph
CO2Et
202

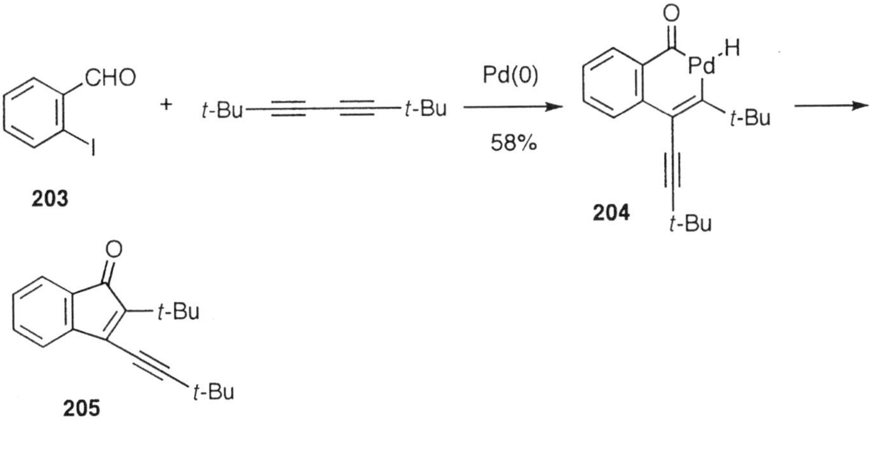

CHO
I
203
t-Bu
t-Bu
Pd(0)
58%
O
Pd
H
t-Bu
t-Bu
204
O
t-Bu
t-Bu
205

3.4 Cross-coupling via Transmetallation

Organometallic chemistry is classified into main group metal chemistry and transition metal chemistry. Since the discovery of Ni-catalysed cross-coupling of aryl and alkenyl halides with Grignard reagents [87], many useful synthetic methods have been developed by combining both main group metal compounds and transition metal complexes. In these reactions, after oxidative addition of halides to transition metal complexes, transmetallation with organometallic compounds of main group metals such as Mg, Zn, B, Al, Sn, Si or Hg occurs, and subsequent reductive elimination gives the coupled products. The transmetallation can be regarded as alkylation of transition metal compounds. Aryl, alkenyl and alkyl halides, and pseudohalides, are used for the coupling with aryl-, alkenyl- and alkylmetal compounds [88]. Couplings involving allylic compounds are treated in Chapter 4.

Three transmetallation reactions are known. The reaction starts by the oxidative addition of halides to transition metal complexes to form **206**. (In this scheme, all ligands are omitted.) (i) The C−C bonds **208** are formed by transmetallation of **206** with **207** and reductive elimination. Mainly Pd and Ni complexes are used as efficient catalysts. Aryl–aryl, aryl–alkenyl, alkenyl–alkenyl bonds, and some alkenyl–alkyl and aryl–alkyl bonds, are formed by the cross-coupling. (ii) Metal hydrides **209** are another partner of the transmetallation, and hydrogenolysis of halides occurs to give **210**. This reaction is discussed in Section 3.8. (iii) C−M′ bonds **212** are formed by the reaction of dimetallic compounds **211** with **206**. These reactions are summarized in Schemes 3.3–3.6.

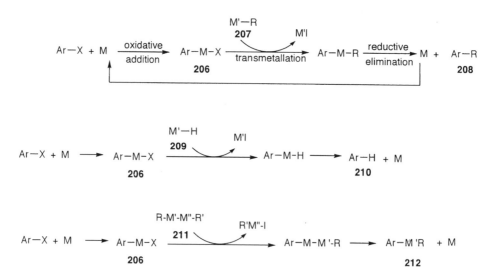

3.4.1 Magnesium Compounds

Phosphine complexes of Ni and Pd show high catalytic activity for the coupling of aryl and alkenyl halides with Grignard reagents [88,89]. The coupling of Grignard reagents with halides is called the Tamao–Kumada–Corriu reaction [87]. The Ni catalyst is used more often than the Pd catalyst, because the Ni catalyst is active for chlorides. In addition, elimination of β-hydrogen from the intermediates is less extensive. The coupling using NiX_2L_2 as a catalyst precursor is explained by the following mechanism. At first the complex **214** is generated *in situ* by the transmetallation of the precursor $NiCl_2(PR_3)_2$ **213** with a Grignard reagent, and converted to the Ni(0) complex **215** as a real catalyst by reductive elimination. The first step of the catalytic cycle is the oxidative addition of aryl halides to the Ni(0) complex **215** to generate the organonickel complex **216**. The key step in the catalyic reaction is transmetallation of the organonickel complex **216** with the Grignard reagent to generate the alkyl-aryl nickel complex **217**. The final step is the reductive elimination of **217** to produce the coupled product **218** with regeneration of the Ni(0) complex **215**.

Aryl and alkenyl chlorides react smoothly with the Ni(0) catalyst. This is a characteristic feature of the Ni(0)-catalysed reaction. Bidentate phosphines such as DPPE (**XLIII**) and DPPP (**XLIV**) give better results than Ph$_3$P, although it was reported that the coupling of aryl Grignard reagents with aryl halides proceeds by using NiCl$_2$ without phosphine [90].

High percentage ee was achieved in the synthesis of the binaphthyl derivative **221** by the coupling of Grignard reagent **219** with bromide **220** using PPFA (**XXX**) [91]. The cross-coupling of aryl triflates with Griganrd reagents is catalyzed by Pd(0). The axially chiral (*S*)-biaryl **224** was prepared with 93% ee in 87% yield by the reaction of the bistriflate, 1-[2,6-bis(trifluoromethylsulfonyloxy)phenyl]naphthalene (**222**) with PhMgBr. PdCl$_2$[(*S*)-phephos] (**XXII**) is the best chiral catalyst [92]. The diphenylated biaryl **223** was obtained in 13% yield. Interesting kinetic resolution occurs in the second cross-coupling to give **223**. The (*R*) isomer of **224** undergoes the second phenylation about five times faster than its (*S*) isomer, indicating that the minor (*R*) isomer of **224** is consumed preferentially at the second asymmetric cross-coupling, which causes an increase of the enantiomeric purity of (*S*)-**224** as the amount of the diphenylation product **223** increased. The remaining triflate in **224** is converted to **225** by the Pd-catalysed diphenylphosphinylation, which is reduced to the phosphine **226** with HSiCl$_3$.

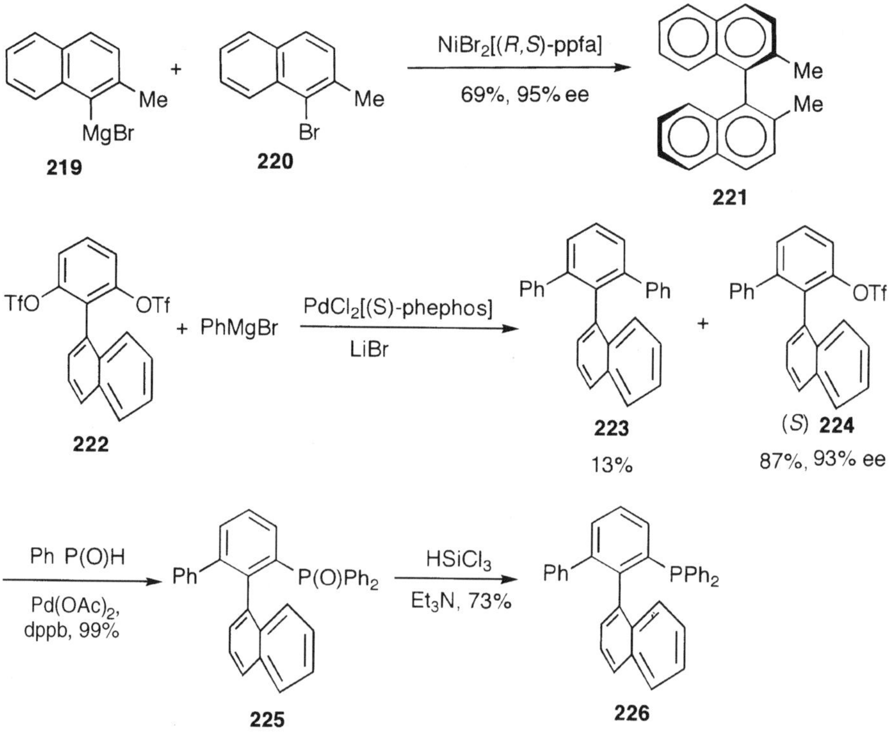

The Ni-catalyzed reaction is utilized for the coupling of halides of heterocyclic compounds such as pyridine and pyrimidine, offering a new substitution route to these heterocyclic compounds [93]. The reaction can be extended to pseudohalides. Even C−O bonds attached to sp^2 carbons such as aryl and alkenyl ethers, thioethers, silyl enol ethers and enol phosphates react with Ni(0) [94,95]. The ether bonds of benzofuran (**227**) [94], and C−S bonds in the benzothiazole **228** are displaced with Grignard reagents [95,96]. Thiophene (**229**) reacts with two equivalents of PhMgCl to yield 1,4-diphenyl-1,3-butadiene (**230**) [97]. Ni-catalysed alkenation of the benzylic dithioacetal **231** with MeMgBr gives **233** via **232**. Dimethylation of **234** to give **235** is achieved by coupling with MeMgBr [98,99]. In these reactions, the oxidative addition of C−O and C−S bonds to Ni(0) takes place, followed by transmetallation due to the strong affinity of Mg with O and S atoms. These bond cleavages are possible only with the Ni catalyst.

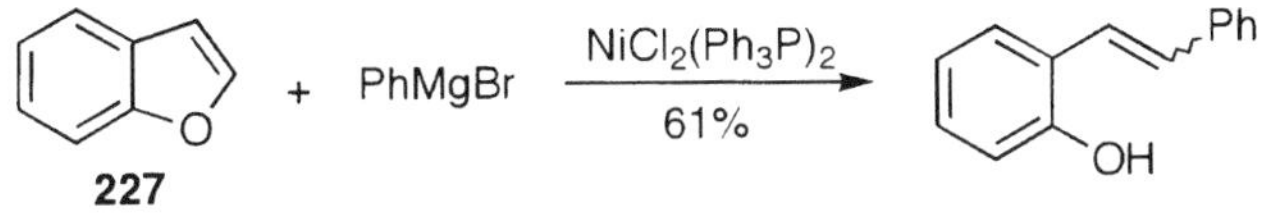

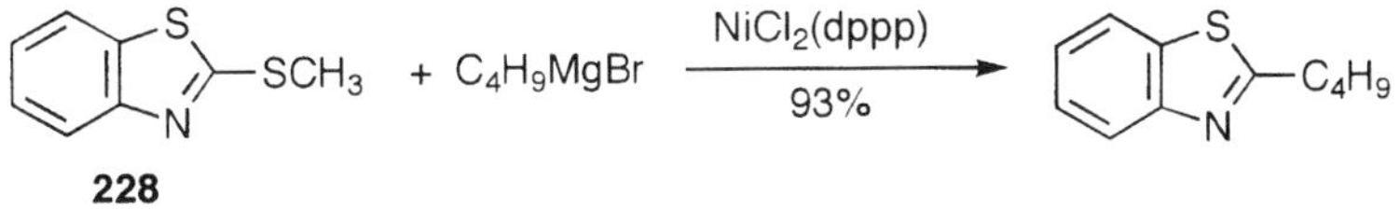

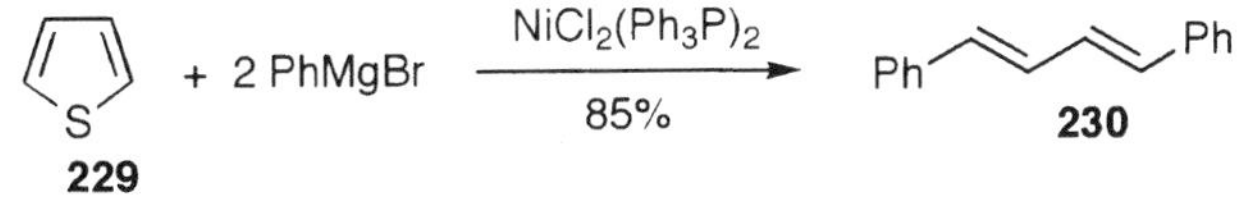

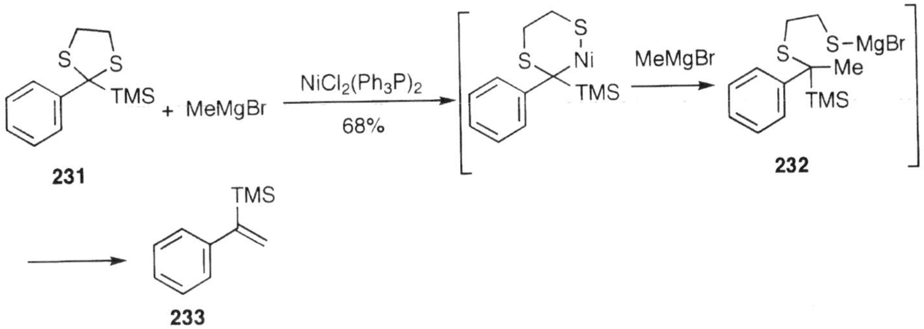

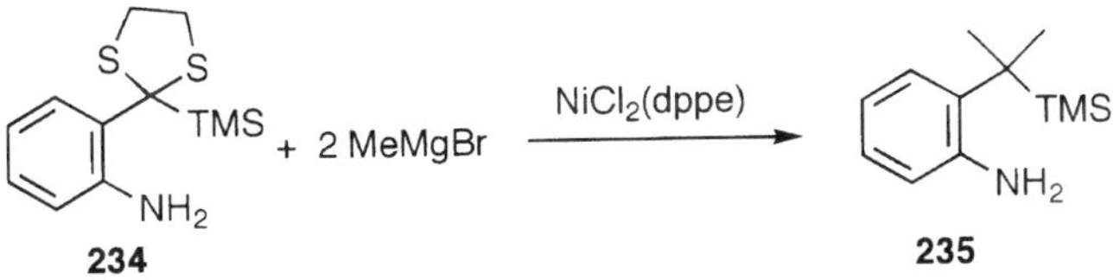

Aryl mesylates are used for the Ni-catalysed cross-coupling with Grignard, organozinc and organoboron reagents [100].

3.4.2 Zinc Compounds

Lithium reagents are not suitable for transmetallation, and they are used after converting to zinc reagents. Cross-coupling of organozinc reagents with alkenyl and aryl halides proceeds generally with high yields and tolerates a wide range of functionality. A smooth reaction occurs using Pd–phosphine complexes as the catalyst and is called the Negishi reaction [101,102]. Organozinc reagents are prepared most conveniently *in situ* from organolithium, magnesium or aluminum compounds and

ZnCl$_2$, and used for the coupling with alkenyl and aryl halides [103]. The reaction of reactive halides with Zn–Cu couple is another method. Alkyl groups can be introduced without β-elimination by using alkylzinc reagents. Organozinc reagents are inert to ketones, esters, amino and cyano groups, and are used without their protection. This feature is an advantage of organozinc reagents. A variety of zinc reagents, such as *n*-alkyl, benzyl, homoallyl and homopropargylzinc reagents, have been shown to couple with aryl and alkenyl halides satisfactorily.

The coupling of aryl- and alkenylzinc reagents with various halides has widespread use in the cross-coupling of aromatic rings [101]. The reactions of zinc derivatives of aromatic and heteroaromatic compounds with aryl and heterocyclic halides have wide synthetic applications [103,104].

The reaction of aryl and alkenyl halides with the Reformatsky reagent **237** in polar solvents affords α-arylacetates [105]. Iodides of heterocyclic compounds such as pyridine, quinoline and pyrimidine react smoothly with Reformatsky reagents. The cross-coupling of **237** with 4-iodo-2,6-dimethyl- and 2-iodo-4,6-dimethylpyrimidine (**236**) occurs smoothly to form **238**. But no reaction of 5-iodo-2,4-dimethylpyrimidine (**239**) takes place [106].

Triflates are used in this reaction [107]. 5-Phenyltropone **241** is prepared by the coupling of triflate **240** with PhZnCl [108]. Instead of rather expensive triflates, triflate equivalents such as phenyl mesylates [100], fluoroalkanesulfonates [109] and the fluorosulfonate **242** [110] are used for the coupling.

As another alternative to triflates, aryl nonaflates (ONf) are easily prepared from phenol and commerically available perfluorobutane sulfonyl fluoride **243**, and used for Pd-catalysed coupling with halides [111]. Aryl nonaflates are more reactive than aryl triflates. *p*-Iodophenyl nonaflate (**244**) gives **245** by chemoselective reaction of the iodide with a Zn reagent, and the remaining nonaflate in **245** is displaced with another Zn reagent to give **246** under different conditions. Reaction of **244** with activated Zn affords the Zn reagent **247**, which is regarded as **248** because the different groups can be introduced stepwise. For example, the arylzinc function in reagent **247** reacts with aryl haldies, and then undergoes Suzuk–Miyaura coupling (Section 3.4.3) with the nonaflate moiety in a stepwise mannner.

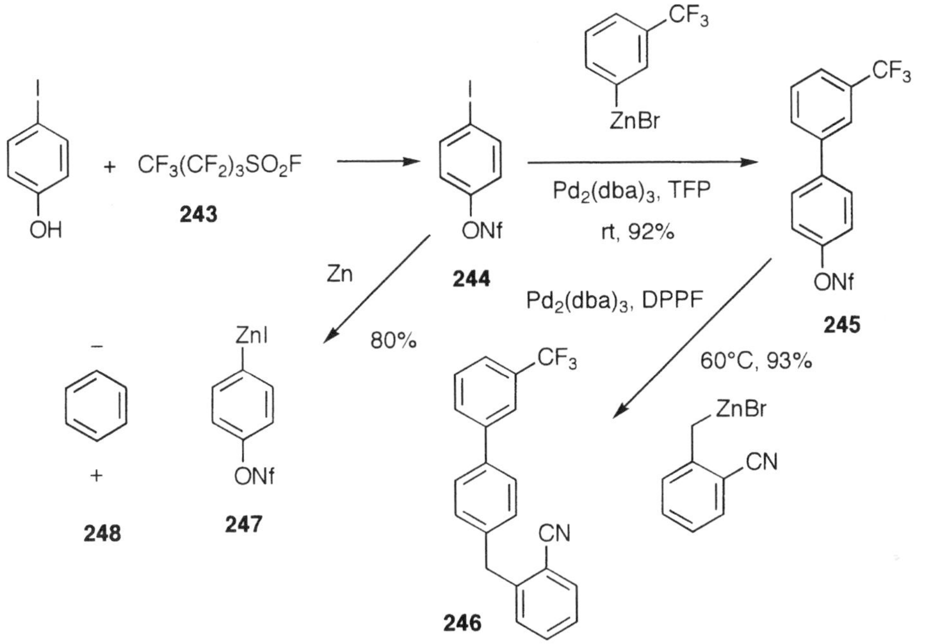

Interestingly, alkylzinc reagents which have β-hydrogens undergo the coupling smoothly to give alkylarenes or alkylalkenes without elimination of the β-hydrogen. For example, some iodozinc derivatives of 1-phenyl-1-heptanone, as represented by the 7-substituted compound **249**, react with various halides and the triflates **250** to give **251** [107]. The cross-coupling of Grignard reagent **252** with vinyl bromide (**253**) gives **254** with 93% ee in the presence of ZnCl$_2$ using BPPFA (**XXV**) as a chiral ligand [112].

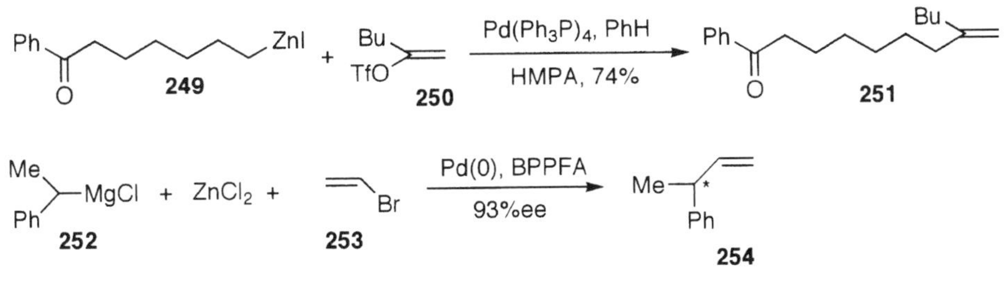

Ni(0) complexes also catalyze the coupling of Zn reagents. The coupling of alkyl halides with alkylzinc reagents is more difficult. It is known that the presence of a carbonyl group or an alkenic bond in alkyl groups sometimes has a favourable effect on the Ni-catalysed alkyl–alkyl coupling. Interestingly it was found that the Ni-catalysed couplings of alkyl iodides **255** and **257** with dialkylzinc reagents proceeds smoothly in the presence of a catalytic amount of acetophenone (**256**), or *m*-trifluoromethylstyrene (**258**) as a cocatalyst [113]. In the absence of the cocatalyst, iodine–Zn exchange occurs mainly to give **259**.

The coupling of alkenyl iodides or bromides with alkylzinc reagents is also catalyzed by Co complexes. Using Co(acac)$_3$ as a catalyst precursor, the coupling product **260** is obtained in THF–NMP in good yield at room temperature [114].

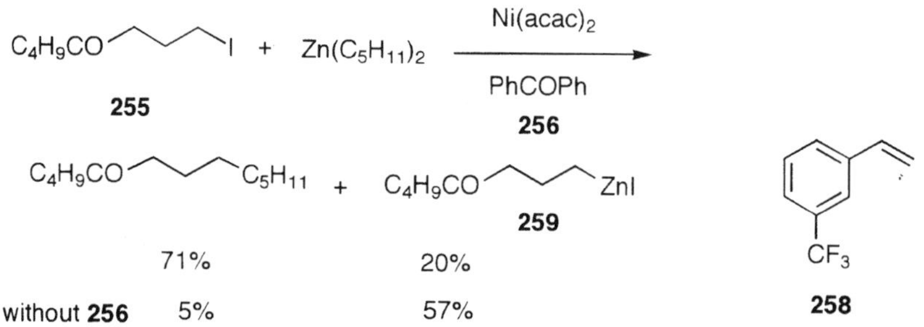

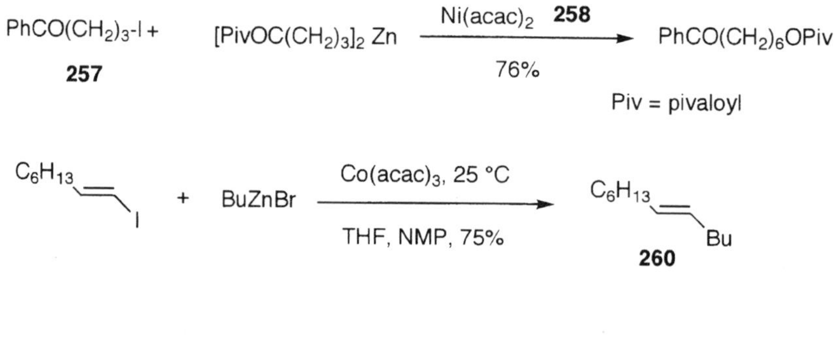

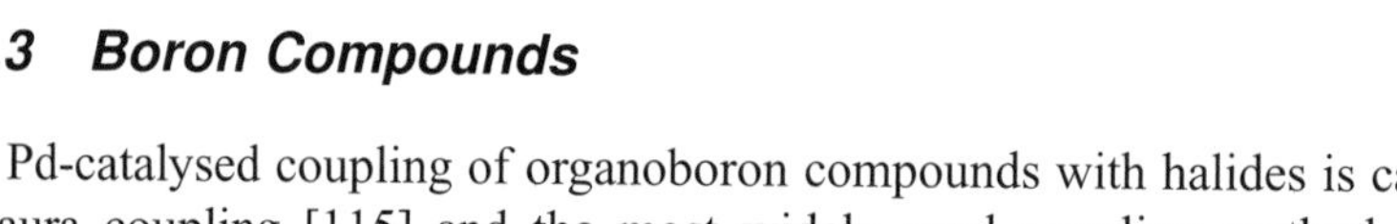

3.4.3 Boron Compounds

The Pd-catalysed coupling of organoboron compounds with halides is called Suzuki–Miyaura coupling [115] and the most widely used coupling method. Tolerance to water, a broad range of functionality, and easier disposal of boron compounds after the coupling than that of Zn and Sn compounds used for similar cross-couplings, have stimulated the application of Suzuki–Miyaura coupling even to commercial processes. The coupling proceeds only in the presence of a base. No transmetallation of boron compounds occurs under neutral conditions [116]. This is a characteristic feature of boron compounds, that differs from other organometallic reagents [117]. The role of the base is said to activate either Pd or boranes. Most probably, the formation of Ar−Pd−OR **261** from Ar−Pd−X and the tetracoordiated ate complex of boron facilitates the transmetallation with organoboranes. Various bases including weak bases such as K_2CO_3 can be used, depending on the reactants. Sometimes the coupling proceeds without phosphine.

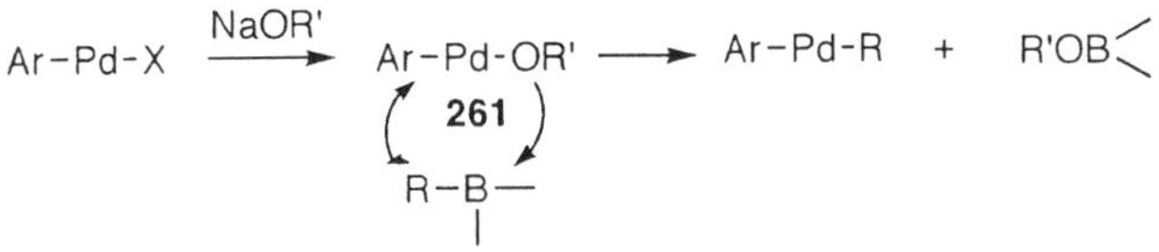

The conjugated dienone **266** is prepared by two-step reactions of the α-methoxyvinylzinc reagent **262** as a methyl ketone precursor. At first the chemoselective reaction with the bromide in **263** affords **264** under neutral conditions. Then coupling of the organoborane moiety in **264** with the bromide **265** under basic conditions gives **266** [117]. When both alkylborane and alkylstannane functions are present in the same molecule, the borane moiety reacts chemoselectively under basic conditions [118]. Under basic conditions, the organoborane in **268** is more reactive than the organostannane, and reacts chemoselectively with the bromide **267** to afford the alkylstannane **269** without attacking the organotin moiety in the same molecule **268**.

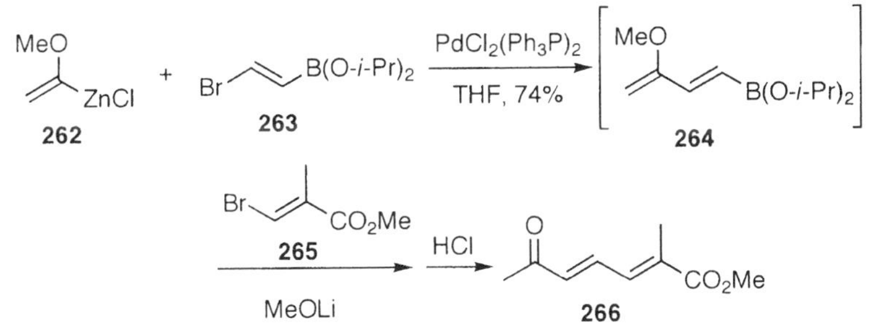

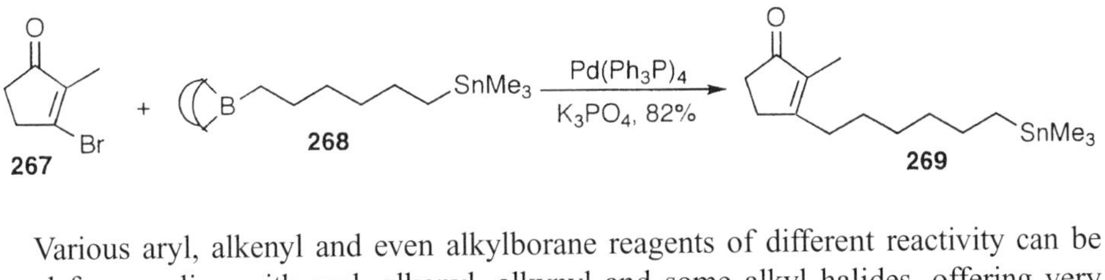

Various aryl, alkenyl and even alkylborane reagents of different reactivity can be used for coupling with aryl, alkenyl, alkynyl and some alkyl halides, offering very useful synthetic methods. The cross-coupling of aryl and heteroarylboronic acids with aryl and heterocyclic halides and triflates provide useful synthetic routes to various aromatic and heteroaromatic derivatives. Sometimes, the reaction proceeds in the absence of a ligand [119]. The coupling of aryl bromides proceeds smoothly in water without organic cosolvent in the presence of Bu$_4$NBr [120]. The coupling of the arylboronate **270** can be carried out smoothly in water, aqueous DMF or acetone as solvents using Pd(OAc)$_2$ without Ph$_3$P [121]. A similar coupling can be carried out using Pd on charcoal as a catalyst in EtOH in the presence of Na$_2$CO$_3$ without ligand in good yields [119]. Phenylboronic acid (**275**) is prepared by the treatment of PhLi with P(OMe)$_3$ and used for the coupling without isolation to give the coupled products in good yields in a one-pot reaction. The *ortho*-palladated tri-*o*-tolylphosphine complex **20** was found to be the very active catalyst for the Suzuki–Miyaura reaction [122]. The reaction has been successfully applied to the commercial production of medicinal compounds such as substituted *β*-lactam **273** by the coupling of the enol triflate **271** [123]. Arenediazonium tetrafluoroborates **274** react smoothly with phenylboronic acid (**275**) in dioxane in the absence of ligand without the addition of a base [124].

Nickel complexes are active for the coupling of aryl chlorides **276** [125], triflates and mesylates **277** with phenylboronic acid (**275**) [100]. The catalyst is generated *in situ* by the treatment of NiCl$_2$(dppf) with BuLi in dioxane or Zn [126], and NiCl$_2$ (dppe)$_2$, TMSPP (**XLVII**) with Zn [126a]. Pd is effective for coupling the chlorides using *t*-Bu$_3$P as a ligand and Cs$_2$CO$_3$ as a base [126b].

The coupling of alkenylboranes with alkenyl halides is particularly useful in the stereoselective synthesis of 1,3-diene systems of four possible double-bond isomers [127]. The (*E*) and (*Z*) forms of the vinylboron compounds **278** and **279** can be prepared by hydroboration of alkynes and haloalkynes, and their reactions with the (*E*)- or (*Z*)-vinyl iodides or bromides **280** and **281** proceed without isomerization; the four possible isomers **282–285** can be prepared in high purity. However, for the efficient preparation of the (*Z,Z*)-dienes **287**, the diisopropyl ester of (*Z*)-alkenylboronic acid **286** should be used. Other boron compounds give poor yields [128].

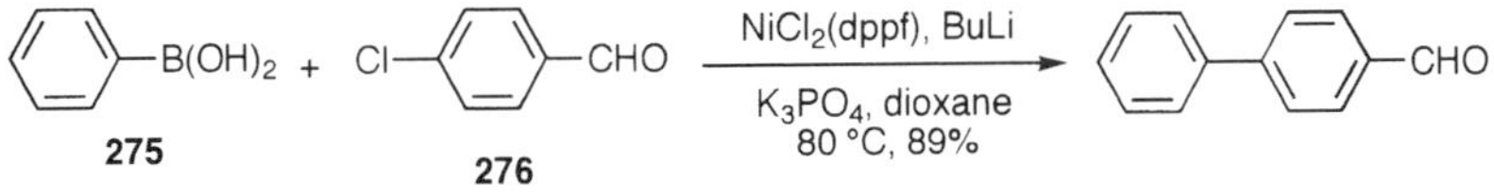

275 + **276** →

$$\text{NiCl}_2(\text{dppf}), \text{BuLi}$$
$$\text{K}_3\text{PO}_4, \text{dioxane}$$
$$80\,°C, 89\%$$

277 + **275** →

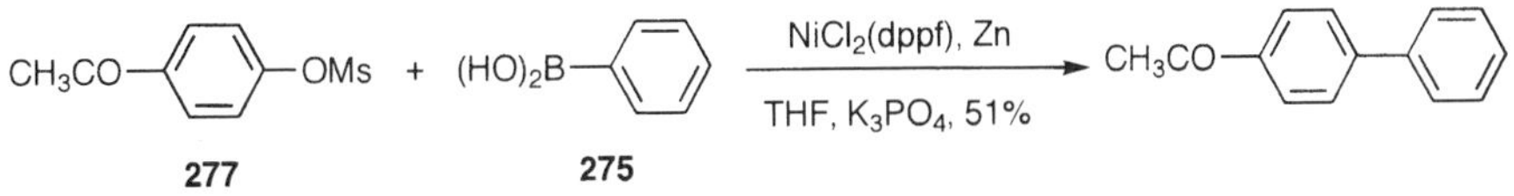

$$\text{NiCl}_2(\text{dppf}), \text{Zn}$$
$$\text{THF}, \text{K}_3\text{PO}_4, 51\%$$

Me—C$_6$H$_4$—Cl + (HO)$_2$B—Ph →

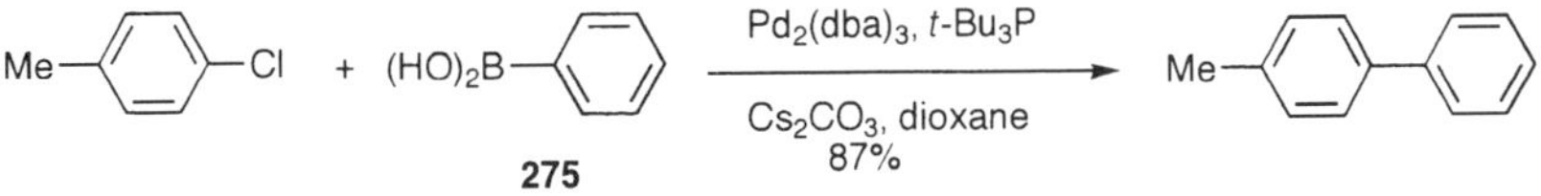

$$\text{Pd}_2(\text{dba})_3, t\text{-Bu}_3\text{P}$$
$$\text{Cs}_2\text{CO}_3, \text{dioxane}$$
$$87\%$$

R—≡ + HBY$_2$ → **278** (E) = 99%

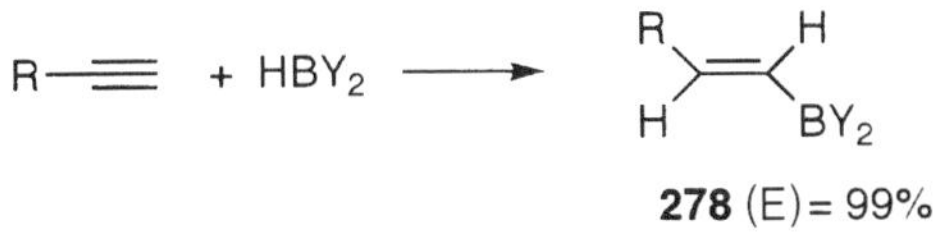

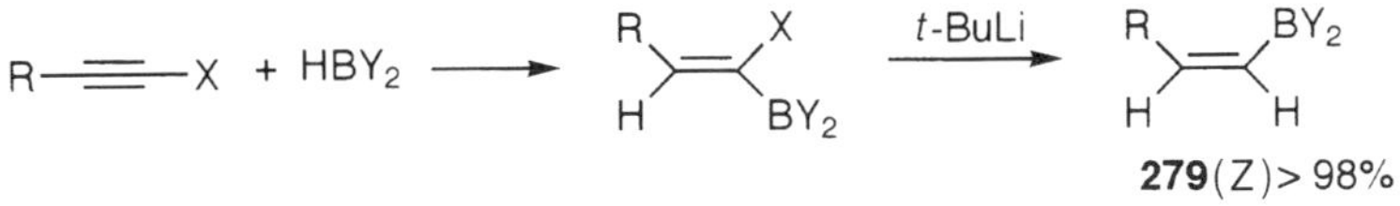

R—≡—X + HBY$_2$ → →t-BuLi→ **279**(Z) > 98%

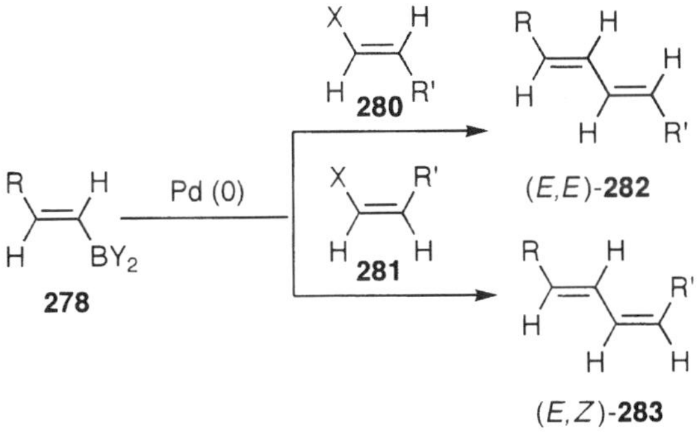

278 Pd (0) →

280 → (E,E)-**282**

281 → (E,Z)-**283**

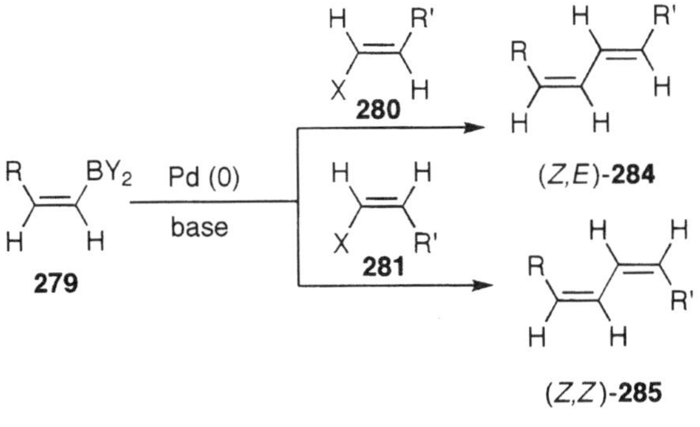

279 Pd (0), base →

280 → (Z,E)-**284**

281 → (Z,Z)-**285**

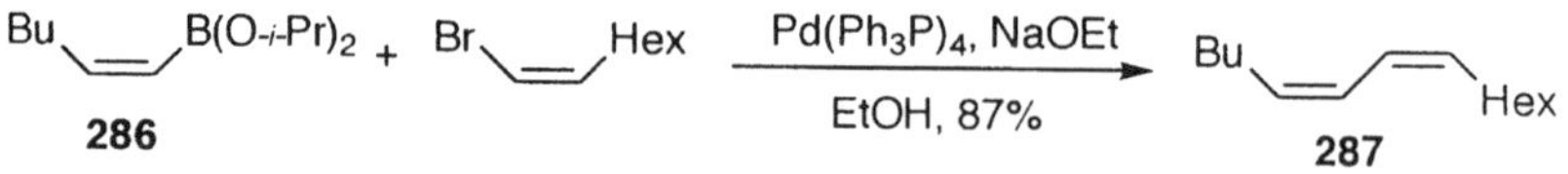

The (*Z,E,E*)-triene systems in leukotriene and DiHETE were constructed by the coupling of the (*Z,E*)-dienylborane with the (*E*)-alkenyl iodide [129,130]. In the total synthesis of the naturally occurring large molecule palytoxin, which has numerous labile functional groups, Suzuki coupling gives the best results for the creation of the (*E,Z*)-1,3-diene part (**290**) by the coupling of the alkenylborane **288** with the (*Z*)-alkenyl iodide **289**. In this case, thallium hydroxide as the base accelerates the reaction 1000 times more than KOH [131].

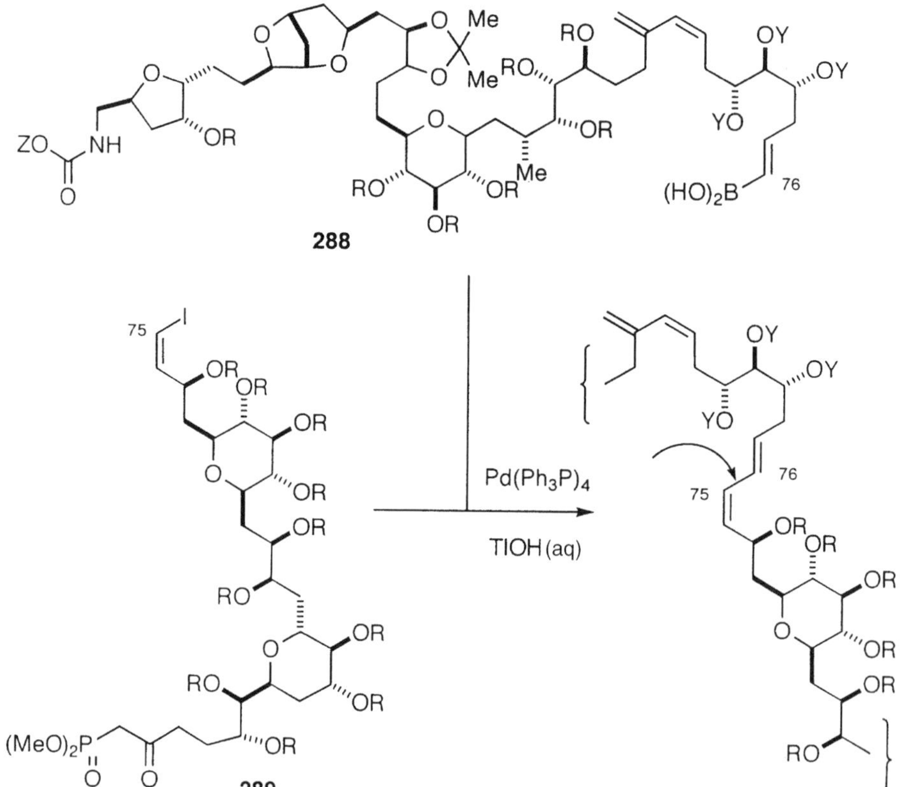

Primary alkylboranes derived by hydroboration of terminal alkenes with 9-BBN-H are coupled with aryl and alkenyl triflates and halides under properly selected conditions. The reaction proceeds smoothly without elimination of β-hydrogen using PdCl$_2$(dppf) or Pd(Ph$_3$P)$_4$ and K$_3$PO$_4$ in dioxane or DMF [132]. The intramolecular cross-coupling of the alkenyl triflate with the alkylborane in **292**, prepared by *in situ* hydroboration of the double bond in **291** with 9-BBN-H, is applied to the annulation to give **293** [133]. The most difficult reaction is the coupling of alkyl–alkyl groups which have β-hydrogens. The coupling of 1-octylborane (**294**) with 1-iodohexane (**295**) gave tetradecane (**296**) successfully in 64% yield [134].

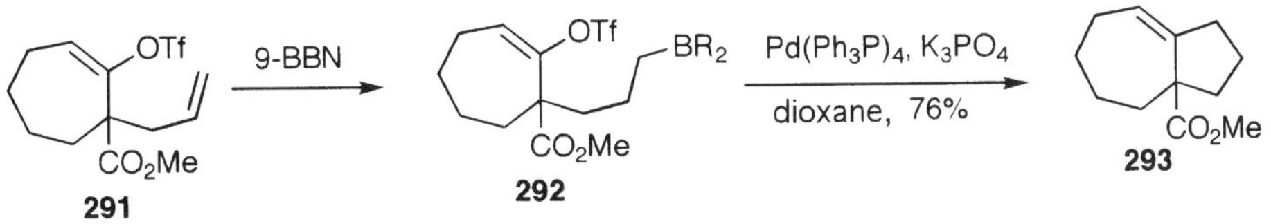

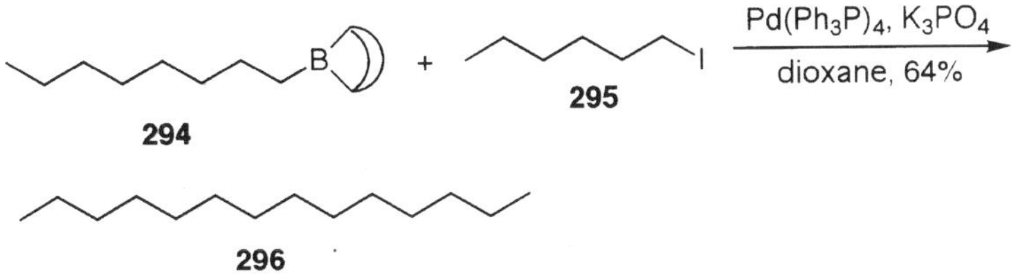

3.4.4 Aluminium and Zirconium Compounds

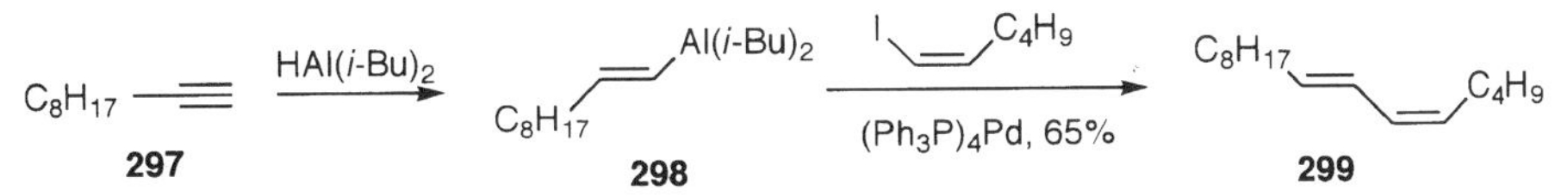

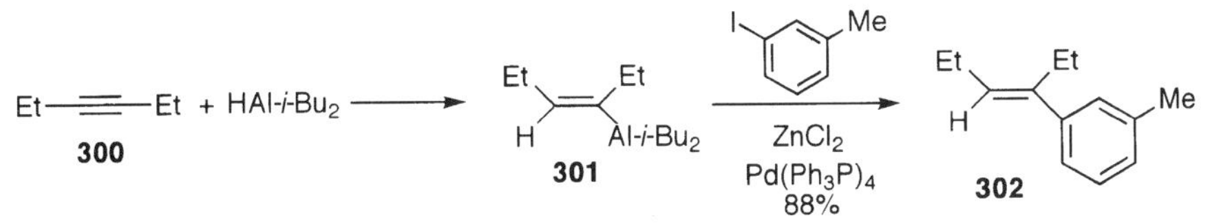

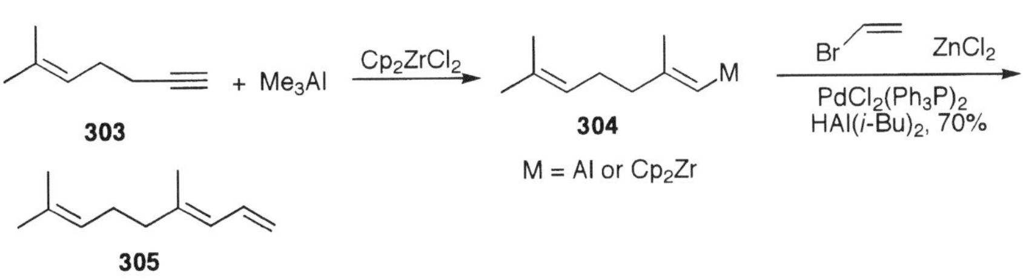

Palladium and Ni complexes are active catalysts for the coupling of Al and Zr reagnts [101,103]. The alkenylalanes **298** are prepared by hydroalumination of terminal alkynes **297**, and their coupling with aryl and alkenyl halides gives the disubstituted alkenes and (*E,E*) or (*E,Z*) conjugated diene **299** stereospecifically [101,103]. The disubstituted alkenylalanes **301** and **304** are obtained by hydroalumination of the internal alkynes **300** or by the *syn* carboalumination of the terminal alkynes **303** in the presence of a stoichiometric amount of Cp_2ZrCl_2. Their coupling with aryl and alkenyl halides affords the trisubstituted alkene **302** or the 1,3-diene **305** using Pd or Ni catalyst in the presence of a stoichiometric amount of $ZnCl_2$. As no reaction takes place without $ZnCl_2$, the reaction seems to proceed by the double transmetallation from Al to Zn and then to Pd (or Ni) [135].

The aryl and enol triflates **306** and **307** couple with Me_3Al, Et_3Al and Bu_3Al [136]. The enol phosphate **309**, derived from ketone **308**, is displaced with methyl group of Me_3Al using Pd catalyst in dichloroethane. Based on this reaction, 4-*tert*-butylcyclohexanone (**308**) is converted to 2-methyl-5-*tert*-butylcyclohexanone (**311**) via **310** [137].

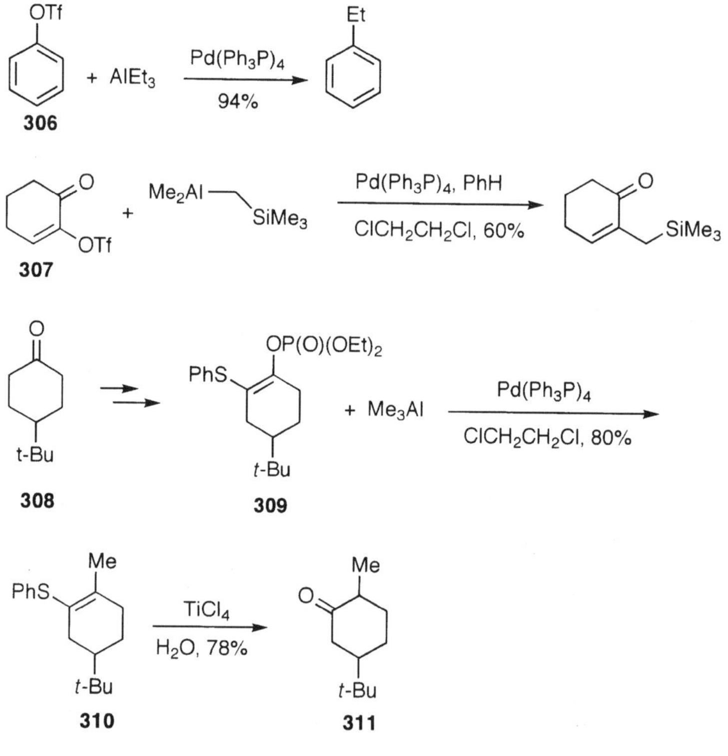

The alkenylzirconium compound **314** is prepared by hydrozirconation of 1-heptyne (**312**) with hydrozirconocene chloride (**313**), and reacts with alkenyl iodide to afford the 1,3-diene **315** [138]. The Zr reagent can be used even in the presence of carbonyl group, which is sensitive to Al or Mg reagents.

3.4.5 Tin Compounds

As organotin compounds (organostannanes) undergo smooth Pd-catalysed transmetallation, aryl halides react with a wide variety of aryl-, alkenyl- and alkylstannanes [139]. Coupling with allylstannane is the first example [140]. The reaction is called the Migita–Kosugi–Stille or Stille coupling. Aryl, alkenyl, allyl, alkynyl and benzyl

312 **313** **314** **315**

groups of organostannanes undergo the transmetallation, and are used for the coupling. Generally only one of four groups on the tin enters into the coupling reaction.

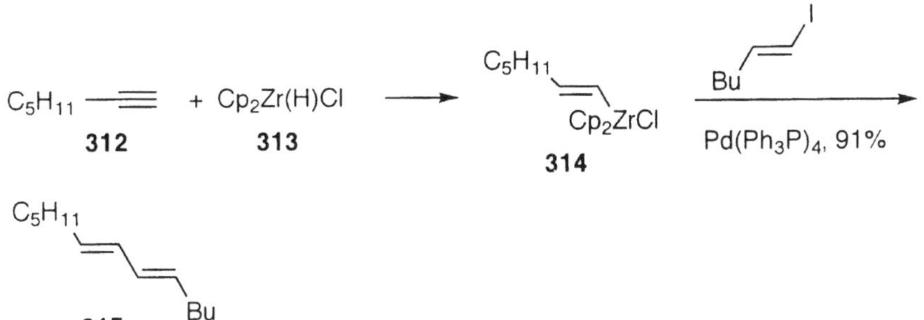

Different groups are transferred with different selectivities from tin. A simple alkyl group has the lowest transfer rate. Transfer of a methyl group by the transmetallation of methylstannanes is slow. Thus unsymmetrical organotin reagents containing three simple alkyl groups (usually methyl or butyl) are chosen, and the fourth group, which is usually alkynyl, alkenyl, aryl, allyl or benzyl, undergoes the transmetallation with aryl and alkenyl halides, pseudohalides and arenediazonium salts. The cross-coupling of these tin groups with aryl, alkenyl, alkynyl and benzyl halides affords a wide variety of the cross-coupled products that are difficult to prepare by the uncatalysed reactions. Usually, Ph_3P is used as a ligand. A large acceleration is observed in the rate of coupling for some organostannanes when tri-2-furylphosphine (TFP) or Ph_3As is used [141]. Although Ph_3As is a good ligand for the coupling of triflates, Ph_3P retards the reaction [141,143]. However, it is claimed that ligandless Pd complexes are active for the cross-coupling at room temperature [142].

The cross-coupling of aromatic and heteroaromatic stannanes has been carried out extensively [144]. Tin compounds of heterocycles, such as the oxazole **316** [145], the thiophene (**318**) [146], furans and pyridines [147], can be coupled with aryl halides.

$$\text{R-X} + \text{R'SnR''}_3 \xrightarrow{\text{Pd(0)}} \text{R-R'} + \text{X-SnR''}_3$$

R' = aryl, alkenyl, allyl, alkynyl, benzyl

316 **317**

318 **319**

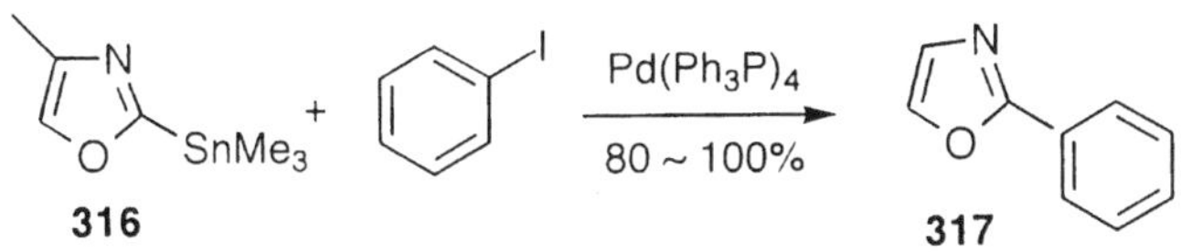

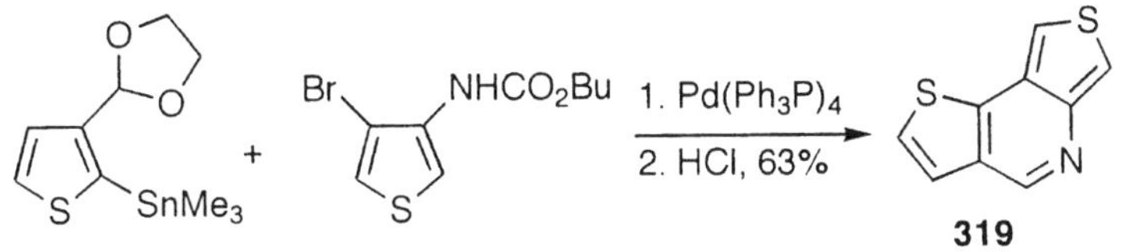

The syntheses of the phenyloxazole **317** [148], and the dithiophenopyridine **319** are typical examples [149].

Coupling of the (*E*)-alkenylstannane **320** with the (*Z*)-alkenyl iodide **321** using a ligandless Pd catalyst gives the (*E,Z*)-diene **322** with retention of the stereochemistry [150]. Arenediazonium salts are also used for the coupling without phosphine [151]. (*Z*)-stilbene (**324**) is obtained unexpectedly by the *cine* substitution of the α-stannylstyrene **323** by the addition–elimination mechanism. This is a good preparative method for *cis*-stilbene (**324**) [152]. The cyclic triflate **325** is converted to the alkenylstannane **326** and its reaction with aryl iodide affords the *cine* product **329**. The reaction is explained via the Pd–carbene complex **328**, which is formed by insertion of the double bond of **326** to Ar−Pd−X to give **327**, rather than transmetallation, and α-elimination of Me₃SnX from **327**. Finally the *cine* product **329** is formed by 1,2-hydride shift of **328** [153,154].

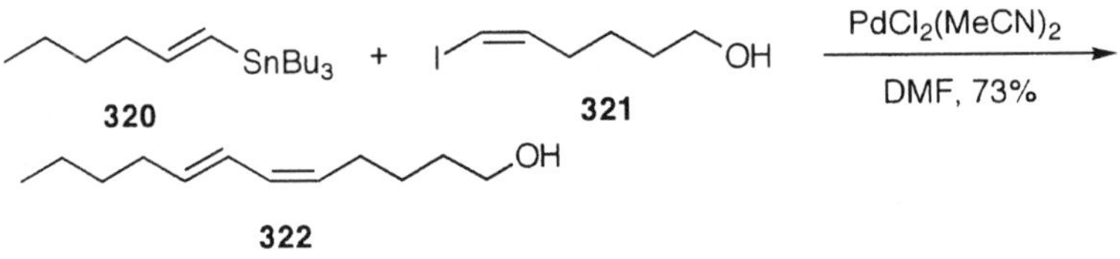

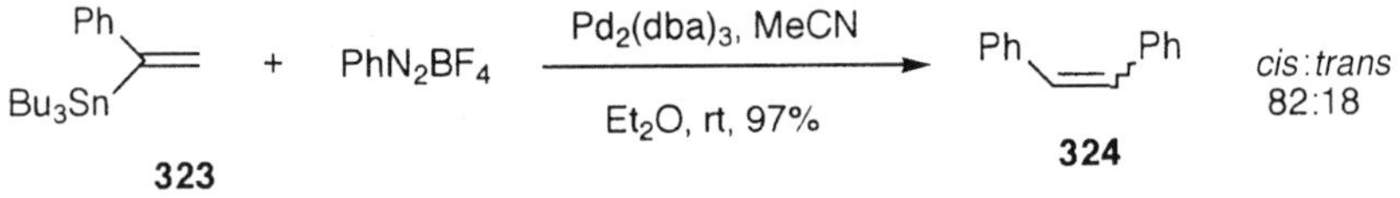

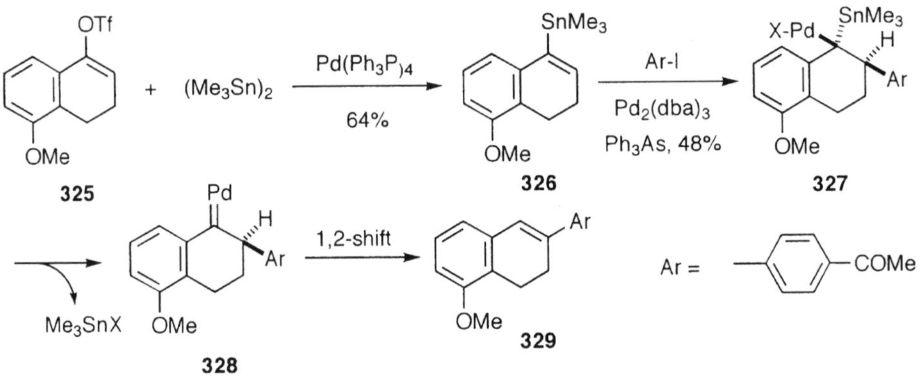

Optimum conditions for the coupling of the alkenyl triflates **330** with the arylstannanes **331** have been studied. Ligandless Pd complexes such as Pd(dba)₂ are most active in the reaction of the enol triflate. Ph₃P inhibits the reaction. NMP as a polar solvent gives the best results. The use of tri(2-furyl)phosphine and Ph₃As in the coupling of stannanes with the halides and triflates increases the rate of the transmetallation of the stannanes to Pd, which is thought to be the rate-determining step of the catalytic cycle. Relative rates are $Ph_3P:(fur)_3P:Ph_3As:Ph_3As + ZnCl_2 = 1:3.5:95:151$ [141,155]. The iminophosphine **332** was shown to be an excellent ligand for the cross-coupling of aryl halides with organostannanes [156].

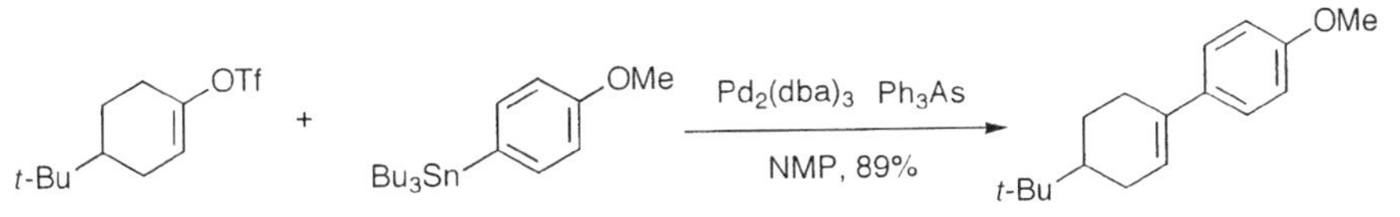

Instead of enol triflate, the enol phosphate **334**, derived from lactone **333**, is used for the coupling to afford the cyclic enol ether **335**. No coupling of phenyl phosphate in **334** takes place [157].

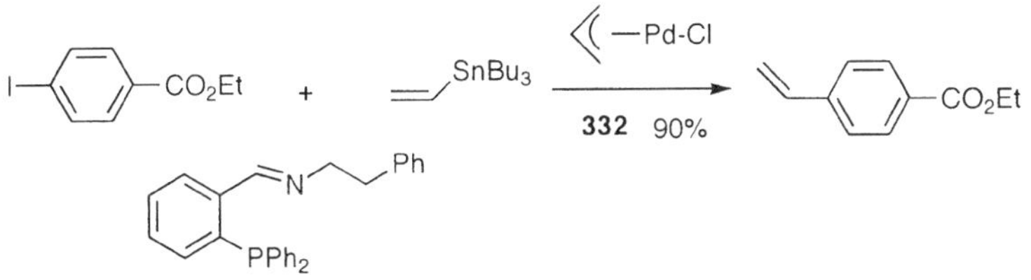

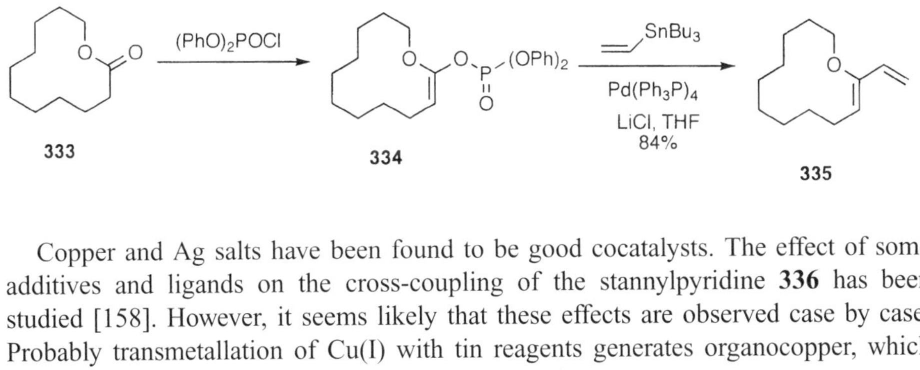

Copper and Ag salts have been found to be good cocatalysts. The effect of some additives and ligands on the cross-coupling of the stannylpyridine **336** has been studied [158]. However, it seems likely that these effects are observed case by case. Probably transmetallation of Cu(I) with tin reagents generates organocopper, which undergoes facile transmetallation with the Pd species.

Coupling of aryl chlorides with vinylstannane can be carried out using Ni(0)–Ph$_3$P as a catalyst [158a].

The construction of the highly functionalized cyclic polyether framework **339** in maitotoxin has been achieved by efficient coupling of the alkenylstannane **338** with the enol triflate **337** [159]. The smooth coupling occurs in the presence of CuCl (2 equivalents) and K_2CO_3 in THF.

The conjugated (*E,E,E*) triene part of rapamycin **341** has also been constructed by the intramolecular coupling of the (*E,E*) iododiene with the (*E*) vinylstannane in **340** in 75% yield [160].

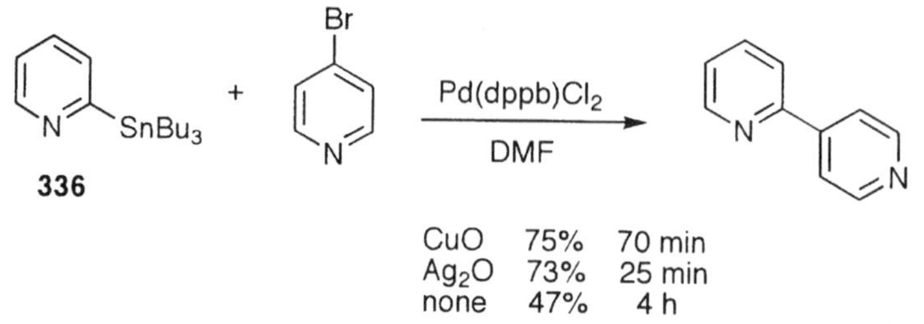

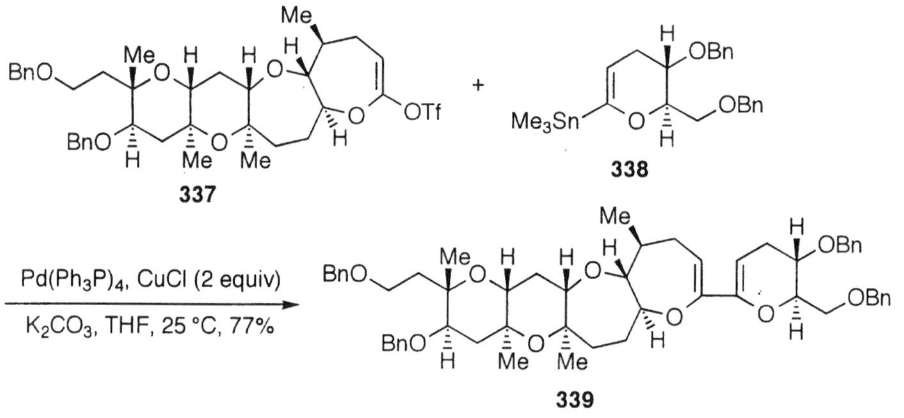

CuO   75%   70 min
Ag₂O   73%   25 min
none   47%   4 h

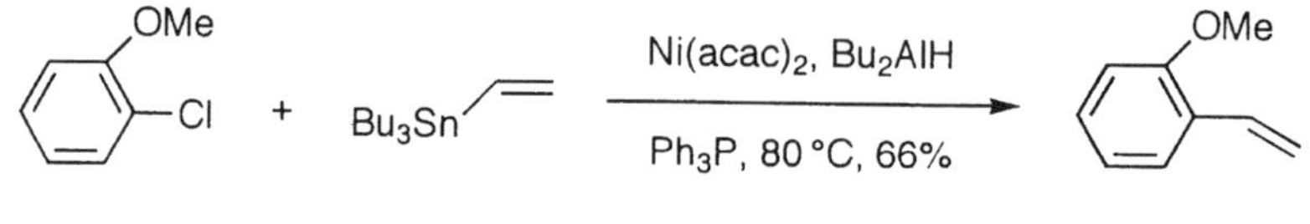

3.4.6 Silicon Compounds

The transmetallation of Si attached to sp² carbons (aryl and alkenyl) is slow, but accelerated by the addition of TASF [161]. Trimethylsilylethylene (**342**), activated by TASF, is an ethylene equivalent because it reacts with aryl and alkenyl halides to afford the styrene derivatives **343** and diene [162]. The rate of the transmetallation is enhanced by using Si compounds attached to alkoxy of fluoride groups [163]. The facile reaction by the action of F⁻ supplied from TASF, TBAF and KF is explained by the formation of five-coordinated silicate compounds [161]. The cross-coupling of the aryl and alkenylsilyl compounds **344** and **346** with aryl or alkenyl halides and triflates offers good synthetic routes for the biaryls **345**, the alkenyl arenes **348** and 1,3-dienes [163]. The aryl chlorides **347**, activated by an electron-withdrawing group (EWG), react with the vinylsilane **346** using triethylphosphine as a ligand more smoothly using NaOH, instead of TBAF [164,165]. The ketene silyl acetal **349** reacts with aryl triflates in the presence of AcOLi using DPPF to give the α-arylcarboxylate **350** [166].

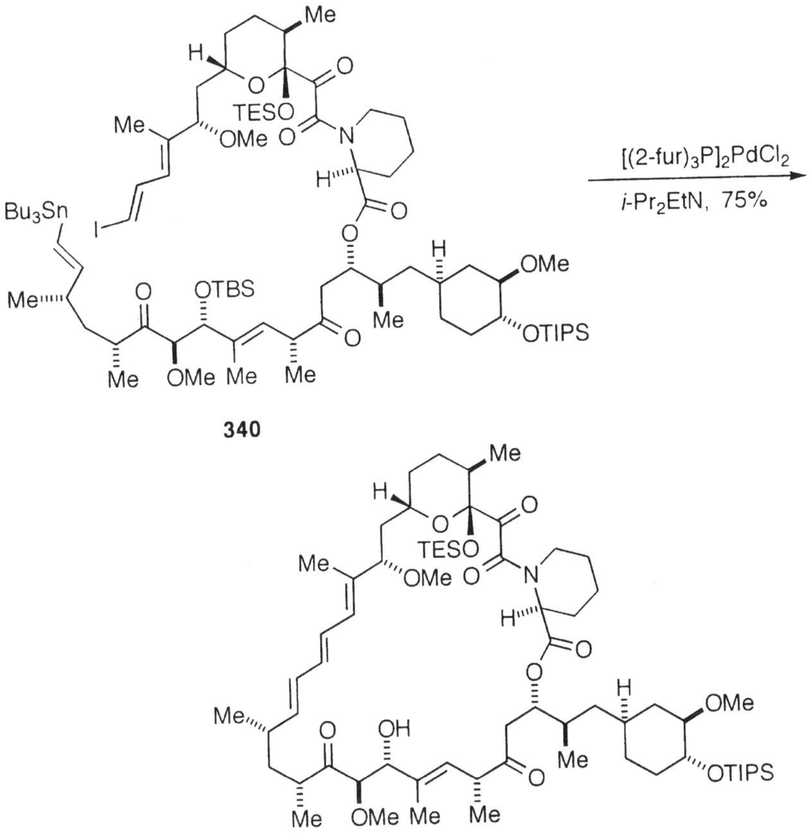

$[(2\text{-fur})_3P]_2PdCl_2$

$i\text{-}Pr_2EtN$, 75%

340

341

3.4.7 Chromium Compounds

Organochromium compounds are useful reagents because they react with aldehydes selectively without attacking ketones [167,168]. Although the chemistry involved is somewhat different from that of the usual 'oxidative addition–transmetallation–reductive elimination' type of reaction, the reaction of Cr(II) with halides is treated here. $CrCl_2$ is commercially available as a moisture- and air-sensitive reagent. Reduction of $CrCl_3$ with $LiAlH_4$ generates $CrCl_2$ *in situ,* which seems to give less reliable results than the commercial product in the case of the coupling of aryl and alkenyl halides [168a]. The arylchromium or alkenylchromium compound **351** is formed by the treatment of aryl or alkenyl halides with a stoichiometric amount of $CrCl_2$. The addition of a catalytic amount of $NiCl_2$ is essential [169,170]. The Ni-catalysed reaction is called the Nozaki–Hiyama–Kishi reaction. Ni(II) is reduced to Ni(0) with Cr(II) and forms alkenylnickel **352** by oxidative addition. Then the alkenylchromium **351** is generated by the transmetallation of **352** with Cr(III) and reacts with aldehyde selectively to give **353** without reacting with ketone. Use of excess $NiCl_2$ tends to induce homocoupling of the halides. The synthetic method with

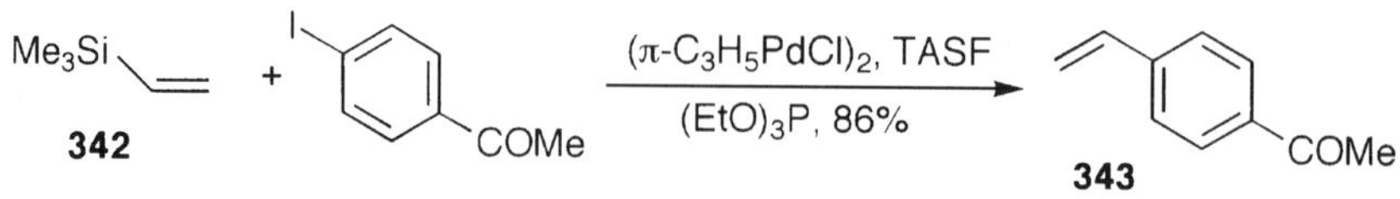

Me₃Si—CH=CH₂ (**342**) + 4-iodo-acetophenone $\xrightarrow[\text{(EtO)}_3\text{P, 86\%}]{(\pi\text{-C}_3\text{H}_5\text{PdCl})_2,\ \text{TASF}}$ vinyl-acetophenone (**343**)

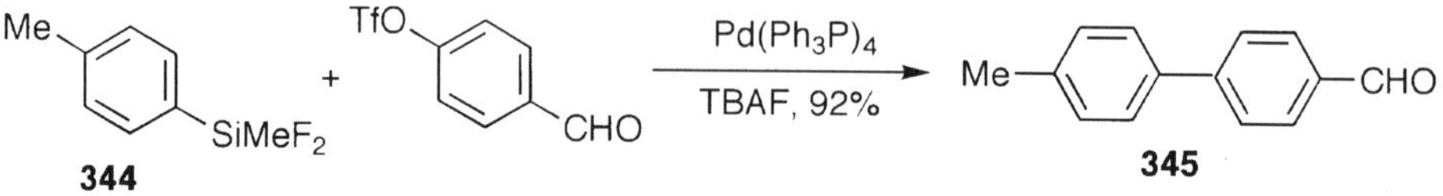

344 (Me–C₆H₄–SiMeF₂) + TfO–C₆H₄–CHO $\xrightarrow[\text{TBAF, 92\%}]{\text{Pd(Ph}_3\text{P)}_4}$ Me–biphenyl–CHO (**345**)

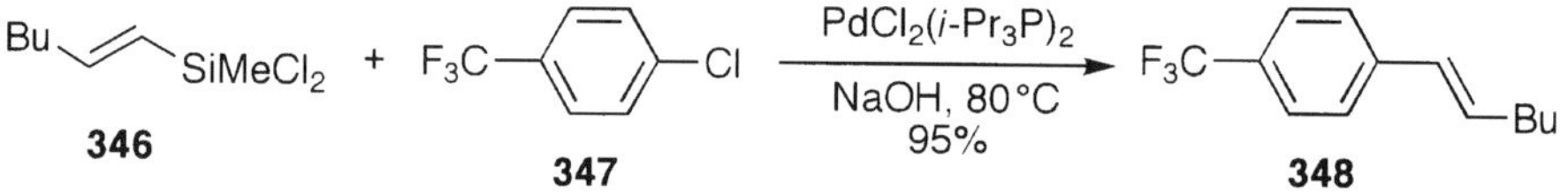

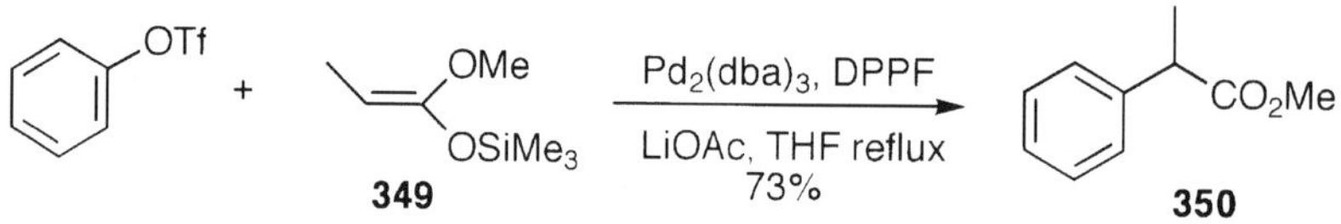

Bu–CH=CH–SiMeCl₂ (**346**) + F₃C–C₆H₄–Cl (**347**) $\xrightarrow[\substack{\text{NaOH, 80\,°C}\\95\%}]{\text{PdCl}_2(i\text{-Pr}_3\text{P})_2}$ F₃C–C₆H₄–CH=CH–Bu (**348**)

Ph–OTf + CH₃–C(OMe)=CH–OSiMe₃ (**349**) $\xrightarrow[\substack{\text{LiOAc, THF reflux}\\73\%}]{\text{Pd}_2(\text{dba})_3,\ \text{DPPF}}$ Ph–CH(CH₃)–CO₂Me (**350**)

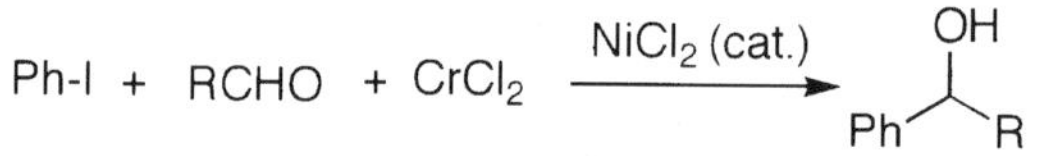

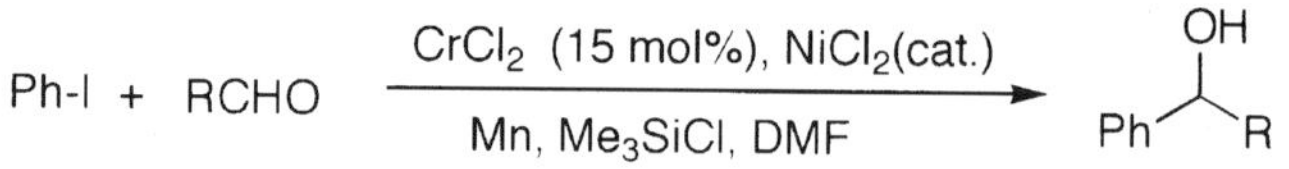

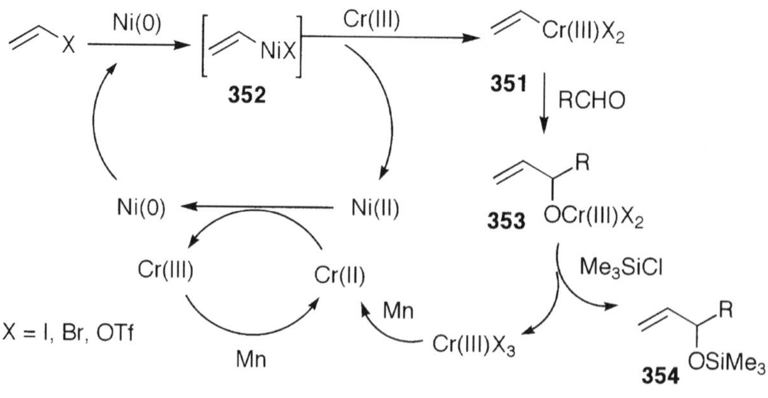

Ph–I + RCHO + CrCl₂ $\xrightarrow{\text{NiCl}_2\ (\text{cat.})}$ Ph–CH(OH)–R

Ph–I + RCHO $\xrightarrow[\text{Mn, Me}_3\text{SiCl, DMF}]{\text{CrCl}_2\ (15\ \text{mol\%}),\ \text{NiCl}_2(\text{cat.})}$ Ph–CH(OH)–R

a catalytic amount of the Cr reagent has been developed by the addition of Mn powder in the presence of Me_3SiCl. The Mn powder reduces Cr(III) to Cr(II), and Me_3SiCl cleaves Cr alkoxide **353** to liberate Cr(III) halide and the silyl ether **354** [171].

The Nozaki–Hiyama–Kishi reaction has been successfully utilized in the total synthesis of palytoxin (**355**) [170] brefeldin (**356**) [172], halichondrin B [173], brevetoxin [174], pinnatoxin A [174a] and others. The coupling of alkynyl iodides with aldehydes can be carried out smoothly using $CrCl_2$ and 0.01% of $NiCl_2$ [175] in THF. The taxane framework **357** was constructed by the intramolecular version. In this reaction, use of a stoichiometric amount of $NiCl_2$ gave better results [176].

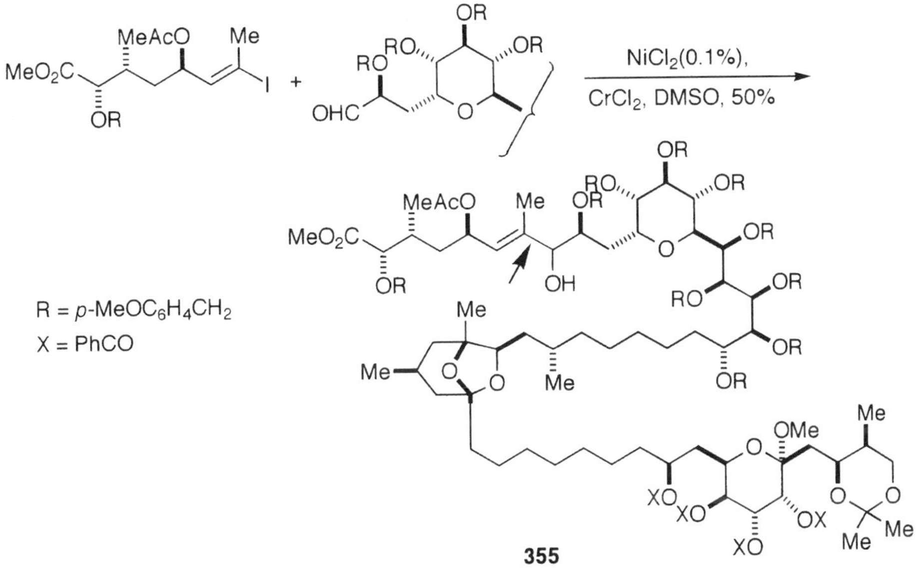

R = *p*-MeOC$_6$H$_4$CH$_2$
X = PhCO

355

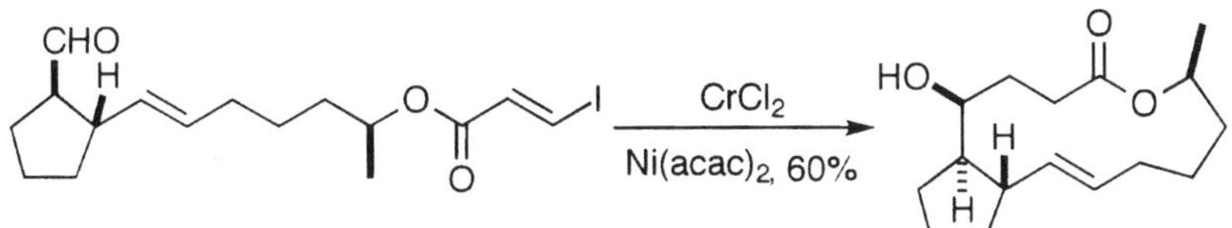

356

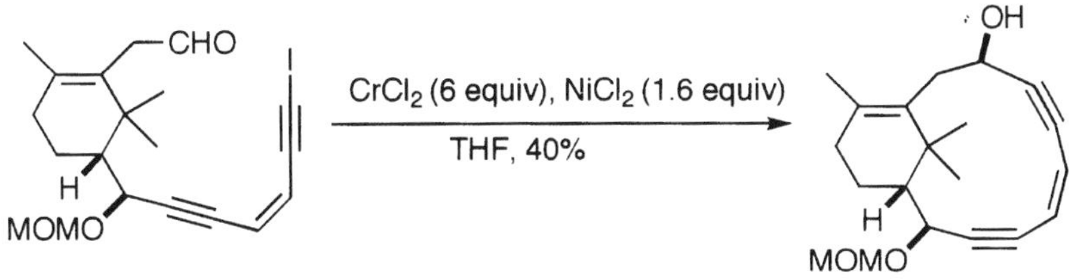

357

3.4.8   Reactions with Dimetallic Compounds

Palladium-catalysed reactions of dimetallic compounds **358** such as X_2B-BX_2, $R_3Sn-SnR_3$, $R_3Sn-SiR_3$ or $R_3Si-SiR_3$ with halides via oxidative addition and transmetallation are useful for the preparation of carbon–main group metal bonds **359**.

Palladium-catalyzed cross-coupling of the stable pinacol ester of diboron **360** or tetra(alkoxy)diborons with halides or triflates affords the arylboronate derivatives **361** which are difficult to prepare by conventional methods using ArLi or Grignard reagents and $B(OMe)_3$. As a ligand, DPPF is the most suitable [177]. It was found that the arylboronate **363** is prepared unexpectedly by the Pd-catalyzed reaction of pinacolborane (**362**) without undergoing expected hydrogenolysis [178].

Palladium-catalysed coupling of $Me_3SnSnMe_3$ (**365**) with halides provides a good synthetic route to organotin reagents. The oxidative addition of halides, transmetallation (stannylation), and reductive elimination sequences afford the organostannane. Aryl, alkenyl, benzyl and allyl halides react with **365** to afford aryl, alkenyl, benzyl and allylstannanes [179]. $(Bu_3Sn)_2$ is unreactive. In many cases, the products of the reaction are used without isolation. The alkenyl chloride **364** is converted to the alkenylstannane **366**. As a synthetic application, the intramolecular coupling between aryl iodides in **367** was carried out in the presence of $(Me_3Sn)_2$. The oxidative addition of one of the iodides and subsequent alkene insertion generate the neopentylpalladium **368**, and its transmetallation with the distannane **365** gives **369**. Oxidative addition of the remaining iodide in **369**, intramolecular transmetallation and reductive elimination, gave the coupling product **370** [180].

The trialkylsilyl group can be introduced to aryl or alkenyl groups using hexaalkyldisilanes. The oxidative addition of alkenyl iodide and transmetallation, followed by reductive elimination, afford the silylated products [181]. The facile Pd-catalysed reaction of Me$_3$SiSiMe$_3$ (**371**) to give **372** proceeds at room temperature in the presence of fluoride anion [182]. Alkenyl and arylsilanes are prepared by the reaction of tris(trimethylsilyl)aluminium (**373**) with halides [183]. Benzyldimethyl-chlorosilane (**375**) is obtained by the coupling of benzyl chloride with dichlorotetramethyldisilane (**374**) [184].

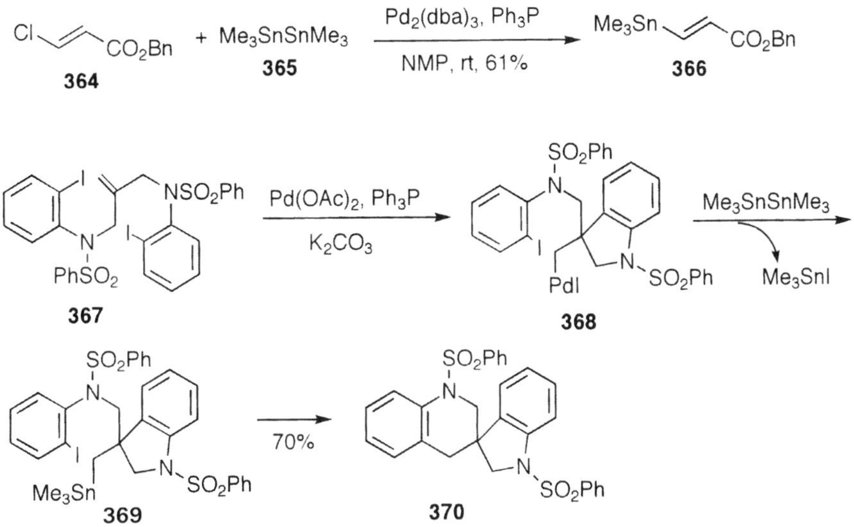

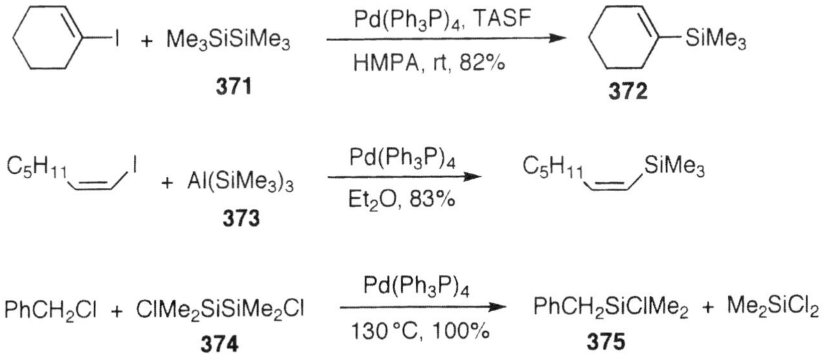

Under certain conditions, aroyl chlorides are converted to arylsilanes by the reaction with disilanes. The oxidative addition of aroyl chloride and decarbonylation are followed by transmetallation and reductive elimination to give aryl silanes. Neat trimellitic anhydride acid chloride (**377**) reacts with dichlorotetramethyldisilane (**376**) at 145 °C to generate **378**, which affords 4-chlorodimethylsilylphthalic anhydride (**379**) by reductive elimination. Finally it was converted to **380** and used for polyimide formation [185]. Biphenyltetracarboxylic anhydride **381** is obtained at a higher temperature (165 °C) in refluxing mesitylene, offering a new synthetic route to biphenyls from acid chloride [185a].

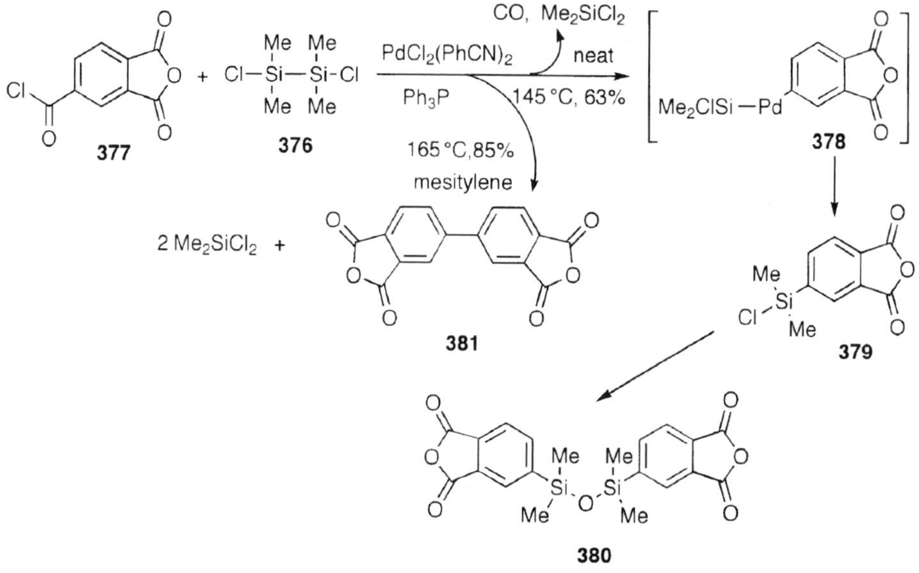

Unexpectedly, the arylsilane **383** can be prepared by the Pd-catalyzed reaction of aryl halides with the hydrosilanes **382**, without giving the expected hydrogenolysis product **384** [186].

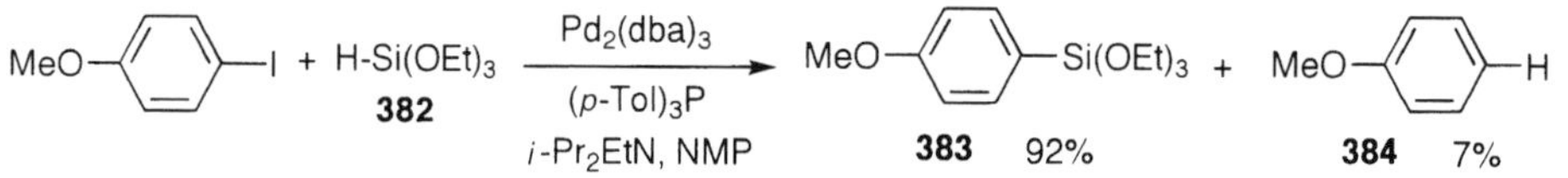

Acyl halides react with dimetal compounds without undergoing decarbonylation. The acetylstannane **385** is prepared by the reaction of acetyl chloride with $(Me_3Sn)_2$ **(365)**. The symmetric 1,2-diketones **388** can be prepared by the reaction of an excess of benzoyl chloride with $(Et_3Sn)_2$ **(386)**. Half of the benzoyl chloride is converted to the benzoylstannane **387**, which is then coupled with the remaining benzoyl chloride under CO atmosphere to afford the α-diketone **388** [187]. Triethyl phosphite is used as a ligand. The reaction of benzoyl chloride with $(Me_3Si)_2$ **(371)** affords benzoyl-trimethylsilane **(389)** [188].

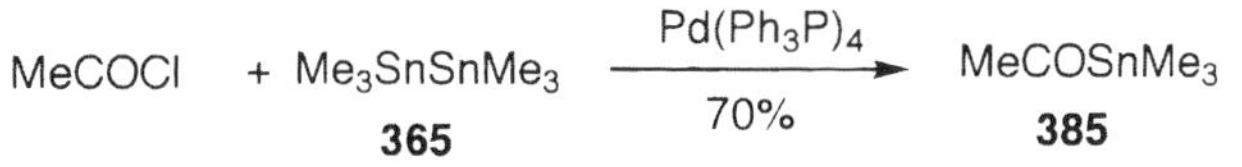

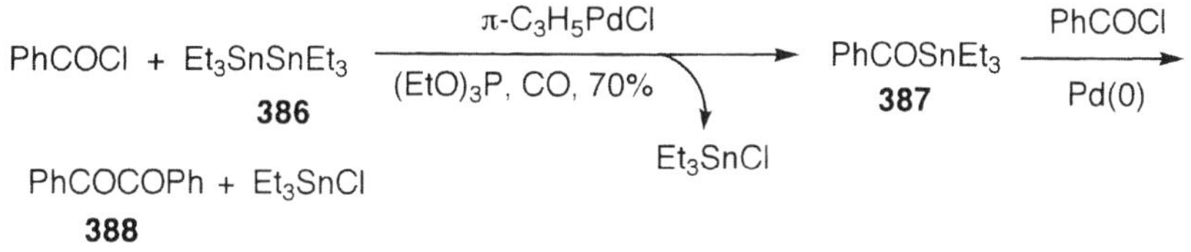

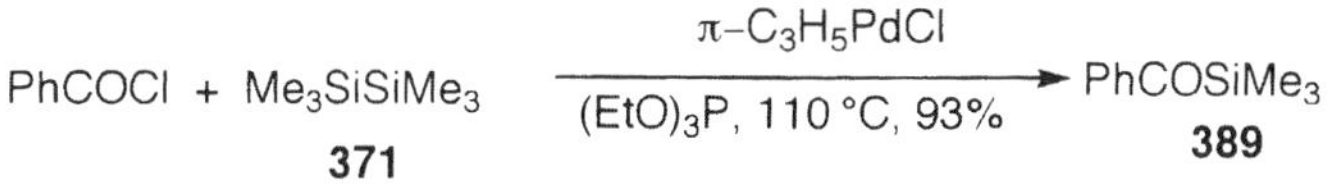

MeCOCl + Me₃SnSnMe₃ $\xrightarrow[\text{70\%}]{\text{Pd(Ph}_3\text{P)}_4}$ MeCOSnMe₃

            365   **385**

PhCOCl + Et₃SnSnEt₃ $\xrightarrow[\text{(EtO)}_3\text{P, CO, 70\%}]{\pi\text{-C}_3\text{H}_5\text{PdCl}}$ PhCOSnEt₃ $\xrightarrow[\text{Pd(0)}]{\text{PhCOCl}}$

         386   **387**

PhCOCOPh + Et₃SnCl   Et₃SnCl

  388

PhCOCl + Me₃SiSiMe₃ $\xrightarrow[\text{(EtO)}_3\text{P, 110 °C, 93\%}]{\pi\text{-C}_3\text{H}_5\text{PdCl}}$ PhCOSiMe₃

          371   **389**

3.5 Reactions with C, N, O, S and P Nucleophiles

Although arylation or alkenylation of active methylene compounds can be carried out using a Cu catalyst, the reaction is sluggish. However, the arylation of malononitrile (**390**) or cyanoacetate proceeds smoothly in the presence of a base and Pd catalysts [189]. Tetracyanoquinodimethane (**392**) is prepared by the coupling of *p*-diiodobenzene with malononitrile (**390**) to give **391**, followed by oxidation [190]. Presence of the cyano group seems to be essential for intermolecular reactions. However, the intramolecular arylation of malonates, *β*-keto esters and *β*-diketones proceeds smoothly [191]. The bromoxazole **393** reacts with phenylsulphonylacetonitrile (**394**)

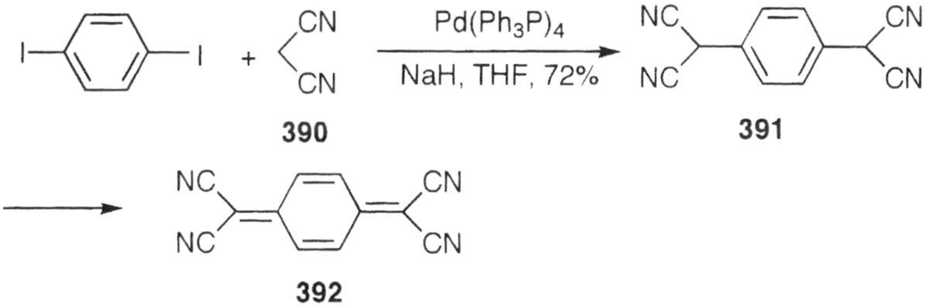

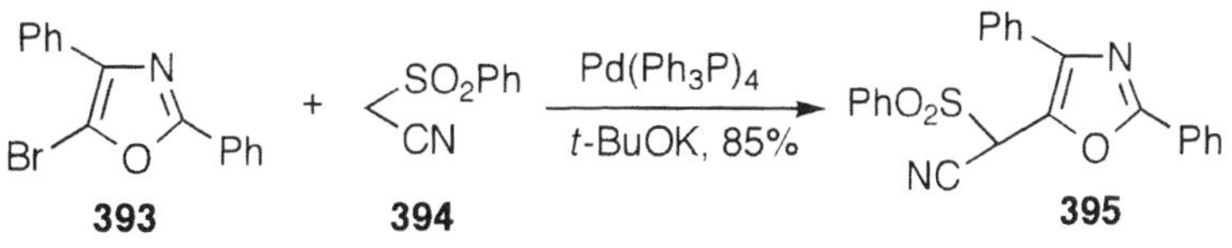

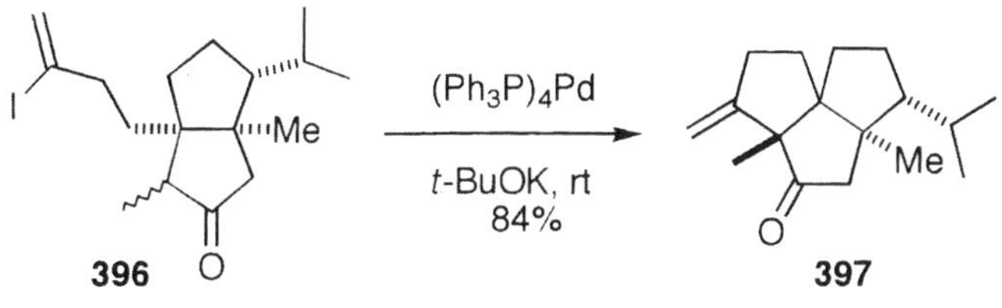

to give **395** [192]. Intramolecular alkenylation of the simple ketone **396** gave **397** by slow addition of *t*-BuOK in *t*-BuOH-THF at room temperature to avoid competing alkyne formation [193].

Intermolecular α-arylation of the ketone **399** with o-tolyl bromide (**398**) gives **400** under selected conditions using *t*-BuONa or KN(SiMe₂)₂ as suitable bases, and BINAP or DPPF (**XLIX**) as a bulky ligand [194]. Furthermore, asymmetric arylation of the ketone **402** with the bromide **401** gave **403** with 98% ee efficiently by using chiral BINAP [195].

Displacement of aryl and alkenyl iodides with KCN, or aryl triflates with $Zn(CN)_2$ [196] to afford the aryl nitrile **404** is catalyzed by a Pd complex. Similar displacement of aryl chlorides is catalyzed by a Ni complex [197].

Metal-catalyzed displacement of aryl iodides and bromides with primary and secondary alkyl- and arylamines, to afford secondary and tertiary arylamines, has been developed rapidly only recently as a useful synthetic method, and extensive studies have been carried out under various conditions [198]. The selection of ligands is important. Bulky ligands such as $(o\text{-Tol})_3P$, BINAP, and DPPF (**XLIX**) [199] were found to give best results. Primary and secondary alkyl and arylamines react smoothly in THF or toluene at 80–100 °C in the presence of *t*-BuONa as a base, as shown by the reaction of aryl bromides **405** and **406**. *t*-BuONa is used because primary and secondary alkoxides are oxidized to aldehydes and ketones. Cs_2CO_3 as a base is used for electronically neutral aryl bromides [200]. Furthermore, $t\text{-Bu}_3P$, a bulky and rarely used phosphine in catalytic reactions, was found to be a good ligand. In particular, the amination of aryl bromide with diarylamine **407** to give the triarylamine **408** proceeds satisfactorily with this ligand [201]. The amination proceeds at room temperature in the presence of crown ethers [202]. Aryl triflates **409** and **410** are similarly aminated even when they have electron-donating groups [203]. The amination of the aryl chloride **411** is catalysed by Ni(O)–DPPF [203a] and $Pd(Cy_3P)_2Cl_2$ [204].

The N-aryl imines **412** as protected anilines can be prepared by Pd-catalysed arylation of benzophenone imine with aryl halides using DPPF and BINAP as ligands, and aniline derivatives are obtained by deprotection [204a].

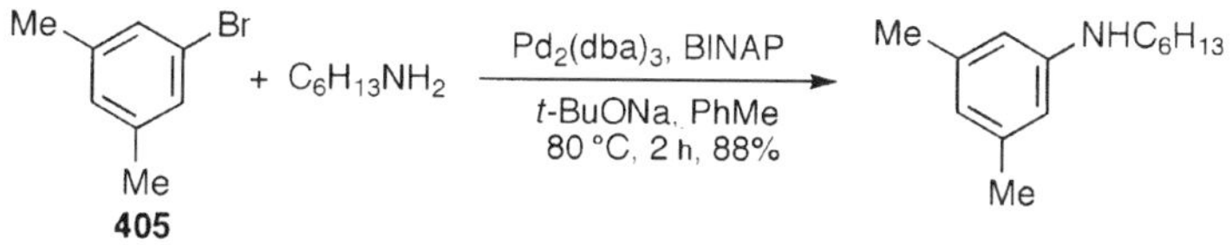

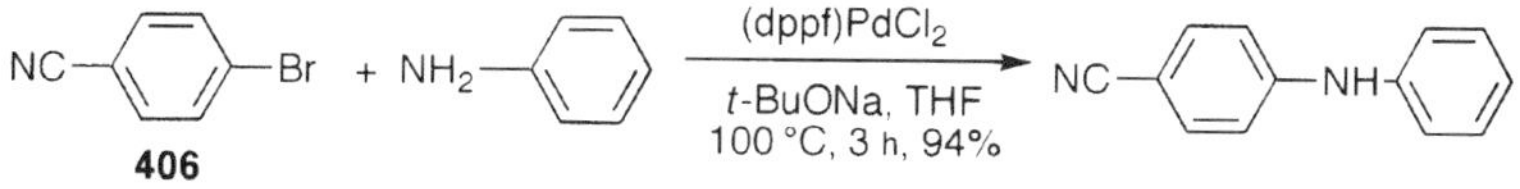

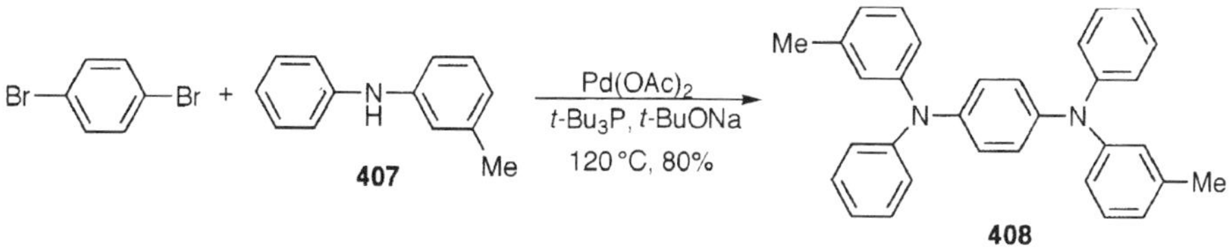

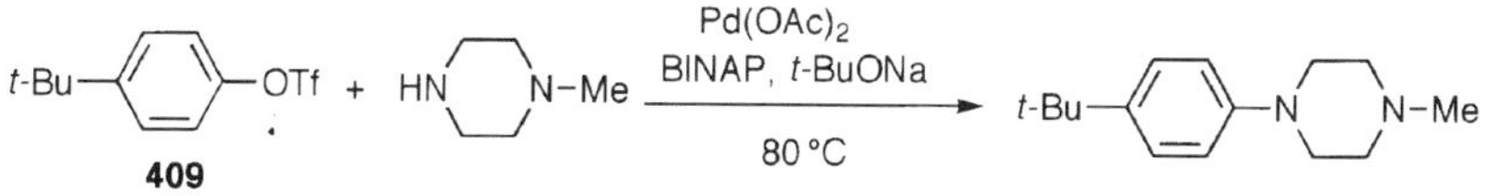

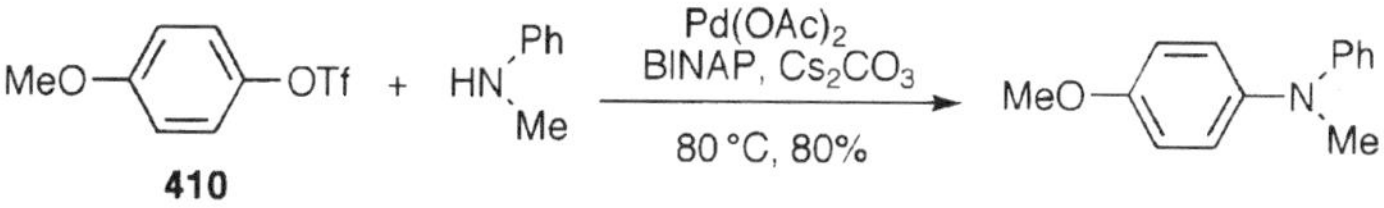

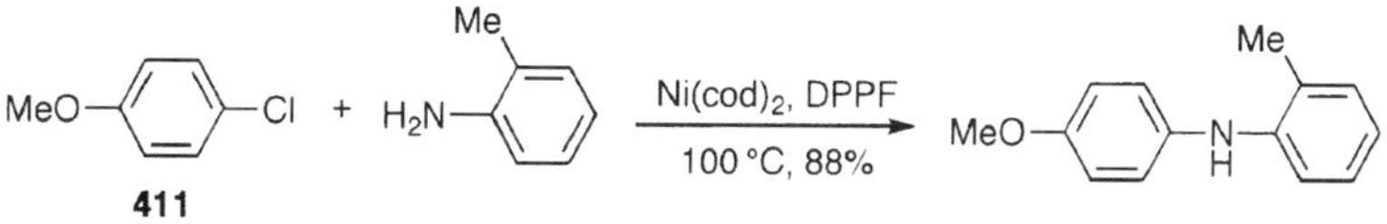

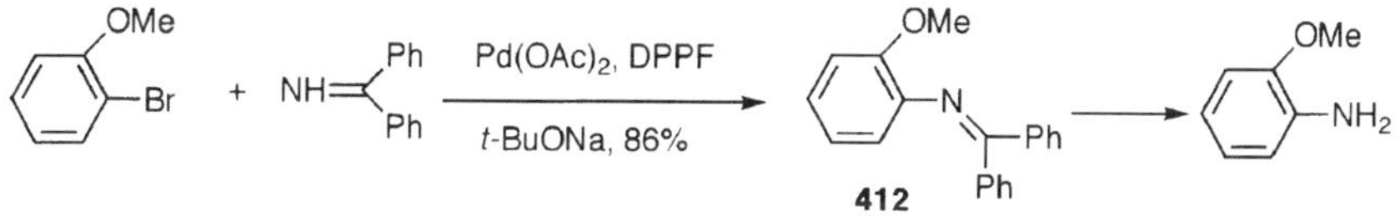

The coupling of chiral amines with aryl bromides proceeds without racemization by proper choice of ligands. Intermolecular amination of a chiral amine proceeds without loss of enantiomeric purity with Pd(0)–(o-Tol)$_3$P. Synthesis of the optically pure indole **415**, an intermediate for the synthesis of a potent ACE inhibitor, has been achieved by the Pd-catalyzed amination of **414**, which is prepared by the Heck reaction of bromide **413** and Rh-catalyzed aymmetric hydrogenation [205].

The Pd-catalysed intermolecular reaction of aryl bromides containing electron-withdrawing substituents with a wide variety of alcohols including MeOH, 2-propanol, benzyl alcohol and t-butyl alcohol gives the aryl ethers **416** under milder conditions than uncatalysed reactions. Bidentate ligands such as BINAP and DPPF (**XLIX**) are effective [206,207]. The aryl Pd alkoxide **417** was isolated as an intermediate, and the formation of the aryl ethers **418** by reductive elimination of **417** was confirmed.

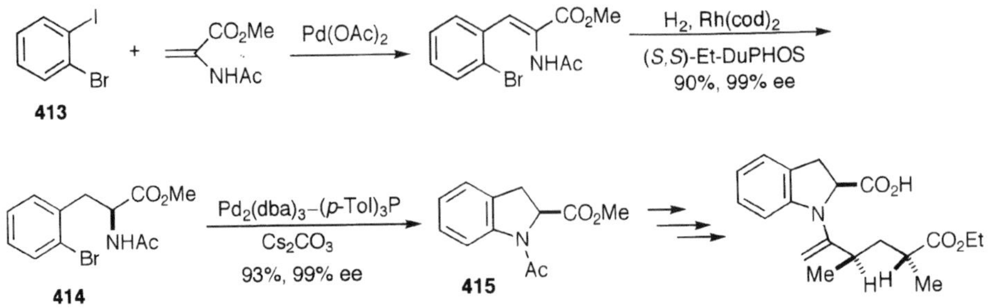

413

414   **415**

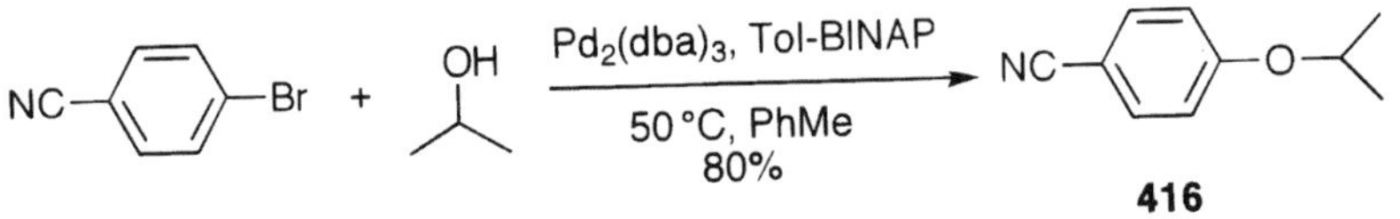

416

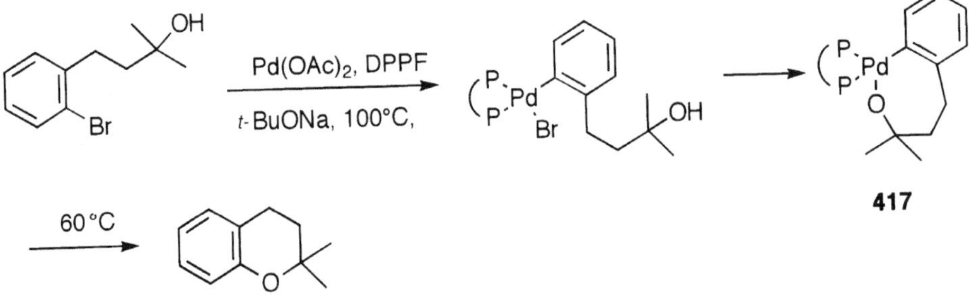

417

418   91%

Ni(cod)$_2$, coordinated by DPPF, is superior to the Pd catalyst in some cases. Ni-catalysed preparation of the aryl *t*-butyl ethers **419** and the silyl ethers **421**, and their facile deprotection offer a good synthetic method for phenols **420** and **422** from aryl bromides [208].

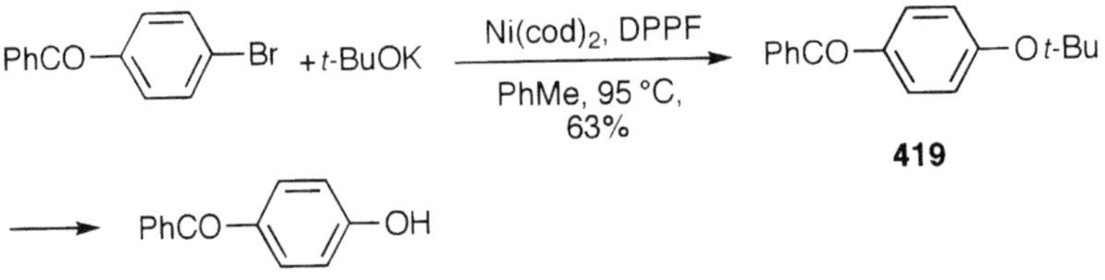

419

420

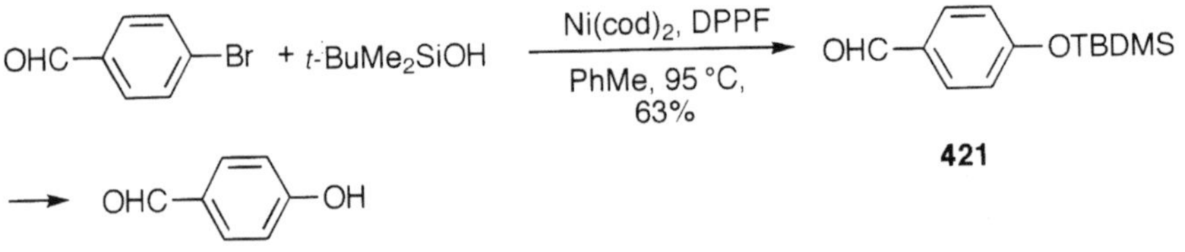

421

422

The aryl sulfides **424** are prepared by the Pd-catalyzed reaction of aryl halides with the mercaptans **423** or thiophenol **425** in DMSO [209]. Pd(OAc)$_2$–BINAP is active for the reaction of aryl triflates in the presence of *t*-BuONa [209a]. The NiBr$_2$ complex of bipyridine is an active catalyst for thioarylation of aryl iodides with the thiophenol **425**, to give diaryl thioether **426** [210].

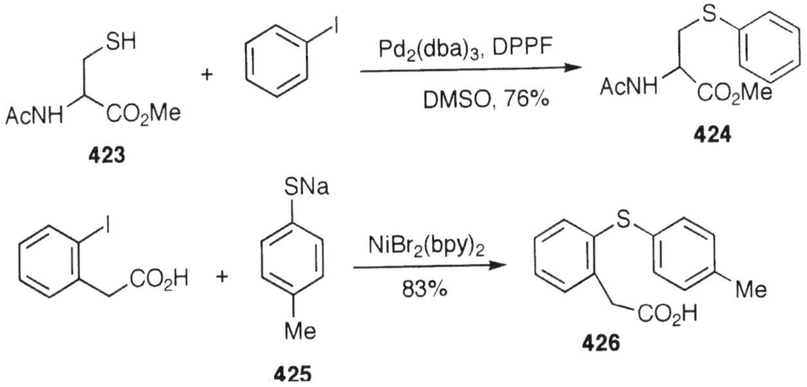

Formation of P—C bonds occurs by the reaction of various phosphorus compounds containing a P—H bond with halides or triflates. Monoaryl, symmetric and unsymmetric diarylphosphinates **428–430** are prepared by the reaction of the rather unstable methyl phosphinate **427** with different amounts of aryl iodides. Trimethyl orthoformate is added to stabilize methyl phosphinate [211].

The coupling of aryl triflates with diphenylphosphine (**432**) catalysed by Ni and Pd complexes is an important synthetic method for optically active bidentate phosphines. Bis-substitution of optically active 2,2′-bis[(trifluoromethanesulfonyl)-oxy]-1,1′-binaphthyl (**431**) with Ph$_2$PH (**432**) to give **433** has been achieved using NiCl$_2$(dppe)

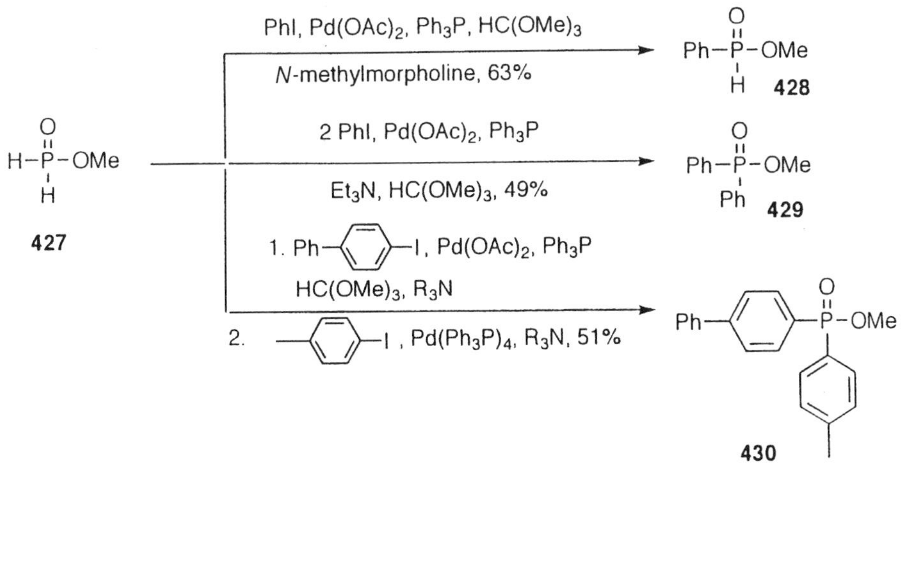

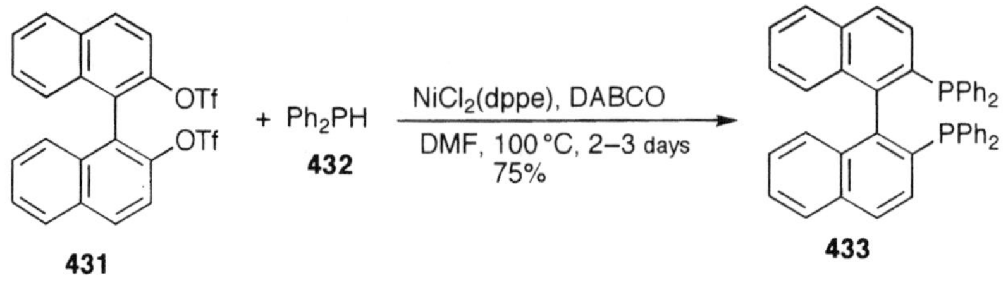
OTf
OTf
+ Ph2PH
432
NiCl2(dppe), DABCO
DMF, 100 °C, 2–3 days
75%
PPh2
PPh2
431
433

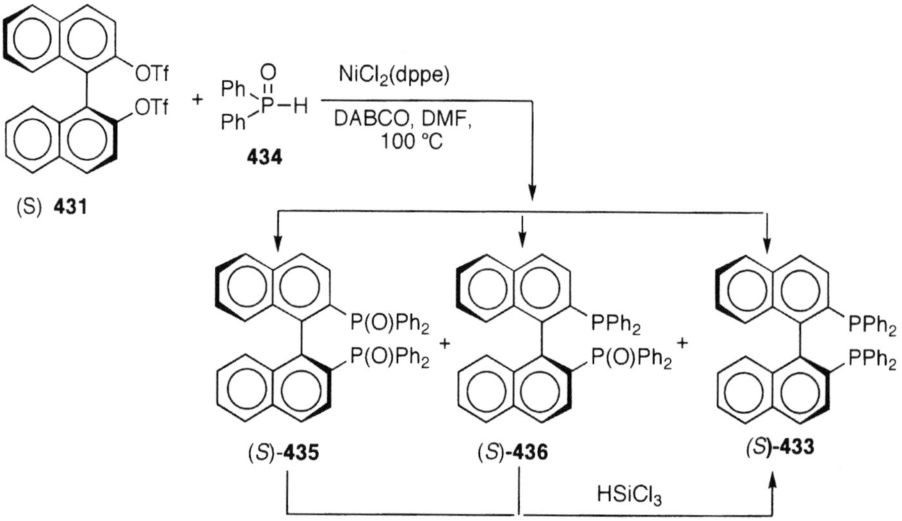
OTf
OTf
+
Ph
Ph
O
P–H
434
NiCl2(dppe)
DABCO, DMF,
100 °C
(S) 431
P(O)Ph2
P(O)Ph2
+
PPh2
P(O)Ph2
+
PPh2
PPh2
(S)-435
(S)-436
(S)-433
HSiCl3

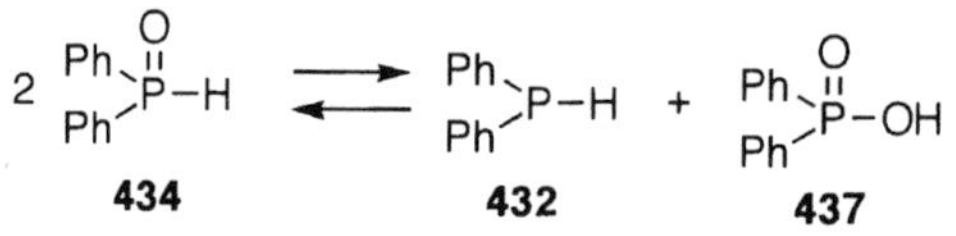
2
Ph
Ph
O
P–H
434
Ph
Ph
P–H
432
+
Ph
Ph
O
P–OH
437

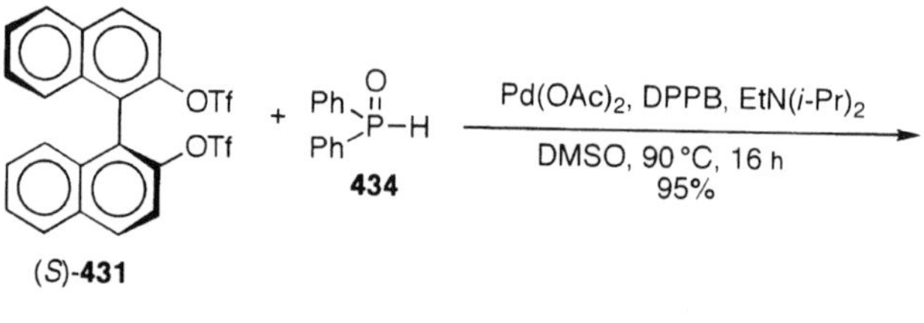
OTf
OTf
+
Ph
Ph
O
P–H
434
Pd(OAc)2, DPPB, EtN(i-Pr)2
DMSO, 90 °C, 16 h
95%
(S)-431

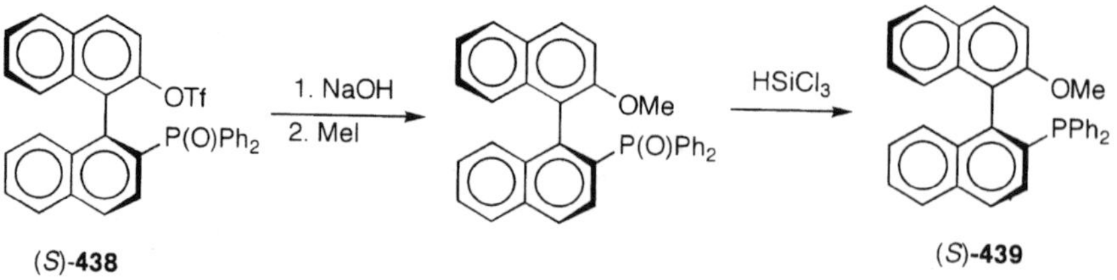
OTf
P(O)Ph2
1. NaOH
2. MeI
OMe
P(O)Ph2
HSiCl3
OMe
PPh2
(S)-438
(S)-439

e) as a catalyst, offering a convenient preparative method of BINAP. For the effective reaction, the use of DABCO as a base is essential. Other amines are ineffective [212]. Diphenylphosphine oxide (**434**) is more easily handled than Ph_2PH (**432**). Reaction of **431** with diphenylphosphine oxide (**434**) catalysed by Ni–dppe complex gives the monooxide **436** and diphosphine **433** as major products, and the dioxide **435** as a byproduct [213]. These products are formed because **434** is in equilibrium with **432** and **437**. The phosphine oxides **435** and **436** are reduced to the phosphine **433** with $HSiCl_3$. However, selective monophosphinylation of **431** with diphenylphosphine oxide (**434**) is catalysed by the Pd complex of DPPB (**XLV**) or DPPP (**XLIV**) to give optically active 2-(diarylphosphino)-1,1′-binaphthyl (**438**). No bis-substitution was observed [214]. The monodentate phosphine (MeO–MOP) (**XXXVI, 439**) can be prepared from **438**.

3.6 Carbonylation and Reactions of Acyl Chlorides

Aromatic carboxylic acids, α,β-unsaturated carboxylic acids, their esters, amides, aldehydes and ketones, are prepared by the carbonylation of aryl halides and alkenyl halides. Pd, Rh, Fe, Ni and Co catalysts are used under different conditions. Among them, the Pd-catalysed carbonylations proceed conveniently under mild conditions in the presence of bases such as K_2CO_3 and Et_3N. The extremely high toxicity of $Ni(CO)_4$ almost prohibits the use of Ni catalysts in laboratories. The Pd-catalysed carbonylations are summarized in Scheme 3.9 [215]. The reaction is explained by the oxidative addition of halides, and insertion of CO to form acylpalladium halides **440**. Acids, esters, and amides are formed by the nucleophilic attack of water, alcohols and amines to **440**. Transmetallation with hydrides and reductive elimination afford aldehydes **441**. Ketones **442** are produced by transmetallation with alkylmetal reagents and reductive elimination.

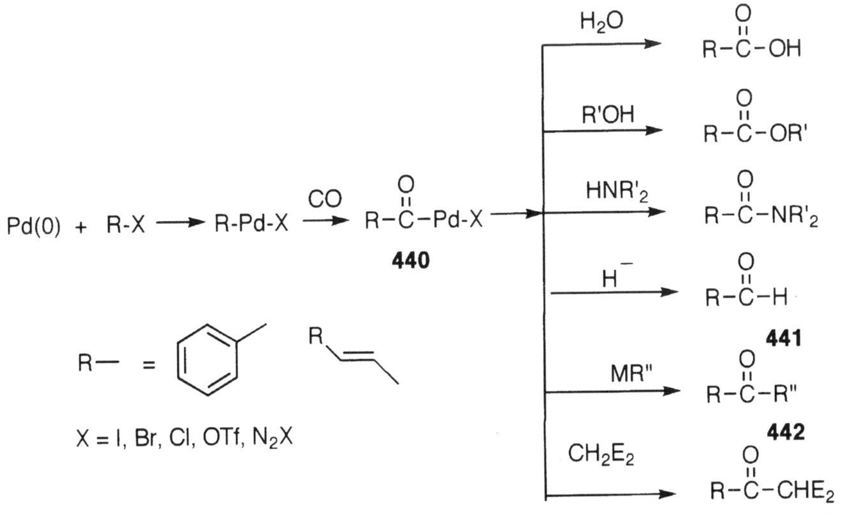

Scheme 5.9 Pd-catalysed carbonylation

3.6.1   Preparation of Carboxylic Acids and Their Derivatives

Even the sterically hindered 2,6-disubstituted aryl iodide **443** is carbonylated smoothly to give **445**. Alkyl iodide present in the alcoholic component **444** remains intact under the carbonylation conditions. This carbonylation reaction is a key reaction in the synthesis of zearalenone (**446**) [216]. Optimal conditions for technical synthesis of the anthranilic acid derivative **448** from bromide **447** has been studied, and it has been found that *N*-acetyl protection of **447**, which has a chelating effect, is important [217]. Cheaply available chlorides are rather inert [13]. The carbonylation of chloride **449** in the presence of DBU and NaI gives the amide **450** [218].

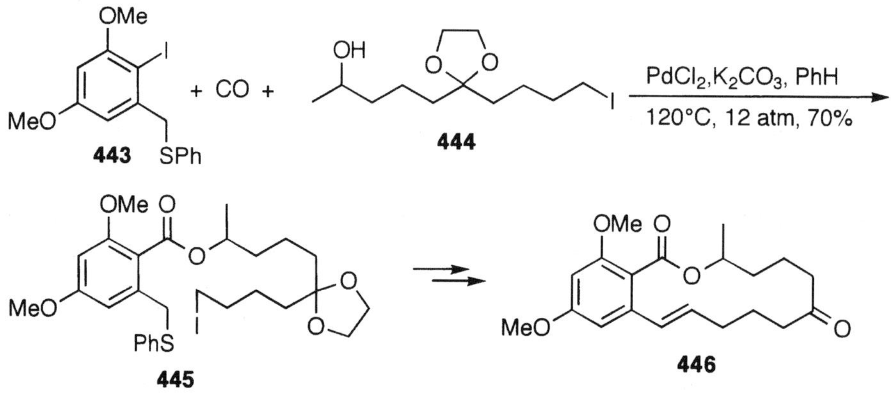

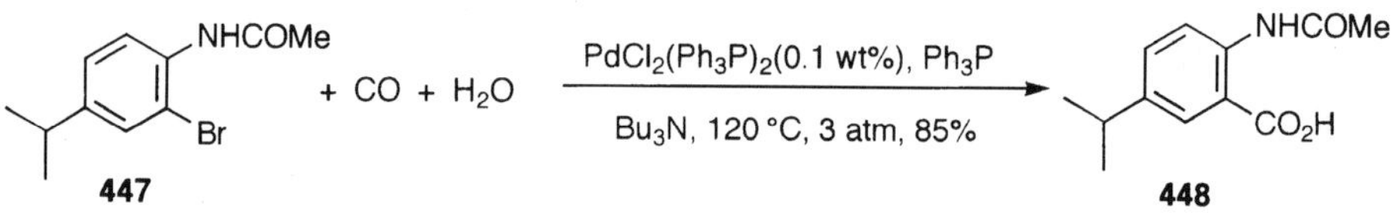

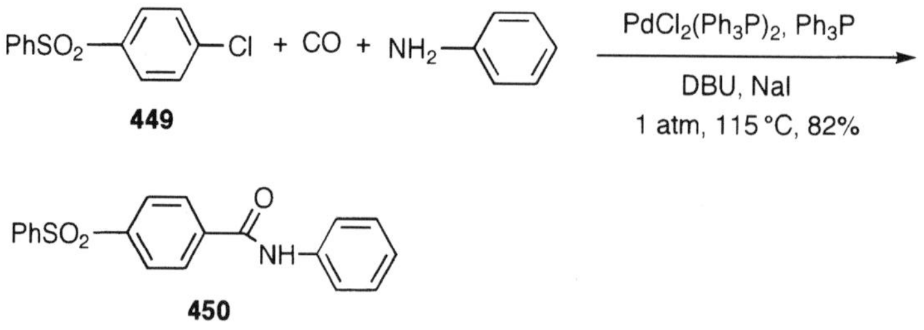

Triflates derived from phenols are carbonylated to form aromatic esters by using Pd-Ph$_3$P. The carbonylation of triflates is 500 times faster if DPPP (**XLIV**) is used [219]. This reaction is a good preparative method of benzoates from phenols and naphthoates from naphthols [220]. Carbonylation of enol triflates derived from ketones and aldehydes affords α,β-unsaturated esters. The enol triflate in **451** is more reactive than the aryl triflate and the carbonylation proceeds stepwise chemoselectively. At first, carbonylation of the enol triflate affords the amide **452**, and then the ester **453** is obtained in DMSO using DPPP [221]. The carbonylation of the enol triflate **454** to form the α,β-unsaturated acid **455** using DPPF as a ligand in aqueous DMF has been applied in the total synthesis of multifunctionalized glycinoeclepin [222].

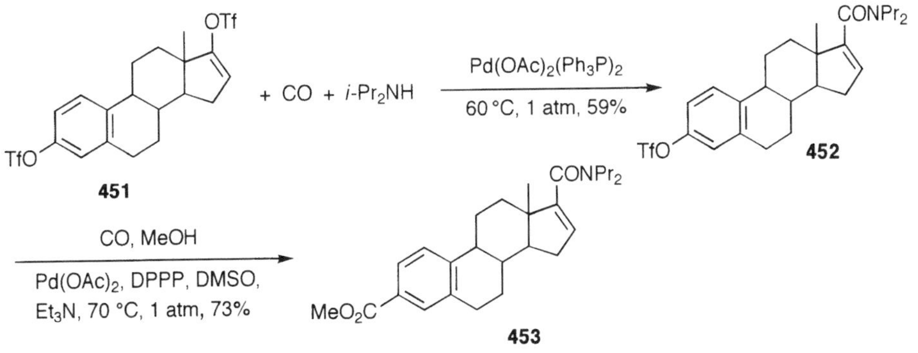

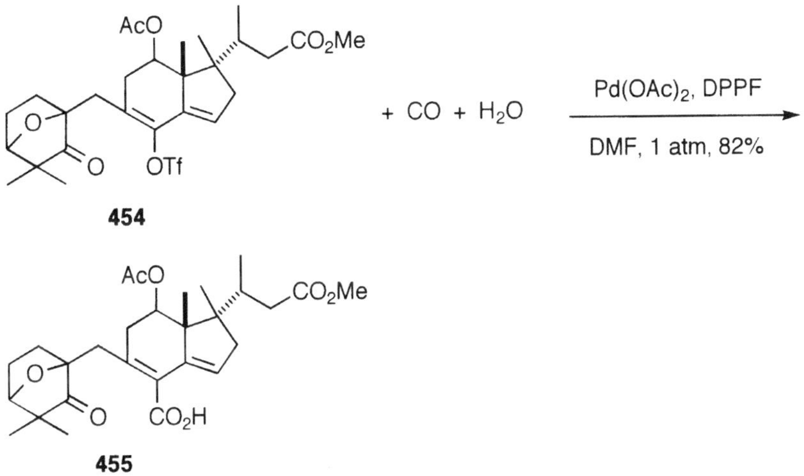

Other pseudohalides are carbonylated. Benzoic acid derivatives are prepared from arenediazonium salts at room temperature without addition of a phosphine ligand [223]. For example, the acid anhydride **457** is prepared by the carbonylation of the benzenediazonium salt **456** in the presence of AcONa. By this method, nitrobenzene can be converted to benzoic acid indirectly.

Carbonylation of halides in the presence of primary and secondary amines at 1 atm affords amides. The intramolecular carbonylation of an aryl bromide which has an amino group affords a lactam. The seven-membered lactam **459** (tomaymycin, neothramycin skeletons) is prepared from **458** by this method [224].

By careful selection of reactions conditions, double carbonylation occurs, which is competitive with monocarbonylation. Utilizing alkylphosphines or DPPB, and secondary amines, the α-keto amide **460** is obtained with high chemoselectivity [225,226].

456

457

458   **459**

$$Ph\text{-}I + CO + Et_2NH \xrightarrow[100\,°C,\ 95\%]{PdCl_2(MePh_2P)_2} PhCOCONEt_2 + PhCONEt_2$$

460

86 : 14

The Pd(0)-catalysed carbonylation of benzyl chloride proceeds under atmospheric pressure in a two-phase system utilizing the water-soluble phosphine to give phenylacetic acid (**461**) [227]. In the total synthesis of the macrolide curvularin (**465**), the Pd-catalysed carbonylation of the benzyl chloride **462** in the presence of the alcohol **463** to give the ester **464** has been applied [228]. Phenylacetic acid (**461**) is produced commercially by $Co_2(CO)_8$-catalysed carbonylation of benzyl chloride. The $Co_2(CO)_8$-catalysed double carbonylation of benzyl chloride using $Ca(OH)_2$ as a base gives phenylpyruvic acid (**466**) with high selectivity [229]. Selection of solvents is important. DME is a good solvent, whereas MeOH is not satisfactory.

Carbonylation of alkyl halides is rare. As an exception, AcOH is produced commercially by the Monsanto process from MeOH and CO using Rh as a catalyst in the presence of HI. In this process (Scheme 3.10), MeI is generated *in situ* from MeOH and HI and undergoes oxidative addition. Insertion of CO generates an acetylrhodium intermediate, and nucleophilic attack of water produces AcOH, regenerating the Rh catalyst and HI (or reductive elimination to give acetyl iodide and hydrolysis).

Acetic anhydride is also produced by the Rh-catalyzed carbonylation of methyl acetate. The method is called the Eastman process (Scheme 3.11). The Rh-catalysed production of acetic anhydride from methyl acetate can be understood by the formation of MeI and acetic acid by the reaction of methyl acetate with HI. Finally, attack of AcOH on the acetylrhodium affords the anhydride and HI, or acetyl iodide reacts with AcOH to give acetic anhydride and HI.

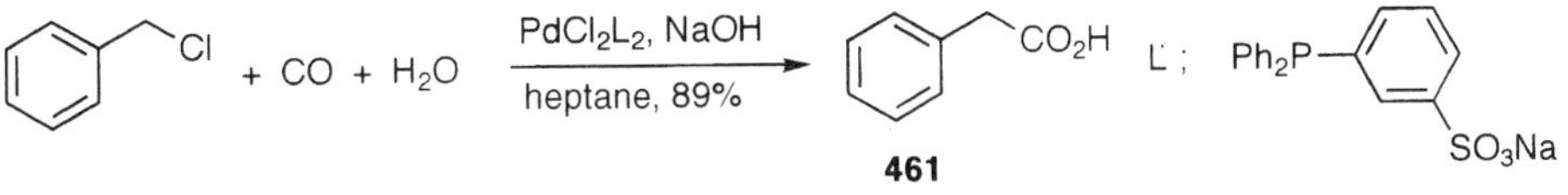

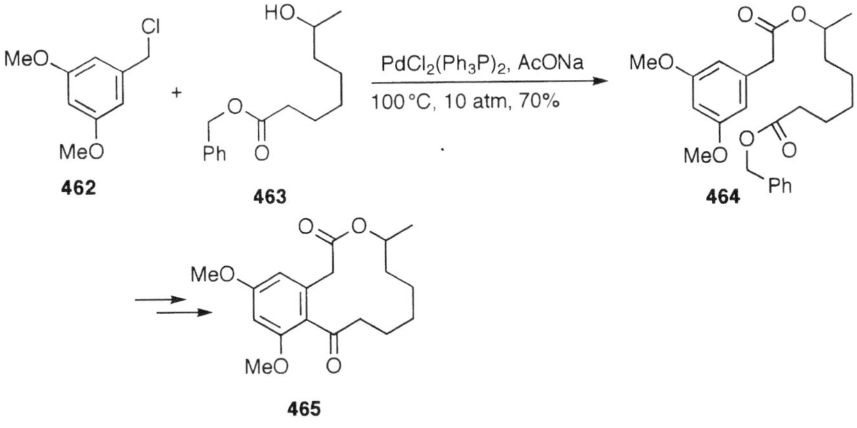

461

462   **463**   **464**

465

$$PhCH_2Cl + CO + H_2O \xrightarrow[\text{Ca(OH)}_2]{\text{Co}_2\text{(CO)}_8} PhCH_2COCO_2H + PhCH_2CO_2H$$

466   **461**

$$CH_3OH + HI \longrightarrow CH_3I + H_2O$$

$$CH_3I + Rh \longrightarrow CH_3-Rh-I$$

$$CH_3-Rh-I + CO \longrightarrow CH_3-\overset{\overset{\displaystyle O}{\|}}{C}-Rh-I$$

$$CH_3-\overset{\overset{\displaystyle O}{\|}}{C}-Rh-I + H_2O \longrightarrow CH_3CO_2H + HI + Rh$$

$$\overline{CH_3OH + CO \longrightarrow CH_3CO_2H}$$

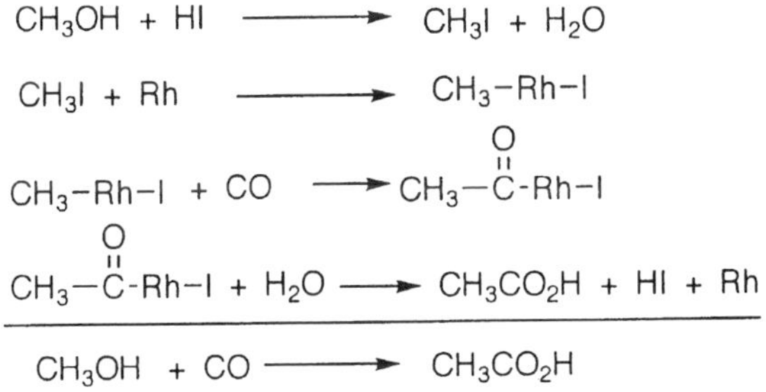

Scheme 3.10   Monsanto process

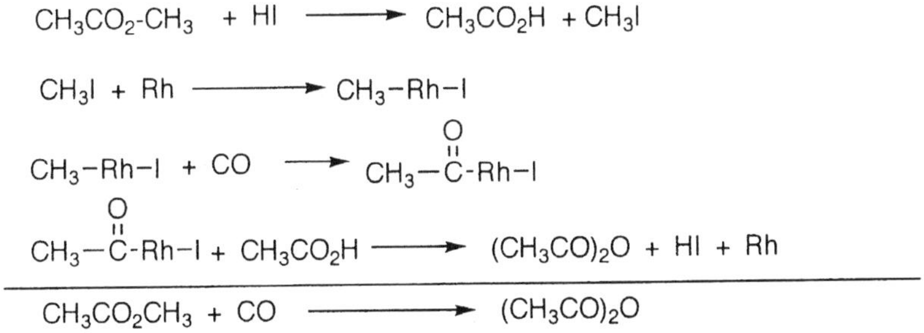

$$CH_3CO_2-CH_3 + HI \longrightarrow CH_3CO_2H + CH_3I$$

$$CH_3I + Rh \longrightarrow CH_3-Rh-I$$

$$CH_3-Rh-I + CO \longrightarrow CH_3-\overset{\overset{\displaystyle O}{\|}}{C}-Rh-I$$

$$CH_3-\overset{\overset{\displaystyle O}{\|}}{C}-Rh-I + CH_3CO_2H \longrightarrow (CH_3CO)_2O + HI + Rh$$

$$\overline{CH_3CO_2CH_3 + CO \longrightarrow (CH_3CO)_2O}$$

Scheme 3.11   Eastman process

3.6.2   Preparation of Aldehydes and Ketones

Aldehydes and ketones are prepared by trapping generated acyl complexes with hydride and carbon nucleophiles. Aldehydes can be prepared by the carbonylation of halides in the presence of various hydrides. The Pd(0)-catalysed carbonylation of aryl, alkenyl iodides and bromides with CO and H_2 (1:1) in aprotic solvents in the presence of tertiary amines affords aldehydes [230]. Sodium formate can be used as a hydride source to produce aldehydes. The α,β-unsaturated aldehydes **468** can be prepared by the carbonylation of aryl, alkenyl halides and triflate **467**, benzyl and allyl chlorides, using tin hydride as the hydride source and Pd(Ph$_3$P)$_4$ as a catalyst [231]. Et$_3$SiH is used as another hydride source [232]. The arenediazonium tetrafluoroborate **469** is converted to the benzaldehyde derivative **470** rapidly in good yield using Et$_3$SiH or PHMS as the hydride [233].

Ketones can be prepared by the carbonylation of halides and pseudohalides in the presence of various organometallic compounds of Zn, B, Al, Sn, Si and Hg, and other carbon nucleophiles, which attack acylmetal intermediates (transmetallation to generate acyl alkylmetal) using Pd or Ni catalyst. The carbonylation of phenyl iodide in the presence of propyl iodide (**471**) and Zn–Cu couple affords phenyl propyl ketone (**476**). The propylzinc iodide **474** is formed from propyl iodide, which gives **475** by transmetallation with **473**, and its reductive elimination gives the ketone **476** [234]. Insertion of CO to phenylpalladium intermediate **472** to form benzoylpalladium **473** is faster than the transmetallation of phenylpalladium **472** with propylzinc (**474**), and hence propylbenzene is not formed. As a simple synthetic route to dialkyl ketones, the diiodide **477** is converted to the dialkylketone **478** in the presence of Zn and CoBr$_2$ (1.5 equivalents) in THF–NMP at room temeprature by bubbling CO into the reaction mixture [235].

Organoboranes are used for ketone synthesis under basic conditions. The cyclic ketone **482** is prepared from alkenyl iodode **479**. Hydroboration of terminal double bond, followed by carbonylation generates **480**, and the cyclic ketone **482** is formed by intramolecular transmetallation of **480** to afford **481** [236].

Organostannanes, such as aryl-, alkenyl- and alkynylstannanes, are useful for the ketone synthesis via transmetallation of acylpalladium with organostannanes and reductive elimination [237]. Carbonylation of the alkenyl triflate and intramolecular trapping with alkenylstannane, using **483**, afford the macrocyclic divinyl ketone **484** [238].

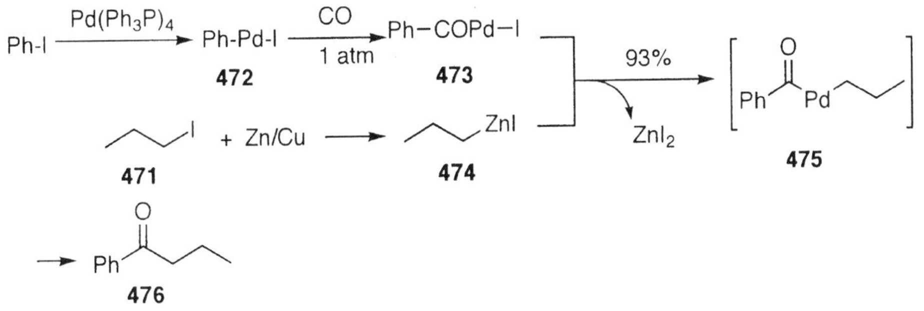

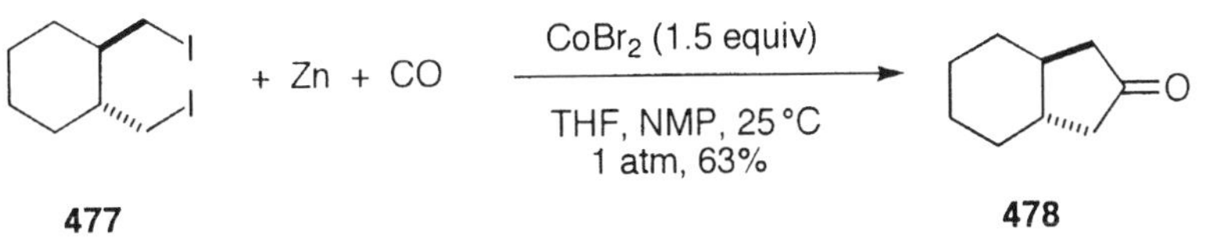

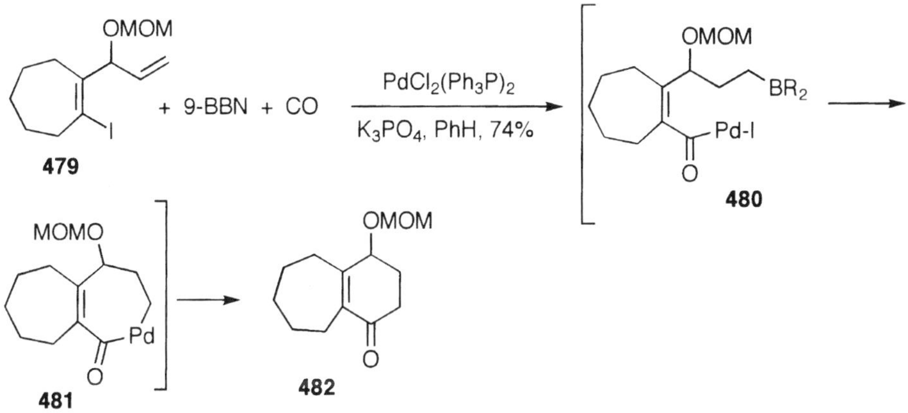

α-Acylmalonates or β-keto esters can be prepared by the carbonylation of aryl iodides in the presence of malonates or β-keto esters. A good ligand is DPPF [239]. In addition to Pd(Ph₃P)₄ (90% yield), other metal complexes such as Li₂CuCl₄ (92%),

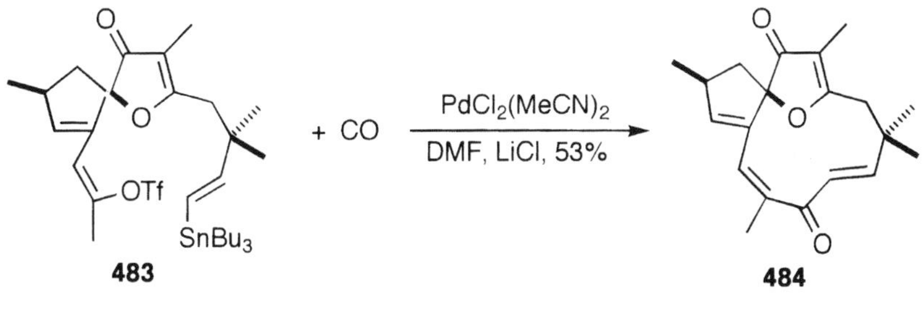

NiBr$_2$ (90%) and NiCl$_2$(Ph$_3$P)$_2$ (92%) can be used as the catalysts for the intramolecular carbonylation of **485** to give **486** [240].

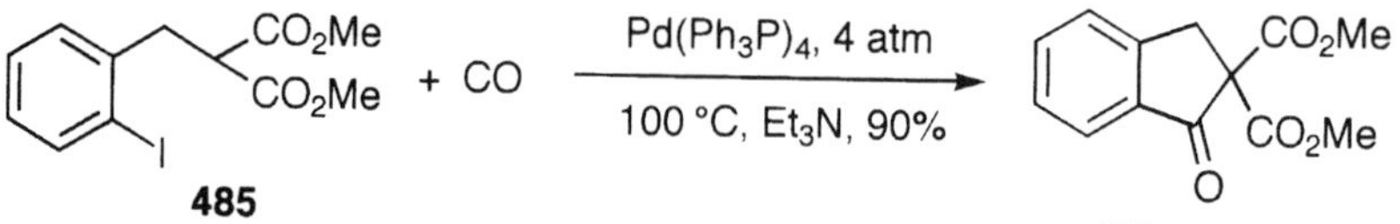

The carbonylation of aryl iodide in the presence of terminal alkynes affords acyl alkynes. Bidentate ligands such as DPPF give good results [241]. When Ph$_3$P is used, phenylacetylene is mainly converted to diphenylacetylene. The alkynyl ketones **488** are prepared by the reaction of the alkenyl triflate **487** with phenylacetylene and CO [242].

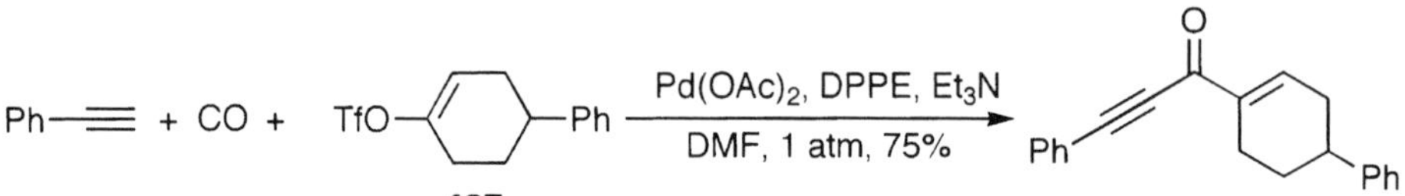

The acylpalladium is formed by CO insertion as the intermediate of the carbonylation. They can be prepared directly by the oxidative addition of acyl chlorides to Pd(0). Thus ketones can be prepared by the reaction of acyl halides with organozinc reagents and organostannanes. Benzoacetate (**490**) is obtained by the reaction of benzoyl chloride with the Reformatsky reagent **489** [243]. The macrocyclic keto lactone **492** is obtained by intramolecular reaction of the alkenylstannane with acyl chloride in **491** [244].

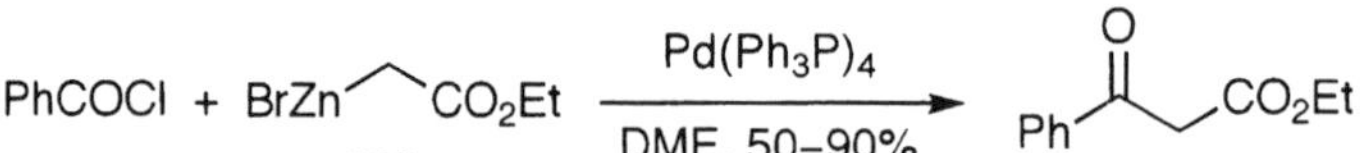

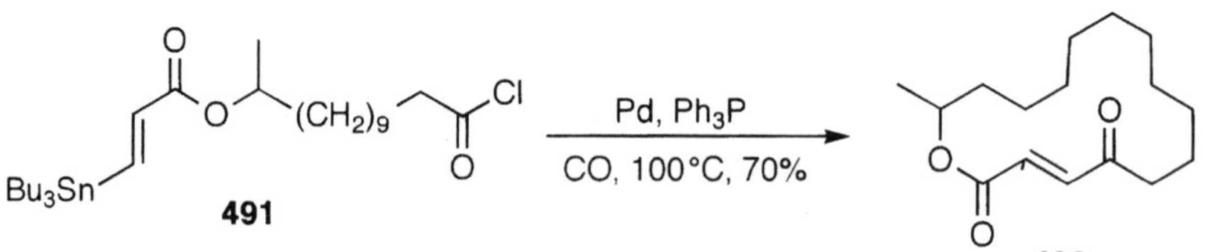

3.6.3 Decarbonylation of Acyl Halides and Aldehydes

Formation of the aroylpalladium **495** by insertion of CO to the arylpalladium **494** (M = Pd) is reversible. Aroylpalladium complexes **495**, prepared directly by the oxidative addition of aroyl halides and aldehydes **496**, undergo decarbonylation by heating in the absence of CO to form **493** [245]. Heating of the aroyl chlorides **497** up to 200 °C with a catalytic amount of $RhCl(Ph_3P)_3$ or $RhCl(CO)(Ph_3P)_2$ affords the chloroarenes **498** [246,247]. The aryl and alkenyl aldehydes **499** and **501** are

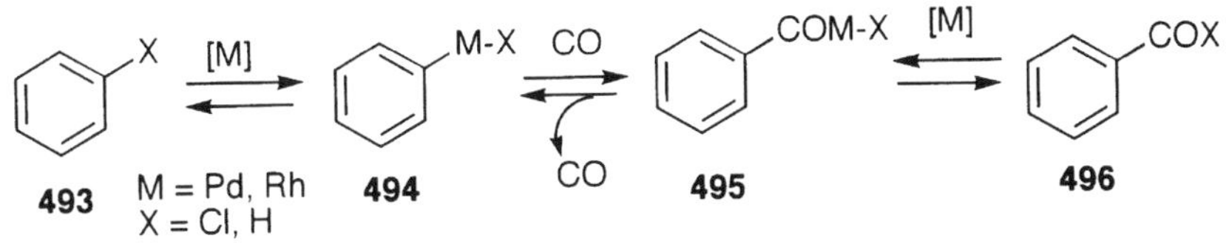

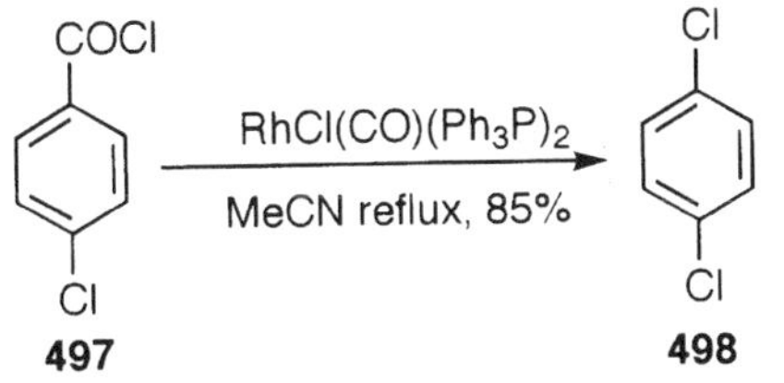

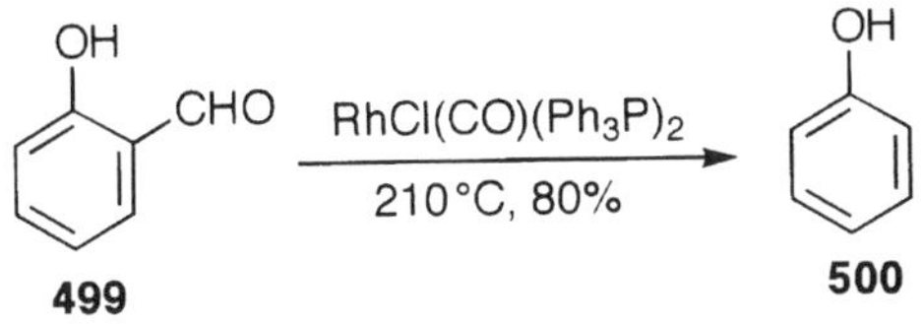

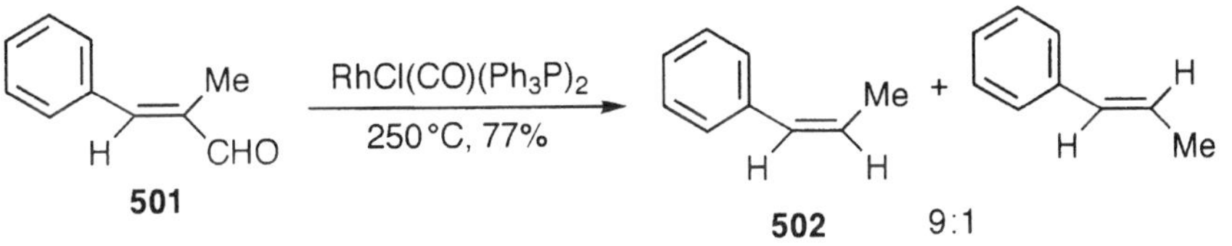

decarbonylated at high temperature with Pd and Rh catalysts to give **500** and **502**.

The decarbonylation of acyl halides and aldehydes proceeds under mild conditions to give aryl halides ArX and arenes ArH with a stoichiometric amount of $RhCl(Ph_3P)_3$. At the same time, $RhCl(CO)(Ph_3P)_2$ is formed which is inactive at moderate temperatures, and the reaction is stoichiometric [245–248].

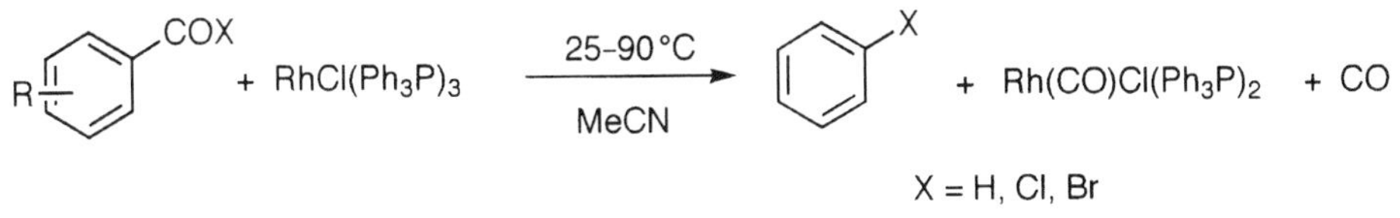

3.7 Preparation of Biaryls by the Coupling of Arenes with Aryl Halides

Biaryls can be prepared from aryl halides in three ways: (i) cross-coupling of aryl halides with aromatic rings, (ii) Reductive homocoupling of aryl halides using Zn and Ni, (iii) cross-coupling of aryl halides with aryltin, zinc and boron compounds (this reaction is treated in Section 3.4). Also, the oxidative coupling of aromatic rings with Pd(II) salts gives biaryls (see Section 11.2).

The Pd-catalysed coupling of aryl halides or triflates with aromatic rings to give biaryls proceeds only intramolecularly [249]. The reaction is carried out with $PdCl_2$ or $Pd(OAc)_2$ as a catalyst in the presence of bases. The oxidative addition of the halide to Pd(0) is followed by the palladation of aromatic ring with Pd(II) species to generate the diarylpalladium **504**. Finally its reductive elimination gives the coupled product. The reaction has been applied to the synthesis of gilvocarcin M (**505**) from **503** via the diarylpalladium **504** [250]. As another example, the alkenylpalladium intermediate **507**, formed from **506** by insertion of two alkyne bonds, attacks the aromatic ring intramolecularly to give the naphthalene ring **508** [251]. Aroyl chlorides are converted to biaryls by the Pd-catalyzed reaction with $(ClMe_2Si)_2$ [185a].

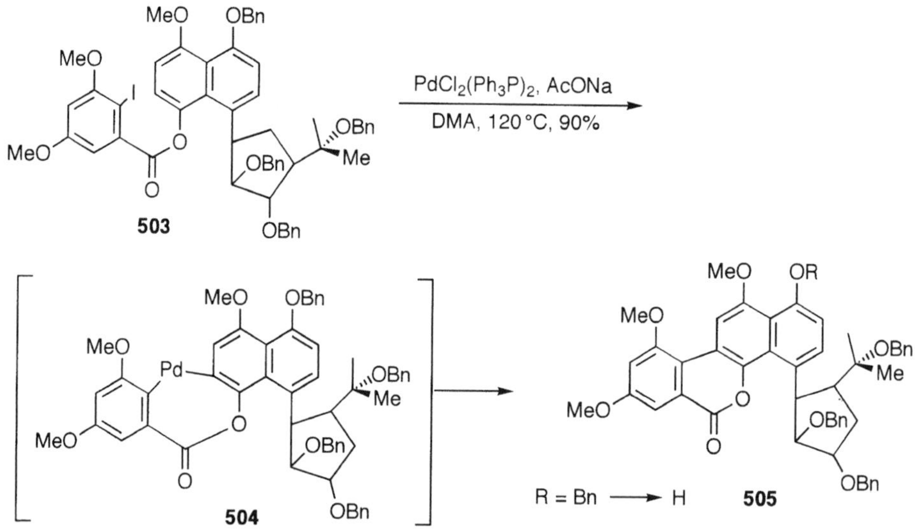

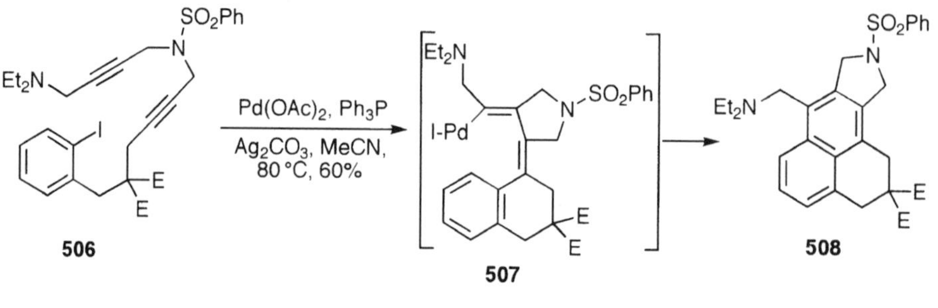

A new and related carboannulation reaction is possible, that involves insertion of alkynes to aryl or alkenyl palladium intermediates, and subsequent cyclization onto an aromatic ring already present in the substrate [252]. The cyclization of the enol triflate **509** to give **510** is an example. The reaction was applied to the synthesis of the indolocarbazole derivative **512** from the iodobisindole **511** [253].

509 **510**

511 **512**

513

514 **515**

516 **517**

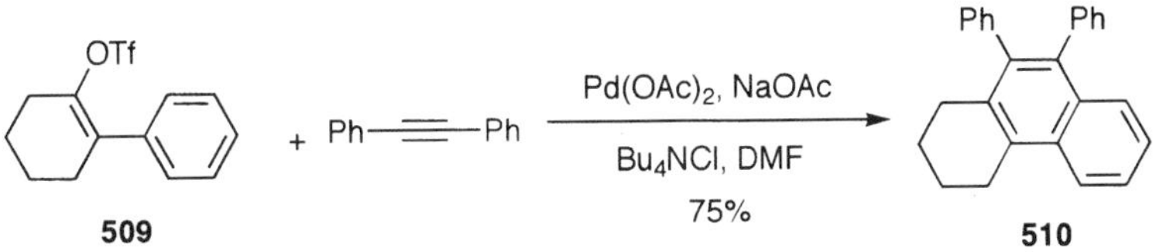

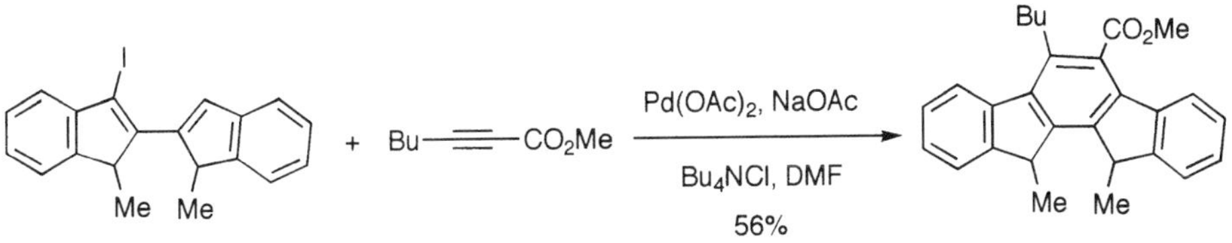

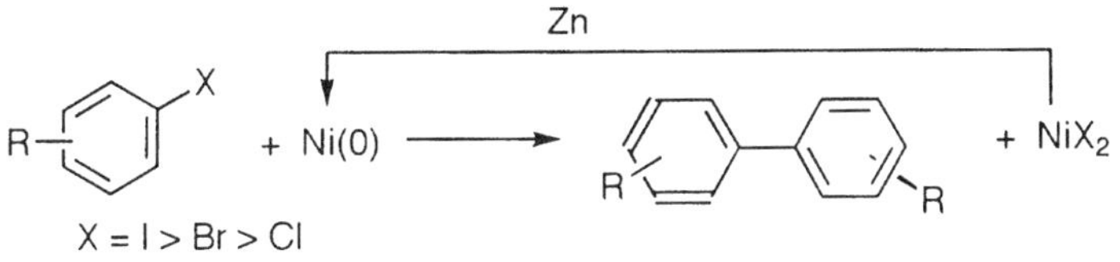

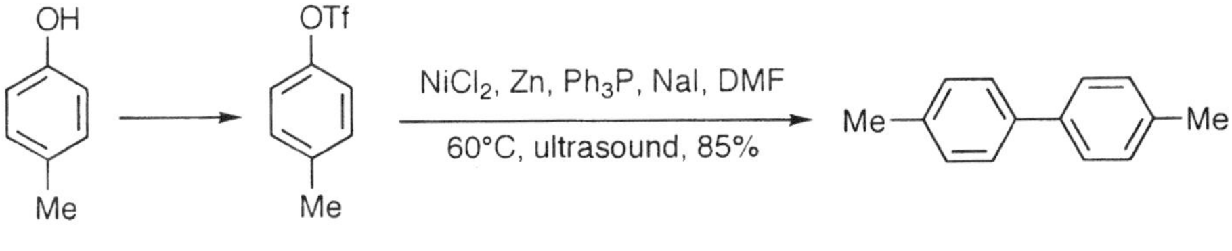

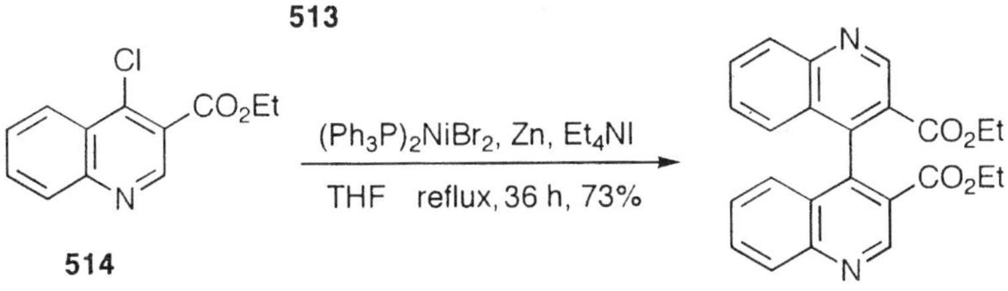

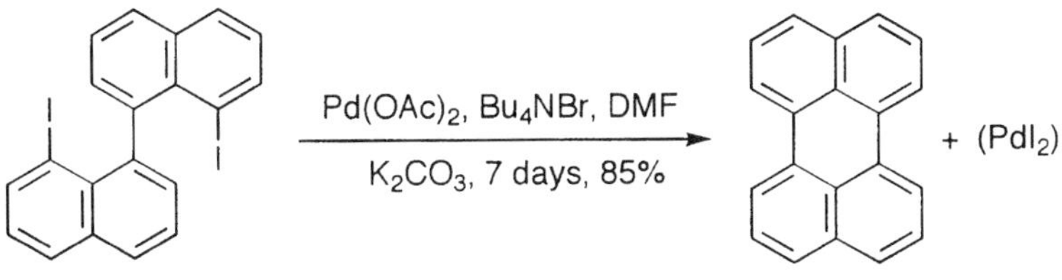

The reductive homocoupling of aryl halides promoted by Cu metal is called the Ullmann reaction. The Ni and Pd versions of Ullmann coupling of aryl and alkenyl halides to give symmetric biaryls and 1,3-dienes proceed smoothly using Ni(0) and Pd(0) species, which are oxidized to NiX_2 and PdX_2. Therefore, reducing agents are required to regenerate Ni(0) and Pd(0) from Ni(II) and Pd(II), in order to make the reaction catalytic. After the reductive coupling, NiX_2 is generated, which is reduced to Ni(0) with Zn. In this way, the reaction can be carried out catalytically by *in situ* regeneration of Ni(0) with Zn [255]. DMF and HMPA are suitable solvents. More smooth reductive coupling is possible using a stoichiometric amount of Ni(0) complex, such as $Ni(cod)_2$ or activated Ni [254]. The homocoupling of the aryl triflates **513** and mesylates derived from phenols is possible using $PdCl_2(Ph_3P)_2/Zn$ and $NiCl_2/Zn$ systems in DMF [256]. Coupling of the chloroquinoline **514** in THF in the presence of a quaternary salt gives **515** [257]. Although the reaction took 7 days, perylene (**517**) was obtained in 85% yield by the intramolecular coupling of the aryl diiodide **516** with $Pd(OAc)_2$ [258]. PdI_2, generated by the coupling, may be reduced to Pd(0) with amines derived from DMF or Bu_4NBr over a long period of time.

3.8 Hydrogenolysis with Hydrides

Oxidative addition of aryl and alkenyl halides, and pseudohalides, followed by transmetallation with various metal hydrides generates Ar−M−H species, reductive elimination of which results in hydrogenolysis of halides. In the main, Pd is used as an efficient catalyst for the hydrogenolysis.

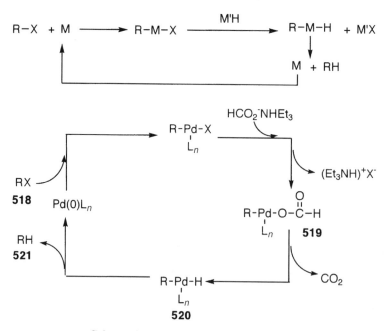

Scheme 3.12   Hydrogenolysis of halides

Halogens of aryl and alkenyl halides are removed easily by the Pd-catalyzed treatment of halides with various hydrides such as $NaBH_4$, $LiAlH_4$, R_3SiH, Bu_3SnH and formic acid. Although aryl halides are reduced easily with sodium formate, triethylammonium formate is a better choice, because it is soluble in organic solvents [259]. After the oxidative addition of the halides **518**, Pd$-$X is displaced by formate to give the Pd$-$formate **519**, which undergoes facile decarboxylation to generate the Pd-H species **520**, and give the hydrogenolyzed products **521** by reductive elimination (Scheme 3.12). Dechlorination of the chloroarene is carried out with HCO_2K or HCO_2NH_4 using Pdcharcoal as the catalyst [260].

The phenolic OH group can be removed by Pd-catalysed hydrogenolysis of its triflate **522** with triethylammonium formate [261]. Naphthol can be converted to naphthalene by the hydrogenolysis of its triflate. The Ni-catalysed reduction of aryl mesylates **523** is possible using MeOH and Zn as the hydrogen donor [262]. Smooth removal of phenol groups as triflates and mesylates is not possible by any other means.

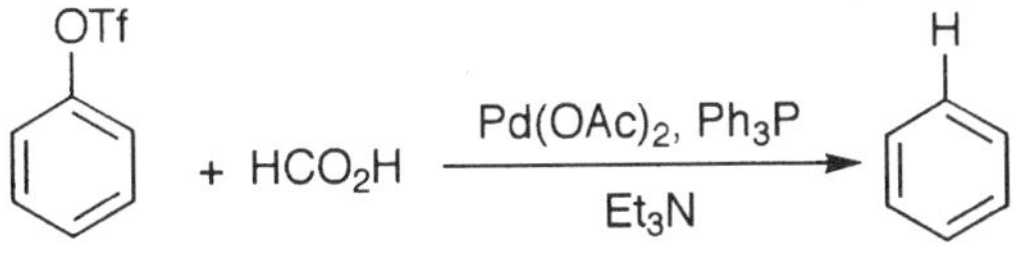

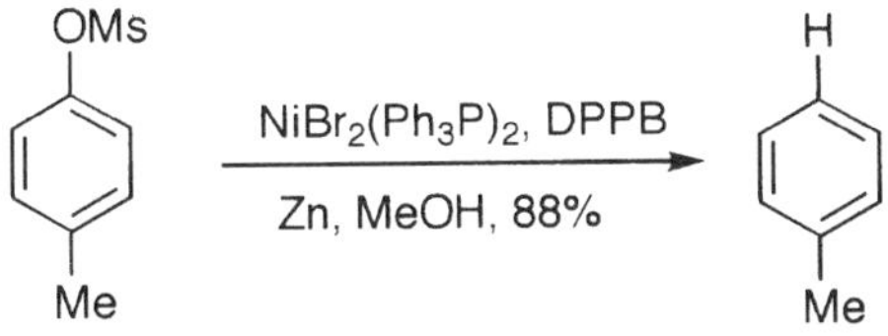

Ketones and aldehydes are converted to alkenes by the hydrogenolysis of their enol triflates with formate. The steroidal enone **524** is converted to the dienol triflate **525** and then to the 1,3-diene **526** by the hydrogenolysis with tributylammonium formate [263,264].

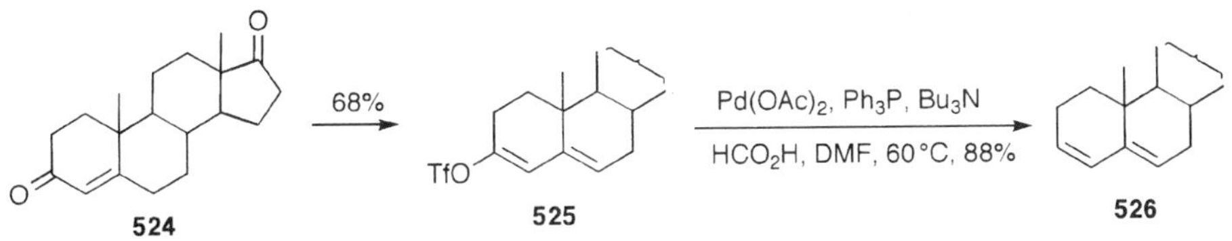

Grignard reagents or alkylaluminum compounds bearing β-hydrogen can be used for the hydrogenolysis. The hydrogenolysis of the alkenyl sulfones **527** with isopropylmagnesium bromide to gives the alkenes **531** is an example [265]. The reaction is explained by transmetallation of **528** to give **529**. β-Elimination of **529** affords the Pd–H species **530**, and its reductive elimination affords the alkene **531**.

527 528 529 530 531

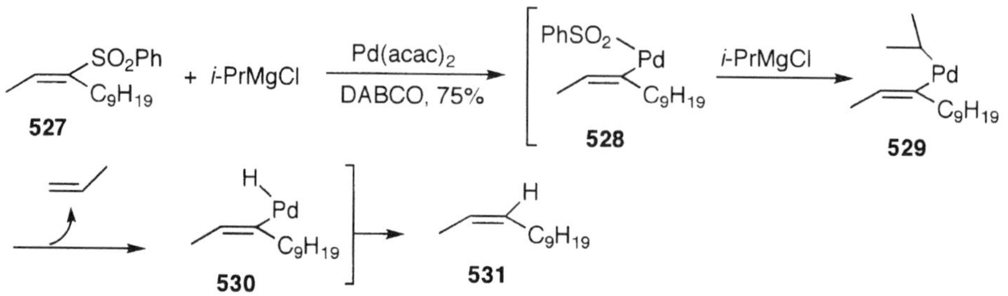

(*Z*)-1-Bromo-1-alkenes **533** can be prepared in high purity based on the stereoselective monohydrogenolysis of the 1,1-dibromo-1-alkene **532** with $HSnBu_3$ [266]. The conjugatd enediyne **535** is prepared in one pot by the treatment of the 1,1-dibromoalka-1-en-3-yne **534** with $HSnBu_3$, followed by the addition of alkyne using CuI as a cocatalyst.

532 533 534 535

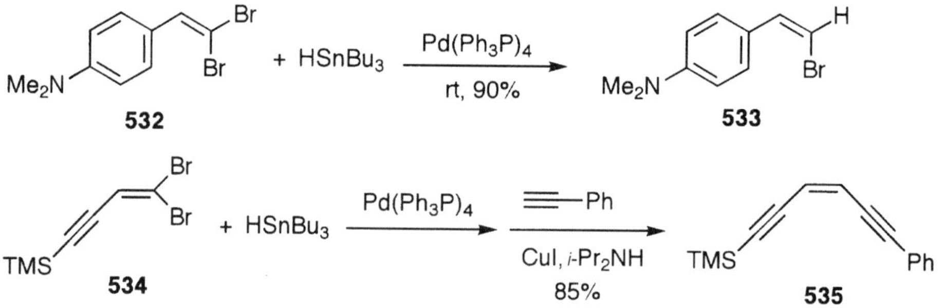

The Pd-catalyzed hydrogenolysis of acyl chlorides with hydrogen to give aldehydes is called the Rosenmund reduction. Rosenmund reduction catalyzed by supporting Pd, is explained by the formation of an acylpalladium complex and its hydrogenolysis [267]. The Pd-catalysed reaction of acyl halides with tin hydride gives aldehydes. This is the tin-assisted version of the Rosenmund reduction [268]. It should be noted that coupling of $HSnBu_3$ to give the ditin **536** is also catalyzed by Pd.

536

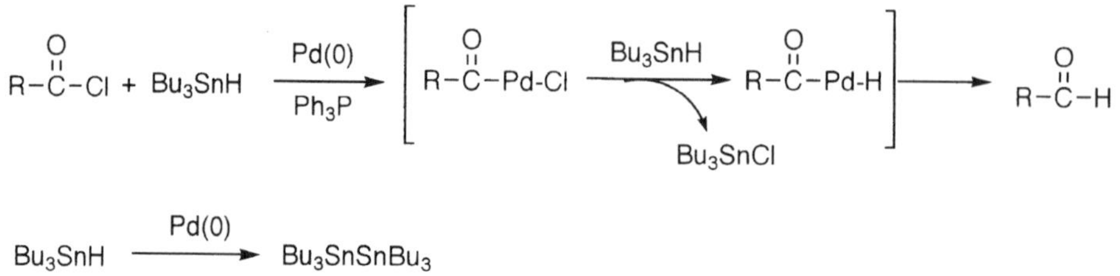

References

1. M. Aizenberg and D. Milstein, *J. Am. Chem. Soc.*, **711**, 8674 (1995).
2. J. Tsuji, *Palladium Reagents and Catalysts*, pp. 125–290, John Wiley, 1995.
2a. J. Tsuji (Ed), *Perspectives in Organopalladium Chemistry for the XXI Century*, pp. 1–124, Elsevier, 1999.
3. T. Mizoroki, K. Mori and A. Ozaki, *Bull. Chem. Soc. Jpn.*, **44**, 581 (1971).
4. R. F. Heck and J. P. Nolley, Jr., *J. Org. Chem.*, **37**, 2320 (1972).
5. R. F. Heck, *Org. React.* **27**, 345 (1982); *Adv. Catal.* **26**, 323 (1977); *Acc. Chem. Res.*, **12**, 146 (1979); A. de Meijere and F. E. Meyer, *Angew. Chem., Int. Ed. Engl.*, **33**, 2379 (1994); S. Bräse and A. de Meijere, in F. Diederich and P. J. Stang, (Eds), *Metal-catalyzed Cross-coupling Reactions*, Wiley-VCH, 1998.
6. C. B. Ziegler and R. F. Heck, *J. Org. Chem.*, **43**, 2941 (1978).
7. J. E. Plevyak, J. E. Dickerson and R. F. Heck, *J. Org. Chem.*, **44**, 4078 (1979).
8. T. Jeffery, *Tetrahedron Lett.*, **26**, 2667 (1985); *Tetrahedron*, **52**, 10113 (1996).
8a. T. Ukai, H. Kawazura, Y. Ishii, J. J. Bennet and J. A. Ibers, *J. Organometal. Chem.*, **65**, 253 (1974).
9. W. A. Herrmann, C. Brossmer, K. öfele, C. P. Reisinger, T. Priermeier, M. Beller and H. Fischer, *Angew. Chem., Int. Ed. Engl.*, **34**, 1844 (1995); M. Beller, T. H. Riermeier, S. Haber, H. J. Kleiner and W. A. Herrmann, *Chem. Ber.*, **129**, 1259 (1996).
10. M. T. Reetz, G. Lohmer and R. Schwickardi, *Angew. Chem., Int. Ed. Engl.* **37**, 481 (1998).
11. M. Beller and A. Zapf, *Synlett*, 792 (1998).
12. S. Hillers, S. Sartori and O. Reiser, *J. Am. Chem. Soc.*, **118**, 2087 (1996).
13. Review, V. V. Grushin and H. Alper, *Chem. Rev.*, **94**, 1047 (1994).
14. Y. Ben-David, M. Portnoy, M. Gozin and D. Milstein, *Organometallics*, **11**, 1995 (1992).
14a. A. F. Littke and G. C. Fu, *J. Org. Chem.*, **64**, 10 (1999).
15. N. A. Bumagin, P. G. More and I. P. Beletskaya, *J. Organometal. Chem.*, **371**, 397 (1989).
16. J. P. Genet, E. Blart and M. Savignac, *Synlett*, **715** (1992); C. Amatore, E. Blart, J. P. Genet, A. Jutand, S. Lemaire-Audoire and M. Savignac, *J. Org. Chem.*, **60**, 6829 (1995).
16a. K. Ritter, *Synthesis*, 735 (1993).
17. A. Arcadi, E. Benocchi, S. Cacchi, L. Caglioti and F. Martinelli, *Tetrahedron Lett.*, **31**, 2463 (1990).
18. K. Kikukawa, K. Nagira, F. Wada and T. Matsuda, *Tetrahedron*, **37**, 31 (1981); K. Kikukawa, K. Maemura, Y. Kiseki, F. Wada and T. Matsuda, *J. Org. Chem.*, **46**, 4885 (1981).
19. M. Beller, H. Fischer and K. Kühlen, *Tetrahedron Lett.*, **35**, 8773 (1994).
20. J. Tsuji and K. Ohno, *J. Am. Chem. Soc.*, **90**, 94 (1968).
21. H. U. Blaser and A. Spencer, *J. Organometal. Chem.*, **233**, 267 (1982); **265**, 323 (1984).
22. A. Kasahara, T. Izumi and K. Kudou, *Chem. Ind.* (London), 467 (1988); A. Kasahara, T. Izumi and K. Kudou, *Synthesis*, 704 (1988).
23. M. S. Stephan, A. J. J. M. Teunissen, G. K. M. Verzijl and J. G. de Vries, *Angew. Chem., Int. Ed. Engl.*, **37**, 662 (1998).
24. F. Ozawa, A. Kubo and T. Hayashi, *J. Am. Chem. Soc.*, **113**, 1417 (1991); *Tetrahedron Lett.*, **34**, 2505 (1993).
25. J. B. Melpolder and R. F. Heck, *J. Org. Chem.*, **41**, 265 (1976); S. A. Buntin and R. F. Heck, *Org. Syn. Coll.*, **7**, 361 (1990).
26. A. J. Chalk and S. A. Magennis, *J. Org. Chem.*, **41**, 1206 (1976).

27. S. Torii, H. Okumoto, F. Akahoshi and T. Kotani, *J. Am. Chem. Soc.*, **111**, 8932 (1989); *Chem. Lett.*, 1971 (1989).

28. R. C. Larock, F. Kondo, K. Narayanan, L. K. Sydnes and M. F. H. Hsu, *Tetrahedron Lett.*, **30**, 5737 (1989).

29. R. C. Larock, W. Y. Leung and S. Stolz-Dunn, *Tetrahedron Lett.*, **30**, 6629 (1989).

30. I. Shimizu and J. Tsuji, *Chem. Lett.*, 233 (1984).

31. V. Gauthier, B. Cazes and J. Gore, *Tetrahedron Lett.*, **32**, 915 (1991).

32. W. Fischetti, K. T. Mak, F. G. Stakem, J. I. Kim, A. L. Reingold and R. F. Heck, *J. Org. Chem.*, **48**, 948 (1983).

33. R. C. Larock, N. Berrios-Pena and K. Narayanan, *J. Org. Chem.*, **55**, 3447 (1990).

33a. R. C. Larock and L. Guo, *Synlett*, 465 (1995).

34. R. C. Larock, C. Tu and P. Pace, *J. Org. Chem.*, **63**, 6859 (1998).

35. R. C. Larock, Y. He, W. W. Leong, X. Han, M. D. Refvik and J. M. Zenner, *J. Org. Chem.*, **63**, 2154 (1998).

36. R. C. Larock, N. G. Berrios-Pena, C. A. Fried, E. K. Yum, C. Tu and W. Leong, *J. Org. Chem.*, **58**, 4509 (1993).

37. Review, E. Negishi, C. Coperet, S. Ma, S. Y. Liou and F. Liu, *Chem. Rev.*, **96**, 365 (1996).

38. K. Nagasawa, Y. Zako, H. Ishihara and I. Shimizu, *J. Org. Chem.*, **58**, 2523 (1993); J. L. Mascarenas, A. M. Garcia, L. Castedo and A. Mourino, *Tetrahedron Lett.*, **33**, 4365 (1992).

39. Review, M. Shibasaki, C. D. J. Boden and A. Kojima, *Tetrahedron*, **53**, 7371 (1997).

40. Y. Sato, M. Sodeoka and M. Shibasaki, *J. Org. Chem.*, **54**, 4738 (1989); Y. Sato, M. Sodeoka and M. Shibasaki, *Chem. Lett.*, 1953 (1990).

41. N. E. Carpenter, Î. J. Kucera and L. E. Overman, *J. Org. Chem.*, **54**, 5846 (1989).

42. R. C. Larock, H. Song, B. E. Baker and W. H. Gong, *Tetrahedron Lett.*, **29**, 2919 (1988); R. C. Larock, W. H. Gong and B. E. Baker, *Tetrahedron Lett.*, **30**, 2603 (1989); *J. Org. Chem.*, **54**, 2047 (1989).

43. Y. Sato, S. Watanabe and M. Shibasaki, *Tetrahedron Lett.*, **33**, 2589 (1992); Y. Sato, S. Nukui, M. Sodeoka and M. Shibasaki, *Tetrahedron*, **50**, 371 (1994).

44. K. Ohrai, K. Kondo, M. Sodeoka and M. Shibasaki, *J. Am. Chem. Soc.*, **116**, 11737 (1994).

45. T. Takemoto, M. Sodeoka, H. Sasai and M. Shibasaki, *J. Am. Chem. Soc.*, **115**, 8477 (1993), M. Shibasaki, C. D. J. Boden and A. Kojima, *Tetrahedron*, **53**, 7371 (1997).

46. J. J. Masters, J. T. Link, L. B. Lawrence, B. Snyder, W. B. Young and S. J. Danishefsky, *J. Am. Chem. Soc.*, **118**, 2843 (1996).

47. V. H. Rawal and S. Iwasa, *J. Org. Chem.*, **59**, 2685 (1994).

48. J. Bonjoch, D. Sole, and J. Bosch, *J. Am. Chem. Soc.*, **117**, 11017 (1995).

49. S. Iyer, C. Ramesh and A. Ramani, *Tetrahedron Lett.*, **38**, 8533 (1997).

50. C. Y. Hong, N. Kado and L. E. Overman, *J. Am. Chem. Soc.*, **119**, 12031 (1997); C. Y. Hong and L. E. Overman, *Tetrahedron Lett.*, **35**, 3453 (1994).

51. T. Ohshima, K. Kagechika, M. Adachi, M. Sodeoka and M. Shibasaki, *J. Am. Chem. Soc.*, **118**, 7108 (1996).

52. L. E. Overman, D. J. Ricca and V. D. Tran, *J. Am. Chem. Soc.*, **115**, 2042 (1993); **119**, 12031 (1997); Reviews L. E. Overman, *Pure Appl. Chem.*, **66**, 1423 (1994); J. T Link and L. E. Overman, in F. Diederich and P. J. Stang (Ed.), *Metal-catalyzed Cross-coupling Reactions*, Wiley-VCH, p. 231 (1998).

53. R. D. Stephens and C. E. Castro, *J. Org. Chem.*, **28**, 3313 (1963).

54. L. Cassar, *J. Organometal. Chem.*, **93**, 259 (1975); H. A. Dieck and R. F. Heck, *J. Organometal. Chem.*, **93**, 259 (1975).

55. K. Sonogashira, Y. Tohda and N. Hagihara, *Tetrahedron Lett.*, 4467 (1975); S. Takahashi, Y. Kuroyama, K. Sonogashira and N. Hagihara, *Synthesis*, 627 (1980).

56. K. Sonogashira, *Comprehensive Organic Synthesis*, **3**, 521 (1991), Pergamon Press: K. Sonogashira in F. Diederich and P. J. Stang (Eds), *Metal-catalyzed Cross-coupling Reactions*, p. 203 (1998), Wiley-VCH.

57. J. P. Genet, E. Blart and M. Savignac, *Synlett*, 715 (1992); C. Amatore, E. Blart, J. P. Genet, A. Jutand, S. Lemaire-Audoire and M. Savignac, *J. Org. Chem.*, **60**, 6829 (1995).

58. M. Alami, F. Ferri and G. Linstrumelle, *Tetrahedron Lett.*, **34**, 6403 (1993).

59. J. F. Nguefack, V. Bolitt and D. Sinou, *Tetrahedron Lett.*, **37**, 5527 (1996).

60. R. Diercks and K. P. C. Vollhardt, *J. Am. Chem. Soc.*, **108**, 3150 (1986); H. Schwager, S. Spyroudis and K. P. C. Vollhardt, *J. Organometal. Chem.*, **382**, 191 (1990); R. Boese, J. R. Green, J. Mittendorf, D. L. Mohler and K. P. C. Vollhardt, *Angew. Chem., Int. Ed. Engl.*, **31**, 1643 (1992).

61. E. Negishi, M. Kotora and C. Xu, *J. Org. Chem.*, **62**, 8957 (1997).

62. K. C. Nicolaou and W. M. Dai, *Angew. Chem., Int. Ed. Engl.*, **30**, 1387 (1991).

63. V. Ratovelomanana, D. Guillerm and G. Limstrumelle, *Tetrahedron Lett.*, **25**, 6001 (1984); **26**, 3811 (1985).

64. P. Magnus and P. A. Carter, *J. Am. Chem. Soc.*, **110**, 1626 (1988); P. Magnus, H. Annoura and J. Harling, *J. Org. Chem.*, **55**, 1709 (1990); K. Tomioka, H. Fujita and K. Koga, *Tetrahedron Lett.*, **30**, 851 (1989).

65. J. A. Porco, F. J. Schoenen, T. J. Stout, J. Clardy and S. L. Schreiber, *J. Am. Chem. Soc.*, **112**, 7410 (1990).

66. A. O. King, N. Okukado and E. Negishi, *Chem. Commun.*, 683 (1977); A. O. King, E. Negishi, F. J. Villani and A. Silveira, *J. Org. Chem.*, **43**, 358 (1978); E. Negishi, *Acc. Chem. Res.*, **15**, 340 (1982).

67. J. K. Stille and J. H. Simpson, *J. Am. Chem. Soc.*, **109**, 2138 (1987); D. E. Rudisill, L. A. Castonguay and J. K. Stille, *Tetrahedron Lett.*, **29**, 1509 (1988).

68. F. Tellier, R. Sauvetre and J. F. Normant, *Tetrahedron Lett.*, **27**, 3147 (1986).

69. R. Rossi, A. Carpita and A. Lezzi, *Tetrahedron*, **40**, 2773 (1984).

70. A. E. Graham, D. McKerrecher, D. H. Davies and R. J. Taylor, *Tetrahedron Lett.*, **37**, 7445 (1996).

71. A. M. W. Miller and C. R. Johnson, *J. Org. Chem.*, **62**, 1582 (1997).

72. D. Madec, S. Pujol, V. Henryon and J. P. Ferezou, *Synlett*, 435 (1995).

73. D. A. Siesel and S. W. Staley, *Tetrahedron Lett.*, **34**, 3679 (1993).

74. Y. Nishihara, K. Ikegashira, A. Mori and T. Hiyama, *Chem. Lett.*, 1233 (1997).

75. M. D. Shair, T. Yoon, K. K. Mosny, T. C. Chou and S. J. Danishefsky, *J. Org. Chem.*, **59**, 3755 (1994); *J. Am. Chem. Soc.*, **118**, 9509 (1996); M. D. Shair, T. Yoon, T. C. Chou and S. J. Danishefsky, *Angew. Chem., Int. Ed. Engl.*, **33**, 2477 (1994).

76. J. Wityak and J. B. Chan, *Synth. Commun.*, **21**, 977 (1991); T. R. Hoye and P. R. Hanson, *Tetrahedron Lett.*, **34**, 5043 (1993).

77. B. M. Trost, J. Dumas and M. Villa, *J. Am. Chem. Soc.*, **114**, 9836 (1992).

78. Y. Zhang, G. Wu, G. Angel and E. Negishi, *J. Am. Chem. Soc.*, **112**, 8590 (1990); **111**, 3454 (1989).

79. F. E. Meyer, P. J. Parsons and A. de Meijere, *J. Org. Chem.*, **56**, 6487 (1991); F. E. Meyer, J. Brandenburg, P. J. Parsons and A. de Meijere, *Chem. Commun.*, 390 (1992).

80. S. Pivsa-Art, T. Satoh, M. Miura and M. Nomura, *Chem. Lett.*, 823 (1997).

81. B. Burns, R. Grigg, V. Sridharan and T. Worakun, *Tetrahedron Lett.*, **29**, 4325 (1988).
82. J. M. Nuss, B. H. Levine, R. A. Rennels and M. M. Heravi, *Tetrahedron Lett.*, **32**, 5243 (1991).
83. R. C. Larock, E. K. Yum and M. D. Refvik, *J. Org. Chem.*, **63**, 7652 (1998).
84. A. Kojima, T. Takemoto, M. Sodeoka and M. Shibasaki, *J. Org. Chem.*, **61**, 4856 (1996).
85. K. R. Roesch and R. C. Larock, *J. Org. Chem.*, **63**, 5306 (1998).
86. W. Tao, L. J. Silverberg, A. L. Rheingold and R. F. Heck, *Organometallics*, **8**, 2550 (1989); R. C. Larock, M. J. Doty and S. Cacchi, *J. Org. Chem.*, **58**, 4579 (1993).
87. K. Tamao, K. Sumitani and M. Kumada, *J. Am. Chem. Soc.*, **94**, 4374 (1972); R. J. Corriu and J. P. Masse, *Chem. Commun.*, 144 (1972).
88. F. Diederich and P. J. Stang (Eds) *Metal-catalyzed Cross-coupling Reactions*, Wiley-VCH, 1998; K. Tamao, *Comprehensive Organic Synthesis*, Vol. 3, p. 435, Pergamon Press, 1991; V. N. Kalinin, *Synthesis*, 413 (1992).
89. M. Kumada, *Pure Appl. Chem.*, **52**, 669 (1980); K. Tamao, K. Sumitani, Y. Kiso, M. Zembayashi, A. Fujioka, S. Kodama, I. Nakajima, A. Minato and M. Kumada, *Bull. Chem. Soc. Jpn.*, **49**, 1958 (1976).
90. Y. Ikoma, F. Taya, E. Ozaki, S. Higuchi, Y. Naoi and K. Fujii, *Synthesis*, 147 (1990).
91. T. Hayashi, K. Hayashizaki, T. Kiyoi and Y. Ito, *J. Am. Chem. Soc.*, **110**, 8153 (1988).
92. T. Hayashi, S. Niizuma, T. Kamikawa, N. Suzuki and Y. Uozumi, *J. Am. Chem. Soc.*, **117**, 9101 (1995).
93. K. Tamao, S. Kodama, T. Nakatsuka, Y. Kiso and M. Kumada, *J. Am. Chem. Soc.*, **97**, 4405 (1975); H. Yamanaka, K. Edo, F. Shoji, S. Konno, T. Sakamoto and M. Mizugami, *Chem. Pharm. Bull.*, **26**, 2160 (1978).
94. E. Wenkert, E. L. Michelotti, C. S. Swindell and M. Tingoli, *J. Org. Chem.*, **49**, 4894 (1984); J. P. Ducoux, P. LeMenez, N. Kunesch, G. Kunesch and E. Wenkert, *Tetrahedron Lett.*, **31**, 2595 (1990).
95. T. Hayashi, Y. Katsuro and M. Kumada, *Tetrahedron Lett.*, **21**, 3915 (1980).
96. H. Takei, M. Miura, H. Sugimura and H. Okamura, *Chem. Lett.*, 1447 (1979); 1209 (1980); H. Okamura, M. Miura and H. Takei, *Tetrahedron Lett.*, 43 (1979).
97. E. Wenkert, T. W. Ferreira and E. L. Michelotti, *Chem. Commun.*, 637 (1979).
98. T. Y. Luh and Z. J. Ni, *Synthesis*, 89 (1990); T. Y. Luh, *Acc. Chem. Res.*, **24**, 257 (1991); *Synlett*, 201 (1996).
99. Z. J. Ni, P. F. Yang, D. K. P. Ng, Y. L. Tzeng and T. Y. Luh, *J. Am. Chem. Soc.*, **112**, 9356 (1990); T. M. Yang and T. Y. Luh, *J. Org. Chem.*, **57**, 4550 (1992).
100. V. Percec, J. Y. Bae and D. H. Hill, *J. Org. Chem.* **60**, 176, 1060, 1066 and 6895 (1995).
101. E. Negishi, *Acc. Chem. Res.*, **15**, 340 (1982).
102. E. Erdik, *Tetrahedron*, **48**, 9577 (1992).
103. S. Baba and E. Negishi, *J. Am. Chem. Soc.*, **98**, 6729 (1976); E. Negishi, A. O. King and N. Okukado, *J. Org. Chem.*, **42**, 1821 (1977); E. Negishi, T. Takahashi, S. Baba, D. E. VanHorn and N. Okukado, *J. Org. Chem.*, **109**, 2393 (1987); *Org. Syn.*, **66**, 60 (1988).
104. A. Pelter, M. Rowlands and I. H. Jenkins, *Tetrahedron Lett.*, **28**, 5213 (1987).
105. J. F. Fauvarque and A. Jutand, *J. Organometal. Chem.*, **132**, C17 (1977), **177**, 273 (1979), **209**, 109 (1981).
106. H. Yamanaka, M. An-naka, Y. Kondo and T. Sakamoto, *Chem. Pharm. Bull.*, **33**, 4309 (1985).
107. Y. Tamaru, H. Ochiai, T. Nakamura and Z. Yoshida, *Angew. Chem., Int. Ed. Engl.*, **26**, 1157 (1987).
108. R. M. Keenan and L. I. Kruse, *Synth. Commun.*, **19**, 793 (1989).

109. Q. Y. Chen and Y. B. He, *Tetrahedron Lett.*, **28**, 2387 (1987).

110. G. P. Roth and C. Sapino, *Tetrahedron Lett.*, **32**, 4073 (1991); G. P. Roth and C. E. Fuller *J. Org. Chem.*, **56**, 3493 (1991).

111. M. Rottländer and P. Knochel, *J. Org. Chem.*, **63**, 3203 (1998).

112. T. Hayashi, A. Yamamoto, M. Hojo and Y. Ito, *Chem. Commun.*, 495 (1989).

113. R. Giovannini, T. Stüdemann, G. Dussin and P. Knochel, *Angew. Chem., Int. Ed. Engl.*, **37**, 2387 (1998).

114. H. Avedissian, L. Berillon, G. Cahiez and P. Knochel, *Tetrahedron Lett.*, **39**, 6163 (1998).

115. A. Suzuki, *Acc. Chem. Res.*, **15**, 178 (1982); *Pure Appl. Chem.*, **57**, 1749 (1985); **63**, 419 (1991); **66**, 213 (1994); N. Miyaura and A. Suzuki, *Chem. Rev.*, **95**, 2457 (1995); V. Snieckus, *Chem. Rev.* **90**, 879 (1990); D. S. Matteson, *Tetrahedron*, **45**, 1859 (1989).

116. N. Miyaura, K. Yamada and A. Suzuki, *Tetrahedron Lett.*, 3437 (1979); N. Miyaura, T. Yanagi and A. Suzuki, *Syn. Commun.*, **11**, 513 (1981).

117. M. Ogima, S. Hyuga, S. Hara and A. Suzuki, *Chem. Lett.*, 1959 (1989).

118. T. Ishiyama, N. Miyaura and A. Suzuki, *Synlett*, 687 (1991).

119. G. Marck, A. Villiger and R. Buchecker, *Tetrahedron Lett.*, **53**, 3277 (1994).

120. D. Badone, M. Baroni, R. Cardamone, A., Ielmini and U. Guzzi, *J. Org. Chem.*, **62**, 7170 (1997).

121. T. I. Wallow and B. M. Novak, *J. Org. Chem.*, **59**, 5034 (1994).

122. M. Beller, H. Fischer, W. A. Herrmann, K. Öfele and C. Brossmer, *Angew. Chem., Int. Ed. Engl.*, **34**, 1848 (1995).

123. N. Yasuda, L. Xavier, D. L. Rieger, Y. Li, A. E. DeDamp and U. H. Dolling, *Tetrahedron Lett.*, **34**, 3211 (1993).

124. S. Darres, T. Jeffery, J. P. Genet, J. L. Brayer and J. P. Demoute, *Tetrahedron Lett.*, **37**, 3857 (1996).

125. S. Saito, S. Oh-taniz and N. Miyaura, *Tetrahedron Lett.*, **37**, 2993 (1996); *J. Org. Chem.*, **62**, 8024 (1997).

126. A. F. Indolese, *Tetrahedron Lett.*, **38**, 3513 (1997).

126a. J. C. Galland, M. Savignac and J. P. Genet, *Tetrahedron Lett.*, **40**, 2323 (1999).

126b. A. F. Littke and G. C. Fu, *Angew. Chem., Int. Ed. Engl.*, **37**, 3387 (1998).

127. N. Miyaura, K. Yamada, H. Suginome and A. Suzuki, *J. Am. Chem. Soc.*, **107**, 972 (1985).

128. N. Miyaura, M. Satoh and A. Suzuki, *Tetrahedron Lett.*, **27**, 3745 (1986).

129. K. C. Nicolaou, J. Y. Ramphal, N. A. Petasis and C. N. Serhan, *Angew. Chem., Int. Ed. Engl.*, **30**, 1100 (1991); K. C. Nicolaou, J. Y. Ramphal, J. M. Palazon and R. A. Spanevello, *Angew. Chem., Int. Ed. Engl.*, **28**, 587 (1989).

130. Y. Kobayashi, T. Shimazaki, H. Taguchi and F. Sato, *J. Org. Chem.*, **55**, 5324 (1990); *Tetrahedron Lett.*, **28**, 5849 (1987).

131. J. Uenishi, J. M. Beau, R. W. Armstrong and Y. Kishi, *J. Am. Chem. Soc.*, **109**, 4756 (1987); **111**, 7525 (1989).

132. N. Miyaura, T. Ishiyama, H. Sasaki, M. Ishikawa, M. Satoh and A. Suzuki, *J. Am. Chem. Soc.*, **111**, 314 (1989); T. Ohe, N, Miyaura and A. Suzuki, *Synlett*, 221 (1990).

133. T. Ohe, N. Miyaura and A. Suzuki, *J. Org. Chem.*, **58**, 2201 (1993); N. Miyaura, M. Ishikawa and A. Suzuki, *Tetrahedron Lett.*, **33**, 2571 (1992).

134. T. Ishiyama, S. Abe, N. Miyaura and A. Suzuki, *Chem. Lett.*, 691 (1992); S. Hyuga, Y. Chiba, N. Yamashita, S. Hara and A. Suzuki, *Chem. Lett.*, 1757 (1987).

135. E. Negishi, N. Okukado, A. O. King, D. E. Van Horn and B. I. Spiegel, *J. Am. Chem. Soc.*, **100**, 2254 (1978).

136. K. Asao, H. Iio and T. Tokoroyama, *Synthesis*, 382 (1990); M. G. Saulnier, J. F. Kadow, M. M. Tun, D. R. Langley and D. M. Vyas, *J. Am. Chem. Soc.*, **111**, 8320 (1989).

137. K. Takai, K. Oshima and H. Nozaki, *Tetrahedron Lett.*, **21**, 2531 (1980); M. Sato, K. Takai, K. Oshima and H. Nozaki, *Tetrahedron Lett.*, **22**, 1609 (1981); *Bull. Chem. Soc. Jpn.*, **57**, 108 (1984).

138. N. Okukado, D. E. Van Horn, W. L. Klima and E. Negishi, *Tetrahedron Lett.*, 1027 (1978); E. Negishi and D. E. van Horn, *J. Am. Chem. Soc.*, **99**, 3168 (1977).

139. J. K. Stille, *Angew. Chem., Int. Ed. Engl.*, **25**, 508 (1986); *Pure Appl. Chem.*, **57**, 1771 (1985); T. N. Mitchell, *Synthesis*, 803 (1992); V. Farina, V. Krishnamurthi and W. J. Scott, *Org. Reaction*, **50**, 1 (1997).

140. M. Kosugi, K. Sasazawa, Y. Shimizu and T. Migita, *Chem. Lett.*, 301 (1977).

141. V. Farina and B. Krishnan, *J. Am. Chem. Soc.*, **113**, 9585 (1991).

142. I. P. Beletzkaya, *J. Organometal. Chem.*, **250**, 551 (1983).

143. V. Farina, B. Krishnan, D. R. Marshall and G. P. Roth, *J. Org. Chem.*, **58**, 5434 (1993).

144. V. N. Kalinin, *Synthesis*, 413 (1992).

145. A. Dondoni, G. Fantin, M. Fogagnolo, A. Medici and P. Pedrini, *Synthsis*, 693 (1987).

146. G. T. Crisp, *Syn. Commun.*, **19**, 307 (1989).

147. T. R. Bailey, *Tetrahedron Lett.*, **27**, 4407 (1986); J. M. Clough, I. S. Mann and D. A. Widdowson, *Tetrahedron Lett.*, **28**, 2645 (1987).

148. C. Chan, P. B. Cox and S. M. Roberts, *Chem. Commun.*, 971 (1988).

149. Y. Yang, S. B. Hörnfeldt and S. Gronowitz, *Synthesis*, 130 (1989); J. Malm, P. Bjork, S. Gronowitz and A. B. Hörnfeldt, *Tetrahedron Lett.*, **33**, 2199 (1992).

150. J. K. Stille and B. L. Groh, *J. Am. Chem. Soc.*, **109**, 813 (1987).

151. K. Kikukawa, K. Kono, F. Wada and T. Matsuda, *J. Org. Chem.*, **84**, 1333 (1983).

152. K. Kikukawa, H. Umekawa and T. Matsuda, *J. Organometal. Chem.*, **311**, C44 (1986).

153. C. A. Busacca, J. Swestock, R. E. Johnson, T. R. Bailey, L. Misza and C. A. Rodger, *J. Org. Chem.*, **59**, 7553 (1994).

154. V. Farina and M. A. Hossain, *Tetrahedron Lett.*, **37**, 6997 (1996).

155. V. Farina and G. P. Roth, *Tetrahedron Lett.*, **32**, 4243 (1991); V. Farina, B. Krishnan, D. R. Marshall and G. P. Roth, *J. Org. Chem.*, **58**, 5434 (1993).

156. E. Shirakawa, H. Yoshida and H. Takaya, *Tetrahedron Lett.*, **38**, 3759 (1997).

157. K. C. Nicolaou, G. Q. Shi, J. L. Gunzner and Z. Yang, *J. Am. Chem. Soc.*, **119**, 5467 (1997).

158. S. Gronowitz, P. Björk, J. Malm and A. B. Hörnfeldt, *J. Organometal. Chem.*, **460**, 127 (1993); J. Malm, P. Björk, S. Gronowitz and A. B. Hörnfeldt, *Tetrahedron Lett.*, **33**, 2199 (1992). **35**, 3195 (1994).

158a. E. Shirakawa, K. Yamasaki and T. Hiyama, *Angew. Chem., Int. Ed. Engl.*, **37**, 1544 (1998).

159. K. C. Nicolaou, M. Sato, N. D. Miller, J. L. Gunzner, J. Renaud and E. Untersteller, *Angew. Chem., Int. Ed. Engl.*, **35**, 889 (1996).

160. A. B. Smith, S. M. Condon, J. A. McCauley, J. L. Leazer, J. W. Leahy and R. E. Maleczka, *J. Am. Chem. Soc.*, **117**, 5407 (1995).

161. Reviews, T. Hiyama and Y. Hatanaka, *Pure Appl. Chem.*, **66**, 1471 (1994); Y. Hatanaka and T. Hiyama, *Synlett*, 845 (1991).

162. Y. Hatanaka and T. Hiyama, *J. Org. Chem.*, **53**, 918 (1988).

163. Y. Hatanaka and T. Hiyama, *J. Org. Chem.*, **54**, 268 (1989); *Tetrahedron Lett.*, **31**, 2719 (1990); K. Tamao, K. Kobayashi and Y. Ito, *Tetrahedron Lett.*, **30**, 6051 (1989).

164. K. Gouda, E. Hagiwara, Y. Hatanaka and T. Hiyama, *J. Org. Chem.*, **61**, 7232 (1996).

165. E. Hagiwara, K. Gouda, Y. Hatanaka and T. Hiyama, *Tetrahedron Lett.*, **38**, 439 (1997).

166. C. Carfagna, A. Musco, G. Sallese, R. Santi and T. Fiorani, *J. Org. Chem.*, **56**, 261 (1991).

167. Y. Okude, S. Hirano, T. Hiyama and H. Nozaki, *J. Am. Chem. Soc.*, **99**, 3179 (1977); K. Takai, K. Kimura, T. Kuroda, T. Hiyama and H. Nozaki, *Tetrahedron Lett.*, **24**, 5281 (1983).

168. P. Cintas, *Synthesis*, 248 (1992).

168a. L. A. Paquette, I. Collado and M. Purdie, *J. Am. Chem. Soc.*, **120**, 2553 (1998).

169. K. Takai, M. Tagashira, T. Kuroda, K. Oshima, K. Utimoto and H. Nozaki, *J. Am. Chem. Soc.*, **108**, 6048 (1986).

170. H. Jin, J. Uenishi, W. J. Christ and Y. Kishi, *J. Am. Chem. Soc.*, **108**, 5644 (1986).

171. A. Fürstner and N. Shi, *J. Am. Chem. Soc.*, **118**, 2533, 12349 (1996).

172. S. L. Schreiber and H. V. Meyers, *J. Am. Chem. Soc.*, **110**, 5198 (1988).

173. T. D. Aicher, K. R. Buszek, F. G. Fang, C. J. Forsyth, S. H. Jung, Y. Kishi, M. C. Matelich, P. M. Scola, D. M. Spero and S. K. Yoon, *J. Am. Chem. Soc.*, **114**, 3162 (1992).

174. K. C. Nicolaou, E. A. Theodrakis, F. P. J. T. Rutjes, J. Tiebes, M. Sato, E. Untersteller and X. Y. Xiao, *J. Am. Chem. Soc.*, **117**, 1171, 1173 (1995).

174a. J. A. McMauley, K. Nagasawa, P. A. Lander, S. G. Mischke, M. A. Semones and Y. Kishi, *J. Am. Chem. Soc.*, **120**, 7647 (1998).

175. C. Crevisy and J. M. Beau, *Tetrahedron Lett.*, **32**, 3171 (1991); T. D. Aicher and Y. Kishi, *Tetrahedron Lett.*, **28**, 3463 (1987).

176. C. W. Harwig, S. Py and A. G. Fallis, *J. Org. Chem.*, **62**, 7902 (1997).

177. T. Ishiyama, M. Murata and N. Miyaura, *J. Org. Chem.*, **60**, 7508 (1995), T. Ishiyama, Y. Itoh, T. Kitano and N. Miyaura, *Tetraheron Lett.*, **38**, 3447 (1997).

178. M. Murata, S. Watanabe and Y. Masuda, *J. Org. Chem.*, **62**, 6458 (1997).

179. M. Kosugi, K. Shimizu, A. Ohtani and T. Migita, *Chem. Lett.*, 829 (1981).

180. T. R. Kelly, O. Li and V. Bhushan, *Tetrahedron Lett.*, **31**, 161 (1990). R. Grigg, A. Teasdale and V. Sridharan, *Tetrahedron Lett.*, **32**, 3859 (1991).

181. H. Matsumoto, S. Nagashima, T. Sato and Y. Nagai, *Angew. Chem., Int. Ed. Engl.*, **17**, 279 (1978).

182. Y. Hatanaka and T. Hiyama, *Tetrahedron Lett.*, **28**, 4715 (1987).

183. B. M. Trost and J. Yoshida, *Tetrahedron Lett.*, **24**, 4895 (1983).

184. H. Matsumoto, M. Kasahara, M. Takahashi, T. Arai, M. Hasegawa, T. Nakano and Y. Nagai, *J. Organometal. Chem.*, **250**, 99 (1983); *Chem. Lett.*, 613 (1982).

185. J. D. Rich, *J. Am. Chem. Soc.*, **111**, 5886 (1989).

185a. T. E. Krafft, J. D. Rich and P. J. McDermott, *J. Org. Chem.*, **55**, 5430 (1990).

186. M. Murata, K. Suzuki, S. Watanabe and Y. Masuda, *J. Org. Chem.*, **62**, 8569 (1997).

187. N. A. Bumagin, Yu. V. Gulevich and I. P. Beletskaya, *J. Organometal. Chem.*, **282**, 421 (1985).

188. A. Ricci, A. Degl'Innoceti, S. Chimichi, M. Fiorenza, G. Rossini and H. Bestmann, *J. Org. Chem.*, **50**, 130 (1985); K. Yamamoto, A. Hayashi, S. Suzuki and J. Tsuji, *Organometallics*, **6**, 974 (1987).

189. M. Uno, K. Seto, W. Ueda, M. Masuda and S. Takahashi, *Synthesis*, 506 (1985).

190. M. Uno, K. Seto, M. Masuda, W. Ueda and S. Takahashi, *Tetrahedron Lett.*, **26**, 1553 (1985).

191. M. A. Ciufolini, H. B. Qi and M. E. Browne, *J. Org. Chem.*, **53**, 4149 (1988); *Tetrahedron Let.*, **28**, 171 (1987).

192. T. Sakamoto, Y. Kondo, T. Suginome, S. Ohta and H. Yamanaka, *Synthesis*, 552 (1992).

193. E. Piers, J. Renaud and S. J. Rettig, *J. Org. Chem.*, **58**, 11 (1993); *Synthesis*, 590 (1998).

194. M. Palucki and S. L. Buchwald, *J. Am. Chem. Soc.*, **119**, 11108 (1997); B. C. Hamann and J. F. Hartwig, *J. Am. Chem. Soc.*, **119**, 12382 (1997); T. Satoh, Y. Kawamura, M. Miura and M. Nomura, *Angew. Chem., Int. Ed. Engl.*, **36**, 1740 (1997).

195. J. Ahman, J. P. Wolfe, M. V. Troutman, M. Palucki and S. L. Buchwald, *J. Am. Chem. Soc.*, **120**, 1918 (1998).

196. K. Takagi, T. Okamoto, Y. Sakakibara, A. Ohno, S. Oka and N. Hayama, *Chem. Lett.*, 471(1973); *Bull. Chem., Soc. Jpn.*, **49**, 3177 (1976); H. G. Selnick, G. R. Smith and A. J. Tebben, *Syn. Commun.*, **25**, 3255 (1995).

197. L. Cassar, M. Foa, F. Montanari and G. P. Marinelli, *J. Organometal. Chem.*, **173**, 335 (1979). Y. Sakakibara, N. Yadai, I. Ibuki, M. Sakai and N. Uchino, *Chem. Lett.*, 1565 (1982).

198. J. F. Hartwig, *Synlett.*, 329 (1997), *Angew. Chem., Int. Ed. Engl.*, **37**, 2047 (1998).

199. J. P. Wolfe, S. Wagaw and S. L. Buchwald, *J. Am. Chem. Soc.*, **118**, 7215 (1996); J. P. Wolfe and S. L. Buchwald, *J. Org. Chem.*, **61**, 1133 (1996); J. Louie and J. F. Hartwig, *Tetrahedron Lett.*, **36**, 3609 (1995); M. S. Driver and J. F. Hartwig, *J. Am. Chem. Soc.*, **118**, 7217 (1996).

200. J. P. Wolfe and S. L. Buchwald, *Tetrahedron Lett.*, **38**, 6359 (1997); J. Ahman and S. L. Buchwald, *Tetrahedron Lett.*, **38**, 6363 (1997).

201. M. Nishiyama, T. Yamamoto and Y. Koie, *Tetrahedron Lett.*, **39**, 617, 2367 (1998).

202. J. P. Wolfe and S. L. Buchwald, *J. Org. Chem.*, **62**, 6066 (1997).

203. J. P. Wolfe and S. L. Buchwald, *J. Org. Chem.*, **62**, 1264 (1997); J. Louie, M. S. Driver, B. C. Hamann and J. F. Hartwig, *J. Org. Chem.*, **62**, 1268 (1997).

203a. J. P. Wolfe and S. L. Buchwald, *J. Am. Chem. Soc.*, **119**, 6054 (1997).

204. N. P. Reddy and M. Tanaka, *Tetrahedron Lett.*, **38**, 4807 (1997).

204a. J. P. Wolfe, J. Ahman, J. P. Sadighi, R. A. Singer and S. L. Buchwald, *Tetrahedron Lett.*, **38**, 6367 (1997); G. Mann, J. F. Hartwig, M. S. Driver and C. Fernandez-Rivas, *J. Am. Chem. Soc.*, **120**, 827 (1998).

205. S. Wagaw, R. A. Rennels and S. L. Buchwald, *J. Am. Chem. Soc.*, **119**, 8451 (1997).

206. J. F. Hartwig, *J. Am. Chem. Soc.*, **118**, 13109 (1996); S. L. Buchwald, *J. Am. Chem. Soc.*, **118**, 10333 (1996).

207. M. Palucki, J. P. Wolfe and S. L. Buchwald, *J. Am. Chem. Soc.*, **119**, 3395 (1997); G. Mann and J. F. Hartwig, *Tetrahedron Lett.*, **38**, 8005 (1997).

208. G. Mann and J. F. Hartwig, *J. Org. Chem.*, **62**, 5413 (1997).

209. M. Kosugi, T. Shimizu and T. Migita, *Chem. Lett.*, 13 (1978); T. Migita, T. Shimizu, Y. Asami, J. Shiobara and M. Kosugi, *Bull. Chem. Soc. Jpn.*, **53**, 1385 (1980); P. G. Ciattini, E. Morera and G. Ortar, *Tetrahedron Lett.*, **36**, 4133 (1995).

209a. N. Zheng, J. C. McWilliams, F. J. Fleitz, J. D. Armstrong III and R. P. Volante, *J. Org. Chem.*, **63**, 9606 (1998).

210. J. T. Kuethe, J. E. Cochran and A. Padwa, *J. Org. Chem.*, **60**, 7082 (1995); H. J. Cristau, B. Chabaud, A. Chene and H. Christol, *Synthesis*, 892 (1981).

211. H. Lei, M. S. Stoakes and A. W. Schwabacher, *Synthesis*, 1255 (1992).

212. D. Cai, J. F. Payack, D. R. Bender, D. L. Hughes, T. R. Verhoeven and P. J. Reider, *J. Org. Chem.*, **59**, 7180 (1994).

213. H. Kumobayashi, private communication.

214. L. Kurz, G. Lee, D. Morgans, Jr., M. J. Waldyke and T. Ward, *Tetrahedron Lett.*, **31**, 6321 (1990); Y. Uozumi, A. Tanahashi, S. Y. Lee and T. Hayashi, *J. Org. Chem.*, **58**, 1945 (1993).

215. A. Schoenberg, I. Bartoletti and R. F. Heck, *J. Org. Chem.*, **39**, 3318 (1974); M. Hidai, T. Hikita, Y. Wada, Y. Fujikura and Y. Uchida, *Bull. Chem. Soc. Jpn.*, **48**, 2075 (1975); T. Ito, K. Mori, T. Mizoroki and A. Ozaki, *Bull. Chem. Soc. Jpn.*, **48**, 2091 (1975); J. K. Stille and P. K. Wong, *J. Org. Chem.*, **40**, 532 (1975).
216. T. Takahashi, T. Nagashima and J. Tsuji, *Chem. Lett.*, 369 (1980).
217. D. Valentine, J. W. Tilley and R. A. LeMahieu, *J. Org. Chem.*, **46**, 4614 (1981).
218. R. J. Perry and B. D. Wilson, *J. Org. Chem.*, **61**, 7482 (1996).
219. R. E. Dolle, S. J. Schmidt and L. I. Kruse, *Chem. Commun.*, 904 (1987).
220. S. Cacchi, E. Morera and G. Ortar, *Tetrahedron Lett.*, **26**, 1109 (1985).
221. D. A. Holt, M. A. Levy, D. L. Ladd, H. J. Oh, J. M. Erb, J. I. Heaslop and B. W. Metcalf, *J. Med. Chem.*, **33**, 937, 943 (1990).
222. A. Murai, N. Tanimoto, N. Sakamoto and T. Masamune, *J. Am. Chem. Soc.*, **110**, 985 (1988).
223. K. Kikukawa, K. Kono, K. Nagira, F. Wada and T. Matsuda, *J. Org. Chem.*, **46**, 4413 (1981).
224. M. Mori, Y. Uozumi, M. Kimura and Y. Ban, *Tetrahedron*, **42**, 3793 (1986).
225. F. Ozawa, T. Sugimoto, Y. Yuasa, M. Santra, T. Yamamoto and A. Yamamoto, *Organometallics*, **3**, 683 (1984); T. Kobayashi and M. Tanaka, *J. Organometal. Chem.*, **233**, C64 (1982).
226. T. Kobayashi and M. Tanaka, *J. Mol. Catal.*, **47**, 41 (1988); T. Kobayashi, F. Abe and M. Tanaka, *J. Mol. Catal.*, **45**, 91 (1988).
227. T. Okano, T. Nakagaki, H. Konishi and J. Kiji, *J. Mol. Catal.*, **54**, 65 (1989).
228. T. Takahashi, H. Ikeda and J. Tsuji, *Tetrahedron Lett.*, **21**, 3885 (1980).
229. R. C. Perron, H. des Abbayes and A. Buloup, *Chem. Commun.*, 1090 (1978); L. Cassar and M. Foa, *J. Organometal. Chem.*, **134**, C15 (1977); M. Foa and F. Francalanci, *J. Mol. Catal.*, **41**, 89 (1987).
230. A. Schoenberg and R. F. Heck, *J. Am. Chem. Soc.*, **96**, 7761 (1974); H. Yoshida, N. Sugita, K. Kudo and Y. Takezaki, *Bull. Chem. Soc. Jpn.*, **49**, 1681 (1976).
231. V. P. Baillargeon and J. K. Stille, *J. Am. Chem. Soc.*, **105**, 7175 (1983); **108**, 452 (1986).
232. I. Pri-Bar and O. Buchman, *J. Org. Chem.*, **49**, 4009 (1984).
233. K. Kikukawa, T. Totoki, F. Wada and T. Matsuda, *J. Organometal. Chem.*, **270**, 283 (1984).
234. Y. Tamaru, H. Ochiai, Y. Yamada and Z. Yoshida, *Tetrahedron Lett.*, **24**, 3869 (1983).
235. A. Devasagayaraj and P. Knochel, *Tetrahedron Lett.*, **36**, 8411 (1995).
236. T. Ishiyama, N. Miyaura and A. Suzuki, *Bull. Chem. Soc. Jpn.*, **64**, 1999 (1991).
237. A. M. Echavarren and J. K. Stille, *J. Am. Chem. Soc.*, **110**, 1557 (1988).
238. A. C. Gyorkos, J. K. Stille and L. S. Hegedus, *J. Am. Chem. Soc.*, **112**, 8465 (1990).
239. T. Kobayashi and M. Tanaka, *Tetrahedron Lett.*, **27**, 4745 (1986).
240. E. Negishi, Y. Zhang, I. Shimoyama and G. Wu, *J. Am. Chem. Soc.*, **111**, 8018 (1989).
241. K. Kobayashi and M. Tanaka, *Chem. Commun.*, 333 (1981).
242. P. G. Ciattini, E. Morera and G. Ortar, *Tetrahedron Lett.*, **32**, 6449 (1991).
243. T. Sato, T. Itoh and T. Fujisawa, *Chem. Lett.*, 1559 (1982).
244. J. E. Baldwin, R. M. Adlington and S. H. Ramcharitar, *Chem. Commun.*, 940 (1991).
245. J. Tsuji and K. Ohno, *Synthesis*, 157 (1969).
246. K. Ohno and J. Tsuji, *J. Am. Chem. Soc.*, **88**, 3452 (1966); **90**, 99 (1968).
247. J. Blum, H. Rosenman and E. D. Bergmann, *J. Org. Chem.*, **33**, 1928 (1968).
248. J. Tsuji and K. Ohno, *J. Am. Chem. Soc.*, **90**, 94 (1968).
249. D. E. Ames and A. Opalko, *Tetrahedron*, **40**, 1919 (1984); *Synthesis*, 234 (1983).

250. P. P. Deshpande and O. R. Martin, *Tetrahedron Lett.*, **31**, 6313 (1990); T. Hosoya, E. Takashiro, T. Matsumoto and K. Suzuki, *J. Am. Chem. Soc.*, **116**, 1004 (1994).
251. R. Grigg, V. Loganathan and V. Sridharan, *Tetrahedron Lett.*, **37**, 3399 (1996).
252. R. C. Larock, M. J. Doty, Q. Tian and J. Zenner, *J. Org. Chem.*, **62**, 7536 (1997); **63**, 2002 (1998).
253. C. A. Merlic and D. M. McInnes, *Tetrahedron Lett.*, **38**, 7661 (1997).
254. M. F. Semmelhack, P. M. Helquist and L. D. Jones, *J. Am. Chem. Soc.*, **93**, 5908 (1971); T. Yamamoto, Z. H. Zhou, T. Kanbara and T. Maruyama, *Chem. Lett.*, 223 (1990).
255. T. Takagi, N. Hayama and S. Inokawa, *Chem. Lett.*, 917 (1979); A. Kende, L. S. Liebeskind and D. M. Braitsch, *Tetrahedron Lett.*, 3375 (1975); M. Zembayashi, K. Tamao, J. Yoshida and M. Kumada, *Tetrahedron Lett.*, 4089 (1977); M. Iyoda, K. Sato and M. Oda, *Tetrahedron Lett.*, **26**, 3829 (1985); M. Iyoda, H. Otsuka, K. Sato, N. Nisato and M. Oda, *Bull. Chem. Soc. Jpn.*, **63**, 80 (1990).
256. J. Yamashita, Y. Inoue, T. Kondo and H. Hashimoto, *Chem. Lett.*, 407 (1986); V. Percec, J. Y. Bae, M. Zhao and D. H. Hill, *J. Org. Chem.*, **60**, 176 (1994).
257. M. Slany and P. J. Stang, *Synthesis*, 1019 (1996).
258. G. Dyker, *Tetrahedron Lett.*, **32**, 7241 (1991); *J. Org. Chem.*, **58**, 234 (1993).
259. N. A. Cortese and R. F. Heck, *J. Org. Chem.*, **42**, 3491 (1977).
260. M. K. Anwer, D. B. Sherman, J. G. Roney and A. F. Spatola, *J. Org. Chem.*, **54**, 1284 (1989).
261. S. Cacchi, P. G. Ciattini, E. Morera and G. Ortar, *Tetrahedron Lett.*, **27**, 5541 (1986).
262. K. Sasaki, T. Kubo, M. Sakai and Y. Kuroda, *Chem. Lett.*, 617 (1997).
263. S. Cacchi, E. Morera and G. Ortar, *Org. Synth.*, **68**, 138 (1990); *Tetrahedron Lett.*, **25**, 4821 (1984).
264. R. E. Dolle, S. J. Schmidt and L. I. Kruse, *Tetrahedron Lett.*, **29**, 1581 (1988).
265. J. L. Fabre, M. Julia and J. N. Verpeaux, *Tetrahedron Lett.*, **23**, 2469 (1982), J. L. Fabre and M. Julia, *Tetrahedron Lett.*, **24**, 4311 (1983).
266. J. Uenishi, R. Kawahama, O. Yonemitsu and J. Tsuji, *J. Org. Chem.*, **61**, 5716 (1996); *Tetrahedron Lett.*, **37**, 6759 (1996).
267. J. Tsuji, K. Ohno and T. Kajimoto, *Tetrahedron Lett.*, 4565 (1965).
268. F. Guibe, P. Four and H. Riviere, *Chem. Commun.*, 432 (1980), *J. Org. Chem.*, **46**, 4439 (1981).

4

REACTIONS OF ALLYLIC COMPOUNDS

4.1 Catalytic and Stoichiometric Reactions of Allylic Compounds

π-Allyl complexes are prepared from various allylic compounds, conjugated dienes and alkenes. Oxidative addition of various allylic halides and esters to low-valent metal complexes provides a general synthetic method for π-allyl complexes [1]. Some of them can be isolated. Both electrophilic and nucleophilic π-allyl complexes are known. These π-allyl complexes react with either nucleophiles or electrophiles depending on metal species, and offer useful synthetic methods. Some bis-π-allyl complexes **1** are known to show amphiphilicity.

Both stoichiometric and catalytic reactions of allylic compounds via π-allyl complexes are known. Reactions of nucleophilic π-allyl complexes with electrophiles involve oxidation of metals and hence constitutes stoichiometric reactions. π-Allyl complexes of Ni, Fe, Mo, Co and others are nucleophilic and undergo the stoichiometric reaction with electrophiles. However, electrophilic π-allyl complexes react with nucleophiles, accompanying reduction of metals. For example, π-allylnickel chloride (**2**) reacts with electrophiles such as aldehydes, generating Ni(II), and hence the reaction is stoichiometric. In contrast, electrophilic π-allylpalladium chloride (**3**) reacts with nucleophiles such as malonate and Pd(0) is generated. Thus repeated oxidative addition of allylic compounds to Pd(0) constitutes a catalytic reaction.

Some metals are amphiphilic, reacting with both electrophiles and nucleophiles. For example, the Ru complex **4** reacts with both aldehyde as the electrophile and malonate as the nucleophile under different conditions [2].

Among π-allyl complexes of several transition metals, the chemistry of π-allylpalladium has been studied most extensively. From the standpoint of organic synthesis, reactions involving π-allylpalladium complexes are by far the most important; therefore, their synthetic applications are mainly treated in this chapter.

The reaction of π-allylpalladium chloride (**3**) with carbon nucleophiles such as malonate, acetoacetate and enamines was discovered in 1962 [3]. This reaction constitutes the basis of stoichiometric as well as catalytic π-allylpalladium chemistry.

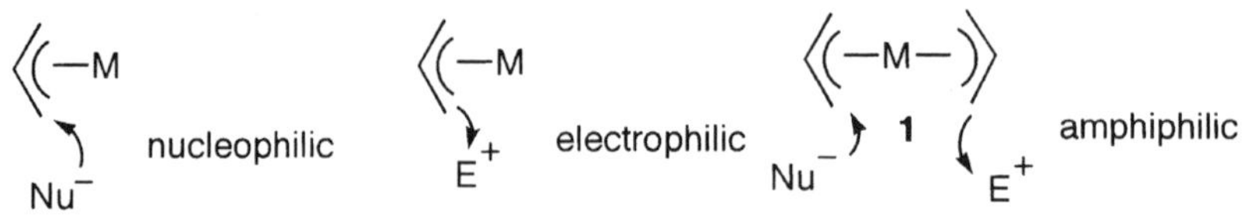

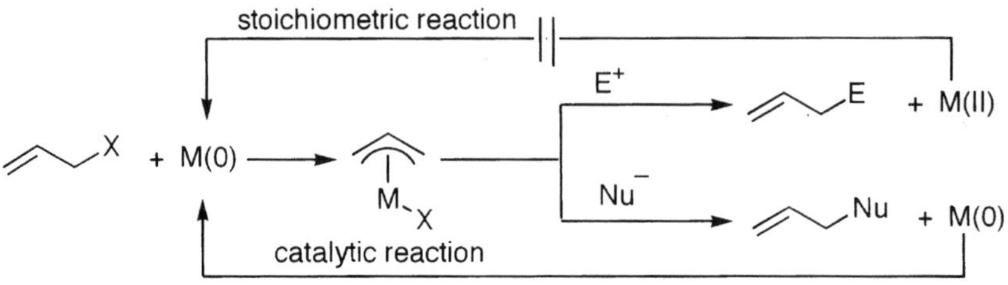

Both stoichiometric and catalytic reactions involving π-allylpalladium complexes are known. Reactions involving π-allylpalladium complexes become stoichiometric or catalytic depending on the preparative methods of the π-allylpalladium complex. Preparation of the π-allylpalladium complexes **6** by the oxidative addition of various allylic compounds **5**, mainly esters to Pd(0), and their reactions with nucleophiles are catalytic. This is because Pd(0) is regenerated after the reaction with the nucleophile, and the Pd(0) reacts again with allylic compounds to form the complex **6**. These catalytic reactions are treated in Section 4.3. However, the preparation of π-allyl complexes **6** from alkenes **7** requires Pd(II) salts. Subsequent reaction with nucleophiles generates Pd(0). As a whole, Pd(II) is consumed, and the reaction ends as the stoichiometric process, because *in situ* reoxidation of Pd(0) to Pd(II) is not attainable in this case. Also, π-allylpalladium complex **9** is formed by the reaction of conjugated dienes **8** with Pd(II), and the reaction of **9** with nucleophiles is stoichiometric.

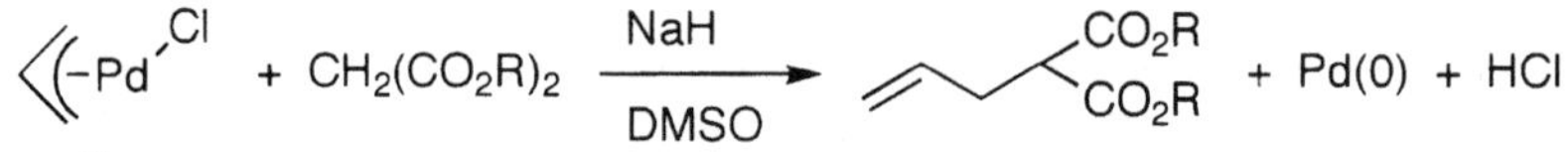

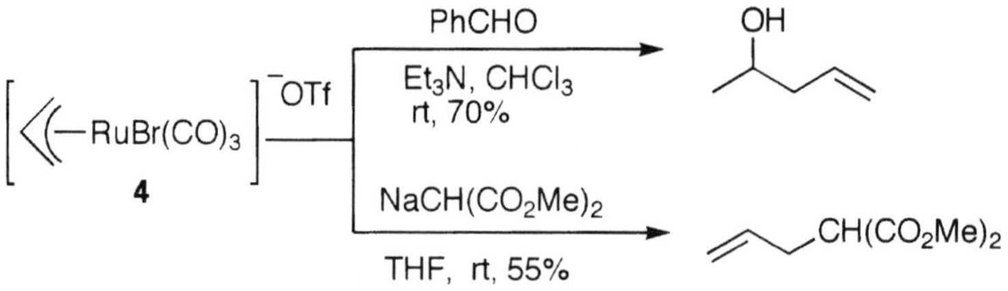

The stoichiometric reactions of allylic compounds is treated in Section 4.2, and the more useful catalytic reactions are treated in Section 4.3.

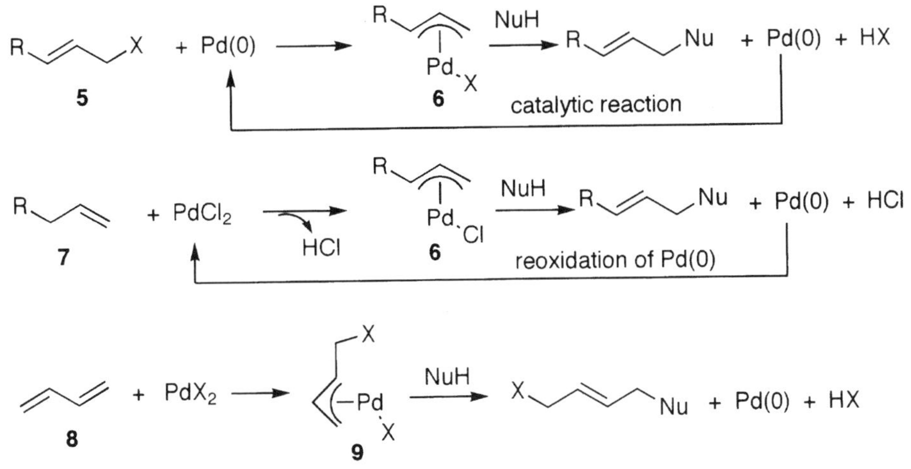

4.2 Stoichiometric Reactions of π-Allyl Complexes

4.2.1 Reactions of Electrophilic π-Allyl Complexes

π-Allylpalladium complexes are electrophilic. Their reactions are stoichiometric when the complexes are prepared from alkenes and $PdCl_2$ under basic reactions. The complex formation occurs in DMF in the presence of bases [4], or in AcOH in the presence of $CuCl_2$ and sodium acetate. It is well-known that π-allylpalladium complex reacts with soft carbon nucleophiles [3]. Combination of these two reactions enables alkylation of alkenes with carbon nucleophiles via π-allylpalladium complexes as a stoichiometric reaction. This reaction offers a method for the oxidative functionalization of alkenes, and has been applied to synthesis of a number of natural products [5]. Conversion of farnesoate (**10**) to geranylgeraniol (**13**) via regioselective formation of the π-allylpalladium complex **11** using methyl 4-methyl-2-phenylsulfonyl-3-penteno-ate (**12**) as a nucleophile is an example [6].

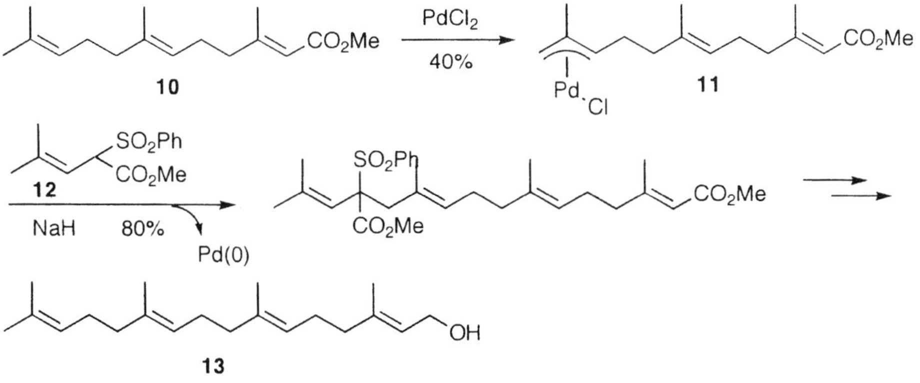

π-Allylic palladium complex **16** is formed particularly easily from the α,β- or β,γ-unsaturated carbonyl compounds **14** or **15**, because elimination of α- or γ-hydrogen is facilitated by their high acidity [7]. The reaction of complex **16**, prepared from the enones **14** or **15**, with a carbon nucleophile leads to **17**. The conversion of **14** and **15** to **17** constitutes γ-alkylation of α,β-unsaturated ketones or esters. The regioselective reaction of the complex **18** with malonate at the 6β-position (γ-alkylation of the enone) in DMSO gave **19** stereoselectively [8]. Thus umpolung is possible via the complex formation, because usually γ-alkylation is possible with an electrophile. Complex **20** is formed from butadiene and $PdCl_2$, and its carbonylation affords 3-hexenedioate (**21**), which is converted to muconate (**23**) by β-elimination via the complex **22** [9].

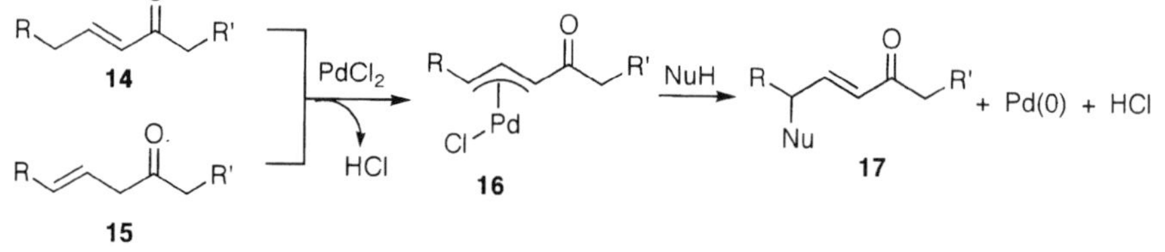

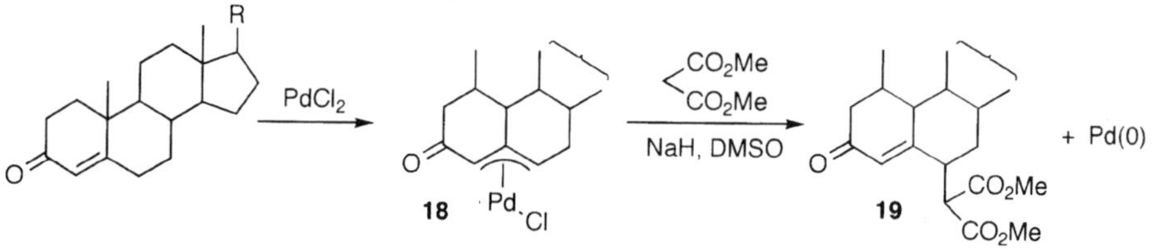

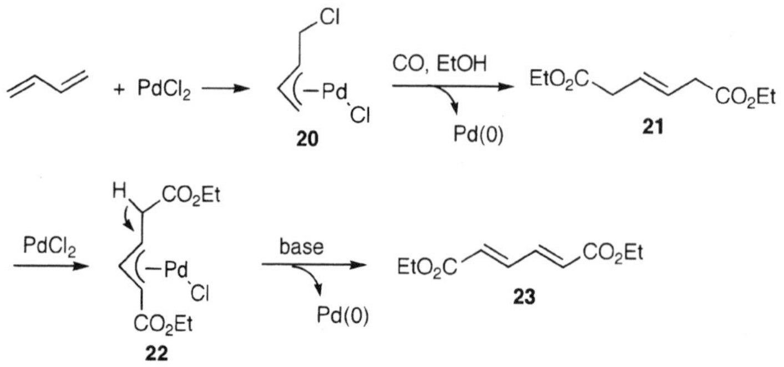

Hard carbon nucleophiles of organometallic compounds react with π-allylpalladium complexes. A steroidal side chain was introduced to **24** regio- and stereoselectively by the reaction of the alkenylzirconium compound **26** with the steroidal π-allylpalladium complex **25**, which was derived from **24** to afford **27** [10].

Thus activation and functionalization of alkenes, enones and conjugated dienes are possible based on the π-allylpalladium complex formation from these unsaturated compounds.

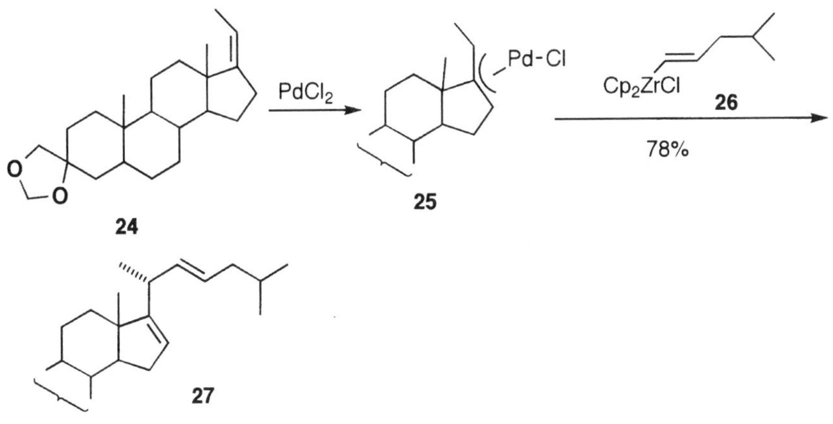

4.2.2 Reactions of Nucleophilic π-Allyl Complexes

π-Allylnickel complexes are prepared by the oxidative addition of allylic halides to Ni(0) complexes, such as Ni(CO)$_4$ or Ni(cod)$_2$, and they attack an electrophilic carbon. Reaction of alkyl iodide **30** with π-allylnickel **29**, prepared from prenyl bromide (**28**), affords **31** [11].

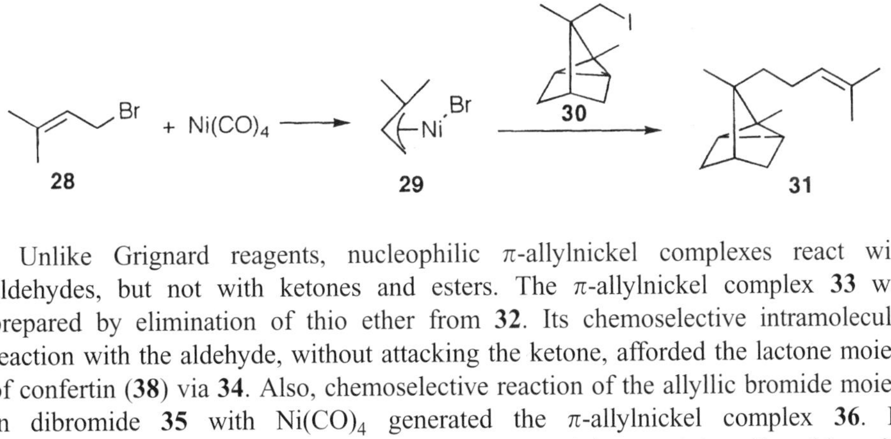

Unlike Grignard reagents, nucleophilic π-allylnickel complexes react with aldehydes, but not with ketones and esters. The π-allylnickel complex **33** was prepared by elimination of thio ether from **32**. Its chemoselective intramolecular reaction with the aldehyde, without attacking the ketone, afforded the lactone moiety of confertin (**38**) via **34**. Also, chemoselective reaction of the allyllic bromide moiety in dibromide **35** with Ni(CO)$_4$ generated the π-allylnickel complex **36**. Its intramolecular reaction with the aldehyde gave **37**, and the remaining alkenyl bromide in **37** was carbonylated via alkenylnickel complex to produce the α-methylene lactone **38**. Frullanolide was synthesized similarly [12].

Allyl complexes of some other transition metals, such as Ru, Fe, Cr and Ti, are also used for synthetic purposes. Reaction of allylic acetates with aldehydes catalyzed by a Ru complex in the presence of Et$_3$N and CO gives the alcohol **39**. It seems likely that the reaction itself is stoichiometric. But a high-valent Ru complex, formed by the reaction with the electrophile, is reduced to a lower valences with Et$_3$N or CO, making whole reaction catalytic [2,13].

$$\text{PhCHO} + \underset{}{\text{OAc}} \xrightarrow[\text{CO, 120°C, 87\%}]{\text{Ru}_3(\text{CO})_{12},\ \text{Et}_3\text{N}} \underset{\mathbf{39}}{\text{Ph}\overset{\text{OH}}{}}$$

Vinyloxiranes are used for facile π-allyl complex formation [14]. The π-allylic ferralactone complex **41** was prepared by oxidative addition of $Fe_2(CO)_9$ to the functionalized vinyloxirane **40** and CO insertion. Treatment of the ferralactone complex **41** with optically active α-methylbenzylamine (**42**) in the presence of $ZnCl_2$ gave the π-allylic ferralactam complex **45** via **44**. In this case, as shown by **43**, the amine attacks the terminal carbon of the allylic system and then the lactone carbonyl. Then, elimination of OH group generates the π-allylic ferralactam complex **45**. Finally the β-lactam **46** was obtained in 64% yield by oxidative decomplexation with Ce(IV) salt. The δ-lactam **47** was a minor product (24%). The precursor of the thienamycin **48** was prepared from **46** [15,16]. This mechanistic explanation is supported by the formation of both π-allyllactone and lactam complexes (**49** and **51**) from the allylic amino alcohol **50** [17].

Allylchromium complexes may be generated by the reaction of allylic bromides **54**, allylic phosphates **56** [18], and vinyloxiranes **59** [19] with $CrCl_2$. They react with the aldehyde group in **53** to give **55** selectively. The reaction with ketones is slow. The reaction takes place regioselectively at the more substituted side of the allylic system of **56** and **59** to give **57, 58, 60** and **61** as the main products. In the synthesis of the highly strained nine-membered ring **64** in neocartinostatin, the best result was obtained by intramolecular $CrCl_2$-promoted regioselective reaction of the aldehyde at the more substituted side of the allylic system in **63** [20].

The nucleophilic π-allyltitanium complex **67** is prepared by the reaction of the conjugated diene **65** with titanocene hydride **66**, generated *in situ* by the treatment of titanocene dichloride with 2 moles of *i*-PrMgCl [21]. The complex is nucleophilic and reacts with aldehydes regio- and stereoselectively to give homoallylic alcohols [22].

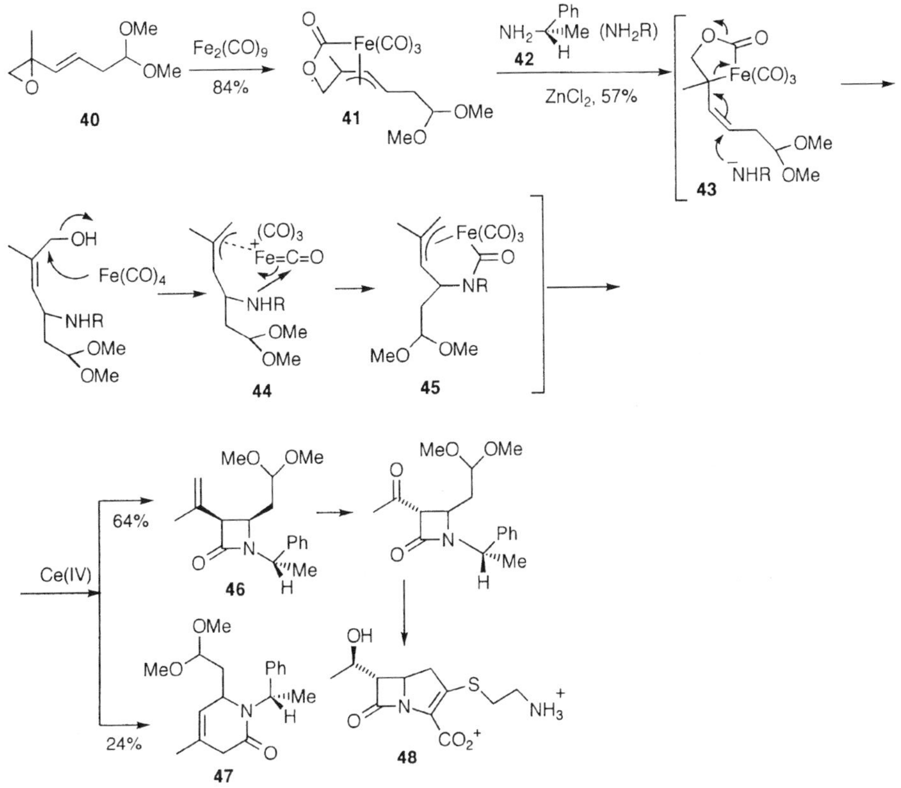
OMe
OMe
Fe2(CO)9
84%
40
Fe(CO)3
41
MeO
OMe
NH2 Ph Me (NH2R)
H
42
ZnCl2, 57%
Fe(CO)3
OMe
NHR OMe
43
OH
Fe(CO)4
NHR
OMe
OMe
(CO)3
Fe=C=O
NHR
44
Fe(CO)3
NR
MeO OMe
45
Ce(IV)
64%
24%
MeO OMe
Ph
Me
H
46
47
OH
S
NH3+
CO2+
48

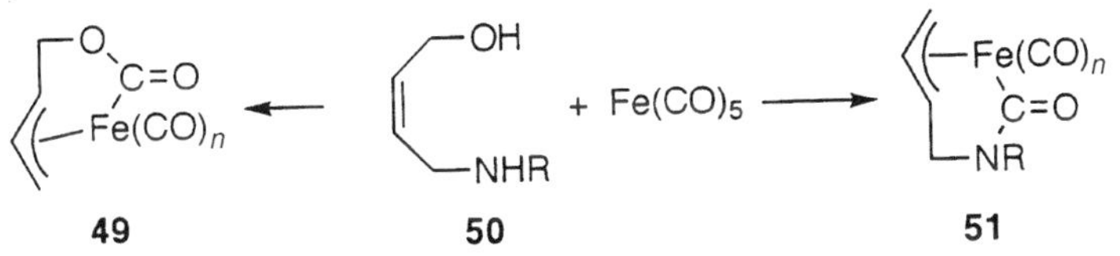
C=O
Fe(CO)n
49
OH
NHR
50
+ Fe(CO)5
Fe(CO)n
C=O
NR
51

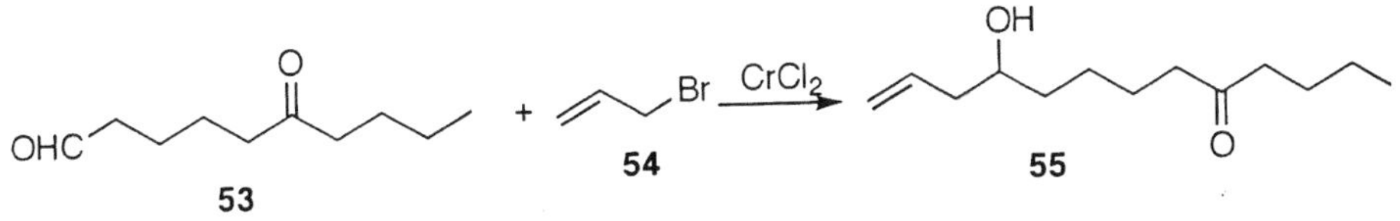
OHC
53
+
Br
54
CrCl2
OH
55

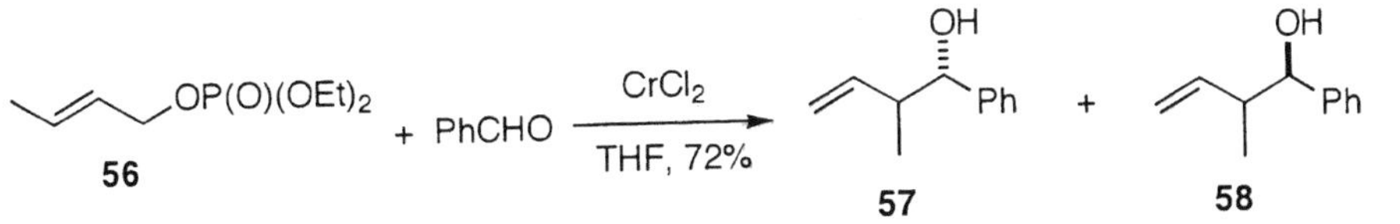
OP(O)(OEt)2
56
+ PhCHO
CrCl2
THF, 72%
OH
Ph
57
+
58

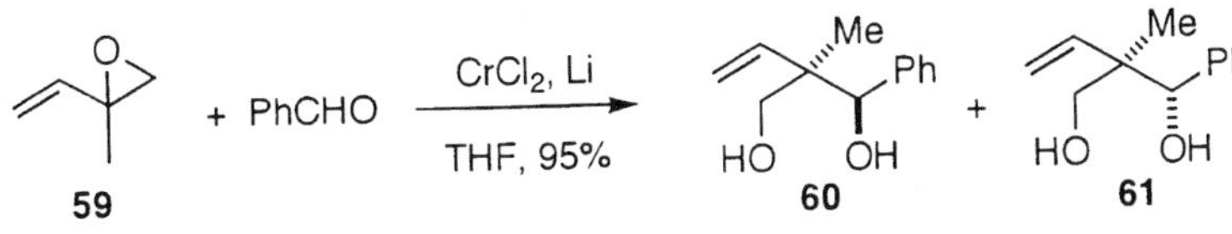
59
+ PhCHO
CrCl2, Li
THF, 95%
Me
Ph
HO
OH
60
+
61

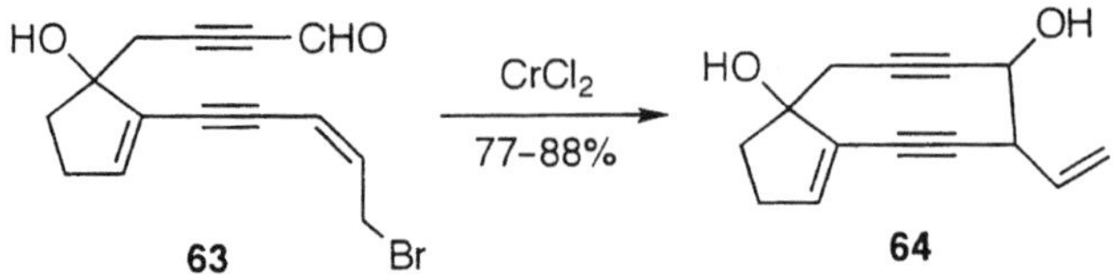

The β,γ-unsaturated ester **68** was prepared by the reaction of the π-allyltitanium complex **67** with methyl chloroformate [23].

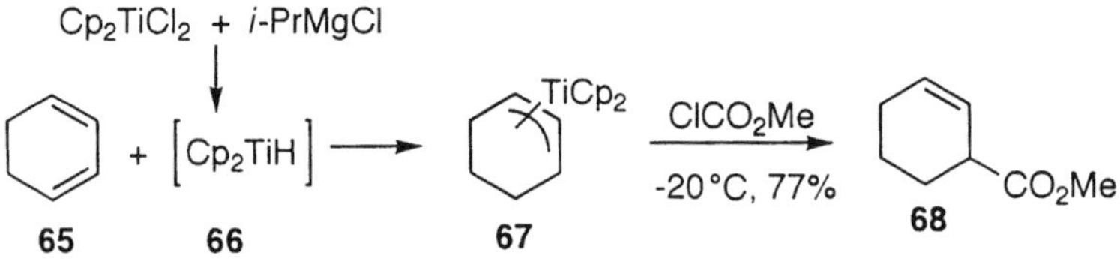

4.3 Catalytic Reactions of Allylic Compounds

4.3.1 *Allylation of Nucleophiles*

Transition metal complexes, particularly Pd(0)-catalysed reactions of allylic compounds via π-allyl intermediates, are useful reactions [24]. Formation of the π-allylpalladium complexes **69** by the oxidative addition of various allylic compounds to Pd(0), and subsequent reaction of the complexes with soft carbon nucleophiles to give **70** are the basis of the catalytic allylation. After the reaction, Pd(0) is regenerated, which undergoes the oxidative addition to the allylic compounds again, making the reaction catalytic. The catalytic reaction was first reported by two groups [25]. Similarly, hard carbon nucleophiles of organometallic compounds of main group metals are allylated with π-allylpalladium intermediates. The reaction proceeds via transmetallation.

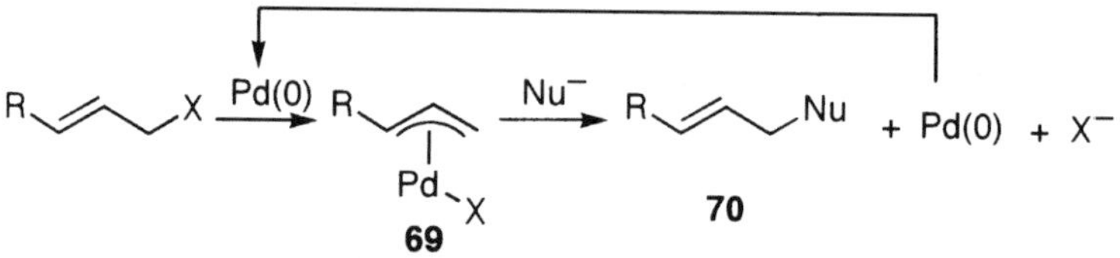

The efficient catalytic cycle is ascribed to the characteristic feature that Pd(0) is more stable than Pd(II). Reactions of π-allylpalladium complexes with carbon nucleophiles are called Tsuji–Trost reactions. In addition to Pd, other transition metal complexes, such as those of Mo [26], Rh [27] and other metals, are used for catalytic allylation.

In addition to the formation of **70** by the usual nucleophilic attack at the terminal carbon of allylic system, the substituted cyclopropanes **72** are formed by the attack at the central sp^2 carbon of the allylic system via the palladacyclobutane **71** and its reductive elimination under certain conditions [28].

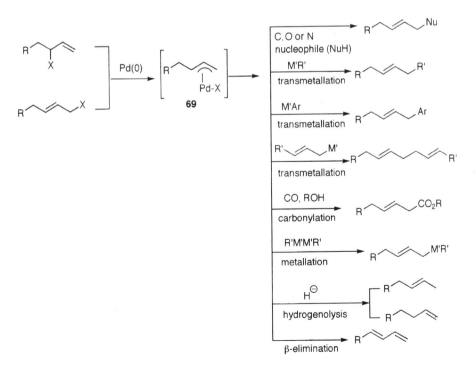

In addition to the catalytic allylation of carbon nucleophiles, several other catalytic transformations of allylic compounds via π-allylpalladium intermediates **69** are possible and they are summarized in Scheme 4.1. Sometimes these reactions are competitive with each other, and the chemoselectivity is dependent on the reactants and the reaction conditions.

A number of allylic leaving groups shown in Scheme 4.2 are cleaved by Pd catalysts. Mainly allylic esters are used as substrates for the catalytic reactions. In addition, even allylic nitro compounds [29,30] and sulfones [31–33] are known to form π-allylpalladium complexes.

The stereochemistry of the Pd-mediated or -catalysed allylation of nucleophiles has been studied extensively [34–36]. In the first step, formation of π-allylpalladium complex **74** by the attack of Pd(0) on an allylic acetate moiety of **73** proceeds by inversion (*anti* attack). Subsequent reaction of soft carbon nucleophiles, N– and O– nucleophiles gives **75** by inversion. Thus the overall retention is observed. However,

Scheme 4.1 Pd-catalysed reactions of allylic compounds

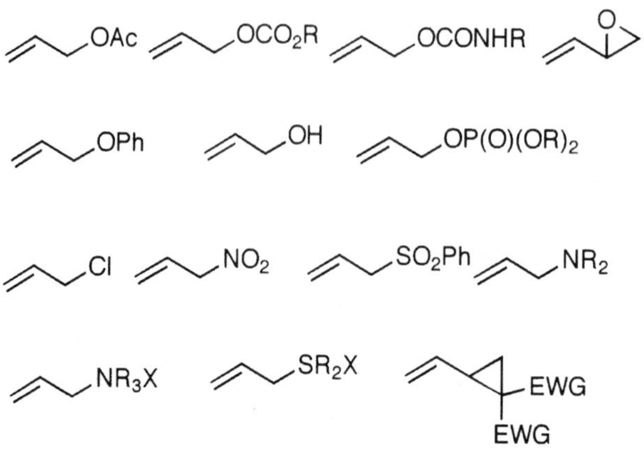

Scheme 4.2   Allylic compounds used for Pd-catalysed reactions

transmetallation of **74** with hard carbon nucleophiles of organometallic compounds affords **76** by retention, and the final product **77** is obtained by reductive elimination, which is retention. Thus an overall inversion is observed in this case [37,38].

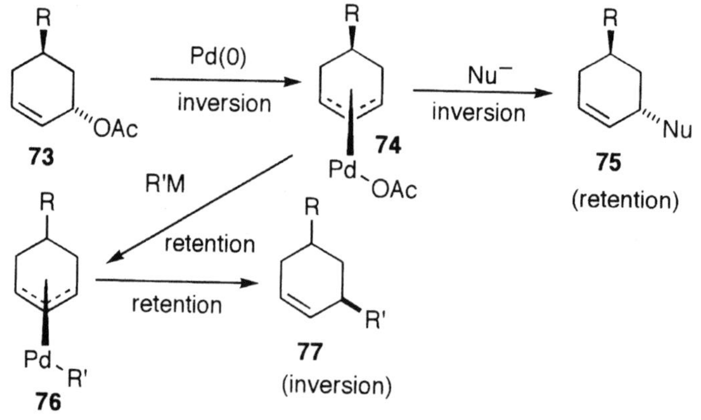

Pd-catalyzed reaction of nucleophiles with substituted π-allyl systems usually occurs at the less substituted side with high regioselectivity, although some exceptions are known. For example, when bulky (*R*)-MeO-MOP (**XXXVI**) is used as a ligand, reaction of methyl methylmalonate at the more substituted side of 1-ary-2-propenyl acetate occurs to give **78** with high branch-selectivity (9:1) as well as high enantioselectivity (87% ee) [39]. Rh complexes are active for catalytic allylation [27]. Using the Wilkinson complex modified by $(MeO)_3P$, the quaternary substituted product **79** was obtained as a main product from the tertiary allylic carbonate [40]. Also, the reaction of chiral allylic carbonate **80** with malonate gave **81** with almost complete retention of the absolute configuration [41]. Products from more substituted side are also obtained using $Mo(CO)_nL_m$ [42].

4.3.2 Allylation of C, N and O Nucleophiles

Reactions under neutral conditions are highly desirable. It was believed for long time that allylation with allylic acetates proceeds only in the presence of bases. An important advance in π-allylpalladium chemistry has been achieved by the introduction of highly reactive allylic carbonates. Their reactions can be carried out under mild neutral conditions [43,44]. Reactions of allylic carbamates [44], allyl aryl ethers [45,46] and vinyl epoxides [47,48] also proceed under neutral conditions without addition of bases. As shown by the following mechanism, the oxidative addition of allyl carbonates is followed by decarboxylation as an irreversible process to afford π-allylpalladium alkoxide **82**, and then the generated alkoxide picks up proton from active methylene compounds (NuH), yielding **83**. This *in situ* formation of the alkoxide is the reason why the reaction of allyl carbonates can be carried out without addition of bases from outside. Alkoxides are rather poor nucleophiles and allyl alkyl ethers are not formed. In addition, the formation of π-allylpalladium complexes from allylic carbonates involving decarboxylation is irreversible. Allylic carbamates behave similarly to carbonates [44]. The chemoselective C-allylation of nitroacetate with the carbonate side of the bis-allylic compound **84** affords **85**, while the acetate group remains unattacked. The chemoselective reaction clearly shows the higher reactivity of allylic carbonates than allylic acetates [49]. No O-alkylation of the nitro compounds, usually observed under basic conditions, occurs.

Neutral allylation with allylic carbonates has wide application in the alkylation of rather labile compounds that are sensitive to acids or bases. For example, successful C-allylation of the sensitive molecule of ascorbic acid (**86**) to give **87** is possible only with allyl carbonate under neutral conditions [50]. However, it was shown recently that allylic acetates react with soft carbon nucleophiles except malonate under neutral conditions [51].

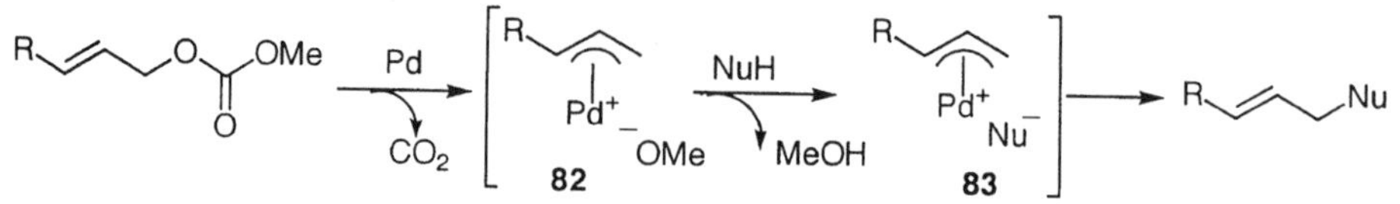

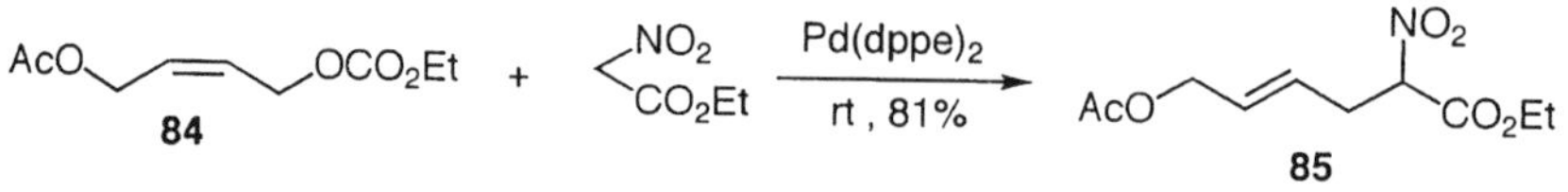

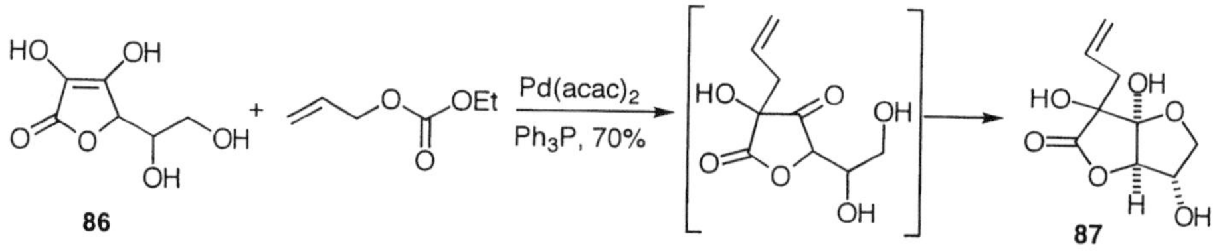

Allylic phosphates are more reactive than allylic acetates. Chemoselective reaction of the allylic phosphate moiety of the bis-allylic compound **88** with one equivalent of malonate, without attacking the allylic acetate moiety, takes place to give **89**. Then the aminated product **90** was obtained by the addition of amine [52].

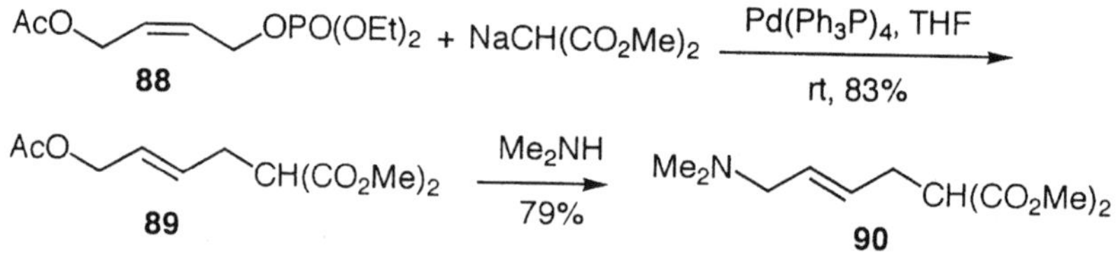

Various cyclic compounds from three-membered rings to macrocycles have been prepared by intramolecular allylation. A typical example of this cyclization is the reaction of the monoacetate of 1,4-butenediol derivative **91** with the active methylene compound **92**, which afforded the allylic alcohol **93**. The three-membered chrysantemic acid derivatives **94** and **95** were then prepared after acetylation of **93**, followed by Pd-catalysed intramolecular allylation [53].

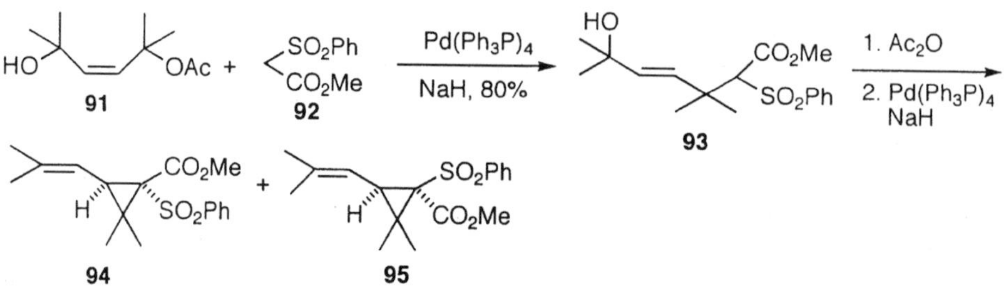

Cyclization of the allyl phenyl ether **96** under neutral conditions gives the cyclopentanone **97** as a main product and the cycloheptanone **98** as a minor product when Ph₃P is used. However, the furan **99** is obtained by O-alkylation using (PhO)₃P as the ligand. Rearrangement of the furan **99** to **97** and **98** takes place with Ph₃P ligand. However, Claisen rearrangement of **99** at 195 °C gives the cycloheptenone **98**, as expected [54].

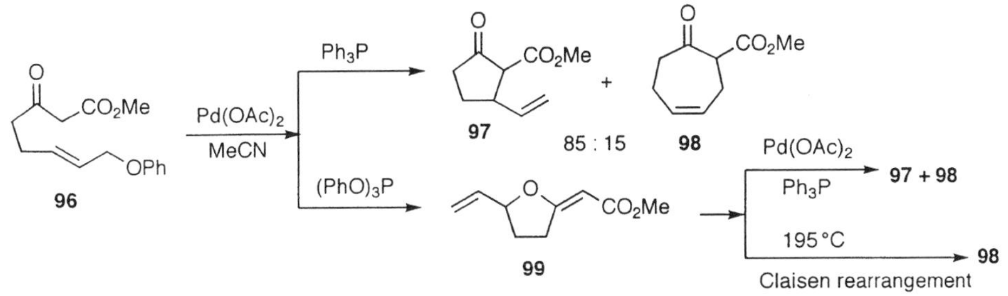

The six-membered ring **101** is formed from **100** without forming eight-membered ring [55], whereas the eight-membered ring **103** is obtained as the main product and the six-membered ring **104** as the minor product from the allyl carbonate **102** using dppe as a ligand [56].

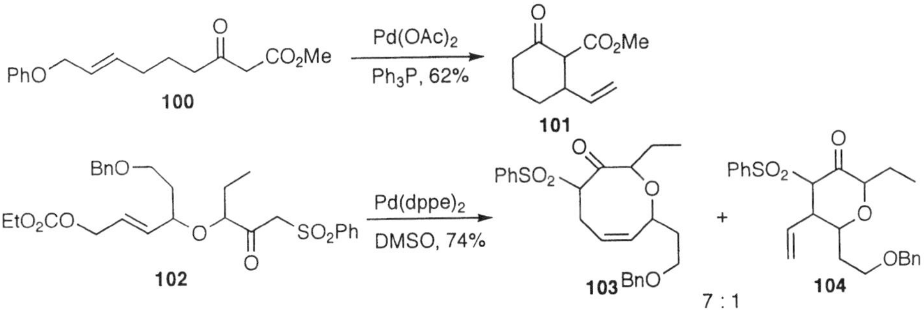

Intramolecular allylation has wide application in the synthesis of macrocycles [57]. Synthesis of humulene (**107**) by the cyclization of allyl acetate **105** to give **106** is an early example [58]. The 14-membered ring **109** was obtained from **108** and converted to cembranolide **110** [59].

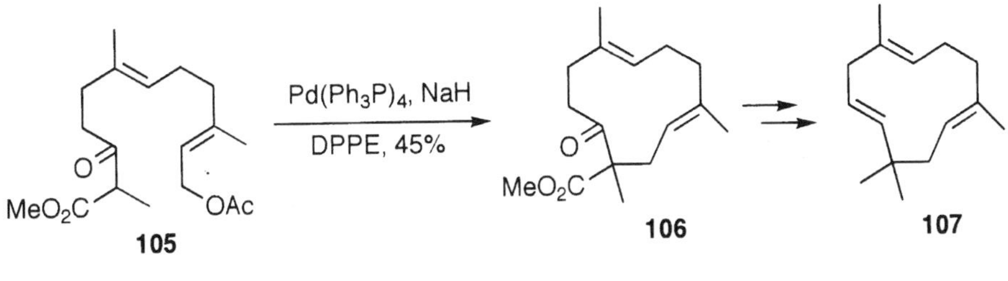

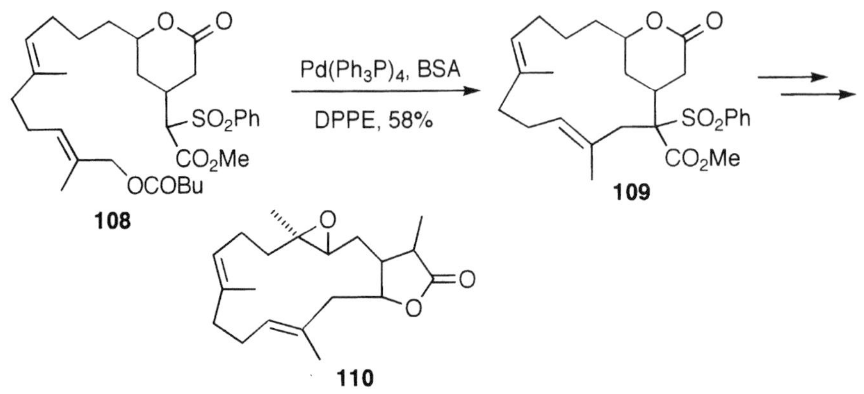

The trichothecene framework **114** was constructed by interesting Pd-catalysed skeletal reorganization via pinacolone rearrangement of the allylic lactone **111**. The *exo*-π-allylpalladium complex **112** is attacked intramolecularly by a carbon bond (pinacolone rearrangement) to give **113** [60].

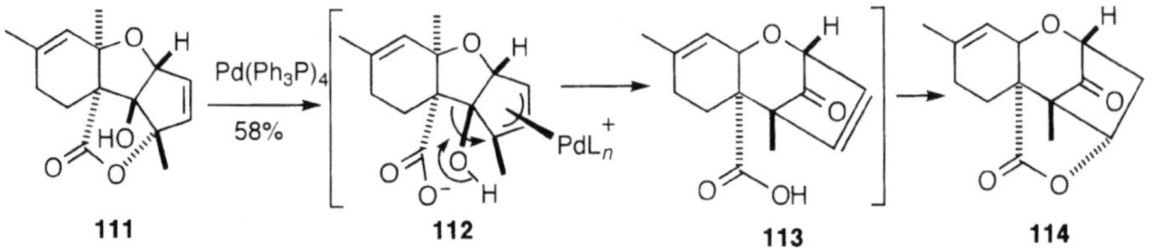

Interestingly, the allylation of stabilized carbon nucleophiles has been found to be reversible. Complete transfer of dimethyl methylmalonate moiety, involving C–C bond cleavage, from the secondary carbon **115** to the primary carbon **116** was observed by treatment with a Pd catalyst for 24 h, showing that the allylic C–C bond cleavage proceeds slowly [61]. Ni(dppb)$_2$ was found a more efficient catalyst for the rearrangement [62].

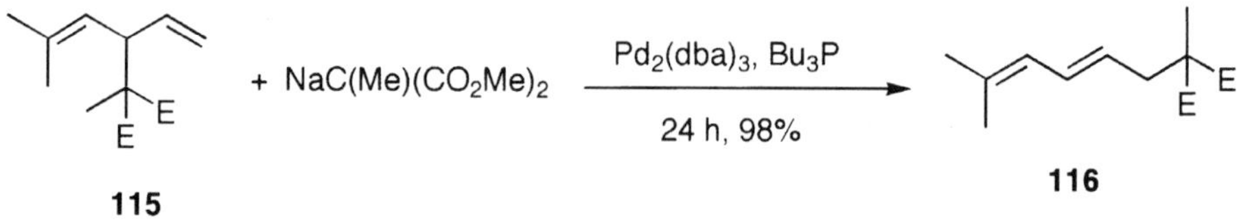

Vinyl epoxides (vinyloxiranes) **117** are reactive allylating agents. The epoxy ring is opened by the oxidative addition of Pd(0) and the π-allylpalladium complex **118** is formed. At the same time, alkoxide is generated, which abstracts a proton from NuH to form **119**, and hence the allylation of Nu⁻ proceeds under neutral conditions. In addition, the 1,4-adduct **120** is formed mainly, rather than the 1,2-adduct **121** [63]. The reaction has been utilized for the introduction of a 15-hydroxy group in a steroid related to oogoniol **124** using TMPP (**XLVI**) as a ligand [64]. The oxirane **122** is the β-form, and Pd(0) attacks from the α-side by inversion. Then the nucleophile comes from the β-side to give **123**. Thus the overall reaction is *syn*-S$_N$2′ in type. The 26-

membered ring **126** was prepared efficiently by the cyclization of the vinyl epoxide **125** [65].

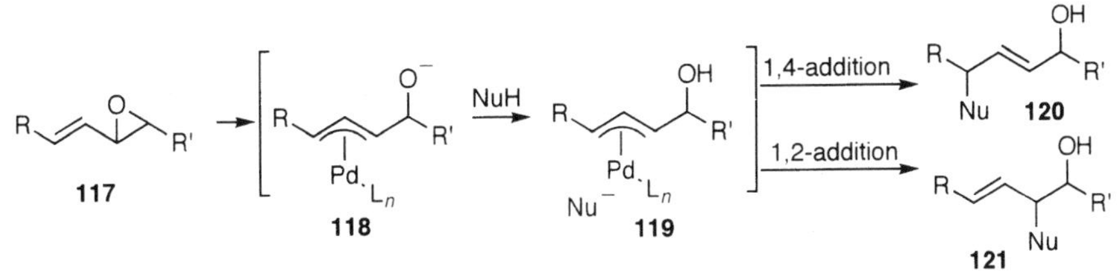

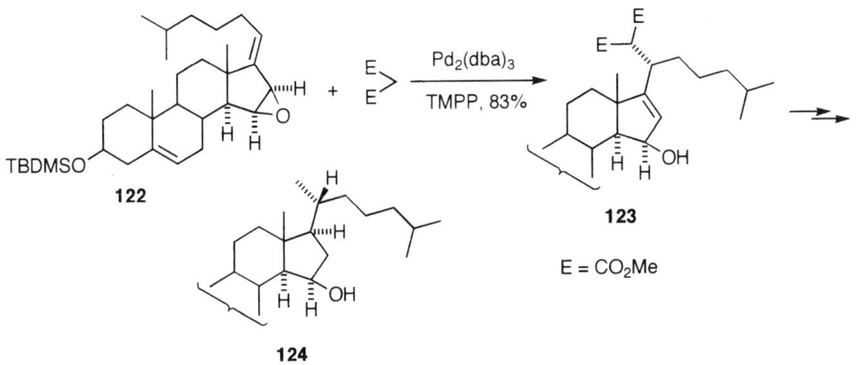

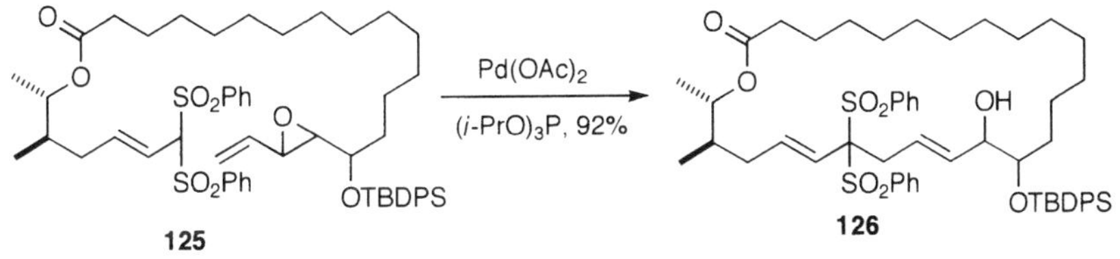

Various unsaturated compounds, such as CO_2, isocyanates and aldehydes, undergo Pd-catalysed cycloaddition with vinyl epoxides. Reaction of CO_2 with **127** affords cyclic carbonates **128** with retention of the configuration at C(3), offering a method of *cis* hydroxylation of epoxides [66], and has been used for the synthesis of the side-chain unsaturated (−)-*exo*-brevicomin (**129**) [67]. The tetrahydrofuran **131** was prepared by [3+2] cycloaddition of activated alkenes such as benzylidene malononitrile (**130**) with vinyl epoxide via Michael addition and allylation [68].

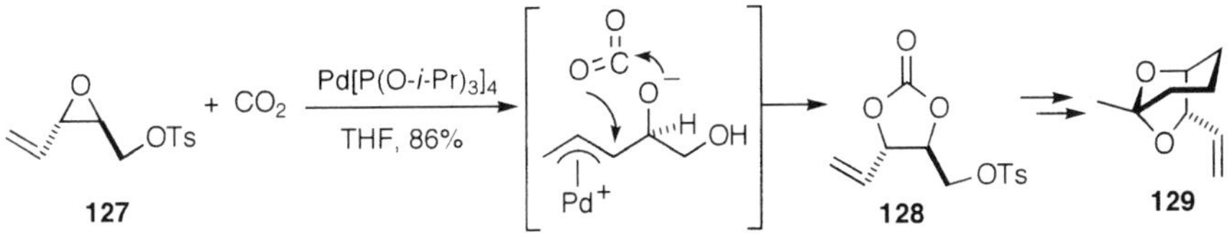

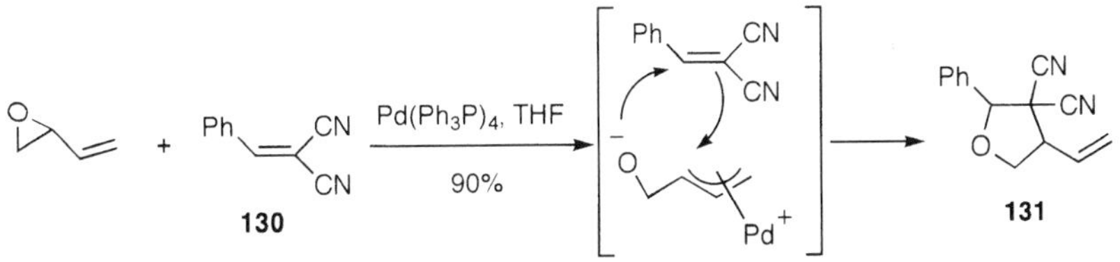

Pd-catalysed allylation of amines proceeds smoothly. Allylamine (**132**) and di- and triallylamines are produced commercially by the Pd-catalysed reaction of ammonia with allyl alcohol using DPPB as a suitable ligand [69]. Allylic alcohols are rather unreactive substrates for π-allylpalladium complex formation under usual conditions. The intramolecular amination of **133** afforded the azaspiro ring **134** and the reaction was applied to the synthesis of perhydrohistrionicotoxin (**135**) [70]. Smooth Pd-catalyzed allylation of the purine base **136** gives **137**, which is utilized for the synthesis of nucleosides [71].

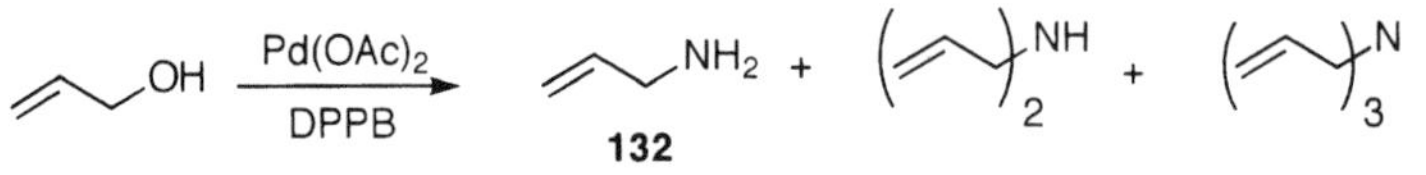

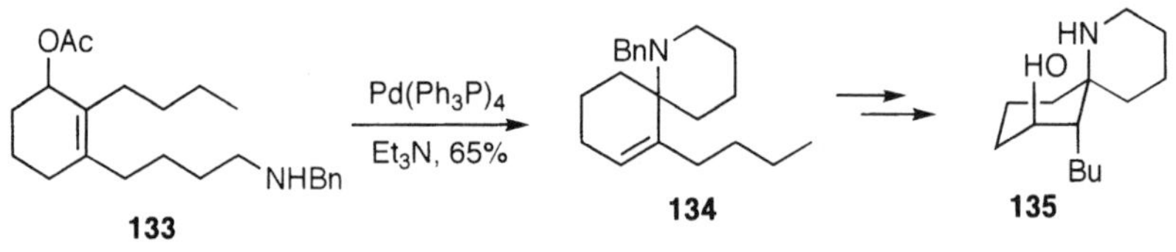

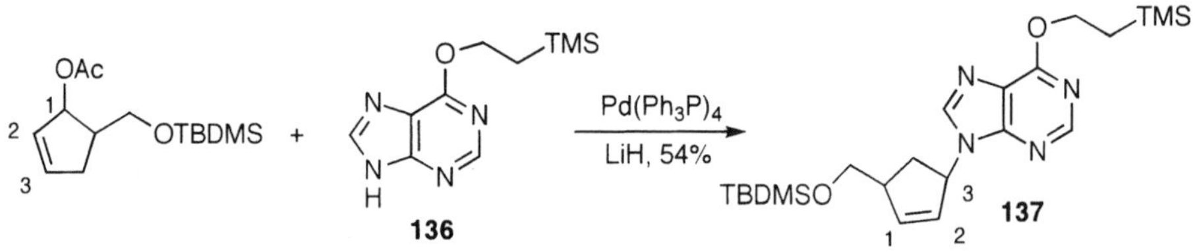

Although the C−N bond of allylamines is difficult to cleave, it can be cleaved in AcOH, probably by forming amine salt. However, the allylation of nucleophiles with the allylamine **138** is catalysed by Ni–DPPB [72]. Removal of the allyl group from allylamines is possible with Ni(dppp)Cl$_2$–DIBAL and used for deprotection of amines, which are protected as allylamines [73]. Pd–DPPB is less active.

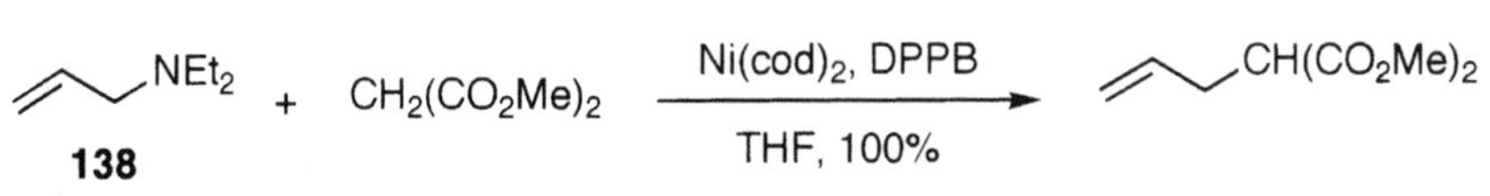

Carboxylates behave as an O−nucleophile and are allylated. Reaction of AcONa with cyclopentadiene monoxide (**139**) proceeds with retention of stereochemistry as shown by **140**, to give the 3,5-*cis*-disubstituted cyclopentene **141** [74]. Although alcohols are rather unreactive nucleophiles for the Pd-catalysed allylation, the alkoxide anions generated by the treatment of silyl ethers with TBSF are easily allylated. Desilylated alcohol from **142** reacts intramolecularly with the *cis* and *trans* vinyl epoxides to give the *cis* and *trans* pyrans **143** and **144** regio- and stereoselectively, and

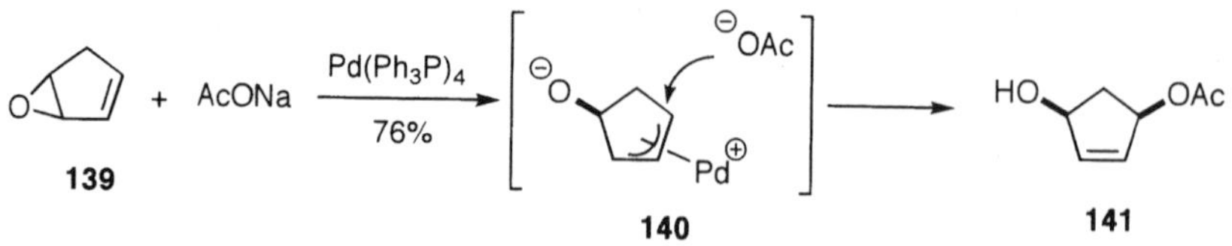

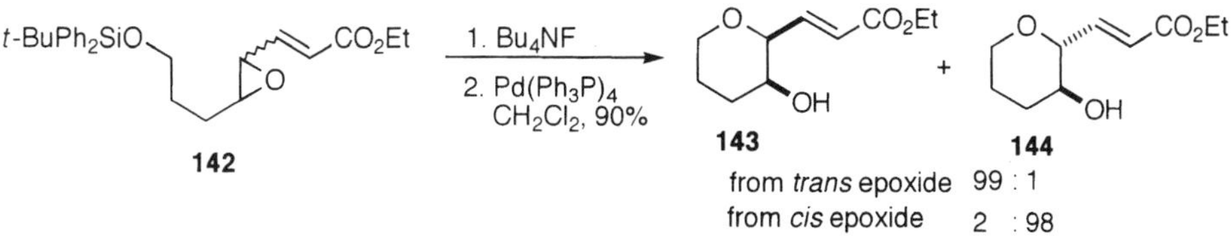

142

143 + **144**

from *trans* epoxide 99 : 1
from *cis* epoxide 2 : 98

the reaction was applied to the synthesis of AB rings of gambiertoxin [75]. In this case, rather unusual 1,2-addition to the vinyl epoxide **142** occurs.

(+)-γ-Lycorane (**150**) has been synthesized by applying allylation of C and N nucleophiles and the Heck reaction. Asymmetric allylation of the malonate derivative **146** with the benzoates of cyclohexene-1,4-diol (**145**) using (*S*)-BINAPO (**XXXIV**) afforded **147** with 40% ee. The intramolecular allylation of the amide using DPPB gave rise to **148**. Without isolation, the intramolecular Heck reaction of **148** occurred by the addition of a tertiary amine to give **149**, which was converted to (+)-γ-lycorane (**150**) [76].

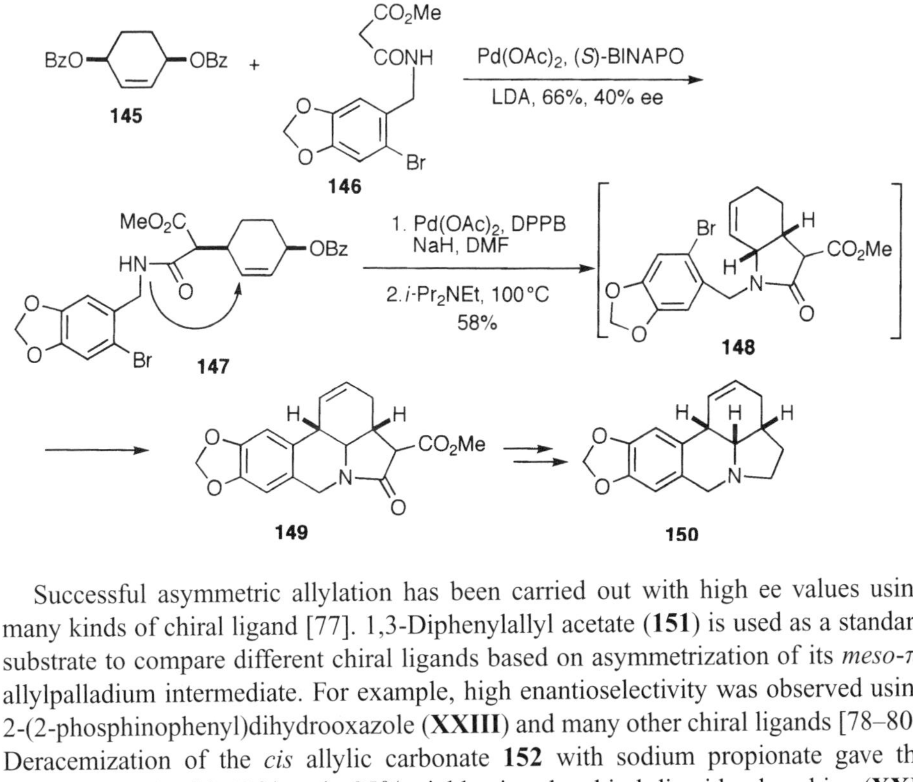

145 + **146**

147

148

149 → **150**

Successful asymmetric allylation has been carried out with high ee values using many kinds of chiral ligand [77]. 1,3-Diphenylallyl acetate (**151**) is used as a standard substrate to compare different chiral ligands based on asymmetrization of its *meso*-π-allylpalladium intermediate. For example, high enantioselectivity was observed using 2-(2-phosphinophenyl)dihydrooxazole (**XXIII**) and many other chiral ligands [78–80]. Deracemization of the *cis* allylic carbonate **152** with sodium propionate gave the propionate **153** with 98% ee in 95% yield using the chiral diamide phosphine (**XXI**) [81]. Reaction of cinnamyl carbonate with dimethyl methylmalonate, catalysed by the Mo complex coordinated by the chiral tetramine ligand **154**, proceeded with high

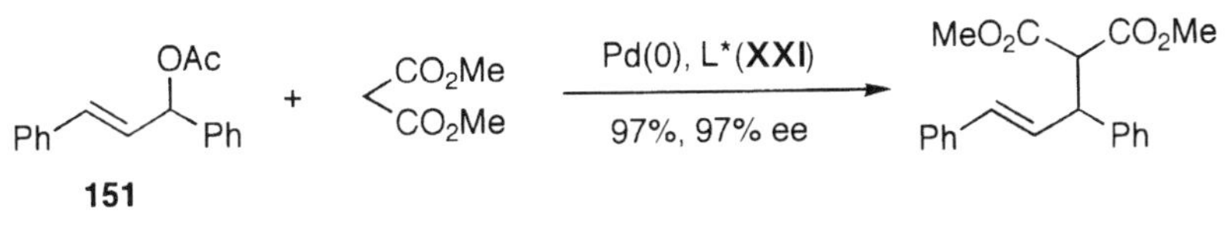

151

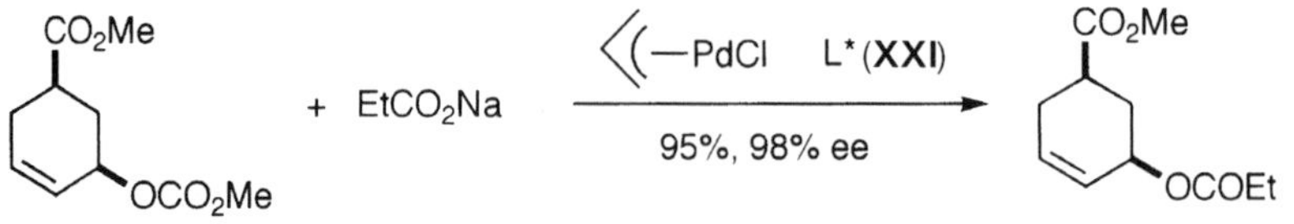

L* =

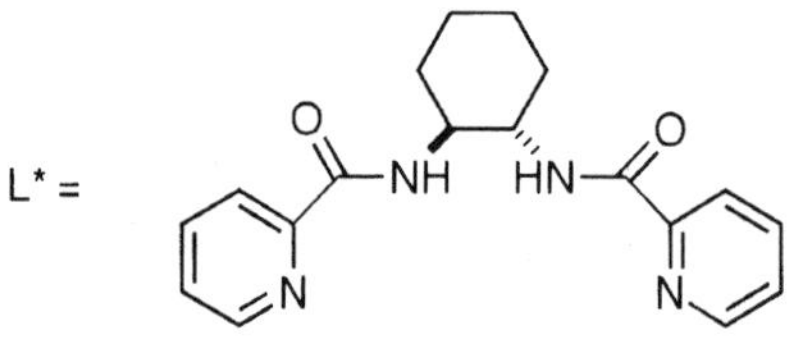

154

regioselectivity and enantioselectivity (98% ee). The product **155** from attack at the more substituted terminus was obtained [42].

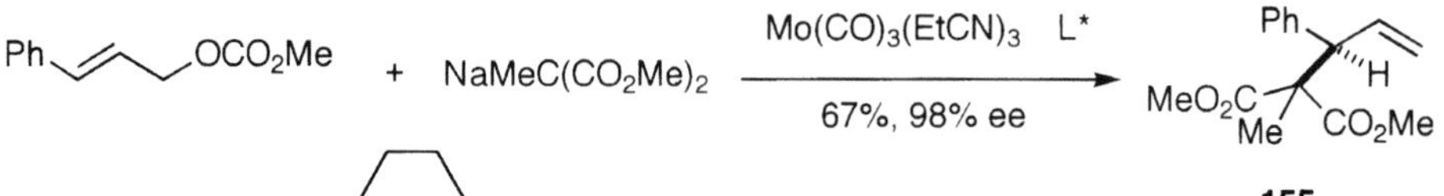

The Pd-catalysed reaction of 2-(trimethylsilylmethyl)allyl acetate (**156**) under basic conditions generates π-allylpalladium **157**. Elimination of the allylic TMS group from **157** generates the dipolar complex **158** (trimethylenemethane complex). The [3+2] cycloaddtion of this species with alkenic bonds **159** bearing an electron-withdrawing group (EWG) gives **160** [82]. For example, the smooth reaction of the cyclohexenonecarboxylate **161** with **156** afforded **162**. The kampane-type diterpene **163** was synthesized by this reaction [83].

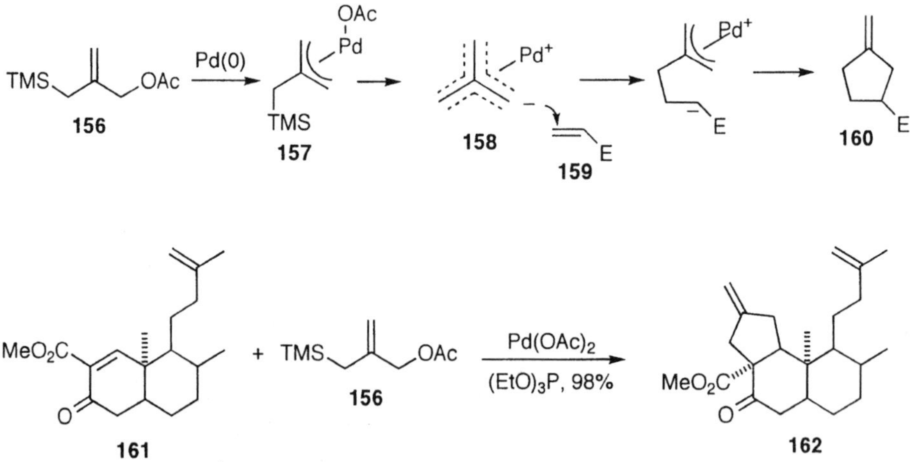

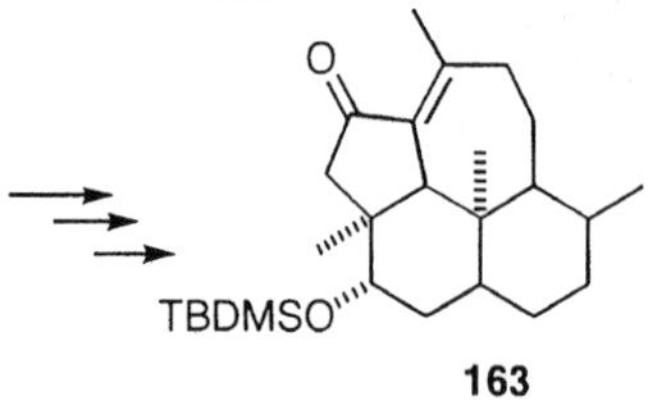

The π-allyl complex **165** is formed from the allyl carbonate **164** bearing an EWG. The dipolar molecule **166** is generated by deprotonation with the alkoxide. The five-membered ring with exomethylene **169** is formed via **168** by the [3+2] cycloaddition with the alkene **167** [84]. This cycloaddition proceeds under neutral conditions. For example, the reaction of the reagent **170** with 5,5-dimethylcyclopentenone gave rise to the bicyclic ketone **171** [85].

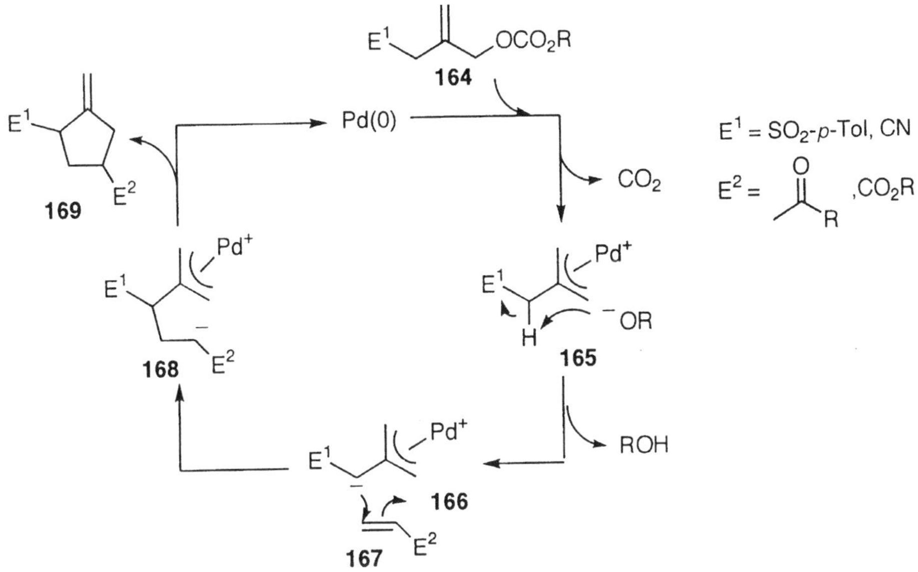

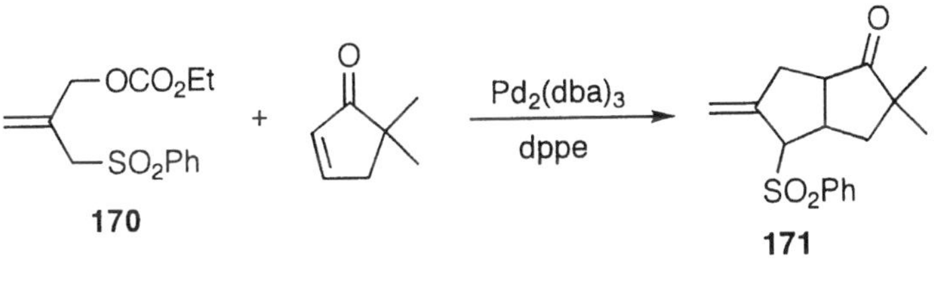

4.3.3 Amphiphilic Bis-π-allylpalladium

It is well-established that π-allylpalladium is electrophilic, and no reaction with electrophiles has been observed. However, there is an evidence that bis-π-allylpalladium (**172**), generated *in situ,* could be amphiphilic. Typically, formation of the 2-substituted 3,6-divinylpyran (**175**) by the reaction of butadiene with aldehyde can be explained by the amphiphilic nature of the bis-π-allylpalladium **173** generated *in situ* as an intermediate, which reacts with the electrophilic carbon and the nucleophilic oxygen in the aldehyde as shown by **174** [86]. As a similar reaction, piperidone is obtained by the reaction of butadiene with isocyanate [87]. The reaction of allyltributylstannane (**176**), allyl chloride and benzalmalononitrile (**177**) in the presence of $PdCl_2(Ph_3P)_2$ (3 mol %) afforded the diallylated product **178** in high yield.

The reaction is explained by the *in situ* formation of bis-π-allylpalladium (**172**), which is amphiphilic and reacts with **177** [88]. It is known that the reaction of the allylstannane **176** with aldehydes proceeds under rather severe conditions. Facile formation of the homoallyl alcohol **179** at room temperature in the presence of a Pt or Pd catalyst is explained by the nucleophilic nature of bis-π-allylplatinum or -palladium generated *in situ* as an intermediate [89].

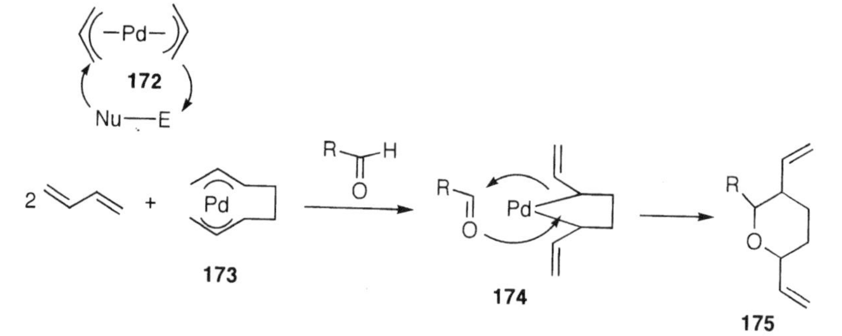

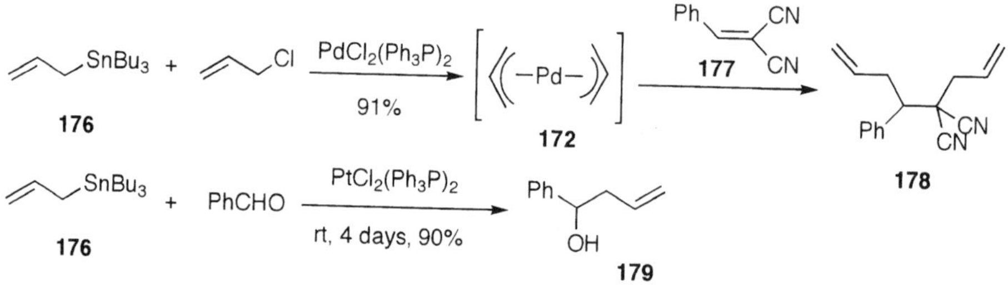

4.3.4   Reactions via Transmetallation

Cross-coupling of allylic compounds occurs by transmetallation between π-allyl intermediates and organometallic compounds of Mg, Zn, B, Al, Si and Sn, and subsequent reductive elimination. Reaction of the allylic dithioacetal **180** with MeMgBr in the presence of an Ni catalyst affords alkenes **184** bearing a *tert*-butyl group [90]. In this reaction, generation of the π-allylnickel **181** by oxidative addition and subsequent transmetallation with MeMgBr afford **182**. Then the methylated product **183** is formed by reductive elimination, and finally the dimethylated product **184** is formed by the sequence of similar reactions.

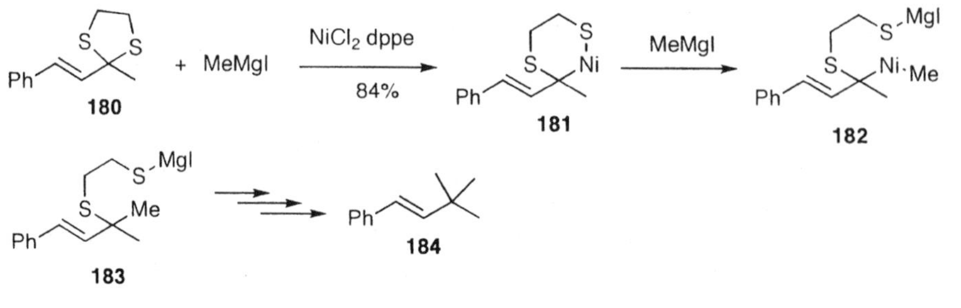

Coupling of the allyl bromide **185** with tolylboronic acid proceeds smoothly with a ligandless Pd catalyst [91]. The allylic carbonate **186** has much higher reactivity toward alkenyl and arylstannanes than the corresponding allylic acetate. The reaction with vinylstannane **187** in aqueous DMF at room temperature with ligandless Pd catalyst gives the 1,4-diene **188** in high yield [92]. The 1,4-diene **191** is obtained by the reaction of allylic chloride **189** with alkenylstannane **190**, involving an inversion of stereochemistry [93,94]. Also, the reaction of PhZnCl with the allylic lactone **192** produced the 3-phenylcyclohexene **193** with inversion as expected [37].

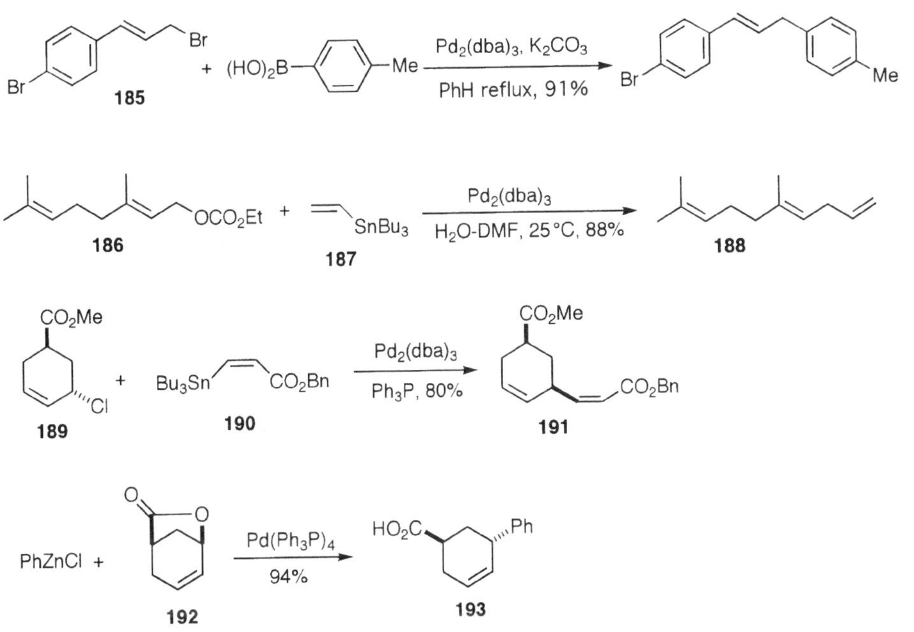

Allylmetal compounds can be prepared by the Pd-catalysed allylation of dimetal compounds. The phenylallylboronate **195** is prepared by the reaction of 2-phenylallyl acetate with bis(pinacolate)diboron (**194**) catalysed by ligandless Pd in DMSO [95]. A good synthetic route to the allylsilane **197** is the reaction of Me$_3$SiSiMe$_3$ with geranyl trifluoroacetate (**196**) catalysed by ligandless Pd via transmetallation of the π-allylpalladium intermediate at room temperature in DMF [96].

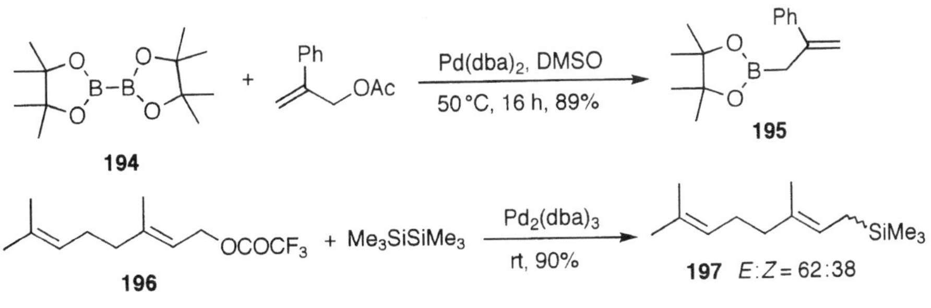

Allylstannane **176** is formed by the reaction of allyl acetate with distannanes [97,98]. In this reaction, umpolung of the electrophilic π-allylpalladium to the nucleophilic allylstannane occurs. Allylation of bromoindole **198** to give allylindole **199** involves the oxidative addition of **198** to Pd, transmetallation with the allylstannane **176**, and final reductive elimination [99].

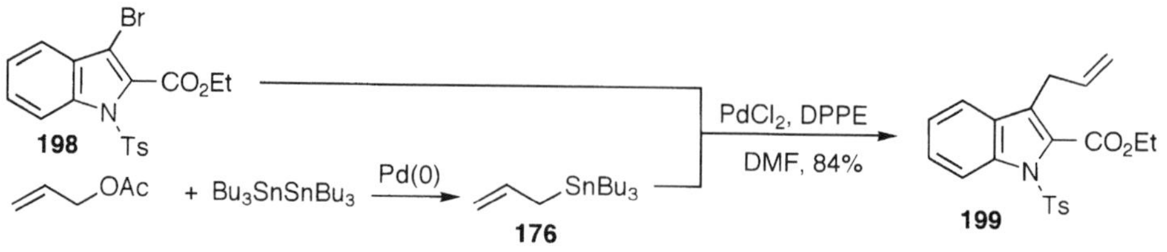

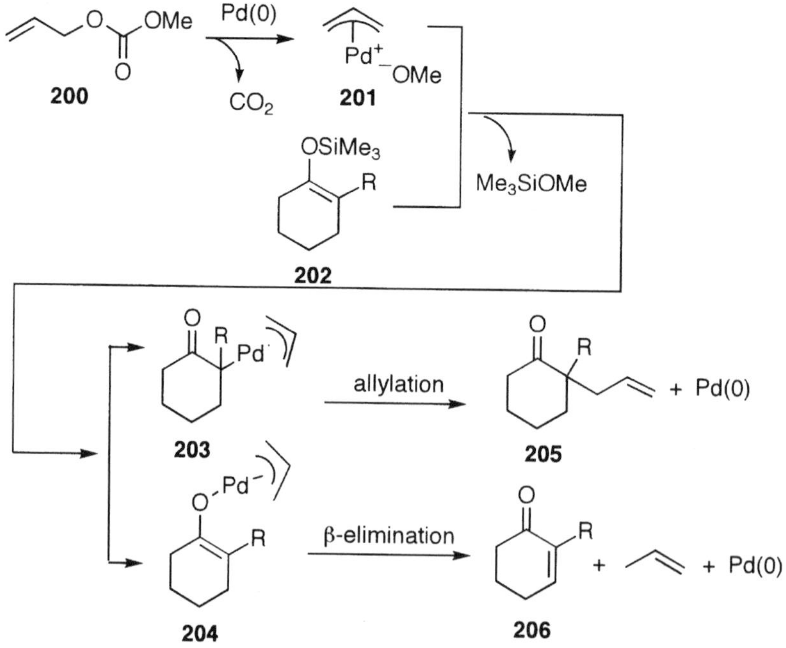

Although simple ketones and esters can not be allylated by Pd catalysts, they are allylated with allyl carbonates via their enol ethers of Si and Sn. In the allylation of the silyl enol ether **202** with allyl carbonate **200**, transmetallation of **202** with the π-allylpalladium methoxide **201**, generated from allyl methyl carbonate (**200**), takes place to generate the Pd enolates **203** and **204**. Depending on the reaction conditions, allyl ketone **205** is formed by the reductive elimination of **203** [100]. When the ratio of Pd:Ph$_3$P is small, the α,β-unsaturated ketone **206** is obtained by β-elimination [101]. For example, the silyl enol ether **208** of aldehyde **207** is allylated with allyl carbonate (**200**) to give α-allylaldehyde **210** via **209**. The α-allyl carboxylate **213** is obtained by allylation of ester **211** with allyl carbonate (**200**), after conversion of ester **211** to the ketene silyl acetal **212** [102]. As the silyl group is trapped in these reactions by the methoxy group, no other trapping agent is necessary. The reaction was applied to the construction of a building block of vitamin D hydrindan ring-side chain. Pd-catalyzed reaction of the silyl enol ether **214** with the *cis*-allyl carbonate **215** afforded the allylated ketone **216** in an overall yield of 87% from 2-methylcyclopentenone [103].

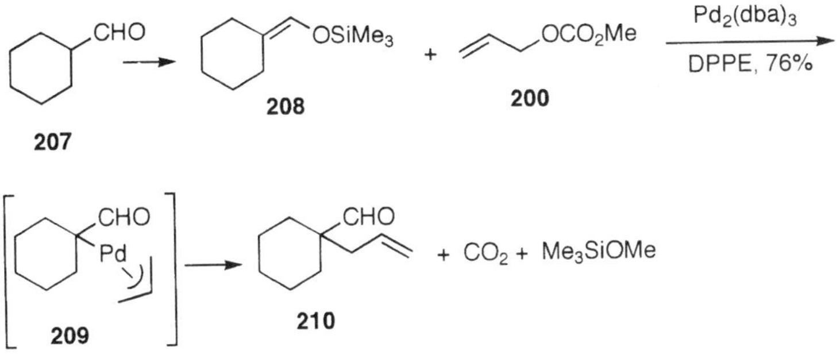

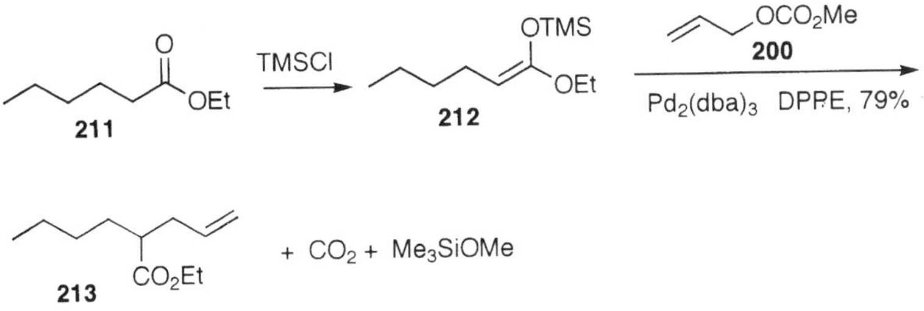

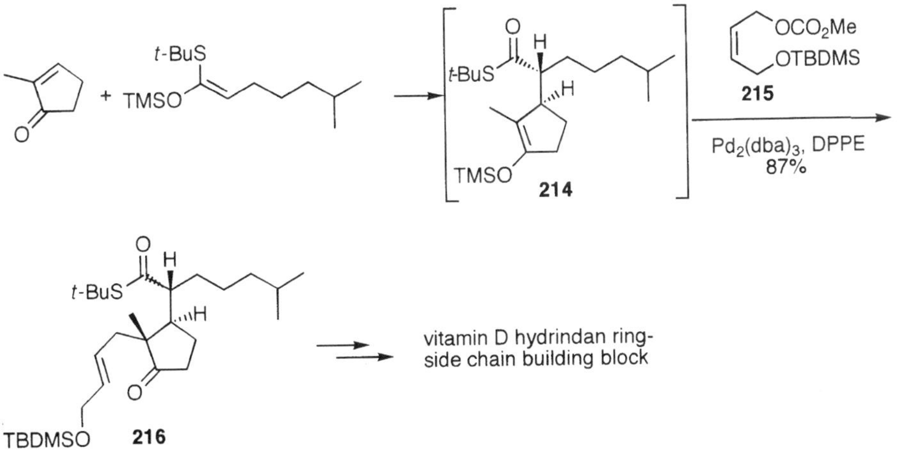

In the reaction of allyl acetate **217** with ketene silyl acetal **218** of methyl acetate, using a Pd catalyst coordinated to DPPP, cyclopropane **220** is formed in addition to the expected allylacetate **219** [104]. The cyclopropanation becomes main reaction when TMEDA, as a ligand, and thallium acetate are added [105]. The cyclopropanation can be understood by the attack of the enolate ion at the central carbon of π-allylpalladium to form the palladacyclobutane **221**, followed by reductive elimination.

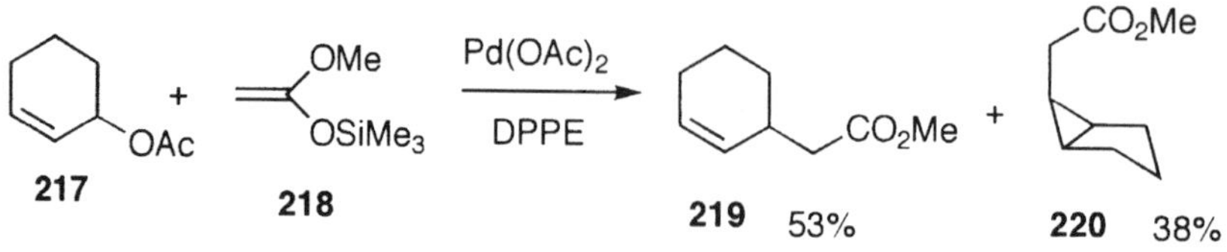

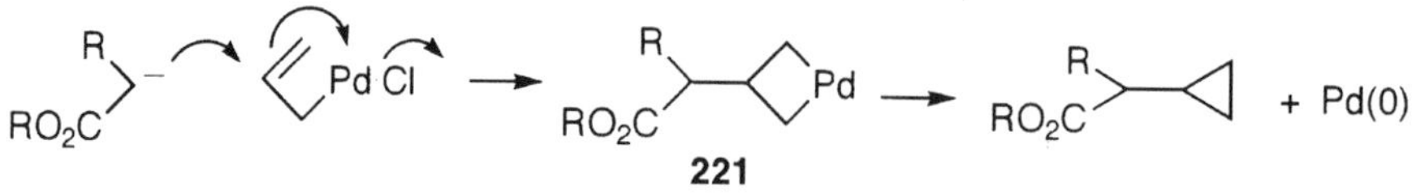

A new preparative method for α,β-unsaturated aldehydes, ketones and esters has been developed, based on the formation of Pd enolates from silyl enol ethers. Allylation of ketones is possible by the Pd-catalysed reaction of silyl enol ethers with allyl carbonates, as described above. However, no allylation occurs in the reaction of the silyl enol ether **222** with allyl carbonate **200** in refluxing MeCN. Instead, the enone **224** is formed regioselectively by the elimination of β-hydrogen without forming the isomeric enone **225** [101]. In this reaction, the allyl group is a hydrogen acceptor, and converted to propylene. The allylation and dehydrogenation are the competitive reactions of the Pd enolate **203** and **204** or **223**. The allylation by reductive elimination of **203** becomes a main path when the ratio of Ph₃P:Pd is larger than 4. When Pd is made more coordinatively unsaturated, by keeping the ratio at 1:1–2, β-elimination occurs mainly at higher temperature to afford the enones **206** or **224**. Enone **224** is obtained regioselectively from **222** without forming the isomeric enone **225**. Lactones can be converted to unsaturated lactones via their ketene silyl acetals [102]. For example, the eight-membered unsaturated lactone **228** was prepared from the lactone **226** via silyl ether **227**, and converted to lanthisan [106].

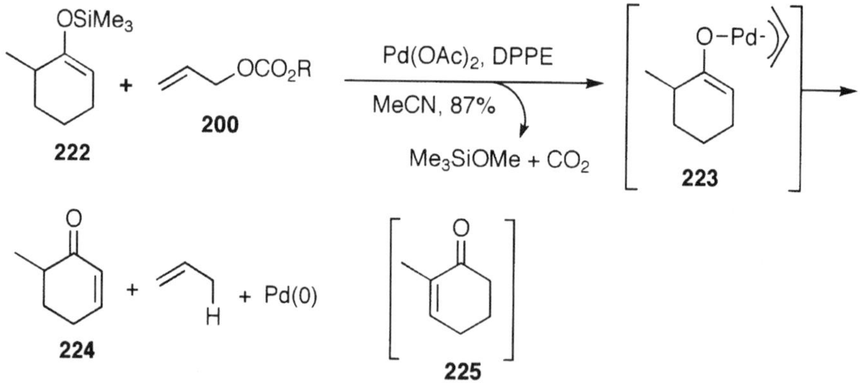

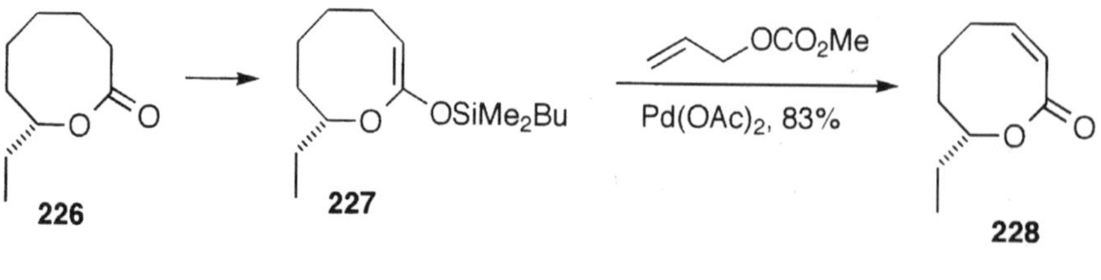

It is known that Sn enolates **232** can be generated by the reaction of enol acetates of ketones **230** and aldehydes with MeOSnBu$_3$ (**231**) [107]. Based on this reaction, the Pd enolates **234** and **223** can be generated by the reaction of the enol acetate **230** with allyl carbonate (**200**), using Pd(0) and MeOSnBu$_3$ as catalysts. In this case, transmetallation of π-allylpalladium methoxide (**229**) with the generated Sn enolate **232** gives the Pd enolates **234** and **223**. At the same time the catalyst, MeOSnBu$_3$ (**231**), is regenerated. Allyl ketone **235** is formed by the reductive elimination of the Pd enolate **234** [108], and the enone **224** is formed by β-elimination [109]. As an example of the allylation, the enol ester **230** is allylated regioselectively with **200** to give the allyl ketone **235** using Pd(0) and MeOSnBu$_3$ as the catalysts. The two isomeric steroidal dienol acetates **237** and **238** are prepared from the steroidal enone **236**. The dienones **239** and **240** are obtained regioselectively by the Pd and Sn-catalyzed reaction of **237** and **238** with allyl carbonate **200** [109].

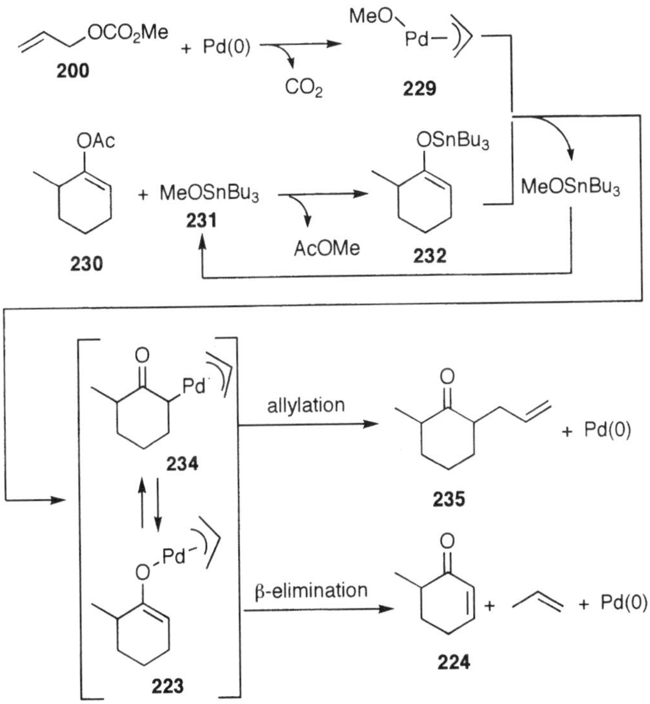

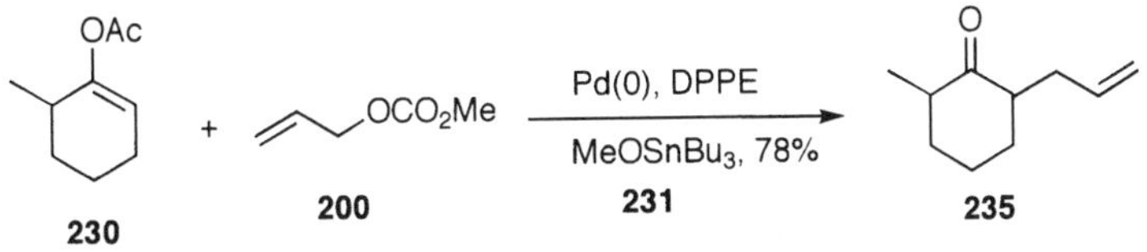

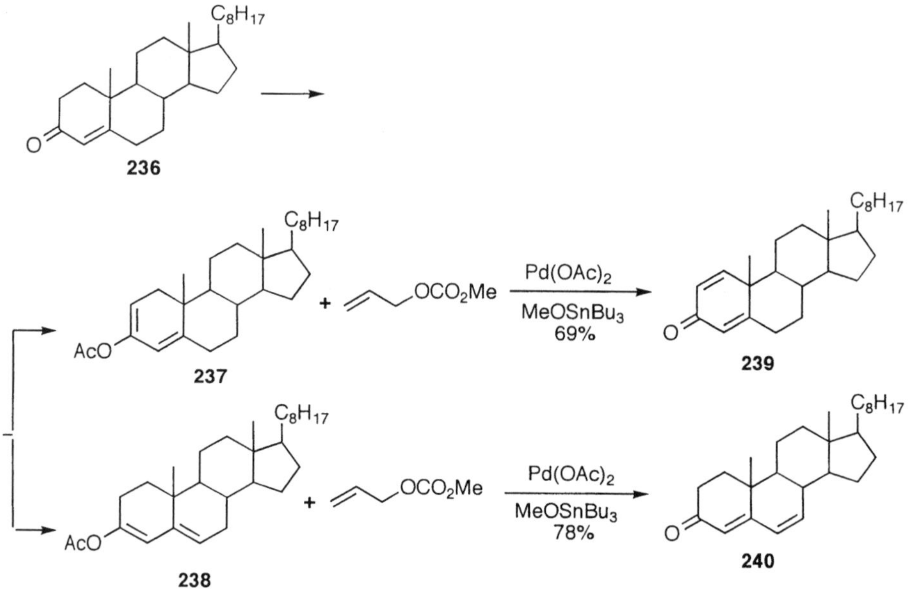

236

237 + allyl-OCO₂Me → **239**

$$\text{Pd(OAc)}_2$$
$$\text{MeOSnBu}_3$$
$$69\%$$

238 + allyl-OCO₂Me → **240**

$$\text{Pd(OAc)}_2$$
$$\text{MeOSnBu}_3$$
$$78\%$$

4.3.5   Carbonylation

β,γ-Unsaturated esters are formed by the carbonylation of allylic compounds. 3-butenoate is formed by facile carbonylation of allyl carbonates **241** at 50 °C under 1–10 atm [110]. Carbonylation of the allyl carbonate **242**, prepared from acrylate and aldehyde, gave the alkylidenesuccinate **243** under mild conditions [111]. Allyl acetates are carbonylated in the presence of NaBr and amine. The carbonylation of **244** is an example [112]. Most conveniently, the carbonylation of allyl chloride **245** proceeds at room temperature and 1 atm in alcohol in the presence of K_2CO_3 using a ligandless Pd catalyst [113]. Carbonylation of allyl amine **246** using dppp at 110 °C and 50 atm gives the β,γ-unsaturated amide **247** [114].

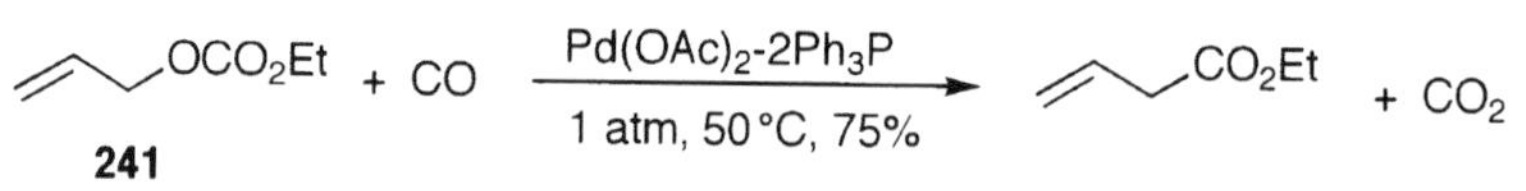

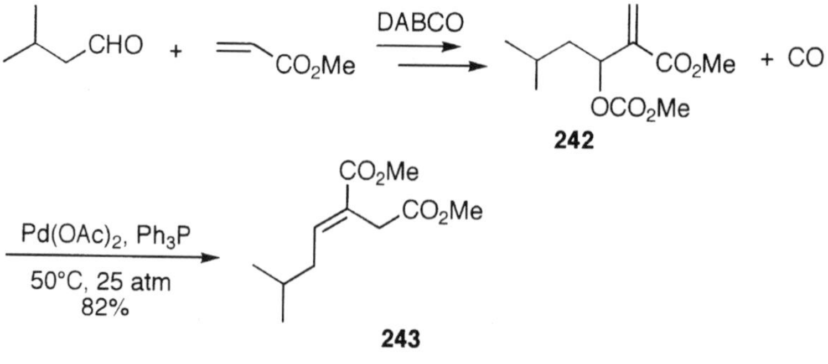

241   + CO $\xrightarrow[\text{1 atm, 50°C, 75\%}]{\text{Pd(OAc)}_2\text{-2Ph}_3\text{P}}$ allyl-CO₂Et + CO₂

CHO + CO₂Me $\xrightarrow{\text{DABCO}}$ **242** + CO

OCO₂Me

$\xrightarrow[\substack{50°C, 25\ atm \\ 82\%}]{\text{Pd(OAc)}_2,\ \text{Ph}_3\text{P}}$ **243**

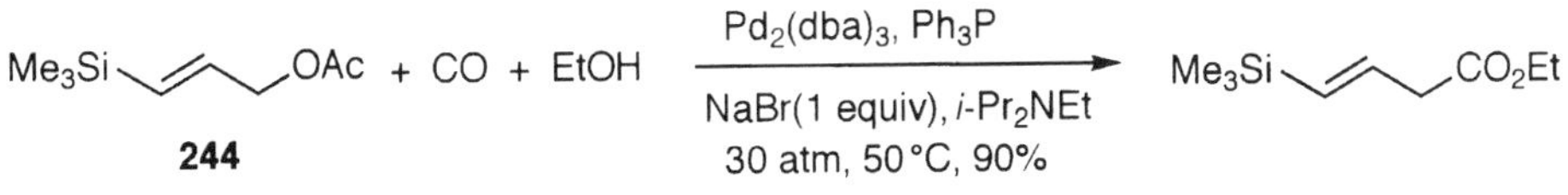

Me₃Si⏜⏜OAc + CO + EtOH →[Pd₂(dba)₃, Ph₃P / NaBr(1 equiv), *i*-Pr₂NEt / 30 atm, 50 °C, 90%] Me₃Si⏜⏜CO₂Et

244

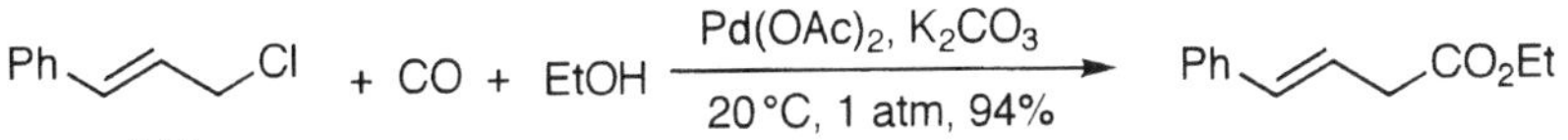

Ph⏜⏜Cl + CO + EtOH →[Pd(OAc)₂, K₂CO₃ / 20 °C, 1 atm, 94%] Ph⏜⏜CO₂Et

245

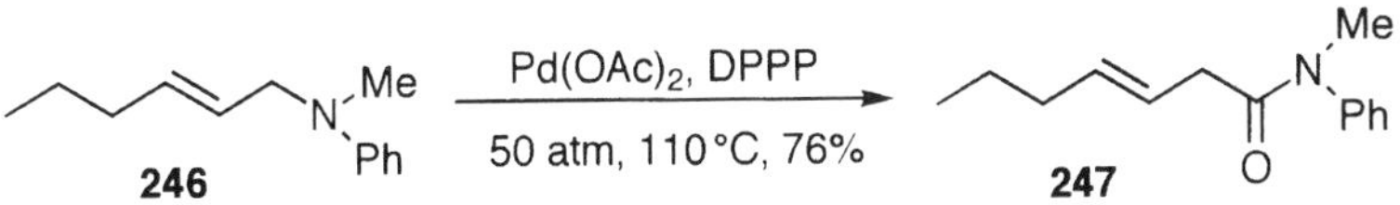

⏜⏜⏜N(Me)(Ph) →[Pd(OAc)₂, DPPP / 50 atm, 110 °C, 76%] **247**

246

Allyl ketones can be prepared by trapping acylpalladium intermediates with main group metal compounds. The allyl ketone **251** can be prepared from the allyl ester **248** by trapping acylpalladium **249** with the alkylzinc compound **250** at room temperature and 1 atm [115]. Reaction of geranyl chloride (**252**) with the furylstannane **253** under CO pressure afforded the ketone **254**, which was converted to dendrolasin (**255**) [93]. Aldehydes are prepared by trapping with a metal hydride. The β,γ-unsaturated aldehyde **257** is prepared by the carbonylation of the allyl chloride **256** in the presence of Bu₃SnH [116,117].

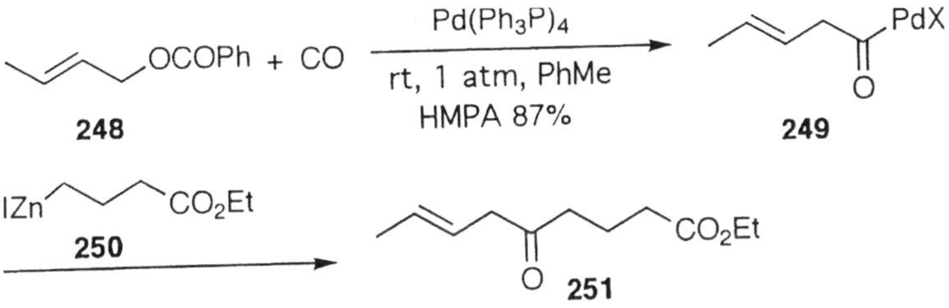

⏜⏜OCOPh + CO →[Pd(Ph₃P)₄ / rt, 1 atm, PhMe / HMPA 87%] **249** PdX

248

IZn⏜⏜CO₂Et

250

→ ⏜⏜⏜CO₂Et **251**

252 + CO + **253** (Bu₃Sn-furyl) →[Pd(dba)₂, Ph₃P / 50 psi, 50 °C, 69%]

254 → **255**

EtO₂C⏜(OMe)⏜Cl + CO + Bu₃SnH →[Pd(Ph₃P)₄ / 50 °C, 86%] EtO₂C⏜(OMe)⏜CHO

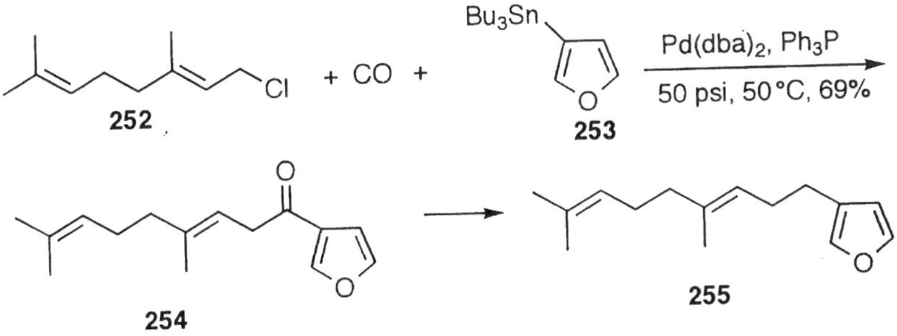

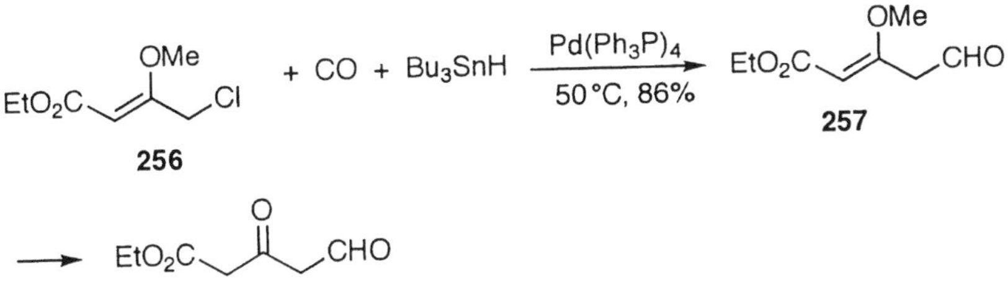

256 → **257**

→ EtO₂C⏜(C=O)⏜CHO

Carbonylation of the 3-phenylallyl acetate **258** under somewhat severe conditions in the presence of tertiary amine and acetic anhydride affords the naphthyl acetate derivative **260**. This interesting cyclocarbonylation is explained by the Friedel–Crafts-type cyclization of the acylpalladium **259** as an intermediate [118,119]. Even 5-phenyl-2,4-pentadienyl acetate (**261**) is cyclocarbonylated to afford 2-phenylphenyl acetate (**262**) [120].

Methyl (*Z*)-2,5-hexadienoate (**263**) is prepared by the treatment of allyl chloride and acetylene with Ni(CO)$_4$ in MeOH at room temperature under 1 atm of CO [121]. When the reaction is carried out in the presence of aqueous acetone (0.4% of water), the lactone **271** is a main product. Lower concentrations of water tends to give the lactone **272**. Formation of the lactones **271** and **272** presents a good example of the versatility of insertion reactions as shown by the following mechanism. At first π-allylnickel is formed from allyl chloride, to which acetylene and CO are inserted to generate the acylnickel **264**. The acid **265** is formed by the reaction of water. In aqueous acetone, intramolecular insertion of the double bond in **264** occurs to form the cyclopentenone **266**. Further domino insertions of CO, acetylene, and CO continue, to form the acylnickel **268**. Then, intramolecular insertion of carbonyl group of the ketone gives the lactone **269**, which is stabilized as the π-allylnickel **270**. Protonation of **270** produces the lactone **271**, and another lactone, **272**, is formed by nucleophilic attack of the π-allylnickel **270** on acetone. Altogether seven domino insertions occur efficiently with remarkably high chemoselectivity in the formation of the lactones **271** and **272** as main products.

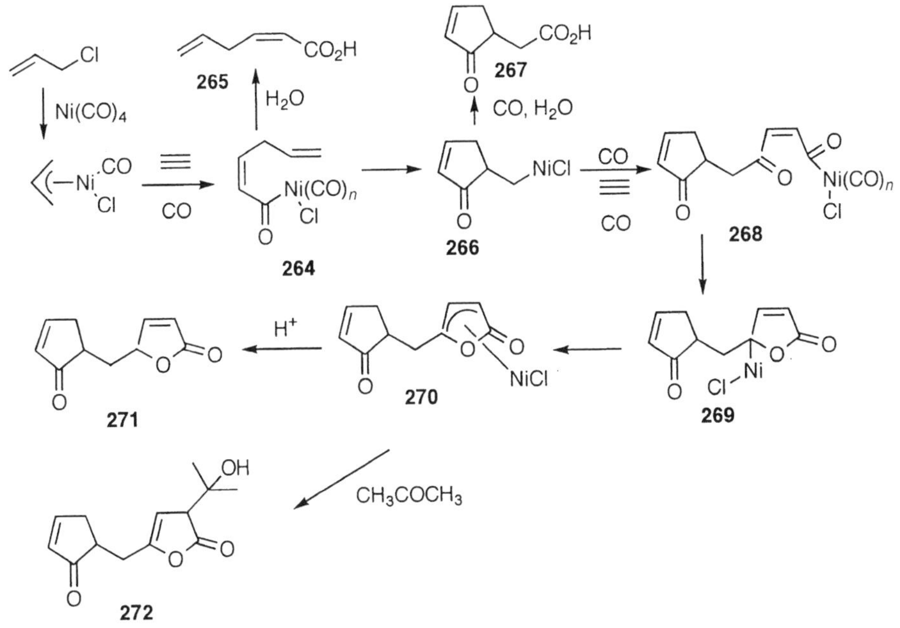

4.3.6 Insertion of Alkenes and Alkynes

Intermolecular insertion of alkenes to π-allyl intermediates is possible with an Ru catalyst. For example, 3,5-dienecarboxamide **274** is formed in high yield by Ru(cod)(cot)-catalysed coupling of 2-butenyl methyl carbonate (**273**) with acrylamide in the presence of *N*-methylpiperidine [122]. Ni-catalysed transformation of allyl 3-butenoate (**275**) to heptadienoic acids **276a** and **276b** proceeds by insertion of the double bond to π-allylnickel intermediate [123].

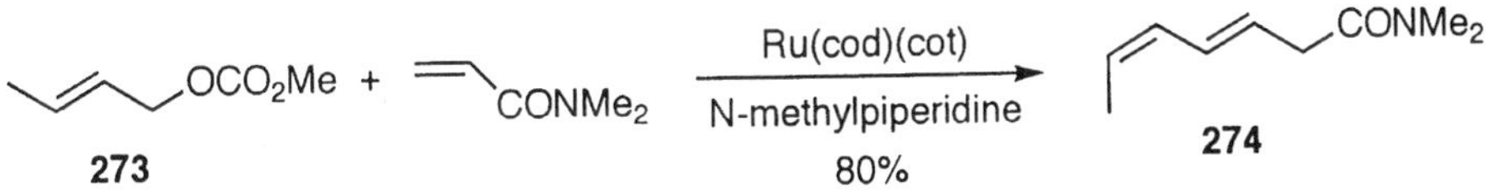

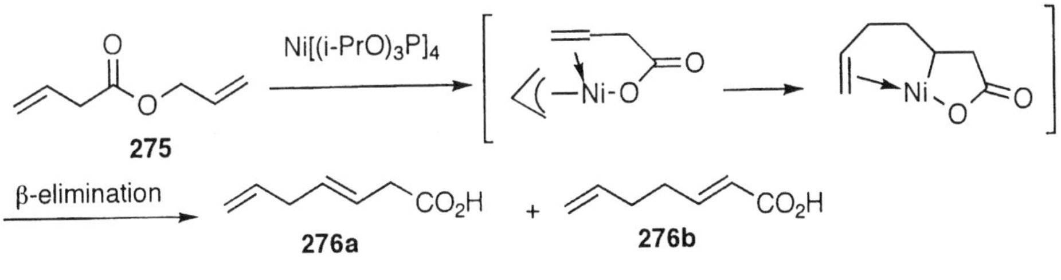

However, whereas intermolecular insertion of alkene to π-allylpalladium is difficult, intramolecular insertion proceeds smoothly. This reaction is known as the metalla-ene (pallada-ene) reaction [124]. Use of AcOH as the solvent is crucial. The Pd-catalysed reaction of **277** under CO atmosphere affords the keto ester **279** by one-pot reaction via domino insertions of triple bond, CO, double bond, and CO to the π-allylpalladium intermediate **278**. The ester **279** was converted to hirsutene (**280**) [125]. The highly strained molecule of [5.5.5.5] fenestrane rings **283** was formed in one pot in 65% yield. The intramolecular insertion of the double bond to the π-allylpalladium generates **282**, and subsequent domino insertions of CO, double bond, and CO produce **283**. Due to the unfavourable stereochemistry of the intermediates, no elimination of β-hydrogen occurs during the reaction [126].

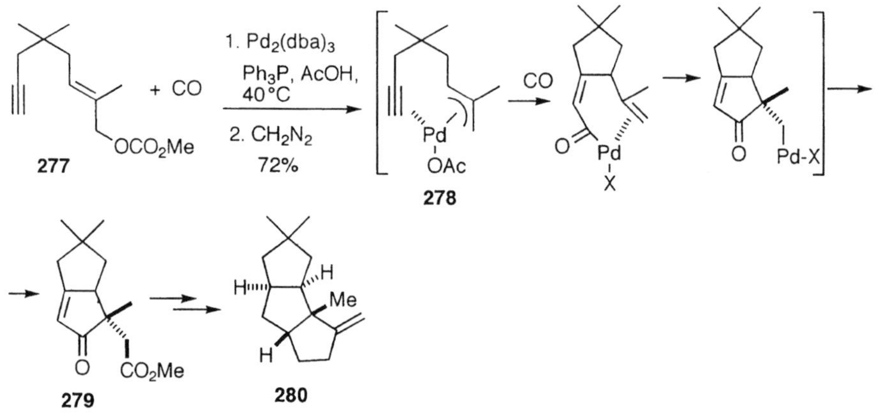

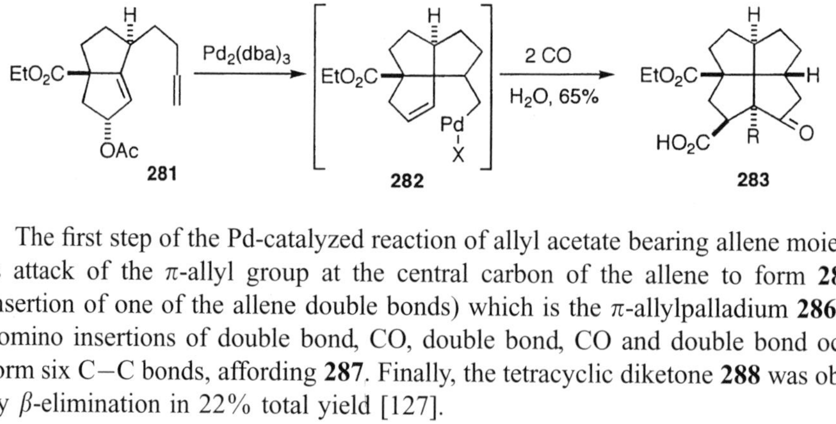

The first step of the Pd-catalyzed reaction of allyl acetate bearing allene moiety **284** is attack of the π-allyl group at the central carbon of the allene to form **285**, (or insertion of one of the allene double bonds) which is the π-allylpalladium **286**. Then domino insertions of double bond, CO, double bond, CO and double bond occur to form six C−C bonds, affording **287**. Finally, the tetracyclic diketone **288** was obtained by β-elimination in 22% total yield [127].

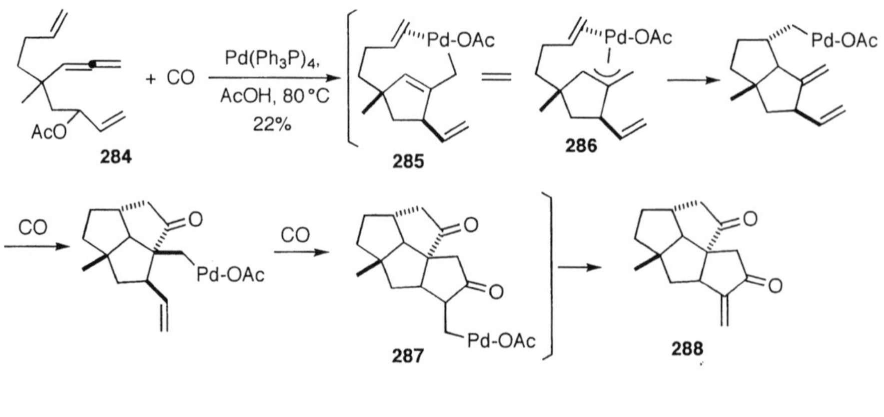

4.3.7 Hydrogenolysis

Allyl compounds are converted to alkenes by the Pd-catalysed hydrogenolysis using various hydride sources [128]. Particularly useful is the regioselective formation of 1-alkenes **291** from terminal allyl compounds **289** and **290** using formic acid [129]. The 2-alkenes **292** are formed mainly when other hydrides such as HSnBu$_3$, HSiR$_3$, LiBHEt$_3$ or SmI$_2$ are used. These regioselectivities can be explained by the following mechanism. Reaction of the π-allylpalladium **293**, generated from allylic compounds **289** and **290**, with formate gives the π-allylpalladium formate **294**, which undergoes decarboxylation and transfer of hydride to the more substituted side of the π-allyl system by the cyclic mechanism as shown by **295** to afford 1-alkenes **291**. π-Allylpalladium formate **294** can be generated directly from the allyl formate **297**. However, reaction of other hydrides generates the π-allylpalladium hydride **296** by transmetallation of **293**, and 2-alkenes **292** are formed by reductive elimination.

Regioselective hydrogenolysis has wide synthetic application. The use of allyl formates is the most convenient for regioselective hydrogenolysis. The hydrogenolysis of other allyl compounds can be carried out using triethylammonium formate. The allylic acetate **298** was converted to the exomethylene **299** [130]. However, the hydrogenolysis of **300** afforded **301** by the isomerization of the internal double bond to the *exo* position [131]. Asymmetric hydrogenolysis of geranyl carbonate (**302**) using (*R*)-MOP-Phen (**XXXVII**) produced the terminal alkene **303** with 85% ee [132]. Regioselective hydrogenolysis is used for the preparation of the 20-keto steroid **307** from the 17-keto steroid **304**. The isopropenyl group is introduced to **304** and converted to the carbonate **305**. Its hydrogenolysis with HCO$_2$H − Et$_3$N affords the 17β-isopropenylsteroid **306** regio- and stereoselectively [133]. Finally, oxidative cleavage of the double bond in **306** gives the 17-keto steroid **307** [134].

High regioselectivity is observed also in the hydrogenolysis of internal allylic formates. The hydrogenolysis of allylic formates **308** and **312** in the A ring of steroids proceeds regio- and stereoselectively [135]. The active Pd(0) catalyst, generated from Pd(OAc)$_2$ and Bu$_3$P in 1:1 ratio, attacks the allyl formate of the β-oriented allyl alcohol **308** from the α-side with inversion, as shown by **309**, and subsequent hydride transfer to the more substituted angular carbon from the α-side produces the A/B *trans* steroid **310**. No regioisomer **311** is formed. However, the α-oriented formate **312** is converted to the A/B *cis* steroid **314** by hydride attack from the β-side as shown by **313**. Thus the AB *cis* and *trans* ring junctions can be constructed from the 3β- and 3α-allylic formates.

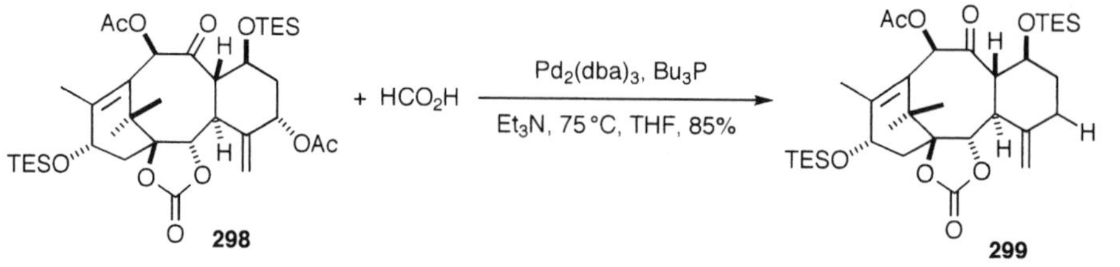

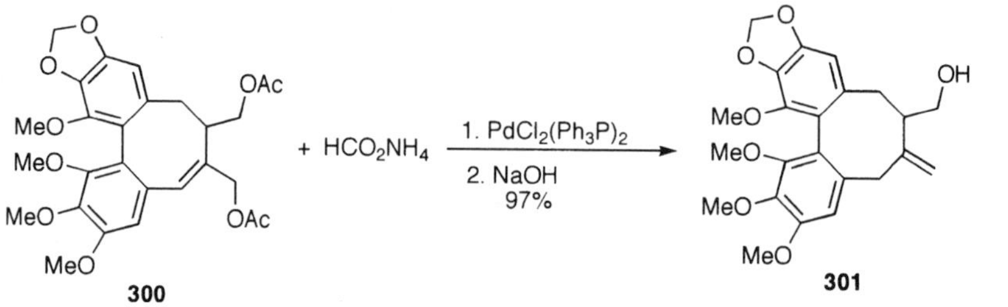

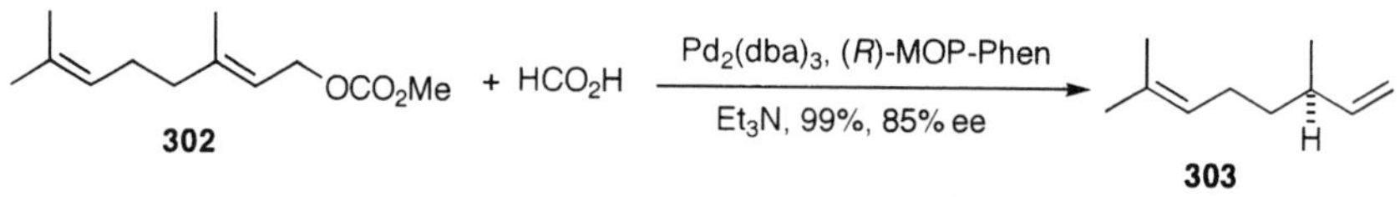

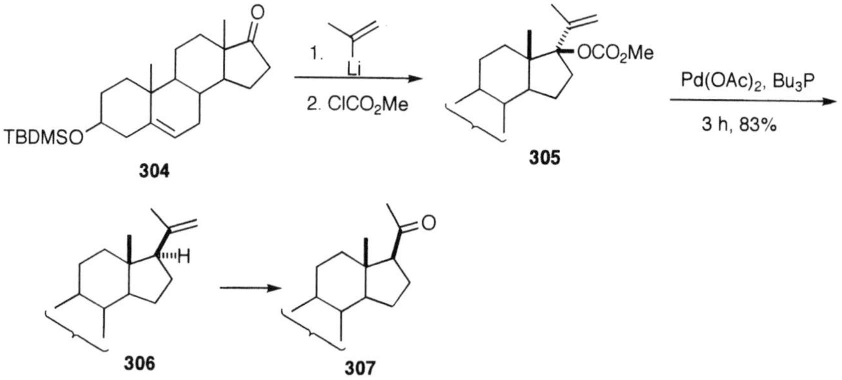

The hydrogenolysis of the allyl ester **315** produced **316** regio- and stereoselectively. The free allyl alcohol in the same molecule was not attacked [136].

Stereocontrolled construction of the natural configuration **327** at C-20 in steroid side-chains is an important problem in steroid synthesis. In addition, preparation of the unnatural epimer **321** has increasing importance, because of its interesting biological activity. Both epimers can be prepared cleanly at will from the common intermediate **317** by Pd-catalysed regio- and stereoselective hydrogenolysis with $HCO_2H–Et_3N$ [137]. The (*E*)-and (*Z*)-allyl alcohols are prepared by the reaction of (*E*)-and (*Z*)-vinyllithiums with the 20-keto steroids **317**, and converted to carbonates **318** and **322**. The Pd-catalyzed hydrogenolysis of these carbonates with $HCO_2H–Et_3N$ proceeds regio- and stereoselectively. The C-20 of the unnatural configuration **321** is constructed from (*E*)-allyl carbonate **318**, and the natural configuration type **327** is obtained from (*Z*)-allyl carbonate **322**. The selectivities are explained by the following mechanism. The stable *syn*-type π-allylpalladium intermediate **319** is generated by the attack of Pd(0) from the α-side of carbonate **318** with inversion, and the hydride is transferred to the more substituted tertiary carbon from the α-side as shown by **320**, producing the unnatural configuration **321**. However, the intermediate **323** formed from the (*Z*)-carbonate **322**, is the sterically congested *anti* form, and isomerized to the stable *syn* form **325** via rotation of the σ-allyl **324**. At the same time, the Pd formate in **323** moves from the α-side to the β-side by this rotation, and hence the hydride transfer occurs as shown by **326** from the β-side to give the natural configuration **327**. In these regio- and stereoselective hydrogenolyses, the active catalyst is prepared by mixing $Pd(OAc)_2$ or $Pd(acac)_2$ and Bu_3P in 1:1 ratio [137a].

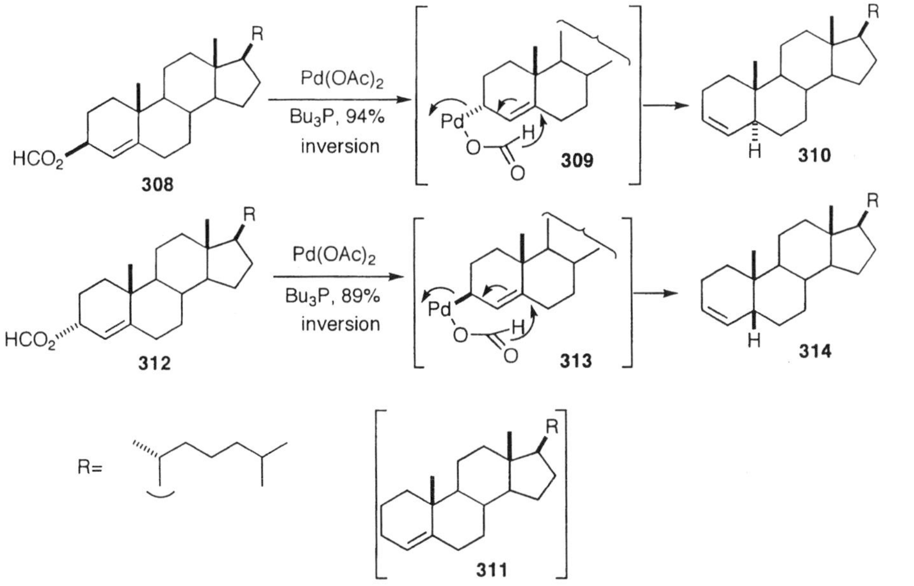

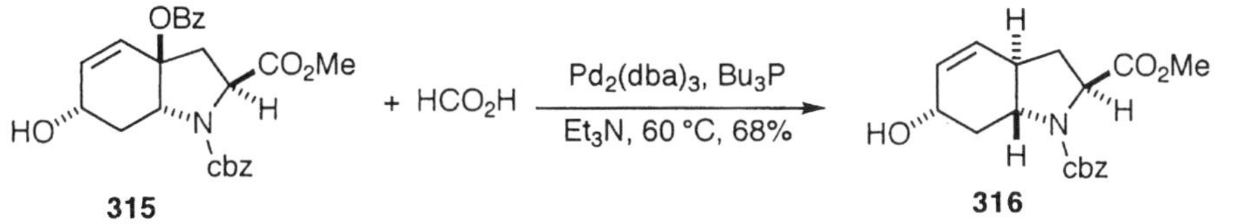

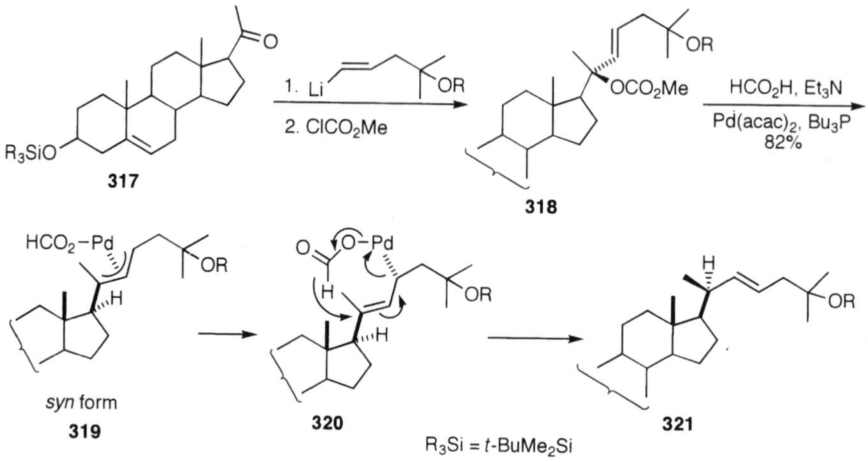

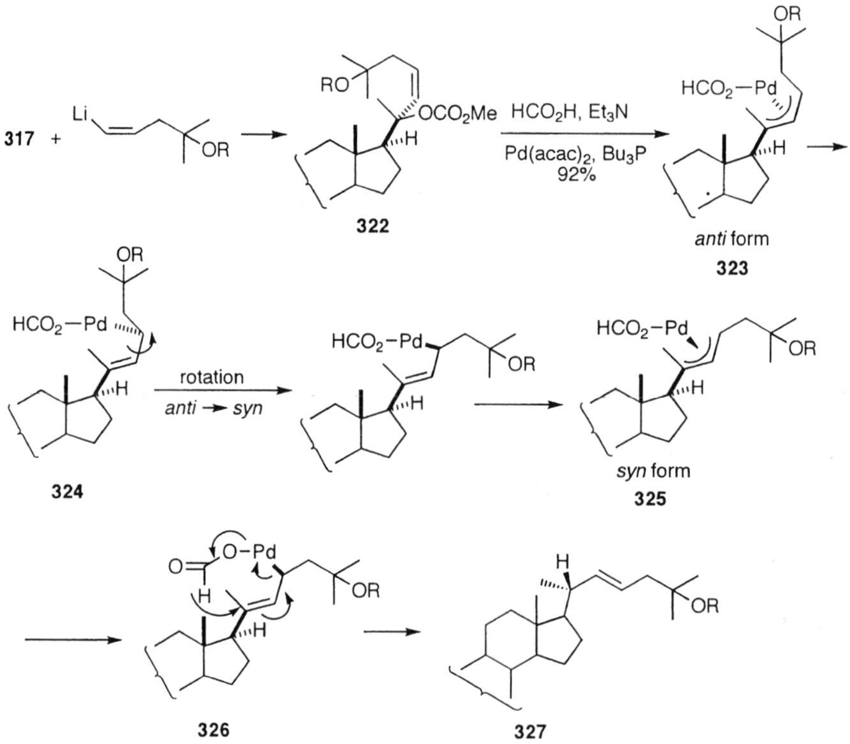

317

318

319   syn form

320

321

R₃Si = *t*-BuMe₂Si

322

323   anti form

324

rotation
anti → syn

325   syn form

326

327

4.3.8 Allyl as a Protecting Group and its Deprotection

Based on facile formation of π-allylpalladium intermediates from various allylic compounds, allyl groups can be used for the protection of carboxylic acids, amines and alcohols. Deprotection can be achieved by two methods using Pd(0) catalysts [128,138]. In one method, the allyl group can be removed as propylene by Pd-catalysed hydrogenolysis; in the other, Pd-catalysed transfer of the allyl group occurs to other nucleophiles, such as amines and active methylene compounds.

The allyl carboxylate **328** can be deprotected to give free carboxylic acid under neutral conditions at room temperature using HCO_2H–Et_3N. Only CO_2 and propylene are formed, which are easily removed [139]. The ester group in a very labile prostaglandin intermediate of a commercial synthetic process is hydrolysed usually under mild conditions using enzymes, because the use of acid or base cannot be tolerated. As an alternative method, deprotection of the allyl ester in the precursor **329** of prostaglandin to the free acid **330** is possible by the Pd-catalysed hydrogenolysis using HCO_2H–Et_3N under neutral conditions without attacking other functional groups in the molecule [140].

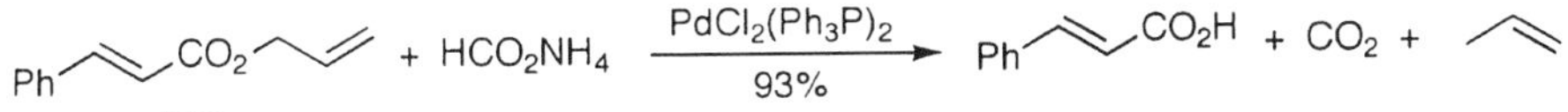

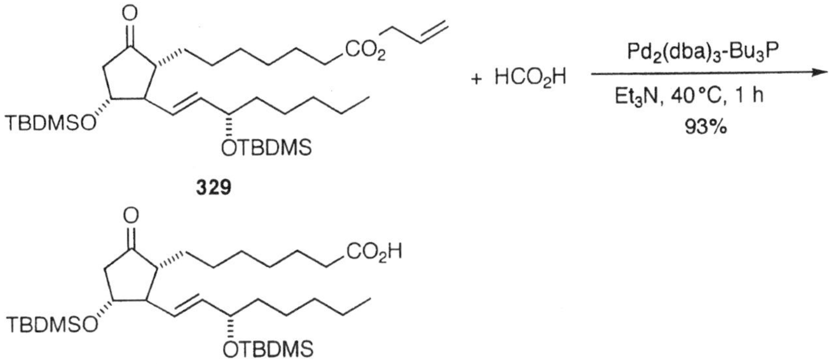

Another method for deallylation is the Pd-catalyzed transfer of the allyl group to reactive nucleophiles. Sodium 2-ethylhexanoate [141], morpholine [142], dimedone [143] and *N,N*-dimethylbarbituric acid [144] are used as allyl scavengers. It is claimed that benzenesulfinic acid **331** or sodium toluenesulfinate are the best allyl scavengers [145].

Allyl phosphates **332** are deprotectd using HCO_2H–amine [146] or $HSnBu_3$ [147]. The method is applied to the protection and deprotection of the amino and phosphate groups in nucleotides synthesized in the solid phase [148].

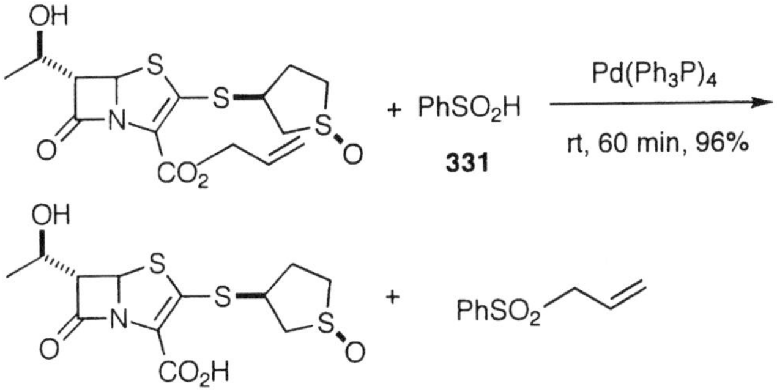

$$\text{Nu-O-P(=O)(O)-O}\diagup\diagdown \quad + \text{ HCO}_2\text{H} \quad \xrightarrow[\text{BuNH}_2]{\text{Pd(Ph}_3\text{P})_4} \quad \text{Nu-O-P(=O)-OH / OH}$$

332

Alcohols are protected as allyl ethers, which are difficult to cleave with the Pd catalyst and deprotected by other methods [149]. Alcohols are conveniently converted to allyl carbonates **334** by treatment with allyl chloroformate (**333**). The allyl carbonates are deprotected using HCO_2H [150], and $HSnBu_3$ [151]. This method is called the AOC (allyloxycarbonyl) method. Phenols are protected as allyl phenyl ethers, which can be cleaved with $HSnBu_3$ [152].

$$\text{ROH} + \text{Cl-C(=O)-O}\diagup\diagdown \quad \longrightarrow \quad \text{R-O-C(=O)-O}\diagup\diagdown \quad + \text{ HCO}_2\text{H} \quad \xrightarrow[\text{Et}_3\text{N}]{\text{Pd(OAc)}_2, \text{Ph}_3\text{P}}$$

333   **334**

$$\text{ROH} + 2\,\text{CO}_2 + \quad \diagup\diagdown$$

One protecting method for ketones and aldehydes is the formation of oximes, but sometimes further protection of the oximes is required. For this purpose, the oximes can be protected as the allyl ethers **335**, which are deprotected by the Pd-catalysed hydrogenolysis with HCO_2H–Et_3N in boiling dioxane under mild conditions without attacking the acetal group in **335** [153].

$$\text{(allyl-O-N=)}\,\text{OMe}\,/\,\text{OMe} \quad + \text{ HCO}_2\text{H} \quad \xrightarrow[\text{Et}_3\text{N, 99\%}]{\text{Pd(OAc)}_2, \text{Ph}_3\text{P}} \quad \text{HO-N=}\,\text{OMe}\,/\,\text{OMe}$$

335

Amines are protected by the AOC method as the allyl carbamates **336**. Deprotection is possible by hydrogenolysis with HCO_2H [154] or with $HSnBu_3$ [155]. Allyl transfer to the dimedone (**337**) is also used [143,156]. Protection of the amino group, as carbamate and phosphoric acid as allyl ester, is applied to protect nucleoside-3-phosphoramidite monomer units such as **338**, and are used in the solid-phase oligonucleotide synthesis. In 60-mer synthesis, 104 allylic protective groups were removed in almost 100% overall yield by a single Pd-catalysed reaction with HCO_2H and $BuNH_2$ [148,157].

Cleavage of allylamines is rather difficult. However, unexpectedly the allylamine **340** is deprotected by allyl transfer to dimethylbarbiturate (**341**) [158] or to 2-thiobenzoic acid using DPPB as a ligand [159]. The 2-thiobenzoic acid method has been applied to indole synthesis [160]. Allylamines are also cleaved with Pd/C in EtOH in the presence of methanesulfonic acid [161] or with $Ni(dppp)Cl_2$ [162]. Thus diallylamine (**339**) can be regarded as protected ammonia.

Pd-catalyzed treatment of the diallyl dicarbamate of hydrazine **342** with $HSnBu_3$ in the absence of proton source produces the tin carbamate **343**, which is converted to the free hydrazine **344** by protonation, and to the amide **345** by the treatment with acetic

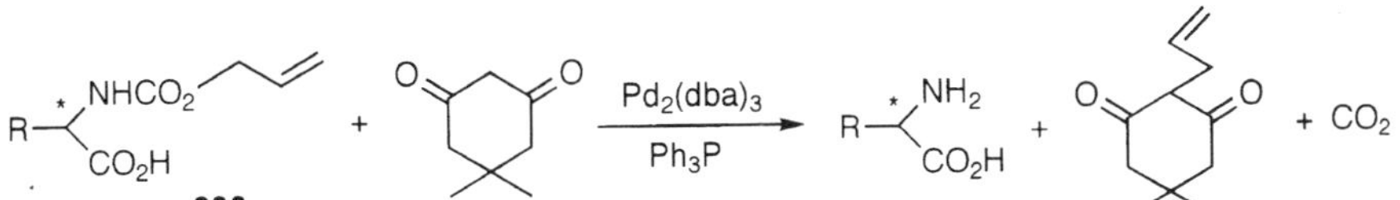

336   **337**

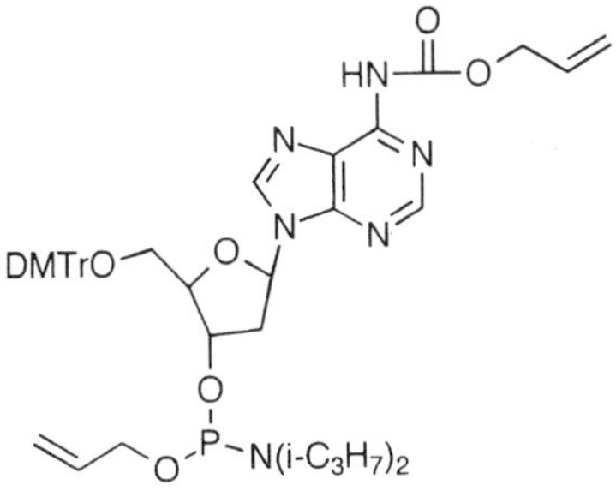

338

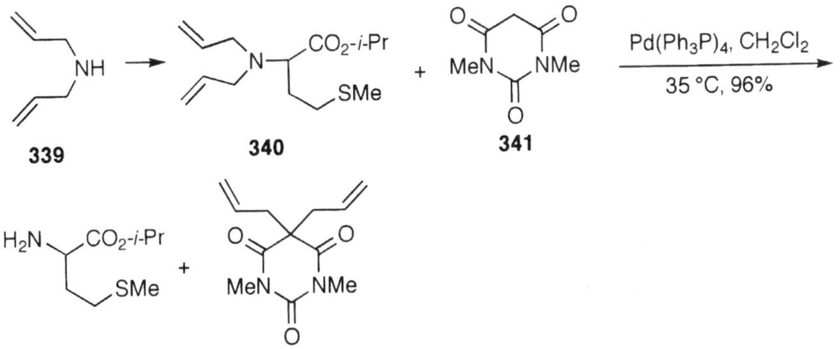

339   **340**   **341**

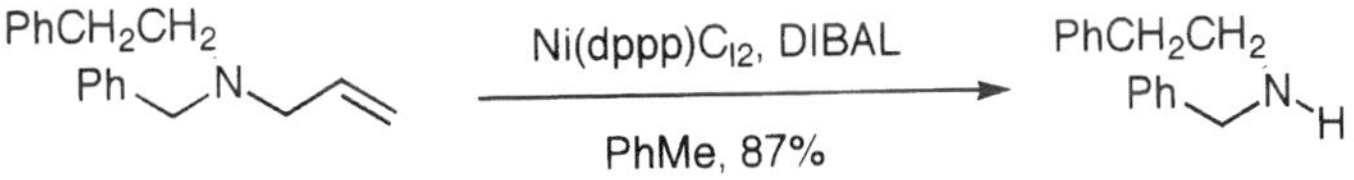

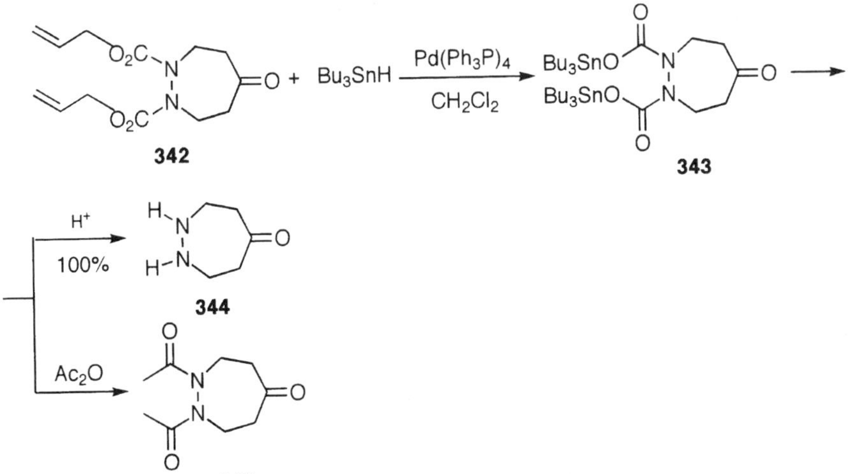

Pd(Ph$_3$P)$_4$, CH$_2$Cl$_2$

35 °C, 96%

PhCH$_2$CH$_2$ Ni(dppp)C$_{l2}$, DIBAL   PhCH$_2$CH$_2$

Ph—N   PhMe, 87%   Ph—N–H

342   Bu$_3$SnH   Pd(Ph$_3$P)$_4$ **343**

CH$_2$Cl$_2$

H$^+$

100%   **344**

Ac$_2$O   **345**

anhydride. By this method, transprotection from allyl carbamates **342** to amides **345** is achieved in one step [163].

4.3.9 *Preparation of Conjugated Dienes by 1,4-Elimination*

When allylic compounds **346** are treated with Pd(0) catalyst in the absence of any nucleophile, 1,4-elimination (dehydropalladation) is a sole reaction path as shown by **347**, and conjugated dienes are formed as a mixture of (*E*) and (*Z*) isomers [164,165]. From terminal allylic compounds, terminal conjugated dienes are formed.

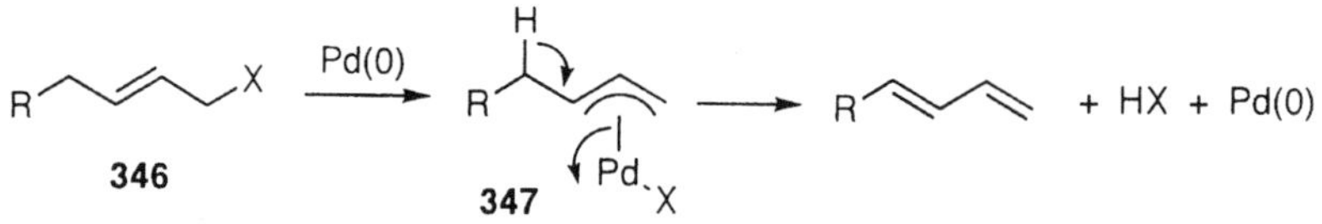

Allylic amines are difficult to cleave, but they can be cleaved by using a cationic Pd complex and DPPB as a ligand. As an example, *N,N*-diethylgeranylamine (**348**) is converted **349**, and its regioselective dehydropalladation affords the myrcene-type conjugated diene **350**, and the reaction is applied to commercial production of the fragrant compound **351**, called kovanol [166]. 1,4-Elimination of allylic amines is also possible as their amine salts using AcOH as a solvent [167].

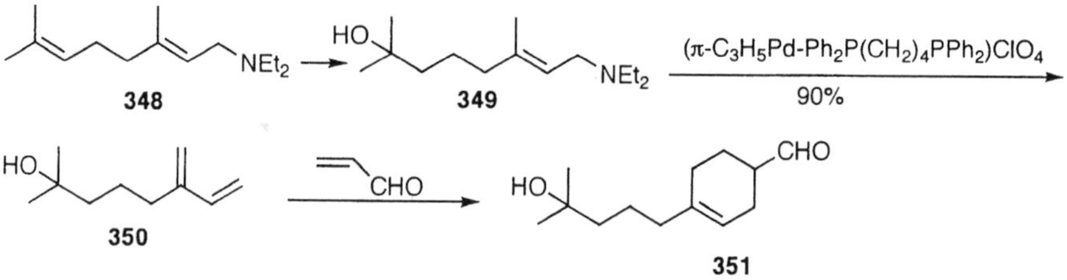

The allyl chloride **353** is prepared from diprenyl ether (**352**), and its Pd-catalysed 1,4-elimination in the presence of AcONa affords the conjugated diene **354**. Citral (**355**) is obtained by the consecutive Claisen and Cope rerrangements of **354** [168].

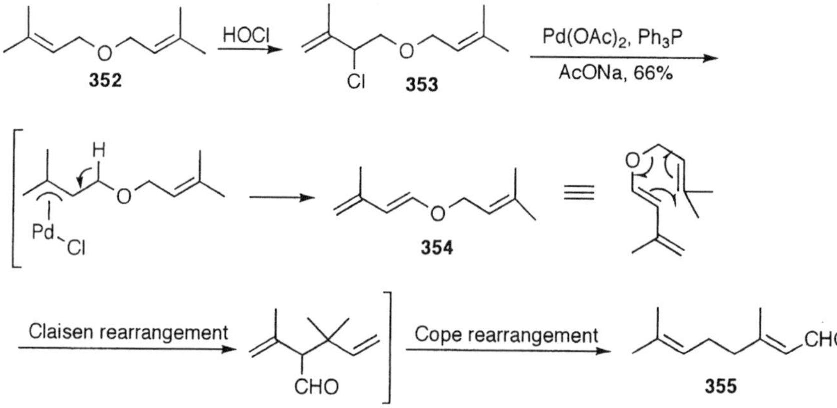

Palladium-catalysed 1,4-elimination proceeds with high regioselectivity in cyclic systems, and is particularly useful for the selective preparation of both homoannular and heteroannular conjugated dienes **362** and **358** in decalin systems [169]. These two conjugated cyclic dienes exist in a number of natural products, and their regioselective preparation is highly desirable. Selective formation of the homoannular diene **362** is considered to be difficult, because acid- or base-promoted elimination of the corresponding allylic alcohols or their derivatives affords only the heteroannular diene **358**. It is possible to prepare both the homo- and heteroannular conjugated dienes at will only by the Pd-catalysed 1,4-elimination of allylic carbonates under mild conditions. The regiospecificity of the elimination is dependent on the stereochemistry of the allylic carbonates in the decalin systems. The heteroannular diene **358** is obtained selectively from the β-oriented 3-allylic carbonate **356** at room temperature. The π-allylpalladium intermediate, in which Pd is α-oriented, is formed by inversion of the stereochemistry, and rapid elimination occurs via the α-oriented angular 5σ-allylpalladium methoxide **357** to give the heteroannular conjugated diene **358**, rather than via 3σ-allylpalladium methoxide **359**. This selectivity may be explained by the different elimination rates of **357** and **359**. The elimination of H−Pd−OAc from the tertiary carbon of **357** seems to be faster than that from **359**. However, the β-oriented π-allylpalladium methoxide is formed from 3α-allylic carbonate **360**, and the homoannular diene **362** is obtained as a major product via the β-oriented 3σ-allylpalladium methoxide **361**. Formation of the β-oriented angular 5σ-allylpalladium **363** is sterically unfavourable. The reaction can be applied to AB rings in steroids.

The reaction has been successfully applied to the synthesis of the precursor **366** of provitamin D which has the homoannular conjugated diene in the B ring [169]. Treatment of 7α-carbonate **364** with the Pd catalyst at 40 °C affords the 5,7-diene **366** regioselectively in good yield. No heteroannular diene **367** is detected. In the intermediate complex **365**, the β-oriented 7σ-allylpalladium undergoes facile *syn* elimination of the 8β-hydrogen to afford **366** exclusively.

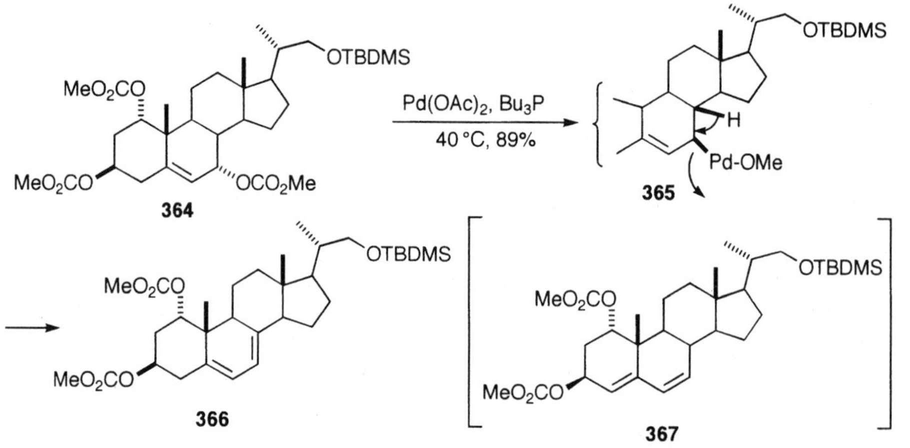

However, a mechanistically interesting result which contradicts the expected *anti* addition–*syn* elimination mechanism of Pd-catalysed 1,4-elimination of allylic compounds has been reported. This is the elimination of the cyclic allylic carbonate **368** which afforded diene **371**, but not diene **370**, as expected from the *anti* addition–*syn* elimination mechanism. The selective formation of **371** is explained by oxidative addition with inversion to genearate **369**, followed by elimination of the *anti* H of **369**, namely *anti* elimination occurs [170].

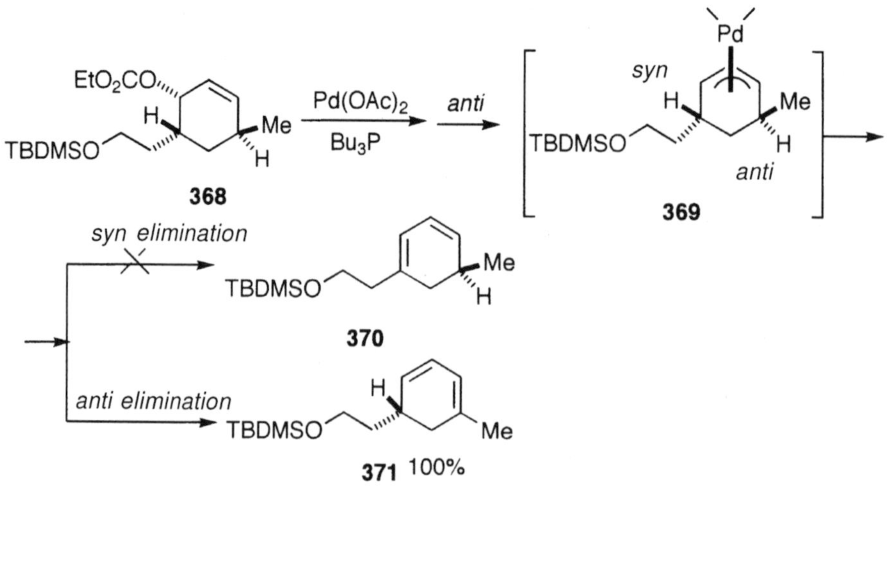

Vinyl epoxides **372** and **373** are converted to conjugated dienes **375** and **378** via the π-allyl intermediates **374** and **377**. In other words, different protons are eliminated, depending on the structure to give either the cyclopentenone **376** or the dienyl alcohol **378**. The unsaturated ketone **376** is an expected product, because elimination of a hydride from the carbon connected to oxygen is common [171].

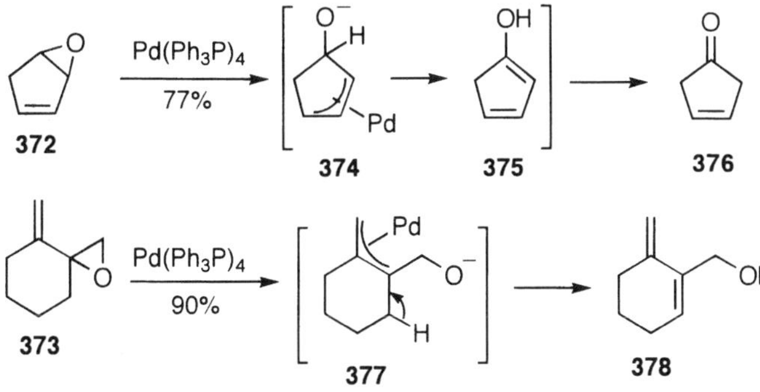

Pd-catalysed decarboxylation-elimination offers a useful method for regioselective generation of conjugated dienes. The polyene system of vitamin A derivative **381** is prepared from the β-acetoxycarboxylic acid **379** by decarboxypalldation, as shown by **380** [172].

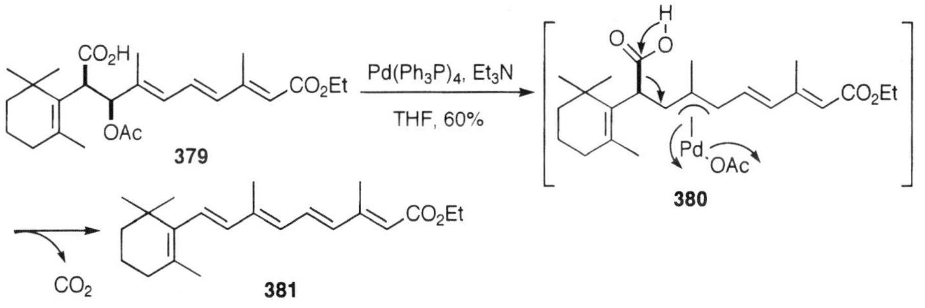

Another interesting reaction is the Pd-catalysed β-decarbopalladation of 4-vinyl cyclic carbonates **382** to afford dienyl aldehydes or ketones **384**. The 2-oxa-1-

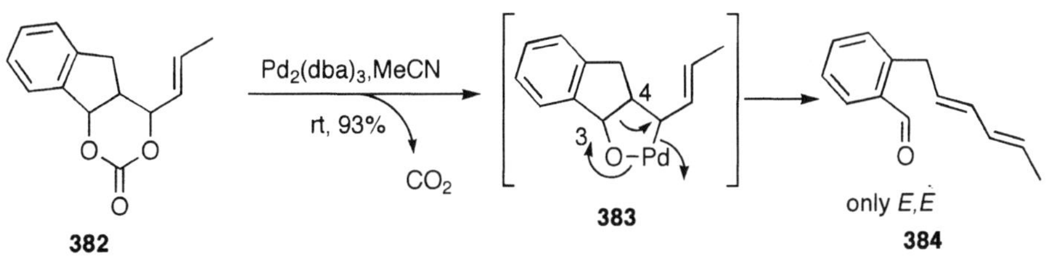

palladacyclopentane **383** is generated by oxidative addition and decarboxylation of the allylic carbonate **382**, and its novel C(3) C(4) bond fission (*cis-β-decarbopalladation*) under mild conditions as shown by **383** affords the dienyl aldehyde **384** [173].

β-Elimination of π-allylpalladium alkoxides produces ketones. Based on this reaction, the secondary alcohol **385** is oxidized to ketone **388** after converting to the allyl carbonate **386**. Treatment of **386** with ligandless Pd in refluxing MeCN affords the ketone **388** by elimination of β-hydrogen as shown by **387** under neutral conditions [174]. Secondary alcohols and allylic alcohols are oxidized easily by this method. Primary alcohols, particularly MeOH, are difficult to oxidize. Therefore, alcohols are oxidized using allyl methyl carbonate (**389**). The π-allylpalladium methoxide **390**, generated from allyl methyl carbonate (**389**), is subjected to exchange reaction with the alcohol **391** to be oxidized to generate the π-allylpalladium alkoxide **392**, which is oxidized to **393** by elimination of β-hydrogen. Using diallyl carbonate (**395**), instead of allyl methyl carbonate, the lactol **394** was oxidized selectively to the lactone **396** in echinosporin synthesis [175]. The oxidation of alcohols using allyl methyl carbonate can be carried out more efficiently using a Ru catalyst. The oxidation of **397** to **398** is an example [176].

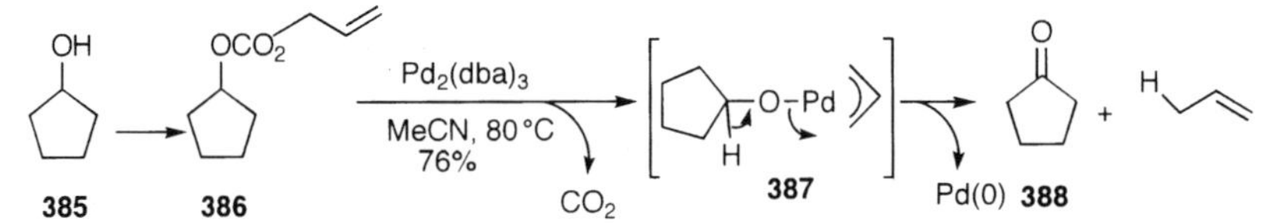

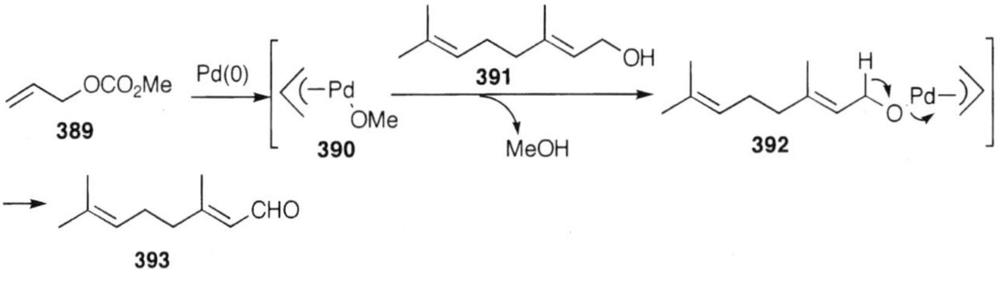

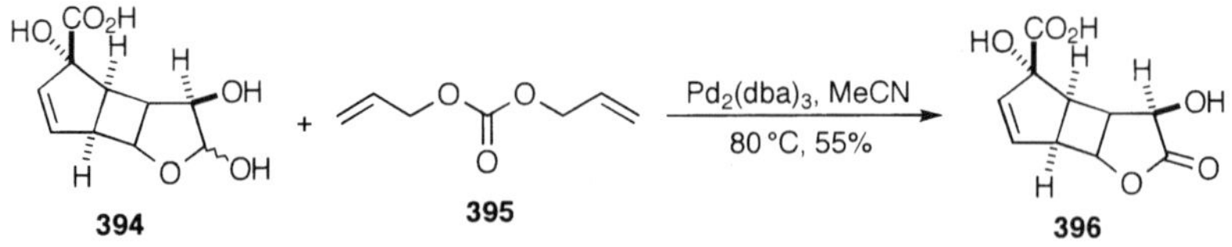

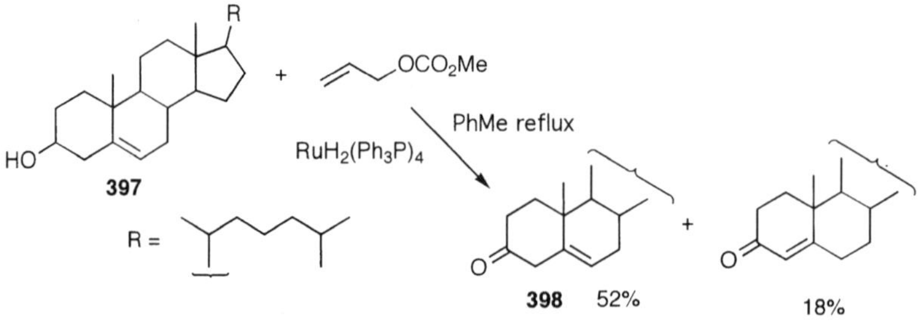

4.3.10 Pd-catalysed Reactions of Allyl β-Keto Carboxylates, Malonates, and Enol Carbonates

Needless to say, β-keto esters are important compounds in organic synthesis. Their usefulness has been considerably expanded, based on Pd-catalysed reactions of allyl β-keto carboxylates **399**. Cleavage of the allylic carbon–oxygen bond and subsequent facile decarboxylation by the treatment of allyl β-keto carboxylates with Pd(0) catalysts generate the π-allylpalladium enolates **400, 401**. These intermediates undergo, depending on the reaction conditions, various transformations which are not possible by conventional methods. Thus new synthetic uses of β-keto esters and malonates based on Pd enolates have been expanded. These reactions proceed under

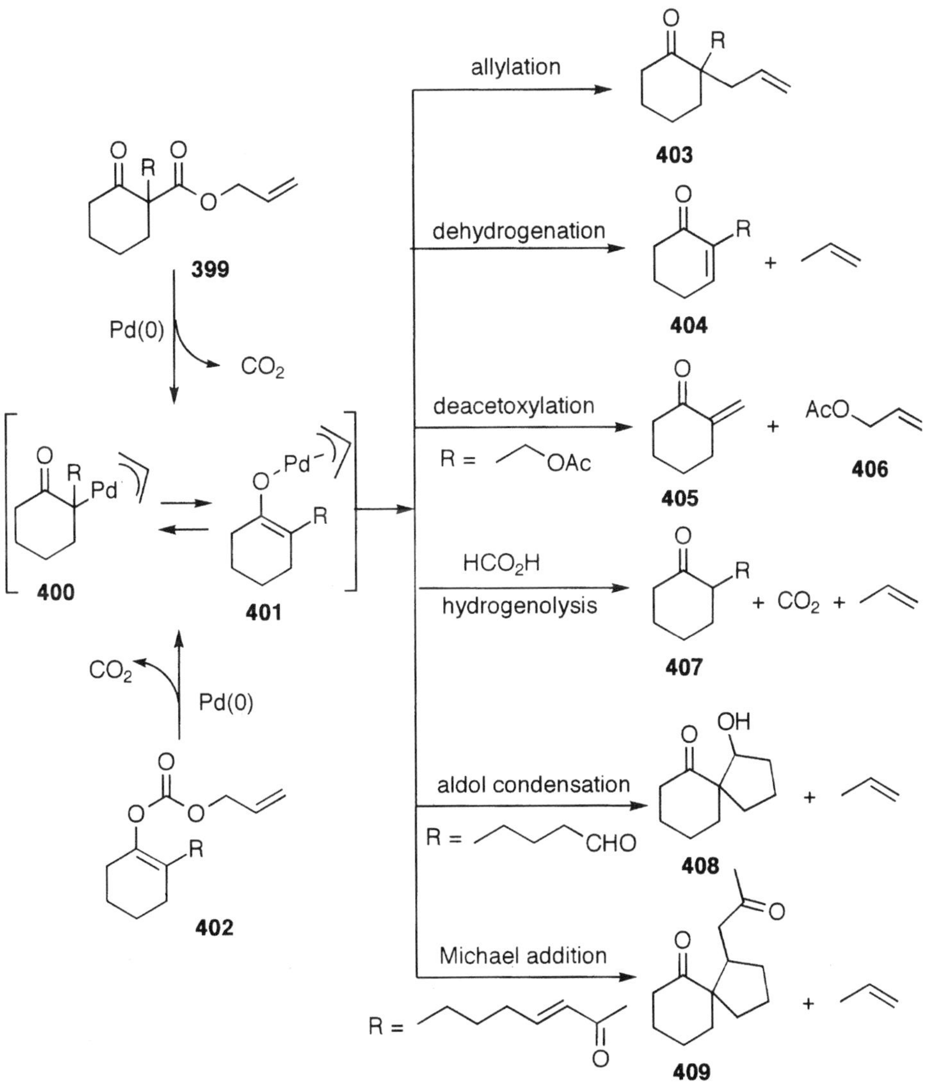

Scheme 4.3 Pd-catalysed transformations of allyl β-keto carboxylates

neutral conditions. The chemistry of β-keto esters of the new generation is summarized in Scheme 4.3 [43]. In addition to allyl β-keto carboxylates, allyl enol carbonates **402**, the structural isomers of allyl β-keto esters, undergo similar reactions via the formation of the π-allylpalladium enolates **400** and **401**. It should be pointed out that similar π-allylpalladium enolates **203, 204, 223** and **234** are formed by the Pd-catalysed reactions of silyl enol ethers and enol acetates with allyl carbonates as described before. Similarly, derivatives of allyl acetate bearing other EWGs such as malonates, nitroacetates, cyanoacetates, and sulfonylacetates undergo Pd-catalysed decarboxylation and further transformations. These reactions are explained below.

4.3.10.1   Decarboxylation and allylation

The reductive coupling of the π-allylpalladium enolates **400** gives the allylated ketones **403**. This reaction is also possible thermally and is called the Carroll reaction. Whereas the Carroll reaction proceeds by heating up to 200 °C, the Pd-catalysed Carroll-type reaction can be carried out under mild conditions (even at room temperature) by reductive elimination of the π-allylpalladium enolate **400** [177,178]. The Pd-catalysed reaction is mechanistically different from the thermal reaction and more versatile, which is explained by the [3,3] sigmatropic rearrangement of the enolate form. For example, thermal Carroll rearrangement of the α,α-disubstituted keto ester **410** is not possible, because there is no possibility of the enolization. However, it rearranges to ketone **411** smoothly with the Pd catalyst, via the π-allylpalladium enolate.

The α,β-disubstituted cyclohexane **414** is prepared by intramolecular allylation of **412** to generate **413**, and subsequent decarboxylation–allylation. The diallylation reactions of **412** are based on the fact that intramolecular allylation of the β-keto ester **412** with the allylic carbonate moiety is faster than decarboxylation of the allyl carboxylate moiety in **412** [179].

4.3.10.2   Decarboxylation and β-elimination (enone synthesis)

When the reaction of allyl β-keto carboxylates is carried out in boiling MeCN, the decarboxylation is followed by elimination of β-hydrogen from the intermediate enolate **400**, affording the α,β-unsaturated ketone **404** [180] (Scheme 4.3). The 1 ~ 1.5:1 ratio of Ph$_3$P:Pd is essential for chemoselective formation of the enones. This means that coordinative unsaturation favours the elimination. Allylation becomes the main reaction path with higher ratios of Ph$_3$P to Pd, which makes the intermediate complex more coordinatively saturated and hence accelerates reductive elimination. The allyl group is the hydride acceptor in the elimination. As supporting evidence, the enone **416** and 1-phenylpropylene (**417**) were obtained in equal amounts by the reaction of the cinnamyl β-keto carboxylate **415**.

As one application of enone formation, α-substituted cyclopentenones can be prepared. Methyl jasmonate (**421**) is produced commercially by applying this as the key reaction. Dieckmann condensation of diallyl adipate (**418**), followed by alkylation gives the α-substituted cyclopentanone carboxylate **419**. The allyl ester **419** undergoes Pd-catalysed decarboxylation-Elimination of β-hydrogen in boiling MeCN by keeping the ratio of Pd:Ph$_3$P as 1:1 ~ 1.5, yielding 2-(2-pentynyl)-2-cyclopentenone (**420**). Methyl jasmonate (**421**) is produced from **420** [181]. This method is also applied to the synthesis of 2-methylcyclopentenone (**422**), which is a useful intermediate for cyclopentanoid synthesis [182]. When the ratio of Ph$_3$P to Pd is higher than 2 in the reaction, decarboxylation–allylation takes place.

The unsaturated allyl β-keto carboxylates **425** and **426**, obtained by the Rh-catalysed reaction of myrcene (**423**) with allyl acetoacetate (**424**) (see 5.2), were subjected to the Pd-catalyzed enone formation. The pseudo-ionone isomers **428** and **429** were obtained at room temperature using dinitriles such as adiponitrile (**427**) or 1,6-dicyanohexane as a solvent and weak ligand without giving an allylated product [183].

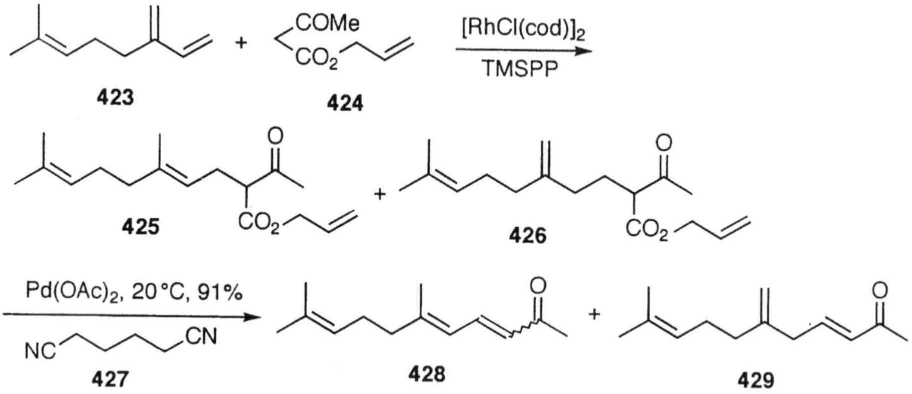

4.3.10.3 Decarboxylation and deacetoxylation (preparation of α-methylene compounds)

The α-acetoxymethyl group can be introduced to allyl β-keto carboxylates by the treatment with formaldehyde, followed by acetylation. When allyl β-keto esters **430** with the acetoxymethyl group at the α-position are treated at room temperature with Pd(0) catalyst, decarboxylation is followed by deacetoxylation as shown by **431** to give the *exo*-methylene ketones **405** and **432** [184]. The allyl group is the acceptor of the acetoxy group and allyl acetate (**433**) is formed, indicating that the acetoxy group is eliminated as allyl acetate more easily than β-hydrogen, and 2-acetoxymethylcyclo-pentenone (**434**) is not formed. As the elimination reaction proceeds even at room temperature under mild neutral conditions, the method offers a good synthetic route for this important functional group. The reaction can be applied to allyl malonates. The alkylation of diallyl malonate (**435**) with bromoacetate and the acetoxymethylation afford the mixed triester **436**, which is converted to allyl ethyl itaconate (**437**).

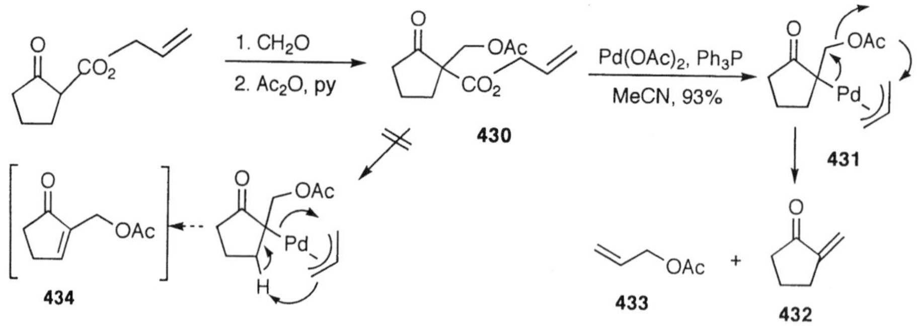

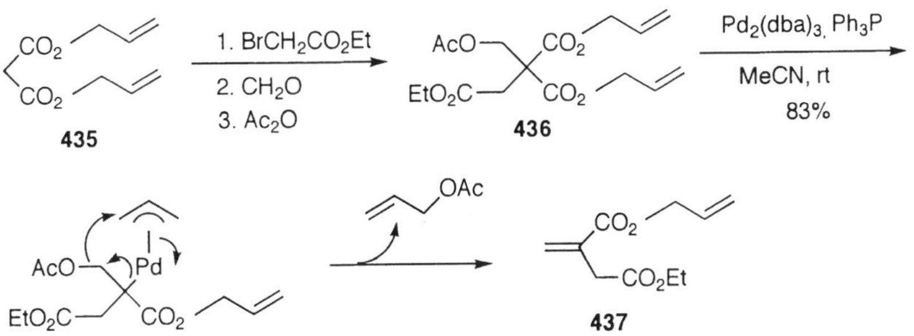

4.3.10.4 Decarboxylation and hydrogenolysis

β-Keto esters and malonates are useful compounds in organic synthesis. After alkylation, they are hydrolysed and decarboxylated to give alkylated ketones or acids. However, rather severe conditions are required for the hydrolysis and decarboxylation of alkylated β-keto esters and malonates. Strongly acidic or basic conditions and high temperatures are necessary. However, allyl β-keto carboxylates and allyl malonates, after the alkylation, can be decarboxylated under extremely mild conditions, namely at room temperature and under neutral conditions by the Pd-catalysed hydrogenolysis with ammonium formate [185]. The hydrogenolysis proceeds by displacement of **439** with HCO_2H, giving only CO_2 and propylene. The reaction can be carried out without attacking acid- or base-sensitive functional groups. The acid-sensitive THP ether in **438** is not cleaved, and no retro-Michael reaction is observed in the decarboxylation of **440**, giving the ketones cleanly.

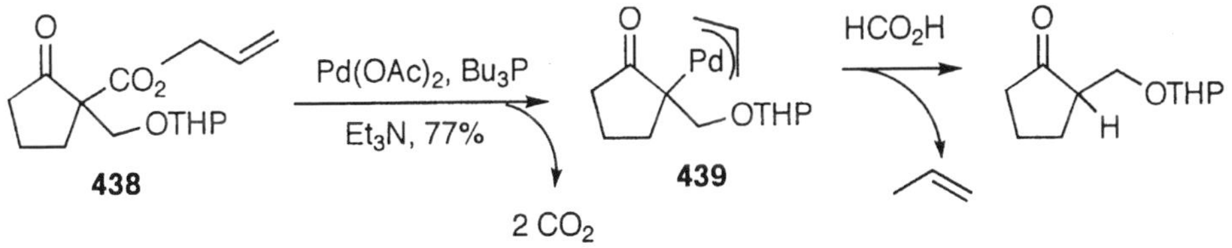

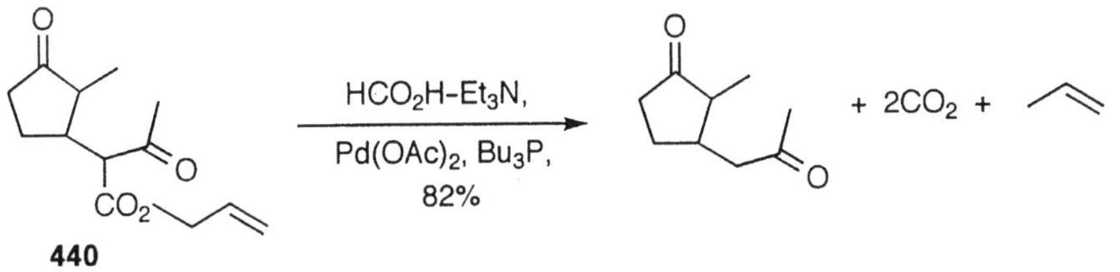

Selective removal of the allyl ester without attacking the methyl ester in **441** to give **442** has been applied to the total synthesis of glycinoeclepin [186]. Hydrogenolysis of diallyl alkylmalonate with formic acid in boiling dioxane affords monocarboxylic acid. Allyl ethyl malonates are converted to ethyl carboxylates [187]. The diallyl α-methylmalonate **443**, attached to a β-lactam ring, undergoes Pd-catalysed stereoselective decarboxylation and hydrogenolysis using an excess of HCO_2H without amine, and the mono acid **444** which has the desired β-oriented methyl group was obtained with high stereoselectivity. Protection of the amide nitrogen with bulky TBDMS is essential, without which no stereoselectivity is observed [188].

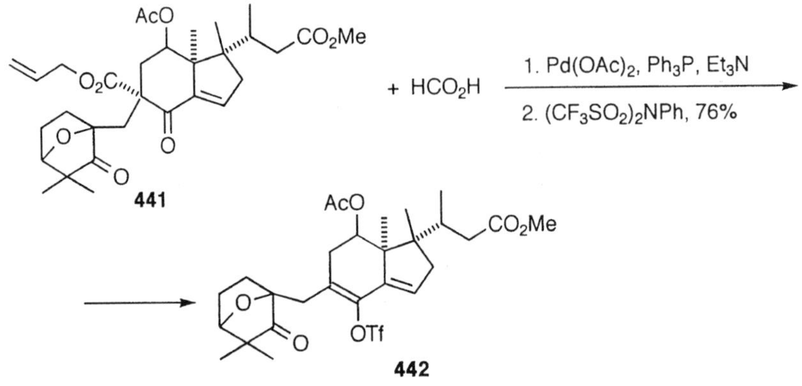

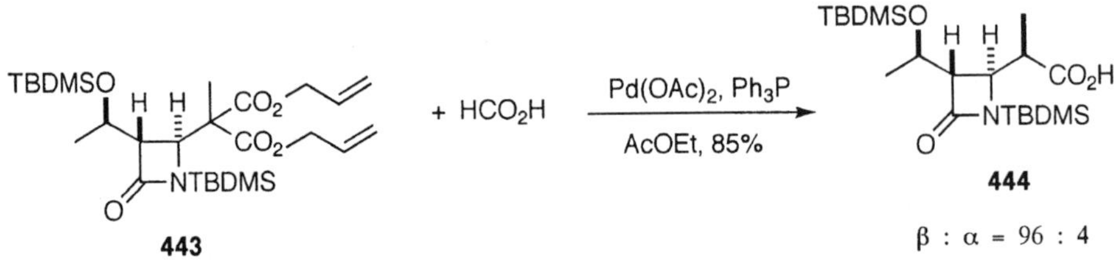

4.3.10.5   *Decarboxylation–aldol condensation and Michael addition*

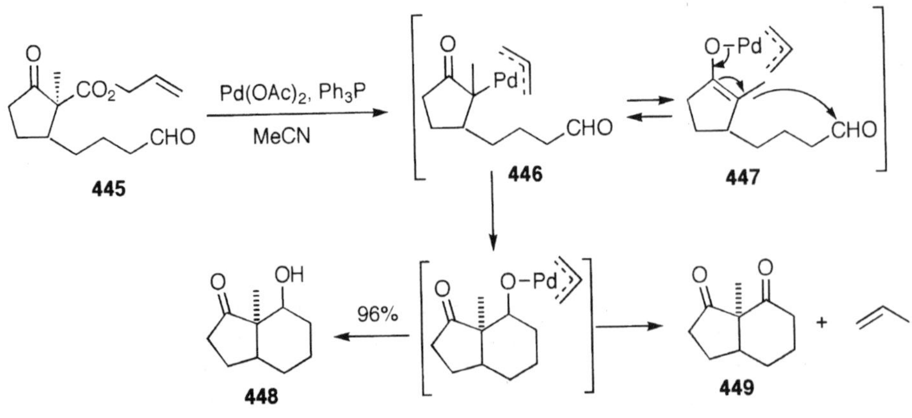

The decarboxylation of allyl β-keto carboxylates generates π-allylpalladium enolates. Aldol condensation and Michael addition are two reactions typical to metal enolates. Actually Pd enolates have been found to undergo intramolecular aldol condensation and Michael addition. When an aldehyde group is present in the allyl β-keto ester **445**, the Pd enolate **446–447** undergoes intramolecular aldol condensation, yielding cyclic aldol **448** as the main product in high yield [189]. At the same time, diketone **449** is formed as a minor product by β-elimination. This is the Pd-catalysed aldol condensation under neutral conditions. The reaction proceeds even in the presence of water, showing that the Pd enolate is not decomposed by water. Allyl acetates bearing other EWGs such as allyl malonate, cyanoacetate and sulfonylacetates undergo similar aldol-type cyclizations.

The Pd enolates also undergo the intramolecular Michael addition when the allyl β-keto carboxylate or diallyl malonate **450**, in which an enone is present at a suitable position, is treated with Pd(0) catalyst [190]. The main product is the saturated ketone **451**, and the allylated product **452** is obtained as a byproduct.

Various Pd-catalysed reactions of allyl β-keto carboxylates, summarized in Scheme 4.3 and shown in the above clearly indicate that synthetic utility of β-keto esters and malonates has been expanded.

4.3.11 Allylic Rearrangement and Isomerization

Rearrangement of allylic esters is catalysed by both Pd(II) and Pd(0) compounds, although their catalyses differ mechanistically. Allylic rearrangement of allylic acetates takes place by Pd(0) formed from Pd(OAc)$_2$–Ph$_3$P as a catalyst [191]. An equilibrium mixture of **453** and **454** in a ratio of 1.9:1.0 was obtained [192]. The Pd(0)–Ph$_3$P-catalysed rearrangement is explained by the formation of a π-allylpalladium complex. The rearrangement of p-tolylsulfone from tertiary **455** to primary **456** occurs [193].

The rearrangement of allylic esters is also catalysed efficiently by PdCl$_2$(MeCN)$_2$ [194]. The Pd(II)-catalysed allylic rearrangement is explained in the following mechanism. After coordination of alkene to PdCl$_2$, the cyclic intermediate **457** is formed by oxypalladation as the rate-determining step. The isomerization ends by the cleavage of **457** [194]. The allylic rearrangement of **458** and **460** in prostaglandin synthesis is catalysed efficiently by PdCl$_2$(MeCN)$_2$ [195,196]. The reaction goes in one direction, irreversibly yielding the thermodynamically stable products **459** and **461**, possibly due to steric reasons. In addition, a complete transfer of the chirality of the carbon–oxygen bond is observed. The rearrangement of the (*E*) and (*Z*) isomers **458** and **460**, respectively, generates the stereochemistry opposite to each other after the rearrangement. The minor product **462** is formed by π-σ-π rearrangement, involving rotation.

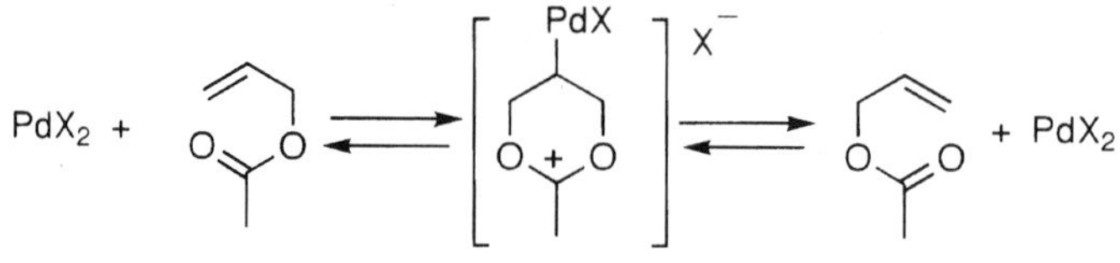

457

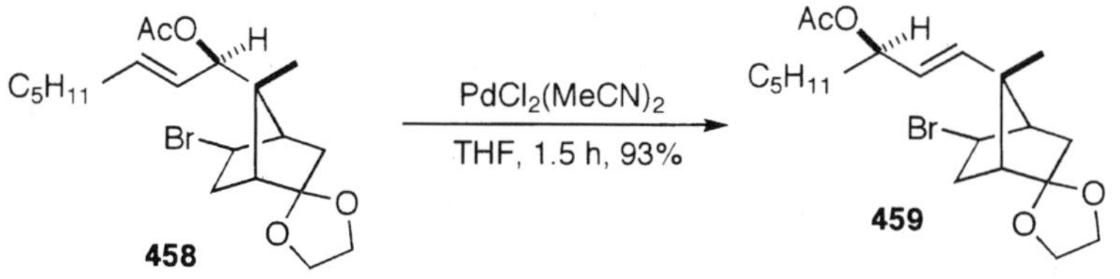

458   PdCl$_2$(MeCN)$_2$ **459**
THF, 1.5 h, 93%

460   PdCl$_2$(MeCN)$_2$ **461**   +   **462**
THF, 1.5 h, 90%

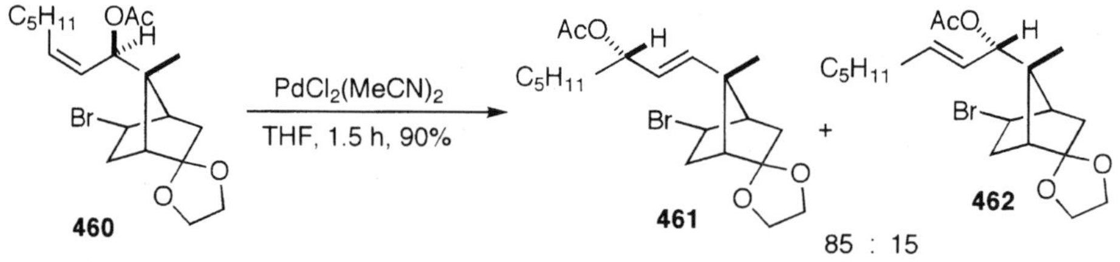

85 : 15

Rearrangement of the acetate of the optically active cyanhydrin **463** proceeds stereoselectively to yield the α,β-unsaturated nitrile **464** with 89% ee [197]. Allyl trichloroacetimidate **465** is rearranged from O to N to give **466** at room temperature with retention of the stereochemistry [198].

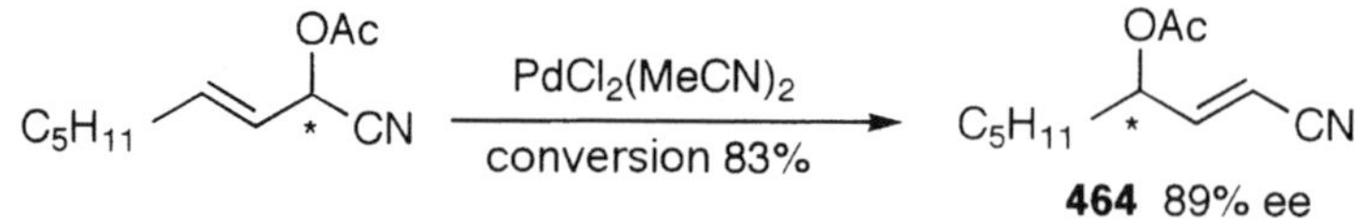

463   PdCl$_2$(MeCN)$_2$ **464** 89% ee
conversion 83%

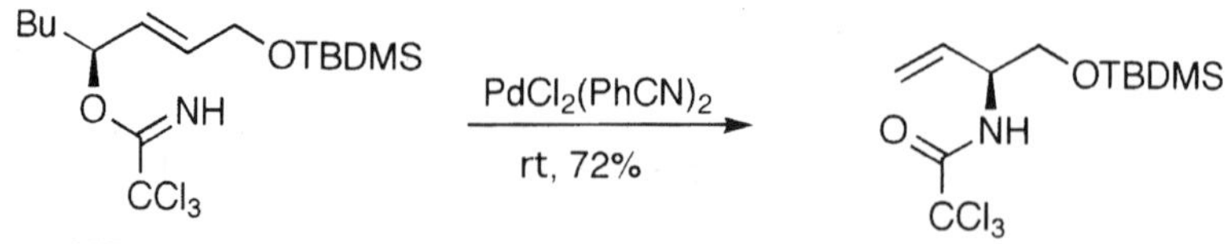

465 88% ee   PdCl$_2$(PhCN)$_2$ **466** 88% ee
rt, 72%

Cope and Claisen rearrangements proceed under milder conditions in the presence of Pd(II) catalyst [194,199]. Cope rearrangement of the linear 1,5-diene **467** to **468** proceeds at room temperature in methylene chloride by the catalysis of $PdCl_2(PhCN)_2$ [200]. Substituents have large effects on the reaction. Oxy-Cope rearrangement of **469** to the ketone **470** proceeds at room temperature in the presence of 10 mol % of $PdCl_2(PhCN)_2$ [201].

Mechanistic studies on the Pd(II)-catalysed Cope rearrangement of the deuterated 1,5-diene **471** has been carried out. After coordination of the alkene to $PdCl_2$, the 4-palladacyclohexyl cation **472** is formed by carbopalladation as the rate determining step. The isomerization ends with the cleavage of **472** to afford **473** [202].

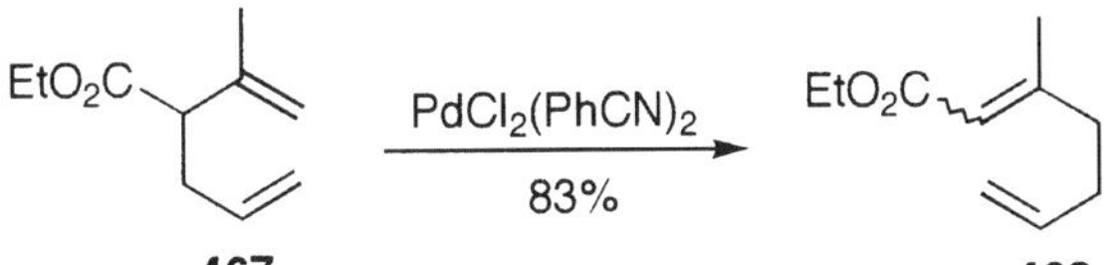

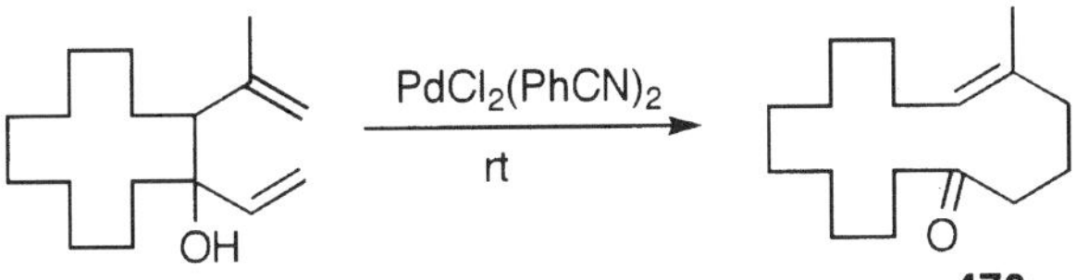

$PdCl_2(PhCN)_2$-catalysed Claisen rearrangement of the allyl vinyl ether **474** derived from cyclic ketone at room temperature affords the *syn* product **475** with high diastereoselectivity [203]. In contrast to thermal Claisen rearrangement, the Pd(II)-catalysed Claisen rearrangement is always stereoselective, irrespective of the geometry of allylic alkenes. The *anti* product is obtained by the thermal rearrangement in the presence of 2,6-dimethylphenol at 100 °C for 10 h.

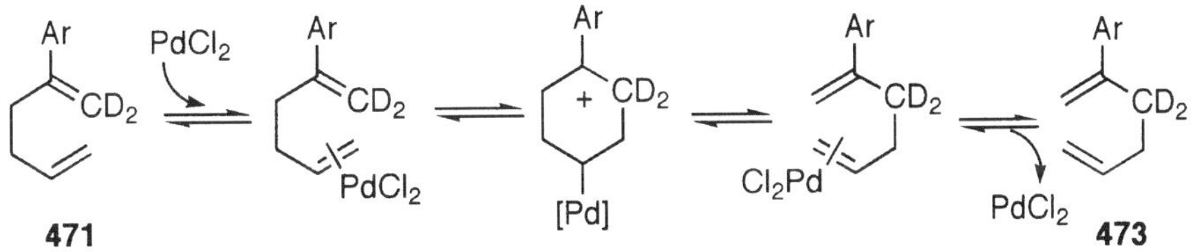

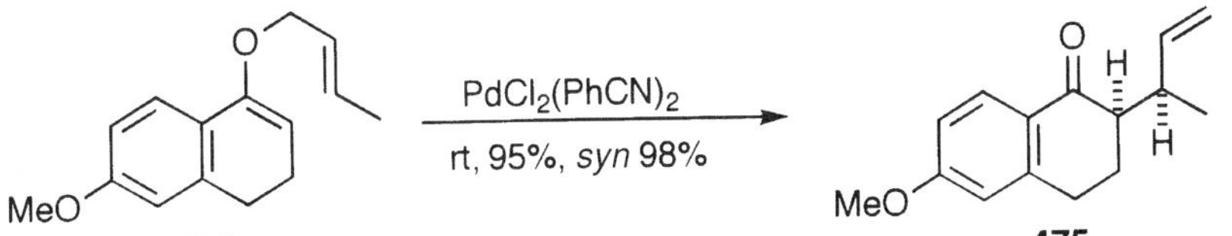

The Pd(II)-catalysed reaction of an allylic alcohol with the ketene acetal **476** at room temperature generates the *ortho* ester **477**. Its Claisen rearrangement via **478** in boiling xylene with a catalytic amount of $PdCl_2(PhCN)_2$ gives the γ,δ-unsaturated ester **479** [204].

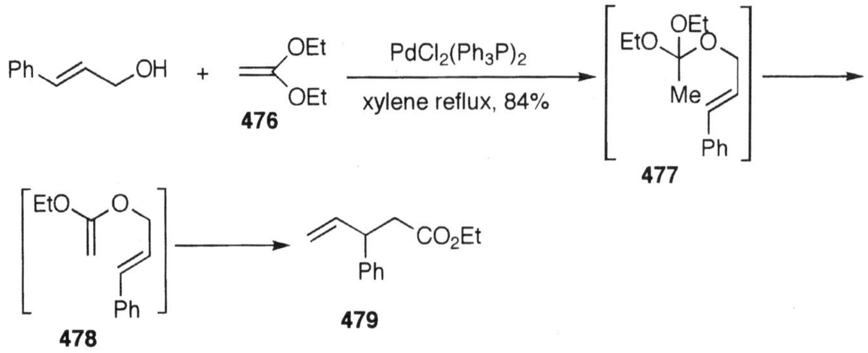

Ph⌒⌒OH + (476 structure with OEt, OEt) →[PdCl₂(Ph₃P)₂ / xylene reflux, 84%] [477 bracketed intermediate] →

[478 bracketed intermediate] → 479 (allyl structure with Ph and CO₂Et)

Allylic double bonds can be isomerized by some transition metal complexes. Isomerization of alkyl allyl ethers **480** to vinyl ethers **481** is catalysed by Pd on carbon [205] and the Wilkinson complex [206], and the vinyl ethers are hydrolysed to aldehydes. Isomerization of the allylic amines to enamines is catalysed by Rh complexes [207]. The asymmetric isomerization of *N,N*-diethylgeranylamine (**483**), catalysed by Rh-(*S*)-BINAP (**XXXI**) complex to produce the (*R*)-enamine **484** with high optical purity, has been achieved with a 300 000 turnover of the Rh catalyst, and citronellal (**485**) with nearly 100% ee is obtained by the hydrolysis of the enamine **484** [208]. Now optically pure *l*-menthol (**486**) is commerically produced in five steps from myrcene (**482**) via citronellal (**485**) by Takasago International Corporation. This is the largest industrial process of asymmetric synthesis in the world [209]. The following stereochemical corelation between the stereochemistries of the chiral Rh catalysts, diethylgeranylamine (**483**), diethylnerylamine (**487**) and the (*R*)- and (*S*)-enamines **484**

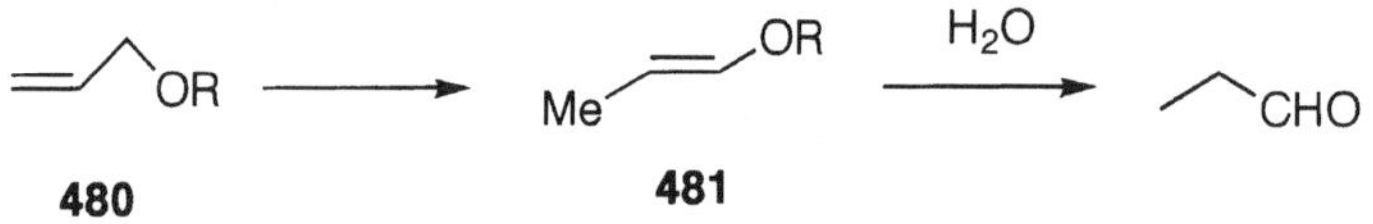

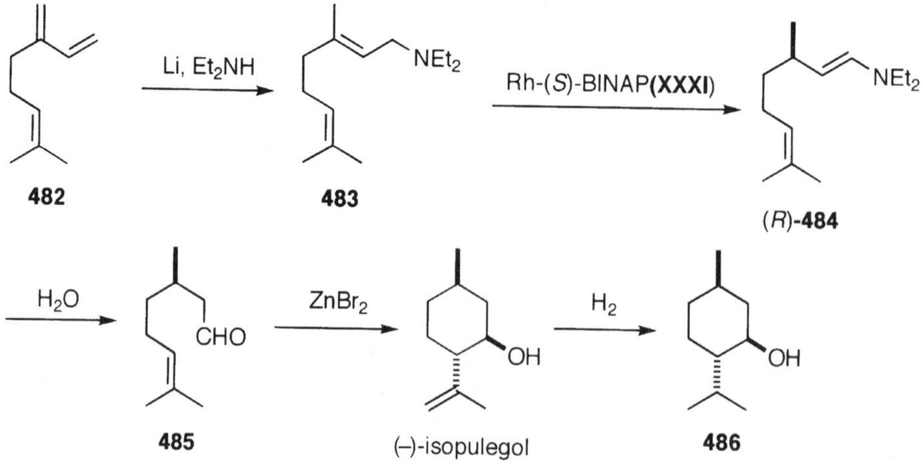

(480 structure)=⌒OR → (481) Me—=⌒OR →[H₂O] ⌒CHO

480   **481**

482 →[Li, Et₂NH] **483** →[Rh-(*S*)-BINAP(**XXXI**)] (*R*)-**484**

→[H₂O] **485** (CHO) →[ZnBr₂] (–)-isopulegol →[H₂] **486**

and **488** was confirmed, and the mechanism of the asymmetric isomerization has been studied [210]. This means that both geranylamine and nerylamine are used for this process. The higher stability of the enamine relative to the allylamine is the driving force of the isomerization.

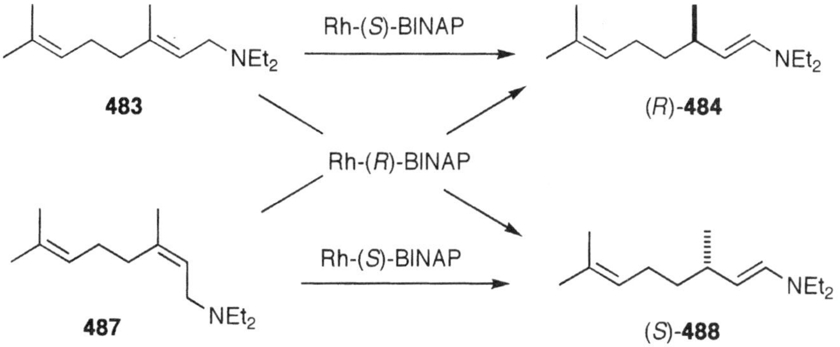

α-Methylene-π-allylpalladiums **490** are generated from the esters of 2,3-alkadienyl alcohols **489**. The complexes are reactive and give either 1,2- or 1,3-dienes, **491** or **492**, depending on reactants. Reaction of the acetate **493** with malonate affords dimethyl 2,3-butadienylmalonate (**494**) [211]. However, hard carbon nucleophiles such as Mg or Zn reagents react with the 2,3-alkadienyl phosphate **495** to give the 2-alkyl-1,3-butadiene **496** [212]. The 3-alkyl-1,3-butadiene-2-carboxylate **498** is obtained in high yield by the carbonylation of the 2-alkyl-2,3-butadienyl carbonate **497** under very mild conditions (room temperature and 1 atm) [213]. The Pd-catalysed carbonylation of 2,3-dienylamine **499** using DPPP as a ligand and TsOH as a promoter under somewhat severe conditions affords the α-vinylacrylamide **500** in high yield [214].

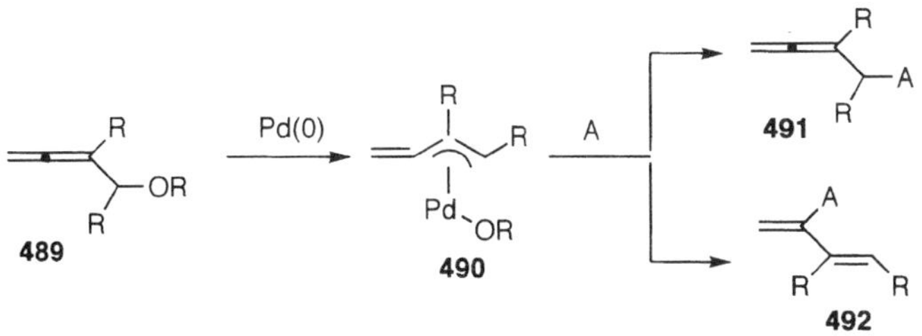

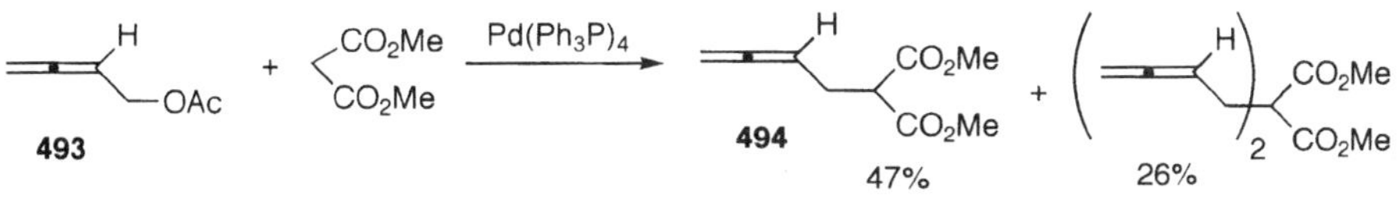

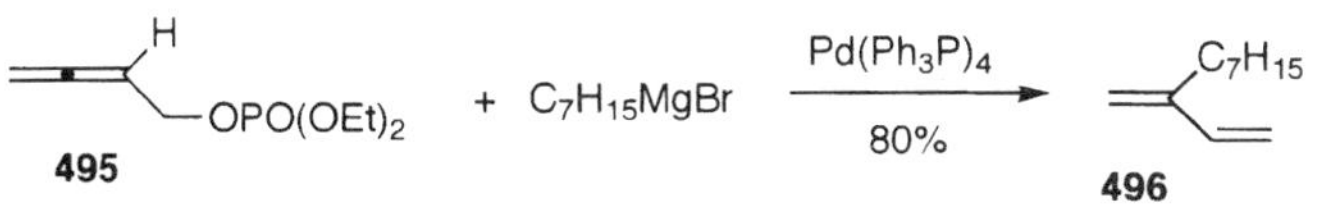

References

1. J. Tsuji, *Palladium Reagents and Catalysts, Innovations in Organic Synthesis*, John Wiley, p. 290, 1995; R. F. Heck, *Palladium Reagents in Organic Synthesis*, Academic Press 1985; B. M. Trost and T. R. Verhoeven, *Comprehensive Organometallic Chemistry*, Vol. 8, p. 799, 1982, Pergamon Press; S. A. Godleski, *Comprehensive Organic Synthesis*, Vol. 4, p. 585 1991, Pergamon Press.
2. T. Kondo, H. Ono, N. Satake, T. Mitsudo and Y. Watanabe, *Organometallics*, **14**, 1945 1995.
3. J. Tsuji, H. Takahashi and M. Morikawa, *Tetrahedron Lett.*, 4387 (1965).
4. R. Hüttel and H. Christ, *Chem. Ber.*, **97**, 1439 (1964); D. Morelli, R. Ugo, F. Conti and M. Donati, *Chem. Commun.*, 801 (1967).
5. B. M. Trost, *Acc. Chem. Res.*, **13**, 385 (1980).
6. B. M. Trost, L. Weber, P. E. Strege, T. J. Fullerton and T. J. Dietsche, *J. Am. Chem. Soc.*, **100**, 3416, 3426 (1978).
7. G. W. Parshall and G. Wilkinson, *Chem. Ind. (London)*, 261 (1962); *In org. Chem.*, **1**, 896 (1962). J. Tsuji, S. Imamura and J. Kiji, *J. Am. Chem. Soc.*, **86**, 4491 (1964).
8. D. J. Collins, W. R. Jackson and R. N. Timms, *Tetrahedron Lett.*, 495 (1976); W. R. Jackson and I. U. Straus, *Tetrahedron Lett.*, 2591 (1975), *Aust. J. Chem.*, **30**, 553 (1977), **31**, 1073 (1978).
9. T. Susuki and J. Tsuji, *Bull. Chem. Soc. Jpn.*, **46**, 655 (1973).
10. J. S. Temple and J. Schwartz, *J. Am. Chem. Soc.*, **102**, 7381 (1980); **104**, 1310 (1982); M. Riediker and J. Schwartz, *Tetrahedron Lett.*, **22**, 4655 (1981).
11. E. J. Corey and M. F. Semmelhack, *J. Am. Chem. Soc.*, **89**, 2755 (1967).
12. M. F. Semmelhack, A. Yamashita, J. C. Tomesch and K. Hirotsu, *J. Am. Chem. Soc.*, **100**, 5565 (1978); M. F. Semmelhack and S. J. Brickner, *J. Am. Chem. Soc.*, **103**, 3945 (1981).
13. Y. Tsuji, T. Mukai, T. Kondo and Y. Watanabe, *J. Organometal. Chem.*, **369**, C51 (1989).
14. G. D. Annis, S. V. Ley, C. R. Self and R. S. Krishnan, *J. Chem. Soc., Perkin I*, 70 (1981).
15. S. T. Hodgson, D. M. Hollinshead and S. V. Ley, *Tetrahedron*, **41**, 5871 (1985); S. V. Ley, *Pure Appl. Chem.*, **66**, 1415 (1994).
16. S. V. Ley, L. R. Cox and G. Meek, *Chem. Rev.*, **96**, 423 (1996).
17. Y. Becker, A. Eisenstadt and Y. Shvo, *Tetrahedron*, **30**, 839 (1974).
18. Y. Okude, S. Hirano, T. Hiyama and H. Nozaki, *J. Am. Chem. Soc.*, **99**, 3179 (1977); K. Takai, K. Nitta and K. Utimoto, *Tetrahedron Lett.*, **29**, 5263 (1988); T. Hiyama, Y. Okude, K. Kimura and H. Nozaki, *Bull. Chem. Soc. Jpn.*, **55**, 561 (1982).
19. O. Fujimura, K. Takai and K. Utimoto, *J. Org. Chem.*, **55**, 1705 (1990).
20. P. A. Wender, J. A. Mckinney and C. Mukai, *J. Am. Chem. Soc.*, **112**, 5369 (1990).

21. H. A. Martin and F. Jelinek, *J. Organometal. Chem.*, **12**, 149 (1968).
22. F. Sato, S. Iijima and M. Sato, *Tetrahedron Lett.*, **22**, 243 (1981).
23. J. Szymoniak, S. Pagneux, D. Felix and C. Moise, *Synlett.*, 46 (1996).
24. J. Tsuji, *Tetrahedron Report* 206; *Tetrahedron*, **42**, 4361 (1986); B. M. Trost, *Tetrahedron Report* 32; *Tetrahedron*, **33**, 2615 (1977): M. F. Semmelhack, *Org. React.*, **19**, 115 (1972); R. Baker, *Chem. Rev.*, **73**, 487 (1973).
25. G. Hata, K. Takahashi and A. Miyake, *Chem. Commun.*, 1392 (1970), K. E. Atkins, W. E. Walker and R. M. Manyik, *Tetrahedron Lett.*, 3821 (1970).
26. B. M. Trost and C. A. Merlic, *J. Am. Chem. Soc.*, **112**, 9590 (1990), Y. D. Ward, L. A. Villanueva, G. D. Allred and L. S. Liebeskind, *J. Am. Chem. Soc.*, **118**, 897 (1996).
27. J. Tsuji, I. Minami and I. Shimizi, *Tetrahdron Lett.*, **25**, 5157 (1984), *J. Organometal. Chem.*, **296**, 269 (1985).
28. A. Aranyos, K. J. Szabo, A. M. Castano and J. E. Bäckvall, *Organometallics*, **16**, 1058 (1997).
29. R. Tamura and L. S. Hegedus, *J. Am. Chem. Soc.*, **104**, 3727 (1982); R. Tamura, Y. Kai, M. Kakihana, K. Hayashi, M. Tsuji, T. Nakamura and D. Oda, *J. Org. Chem.*, **51**, 437 (1986).
30. N. Ono, I. Hamamoto and A. Kaji, *Chem. Commun.*, 821 (1982); *Synthesis*, 950 (1985).
31. B. M. Trost, N. R. Schmuff and M. J. Miller, *J. Am. Chem. Soc.*, **102**, 5979 (1980); B. M. Trost and C. A. Merlic, *J. Org. Chem.*, **55**, 1127 (1290).
32. H. Kotake, T. Yamamoto and H. Kinoshita, *Chem. Lett.*, 1331 (1982).
33. T. Cuvigny, M. Julia and C. Rolando, *J. Organometal. Chem.*, **285**, 395 (1985).
34. B. M. Trost, L. Weber, P. E. Strege, T. J. Fullerton and T. Dietsche, *J. Am. Chem. Soc.*, **100**, 3416 (1978); B. M. Trost and T. R. Verhoeven, *J. Am. Chem. Soc.*, **100**, 3435 (1978); **102**, 4730 (1980).
35. T. Hayashi, T. Hagihara, M. Konishi and M. Kumada, *J. Am. Chem. Soc.*, **105**, 7767 (1983).
36. B. Akermark, J. E. Bäckvall, A. Löwenborg and K. Zetterberg, *J. Organometal. Chem.* **166**, C33 (1979); B. Akermark and A. Jutand, *J. Organometal. Chem.*, **217**, C41 (1981).
37. H. Matsushita and E. Negishi, *Chem. Commun.*, 160 (1982).
38. J. S. Temple, M. Riediker and J. Schwartz, *J. Am. Chem. Soc.*, **104**, 1310 (1982).
39. T. Hayashi, T. Kawatsura and Y. Uozumi, *Chem. Commun.*, 561 (1997).
40. P. A. Evans and J. D. Nelson, *Tetrahedron Lett.*, **39**, 1725 (1998).
41. P. A. Evans and J. D. Nelson, *J. Am. Chem. Soc.*, **120**, 5581 (1998).
42. B. M. Trost and I. Hachiya, *J. Am. Chem. Soc.*, **120**, 1104 (1998).
43. J. Tsuji and I. Minami, *Acc. Chem. Res.*, **20**, 140 (1987); J. Tsuji, *Tetrahedron*, **42**, 4361 (1986).
44. J. Tsuji, I. Shimizu, I. Minami and Y. Ohashi, *Tetrahedron Lett.*, **23**, 4809 (1982). J. Tsuji, I. Shimizu, I. Minami, Y. Ohashi, T. Sugiura and K. Takahashi, *J. Org. Chem.*, **50**, 1523 (1985).
45. J. Tsuji, Y. Kobayashi, H. Kataoka and T. Takahashi, *Tetrahedron Lett.*, **21**, 1475 (1980).
46. G. Hata, K. Takahashi and A. Miyake, *Chem. Commun.*, 1392 (1970); *Bull. Chem. Soc. Jpn.*, **45**, 230 (1972).
47. J. Tsuji, H. Kataoka and Y. Kobayashi, *Tetrahedron Lett.*, **22**, 2575 (1981).
48. B. M. Trost and G. A. Molander, *J. Am. Chem. Soc.*, **103**, 5969 (1981).
49. J. P. Genet and D. Ferroud, *Tetrahedron Lett.*, **25**, 3579 (1984).
50. M. Moreno-Manas, R. Pleixats and M. Villarroya, *J. Org. Chem.*, **55**, 4925 (1990).
51. G. Giambastiani and G. Poli, *J. Org. Chem.*, **63**, 9608 (1998).

52. Y. Tanigawa, K. Nishimura, A. Kawasaki and S. Murahashi, *Tetrahedron Lett.*, **23**, 5549 (1982).
53. J. P. Genet, F. Piau and J. Ficini, *Tetrahedron Lett.*, **21**, 3183 (1980); F. Colobert and J. Genet, *Tetrahedron Lett.*, **26**, 2779 (1985).
54. J. Tsuji, Y. Kobayashi, H. Kataoka and T. Takahashi, *Tetrahedron Lett.*, **21**, 1475, 3393 (1980).
55. K. Yamamoto and J. Tsuji, *Tetrahedron Lett.*, **23**, 3089 (1982).
56. Y. G. Suh, B. A. Koo, E. N. Kim and N. S. Choi, *Tetrahedron Lett.*, **36**, 2089 (1995).
57. B. M. Trost, *Angew. Chem., Int. Ed. Engl.*, **28**, 1173 (1989).
58. Y. Kitagawa, A. Itoh, S. Hashimoto, H. Yamamoto and H. Nozaki, *J. Am. Chem. Soc.*, **99**, 3864 (1977).
59. J. A. Marshall, R. C. Andrews and L. Lebioda, *J. Org. Chem.*, **52**, 2378 (1987).
60. J. D. White, N. S. Kim, D. E. Hill and J. A. Thomas, *Synthesis*, 619 (1998).
61. Y. I. M. Nelsson, P. G. Andersson and J. E. Bäckvall, *J. Am. Chem. Soc.*, **115**, 6609 (1993).
62. H. Bricout, J. F. Carpentier and A. Mortreux, *Tetrahedron Lett.*, **38**, 1053 (1997).
63. J. Tsuji, H. Kataoka and Y. Kobayashi, *Tetrahedron Lett.*, **22**, 2575 (1981); B. M. Trost and G. A. Molander, *J. Am. Chem. Soc.*, **103**, 5969 (1981).
64. T. Takahashi, A. Ootake and J. Tsuji, *Tetrahedron*, **41**, 5747 (1985).
65. B. M. Trost, J. T. Hane and P. Metz, *Tetrahedron Lett.*, **27**, 5695 (1986).
66. B. M. Trost and S. R. Angle, *J. Am. Chem. Soc.*, **107**, 6123 (1985).
67. S. Wershofen and H. D. Scharf, *Synthesis*, 854 (1988).
68. J. G. Shim and Y. Yamamoto, *J. Org. Chem.*, **63**, 3067 (1998).
69. Showa Denko Inc. *Jpn. Pat. Kokai*, 1-165555,1-153660 (1989).
70. S. A. Godleski, D. J. Heathcock, J. D. Meinhart and S. V. Wallendael, *J. Org. Chem.*, **48**, 2101 (1983).
71. L. Gundersen, T. Benneche and K. Undheim, *Tetrahedron Lett.*, **33**, 1085 (1992).
72. H. Bricout, J. F. Carpentier and A. Mortreux, *Chem. Commun.*, 1393 (1997).
73. T. Taniguchi and K. Ogasawara, *Tetrahedron Lett.*, **39**, 4679 (1998).
74. D. R. Deardorf and D. C. Myles, *Org. Syn.*, **67**, 114 (1988).
75. T. Suzuki, O. Sato, H. Hirama, Y. Yamamoto, M. Murata, T. Yamamoto and N. Harada, *Tetrahedron Lett.*, **32**, 4505 (1991).
76. Y. Yoshizaki, H. Satoh, Y. Sato, S. Nukui, M. Shibasaki and M. Mori, *J. Org. Chem.*, **60**, 2016 (1995).
77. B. M. Trost and D. L. Van Vraken, *Chem. Rev.*, **96**, 395 (1996).
78. P. von Matt and A. Pfaltz, *Angew. Chem., Int. Ed. Engl.*, **32**, 566 (1993).
79. G. J. Dawson, C. G. Frost, J. M. J. Williams and S. J. Coote, *Tetrahedron Lett.*, 3149 (1993).
80. J. Spinz, M. Kiefer, G. Helmchen, M. Reggelin, G. Huttner, O. Walter and L. Zsolnai, *Tetrahedron Lett.*, **35**, 1523 (1994).
81. B. M. Trost and M. G. Organ, *J. Am. Chem. Soc.*, **116**, 10320 (1994).
82. B. M. Trost, *Angew. Chem., Int. Ed. Engl.* **25**, 1 (1986); D. M. T. Chan *Comprehensive Organic Synthesis*, Vol. 5, p. 287 Pergamon Press, 1991.
83. L. A. Paquette, D. R. Sauer, D. G. Cleary, M. A. Kinsella, C. K. Blackwell and L. G. Anderson, *J. Am. Chem. Soc.*, **114**, 7375 (1992).
84. I. Shimizu, Y. Ohashi and J. Tsuji, *Tetrahedron Lett.*, **25**, 5157 (1984).
85. P. Breuilles and D. Uguen, *Tetrahedron Lett.*, **28**, 6053 (1987).
86. K. Ohno, T. Mitsuyasu and J. Tsuji, *Tetrahedron Lett.*, 67 (1971); *Tetrahedron*, **28**, 3705 (1972); P. Haynes, *Tetrahedron Lett.*, 3687 (1970); R. M. Manyik, W. E. Walker, K. E. Atkins and E. S. Hammack, *Tetrahedron Lett.*, 3813 (1970).

87. K. Ohno and J. Tsuji, *Chem. Commun.*, 247 (1971).

88. H. Nakamura, J. G. Shim and Y. Yamamoto, *J. Am. Chem. Soc.*, **119**, 8113 (1997).

89. H. Nakamura, N. Asao and Y. Yamamoto, *Chem. Commun.*, 1273 (1995).

90. T. M. Yuan and T. Y. Luh, *J. Org. Chem.*, **57**, 4550 (1992).

91. M. Moreno-Manas, F. Pajuelo and R. Pleixats, *J. Org. Chem.*, **60**, 2396 (1995).

92. A. M. Castano and A. M. Echavarren, *Tetrahedron Lett.*, **37**, 6587 (1996).

93. F. K. Sheffy, J. P. Godschalx and J. K. Stille, *J. Am. Chem. Soc.*, **106**, 4833 (1984).

94. J. P. Godschalx and J. K. Stille, *Tetrahedron Lett.*, **21**, 2599 (1980).

95. T. Ishiyama, T. Abiko and N. Miyaura, *Tetrahedron Lett.*, **37**, 6889 (1996).

96. Y. Tsuji, M. Funato, M. Ozawa, H. Ogiyama and T. Kawamura, *J. Org. Chem.*, **61**, 5779 (1996).

97. N. A. Bumagin, A. N. Kasatkin and I. P. Beletskaya, Izv. Akad. Nauk SSSR, ser. Khim., 636 (1984), *Chem. Abst.*, **101**, 91109 (1984).

98. B. M. Trost and K. M. Pietrusiewicz, *Tetrahedron Lett.*, **26**, 4039 (1985).

99. Y. Yokoyama, S. Ito, Y. Takahashi and Y. Murakami, *Tetrahedron Lett.*, **26**, 6457 (1985).

100. J. Tsuji, I. Minami and I. Shimizu, *Chem. Lett.*, 1325 (1983).

101. J. Tsuji, I. Minami and I. Shimizu, *Tetrahedron Lett.*, **24**, 5635 (1983).

102. J. Tsuji, K. Takahashi, I. Minami and I. Shimizu, *Tetrahedron Lett.*, **25**, 4783 (1984).

103. P. Grzywacz, S. Marczak and J. Wicha, *J. Org. Chem.*, **62**, 5293 (1997).

104. C. Carfagna, L. Mariani, A. Musco, G. Sallese and R. Santi, *J. Org. Chem.*, **56**, 3924 (1991).

105. M. Formica, A. Musco and R. Pontelloni, *J. Mol. Catal.*, **84**, 239 (1993).

106. K. Tsushima and A. Murai, *Chem. Lett.*, 761 (1990).

107. M. Pereyre, B. Bellegrade, J. Mendelsohn and J. Valade, *J. Organometal. Chem.*, **11**, 97 (1968).

108. J. Tsuji, I. Minami and I. Shimizu, *Tetrahedron Lett.*, **24**, 4713 (1983).

109. I. Minami, K. Takahashi, I. Shimizu, T. Kimura and J. Tsuji, *Tetrahedron*, **42**, 2971 (1986); J. Tsuji, I. Minami and I. Shimizu, *Tetrahedron Lett.*, **24**, 5639 (1983).

110. J. Tsuji, K. Sato and H. Okumoto, *Tetrahedron Lett.*, **23**, 5189 (1982); *J. Org. Chem.*, **49**, 1341 (1984).

111. K. Yamamoto, R. Deguchi and J. Tsuji, *Bull. Chem. Soc. Jpn.*, **58**, 3397 (1985); S. Z. Wang, K. Yamamoto, H. Yamada and T. Takahashi, *Tetrahedron*, **48**, 2333 (1992).

112. S. Murahashi, Y. Imada, Y. Taniguchi and S. Higashiura, *J. Org. Chem.*, **58**, 1538 (1993).

113. J. Kiji, T. Okano, Y. Higashimae and Y. Fukui, *Bull. Chem. Soc. Jpn.*, **69**, 1029 (1996).

114. S. Murahashi, Y. Imada and K. Nishimura, *Tetrahedron*, **50**, 453 (1994).

115. K. Yasui, K. Fugami, S. Tanaka and Y. Tamaru, *J. Org. Chem.*, **60**, 1365 (1995).

116. V. P. Baillargeon and J. K. Stille, *J. Am. Chem. Soc.*, **105**, 7175 (1983).

117. V. P. Baillargeon and J. K. Stille, **108**, 452 (1986).

118. Y. Ishii and M. Hidai, *J. Organometal. Chem.*, **428**, 279 (1992).

119. H. Matsuzaka, Y. Hiroe, M. Iwasaki, Y. Ishii, Y. Koyasu and M. Hidai, *J. Org. Chem.*, **53**, 3832 (1988).

120. Y. Ishii, C. Gao, W. X. Xu, M. Iwasaki and M. Hidai, *J. Org. Chem.*, **58**, 6818 (1993).

121. G. P. Chiusoli and L. Cassar, *Angew. Chem., Int. Ed. Engl.*, **6**, 124 (1967); G. P. Chiusoli, *Acc. Chem. Res.*, **6**, 422 (1973); L. Cassar, G. P. Chiusoli and F. Guerrieri, *Synthesis*, 509 (1973).

122. T. Mitsudo, S. W. Zhang, T. Kondo and Y. Watanabe, *Tetrahedron Lett.*, **33**, 341 (1992).

123. G. P. Chiusoli, G. Salerno and F. Dallatomasina, *Chem. Commun.*, 793 (1977); M. Catellani, G. P. Chiusoli, G. Salerno and F. Dallatomasina, *J. Organometal. Chem.*, **146**, C19 (1978).

124. W. Oppolzer, *Pure Appl. Chem.*, **60**, 39 (1988); *Angew. Chem., Int. Ed. Engl.*, **28**, 38 (1989).

125. W. Oppolzer and C. Robyr, *Tetrahedron*, **50**, 415 (1994).

126. R. Keese, R. Guidetti-Grept and B. Herzog, *Tetraheddron Lett.*, **33**, 1207 (1992).

127. T. Doi, A. Yanagisawa, S. Nakanishi, K. Yamamoto and T. Takahashi, *J. Org. Chem.*, **61**, 2602 (1996).

128. J. Tsuji and T. Mandai, *Synthesis*, 1 (1996).

129. J. Tsuji and T. Yamakawa, *Tetrahedron Lett.*, 613 (1979); J. Tsuji, I. Shimizu and I, Minami, *Synthesis*, 623 (1986).

130. T. Mukaiyama, I. Shiina, M. Satoh, K. Nishimura and K. Satoh, *Chem. Lett.*, 223 (1996).

131. M. Tanaka, C. Mukaiyama, H. Mitsuhashi and T. Wakamatsu, *Tetrahedron Lett.*, **33**, 4165 (1992).

132. T. Hayashi, H. Iwamura, Y. Uozumi, Y. Matsumoto and F. Ozawa, *Synthesis*, 526 (1994).

133. T. Mandai, T, Murakami, T. Suzuki, M. Fujita, M. Kawada and J. Tsuji, *Tetrahedron Lett.*, **33**, 2987 (1992).

134. T. Mandai, Y. Kaihara and J. Tsuji, *J. Org. Chem.*, **59**, 5847 (1994).

135. T. Mandai, T. Matsumoto, M. Kawada and J. Tsuji, *J. Org. Chem.*, **57**, 1326 (1992); *Tetrahedron*, **49**, 5483 (1993).

136. P. Wipf, Y. Kim and D. M. Goldstein, *J. Am. Chem. Soc.*, **117**, 11106 (1995).

137. T. Mandai, T. Matsumoto, M. Kawada and J. Tsuji, *J. Org. Chem.*, **57**, 6090 (1992); *Tetrahedron*, **50**, 475 (1994).

137a. T. Mandai, T. Matsumoto, J. Tsuji and S. Saito, *Tetrahedron Lett.*, **34**, 2513 (1993).

138. C. J. Solomon, E. G. Mata and O. A. Mascaretti, *Tetrahedron*, **49**, 3714 (1993).

139. J. Tsuji and T. Yamakawa, *Tetrahedron Lett.*, 613 (1979).

140. N. Ono, M. Tsuboi, S. Okamoto, T. Tanami and F. Sato, *Chem. Lett.*, 2095 (1992).

141. P. D. Jeffery and S. W. McCombie, *J. Org. Chem.*, **47**, 587 (1982).

142. H. Kunz and H. Waldman, *Angew. Chem., Int. Ed. Engl.*, **23**, 71 (1984).

143. H. Kunz and C. Unverzagt, *Angew. Chem., Int. Ed. Engl.*, **23**, 436 (1984).

144. H. Kunz and J. März, *Angew. Chem., Int. Ed. Engl.*, **27**, 1375 (1988).

145. M. Honda, H. Morita and I. Nagakura, *J. Org. Chem.*, **62**, 8932 (1997).

146. Y. Hayakawa, S. Wakabayashi, T. Nobori and R. Noyori, *Tetrahedron Lett.*, **28**, 2259 (1987).

147. H. X. Zhang, F. Guibe and G. Balavoine, *Tetrahedron Lett.*, **29**, 623 (1988).

148. Y. Hayakawa, S. Wakabayashi, H. Kato and R. Noyori, *J. Am. Chem. Soc.*, **112**, 1691 (1990).

149. F. Guibe, *Tetrahedron*, **53**, 13509 (1997).

150. M. Kloosterman, J. H. Boom, P. Chatelard, P. Boullanger and G. Descotes, *Tetrahedron Lett.*, **26**, 5045 (1985).

151. F. Guibe and Y. S. M'Leux, *Tetrahedron Lett.*, **22**, 3591 (1981).

152. F. Guibe, O. Dangles and G. Balavoine, *Tetrahedron Lett.*, **27**, 2368 (1986).

153. T. Yamada, K. Goto, Y. Mitsuda and J. Tsuji, *Tetrahedron Lett.*, **28**, 4557 (1987).

154. I. Minami, Y. Ohashi, I. Shimizu and J. Tsuji, *Tetrahedron Lett.*, **26**, 2449 (1985).

155. O. Dangles, F. Guibe, G. Balavoine, S. Lavielle and A. Marquet, *J. Org. Chem.*, **52**, 4984 (1987).

156. H. Kunz, *Angew. Chem., Int. Ed. Engl.*, **26**, 294 (1987).

157. Y. Hayakawa, H. Kato, M. Uchiyama, H. Kajino and R. Noyori, *J. Org. Chem.*, **51**, 2402 (1986).

158. F. Garro-Helion, A. Merzouk and F. Guibe, *J. Org. Chem.*, **58**, 6109 (1993).

159. S. Lemaire-Audoire, M. Savignac, J. P. Genet and J. M. Bernard, *Tetrahedron Lett.*, **36**, 1267 (1995).

160. D. Zhang and L. S. Liebeskind, *J. Org. Chem.*, **61**, 2594 (1996), W. F. Bailey and X. L. Jiang, *J. Org. Chem.*, **61**, 2596 (1996).

161. S. J-Figueroa, Y. Liu, J. M. Muchowski and D. G. Putman, *Tetrahedron Lett.*, **39**, 1313 (1998).

162. T. Taniguchi and K. Ogasawara, *Tetrahedron Lett.*, **39**, 4679 (1998).

163. E. C. Roos, P. Bernabe, H. Hiemstra, W. N. Speckamp, B. Kaptein and W. H. J. Boesten, *J. Org. Chem.*, **60**, 1733 (1995).

164. J. Tsuji, T. Yamakawa, M. Kaito and T. Mandai, *Tetrahedron Lett.*, 2075 (1978).

165. T. Mandai, H. Yasuda, M. Kaito, J. Tsuji, R. Yamaoka and H. Fukami, *Tetrahedron*, **35**, 309 (1979).

166. H. Kumobayashi, S. Mitsuhashi, S. Akutagawa and S. Otsuka, *Chem. Lett.*, 157 (1986).

167. A. J. Chark, V. Wertheimer and S. A. Magennis, *J. Mol. Catal.*, **19**, 189 (1983).

168. S. Suzuki, Y. Fujita and T. Nishida, *Tetrahedron Lett.*, **24**, 5373 (1983).

169. T. Mandai, T. Matsumoto, Y. Nakao, H. Teramoto, M. Kawada and J. Tsuji, *Tetrahedron Lett.*, **33**, 2549 (1992).

170. J. M. Takacs, E. C. Lawson and F. Clement, *J. Am. Chem. Soc.*, **119**, 5956 (1997).

171. M. Suzuki, Y. Oda and R. Noyori, *J. Am. Chem. Soc.*, **101**, 1623 (1979).

172. B. M. Trost and J. M. Fortunak, *J. Am. Chem. Soc.*, **102**, 2841 (1980), *Tetrahedron Lett.*, **22**, 3459 (1981).

173. H. Harayama, T. Kuroki, M. Kimura, S. Tanaka and Y. Tamaru, *Angew. Chem., Int. Ed. Engl.*, **36**, 2352 (1997).

174. J. Tsuji, I. Minami and I. Shimizu, *Tetrahedron Lett.*, **25**, 2791 (1984).

175. A. S. Smith, G. A. Sulikowski and K. Fujimoto, *J. Am. Chem. Soc.*, **111**, 8039 (1989).

176. I. Minami, M. Yamada and J. Tsuji, *Tetrahedron Lett.*, **27**, 1805 (1986); I. Minami and J. Tsuji, *Tetrahedron*, **43**, 3903 (1987).

177. I. Shimizu, T. Yamada and J. Tsuji, *Tetrahedron Lett.*, **21**, 3199 (1980).

178. T. Tsuda, Y. Chujo, S. Nishi, K. Tawara and T. Saegusa, *J. Am. Chem. Soc.*, **102**, 6381 (1980).

179. I. Shimizu, Y. Ohashi and J. Tsuji, *Tetrahedron Lett.*, **24**, 3865 (1983).

180. I. Shimizu and J. Tsuji, *J. Am. Chem. Soc.*, **104**, 5844 (1982). I. Minami, M. Nisar, M. Yuhara, I. Shimizu and J. Tsuji, *Synthesis*, 992 (1987).

181. H. Kataoka, T. Yamada, K. Goto and J. Tsuji, *Tetrahedron*, **43**, 4107 (1987).

182. J. Tsuji, M. Nisar, I. Shimizu and I. Minami, *Synthesis*, 1009 (1984).

183. C. Mercier, G. Mignani, M. Aufrand and G. Allmang, *Tetrahedron Lett.*, **32**, 1433 (1991).

184. J. Tsuji, M. Nisar and I. Minami, *Tetrahedron Lett.*, **27**, 2483 (1986); *Chem. Lett.*, 23 (1987).

185. J. Tsuji, M. Nisar and I. Shimizu, *J. Org. Chem.*, **50**, 3416 (1985).

186. A. Murai, N. Tanimoto, N. Sakamoto and T. Masamune, *J. Am. Chem. Soc.*, **110**, 1985 (1988).

187. T. Mandai, M. Imaji, H. Takeda, M. Kawada, J. Nokami and J. Tsuji, *J. Org. Chem.*, **54**, 5395 (1989).

188. T. Murayama, A. Yoshida, T. Kobayashi and T. Miura, *Tetrahedron Lett.*, **35**, 2271 (1994).

189. J. Nokami, T. Mandai, H. Watanabe, H. Ohyama and J. Tsuji, *J. Am. Chem. Soc.*, **111**, 4126 (1989).

190. J. Nokami, H. Watanabe, T. Mandai, M. Kawada and J. Tsuji, *Tetrahedron Lett.*, **30**, 4829 (1989).

191. W. E. Walker, R. M. Manyik, K. E. Atkins and M. L. Farmer, *Tetrahedron Lett.*, 3817 (1970).
192. J. Tsuji, K. Tsuruoka and K. Yamamoto, *Bull. Chem. Soc. Jpn.*, **49**, 1701 (1976).
193. K. Inomata, Y. Murata, H. Kato, Y. Tsukahara, H. Kinoshita and H. Kotake, *Chem. Lett.*, 931 (1985).
194. L. E. Overman and F. M. Knoll, *Tetrahedron Lett.*, 321 (1979); L. E. Overman, *Angew. Chem., Int. Ed. Engl.*, **23**, 579 (1984).
195. P. A. Grieco, T. Takigawa, S. L. Bongers and H. Tanaka, *J. Am. Chem. Soc.*, **102**, 7587 (1980).
196. S. J. Danishefsky, M. P. Cabal and K. Chow, *J. Am. Chem. Soc.*, **111**, 3456 (1989).
197. H. Abe, H. Nitta, A. Mori and S. Inoue, *Chem. Lett.*, 2443 (1992).
198. J. Genda, A. C. Helland, B. Ernst and D. Bellus, *Synthesis*, 729 (1993).
199. R. P. Lutz, *Chem. Rev.*, **84**, 205 (1984).
200. L. E. Overman and F. M. Knoll, *J. Am. Chem. Soc.*, **102**, 865 (1980). L. E. Overman and A. F. Lenaldo, *Tetrahedron Lett.*, **24**, 3757 (1983).
201. N. Bluthe, M. Malacria and J. Gore, *Tetrahedron Lettt.;* **24**, 1157 (1983).
202. L. E. Overman and A. F. Renaldo, *J. Am. Chem. Soc.*, **112**, 3945 (1990).
203. K. Mikami, K. Takahashi and T. Nakai, *Tetrahedron Lett.*, **28**, 5879 (1987).
204. M. Oshima, M. Murakami and T. Mukaiyama, *Chem. Lett.*, 1535 (1984).
205. H. A. J. Carless and D. J. Haywood, *Chem. Commun.*, 980 (1980).
206. E. J. Corey and J. W. Suggs, *J. Org. Chem.*, **38**, 3224 (1973); P. A. Gent and R. Grigg, *Chem. Commun.*, 277 (1974).
207. H. Kumobayashi, S. Akutagawa and S. Otsuka, *J. Am. Chem. Soc.*, **100**, 3949 (1978).
208. K. Tani, T. Yamagata, S. Akutagawa, H. Kumobayshi, T. Taketomi, H. Takaya, A. Miyashita, R. Noyori and S. Otsuka, *Chem. Commun.*, 600 (1982); K. Tani, T. Yamagata, S. Akutagawa, H. Kumobayshi, T. Taketomi, H. Takaya, A. Miyashita, R. Noyori and S. Otsuka, *J. Am. Chem. Soc.*, **106**, 5208 (1984); K. Tani, T. Yamagata, Y. Tatsuno, Y. Yamagata, K. Tomita, S. Akutagawa, H. Kumobayashi and S. Otsuka, *Angew. Chem., Int. Ed. Engl.*, **24**, 217 (1985).
209. H. Kumobayashi, *Recl. Trav. Chim. Pays-Bas*, **115**, 201 (1996), S. Akutagawa, *Appl. Catal. A*, **128**, 171 (1995); S. Otsuka and K. Tani, *Synthesis*, 665 (1991).
210. S. Inoue, H. Takaya, K. Tani, S. Otsuka, T. Sato and R. Noyori, *J. Am. Chem. Soc.*, **112**, 4897 (1990); M. Yamakawa and R. Noyori, *Organometallics*, **11**, 3167 (1992).
211. D. Djahanbuni, B. Cazes and J. Gore, *Tetrahedron Lett.*, **25**, 203 (1984), *Tetrahedron*, **43**, 3441 (1987); B. Cazes, D. Djahanbini, J. Gore, J. P. Genet and J. M. Gaudin, *Synthesis*, 983 (1988).
212. H. Kleijn, H. Wertmijze, J. Meijer and P. Vermeer, *Recl. Trav. Chim. Pays-Bas*, **102**, 378 (1983).
213. J. Nokami, A. Maihara and J. Tsuji, *Tetrahedron Lett.*, **31**, 5629 (1990).
214. Y. Imada, G. Vasaqpollo and H. Alper, *J. Org. Chem.*, **61**, 7982 (1996).

5

REACTIONS OF CONJUGATED DIENES

Reaction of conjugated dienes with aryl and alkenyl halides is treated in Section 3.2 and oxidative difunctionalization of conjugated dienes with Pd(II) is treated in Section 11.3. This chapter covers other reactions of conjugated dienes as major reactants.

Polymerizations of butadiene and isoprene catalysed by Ti or Ni complexes are large-scale industrial processes, and polybutadiene and polyisoprene as synthetic rubbers are produced. The catalysts are prepared by the treatment of Ni or Ti compounds with alkylaluminiums, and slight modification of the catalysts induces 1,4-polymerization or 1,2-polymerization. Also, *cis* or *trans* polymer is obtained selectively by changing the catalysts. Although the relationship between nature of the catalysts and structures of the polymers is interesting, only oligomerization of conjugated dienes is treated in this chapter, because oligomers are more useful in organic synthesis.

5.1 Formation of Cyclic Oligomers by Cycloaddition

Several cyclic oligomers **1–5** are prepared from butadiene using transition metal catalysts. The preparation of 1,5-cyclooctadiene (**3**; 1,5-COD) by a catalyst prepared from Ni(CO)$_4$ and phosphine is the first report on cyclooligomerzation of butadiene [1]. However, the activity of this catalyst is low due to strong coordination of CO. Catalyst prepared from TiCl$_4$ and Et$_3$Al has higher catalytic activity for the formation of 1,5-COD and 1,5,9-cyclododecatriene (1,5,9-CDT; **4**). Also Ni(0) catalysts are active for the preparation of COD and CDT. In addition to COD and CDT, the cyclic

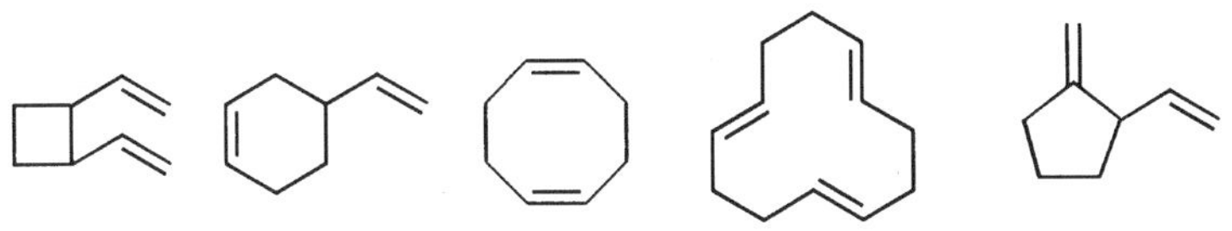

    1   **2**   **3**   **4**   **5**

compounds **1, 2** and **5** are formed [2]. The cyclic compounds **1–5** are prepared with high selectivity using proper catalysts. Also, cocyclizations with alkynes and alkenes give cylic compounds with various ring sizes [2].

Cyclization of butadiene catalysed by Ni(0) catalysts proceeds via π-allylnickel complexes. At first, the metallacyclic bis-π-allylnickel complex **6**, in which Ni is bivalent, is formed by oxidative cyclization. The bis-π-allyl complex **6** may also be represented by σ-allyl structures **7, 8** and **9**. Reductive elimination of **7, 8** and **9** produces the cyclic dimers **1, 2** and **3** by [2+2], [2+4] and [4+4] cycloadditions. Selectivity for **1, 2** and **3** is controlled by phosphine ligands. The catalyst made of a 1:1 ratio of Ni and a phosphine ligand affords the cyclic dimers **1, 2** and **3**. In particular, **1** and **3** are obtained selectively by using the bulky phosphite **11**. 1,2-Divinylcyclobutane (**1**) can be isolated only at a low temperature, because it undergoes facile Cope rearrangement to form 1,5-COD on warming. Use of tricyclohexylphosphine produces 4-vinylcyclohexene (**2**) with high selectivity.

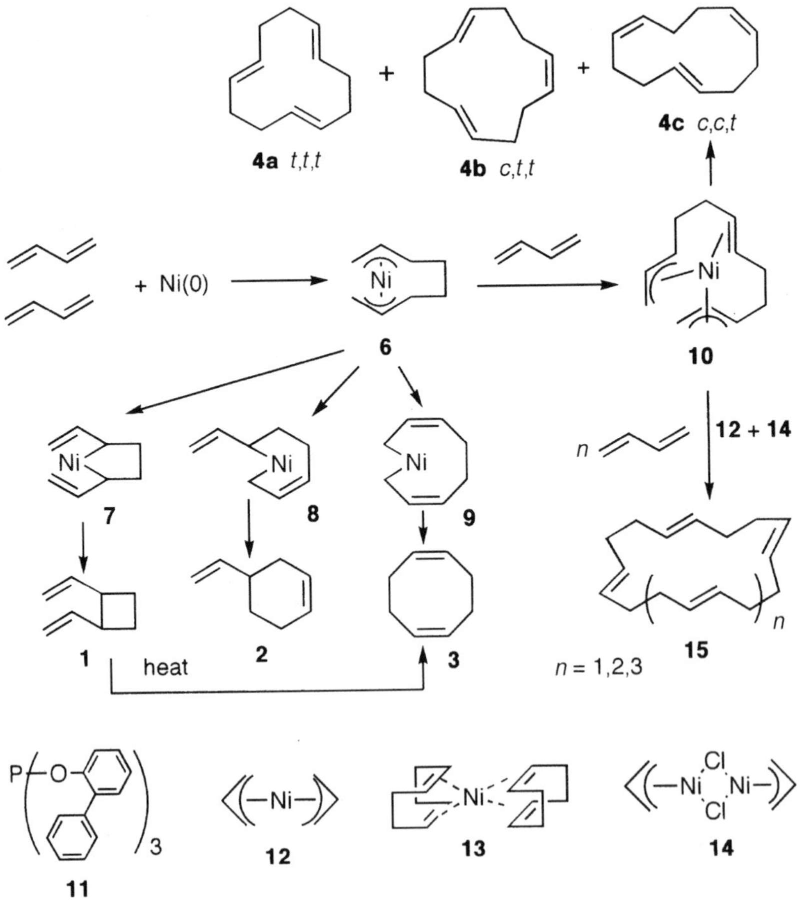

A very active form of Ni(0) is generated from bis-π-allylnickel (**12**) and Ni(cod)₂ (**13**) in the absence of phosphine ligand, and is often called naked Ni(0). Three butadienes coordinate to naked Ni(0) and the 18-electron trimeric complex **10** is formed. 1,5,9-CDT (**4**) is formed by reductive elimination of **10**. However, further ring expansion occurs to give 16, 20, 24 and higher membered macrocyclic polyalkenes **15** when the catalyst prepared by mixing naked Ni(0) **13** and π-allylnickel chloride (**14**) is used. This is an interesting cyclization reaction, although its usefulness is limited because selectivity is not high [3]. Continuing insertion of butadiene finally forms polybutadiene, which may be huge macrocycles. Classical synthesis of macrocyclic compounds is usually carried out under high-dilution conditions, which is not an efficient method. It should be emphasized that transition metal-catalysed cyclization of conjugated dienes, and cocyclization with alkenes and alkynes proceed efficiently without high dilution. This is due to the template effect of metal complexes.

In the cyclization reactions mentioned above, an Ni−H species is not involved and hence no migration of hydrogen occurs. However, 1-methylene-2-vinylcyclopentane (**5**) is formed using the catalyst prepared by the reduction of NiCl$_2$ with NaBH$_4$ in the presenc of Bu$_3$P [4]. Addition of alcohol is essential. In this reaction, the π-allyl alkene complex **16** is formed by insertion of butadiene to the Ni−H bond. Subsequent intramolecular insertion of terminal alkene in **16** gives **17**, and elimination of β-hydrogen affords **5** and regenerates the Ni−H species. This reaction is related to the metalla-ene reaction (see Section 4.3.6), and **5** is prepared by the Pd-catalysed cyclization of 1-acetoxy-2,7-octadiene (**18**).

1,5,9-CDT is produced commercially and used for production of 12-nylon. Interesting synthetic reactions have been reported based on the modification of CDT- and COD-forming reactions. Di- and tetraazacyclododecatrienes **20** and **21** were prepared by cooligomerization of the diazadiene **19** and butadiene [5,6].

Interesting synthetic applications of the [4+4] and [4+4+4] cycloadditions are reported. A novel, short-step synthetic method of muscone (**24**) has been developed using complex **10** as a starting compound [7]. Insertion of allene to the Ni−carbon bond in **10** at low temperature gives the bis-π-allylnickel **22**. Then isonitrile is inserted to **22**. When the reaction mixture is warmed, the 15-membered cyclic compound **23** is formed by reductive elimination, and conversion of **23** to muscone (**24**) is achieved by hydrolysis and subsequent hydrogenation in 43% overall yield.

Intramolecular [4+4] cycloaddition proceeds smoothly and is a useful reaction. The skeleton **26** of asteriscanolide (**27**) was constructed by utilizing Ni-catalysed

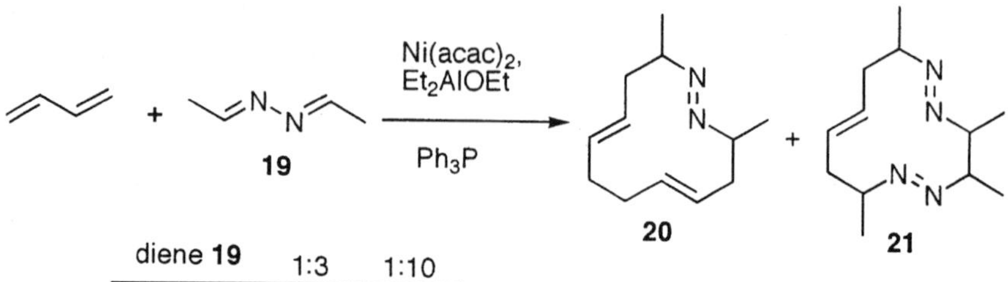

diene 19	1:3	1:10
yield	96%	100%
20:21	68:32	6:94

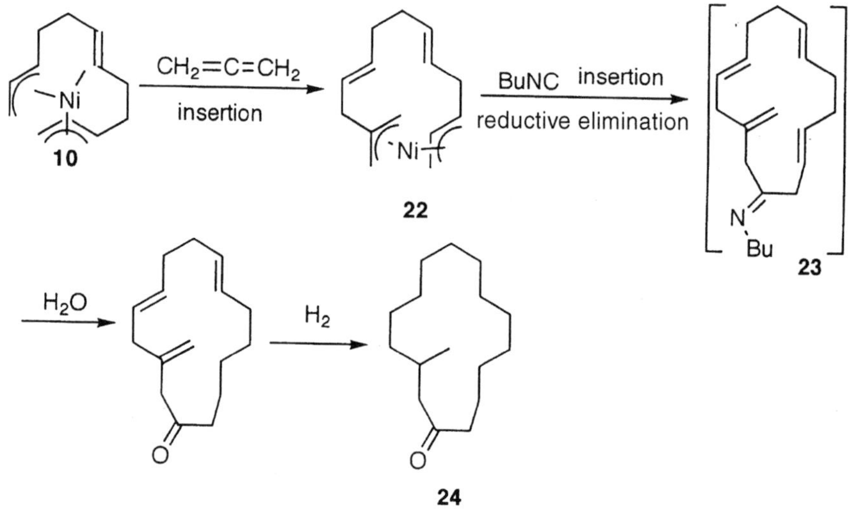

intramolecular [4+4] cycloaddition of the tetraene **25** as a key reaction [8]. The eight-membered ring skeletons **29a** and **29b** as partial structures of the taxane molecule are formed by the intramolecular reaction of **28** using Ni(0) coordinated by tri(o-phenylphenyl) phosphite (**11**) as a catalyst [9].

Using the diazadiene Fe complex **31**, derived from optically active menthol, as a catalyst, enantioselective [4+4] cocycloaddition of isoprene and piperylene (**30**) occurred to give 3,5-dimethyl-1,5-cyclooctadine (**32**) with 61% ee in 89% yield without forming homodimers [10].

Three isomers of the substituted cyclobutanes **33, 34** and **35** are obtained by the [2+2] cycloaddition of isoprene, and separated at low temperature from other cooligomers without undergoing Cope rearrangement. When the mixture was subjected to hydroboration and oxidation, the alcohol **37** was obtained from the isomer **34**, and easily separated from **35** and the diol **36**. The alcohol **37** is a pheromone called grandisol [11]. Although overall yield was 15%, this is the shortest synthetic route to this pheromone.

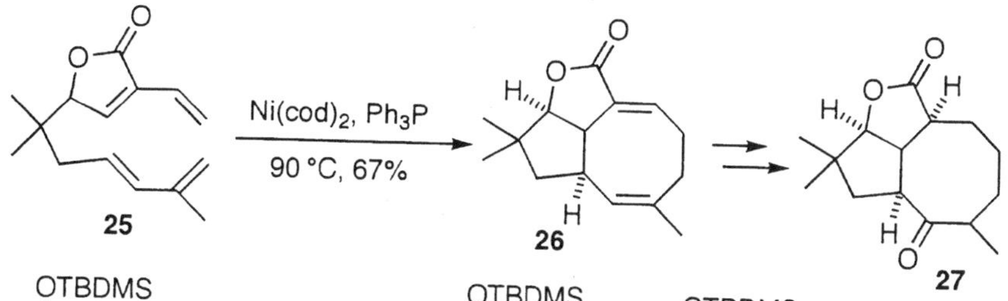

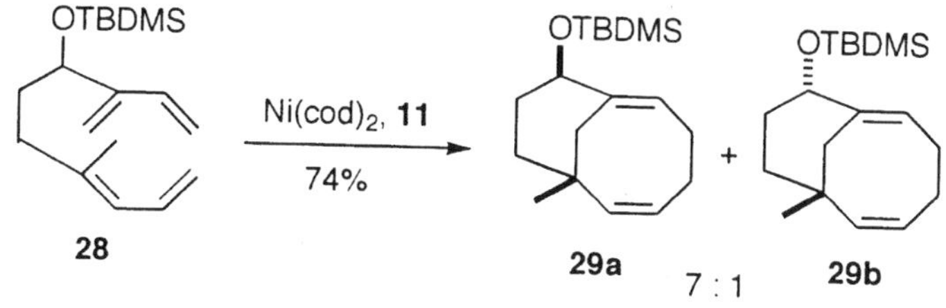

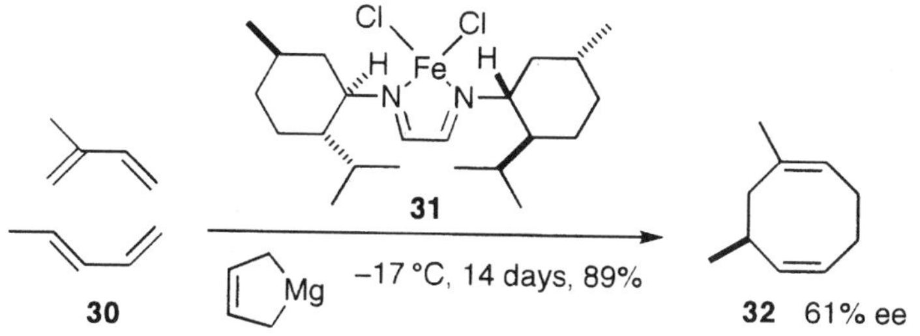

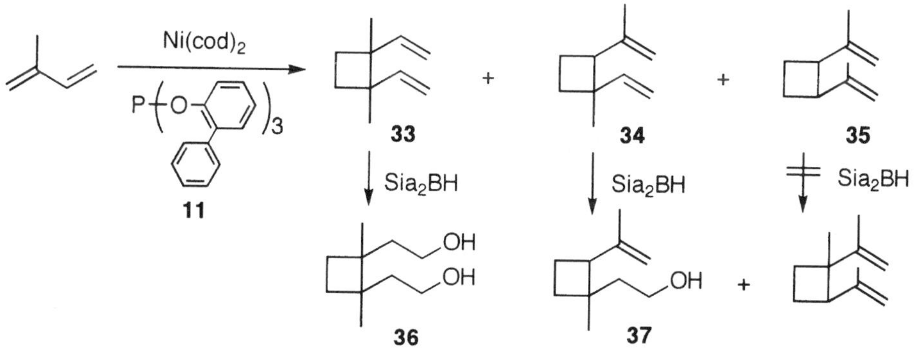

In addition to homooligomerization, butadiene undergoes cooligomerization with unsaturated bonds of alkenes and alkynes offering simple synthetic methods of cyclic compounds of various sizes. Even C=O and C=N bonds participate in the cooligomerization. 1,5-Cyclodecadiene (**38**) is formed by the [4+4+2] cycloaddition of ethylene and butadiene in 1:2 ratio using the Ni catalyst coordinated by $(PhO)_3P$ [2d]. The reaction of ethylene and butadiene in 1:1 ratio is catalysed by a Ti complex having bipyridyl ligand, and produces vinylcyclobutane (**41**) [12]. In this reaction, π-methallyltitanium **39** is formed by insertion of butadiene to Ti−H species. Subsequent insertion of ethylene and intramolecular alkene insertion generate the cyclobutane **40**. β-Elimination gives vinylcyclobutane (**41**) besides the regenerated Ti–H species.

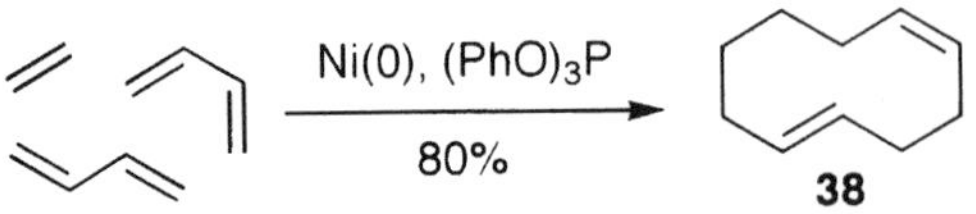

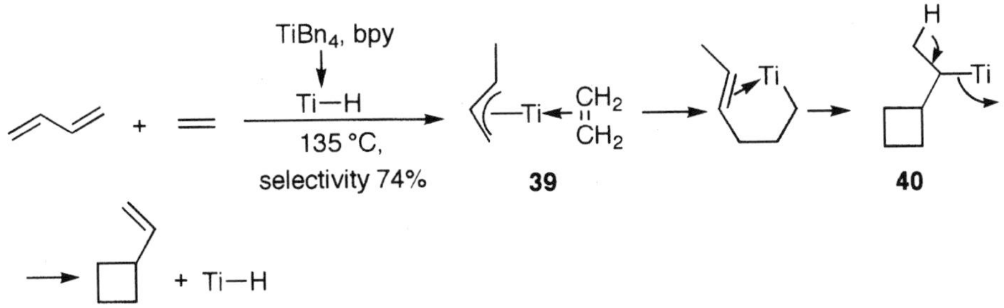

The 10-membered 1,2-disubstituted cyclodecatriene **42** can be prepared in high yield by the cooligomerization of alkyne with butadiene in 1:2 ratio. Selective hydrogenation to give **43** and its ozonization afford the linear diketone **44** [13]. The 20-membered cyclic diketone **47** was prepared from **46**, which was obtained by the cooligomerization of cyclododecyne (**45**) with butadiene. Selective hydrogenation of two disubstituted double bonds in **46** and ozonization of the remaining double bond afford **47**.

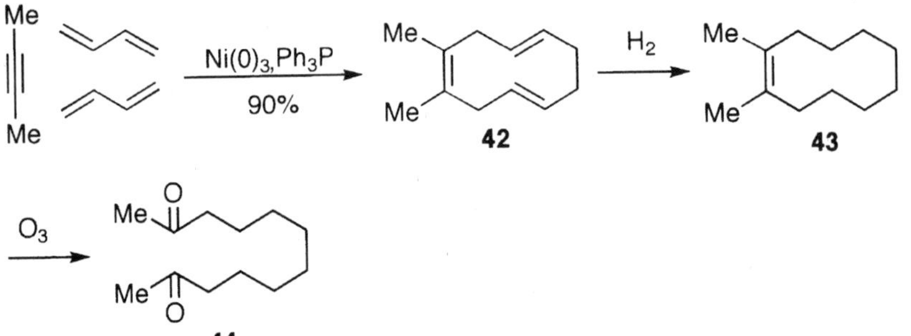

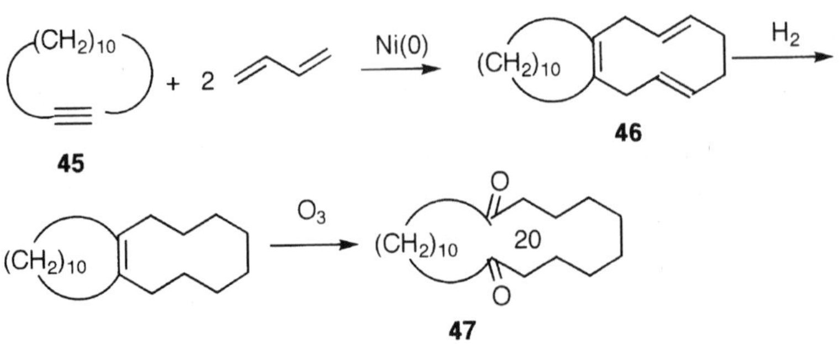

Under high concentration of alkynes, the cooligomerization of alkyne and butadiene in 2:2 ratio takes place using the Ni(0)catalyst, coordinated by $(PhO)_3P$, to give the 12-membered cyclic tetraenes **48** [13]. Acetylene and butadiene react in 2:1 ratio to give 5-vinyl-1,3-cyclohexadiene (**49**) using the Ni complex of Bu_3P [14].

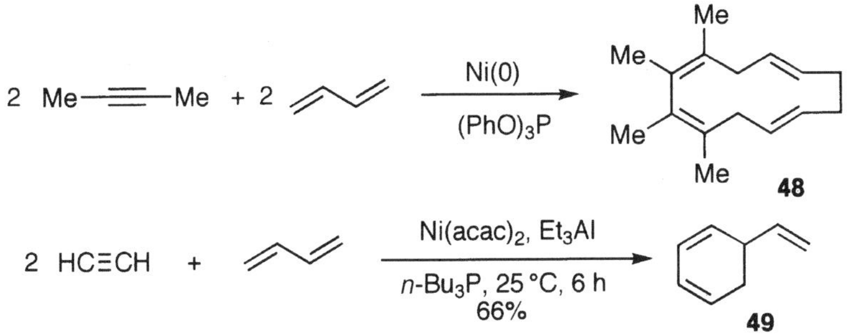

Alkynes are poor dienophiles in the Diels–Alder reaction; decomposition occurs by an attempted thermal intramolecular Diels–Alder reaction of dienynes at 160 °C. In contrast, the Ni-catalysed [4+2] cycloaddition of the dienyne **50** proceeded smoothly at room temperature using tri(hexafluoro)isopropyl phosphite to give **51**, which was converted to the yohimbine skeleton **52** [15]. The same reaction is catalysed by $RhCl(Ph_3P)_3$ in trifluoroethanol [16]. Intramolecular Diels–Alder reactions of the 6,8-dieneyne **53** and the 1,3,8-triene **55**, efficiently catalysed by $[Rh(dppe)(CH_2CH_2)_2]SbF_6$ at room temperature, gave **54** and **56** [17].

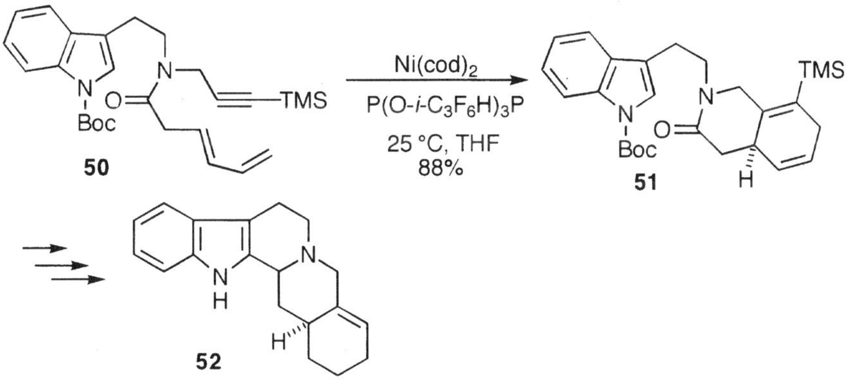

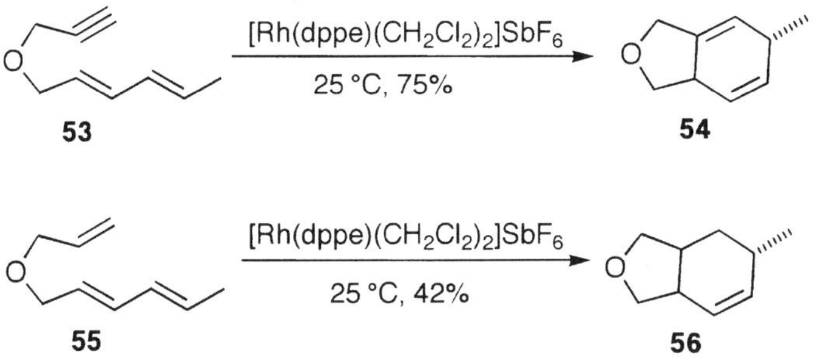

The Rh-catalysed intermolecular [4+2] cycloaddition of a diene and a terminal alkyne affords the 1,4-disubstituted-1,3-cyclohexadiene **58** via **57** [18]. The 1,2,6-trisubstituted-1,4-cyclohexadiene **60** is obtained by the Fe-catalysed [4+2] cycloaddition of dienes and internal alkynes. The catalyst **59** is prepared by the reaction of a 1,4-diaza-1,3-diene complex of FeCl₂ **59** with EtMgBr [19].

As a related reaction, the bicyclo[5.3.0]decane derivative **64** was obtained at 30 °C by the Rh-catalysed intramolecular [5+2] cycloaddition of the alkyne with the vinylcyclopropane moiety in **61**. The latter behaves as a pseudo-1,3-diene in oxidative addition, and generates **62**. This is followed by rearrangement to **63**, whose reductive elimination gives **64** [20]. [Rh(CO)₂Cl]₂ is a better catalyst than RhCl(Ph₃P)₃. The reaction can be extended to alkenes [20a].

Ni- and Rh-catalyzed intramolecular [4+2] cycloadditions of the diene-allene **65** using tri(*o*-phenylphenyl) phosphite gave the different products **66** and **67** [21].

No homocyclization of butadiene is possible with a Pd catalyst, because easy hydrogen transfer occurs with the Pd catalyst to give linear oligomers. However, Pd-catalysed cocyclization occurs with some C=O and C=N bonds to give the heterocycles **68**. The divinyltetrahydropyran **71** and the unsaturated alcohol **72** are obtained by the reaction of aldehydes [22–24]. The unsaturated alcohol **72** is obtained as a main product when a ratio of Ph$_3$P and Pd is 1–2, and pyran **71** is the main product when the ratio is larger than 3. Formation of the pyran **71** and the unsaturated alcohol **72** is interesting, because benzaldehyde, an electrophile, reacts with the bis-π-allylpalladium intermediate **68** formed from buadiene. The reaction can be understood by the amphiphilic nature of the bis-π-allylpalladium **68**. The nucleophilic attack of one of the allyl group in **68** to the carbonyl group of benzaldehyde generates **70**. Subsequent electrophilic attack (or reductive elimination) of the remaining π-allylpalladium in **70** affords the pyran **71**, which is favoured by a higher ratio of Ph$_3$P to Pd, and the alcohol **72** is formed by β-elimination of the allyl group when the ratio of Ph$_3$P is small.

In contrast, Ni(0)-catalysed intramolecular reaction of the diene aldehyde **73** and ketones with hydrosilane proceeds smoothly to give five-, six- and seven-membered alcohols. The π-allylnickel species **75** is formed by the reaction of **73** with Ni hydride **74**, generated by the oxidative addition of hydrosilane. It is well-known that π-allylnickel reacts with carbonyl as an electrophile (or insertion of carbonyl) to afford

Ni alkoxide as a stoichiometric reaction. However, in this reaction the silylnickel alkoxide **76** is formed, and its reductive elimination affords silyl ethers **77** and **78**. At the same time, Ni(0) is regenerated to make the reaction catalytic. The formal total synthesis of elaeokanine C (**79**) was carried out by this reaction [25]. Homoallylic alcohol **81** is obtained by the intermolecular reaction of benzaldehyde with the diene **80** and hydrosilane in high regio- and stereoselective manner [26].

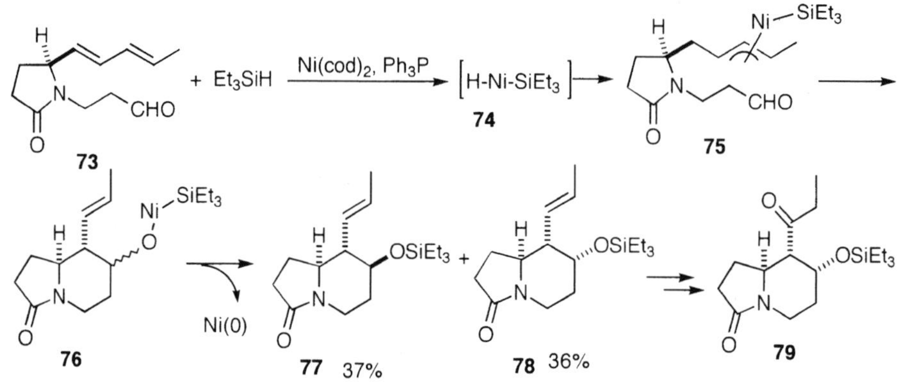

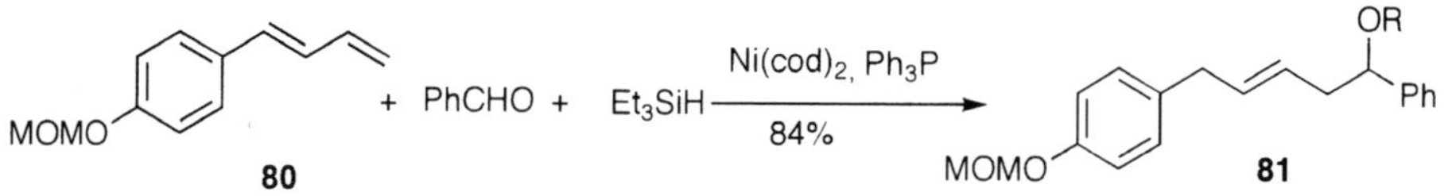

The C=N bonds of isocyanates [27] and Schiff bases **83** [28] react with butadiene to give the piperidone **82** and piperidines **84**. The nucleophilic attack of CO_2 to the amphiphilic bis-π-allylpalladium **68** generates the π-allylpalladium carboxylate **85**, from which the six-membered lactone **86** and five-membered lactone **87** are obtained under certain conditions [29–31]. The unsaturated ester **88** is also formed.

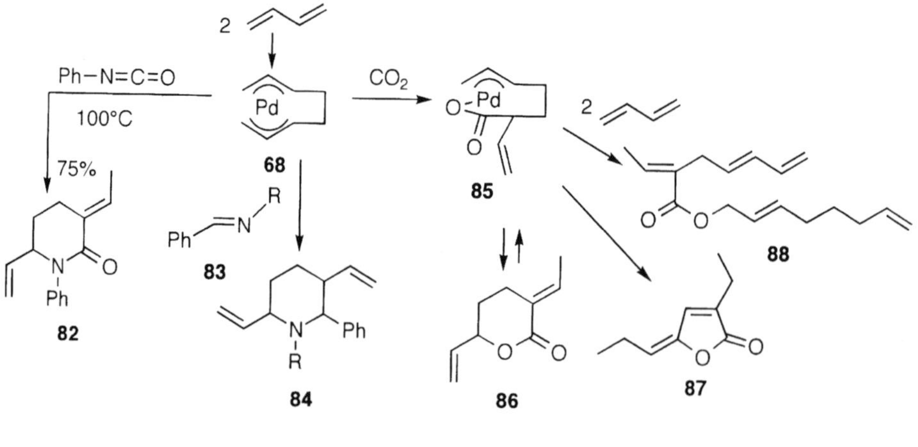

5.2 Formation of Linear Oligomers and Telomers

The linear dimers **89–91** are formed by Ni [32], Co [33], Fe [34] and Pd [35] catalysts. Linear dimers **90** and **91** are produced via the formation of metal M−H, accompanying migration of hydrogen. The formation of **89** is discussed later. The mechanism of the formation of **91** was studied by an experiment using butadiene **92** deuterated at the terminal carbons. In the formation of the branched dimer **91** from the deuterated butadiene **92**, catalysed by Co or Fe complexes, insertion of the second butadiene occurs at the substituted side of the π-allyl complex **93** to give **94**. Finally, the triene **96** is formed from **95** and Fe−H(D) is regenerated.

Dimerization of isoprene, if carried out regioselectively to give the head-to-tail dimer **97**, would be useful for terpene synthesis. So far few reports on successful regioselective dimerization of isoprene have been given, and this remains an unsolved problem. 2,6-Dimethyl-1,3,6-octatriene (**97**), the head-to-tail dimer, can be prepared with high selectivity using Zr [36], Ni [37,38] and Pd [39] catalysts. However, selective functionalization of this dimer to the terpene alcohol **98** is not easy.

89 Pd, Ni cat.   **90** Fe, Co cat.   **91** Co, Fe cat.

92   **93**   **94**

95   **96** + D−Fe

97   **98**

Several linear cooligomers of butadiene are prepared with alkenes and alkynes. Commercially important 1,4-hexadiene (**103**) is prepared by the reaction of ethylene and butadiene catalysed by Ni [40], Fe [41] and Rh [42]. The experiment carried out using deuterated ethylene (**100**) supports the mechanism that the insertion of butadiene to M−H forms the π-allyl complex **99**. Insertion of ethylene (**100**) to **99** gives **101**, and its β-elimination affords the cooligomer **102**, tetradeuterated at C-1,1,2,6 of **103**.

99 100 101 102 103

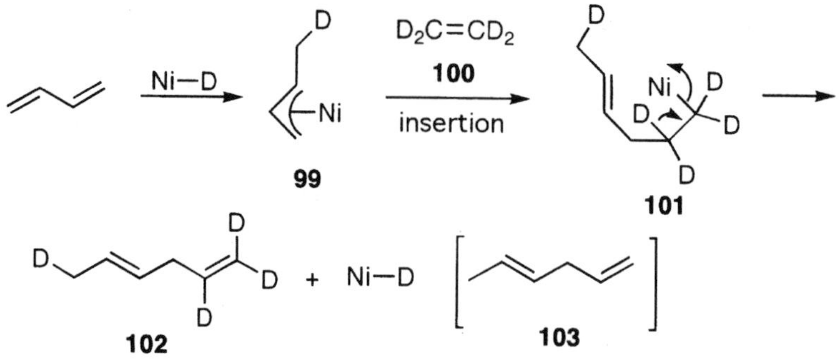

Functionalized alkenes are used for the cooligomerization. Phenyl-1,4,8-decatriene (**104**) is obtained by the Ni-catalysed 1:2 addition of styrene and butadiene [42a]. Pd catalyst affords the 1:1 adduct [43]. Co or Fe catalyst gives the 1:1 adducts **105** and **106** of methyl acrylate and butadiene [44,42a]. The 1:1 adducts **107** and **108** are obtained by the Ru-catalyzed coupling of butadiene and acrylamide [45]. Reaction of methyl methacrylate affords the 1:2 adduct **109** with Ni–Ph_3P catalyst at 0 °C, whereas the oligomer **110** is obtained at higher temperature [46].

104

105 106

107 108

109 110

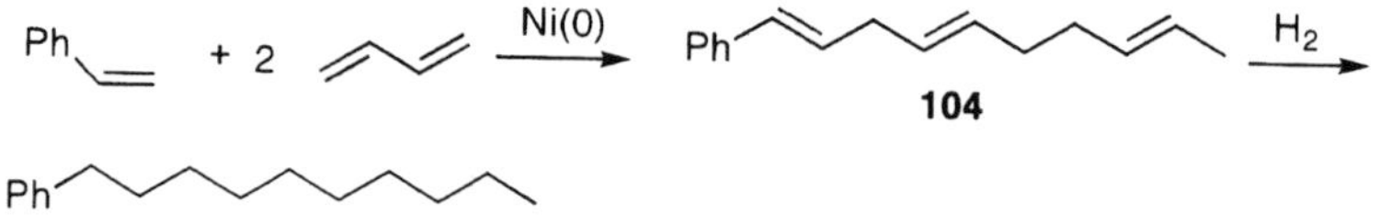

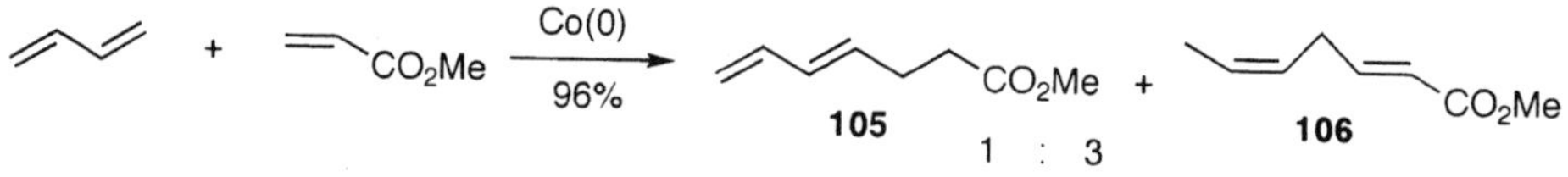

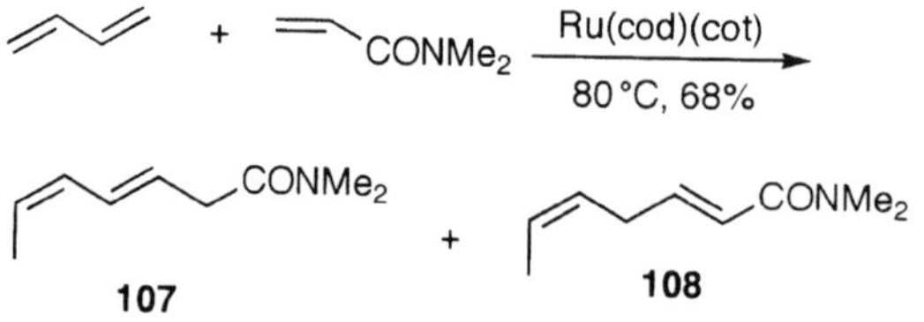

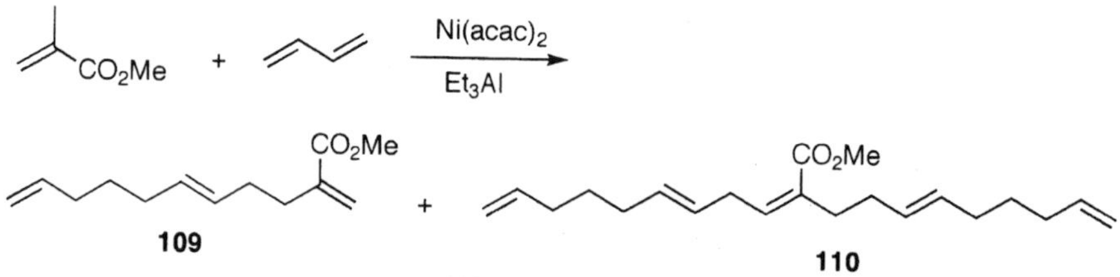

	109	110
Ph_3P, 20 °C, 120 h	6%	62%
Ph_3Sb, 80 °C, 2 h	1%	79%

Ene-type products are obtained by Co- and Fe-catalysed reaction of dienynes. Co-catalysed cyclization of substrate **111** proceeds smoothly with respect to the diene, acetylene and allylic ether moiety to afford **114**. In this cyclization, the π-allyl complex **112** is formed by insertion of the diene to Co–H, followed by domino insertions of the triple and double bonds to give **113**. The final step is the elimination of the β-alkoxide group from **113** to form **114** [47]. The six-membered ene-type products **117** and **118** are obtained from the reaction of **115** catalysed by an Fe bipyridyl complex. The reaction seems to involve oxidative cyclization to form **116**. Subsequent β-elimination and reductive elimination provide **117** and **118**. As another possibility, insertion of the diene to Fe–H gives a π-allyl complex. Then double bond insertion and β-elimination should give **117** and **118** [48].

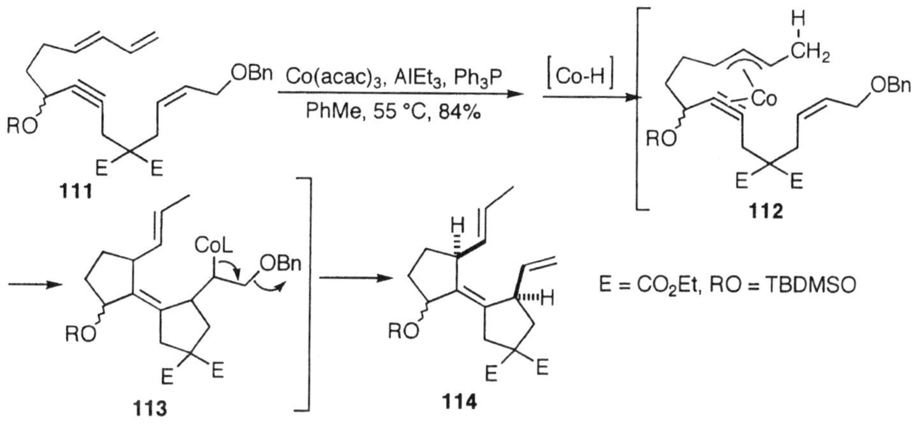

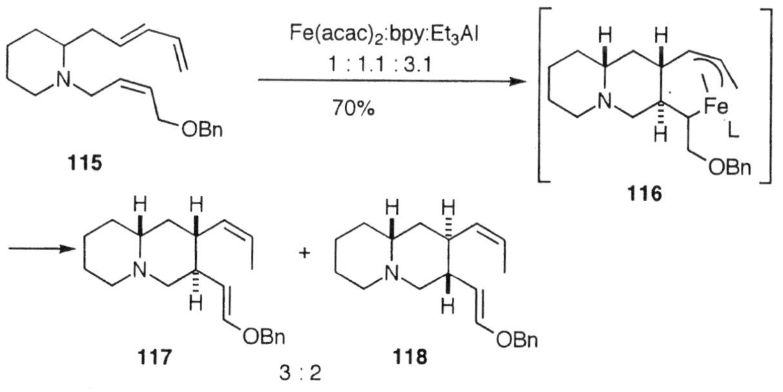

Dienylation of alkynes proceeds in the presence of Co catalysts, and the conjugated diene system is retained in the adducts. The conjugated trienes **120** and **121** are obtained from 2-butyne (**119**) and butadiene [44]. The dienylation occurs to give the conjugated 1,8-diphenyl-1,3,5,7-octatetraene (**122**) by attack of two phenylacetylene molecules to both side of butadiene [42a].

$$\text{Me}\!\!=\!\!\!=\!\!\!\text{Me} + \text{(diene)} \xrightarrow{\text{Co}} \text{120} + \text{121}$$

119   **120** 93%   **121** 7%

$$\text{Ph}\!\!=\!\!\!=\!\!\! + \text{(diene)} \xrightarrow{\text{Co}} \text{Ph}\text{~~~}\text{Ph}$$

122

The Ru−Bu₃P complex-catalysed addition of 1-hexyne to the functionalized diene **123** gives **126** and **127** via the π-allyl complex **125**, which is formed by insertion of the diene to the Ru alkynyl complex **124** [49].

124

125

126   38 : 62   **127**

Unique linear dimerization of butadiene is catalysed by a Pd−Ph₃P complex [50]. This Pd-catalysed linear dimerization affords 1,3,7-octatriene (**89**) by β-elimination of **128** and reductive elimination of **129**. Transfer of H from C(5) to C(3) in the intermediate bis-π-allylpalladium **68** occurs to give 1,3,7-octatriene (**89**).

More importantly the telomers **133** and **134** are formed by the reaction of pronucleophiles (NuH), incorporating the Nu group [51,52,53]. The reaction can be explained by the amphiphilic nature of bis-π-allylpalladium **68**. The first step is the nucleophilic attack of the amphiphilic bis-π-allylpalladium **68** to NuH (or protonation) as shown by **130** to generate π-allylpalladium **131**=**132**, and their electrophilic attacks as shown by **131** and **132** afford **133** and **134** [53a]. The reaction, carried out in deuterated MeOD, affords 1-methoxy-2,7-octadiene (**135**), deuterated at carbon 6 [54]. Pd(Ph₃P)₄ serves as the catalyst. Most conveniently, Pd(OAc)₂ and Ph₃P are used as the catalyst precursor. In this case, Pd(OAc)₂ is reduced *in situ* to Pd(0), which is an active species.

Telomerization with various nucleophiles affords interesting functionalized dimers **133** and **134** [50–54]. The Pd-catalyzed telomerizations of butadiene with various nucleophles gives 1-substituted 2,7-octadienes **133** as major products, and 3-substituted 1,7-octadienes **134** as minor products as summarized in Scheme 5.1. The telomers obtained by the Pd-catalysed reactions are useful building blocks. Natural products, such as steroids and macrolides, are synthesized efficiently using these telomers [55].

Formation of 2,7-octadien-1-ol (**136**) by the reaction of water attracts attention as a new commercial process for *n*-octanol (**155**). The Pd-catalysed reaction of water under usual conditions is very sluggish. The addition of CO_2 facilitates the telomerization of water, affording 2,6-octadien-1-ol (**136**) as the major product [56]. In the absence of CO_2, only 1,3,7-octatriene (**89**) is formed. Octadienyl carbonate is probably formed, which is than easily hydrolyzed to give **136**. A commercial process for the production of 2,7-octadien-1-ol (**136**) has been developed, which is operated in two phases – water and hexane – in the presence of carbonate salts. The water-soluble sulphonated phosphine ligand **XLVIII** or **154** is used [57]. The reaction proceeds in the aqueous phase and the product goes into the organic layer. The Pd catalyst always stays in the aqueous phase, and hence 2,7-octadien-1-ol (**136**) can be separated easily from the catalyst. 2,7-Octadien-1-ol (**136**) is used for the commercial production of 1-octanol (**155**) by hydrogenation. 1,9-Nonanediol (**156**) is also produced by isomerization of the allyl alcohol to the aldehyde, hydroformylation, and reduction.

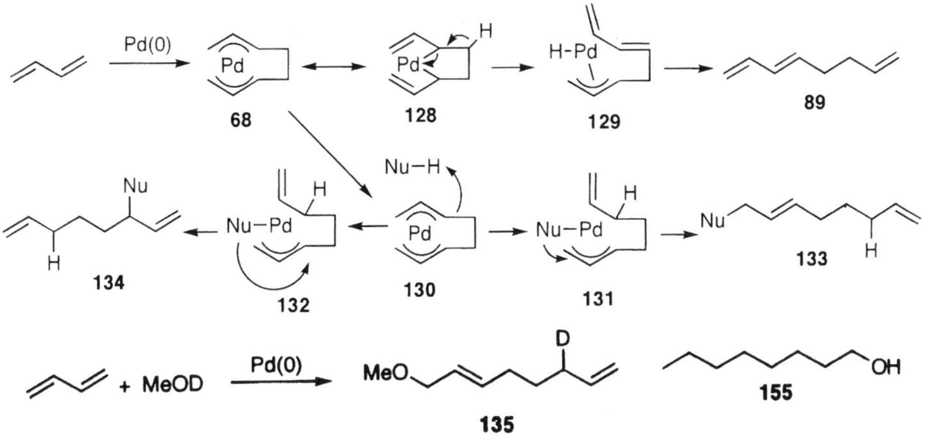

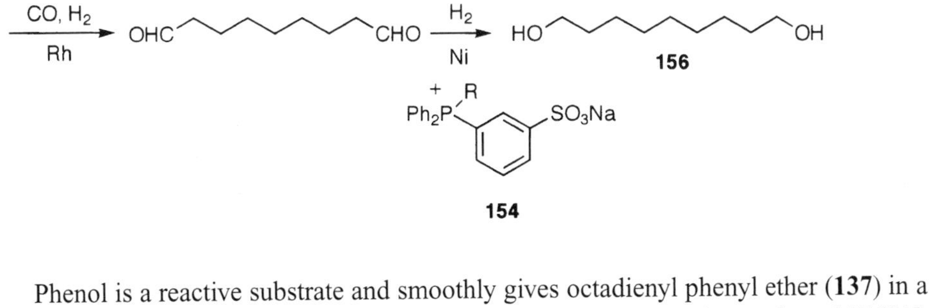

Phenol is a reactive substrate and smoothly gives octadienyl phenyl ether (**137**) in a high yield [52]. Primary alcohols such as MeOH react easily to form ethers **135** [51]. The higher the classes of alcohols, the lower the reactivity [58].

Carboxylic acids react with butadiene as alkali carboxylates. A mixture of isomeric 1- and 3-acetoxyoctadienes **138** and **139** is formed by the reaction of acetic acid [51].

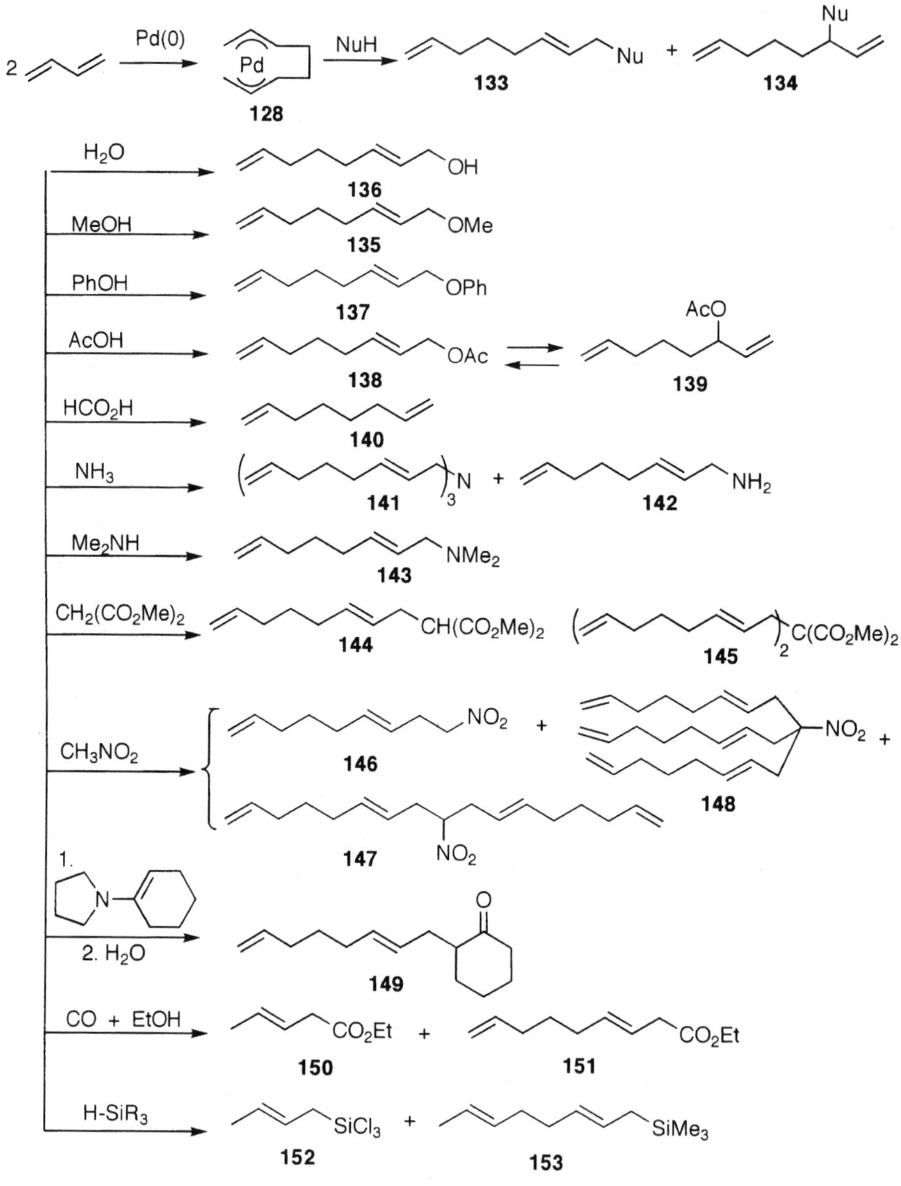

Nu

2 →(Pd(0))→ **128** (Pd) →(NuH)→ **133** Nu + **134**

H₂O → **136** OH

MeOH → **135** OMe

PhOH → **137** OPh

AcOH → **138** OAc ⇌ **139** AcO

HCO₂H → **140**

NH₃ → $\left(\quad\right)_3$N **141** + **142** NH₂

Me₂NH → **143** NMe₂

CH₂(CO₂Me)₂ → **144** CH(CO₂Me)₂ $\left(\quad\right)_2$C(CO₂Me)₂ **145**

CH₃NO₂ → **146** NO₂ + **148** NO₂ + **147** NO₂

1. [pyrrolidine-cyclohexenyl] 2. H₂O → **149**

CO + EtOH → **150** CO₂Et + **151** CO₂Et

H-SiR₃ → **152** SiCl₃ + **153** SiMe₃

Scheme 5.1 Butadiene telomers obtained by Pd-catalysed reactions

The reaction is very slow in acetic acid alone, and accelerated as acetate by the addition of bases [59]. These two isomers undergo Pd-catalysed allylic rearrangement with each other. 3-Acetoxy-1,7-octadiene (**139**) is converted to the allylic alcohol **157** and to the enone **158**, which is used as a bisannulation reagent [60]. Thus Michael addition of **158** to 2-methylcyclopentanedione (**159**) and aldol condensation give **160**. The terminal alkene is oxidized using PdCl₂/CuCl/O₂ to the methyl ketone **161**. After reduction of the double bond in **161**, aldol condensation affords the tricyclic system **162**.

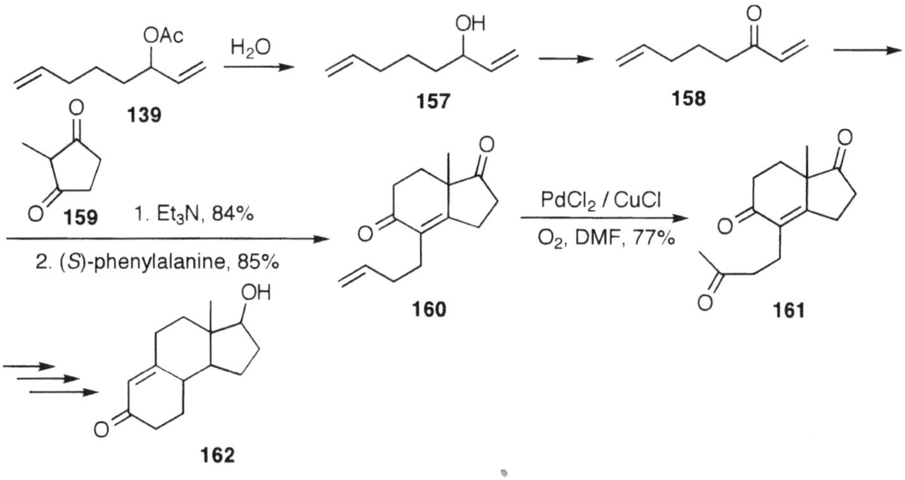

The telomer **138** is a good building block for the 10-membered skeleton of diplodialide (**166**) [61]. The terminal double bond in **138** is oxidized with PdCl$_2$/CuCl/O$_2$ to the methyl ketone **163** and converted to **164**. The phenylthioacetate **165** is prepared and its cyclization gives 9-decanolide (**166**).

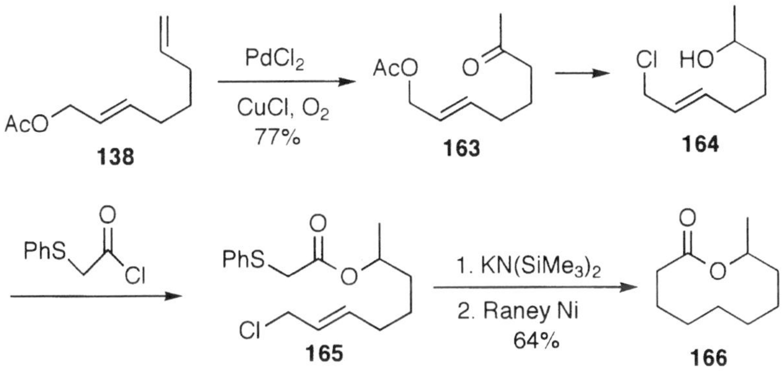

As shown in Scheme 5.1, formic acid behaves differently to other carboxylic acids. Expected octadienyl formate is not formed. The reaction with formic acid in the presence of Et$_3$N affords 1,7-octadiene (**140**) [62–64]. The first step is the protonation of bis-π-allylpalladium **68** with formic acid to generate **167**. Formic acid is a hydride source. It is known that the Pd hydride, formed by the decarboxylation of palladium formate, attacks the substituted side of π-allylpalladium as shown by **167** to form the terminal alkene **140** [65]. The regioselective attack of Pd—H at the more substituted side of π-allyl systems is covered in Section 4.3.7.

Reaction of aqueous ammonia (28%) with butadiene in MeCN in the presence of Pd(OAc)$_2$ and Ph$_3$P at 80 °C gives tri-2,7-octadienylamine (**141**) as the main product [66,67]. The reaction proceeds stepwise, but the primary amine **142** is more reactive than ammonia, and the secondary amine is more reactive than the primary amine. Thus the main product is the trioctadienylamine (**141**), even when the reaction is stopped before completion.

Both aromatic and aliphatic amines react with butadiene to give tertiary octadienylamines **143** [51]. Amines with higher basicity show higher reactivity, and electron-donating substituents on aniline have an accelerating effect [68].

Enamines, as nucleophiles, react with butadiene, and the 2-octadienyl ketones **149** or aldehydes are obtained after hydrolysis [69]. This is a good way of introducing the octadienyl group at the α-carbon of ketones or aldehydes, because butadiene does not react with ketones or aldehydes directly. Active methylene or methine compounds to which two EWGs, such as carbonyl, alkoxycarbonyl, formyl, cyano, nitro or sulfonyl groups are attached, react with butadiene smoothly and the acidic hydrogens are displaced with the 2,7-octadienyl group to give mono- and disubstituted compounds [53]. 3-Substituted 1,7-octadienes are obtained as minor products. β-Keto esters, β-diketones, malonates, α-formyl ketones, cyanoacetates, nitroacetates, cyanoaceta-mide and phenylsulfonylacetates react with butadiene smoothly. Di(octadienyl) malonate (**145**), obtained by this reaction, is converted to the interesting fatty acid **168** which has an acid function at the center of the long carbon chain.

Nitroalkanes react smoothly in *tert*-butyl alcohol as a solvent with butadiene, and their acidic hydrogens are displaced with the octadienyl group. From nitromethane three products, **146**, **147** and **148**, are formed, accompanied by 3-substituted 1,7-octadiene as a minor product. Hydrogenation of **147** affords the fatty amine **169** which has a primary amino function at the center of the long chain [66,70]. The telomer **171**, obtained from nitroethane (**170**) was converted to recifeiolide (**175**) by the following sequence of reactions [61]. The nitro group in **171** was converted to the ketone and protected to give **172**. The terminal double bond was converted to iodide via Ti-catalysed alumination and the ketone was reduced to alcohol to give **173**. Cyclization of the phenylthioacetate **174** afforded recifeiolide (**175**). The nitro compound **147** is a good building block for civetone carboxylate [71].

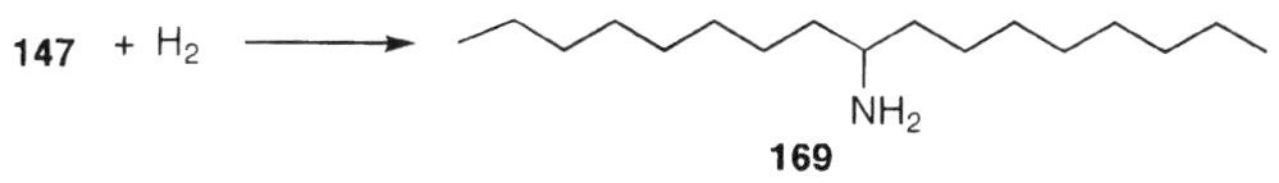

147 + H₂ ⟶

NH₂
169

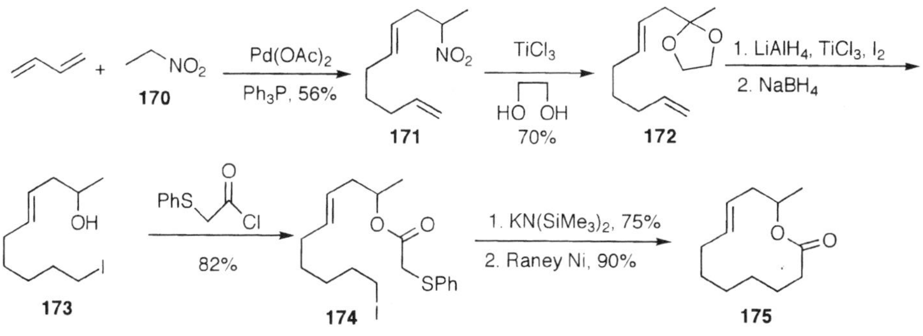

170

Pd(OAc)₂
Ph₃P, 56%

171

TiCl₃

HO OH
70%

172

1. LiAlH₄, TiCl₃, I₂
2. NaBH₄

173
OH

PhS
O
Cl

82%

O O
SPh
174

1. KN(SiMe₃)₂, 75%
2. Raney Ni, 90%

O O
175

The intramolecular reactions of the 1,3,8,10-undecatetraene system in [di(2,4-pentadienyl)malonate] **176** with nucleophiles proceeded smoothly to give the five-membered rings. Reaction of nitromethane gave **177**, in which the nucleophile is introduced at the terminal carbon [72].

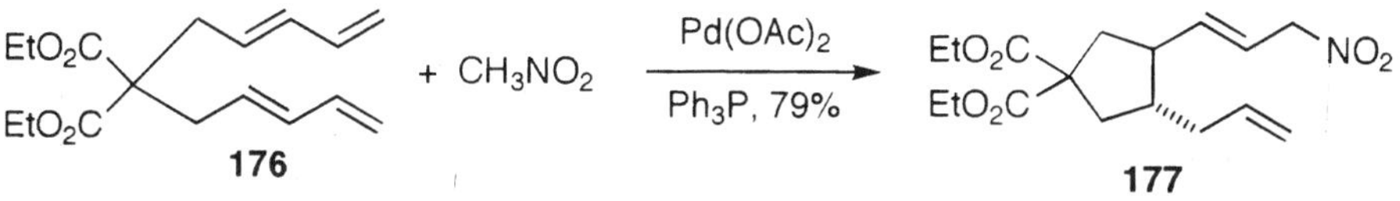

EtO₂C
EtO₂C
176

+ CH₃NO₂

Pd(OAc)₂
Ph₃P, 79%

EtO₂C
EtO₂C
NO₂
177

Carbonylation of butadiene gives two different products, depending on the catalytic species. When PdCl₂ is used in alcohol, 3-pentenoate (**150**) is obtained [73,74]. Further carbonylation of **150**, catalysed by cobalt carbonyl, affords the adipate **178** [75]. However, 3,8-nonadienoate (**151**) is obtained by dimerization and carbonylation when Pd(OAc)₂ and Ph₃P are used [76,77]. The presence of chloride ion firmly attached to Pd makes the difference.

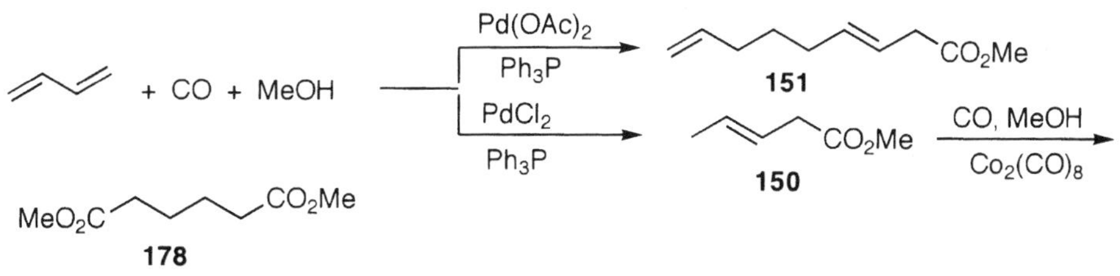

+ CO + MeOH

Pd(OAc)₂
Ph₃P

PdCl₂
Ph₃P

CO₂Me
151

CO₂Me
150

CO, MeOH
Co₂(CO)₈

MeO₂C
CO₂Me
178

A Rh complex coordinated by the water-soluble phosphine TMSPP (**XLVII**) catalyses the 1:1 coupling of isoprene with the malonamide derivative **179** to give **180**, and feprazone (**181**) was prepared after isomerization [78]. Under similar conditions, reaction of myrcene (**182**) with methyl acetoacetate gives **183** and **184** in 97% yield, which are used for pseudo-ionone synthesis [79]. The Ni(0)-catalysed 1:1 coupling of 1,3-cyclohexadiene (**185**) with malonate produces **186** [80].

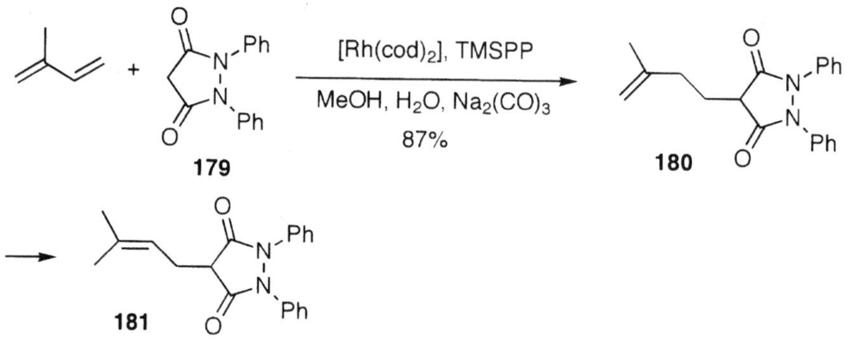

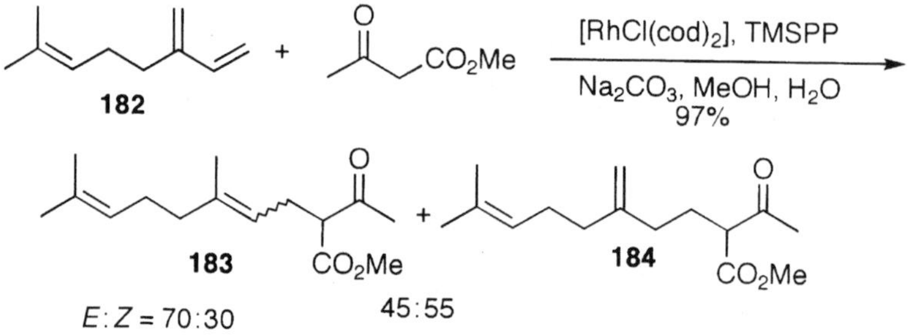

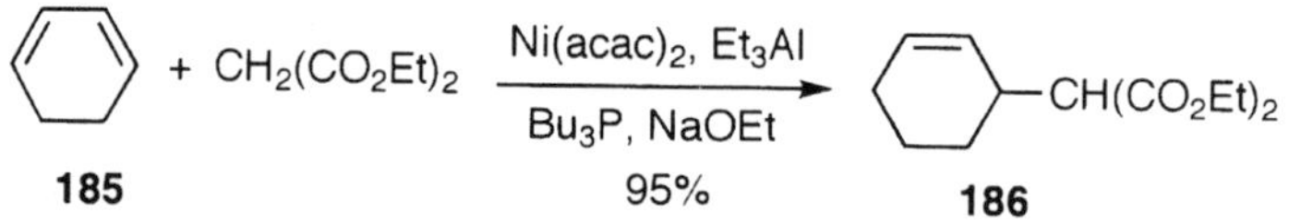

Nickel-catalysed addition of HCN to butadiene was developed by du Pont for adiponitrile production [81]. A Ni(0)–phosphite complex is used as the catalyst in the presence of Lewis acids. Oxidative addition of HCN to Ni(0), followed by insertion of butadiene, generates π-allyl intermediate **187**. Reductive elimination of **187** yields **188** and **189**, and isomerization of the double bond in **189** to the terminal position gives 4-pentenonitrile (**190**). Then, insertion of **190** to H–Ni–CN affords adiponitrile (**191**).

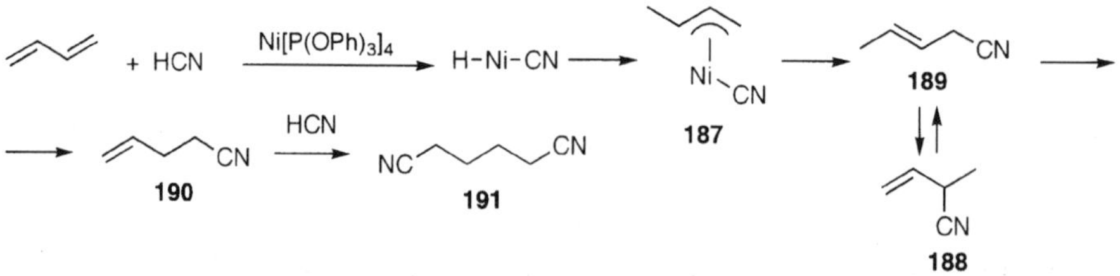

The Co reagent **192**, prepared by the reaction of $Co_2(CO)_8$ with sodium, is reactive, and the acylcobalt complex **193** is formed by the reaction of acyl halides. Insertion of butadiene at the Co–acyl bond generates the π-allylcobalt complex **194**, from which the acylbutadiene **195** is formed by deprotonation with a base [82]. Based on this reaction, various acyldienes are prepared by $Co_2(CO)_8$-catalysed reaction of active alkyl halides, conjugated dienes and CO. The Co-catalysed reaction can be carried out smoothly under phase-transfer conditions. For example, 6-phenyl-3,5-hexadien-2-one (**197**) was prepared in 86% yield by the reaction of MeI, 1-phenylbutadiene (**196**) and CO in the presence of cetyltrimethylammonium bromide [83].

$$Co_2(CO)_8 + 2\,Na \longrightarrow \underset{\mathbf{192}}{2\,NaCo(CO)_4} \xrightarrow{RCOCl} \underset{\mathbf{193}}{R-\overset{O}{\overset{\|}{C}}-Co(CO)_4} \longrightarrow$$

$$R-\overset{O}{\overset{\|}{C}}-Co(CO)_4 \longrightarrow \left[\begin{array}{c} \overset{O}{\overset{\|}{H_2C-C-R}} \\ | \\ CH \\ \| \\ \text{Co(CO)}_3 \end{array}\right] \longrightarrow \underset{\mathbf{194}}{\overset{O}{\overset{\|}{H_2C-C-CH_3}} \quad Co(CO)_3} \xrightarrow[\text{CO}]{\text{base}}$$

$$\left[\begin{array}{c} \overset{O}{\overset{\|}{H\cdot CH-C-CH_3}} \\ \text{Co(CO)}_3 \end{array}\right] \longrightarrow \underset{\mathbf{195}}{\quad} + HCo(CO)_4$$

$$MeI + \underset{\mathbf{196}}{Ph} + CO \xrightarrow[\underset{\mathbf{86\%}}{Me_3(Cet)NCl}]{Co_2(CO)_8} \underset{\mathbf{197}}{Ph}$$

5.3 Bis-metallation, Carbometallation and Hydrometallation

Conjugated dienes undergo metallation to give the 1,4-adduct **198** and the dimerization–1,8-addition product **199** with main group metal compounds. The reaction proceeds by oxidative addition of main group metal compounds to transition metal complexes. Reactive allylmetal compounds **198** and **199** as useful synthetic intermediates are prepared by this methods.

$$R_nM'A + \underset{}{\diagup} \longrightarrow \underset{\mathbf{198}}{R_nM' \diagdown A} + \underset{\mathbf{199}}{R_nM' \diagdown A}$$

A = H, R, $M'R_n$

Disilanes add to conjugated dienes by splitting their Si−Si bond. The dimerization and 1,8-disilylation of conjugated dienes with hexamethyldisilane are catalysed by $PdCl_2(PhCN)_2$ at 90 °C [84]. Using $Pd_3(dba)_3$ without phosphine as an active catalyst, the reaction proceeds in DMF at room temperature to give **200** [85]. Smooth dimerization and 1,8-distannation of butadiene with hexamethyldistannane to give 1,8-distannyl-2,6-octadiene (**201**) is catalysed by $Pd_2(dba)_3$ without phosphine [86]. The cyclopentane **203** is formed by the reaction of di-(2,4-pentadienyl)malonate (**202**) with hexamethyldistannane [87]. However, 1,4-disilylation of 1,3-dienes occurs by Pt-catalysed metallation [88], and the selective 1,4-silylstannation of butadiene with the silylstannane **204** to give **205** is catalysed by $Pt(CO)_2(Ph_3P)_2$ [89].

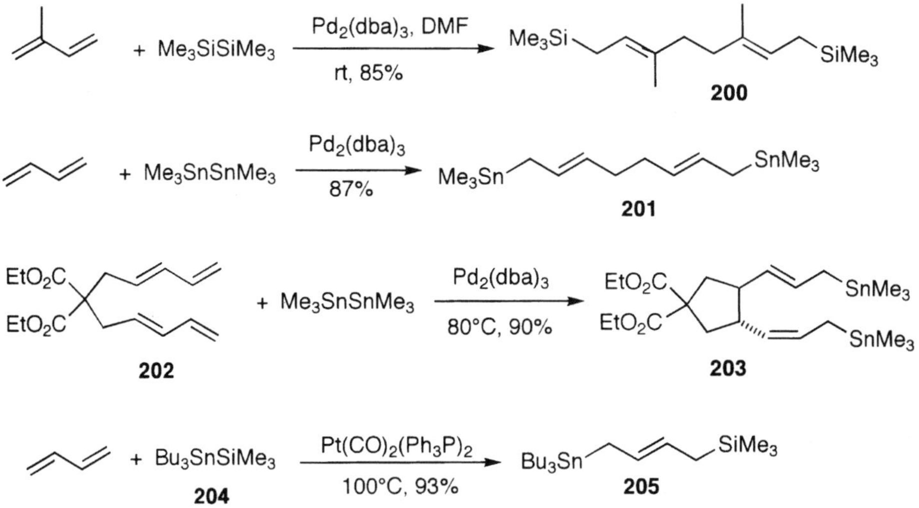

Addition of bis(pinacolate)diboron (**206**) to dienes is catalysed by Pt complexes. The 1:1 adduct **207** is obtained with $Pt(Ph_3P)_4$, whereas $Pt(dba)_2$ without phosphine is a very active catalyst and the 1,8-adduct **208** is obtained at room temperature [90].

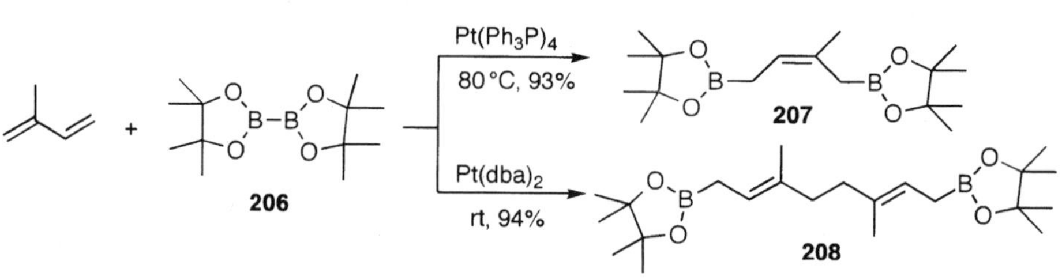

Silaboration of diene with the silylborane **209** is also catalysed by Pt complexes to give the 1,4-adduct **211** as an *E/Z* mixture which reacts with benzaldehyde to afford the homoallyl alcohol **212** having a silyl group [91]. However, the silaborative coupling of the diene **210** with **209** and benzaldehyde proceeds in one step under milder condition to yield the silyl-protected homoallyl alcohol **215** having a boryl group with high stereoselectivity. In this coupling, reaction of the intermediate **213** with aldehyde generates the alkoxysilylplatinum **214**, and finally **215** is obtained by reductive elimination.

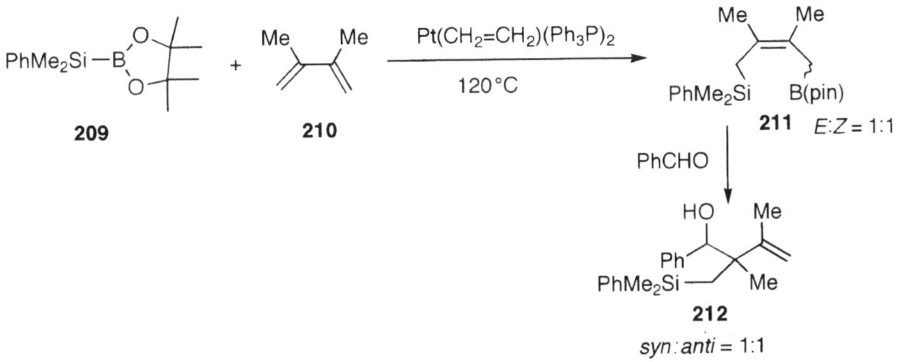

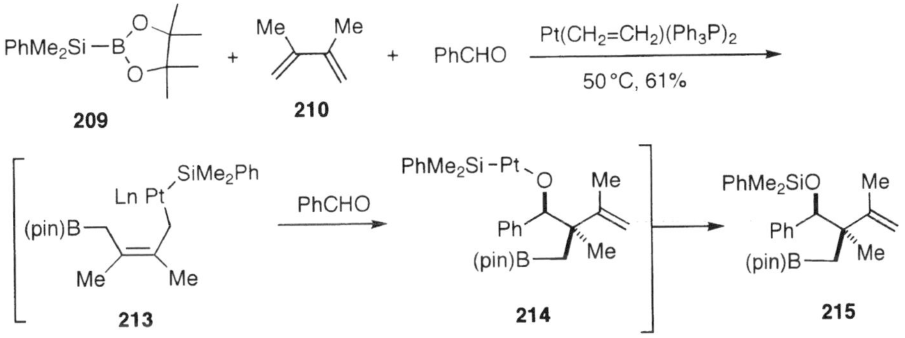

As an example of carbometallation, the 1,4-carbosilylation product **218** is obtained by the reaction of dienes, disilanes and acid chlorides of aromatic and α,β-unsaturated acids at 80 °C. The phenylpalladium **216** is formed by the oxidative addition of benzoyl chloride, followed by facile decarbonylation at 80 °C, and reacts with butadiene to generate the benzyl-π-allylic complex **217**. Then, transmetallation with the disilane and reductive elimination afford 4-silyl-2-butenylbenzene **218** [92]. Regioselective carbomagnesation of isoprene with allylic magnesium bromide **219** catalysed by Cp_2TiCl_2 gives **220**, which is useful for terpene synthesis [93,94].

1,4-Hydrometallation of dienes proceeds smoothly to give the allylmetal compounds **221**. 1,4-Hydromagnesation is catalysed by Cp_2TiCl_2. As the Cp_2TiCl_2-catalysed exchange reaction between Grignard reagents which have β-hydrogen and conjugated dienes, such as butadiene and isoprene, is irreversible, the methallylic Grignard reagent **224** can be prepared in a quantitative yield by the exchange reaction of *i*-PrMgBr and isoprene [95–97]. The reaction can be explained by the following mechanism. Cp_2TiH (**222**) is generated by the transmetallation of Cp_2TiCl_2 with *i*-PrMgBr, followed by β-elimination. Then insertion of isoprene generates the π-allylic

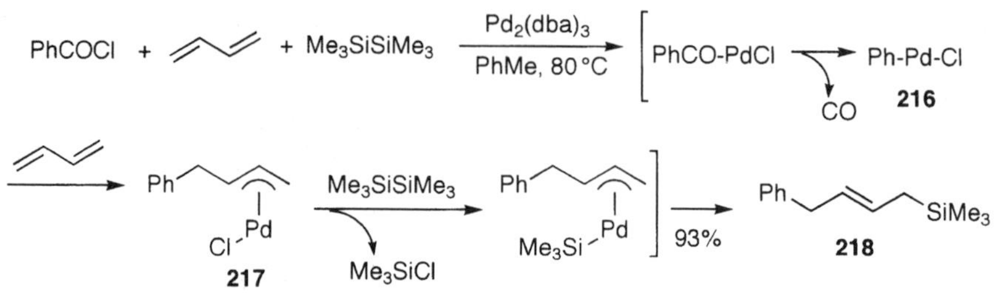

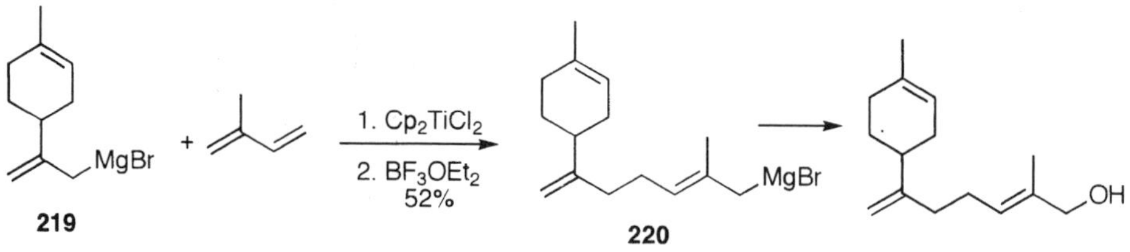

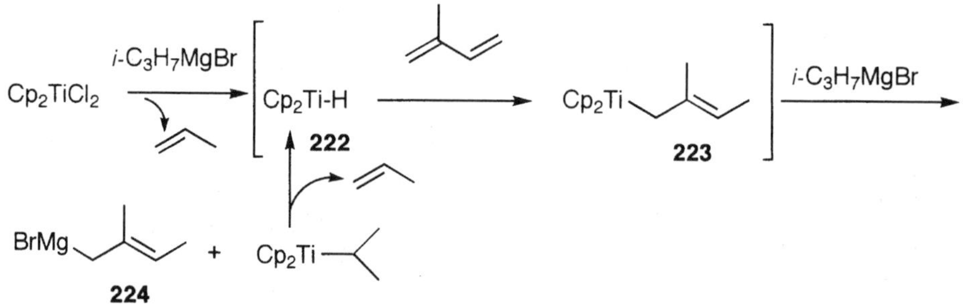

Ti complex **223**, and transmetallation with the Grignard reagent gives the methallylic Grignard **224**, and the Ti-hydride **222** is regenerated.

Hydroboration proceeds without a catalyst, but hydroboration with less active catecholborane (**225**) is accelerated by catalysts. Usually 1,2-addition to conjugated dienes takes place, but the Pd-catalysed reaction of catecholborane (**225**) gives the 1,4-adduct **226**. This reaction is not catalysed by Rh complexes [98]. Hydroalumination of conjugated dienes catalysed by Cp$_2$TiCl$_2$ affords the allylic aluminium compounds **227** by 1,4-addition. The Pd-catalysed hydrostannation of isoprene with HSnBu$_3$ affords the (*Z*)-2-alkenylstannane **228** with high regio- and stereoselectivities [99].

Hydrosilylation of butadiene gives rise to different products depending on the kinds of hydrosilanes and the reaction conditions. Trimethylsilane produces the 1:2 adduct, namely 1-trimethylsilyl-2,6-octadiene (**231**) in high yields [100,101]. Unlike other telomers, which have the 2,7-octadienyl chain, the 2,6-octadienyl chain is formed by the hydrosilylation. However, 1-trichlorosilyl-2-butene (**230**) as the 1:1 adduct is formed selectively with trichlorosilane, which is more reactive than trialkylsilanes. The

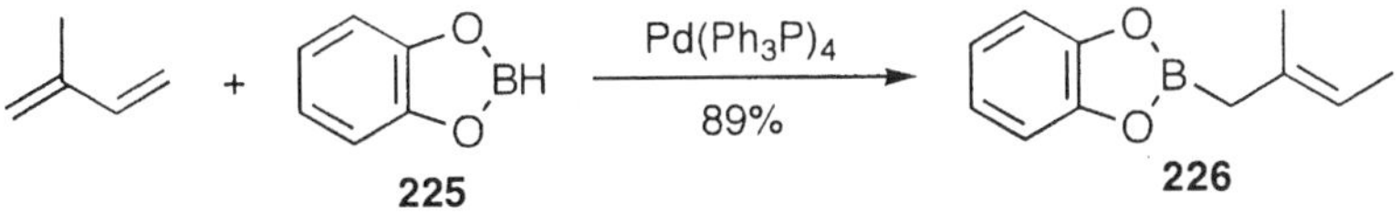

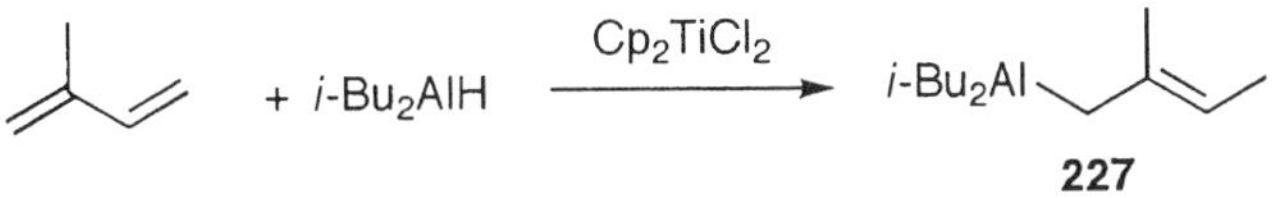

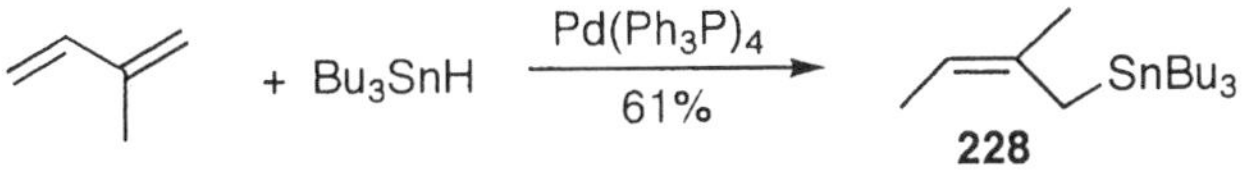

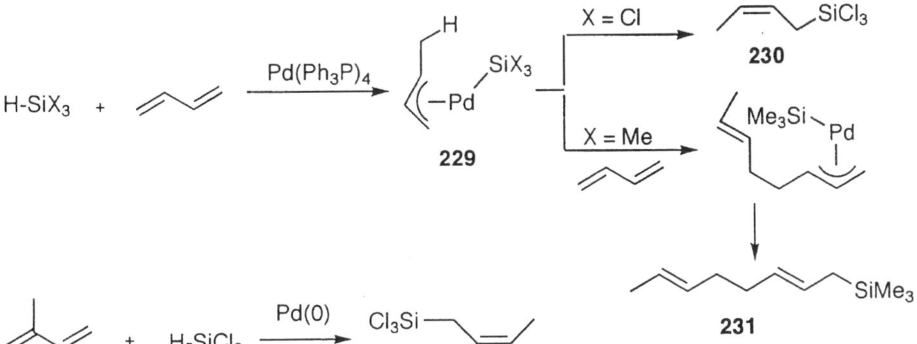

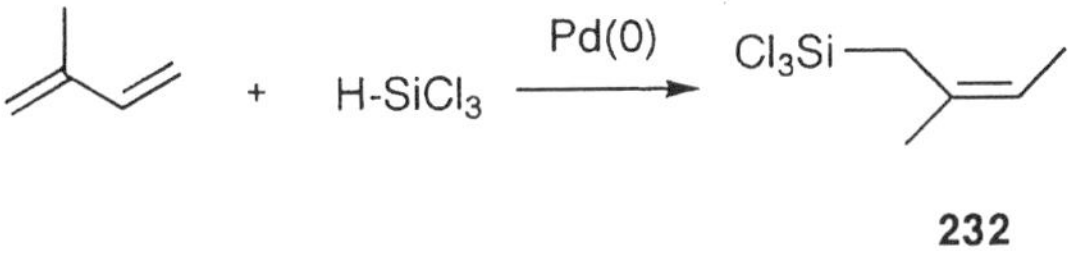

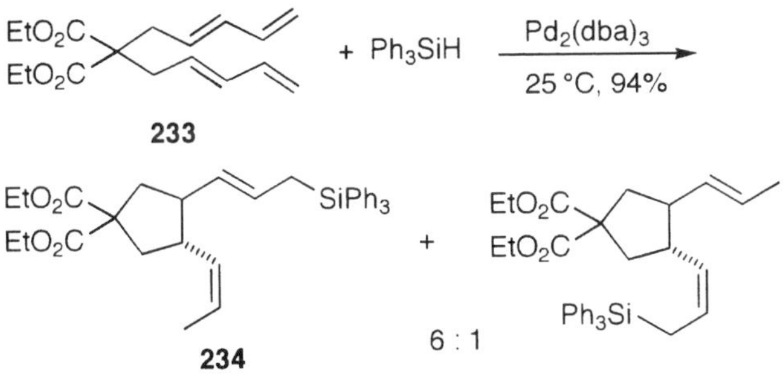

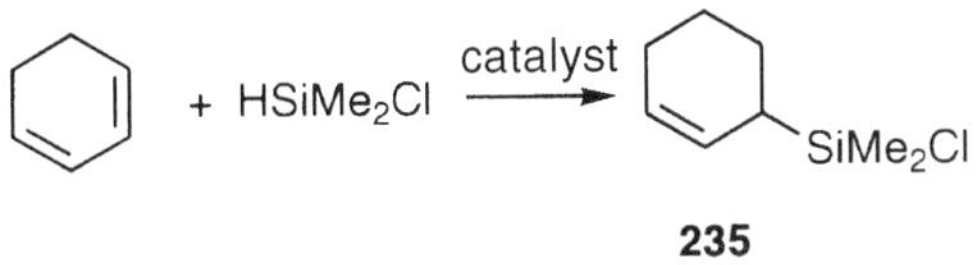

Catalyst	Conditions	Yield of 235
Ni(acac)$_2$/AlEt$_3$	−78–0 °C, 1 h	72%
Pd(Ph$_3$P)$_4$	100 °C, 6 h	82%
RhCl(Ph$_3$P)$_3$	100 °C, 6 h	83%

reaction gives the (Z) form stereoselectively. A mixture of the 1:1 and 1:2 adducts (83.5 and 5.2%) is obtained with dichloromethylsilane [101]. The reaction of chlorosilanes proceeds even in the absence of a phosphine ligand. The reaction of trichlorosilane with isoprene gives (Z)-1-trichlorosilyl-2-methyl-2-butene (**232**) regio- and stereoselectively. No reaction takes place with trialkylsilanes and isoprene [101, 102]. Cyclization of 1,3,8,10-undecatetraene system in di(2,4-pentadienyl) malonate (**233**) via hydrosilyation gave the cyclopentane derivative **234**, which corresponds to 2,6-octadienylsilane [103].

The Ni catalyst is most active for the hydrosilylation of 1,3-cyclohexadiene to afford **235** with $HSiMe_2Cl$, and the reaction proceeds at $-78\,°C$ in 1 h, whereas Pd- and Rh-catalysed reactions proceed at $100\,°C$ in 6 h [104].

References

1. H. W. B. Reed, *J. Chem. Soc.*, 1931 (1954).
2. P. Heimbach, P. W. Jolly and G. Wilke, *Adv. Organometal. Chem.*, **8**, 29 (1970); G. Wilke, B. Bogdanovic, P. Hardt, P. Heimbach, W. Keim, M. Kröner, W. Oberkirch, K. Tanaka, E. Steinrücke, D. Walter and H. Zimmermann, *Angew. Chem., Int. Ed., Engl.*, **5**, 151 (1966); P. Heinbach, *Angew. Chem., Int. Ed., Engl.*, **12**, 975 (1973); G. Wilke, *Angew. Chem., Int. Ed., Engl.*, **27**, 186 (1988); R. Baker, *Chem. Rev.*, **73**, 487 (1973); P. W. Jolly, *Comprehensive Organometallic Chemistry*, vol. 8, p. 671, Pergaman Press, (1982); M. F. Semmelhack, *Org. React.*, **19**, 115 (1972); M. Lautens, W. Klute and W. Tam, *Chem. Rev.*, **96**, 49 (1996).
3. A. Miyake, H. Kondo and M. Nishino, *Angew. Chem., Int. Ed. Engl.*, **10**, 802 (1971).
4. J. Kiji, K. Masui and J. Furukawa, *Bull. Chem. Soc. Jpn.*, **44**, 1956 (1971).
5. P. Heimbach, B. Hugelin, H. Peter, A. Roloff and E. Troxler, *Angew. Chem.*, **88**, 29 (1976).
6. P. Brun, A. Tenaglia and B. Waegell, *Tetrahedron*, **41**, 5019 (1985).
7. R. Baker, R. C. Cookson and J. R. Vinson, *Chem. Commun.*, 515 (1974).
8. P. A. Wender, M. C. Ihle and C. R. D. Correia, *J. Am. Chem. Soc.*, **110**, 5904 (1988).
9. P. A. Wender and M. J. Tebbe, *Synthesis*, 1089 (1991), P. A. Wender and M. L. Snapper, *Tetrahedron Lett.*, **28**, 2221 (1987).
10. K. U. Baldenius, H. tom Dieck, W. A. König, D. Icheln and T. Runge, *Angew. Chem., Int. Ed. Engl.*, **31**, 305 (1992).
11. W. E. Billups, J. H. Crosse and C. V. Smith, *J. Am. Chem. Soc.*, **95**, 3438 (1973).
12. L. G. Cannell, *J. Am. Chem. Soc.*, **94**, 6867 (1972).
13. W. Brenner, P. Heimbach, K. J. Ploner and F. Thömel, *Angew. Chem., Int. Ed., Engl.*, **8**, 753 (1969).
14. D. R. Fahey, *J. Org. Chem.*, **37**, 4471 (1972).
15. P. A. Wender and T. E. Smith, *J. Org. Chem.*, **60**, 2962 (1995), **61**, 824 (1996).
16. R. S. Jolly, G. Luedtke, D. Sheehan and T. Livinghouse, *J. Am. Chem. Soc.*, **112**, 4965 (1990).
17. S. R. Gilbertson and G. S. Hoge, *Tetrahedron Lett.*, **39**, 2075 (1998).
18. I. Matsuda, M. Shibata, S. Sata and Y. Izumi, *Tetrahedron Lett.*, **29**, 3361 (1987).
19. H. tom Dieck and R. Diercks, *Angew. Chem., Int. Ed. Engl.*, **22**, 778 (1983).
20. P. A. Wender, H. Takahashi and B. Witulski, *J. Am. Chem. Soc.*, **117**, 4720 (1995); P. A. Wender and D. Sperandio, *J. Org. Chem.*, **63**, 4164 (1998).

20a. P. A. Wender, C. O. Husfeld, E. Langkopf and J. A. Love, *J. Am. Chem. Soc.*, **120**, 1940 (1998).

21. P. A. Wender, T. E. Jenkins and S. Suzuki, *J. Am. Chem. Soc.*, **117**, 1843 (1995).

22. K. Ohno, T. Mitsuyasu and J. Tsuji, *Tetrahedron Lett.*, 67 (1971); *Tetrahedron*, **28**, 3705 (1972).

23. P. Hayns, *Tetrahedron Lett.*, 3687 (1970).

24. R. M. Manyik, W. E. Walker, K. E. Atkins and E. S. Hammack, *Tetrahedron Lett.*, 3813 (1970).

25. Y. Sato, M. Takimoto, K. Hayashi, T. Katsuhara, K. Takagi and M. Mori, *J. Am. Chem. Soc.*, **116**, 9771 (1994); Y. Sato, N. Saito and M. Mori, *Tetrahedron Lett.*, **38**, 3911 (1997).

26. M. Takimoto, Y. Hiraga, Y. Sato and M. Mori, *Tetrahedron Lett.*, **39**, 4543 (1998).

27. K. Ohno and J. Tsuji, *Chem. Commun.*, 247 (1971).

28. J. Kiji, K. Yamamoto, H. Tomita and J. Furukawa, *Chem. Commun.*, 506 (1974).

29. P. Braunstein, D. Matt and D. Nobel, *Chem. Rev.*, **88**, 747 (1988).

30. Y. Sasaki, Y. Inoue and H. Hashimoto, *Chem. Commun.*, 605 (1976).

31. A. Behr, K. D. Juszak and W. Keim, *Synthesis*, 574 (1983).

32. T. Ohta, K. Ebina and N. Yamazaki, *Bull. Chem. Soc. Jpn.*, **44**, 1321 (1971).

33. S. Otsuka, T. Taketomi and T. Kikuchi, *J. Am. Chem. Soc.*, **85**, 3709 (1963).

34. M. Hidai, Y. Uchida and A. Misono, *Bull. Chem. Soc. Jpn.*, **38**, 1243 (1965).

35. S. Takahashi, T. Shibano and N. Hagihara, *Bull. Chem. Soc. Jpn.*, **41**, 454 (1968).

36. Y. Uchida, K. Furuhata and S. Yoshida, *Bull. Chem. Soc. Jpn.*, **44**, 1966 (1971).

37. S. Watanabe, K. Suga and H. Kikuchi, *Aust. J. Chem.*, **23**, 385 (1970).

38. I. Mochida, S. Yuasa and T. Seiyama, *Chem. Lett.*, 901 (1974).

39. K. Takahashi, G. Hata and A. Miyake, *Bull. Chem. Soc. Jpn.*, **46**, 600 (1973).

40. M. Iwamoto and S. Yuguchi, *Bull. Chem. Soc. Jpn.*, **41**, 150 (1968).

41. G. Hata and A. Miyake, *Bull. Chem. Soc. Jpn.*, **41**, 2762 (1968); *J. Am. Chem. Soc.*, **86**, 3703 (1964).

42. R. Cramer, *Acc. Chem. Res.*, **1**, 186 (1968).

42a. H. Müller, D. Wittenberg, H. Seibt and E. Scharf, *Angew. Chem., Int. Ed. Engl.*, **4**, 327 (1965).

43. T. Ito, K. Takahashi and Y. Takami, *Tetrahedron Lett.*, 5049 (1973).

44. H. Bönnemann, *Angew. Chem., Int. Ed. Engl.*, **12**, 964 (1973).

45. T. Mitsudo, S. W. Zhang, T. Kondo and Y, Watanabe, *Tetrahedron Lett.*, **33**, 341 (1992).

46. H. Singer, W. Umbach and M. Dohr, *Synthesis*, 42 (1972); H. Singer, *Synthesis*, 189 (1974).

47. J. M. Takacs and S. J. Mehrman, *Tetrahedron Lett.*, **37**, 2749 (1996).

48. J. M. Takacs, Y. C. Myoung and L. G. Anderson, *J. Org. Chem.*, **59**, 6928 (1994), J. M. Takacs, J. J. Weidne, P. W. Newsome, B. E. Takacs, R. Chidamdaram and R. Shoemaker, *J. Org. Chem.*, **60**, 3473 (1995).

49. T. Mitsudo, Y. Nakagawa, K. Watanabe, Y. Hori, H. Misawa, H. Watanabe and Y. Watanabe, *J. Org. Chem.*, **50**, 565 (1985).

50. J. Tsuji, *Adv. Organometal. Chem.*, **17**, 141 (1979); J. Tsuji, *Acc. Chem. Res.*, **6**, 8 (1973).

51. S. Takahashi, T. Shibano and N. Hagihara, *Tetrahedron Lett.*, 2451 (1967).

52. E. J. Smutny, *J. Am. Chem. Soc.*, **89**, 6793 (1967).

53. G. Hata, K. Takahashi and A. Miyake, *J. Org. Chem.*, **36**, 2116 (1971).

53a. P. W. Jolly, R. Mynott, B. Raspel and K. P. Schick, *Organometallics*, **5**, 473 (1986).

54. S. Takahashi, H. Yamazaki and N. Hagihara, *Bull. Chem. Soc. Jpn.*, **14**, 254 (1968).

55. J. Tsuji, *Pure Appl. Chem.*, **51**, 1235 (1979); **53**, 2371 (1981).
56. K. E. Atkins, W. E. Walker and R. M. Manyik, *Chem. Commun.*, 330 (1971).
57. N. Yoshimura, Y. Tokitoh, M. Matsumoto and M. Tamura, *J. Chem. Soc. Jpn. Chem. Ind.* (in Japanese), 119 (1993).
58. J. Beger and H. Reichel, *J. Prak. Chem.*, **315**, 1067 (1973).
59. W. E. Walker, R. M. Manyik, K. E. Atkins and M. L. Farmer, *Tetrahedron Lett.*, 3817 (1970).
60. J. Tsuji, I. Shimizu, H. Suzuki and Y. Naito, *J. Am. Chem. Soc.*, **101**, 5070 (1979).
61. T. Takahashi, S. Hashiguchi, K. Kasuga and J. Tsuji, *J. Am. Chem. Soc.*, **100**, 7424 (1978).
62. S. Gardner and D. Wright, *Tetrahedron Lett.*, 163 (1972).
63. P. Riffia, G. Gregorio, F. Conti, G. F. Pregaglia and R. Ugo, *J. Organometal. Chem.*, **55**, 405 (1973).
64. C. U. Pitman, R. M. Hanes and J. J. Yang, *J. Mol. Catal.*, **15**, 377 (1982).
65. J. Tsuji and T. Yamakawa, *Tetrahedron Lett.*, 613 (1979).
66. T. Mitsuyasu, M. Hara and J. Tsuji, *Chem. Commun.*, 345 (1971).
67. J. Tsuji and M. Takahashi, *J. Mol. Catal.*, **10**, 107 (1981).
68. K. Takahashi, A. Miyake and G. Hata, *Bull. Chem. Soc. Jpn.*, **45**, 1183 (1972).
69. J. Tsuji, *Bull. Chem. Soc. Jpn.*, **46**, 1896 (1973).
70. T. Mitsuyasu and J. Tsuji, *Tetrahedron Lett.*, **30**, 831 (1974).
71. J. Tsuji and T. Mandai, *Tetrahedron Lett.*, 3285 (1977).
72. J. M. Takacs and J. Zhu, *J. Org. Chem.*, **54**, 5193 (1989); *Tetrahedron Lett.*, **31**, 1117 (1990); J. M. Takacs and S. V. Chandramouli, *J. Org. Chem.*, **58**, 7315 (1993).
73. S. Hosaka and J. Tsuji, *Tetrahedron Lett.*, **27**, 3821 (1971).
74. S. Brewis and P. R. Hughes, *Chem. Commun.*, 157 (1965).
75. E. Drent and J. van Goch, *Eur. Patent, Appl.*, EP 284170; *Chem. Abst.*, **110**, 39519 (1989).
76. J. Tsuji, Y. Mori and M. Hara, *Tetrahedron*, **28**, 3721 (1972).
77. W. E. Billups, W. E. Walker and T. C. Shield, *Chem. Commun.*, 1067 (1971).
78. G. Mignani, D. Morel and Y. Colleuille, *Tetrahedron Lett.*, **26**, 6337 (1985).
79. C. Mercier, G. Mignani, M. Aufrand and G. Allmang, *Tetrahedron Lett.*, **32**, 1433 (1991).
80. O. S. Andell, J. E. Bäckvall and C. Morberg, *Acta Chem. Scand. Ser. B*, **40**, 184 (1986).
81. J. W. Parshall, *J. Mol. Catal.*, **4**, 244 (1978).
82. R. F. Heck, *J. Am. Chem. Soc.*, **85**, 3381 (1963).
83. S. Gamarotta and H. Alper, *J. Organometal. Chem.*, **194**, C19 (1980); H. Alper and J. K. Currie, *Tetrahedron Lett.*, 2665 (1979).
84. H. Sakurai, Y. Eriyama, Y. Kamiyama and Y. Nakadaira, *J. Organometal. Chem.*, **264**, 229 (1984).
85. Y. Obora, Y. Tsuji and T. Kawamura, *Organometallics*, **12**, 2853 (1993).
86. Y. Tsuji and T. Kakehi, *Chem. Commun.*, 1000 (1992).
87. Y. Obora, Y. Tsuji, T. Kakehi, M. Kobayashi, Y. Shinkai, M. Ebihara and T. Kawamura, *Chem. Soc. Perkin Trans. I*, 599 (1995).
88. Y. Tsuji, R. M. Lago, S. Tomohiro and H. Tsuneishi, *Organometallics*, **11**, 2353 (1992).
89. Y. Tsuji and Y. Obora, *J. Am. Chem. Soc.*, **113**, 9368 (1991).
90. T. Ishiyama, M. Yamamoto and N. Miyaura, *Chem. Commun.*, 2073 (1996).
91. M. Suginome, H. Nakamura, T. Matsuda and Y. Ito, *J. Am. Chem. Soc.*, **120**, 4248 (1998).
92. Y. Obora, Y. Tsuji and T. Kawamura, *J. Am. Chem. Soc.*, **115**, 10414 (1993); **117**, 9814 (1995).
93. S. Akutagawa and S. Otsuka, *J. Am. Chem. Soc.*, **97**, 6870 (1975).
94. F. Barbot and Ph. Miginiac, *J. Organometal. Chem.*, **145**, 269 (1978).

95. F. Sato, H. Ishikawa and M. Sato, *Tetrahedron Lett.*, 365 (1980).
96. H. A. Martin and F. Jellinek, *J. Organomet. Chem.*, **12**, 149 (1968).
97. E. Colomer and R. Corriu, *J. Organomet. Chem.*, **82**, 367 (1974).
98. M. Satoh, Y. Nomoto, N. Miyaura and A. Suzuki, *Tetrahedron Lett.*, **30**, 3789 (1989).
99. H. Miyake and K. Yamamura, *Chem. Lett.*, 507 (1992).
100. S. Takahashi, T. Shibano and N. Hagihara, *Chem. Commun.*, 161 (1969).
101. M. Hara, K. Ohno and J. Tsuji, *Chem. Commun.*, 247 (1971). J. Tsuji, M. Hara and K. Ohno, *Tetrahedron*, **30**, 2143 (1974).
102. I. Ojima, *J. Organometal. Chem.*, **134**, C1 (1977), I. Ojima and M. Kumagai, *J. Organometal. Chem.*, **157**, 359 (1978).
103. J. M. Takacs and S. Chandramouli, *Organometallics*, **9**, 2877 (1990).
104. C. Bismara, R. D. Fabio, D. Donati, T. Rossi and R. J. Thomas, *Tetrahedron Lett.*, **36**, 4283 (1995).

6
REACTIONS OF PROPARGYLIC COMPOUNDS

6.1 Classification of Catalytic Reactions Based on Mechanistic Consideration

Propargylic compounds (2-alkynyl compounds) are derivatives of alkynes and they undergo several types of transformations in the presence of transition metal catalysts. However, catalytic reactions of propargylic compounds, particularly their esters and halides, clearly differ mechanistically from those of simple alkynes, except in a few cases. Therefore, the catalytic reactions of propargylic compounds are treated independently from those of simple alkynes. The most extensive studies have been carried out using Pd catalysts, and mainly Pd-catalysed reactions are treated in this chapter [1].

The Pd(0)-catalysed reactions of propargylic compounds can be understood by the following mechanistic consideration. Complex formation by stoichiometric reaction of the propargylic chlorides **1** and **3** with $Pd(Ph_3P)_4$ have been studied, and the σ-allenylpalladium **2** and the propargylpalladium (or, σ-prop-2-ynylpalladium) **4** are isolated as yellow powders [2]. Allenylpalladium chloride **2** is formed by S_N2' type displacement of the chlorine with Pd(0). Compound **4** is generated by direct oxidative addition, and formed when a bulky group such as trimethylsilyl or *tert*-butyl is attached to the alkyne terminal. It is reasonable to expect that Pd(0)-catalysed reactions of various propargylic compounds should proceed by the formation of either **2** or **4** as intermediates.

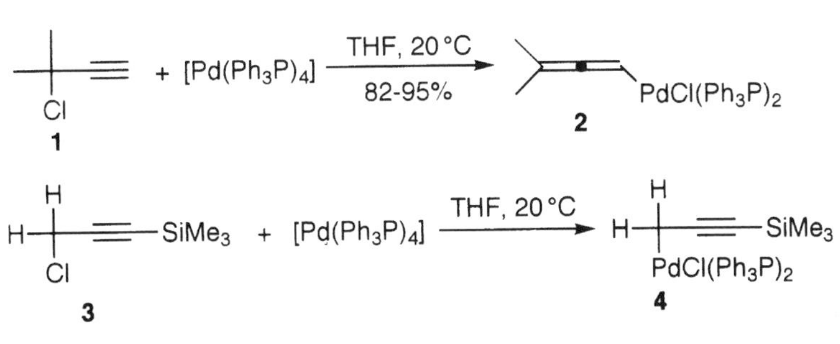

From a mechanistic viewpoint, the Pd(0)-catalysed reactions of propargylic compounds so far discovered can be classified into four types **I–IV**. The allenyl complexes **5** undergo three types of transformations depending on reactants. Type **I** reactions proceed by insertion of unsaturated bonds to the σ-bond between Pd and the sp^2 carbon in **5**. Type **Ia** is the insertion of alkenes to the palladium–carbon σ-bond, and the 1,2,4-alkatrienes are formed by β-elimination. Alkynes insert to form the alkenylpalladium **6**, which undergoes various transformations such as insertion of unsaturated bonds and anion captures.

Insertion of CO also generates the acylpalladium intermediate **7**, and 2,3-alkadienoates are obtained by alcoholysis (Type **Ib**).

Type Ia

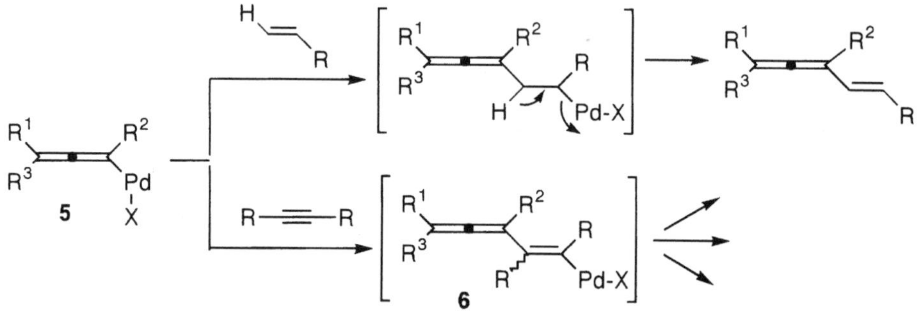

Type I b

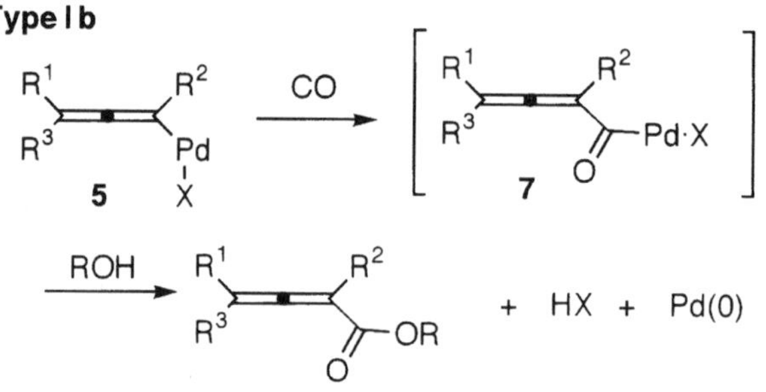

The reactions of type **II** proceed by transmetallation of the complex **5**. The transmetallation of **5** with hard carbon nucleophiles M′R (M′ = main group metals) such as Grignard reagents and metal hydrides MH generates **8**. Subsequent reductive elimination gives rise to an allene derivative as the final product. Coupling reactions of terminal alkynes in the presence of CuI belong to Type **II**.

Type II

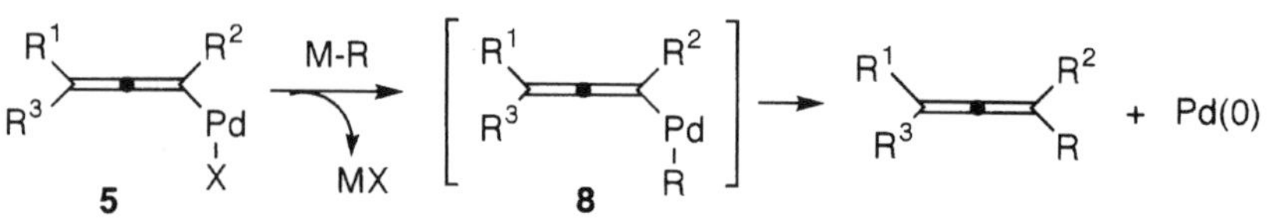

Type **III** reactions proceed by attack of a nucleophile at the central sp carbon of the allenyl system of the complexes **5**. Reactions of soft carbon nucleophiles derived from active methylene compounds, such as β-keto esters or malonates, and oxygen nucleophiles belong to this type. The attack of the nucleophile generates the intermediates **9**, which are regarded as the palladium–carbene complexes **10**. The intermediates **9** pick up a proton from the active methylene compound and π-allylpalladium complexes **11** are formed, which undergo further reaction with the nucleophile, as expected, and hence the alkenes **12** are formed by the introduction of two nucleophiles.

Type III

Two reactions of type **IV** are known. They proceed via the propargylpalladium complexes **13**. Formation of alkynes **14** by direct displacement of Pd in **13** with hydride is one reaction. Elimination of Pd−H species from **13** to form the enyne **15** is another reaction which proceeds via the propargylpalladium intermediate **13**.

Type IV

Several propargylic derivatives can be used for Pd-catalysed reactions, although they have different reactivities. Propargylic carbonates **16** are highly reactive and undergo various Pd-catalysed reactions smoothly, especially under neutral conditions. The allenylpalladium methoxides **17** are generated by facile irreversible oxidative addition of the carbonates with evolution of CO_2. Extensive studies on propargylic compounds have been carried out using propargylic carbonates as convenient substrates. Also, the 2-(1-alkynyl)oxiranes **18** undergo facile reactions using Pd(0) catalysts under neutral conditions by forming alkoxypalladium complexes **19** as intermediates. Examples of reaction types **I–IV** are now discussed.

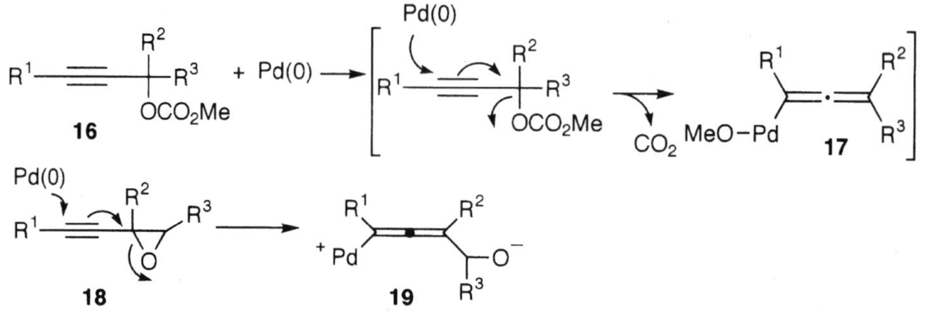

6.2 Reactions via Insertion to the Palladium–sp² Carbon Bond (Type I)

6.2.1 Reaction of Alkenes and Alkynes

Reaction of the propargylic carbonate **20** with alkene **21** offers a good synthetic route to the 1,2,4-alkatriene **23**. Smooth insertion of methyl acrylate (**21**) to the allenylpalladium bond generates **22**, and subsequent elimination of β-hydrogen affords the 2,4,5-alkatrienoate **23** in a good yield [3]. The reaction of allyl alcohol (**25**) as an alkene component with **24** gives aldehyde **27** by the elimination of β-hydrogen from the carbon attached to the oxygen in the intermediate **26**.

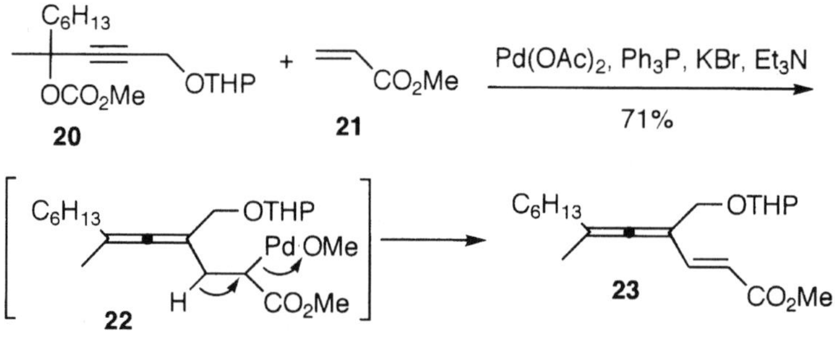

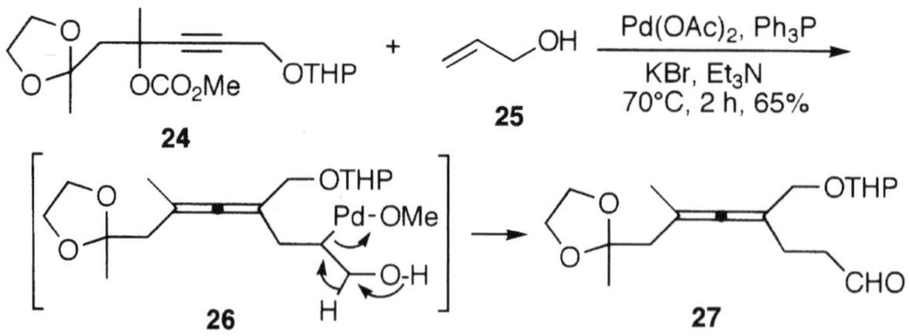

Polycyclic compounds are prepared by domino insertions of alkenes and alkynes to the allenylpalladium intermediates. As examples, 1,6-enyne **28** underwent intramolecular alkene insertion twice via the allenylpalladium **29**, to form the intermediate bearing bicyclo[3.1.0]hexane structure **30**, which was trapped with CO to give ester **31** and with phenyl group of NaBPh$_4$, to give **32** [4]. The triyne **33** generates the allenylpalladium **34**, and triscyclization starting from **34** by domino insertions of two alkyne and one alkene bonds affords **36** via **35** in 82% [5].

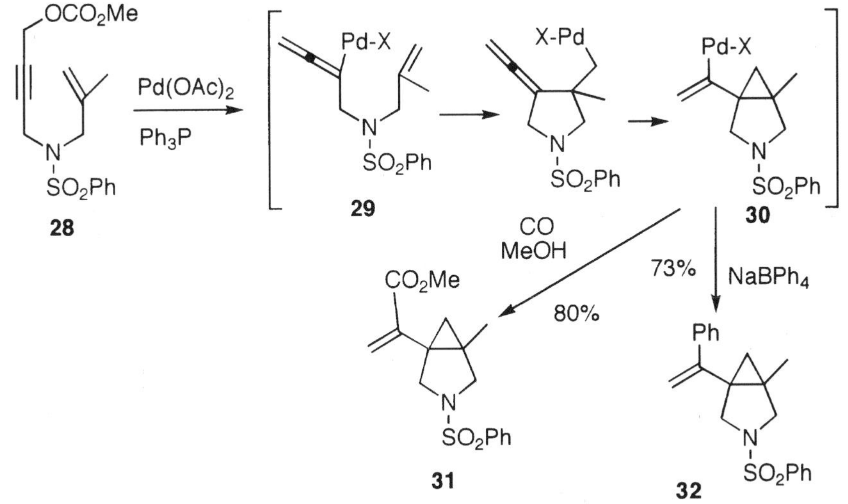

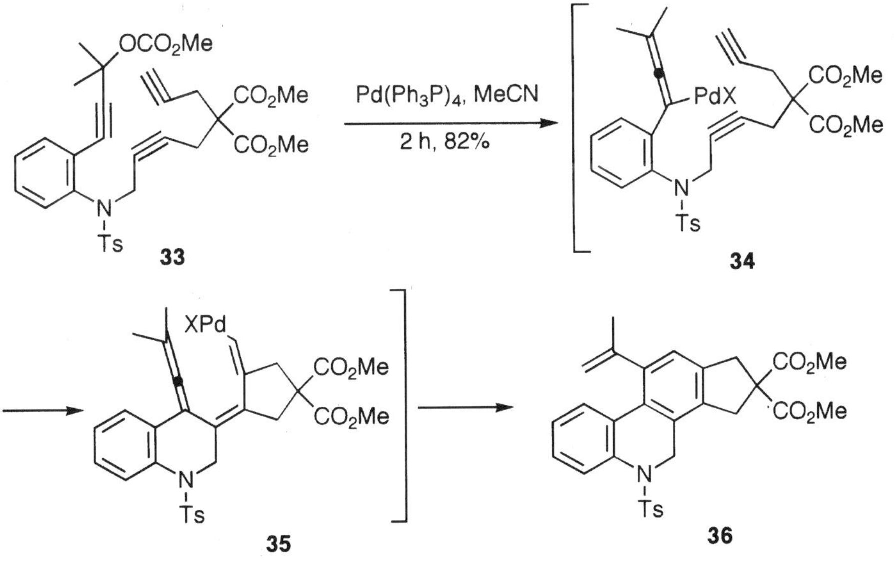

6.2.2 Carbonylation

Propargylic compounds undergo facile Pd-catalysed mono- and di-carbonylations depending on the reaction conditions [6]. The facile monocarbonylation of propargylic carbonates **37** proceeds under mild conditions [7]. The carbonylation of propargylic alcohols carried out under somewhat severe conditions affords mainly dicarbonylation products [8,9].

The mono- and dicarbonylations in an alcohol can be understood by the following mechanism. At first, CO insertion to the allenylpalladium intermediates generates acylpalladium complexes **38** which react with alcohol to give the 2,3-alkadienoates **39**. The carbonylation of propargylic carbonates under mild conditions stops at this stage. Under high pressure of CO, or in the presence of an activating group (for example, $R^3 = CO_2R$), further attack of CO at the central sp carbon of the allenyl system in **39** takes place to give the diesters **40**.

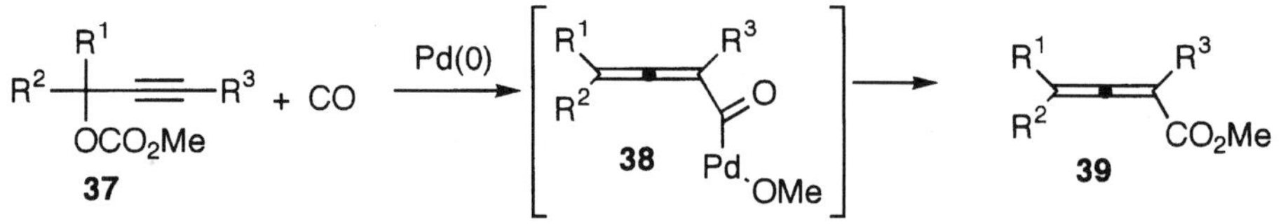

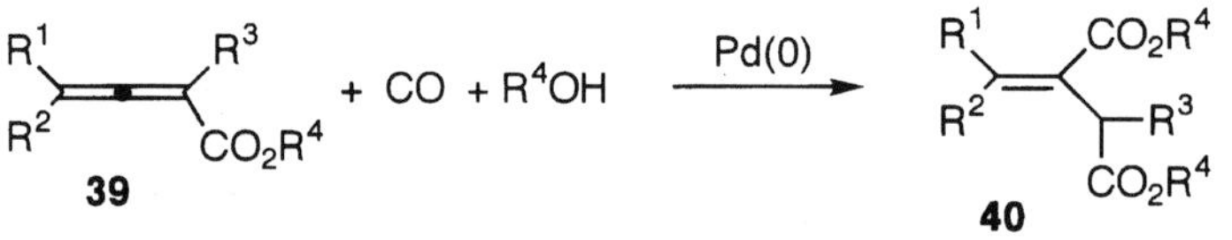

The 2,3-alkadienoate **42** is obtained by smooth decarboxylation–carbonylation of propargylic carbonate **41** in an alcohol under mild neutral conditions [7]. The reaction proceeds at 50 °C under 1–10 atm pressure of CO. The tertiary propargylic carbonates **43**, with a terminal triple bond, are the most reactive and give high yields of esters **44**. Propargylic acetates [10] and propargyl bromides [11] are also carbonylated under more severe conditions than those of carbonates. Carbonylation of propargylic mesylate **45** gave ester **46** with net inversion of the configuration [12].

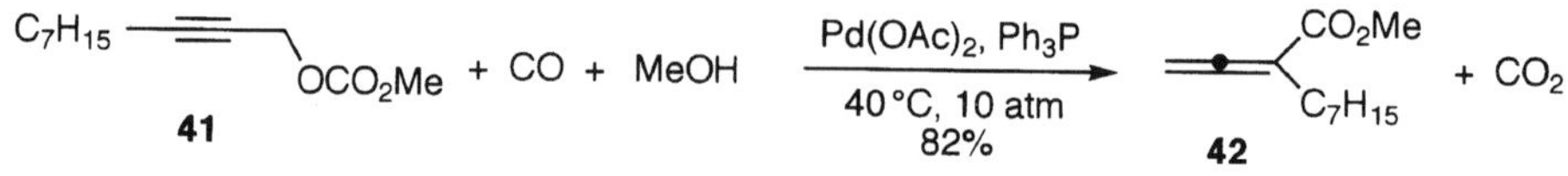

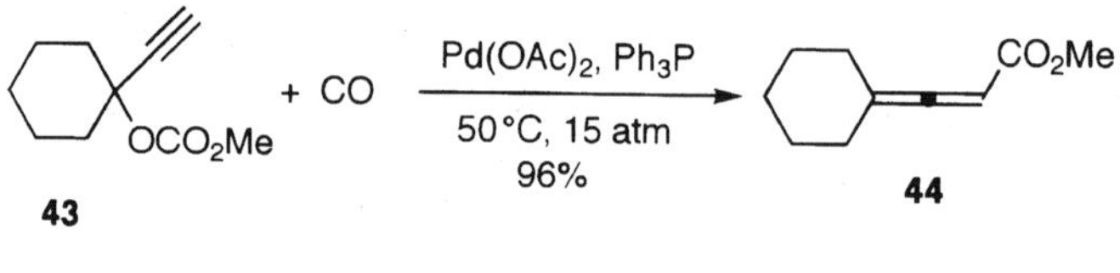

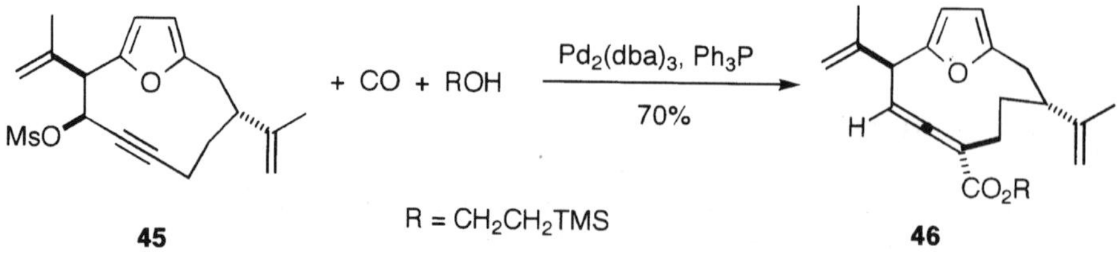

The 5-hydroxy-2,3-dienoate **49** is obtained from 2-alkynyloxirane **47** via the allenylpalladium **48**. In the presence of hydroxymethyl group in **50**, the 2,3-dihydrofuran-4-ol derivative **51** is obtained [13].

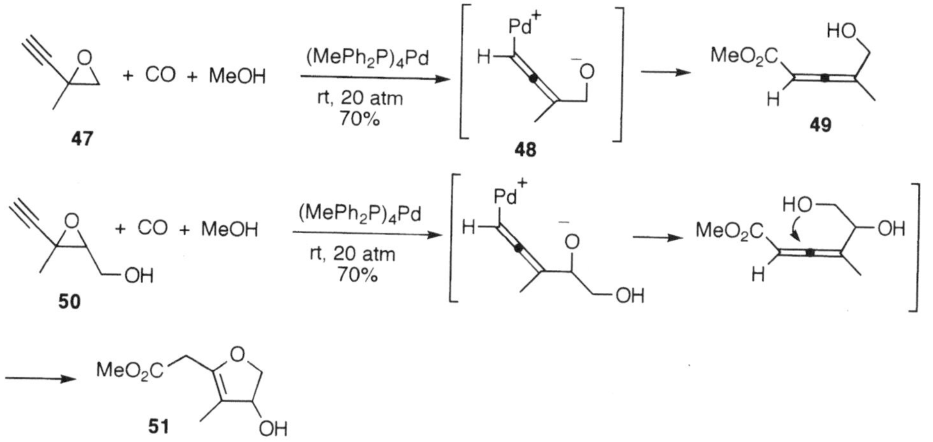

Substituted propargylamines **52** can be carbonylated in the presence of TsOH which may facilitate the oxidative addition of Pd(0) to the C–N bond to generate **53**, and the 2,3-dienecarbamides **54** are obtained by CO insertion using DPPP as a ligand. The terminal alkynes **55** are converted to the 2,4-dienecarbamide **56** [14].

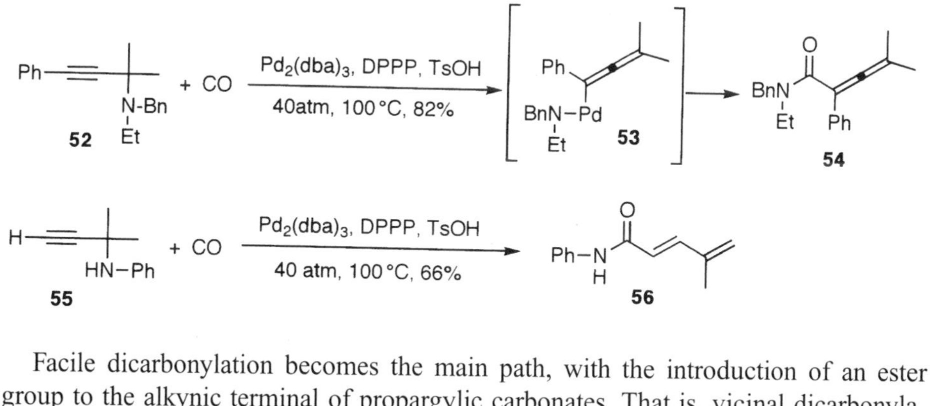

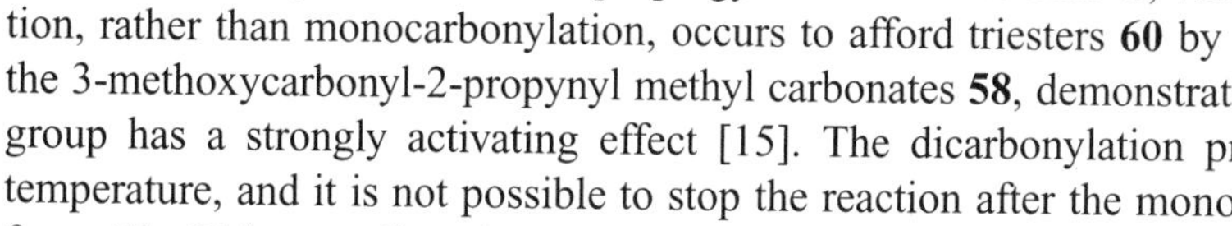

Facile dicarbonylation becomes the main path, with the introduction of an ester group to the alkynic terminal of propargylic carbonates. That is, vicinal dicarbonylation, rather than monocarbonylation, occurs to afford triesters **60** by carbonylation of the 3-methoxycarbonyl-2-propynyl methyl carbonates **58**, demonstrating that the ester group has a strongly activating effect [15]. The dicarbonylation proceeds at room temperature, and it is not possible to stop the reaction after the monocarbonylation to form **59**. Bidentate ligands such as DPPP and DPPF are the most effective for dicarbonylation. The alkynic esters **58** are easily prepared in one step by the treatment of propargylic alcohols **57** with two equivalents of *n*-BuLi, followed by the reaction of two equivalents of methyl chloroformate.

One possible mechanism is the following. The allenyl geminal diester **61**, not isolable, is expected to be susceptible to Michael-type addition of Pd(0)L*ₙ* species to the allenyl sp carbon, resulting in the formation of the palladacyclopropane **62**. Insertion of CO to **62** and methanolysis afford the triester **63**. The alkene geometry of the product **63** is exclusively (E). The high stereoselectivty is rationalized by assuming that a nucleophilic attack of Pd(0) species to the allenyl sp carbon in **61** occurs from the less hindered side of a smaller alkyl substituent (R_s).

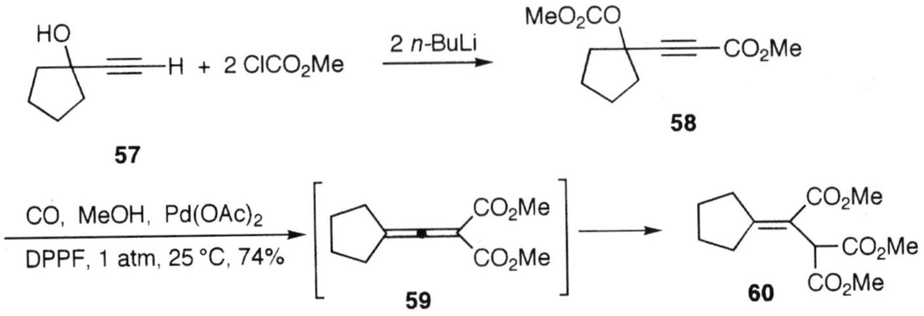

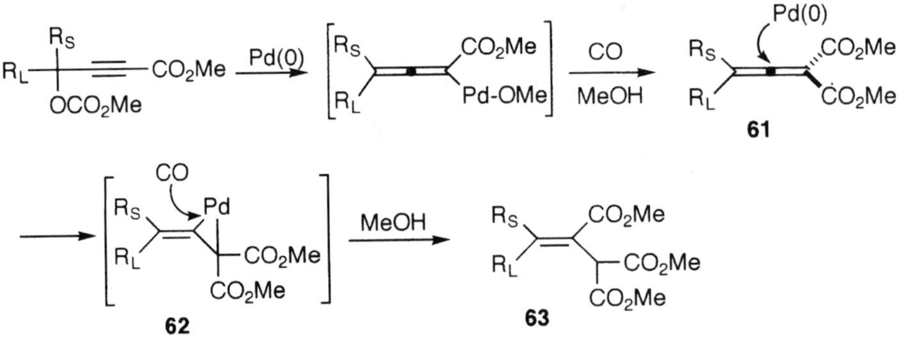

Bis(methylene)butanedioate **68**, which is an interesting derivative of butadiene as well as acrylate, can be prepared by the dicarbonylation of the dicarbonate of butynediol **64** [16]. The monocarbonylation of **64** affords the allylic carbonate **65**, and the diester **68** is obtained by further carbonylation of π- or σ-allyl complexes **66** and **67**.

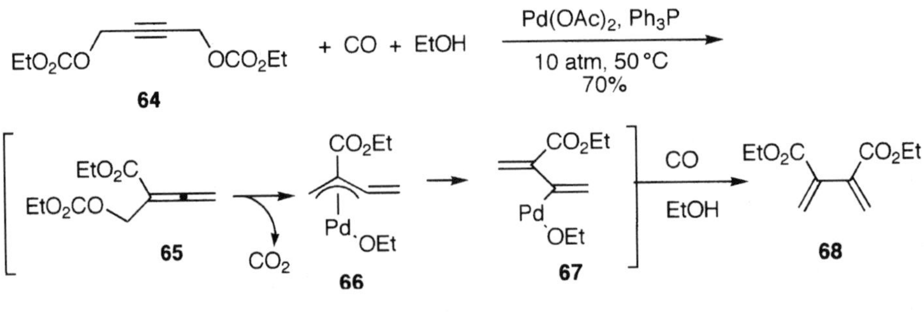

Propargylic alcohols are less reactive than their esters, and their carbonylation has been carried out under somewhat severe conditions (100 °C, 100 atm) [8,9]. Carbonylation of propargyl alcohol (**69**) in MeOH without using a phosphine ligand proceeds in the presence of HCl to afford methyl itaconate (**71**) as the main product and methyl aconitate (**72**) as a minor product. PdCl$_2$ or Pd/charcoal is an active catalyst [9]. 2,3-Butadienate (**70**) is a primary product of the Pd(0)-catalysed carbonylation.

Terminal and internal propargylic alcohols **73** are carbonylated using DPPB under CO-H$_2$ pressure, and 2(5H)-furanones **75** are obtained via the lactonization of 2,3-dienecarboxylic acids **74**. Hydrogen pressure is claimed to be essential [17].

Dicarbonylation in aprotic solvents yields acid anhydrides. Fulgide (dimethylene-succinic anhydride derivative) **78** was obtained in 49% yield by the Pd(0)-catalysed dicarbonylation of 2,5-dimethyl-3-hexyne-2,5-diol (**76**) via **77** in benzene [6]. The fulgide-forming reaction proceeds more smoothly using Pd(OAc)$_2$ as the catalyst in the presence of iodine (Pd:I$_2$ = 1:1) instead of hydrochloric acid [18].

The α-vinylidene-γ-lactones **80** are prepared from 5-hydroxy-2-alkynyl methyl carbonates **79** under mild conditions in good yields [19]. The reaction proceeds at room temperature under 1–10 atm of CO. Bidentate ligands, particularly DPPP and DPPF, are the best ligands.

Carbonylation of propargyl carbonates bearing an amino group yields lactams. The α-vinylidene–β-lactams **82** are prepared by the carbonylation of 4-benzylamino-2-alkynyl methyl carbonates **81** [20]. The best results are obtained by using the cyclic phosphite (4-ethyl-2,6,7-trioxa-1-phosphabicyclo[2,2,2]octane) (**83**). The lactam formation is carried out in THF or MeCN as solvents at 50 °C under 1–10 atm of CO.

Active methylene compounds can be used for trapping the acylpalladium intermediates [21]. The triketone **85** is obtained in the presence of cyclohexane-1,3-dione (**84**). The carbonylation proceeds under 1 atm of CO at 50 °C.

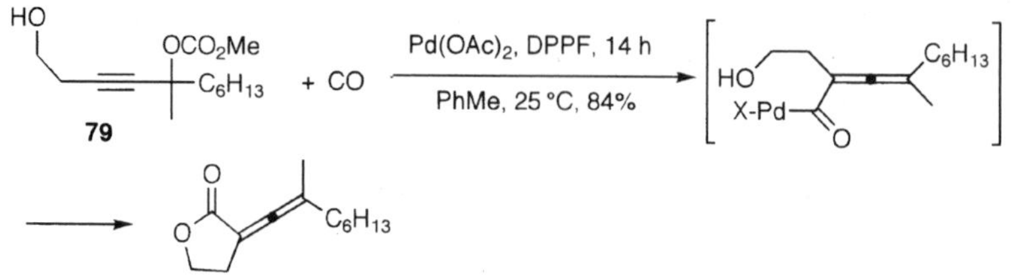

Interesting domino carbonylation, Diels–Alder and ene reactions are possible in the presence of alkene bonds in propargylic carbonates. The allenyl esters formed by the carbonylation of propargylic carbonates are highly reactive, and undergo subsequent intramolecular reaction in the presence of alkenic bonds in the same molecules. The smooth carbonylation, followed by intramolecular Diels–Alder reaction of the dienyne carbonates **86**, proceeds at 50 °C under 1 atm of CO in benzene smoothly without stopping after formation of the allenyl esters **87**, and the cyclized products **88** of the intramolecular Diels-Alder reaction are obtained in good yields [22]. As a ligand, DPPP is the most suitable.

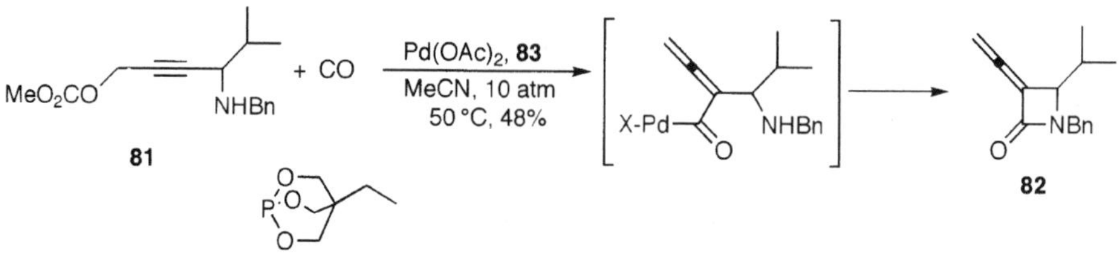

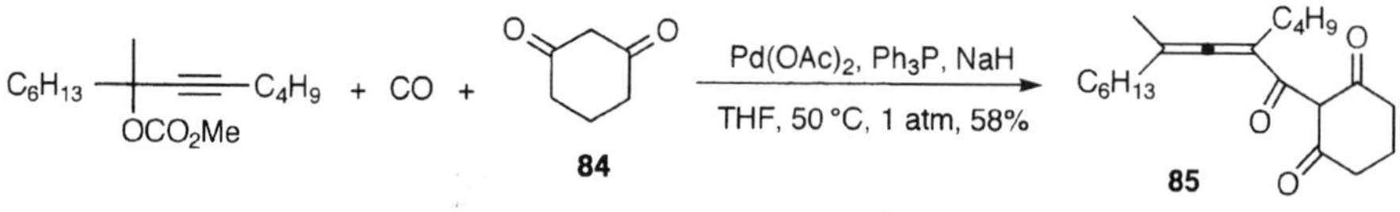

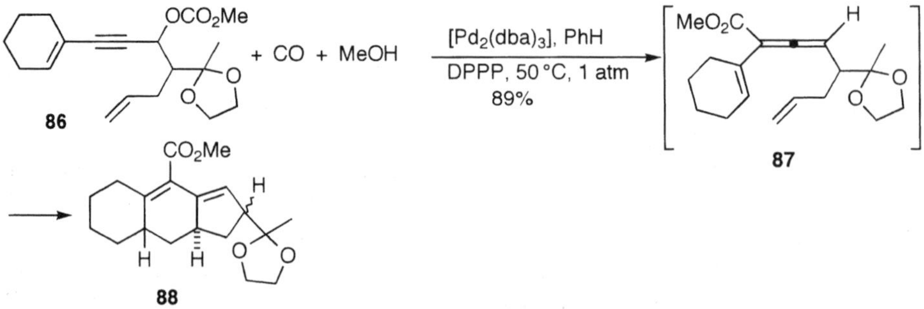

The reaction proceeds even at room temperature when a dienophile has an electron-donating group. The cyclized products **91** and **92** are obtained in good yields from propargylic carbonate **89** at room temperature under 1 atm. This means that the Diels–Alder reaction of the intermediate **90**, in which the diene has an electron-withdrawing group (CO₂Me) and the dienophile has an electron-donating group (OMe), proceeds at room temperature. The electronic effect in this case is opposite to the effect usually observed in Diels–Alder reactions [23]. This reaction offers a good synthetic route to the bicyclic skeleton of glycinoeclepin (**93**).

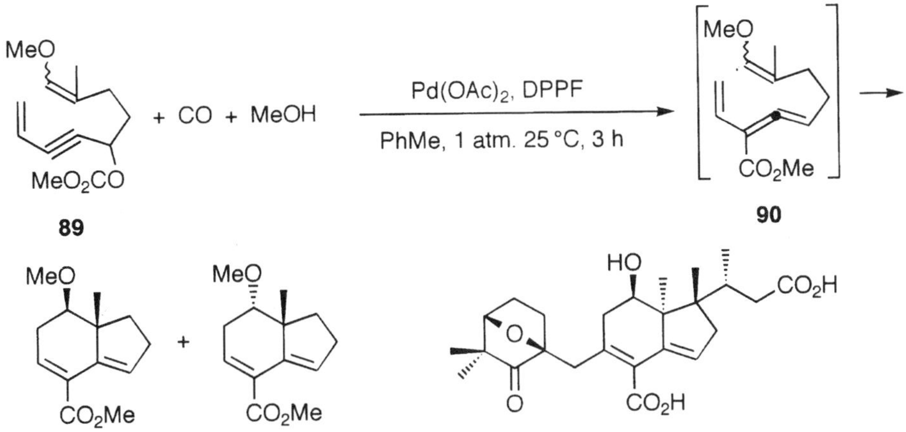

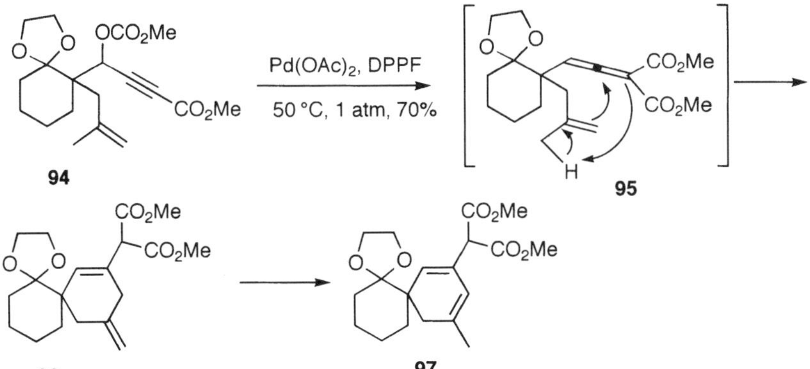

The allenyl esters formed by the carbonylation undergo an intramolecular ene reaction when there exists an alkenic bond at a suitable position. A particularly facile ene reaction is observed with the propargyl carbonates with an ester group attached to the triple bond [24]. The propargyl carbonate **94**, bearing an isopropenyl terminus, undergoes Pd-catalysed monocarbonylation, followed by ene reaction via **95** to yield the six-membered ring **96**. The ensuing thermodynamically controlled alkene isomerization from the 1,4-diene **96** to the 1,3-diene, affords 1,3-cyclohexadiene **97** in high yields. In sharp contrast, the carbonylation of propargyl carbonate **98** with the 2-methyl-1-propenyl terminus yields the five-membered ring **100**, having an isopropenyl group via the intermediate **99**. The triesters corresponding to **60** are not formed by further carbonylation of the intermediary allenyl geminal diesters **95** and **99**. Instead, the intramolecular ene reaction proceeds preferentially. The reaction takes place smoothly at 50 °C under 1 atm of CO. Pd(OAc)$_2$ combined with DPPP or DPPF (1:1 ratio) is used as a suitable catalyst.

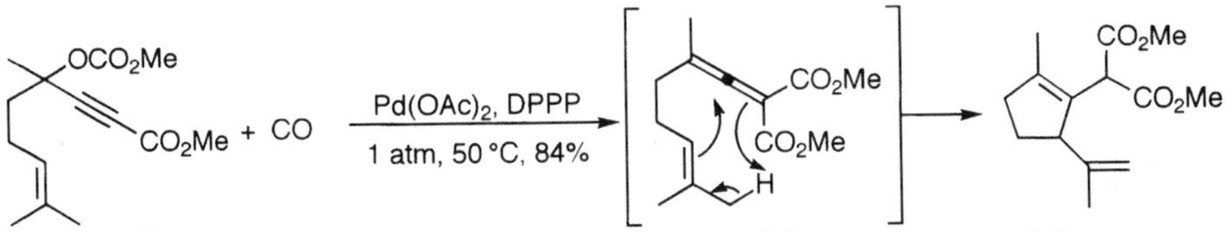

The domino carbonylation and Diels–Alder reaction proceed only as an intramolecular version. Attempted carbonylation and intermolecular Diels–Alder reaction of conjugated 2-yne-4-enyl carbonates **101** in the presence of various alkenes as dienophiles give entirely different carbocyclization products without undergoing the intermolecular Diels–Alder reaction. The 5-alkylidene-2-cyclopenten-4-onecarboxy-lates **102** were obtained unexpectedly by the incorporation of two molecules of CO in 82% yield from **101** at 50 °C under 1 atm [25]. The use of bidentate ligands such as DPPP or DPPE is important. The following mechanism of the carbocyclization of **103** has been proposed. The formation of palladacyclopentene **105** from **104** (oxidative cyclization) is proposed as an intermediate of **108**. Then CO insertion to the palladacycle **105** generates acylpalladium **106**. Subsequent reductive elimination affords the cyclopentenone **107**, which isomerizes to the cyclopentenone **108** as the final product.

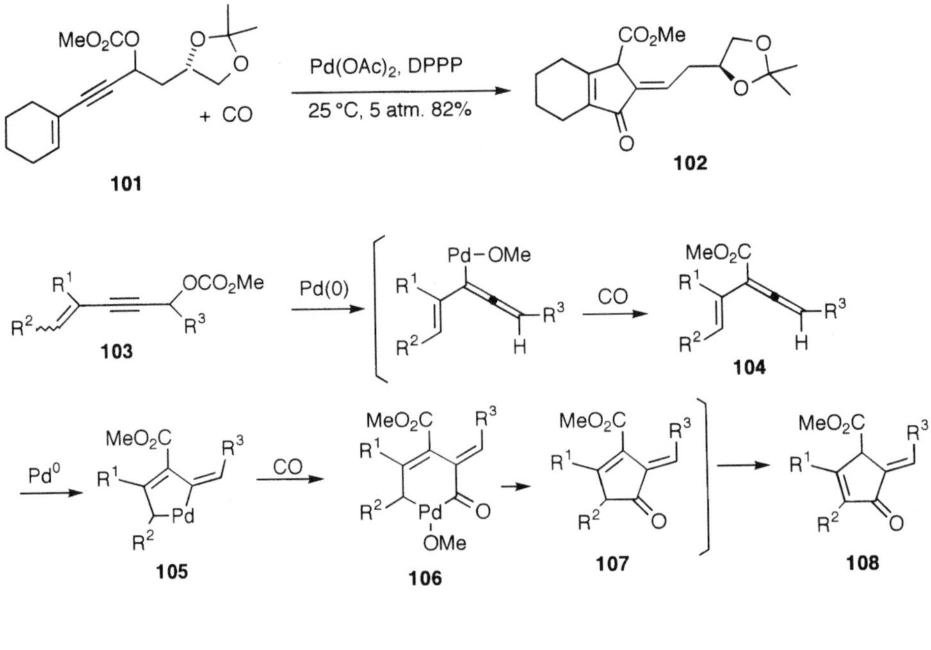

6.3 Reactions via Transmetallation and Related Reactions (Type II)

The reactions of propargylic halides, acetates and phosphates with hard carbon nucleophiles M′R (M′ = main group metals such as Mg, Zn or B) gives allenyl derivatives. The reaction of octylmagnesium chloride with 3-chloro-1-butyne (**109**) gives 2,3-dodecadiene (**110**) [26]. The 1,3-dimethylated allenes **114** are obtained by the Ni-catalysed reaction of propargylic dithioacetals **111** with MeMgI. Based on this reaction, the propargylic dithioacetal **111** can be regarded as 1,3-dication synthons of allene **115** [27]. In this coupling, oxidative addition of Ni(0) to the C−S bonds in **111** and **112** and subsequent transmetallation afford **114**.

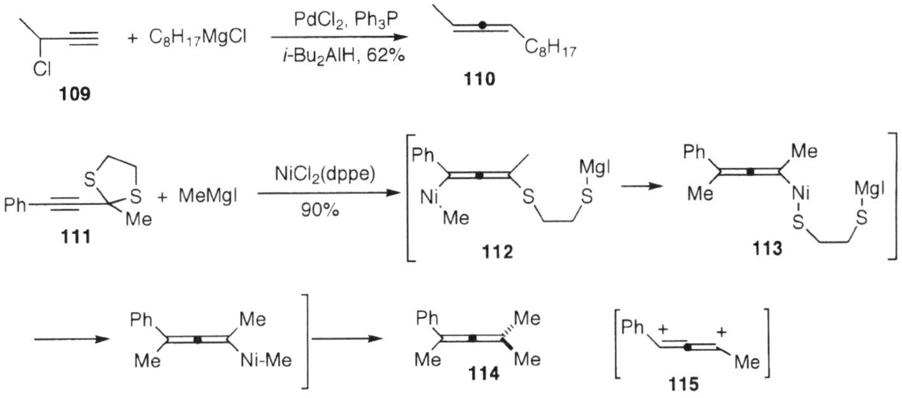

Reaction of PhZnCl with 3-acetoxy-3-methyl-1-butyne (**116**) gives 1-phenyl-3-methyl-1,2-butadiene (**119**) in high yield [28–30]. The reaction can be explained by transmetallation of the allenylpalladium intermediate **117** with PhZnCl to generate the allenyl(phenyl)palladium intermediate **118**, followed by reductive elimination to afford **119**.

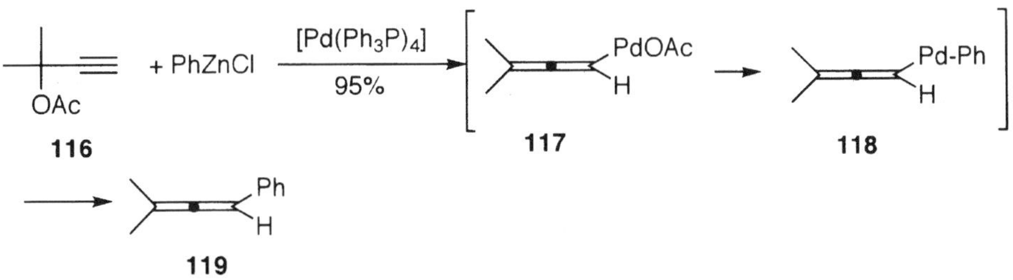

The (R) allene **121** was obtained with high *anti* stereoselectivity in the reaction of (R)-(−)-1-trifluoroacetoxy-1-phenyl-2-propyne (**120**) with PhZnCl. 2-Alkynyloxiranes react smoothly with alkynyl, alkenyl and arylzinc reagents. Reaction of 2-methyl-2-(1-propynyl)oxirane (**122**) with vinylzinc chloride (**123**) yields 2,4-dimethyl-2,3,5-hexatrien-1-ol (**124**) [31].

Organoboranes react with propargylic carbonates. Usually, addition of a base is indispensable for Pd-catalysed reactions of organoboranes with aryl, alkenyl and allyl halides. However, the reaction of organoboranes with methyl propargyl carbonates proceeds without addition of the base, because methoxide as the base is generated *in situ* from the carbonates. For example, the 1,2,4-alkatriene **127** is obtained by the reaction of alkenylboranes **126** with propargylic carbonate **125** under neutral conditions [32].

(R)-**120**

[Pd(Ph₃P)₄]
THF, Et₂O

(R)-**121**

anti : syn
82 : 18

122 + **123** ZnCl

[Pd(Ph₃P)₄], THF
25 °C, 75%

124

C₆H₁₃ ... C₄H₉ **125** + **126**

[Pd(Ph₃P)₄]
THF, 95%

127

The Pd/Cu-catalysed coupling of propargyl compounds with terminal alkynes proceeds via transmetallation. The allenylalkynes **133** can be prepared in good yields by Pd-catalysed coupling of terminal alkynes **130** with propargylic compounds **128**, such as carbonates, acetates and halides, in the presence of a catalytic amount of CuI as a cocatalyst. Addition of CuI is not necessary when metal acetylides are used. The Pd/Cu-catalysed coupling of propargylic carbonates with terminal alkynes proceeds at room temperature rapidly within 30 min [33]. The reaction is explained by the following mechanism. At first, reaction of CuI with the terminal alkyne **130** affords the copper acetylide **131**, and its transmetallation with **129** generates the allenyl(alk-ynyl)palladium intermediate **132**. Finally its reductive elimination gives the allenylalkyne **133** at room temperature in 85% yield in 25 min.

Various metal acetylides are used for smooth coupling with propargylic halides and acetates. 2,3-Alkadien-5-yn-1-ols are obtained by the reaction of 2-(1-alkynyl)oxiranes [28,29]. As a synthetic application, the unstable 2,3-octadiene-5,7-diyn-1-ol (**136**), a fungus metabolite, has been synthesized by the coupling of 4-trimethylsilylbutadiy-nylzinc chloride (**134**) with 2-ethynyloxirane (**135**) followed by desilylation [31].

The allenylalkyne **139**, the primary product of the coupling of **137** and **138**, undergoes further reaction with terminal alkyne. The Pd-catalysed bis-alkynylation of propargylic carbonate **137** occurs to afford enediynes **144**. The reaction is understood by the following mechanism. The insertion of one double bond of the allene in **139** to the alkynylpalladium **141**, formed from the terminal alkyne **140**, generates **142** which is stabilized as the π-allylpalladium **143**. Reductive elimination of **143** affords the enedyne **144** [34].

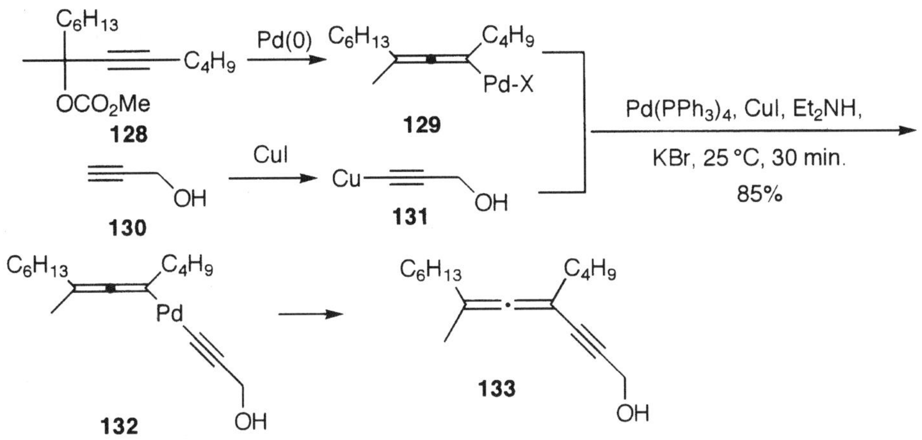

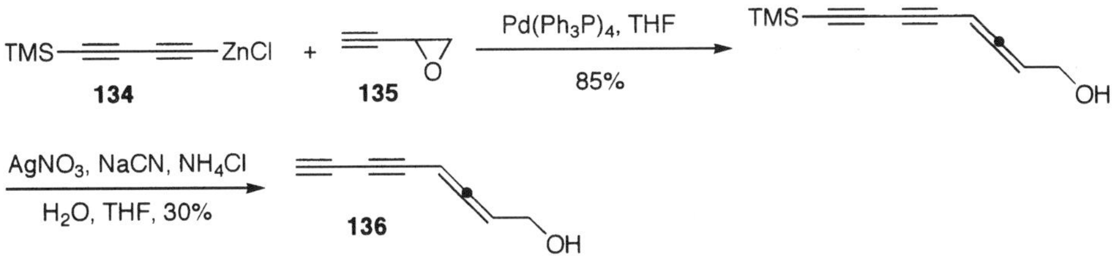

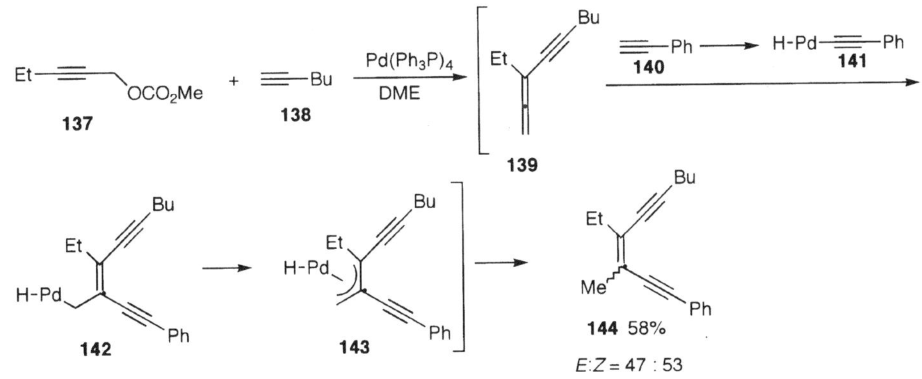

6.4 Reactions with C, O and N Nucleophiles (Type III)

Type **III** reaction proceeds by an attack of a nucleophile at the central sp carbon of the allenylpalladium. Soft carbon nucleophiles such as β-keto esters and malonates react with propargylic carbonates under neutral conditions using DPPE as a ligand [35].

The 2,3-disubstituted propenes **151** and **152** are obtained by the reaction of 2-propynyl carbonate (**145**) with two moles of malonate under neutral conditions in boiling THF. Attack of the carbanion **146** to the central sp carbon of the allenylpalladium **147** generates **148**, which picks up a proton from malonate to form the π-allylpalladium intermediate **150**. The intermediate **148** can be considered as the palladium carbene complex **149**. Again, attack of the malonate anion **146** on the π-allylpalladium **150** gives **151**, as expected. Migration of the double bond in **151** affords **152** [35,36]. Thus propargylic carbonate **145** has two reaction sites for nucleophiles. The best ligand for the reaction is DPPE. The reaction is slow when Ph$_3$P is used.

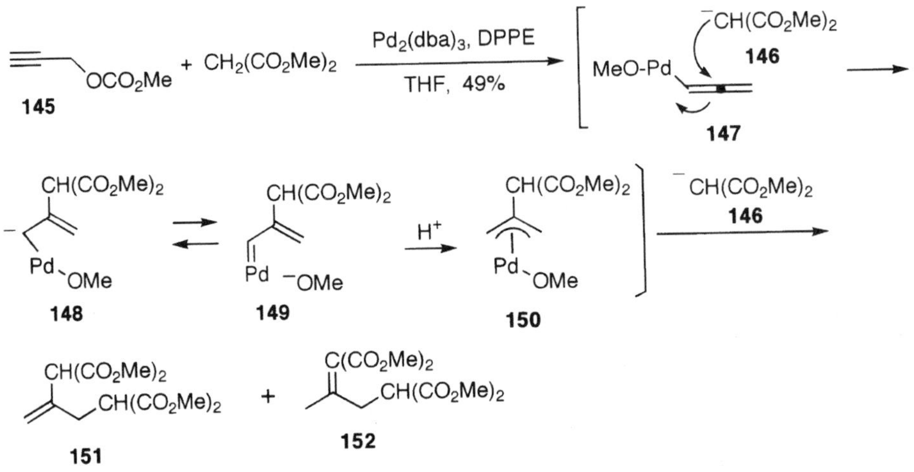

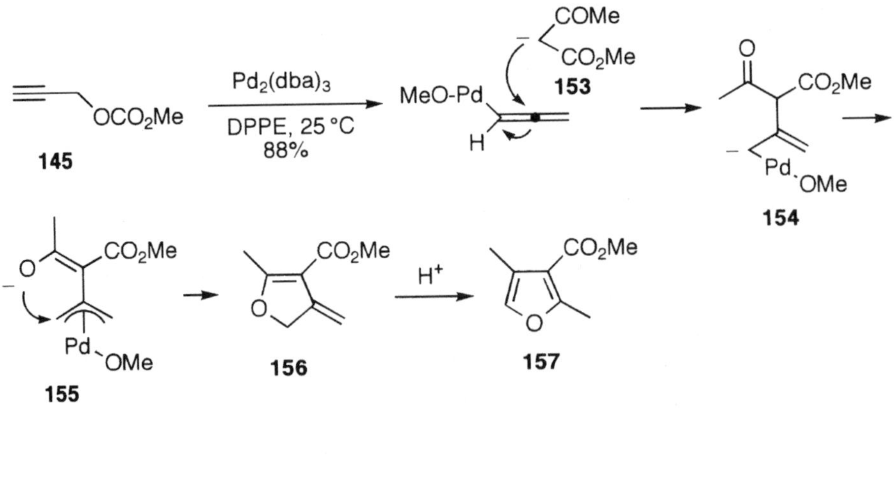

Methyl acetoacetate anion **153**, bearing one active hydrogen, reacts with methyl propargylic carbonate (**145**) in a 1:1 ratio in THF at room temperature, giving an entirely different product. At first, C-alkylation generates the π-allylpalladium intermediate **155** via **154**. The intramolecular attack of the oxygen nucleophile of the enolate **155** gives 4-(methoxycarbonyl)-5-methyl-3-methylene-2,3-dihydrofuran (**156**) in 88% yield under neutral conditions. The methylenefuran **156** is unstable and isomerizes to the stable furan **157** quantitatively under slightly acidic conditions [35, 36]. Thus the C- and O-alkylations of propargylic carbonates offer new synthetic routes to furans. Formation of the furan **159** by the reaction of acetonedicarboxylate (**158**) with **146** is an example.

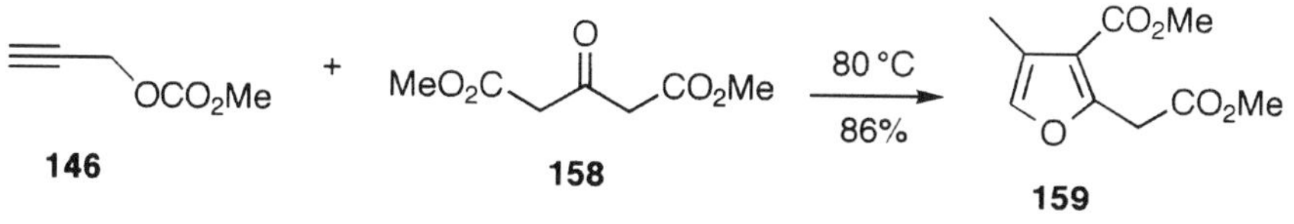

In order to study the mechanism of the furan formation, the following reactions were carried out. The reaction of isomeric 2-butynyl methyl carbonate (**160**) and 1-methylpropynyl methyl carbonate (**161**) with methyl acetoacetate (**162**) gives the same product **163**. This shows that the furan formation from **160** and **161** proceeds via a common intermediate.

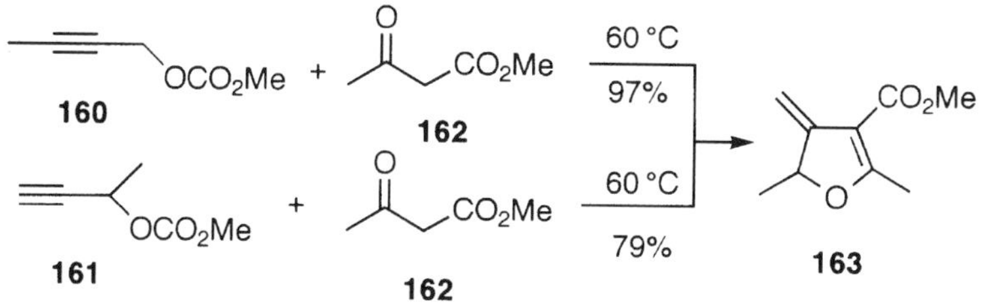

However, reactions of **160** and **161** with methyl 2,2-bisdeuterioacetoacetate (**164**) give the products deuterated at different carbons; namely 2-deuterio-3-hydrofuran **166** is obtained from **160** via **165**. Also, reaction of **161** with **164** affords the furan **168**, deuterated at the exomethylene carbon via **167**. These results are explained by assuming that the attack by the oxygen nucleophile (O-allylation) occurs at the more substituted side of π-allylpalladium systems in **165** and **167**.

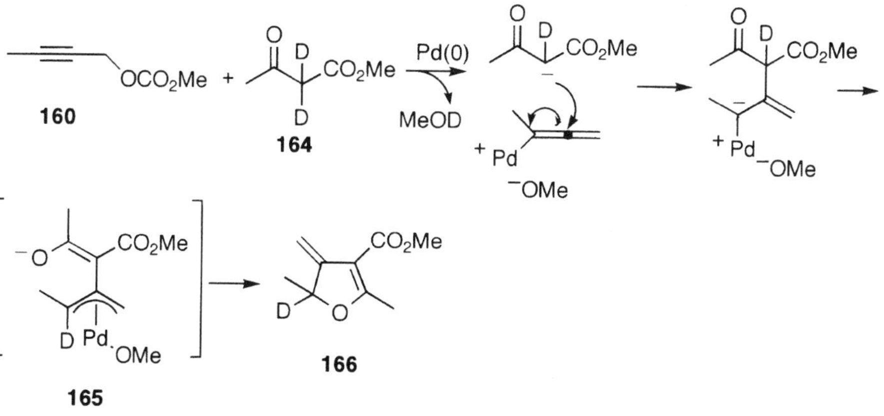

The 2-vinyloxiranes undergo Pd-catalysed furan annelation with soft nucleophiles [36,37]. The allenylpalladium **170** is generated from ethynyloxirane **169**, and the attack of acetoacetate anion **153** at the central carbon of **170** forms **171**. The oxygen anion attacks the more substituted side of π-allyl palladium **171** to give **172**, and the tetrasubstituted furan **173** is formed by elimination of formaldehyde as shown by **172**.

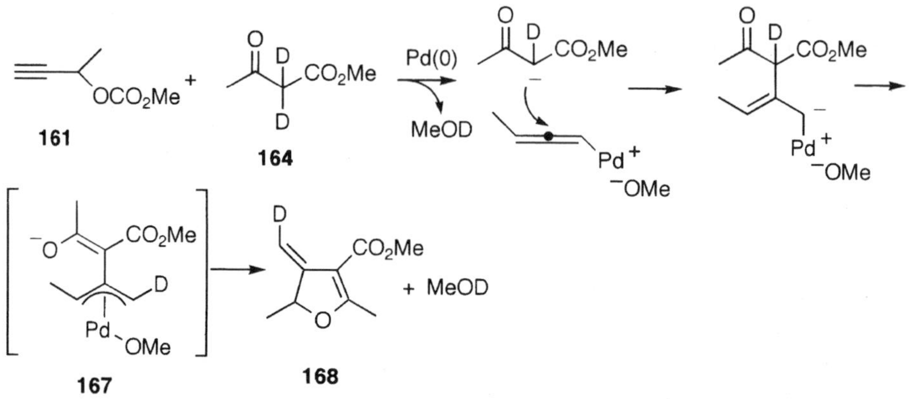

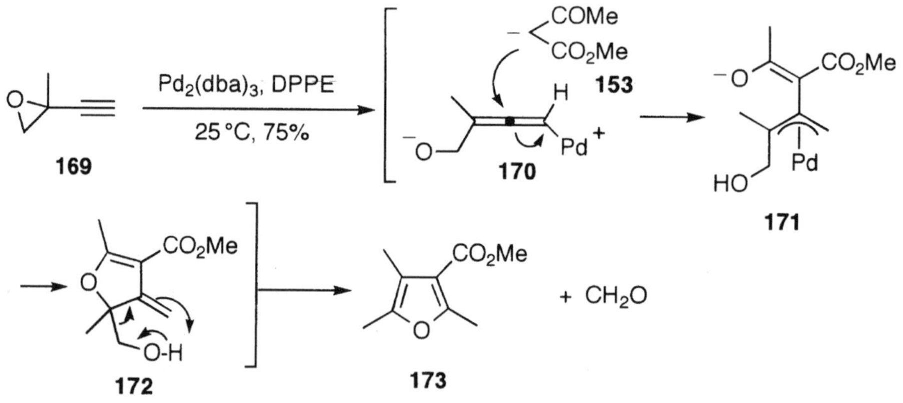

In an intramolecular reaction of the malonate in **174**, the intermediate π-allylpalladium complex **175** undergoes elimination of β-hydrogen to form the diene **176** [23].

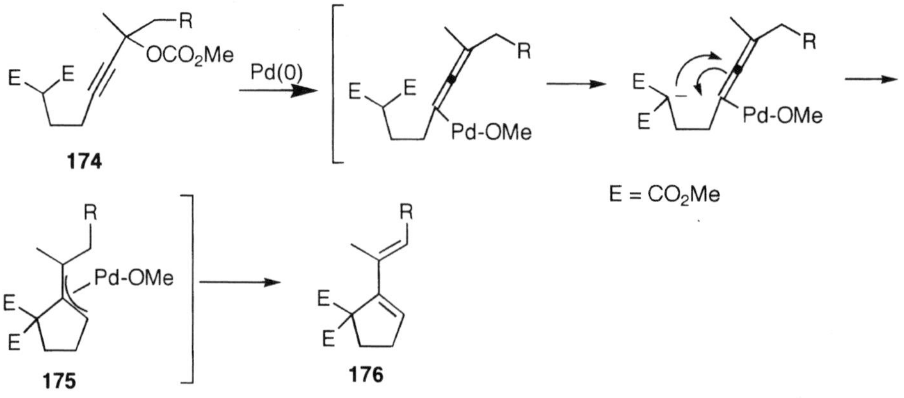

$E = CO_2Me$

Allenyl α,β-unsaturated isoprenoid aldehydes can be prepared in high yields by rearrangement of mixed propargyl enol carbonates. The carbonate **178** was prepared from β-ionone (**177**) and subjected to Pd-catalysed rearrangement, and a mixture of the allenyl aldehyde **180** as an $(E)/(Z)$ mixture and the alkynyl aldehyde **181** was obtained in 54% combined yield, accompanied by the elimination product **182** (25%) using trinaphthylphosphine as a ligand. The allenylpalladium **179** is an intermediate of the rearrangement and its reductive elimination affords **180**. The aldehyde **180** is converted to the conjugated polyenal with hydrogen bromide, and the all-*trans* isomer of retinal (**183**) was obtained by a simple equilibration [38]. The reaction is applicable to syntheses of polyunsaturated aldehydes and ketones. The single product **185** was obtained by the Pd-catalysed rearrangement of propargylic enol carbonate **184** in a high yield [39]. It should be added that the corresponding allyl enol carbonates undergo a similar rearrangement to give α-allyl aldehydes in high yields [40].

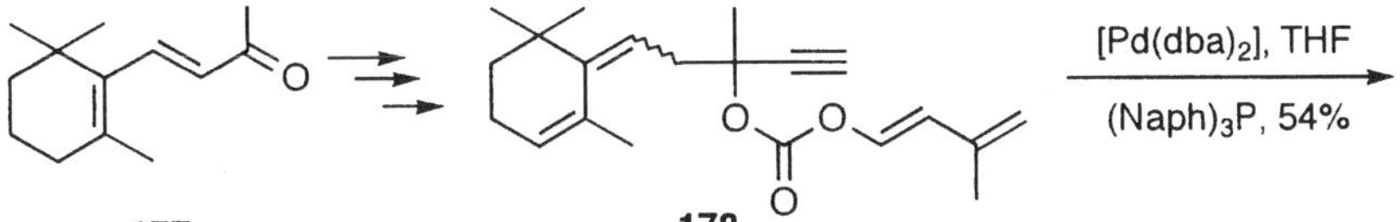

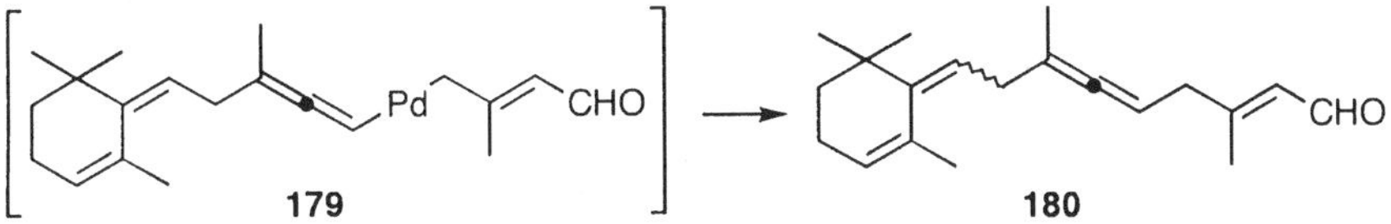

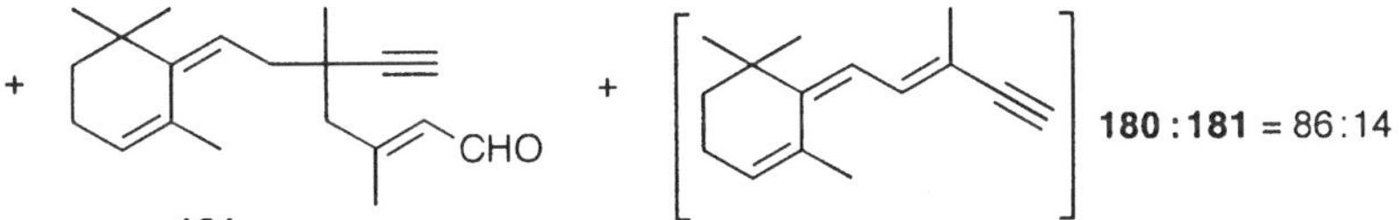

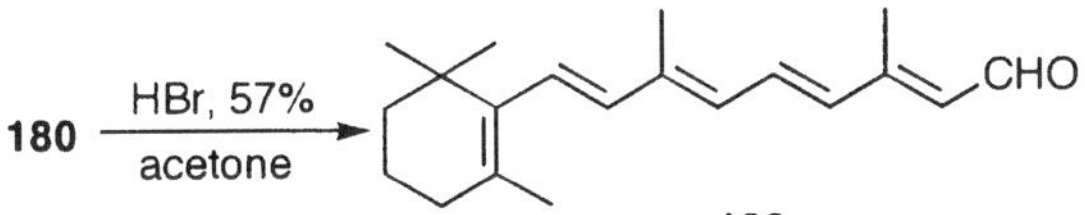

Umpolung of propargyl compounds occurs in the presence of excess Et_2Zn, and homopropargyl alcohols **187** are obtained by the reaction of propargyl benzoates (**186**) with benzaldehyde, although the exact mechanism is not known. Ethylallene, which is expected to be formed by transmetallation of the allenylpalladium with Et_2Zn and reductive elimination, is not formed [41].

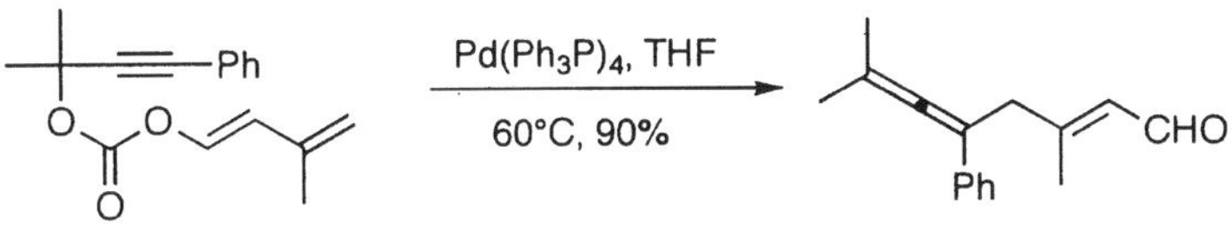

The propargyl carbonates **188**, which have a hydroxy group at C(5), undergo cyclization by an attack of an alkoxy group at the central carbon of the allenyl system **189**. The intermediate σ-allylpalladium complex **190** undergoes elimination of β-hydrogen to give the dienes **191** and **192**, which are converted to the more stable furan **193** as the final product [41,42]. The reaction proceeds in the presence of DBU in 80% yield in refluxing dioxane using DPPP. The primary products **191** and **192** are isomerized easily to furan **193** during column chromatography on silica gel. Similarly, dihydropyrans **196** and **197** are formed from 6-hydroxy carbonates **194** via the intermediate **195** [42,43].

Nitrogen nucleophiles also react smoothly. The reaction of optically active propargyl mesylate **198** with aniline without a catalyst gives **200** with inversion. However, the Pd-catalysed reaction affords **199** with retention of stereochemistry [44]. The 4-ethenylidene-2-oxazolidinone **202** is obtained by intramolecular reaction of the biscarbamate of the 2-butyn-1,4-diol **201** [45].

$$\text{201} \quad \xrightarrow[\text{rt, 73\%}]{\text{Pd}_2(\text{dba})_3,\ \text{Et}_3\text{N}} \quad \text{202}$$

6.5 Hydrogenolysis with Formate and Other Hydrides to Give Allenes and Internal Alkynes (Types II and IV)

Palladium-catalysed reactions via the propargylic palladium intermediate **13** is rather rare. One reaction that does proceed via **13** is decarboxylation–hydrogenolysis of propargylic formates which have internal alkynic bond and various functional groups. Allenes and internal alkynes are prepared by treatment of propargyl formates with Pd catalyst, or the hydrogenolysis of other propargyl compounds with HCO_2NH_4 or HCO_2H–Et_3N. Propargyl formates **203** or **205**, which have terminal alkyne or no substituent on propargylic carbon, are converted selectively to allenes **204** and **206** [46,47]. Allene formation can be carried out most satisfactorily from primary and secondary propargylic formates with a terminal alkyne. Tertiary propargylic formates cannot be prepared, and in that case the corresponding carbonates are hydrogenolysed by treatment with triethylammonium formate. However, substituted propargyl formates **208** are converted to the internal alkyne [47, 48]. Exceptionally, the primary propargylic formate **205** with an internal alkynic bond gives allene **206** as the main product, accompanied by a small amount of alkyne **207**. Allene formation has been applied to the synthesis of the corticoid β-methasone (**214**) [49]. The carbonate **211** was prepared from **210** using methyl phenyl carbonate. This is a better reagent than methyl chloroformate for the preparation of methyl carbonates from alcohols. The Pd-catalysed hydrogenolysis of propargyl carbonate **211** afforded allene **212** cleanly, which was converted to β-methasone (**214**) via **213**.

$$\text{203} \quad \xrightarrow[\text{4 h, 25 °C, 86\%}]{\text{Pd(OAc)}_2,\ n\text{-Bu}_3\text{P, PhH}} \quad \text{204} \ + \quad \quad 99:1$$

$$\text{205} \quad \xrightarrow[\text{25 °C, 93\%}]{\text{Pd(acac)}_2,\ n\text{-Bu}_3\text{P}}$$

$$\text{206} \quad 91:9 \quad + \quad \text{207}$$

$$\text{208} \quad \xrightarrow[\substack{\text{25 °C, 97\%,}\\ \text{selectivity 97\%}}]{\text{Pd(acac)}_2,\ n\text{-Bu}_3\text{P, PhH}} \quad \text{209}$$

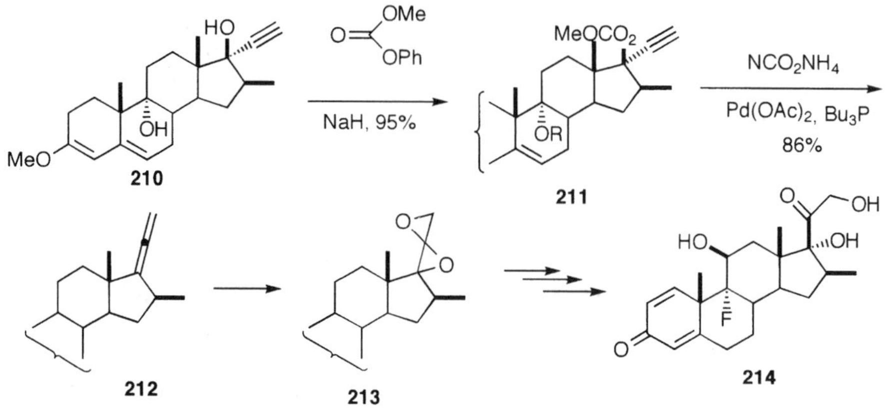

Allenes are reduced further to 1-alkenes with excess ammonium formate at 100 °C. Whereas allene **216** is formed with two equivalents of formate at room temperature in THF, terminal alkene **217** is obtained with excess HCO_2NH_4 directly from **215**. The use of propargyl formates is therefore recommended for clean allene formation without overreduction whenever the formates are available.

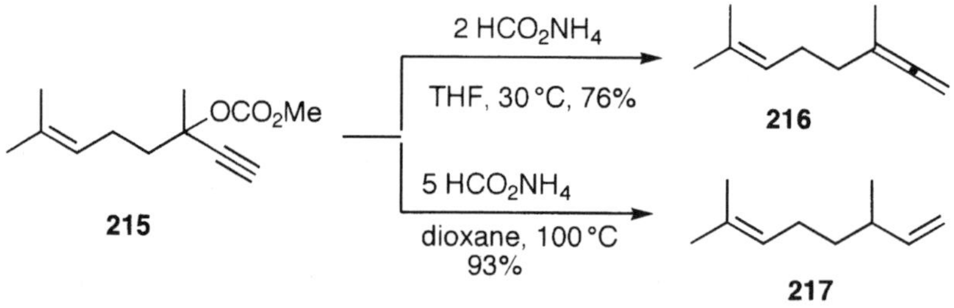

The following mechanism has been proposed for the hydrogenolysis of propargylic formate **219**. The allenylpalladium formate complex **221** is generated by oxidative addition of **219**. The same intermediate is formed by oxidative addition of the corresponding propargylic carbonate **218** via **220** in the presence of triethylammonium formate. Decarboxylation of **221** gives the allenylpalladium hydride complex **224**. Finally, reductive elimination of **224** gives the allene **225**, and the Pd(0) species is regenerated. Also, the propargylic formate complex **223** is formed from **218** depending on the substituent. The alkyne **227** is obtained from the propargylpalladium hydride complex **226** via **223**. Formation of the alkynes **227** as the main product from the hydrogenolysis of propargylic carbonate **218** is discussed here although the reaction belongs to type **IV**.

Palladium(0)-catalysed selective allene formation by hydrogenolysis is a convenient synthetic route to allenes from propargylic alcohols. The method offers a new and efficient preparative method for allenes from easily available propargylic alcohols after converting them to formates or carbonates. The best catalyst for smooth hydrogenolysis is prepared by mixing $Pd(OAc)_2$ or $Pd(acac)_2$ with n-Bu_3P in a ratio of 1:1 [50]. This reaction is also useful for the selective preparation of internal alkynes, particularly those with labile functional groups. It is not always easy to prepare disubstituted alkynes **229** by simple alkylation of the terminal alkynes **228**, particularly in the presence of labile functional groups. However, propargylic alcohols **230** are prepared more easily by reaction of terminal alkynes **228** with ketones or aldehydes, and internal alkynes **232** are prepared by the hydrogenolysis of their formates or carbonates **231** (formates of primary and secondary alcohols, and carbonates of tertiary alcohols).

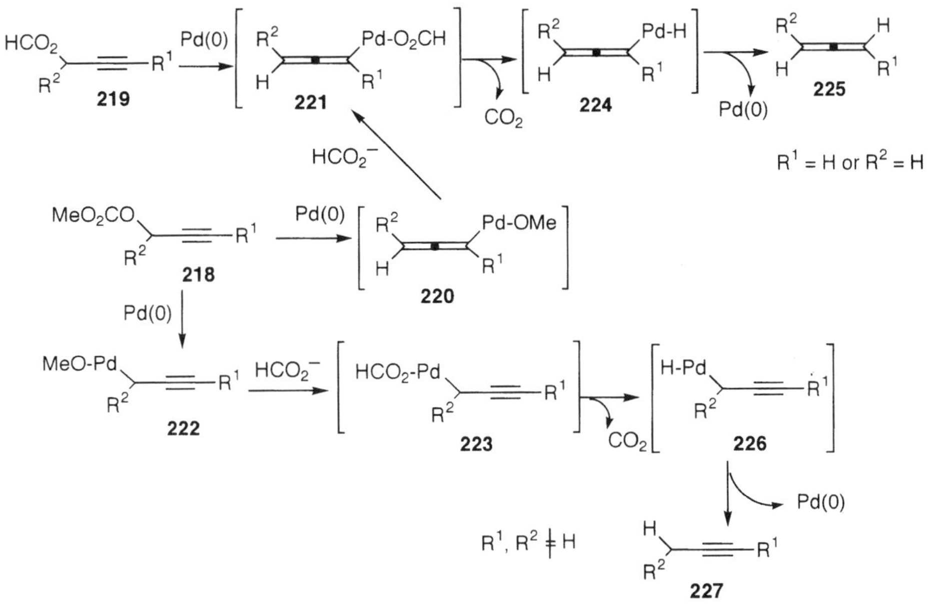

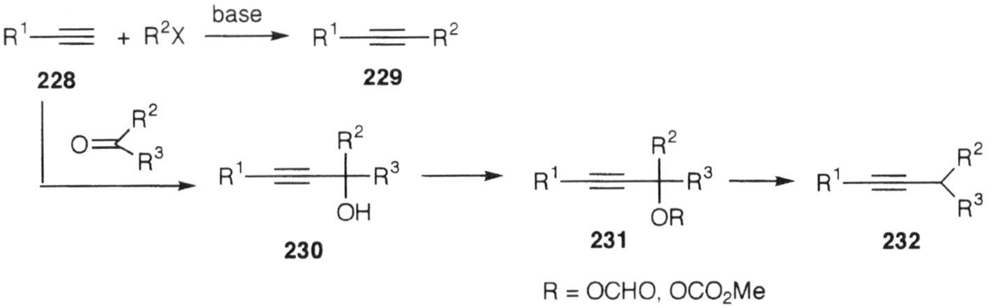

Reduction of propargylic compounds with SmI_2 is possible in the presence of Pd catalysts. Propargylic acetates **233** are converted mainly to the allene **236** by Pd-catalysed reaction with SmI_2 in the presence of a proton source [51]. In this reaction, the allenylpalladium **234** is reduced with Sm(II) to the allenyl anion **235**, which is protonated to give allene **236**. The alkyne **237** is a byproduct. 2,3-Naphthoquinodimethane (**240**) as a reactive intermediate, can be generated by applying this reaction.

Treatment of 1,2-di(1-acetoxypropynyl)benzene (**238**) with SmI_2 generates 1,2-dipropadienylbenzene (**239**), which spontaneously cyclizes to give 2,3-naphthoquino-dimethane (**240**). Then facile Diels–Alder reaction of **240** with fumarate (**241**) gives the tetrahydroanthracene derivative **242** [52].

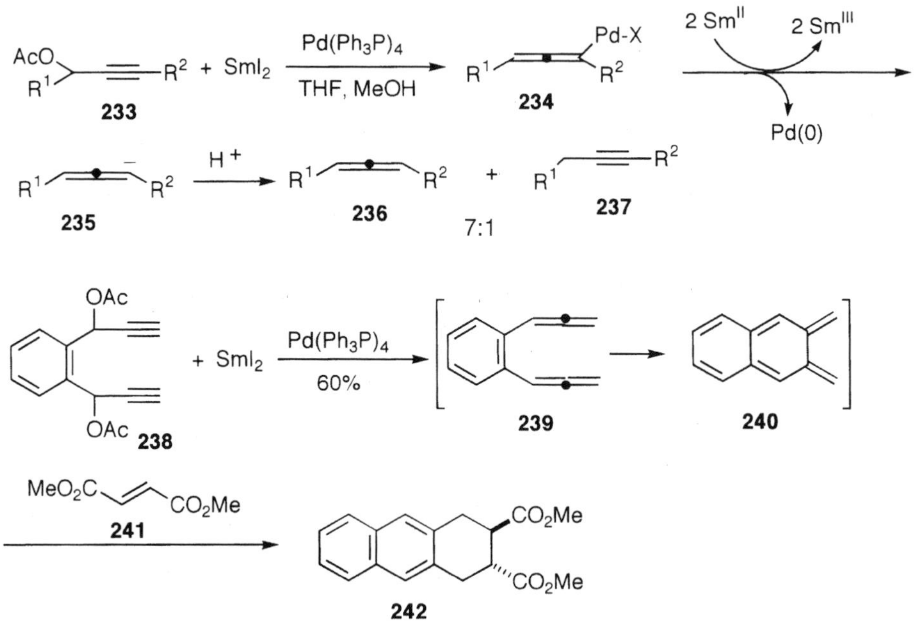

Racemization occurs by the reaction of a chiral propargyl phosphate with SmI_2 and *t*-BuOH. However, asymmetric synthesis of allenic esters by dynamic kinetic protonation of the racemic allenylpalladium species, generated from propargyl phosphates, has been carried out using SmI_2 and chiral proton sources. The racemic allenylmetal intermediates **245** and **246** are formed by the treatment of the propargyl phosphate **243** with SmI_2. Their protonation with (*R*)-pantolactone **244** as a chiral proton source proceeds enantioselectively and the (*R*)-allenic ester **247** with 95% ee was obtained [53].

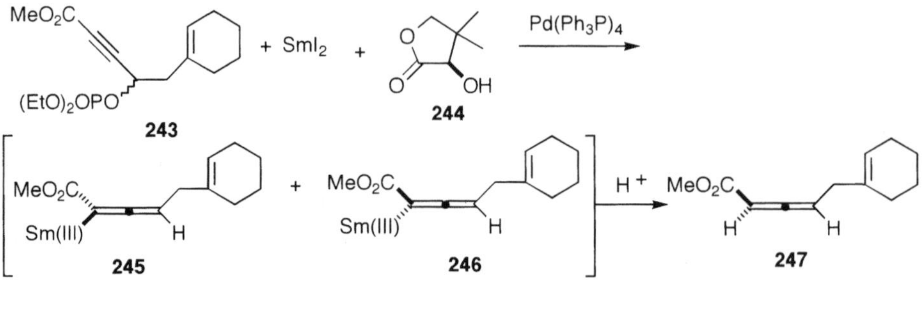

6.6 Elimination Reaction via Propargylpalladium Intermediates (Type IV)

When propargylic carbonates **248** are treated with a Pd catalyst in the absence of other reactants, elimination of β-hydrogen from the propargylpalladium intermediate **249** occurs to afford conjugated enynes **250**. Formation of 1,2,3-alkatriene **252** by elimination of the allenylpalladium intermediate **251** is not observed. The elimination reaction proceeds with a variety of propargylic carbonates under mild and neutral conditions. Bidentate phosphines, particularly DPPF is the best ligand for tertiary carbonates, and Ph_3P is used for secondary carbonates. The reaction of tertiary propargylic carbonate **253**, derived from cyclooctanone, proceeds smoothly to give the conjugated enyne **254** in good yield. The reaction is applicable to other cyclic ketones except cyclohexanone [47,54].

6.7 Miscellaneous Reactions

Some reactions of propargylic compounds which do not belong to the above-mentioned types are known. These reactions are surveyed in this section, although their mechanisms are not always clear.

Propargylic acetates are less reactive than the corresponding carbonates toward Pd(0) and do not form allenylpalladium intermediates easily. Addition reaction of acetic acid to propargylic acetate **255** catalysed by $Pd_2(dba)_3$ and Ph_3P gives the allylic geminal diacetate **258**. The reaction is explained by the isomerization of triple bond to allene **256**, and the insertion of its double bond to H-Pd–OAc to generate π-allylpalladium **257**, which is attacked by acetoxyl anion to give **258**. An intramolecular version of this reaction offers a good synthetic route to acetoxylactones. The 14-membered acetoxylactone **260** was obtained in refluxing benzene in 52% yield from **259** [55].

A Pd-catalysed variant of the Nazarov cyclization of 1-ethynyl-2-propenyl acetate derivatives to form cyclopentenone dierivatives involving 1,2-acetoxy migration is possible by catalysis of the Pd(II) salt. 3-Acetoxty-4-methylene-1-decyne (**261**) was converted to 3-hexyl-2-cyclopentenone (**263**) in 63% yield in the presence of one equivalent of acetic acid. Formation of the acetoxycyclopentadiene **262** as a precursor to **263** was confirmed by trapping with *N*-phenylmaleimide. However, the mechanism of this interesting reaction is not clear [56].

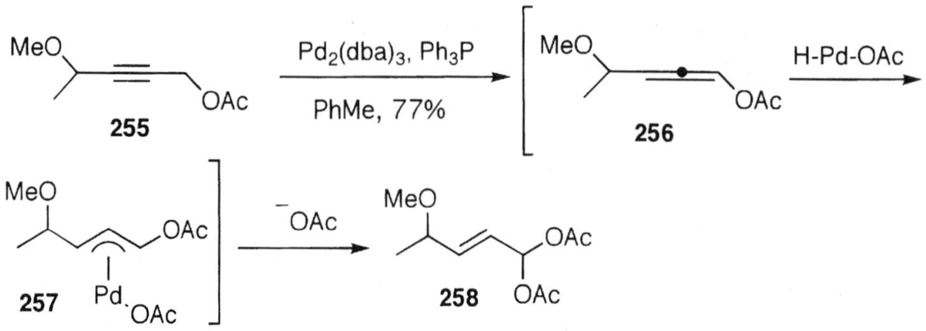

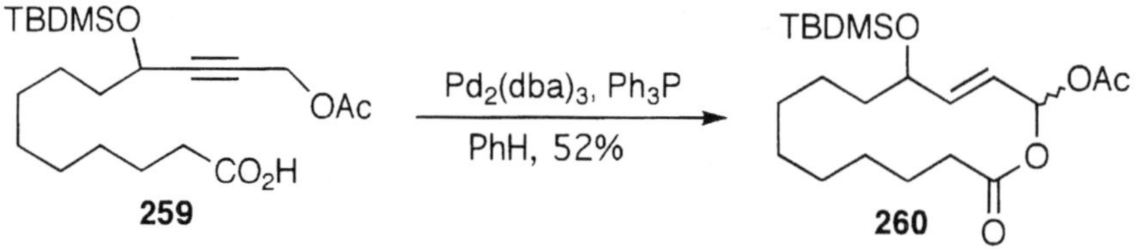

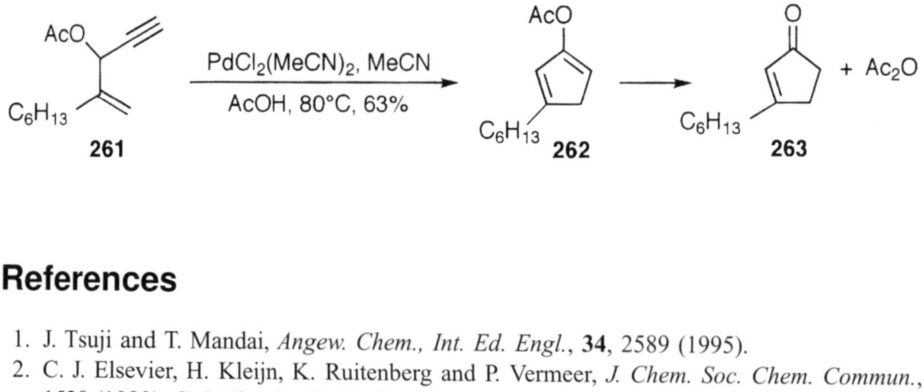

References

1. J. Tsuji and T. Mandai, *Angew. Chem., Int. Ed. Engl.*, **34**, 2589 (1995).
2. C. J. Elsevier, H. Kleijn, K. Ruitenberg and P. Vermeer, *J. Chem. Soc. Chem. Commun.*, 1529 (1983); C. J. Elsevier, H. Kleijn, J. Boersma and P. Vermeer, *Organometallics*, **5**, 716 (1986).
3. T. Mandai, M. Ogawa, H. Yamaoki, T. Nakata, H. Murayama, M. Kawada and J. Tsuji, *Tetrahedron Lett.*, **33**, 3397 (1991).
4. R. Grigg, R. Rasul, J. Redpath and D. Wilson, *Tetrahedron Lett.*, **37**, 4609 (1996).
5. R. Grigg, R. Rasul and V. Savic, *Tetrahedron Lett.*, **38**, 1825 (1997).
6. Review, J. Tsuji and T. Mandai, *J. Organometal. Chem.*, **451**, 15 (1993).
7. J. Tsuji, T. Sugiura and I. Minami, *Tetrahedron Lett.*, **27**, 731 (1986).
8. J. Tsuji and T. Nogi, *Tetrahedron Lett.*, 1801 (1966).
9. T. Nogi and J. Tsuji, *Tetrahedron*, **25**, 4099 (1969).

10. H. Arzoumanian, M. Choukrad and D. Nuel, *J. Mol. Catal.*, **85**, 287 (1993).
11. N. D. Trieu, C. J. Elsevier and K. Vrieze, *J. Organometal. Chem.*, **325**, C23 (1987).
12. J. A. Marshall and E. M. Wallace, *J. Org. Chem.*, **60**, 796 (1995).
13. M. E. Piotti and H. Alper, *J. Org. Chem.*, **62**, 8484 (1997).
14. Y. Imada and H. Alper, *J. Org. Chem.*, **61**, 6766 (1996).
15. T. Mandai, Y. Tsujiguchi, S. Matsuoka, J. Tsuji and S. Saito, *Tetrahedron Lett.*, **35**, 569 (1994).
16. J. Kiji, T. Okano, E. Fujii and J. Tsuji, *Synthesis*, 869 (1997).
17. W. Y. Yu and H. Alper, *J. Org. Chem.*, **62**, 5684 (1997).
18. J. Kiji, H. Konishi, T. Okano, S. Kometani and A. Iwasa, *Chem. Lett.*, 313 (1987).
19. T. Mandai, Y. Tsujiguchi, S. Matsuoka, S. Saito and J. Tsuji, *J. Organometal. Chem.*, **488**, 127 (1995).
20. T. Mandai, K. Ryoden, M. Kawada and J. Tsuji, *Tetrahedron Lett.*, **32**, 7683 (1991).
21. T. Mandai, H. Kunitomi, K. Higashi, M. Kawada and J. Tsuji, *Synlett*, 697 (1991).
22. T. Mandai, S. Suzuki, A. Ikawa, T. Murakami, M. Kawada and J. Tsuji, *Tetrahedron Lett.*, **32**, 7687 (1991).
23. T. Mandai and J. Tsuji, unpublished results.
24. T. Mandai, Y. Tsujiguchi, J. Tsuji and S. Saito, *Tetrahedron Lett.*, **35**, 5701 (1994).
25. T. Mandai, J. Tsuji, Y. Tsujiguchi and S. Saito, *J. Am. Chem. Soc.*, **115**, 5865 (1993).
26. T. Jeffery Luong and G. Linstrumelle, *Tetrahedron Lett.*, **21**, 5019 (1980).
27. H. R. Tseng and T. Y. Luh, *J. Org. Chem.*, **61**, 8685 (1996).
28. K. Ruitenberg, H. Kleijn, C. J. Elsevier, J. Meijer and P. Vermeer, *Tetrahedron Lett.*, **22**, 1451 (1981); K. Ruitenberg, H. Kleijn, H. Westmijze, J. Meijer and P. Vermeer, *Rec. Trav. Chim. Pays-bas*, **101**, 405 (1982).
29. E. Keinan and E. Bosch, *J. Org. Chem.*, **51**, 4006 (1986).
30. J. Elsevier, P. M. Stehouwer and H. Westmijze, *J. Org. Chem.*, **48**, 1103 (1983).
31. H. Kleijn, I. Meijer, G. C. Overbeek and P. Vermeer, *Rec. Trav. Chim. Pays-bas*, **101**, 97 (1982).
32. T. Moriya, N. Miyaura and A. Suzuki, *Synlett*, 149 (1994).
33. T. Mandai, T. Nakata, H. Murayama, H. Yamaoki, M. Ogawa, M., Kawada and J. Tsuji, *Tetrahedron Lett.*, **31**, 7179 (1990); *J. Organometal. Chem.*, **417**, 305 (1991).
34. M. Hayashi and K. Saigo, *Tetrahedron Lett.*, **38**, 6241 (1997).
35. J. Tsuji, H. Watanabe, I. Minami and I. Shimizu, *J. Am. Chem. Soc.*, **107**, 2196 (1985).
36. I. Minami, M. Yuhara, H. Watanabe and J. Tsuji, *J. Organometal. Chem.*, **334**, 225 (1987).
37. I. Minami, M. Yuhara and J. Tsuji, *Tetrahedron Lett.*, **28**, 629 (1987).
38. H. Bienayme, *Tetrahedron Lett.*, **35**, 7383 (1994).
39. H. Bienayme, *Tetrahedron Lett.*, **35**, 7387 (1994).
40. J. Tsuji, I. Minami and I. Shimizu, *Tetrahedron Lett.*, **24**, 1793 (1983).
41. Y. Tamaru, S. Goto, A. Tanaka, M. Shimizu and M. Kimura, *Angew. Chem., Int. Ed. Engl.*, **35**, 878 (1996).
42. T. Mandai and J. Tsuji, *Angew. Chem., Int. Ed. Engl.*, **34**, 2608 (1995).
43. C. Foumier-Nguefack, P. Lhoste and D. Sinou, *Synlett*, 553 (1996).
44. J. A. Marshall and M. A. Wolf, *J. Org. Chem.*, **61**, 3238 (1996).
45. M. Kimura, Y. Wakamiya, Y. Horino and Y. Tamaru, *Tetrahedron Lett.*, **38**, 3963 (1997).
46. J. Tsuji, T. Sugiura, M. Yuhara and I. Minami, *Chem. Commun.*, 922 (1986); J. Tsuji, T. Sugiura and I. Minami, *Synthesis*, 603 (1987).
47. T. Mandai, T. Matsumoto, Y. Tsujiguchi, S. Matsuoka and J. Tsuji, *J. Organometal. Chem.*, **473**, 343 (1994).
48. T. Mandai, T. Matsumoto, M. Kawada and J. Tsuji, *Tetrahedron Lett.*, **34**, 2161 (1993).

49. D. R. Andrews, R. A. Giusto and A. R. Sudhaker, *Tetrahedron Lett.*, **37**, 3417 (1996).
50. T. Mandai, T. Matsumoto, J. Tsuji and S. Saito, *Tetrahedron Lett.*, **34**, 2513 (1993).
51. T. Tabushi, J. Inanaga and M. Yamaguchi, *Tetrahedron Lett.*, **27**, 5237 (1986).
52. J. Inanaga, Y. Sugimoto and T. Hanamoto, *Tetrahedron Lett.*, **33**, 7035 (1992).
53. K. Mikami and A. Yoshida, *Angew. Chem., Int. Ed. Engl.*, **36**, 858 (1997).
54. T. Mandai, Y. Tsujiguchi, S. Matsuoka and J. Tsuji, *Tetrahedron Lett.*, **34**, 7615 (1993).
55. B. M. Trost, W. Brieden and K. H. Baringhaus, *Angew. Chem., Int. Ed. Engl.*, **31**, 1335 (1992).
56. V. Rautenstrauch, *J. Org. Chem.*, **49**, 950 (1984).

7

REACTIONS OF ALKENES AND ALKYNES

Alkenes and alkynes coordinate to transition metals and undergo a variety of reactions, and are very important substrates for transition metal-catalysed reactions. Their reactions with halides, cyclization via carbene complexes, hydrogenation, and oxidative reactions with Pd(II) are treated in Sections 3.2, 8.2, 10.1 and 11.1, respectively. The many other reactions of alkenes and alkynes as main reactants are treated in this chapter.

Transition metal-catalysed reactions involving alkenes proceed mostly by the formation of π-complexes (η^2-complexes). Reactions of alkynes can be understood by the formation of three intermediate complexes, namely π-alkyne complexes (η^2-complexes) **1**, alkynyl complexes **2**, and η^1-vinylidene complexes (carbene complexes) **3**.

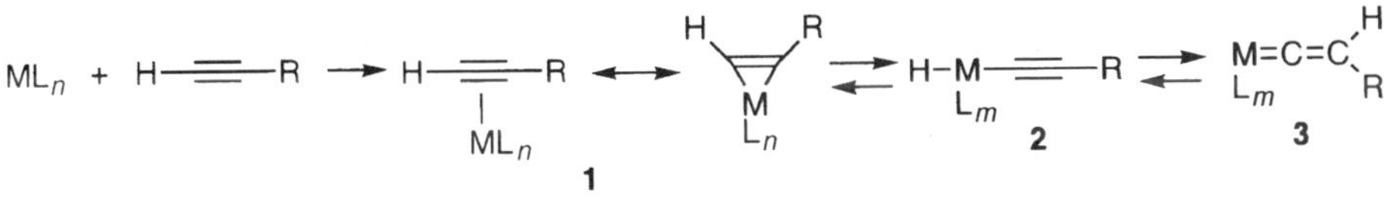

7.1 Carbonylation

1,2-Addition of H and CO to alkenes and alkynes catalysed by transition metal complexes is called hydrocarbonylation, and is useful for the syntheses of carboxylic acids, their esters, aldehydes and ketones [1]. Oxidative carbonylation of alkenes and alkynes with Pd(II), treated in Section 11.1.5, differs mechanistically from hydrocarbonylation. Some carbonylation reactions occur at under 1 atm or low pressures, without using a high-pressure laboratory apparatus. Several commercial processes based on hydrocarbonylation have been developed.

7.1.1 *Preparation of Carboxylic Acids, Esters and Ketones*

The 1,2-addition of H and CO_2H, or CO_2R to alkenes and alkynes is called hydrocarboxylation, or hydroesterification, and proceeds with catalytic amounts of

Pd(0) complexes, $Co_2(CO)_8$ or $Ni(CO)_4$ [1,2]. Most conveniently, Pd(0)-catalysed carbonylations of alkenes can be carried out under mild conditions in a laboratory with or without using a high pressure apparatus. Carbonylation in the presence of a small amount of HCl is explained by the following mechanism. The first step is oxidative addition of HX to Pd(0) to generate **4**. Then insertion of alkenes to H-PdX **4** gives the alkylpalladium bond **5**, and the acylpalladium complex **6** is formed by subsequent CO insertion. The last step is nucleophilic attack of alcohol or water to the acylpalladium complex **6** to give the ester **7** or acid, with regeneration of H-PdX.

Extensive studies on the carbonylation of alkenes have been carried out under different conditions using several types of Pd catalysts. Pd(0), formed from $PdCl_2$, is an active catalyst, and even Pd on carbon is active [3]. $PdCl_2(Ph_3P)_2$ is a suitable catalyst precursor [4]. Addition of a small amount of HCl accelerates the reaction. Usually a mixture of linear and branched saturated esters **8** and **9** is obtained from 1-alkenes. Their ratio changes depending on the reaction conditions, such as temperature, CO pressure, and the kinds and amount of phosphine ligands. For example, the carbonylation of styrene using a bidentate ligand affords the linear ester, and the branched ester with Ph_3P [5]. 6-Methoxy-2-vinylnaphthalene (**10**) is regioselectively carbonylated to give the branched ester **11** in a good yield using cyclohexyldiphenylphosphine as a ligand [6].

A catalyst system of $PdCl_2$ and $CuCl_2$ in aqueous HCl under oxygen is used for carbonylation under mild conditions (1 atm, room temp) to give branched esters with high selectivity [7]. Although oxidative carbonylation is expected under these conditions, no such a reaction occurs. Branched esters are formed as the main product by the carbonylation of alkenes at room temperature and 1 atm using Pd on carbon and an excess of $CuCl_2$, although the reaction is very slow (100% yield after 8 days) [8].

However, the linear ester **12** can be prepared as a major product from 1-alkene using Pd(OAc)$_2$ or even Pd on carbon as a catalyst and DPPB as a ligand in DME in the presence of formate, formic acid or oxalic acid under CO pressure [9]. Linear ester **13** is obtained from 1-octene as a main product using PdCl$_2$(Ph$_3$P)$_2$ coordinated by SnCl$_2$ [10].

$$C_6H_{13}\diagup\!\!= \;+\; CO \quad\xrightarrow[\text{82 atm, 150 °C, 64%}]{\text{Pd(Ph}_3\text{P)}_4,\ \text{DPPB, HCO}_2\text{Bu}}\quad C_6H_{13}\diagdown\!\diagup\diagdown\!\text{CO}_2\text{Bu}$$
$$\textbf{12}$$

$$C_6H_{13}\diagup\!\!= \;+\; CO \;+\; EtOH \quad\xrightarrow[\text{100 atm, 85 °C, 62%}]{\text{PdCl}_2(\text{Ph}_3\text{P})_2,\ [\text{Et}_4\text{N}][\text{SnCl}_3]}\quad C_6H_{13}\diagdown\!\diagup\diagdown\!\text{CO}_2\text{Et}$$
$$\textbf{13}\quad\text{Conv. 96\%}$$

Sometimes, keto esters are formed. Carbonylation of ethylene in alcohol affords propionate (**14**) as a main product, accompanied by a small amount of 4-oxohexanoate (**15**) [3]. Keto esters are obtained by the carbonylation of some dienes via insertion of alkene to an acylpalladium intermediate [11]. Carbonylation of 1,5-COD (**16**) in alcohols affords the mono- and diesters **17** and **18** [12]. On the other hand, bicyclo[3,3,1]-2-nonen-9-one (**20**) is formed in 40% yield in THF by the intramolecular insertion of the alkene to the acylpalladium bond in **19** [13].

$$CH_2\!=\!CH_2 \;+\; CO \;+\; EtOH \quad\xrightarrow[\text{80 °C}]{\text{PdCl}_2}\quad \diagup\diagdown\!\text{CO}_2\text{Et} \;+\; \text{(keto ester)CO}_2\text{Et}$$
$$\textbf{14}\qquad\qquad\qquad\textbf{15}$$

Alternating copolymerization of alkenes and CO is catalyzed by Pd [14]. Formation of polyketone from strained alkenes such as norbornadiene is the first example [15]. Polyketones, named carilon, are commercially produced from CO and ethylene or propylene by Shell. Isotactic polyketone is obtained by copolymerization of propylene and CO [16]. The copolymerization proceeds at room temperature in methylene chloride with high enantioselectivity to give the completely isotactic chiral polyketone **21** having the (*S*) form and high molecular weight (104 400) using a cationic Pd(II) complex coordinated by the unsymmetrical chiral phosphine–phosphite ligand [(*R,S*)-BINAPHOS] (**XXXVIII**) [17]. The high enantioselectivity is explained by assuming that CO always coordinates at the position *trans* to the phosphine ligand, and propylene always coordinates *trans* to the phosphite ligand.

PdL* = [Pd((R,S)-binaphos)-(Me)(MeCN)][BAr₄]
Ar = 3,5-bis(trifluoromethyl)phenyl

21

The carbonylation of acetylene to produce acrylic acid or ester (**22**) catalysed by $Ni(CO)_4$ was a historical industrial process developed by Reppe.

22

Alkynes are reactive compounds for Pd(0)-catalysed carbonylation. Monohydroesterification gives regioisomers of α,β-unsaturated esters, depending on the reaction conditions. The hydroesterification of alkynes is accelerated by the addition of HCl or HI, and explained by the insertion of alkynes to the H−PdX bond, followed by CO insertion. Methyl methacrylate (**23**) is obtained in high yields and high regioselectivity by the carbonylation of propyne using Pd black in the presence of hydrogen iodide [18, 19]. Regioselective carbonylation of the terminal alkyne **24** has been applied to the synthesis of the carbapenem derivative **25** [20]. $Pd(OAc)_2$, coordinated by 2-pyridyldiphenylphosphine is a very selective (99% selective) and active catalyst precursor (40 000 turnover) for the carbonylation of terminal alkynes to give branched unsaturated esters in the presence of methanesulfonic acid or *p*-toluenesulfonic acid [21].

23

24   **25**

In contrast, linear unsaturated esters are formed from terminal alkynes, depending on substrates and reaction conditions. A ratio of linear and branched α,β-unsaturated esters from higher terminal alkynes changes depending on ligands. The linear ester **26** was obtained in 81% selectivity (65% total yield) from 1-heptyne using $PdCl_2$, coordinated by dimethylphenylphosphine and $SnCl_2$, at 80 °C and 240 atm. With $PdCl_2(Ph_3P)_2$ in the absence of $SnCl_2$, the selectivity was 19% (84% yield) [19].

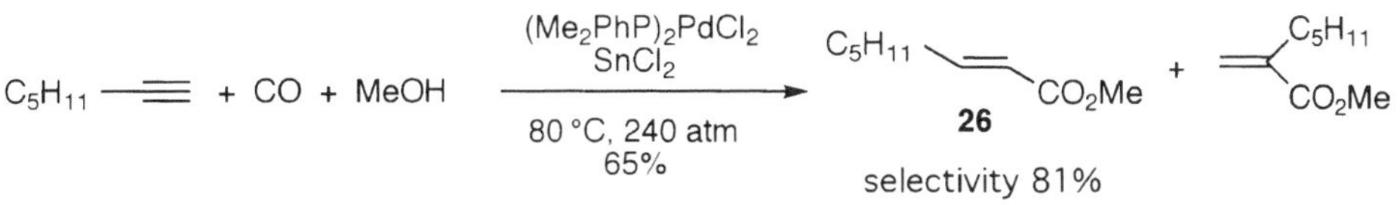
C_5H_{11}
+ CO + MeOH
$(Me_2PhP)_2PdCl_2$
$SnCl_2$
80 °C, 240 atm
65%
C_5H_{11}
CO_2Me
26
+
C_5H_{11}
CO_2Me
selectivity 81%

Monocarbonylation of 3-butyn-1-ol (**27**) in the presence of thiourea as an additive gives α-methylene-γ-lactone (**28**), a structure widely distributed in certain natural products [22]. A derivative of vernolepine **30** is prepared from **29** by this method [23]. Carbonylation of the rigid system of 2-*exo*-ethynyl-7-*syn*-norbornanol (**31**) with $PdCl_2$ in the presence of thiourea at 50 °C afforded the α-methylene-δ-lactone **32** in 47% yield [24].

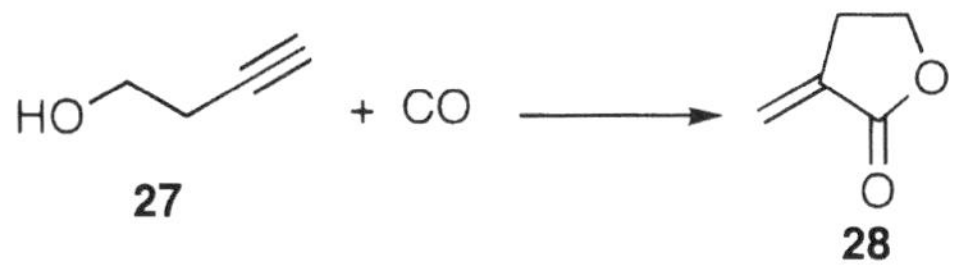
HO
+ CO
27
O
O
28

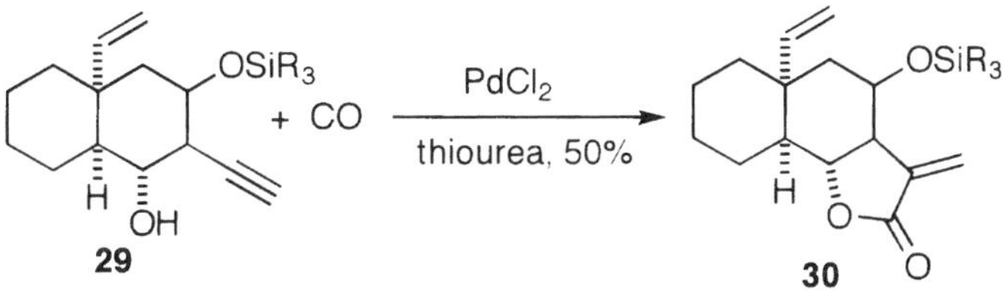
$OSiR_3$
+ CO
H
OH
29
$PdCl_2$
thiourea, 50%
$OSiR_3$
H
O
O
30

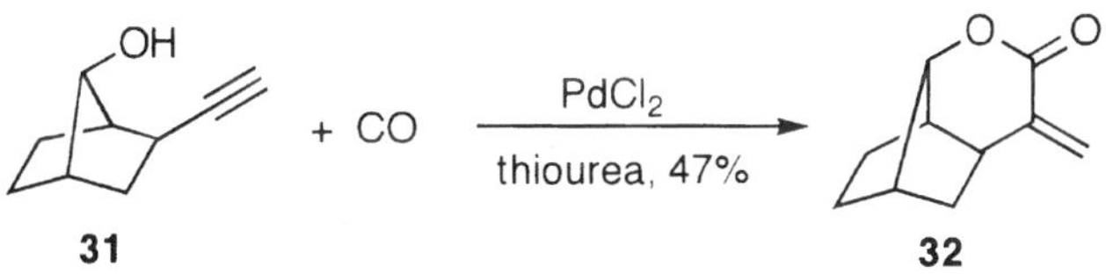
OH
+ CO
31
$PdCl_2$
thiourea, 47%
O
O
32

7.1.2 Preparation of Aldehydes (Hydroformylation) and Alcohols

Formation of aldehydes by the reaction of alkene, CO and H_2 catalysed by $Co_2(CO)_8$ was discovered by Rölen in 1938 [25]. This is the 1,2-addition of H and CHO to alkenes, and hence called hydroformylation or the oxo reaction. Production of butanal, (**33**) from propylene as a main product is an important industrial process. Aldol condensation of butanal, followed by hydrogenation affords 2-ethyl-1-hexanol (**34**), which is converted to phthalate, and used as a plasticizer of poly(vinyl chloride).

Hydrogenolysis of $Co_2(CO)_8$ generates hydrocobalt carbonyl **35** as a catalyst precursor. The hydroformylation is explained by the insertion of propylene to hydrocobalt carbonyl **36** to give propylcobalt **37**. Then CO insertion generates the acylcobalt complex **38**. Finally, attack of hydrogen gives butanal (**33**) and regenerates the hydrocobalt species **36**.

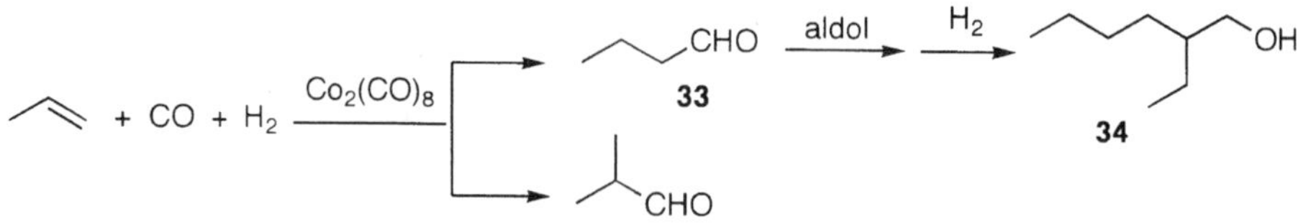

Rh complexes show higher catalytic activity than $Co_2(CO)_8$, and at present the Rh

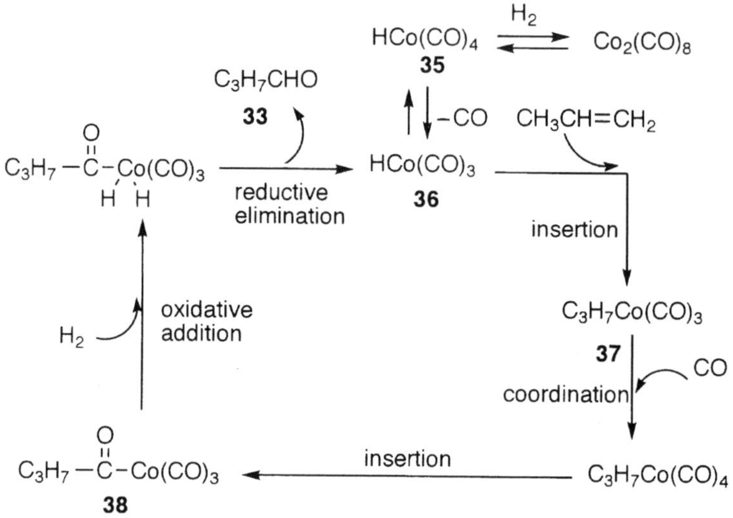

catalyst is used in industrial oxo process. The Rh-catalysed industrial production of butanal in an aqueous phase using TMSPP (**XLVII**) as a ligand has been developed by Ruhrchemie. In this process the catalyst always stays in the aqueous phase and can be recycled after facile separation from organic products [26].

Regioselective formation of linear aldehydes is important in industrial process. The ligand BIPHEPHOS (**L**), developed by Union Carbide, gives the highest ratio of butanal from propylene. This ligand is useful for regioselective formation of linear aldehydes from various functionalized 1-alkenes under mild conditions. The linear aldehyde **40** was obtained from **39** and converted to the indolizidine alkaloid **41** [27].

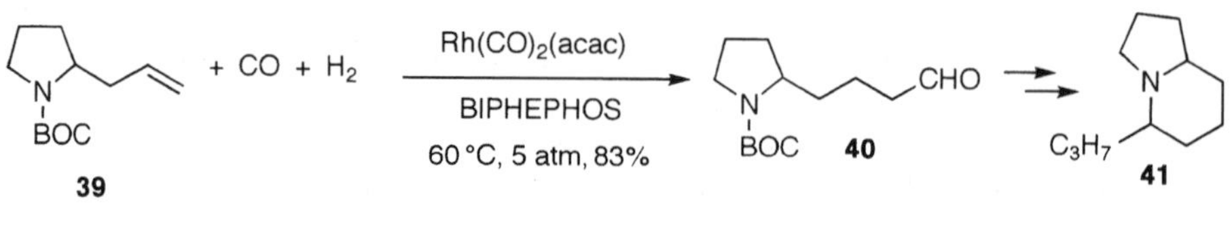

Several interesting synthetic applications of hydroformylation are known. As one example, hydroformylation of enol ethers to afford β-hydroxy aldehydes can be extended to catalytic aldol synthesis. The Rh-catalysed hydroformylation of the cyclic enol ether **42** affords the protected *syn*-3,5-dihydroxyaldehyde **43** without forming the *anti* product **44**. The regioselectivity of attack on the terminal carbon is also very high [28].

Enantioselective hydroformylation has been attempted using various chiral ligands [29]. Highly enantionselective Rh-catalysed hydroformylation of some terminal alkenes has been achieved using the chiral phosphine–phosphite ligand (*R*)-[2-(diphenylphosphino-1,1′-binaphthalen-2′-yl)-(S)-1,1′-binaphthalen-2,2′-yl] phosphite [(*R*,*S*)-BINAPHOS] (**XXXVIII**) and its enantiomer (*S*,*R*)-BINAPHOS. Hydroformylation of styrene with the chiral catalyst afforded the branched aldehyde **45** as the main product with 94% ee. However, the linear aldehyde is obtained as the main product from 1-hexene with this ligand, and there is no possibility of asymmetric hydroformylation [30].

Aldehydes obtained by hydroformylation can be used for further catalytic transformations. Co-catalysed carbonylation of aldehydes in the presence of amides gives rise to acylamino acids. The reaction is called the Wakamatsu reaction [31]. Hydroformylation of acrylonitrile in acetamide affords the linear aldehyde **46**. Then hydroxy amide **47** is formed from the aldehyde **46** and the amide, and converted to the alkylcobalt complex **48** by dehydration. Insertion of CO generates the acylcobalt complex **49**, which reacts with water to give the acylamino acid **50** and regenerates hydrocobaltcarbonyl species **36**. Acylamino acids **50** are prepared in one pot from alkenes by hydroformylation, followed by the Wakamatsu reaction. It should be noticed that amino acids **50** are formed by the reaction of water generated during the reaction with the acylcobalt complex **49**, rather than aldehyde, even though the reaction is carried out under hydrogen pressure. As an intramolecular version, *N*-benzoylpipecorinic acid (**52**) was prepared by the reaction of *N*-benzoyl-3-butenylamine (**51**) [32]. The Wakamatsu reaction is also catalyzed by Pd catalysts such as $PdCl_2(Ph_3P)_2$ and Pd/C. Formation of the amino acid **54** from 3-methylbutanal (**53**) is an example [33].

Studies on the hydroformylation of simple alkynes to give unsaturated aldehydes is less extensive. 2-Phenylcinnamaldehyde is obtained by the oxo reaction of diphenylacetylene catalysed by $Rh_4(CO)_{12}$ [34]. The triple bond in the conjugated enyne **55** was attacked by CO using the same catalyst, and aldehyde **56** and cyclopentenone **57** were obtained. The latter is formed by intramolecular insertion of a double bond to acylrhodium bond followed by hydrogenolysis. Formation of **57** may be regarded as a catalytic intramolecular Pauson–Khand reaction.

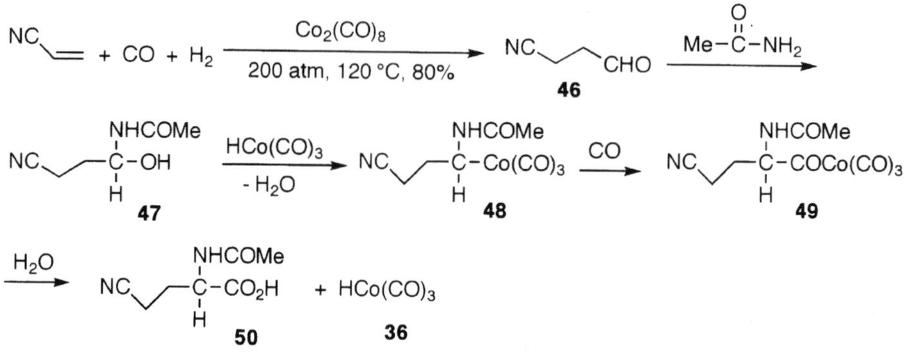

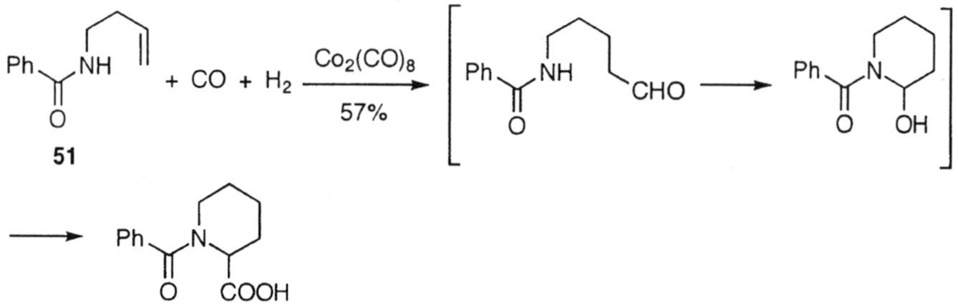

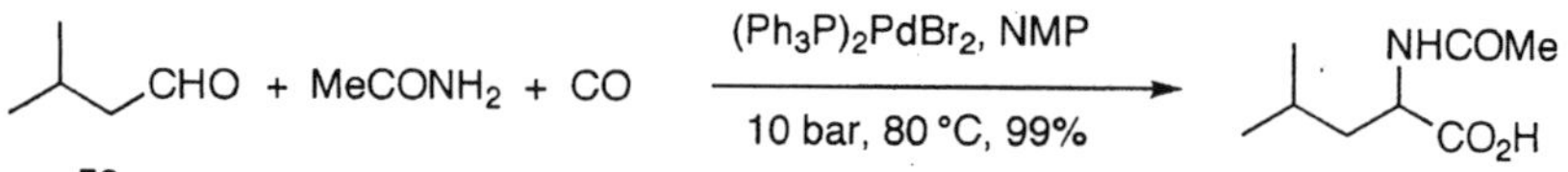

Clean hydroformylation of the internal alkynes **58** to give the α,β-unsaturated aldehydes **59** was achieved using the Rh catalyst coordinated by BIPHEPHOS (**L**) under 1 atm of CO/H_2 at room temperature in CH_2Cl_2 as a solvent [35].

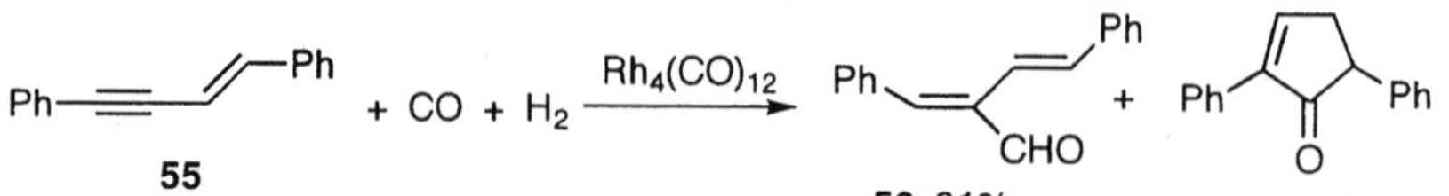

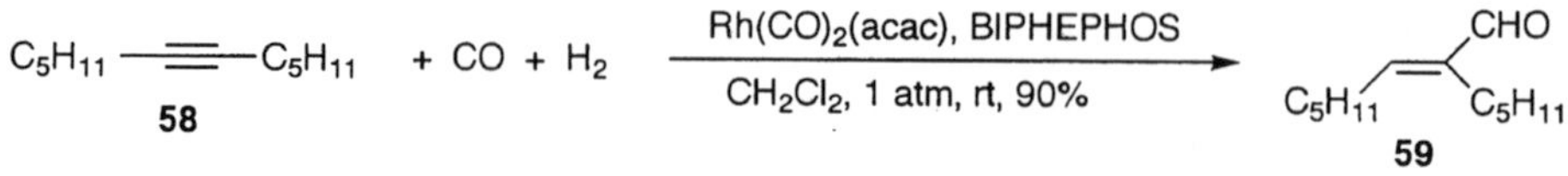

Hydrosilane $HSiR_3$ behaves similar to H_2 toward transition metal complexes in some cases. When $HSiR_3$ is used instead of hydrogen in hydroformylation, two reactions are expected. One is a hydrocarbonylation-type reaction, by which formation of the silyl enol ethers **62** via the acylmetal intermediate **61**, and the acylsilanes **64** via the acyl complex **63**, are expected; in practice both reactions are observed. The other possibility is silylformylation to form **65**, which is unknown, even though silylformylation of alkynes is known. When $Co_2(CO)_8$ is used, the silyl enol ether of aldehyde **66** is obtained [36]. However, the silyl enol ether **67** of acylsilane **68** is obtained when an Ir complex is used, and converted to the acylsilane **68** by hydrolysis [37].

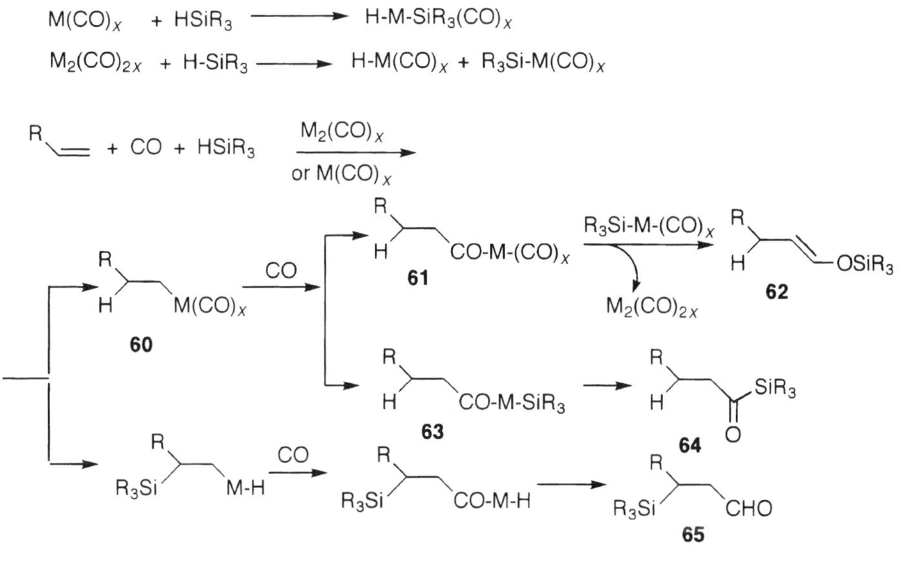

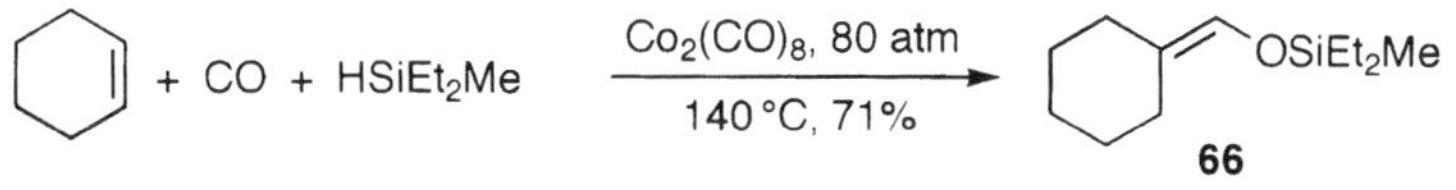

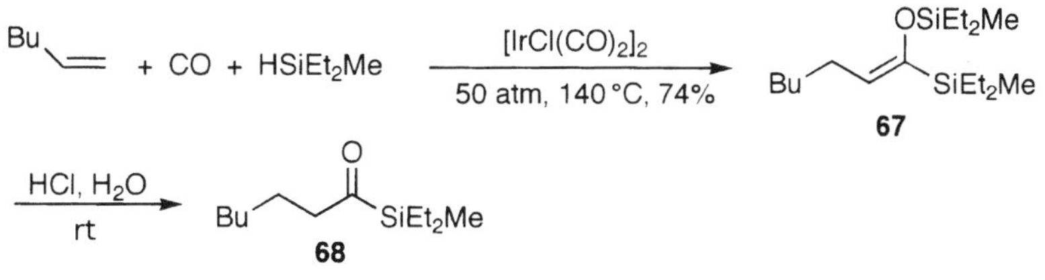

When hydrosilane is used instead of hydrogen in the Rh-catalysed carbonylation of alkynes, the silylformylation of alkynes occurs to give the β-silyl-α,β-unsaturated aldehydes **71** [38,39]. Oxidative addition of silane to Rh generates the silylrhodium hydride **69**. It should be noticed that insertion of alkyne to the $Rh-SiR_3$ bond, but not to the Rh–H bond, occurs to generate **70**. Insertion of CO to **70** and reductive elimination afford **71**. The reaction of the terminal alkyne **72** under similar conditions afforded the γ-lactam **73** as a silylformylation product in the presence of DBU [40]. Conversion of **73** to **74** and subsequent Rh-catalysed hydroformylation afforded the aldehyde **75**. The cyclized product **76** was converted to isoretronecanol (**77**). Interestingly the β-lactam **80** is formed from the acylrhodium **79** in the attempted silylformylation of **78** [41], whereas the terminal aldehyde **82** is obtained by different regiochemical reaction from the terminal alkyne **81** [42].

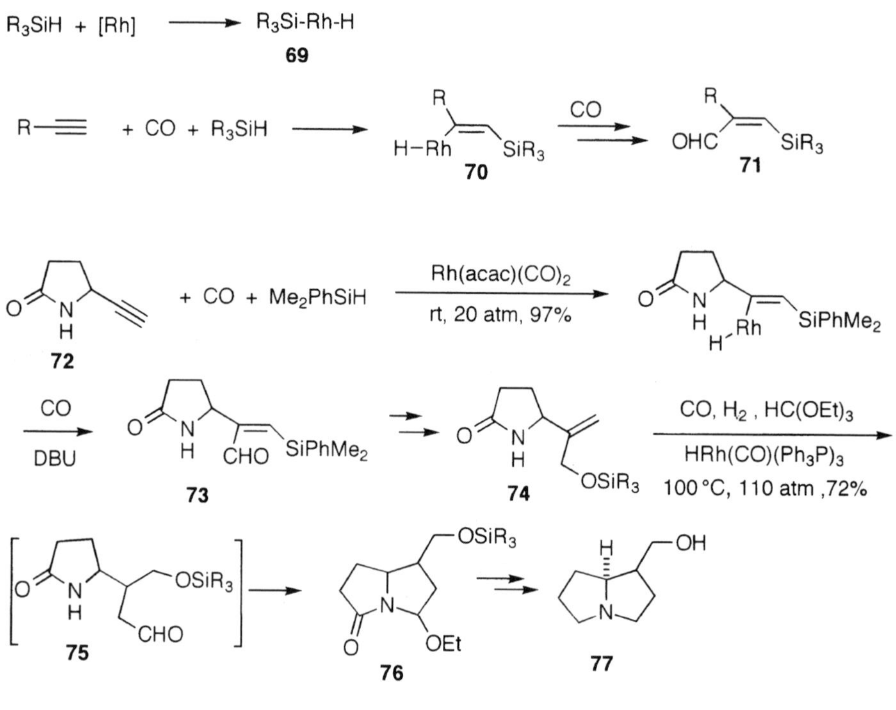

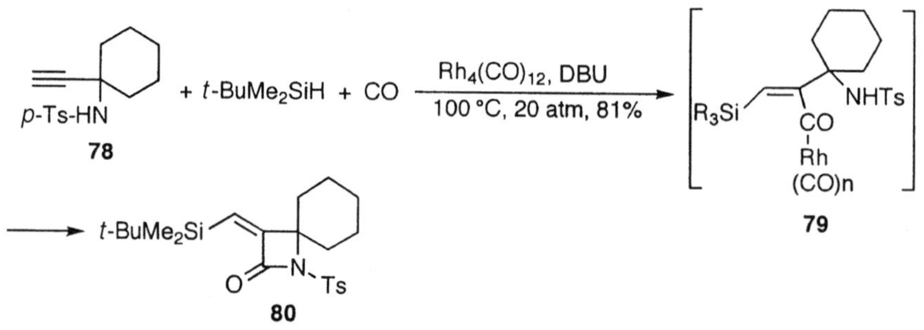

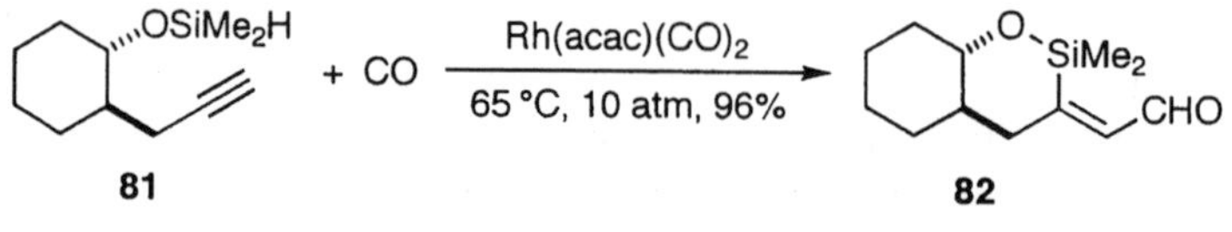

Reductive carbonylation of alkenes occurs with $Fe(CO)_5$. The water-gas-shift reaction catalysed by $Fe(CO)_5$ generates H_2 from water and CO as shown in eq. (7.1). CO is a reducing agent of water. 1-Butanol had been produced once commercially by the reaction of propylene with CO in water catalysed by $Fe(CO)_5$. The process was called Reppe butanol process.

$$\text{propylene} + 3\,CO + 2\,H_2O \xrightarrow{Fe(CO)_5} \text{butanol-OH} + 2\,CO_2$$

$$(CO)_4Fe\text{-}CO + H_2O \longrightarrow \underset{\underset{H}{|}}{(CO)_4Fe\text{-}CO_2H} \xrightarrow{-CO_2} \underset{\underset{H}{|}}{(CO)_4Fe\text{-}H} \xrightarrow[H_2]{CO} Fe(CO)_5 \qquad (7.1)$$

7.1.3 *Decarbonylation of Aldehydes*

The insertion of CO is reversible. In the reverse process, aldehydes and acyl halides are decarbonylated using Pd and Rh catalysts. Both stoichiometric and catalytic decarbonylations are known [43]. Synthetically useful is the decarbonylation of aldehydes under mild conditions mediated by $(Ph_3P)_3RhCl$, the Wilkinson complex (**83**), which is a strong CO accepter and converted to the carbonyl complex **84** [44]. The first step of the decarbonylation is oxidative addition of aldehyde to the coordinatively unsaturated complex **85** to form the acylrhodium complex **86**. Then migration of the alkyl group from the acyl group to Rh occurs to form **87**, and its reductive elimination affords alkane and **84**. The deep red colour of a benzene solution of Wilkinson complex (**83**) turns rapidly to yellow by the decarbonylation when a simple aldehyde is added. Sterically hindered aldehydes are decarbonylated in boiling benzonitrile. Alkanes are obtained from aliphatic aldehydes, and alkenes from α,β-unsaturated aldehydes.

$$RCH_2CH_2\text{-}CHO + \underset{\mathbf{83}}{(Ph_3P)_3RhCl} \longrightarrow \underset{\mathbf{84}}{(Ph_3P)_2Rh(CO)Cl} + RCH_2CH_3 + \left[R\underset{\diagdown}{=}\right]$$

The chiral aldehyde **88** was decarbonylated to give **89** with overall retention of the stereochemistry [45]. The unsaturated aldehyde in the polyfunctionalized molecule **90** was decarbonylated smoothly to afford alkene **91** [46]. The decarbonylation of aldehydes catalysed by a supported Pd or Rh complex is carried out at high temperature [47]. Myrtenal (**92**) is decarbonylated at 195 °C to give apopinene (**93**) using a supported Pd catalys [48].

Although the decarbonylation of aliphatic acyl halides is possible with the Rh complex, a mixture of alkenes is obtained [43,44].

7.2 Cycloaddition Reactions

Cycloadditions are useful for the preparation of cyclic ompounds. Several thermal and photoactivated cycloadditions, typically [4+2] (Diels–Alder reaction), are known. They proceed with functionalized electronically activated dienes and monenes. However, various cycloaddition reactions of alkenes and alkynes without their electronical activation, either mediated or catalysed by transition metal complexes under milder conditions, are known, offering a useful synthetic route to various cyclic compounds in one step. Transition metal complexes are regarded as templates and the reactions proceed with or without forming metallacycles [49].

Many cyclization reactions via formation of metallacycles from alkynes and alkenes are known. Formally these reactions can be considered as oxidative cyclization (coupling) involving oxidation of the central metals. Although confusing, they are also called the reductive cyclization, because alkynes and alkenes are reduced to alkenes and alkanes by the metallacycle formation. Three basic patterns for the intermolecular oxidative coupling to give the metallacyclopentane **94**, metallacyclopentene **95** and metallacyclopentadiene **96** are known. (For simplicity only ethylene and acetylene are used. The reaction can be extended to substituted alkenes and alkynes too). Formation of these metallacycles is not a one-step process, and is understood by initial formation of an η^2 complex, or metallacyclopropene **99**, followed by insertion of the alkyne or alkene to generate the metallacycles **94–96, 100** and **101–103** (Scheme 7.1).

These metallacycles are converted further to various acyclic and cyclic products. Trapping of the metalacycles with X–Y affords acyclic compounds **97**. Cyclic

(a) Formation of metallacycles by intermolecular oxidative cyclization

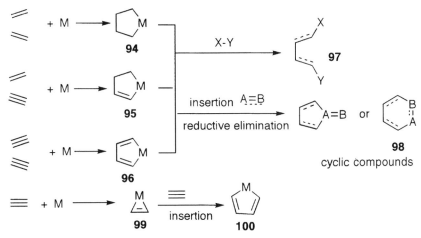

(b) Formation of metallacycles by intermolecular oxidative cyclization

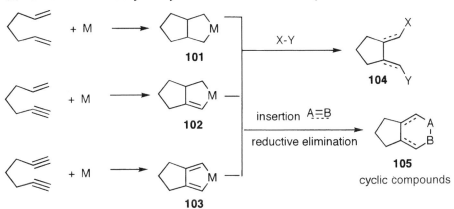

Scheme 7.1   Formation of metallacycles

products **98** are obtained by the insertion of an unsaturated bond A=B and reductive elimination. Intramolecular reactions of dienes, diynes and enynes, to afford the bicyclic metallacycles **101**, **102** and **103**, can be understood as oxidative cyclization, and various monocyclic products **104** and bicyclic products **105** are obtained from them.

7.2.1 Cyclotrimerization and Cyclotetramerization of Alkynes

Benzene and cyclooctatetraene (COT) derivatives are formed by [2+2+2] and [2+2+2+2] cycloadditions of alkynes. At first the metallacyclopropene **107** and metallacyclopentadiene **108** are formed. Benzene and COT (**106**) are formed by reductive elimination of the metallacycloheptatriene **109** and the metallacyclononate-traene **110**. Formation of benzene by the [2+2+2] cycloaddition of acetylene is catalysed by several transition metals. Synthesis of benzene derivatives from substituted alkynes is catalysed by solid catalyst and complex catalysts of Fe, Ni, Co, Pd, Rh, Zr, Ta etc.

The cyclization mechansim has been studied using a Co complex. Oxidative cyclization is explained by initial formation of the η^2-cobaltacyclopropene **107**, and subsequent insertion of alkyne gives the cobaltacyclopentadiene **108**. Ring expansion by the subsequent alkyne insertion to **108** affords cobaltacycloheptatriene **109**, which undergoes reductive elimination to form a benzene derivative [50]. The cobaltacyclopropene **111** and the cobaltacyclopentadiene **112** have been isolated by the reaction of alkynes with $CpCo(Ph_3P)_2$. Highly efficient cyclization of 3-hexyne to hexaethylbenzene occurs using a catalyst of unknown structure prepared by the treatment of Pd on carbon with Me_3SiCl in THF [51].

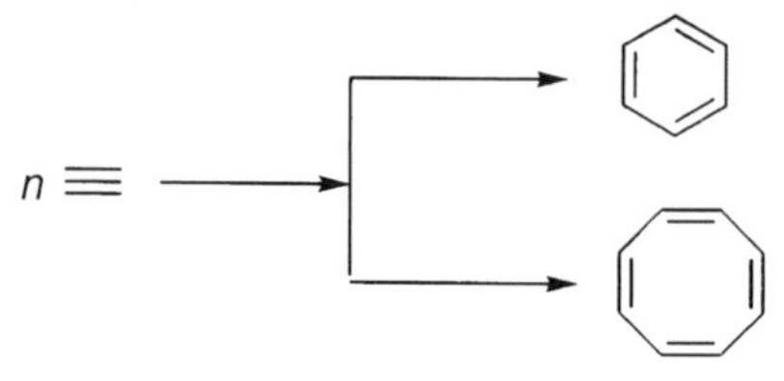

106

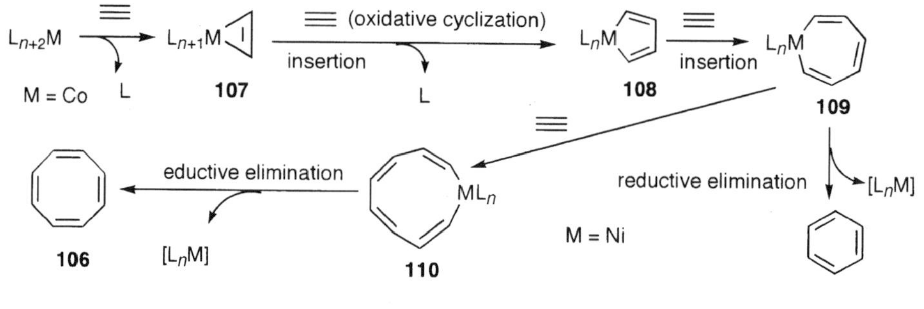

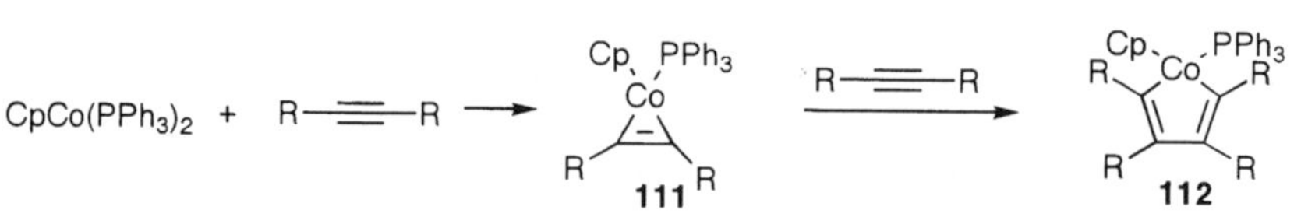

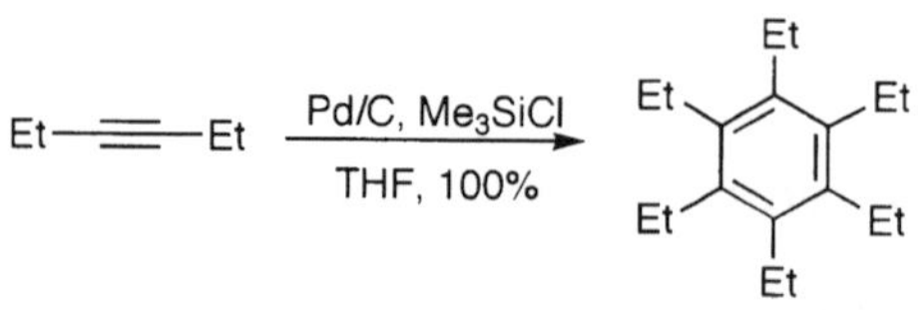

Although some degree of regioselectivity is observed in the formation of cobaltacyclopentadienes **112** from unsymmetrically substituted alkynes, the reaction has little synthetic interest because the cyclization of terminal alkynes and unsymmetric internal alkynes gives a mixture of all possible regioisomers of substituted benzenes. Also, cross-cyclotrimerization of different alkynes gives a mixture of isomers.

As one partial solution, the selective cross-cyclization is possible by stepwise reaction of Cp_2Zr (**113**), generated from zirconocene dichloride and BuLi, with three different symmetric internal alkynes in one pot. The method is based on the conversion of zirconacyclopentadiene **114** to the benzene derivative **115** by the treatment with dimethyl acetylenedicarboxylate and CuCl [52].

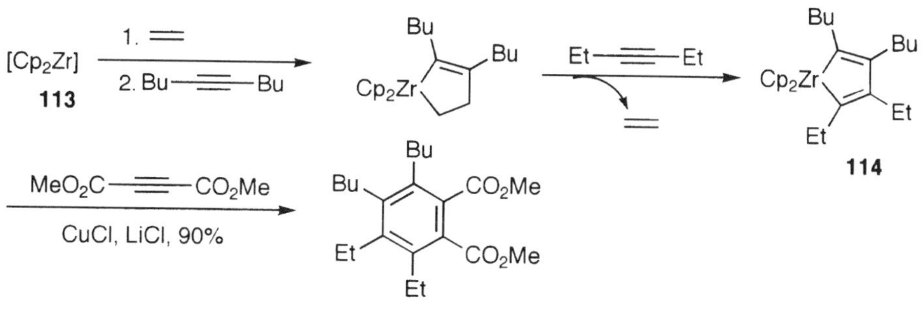

The metallacyclopentadienes **108** are converted to various cyclic compounds by insertion of several unsaturated bonds as summarized in Scheme 7.2.

Control of the regioselectivity in Co-catalysed cross-cyclization has been solved in an ingenious way utilizing 1,5-hexadiyne (**116**) as one component and bis(trimethyl-silyl)acetylene (**118**) as the other. Although bulky bis(trimethylsilyl)acetylene (**118**) itself cannot cyclotrimerize to hexasilylbenzene due to steric hindrance, it reacts with the cobaltacyclopentadiene **117** formed from the less bulky diyne **116** to produce the

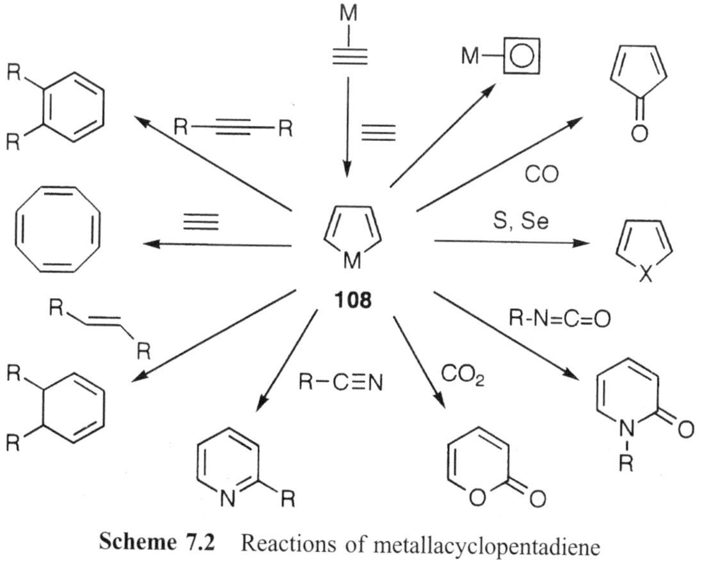

Scheme 7.2   Reactions of metallacyclopentadiene

benzocyclobutene **119** when the reaction catalysed by CpCo(CO)$_2$ is carried out in a large excess of **118**. Benzocyclobutene **119** is converted to quinodimethane **120** on heating, and the subsequent intermolecular Diels–Alder reaction with alkene **121** affords the tetralin **122**. This rather simple domino reaction sequence has high synthetic potential, and is very useful for facile syntheses of many natural products [53]. In an efficient synthesis of estrone, reaction of 1,5-diyne **123** with **118** affords the benzocyclobutene **124**. The ring-opened intermediate **125** undergoes intramolecular Diels–Alder reaction to give **126**, which is converted to estrone [54].

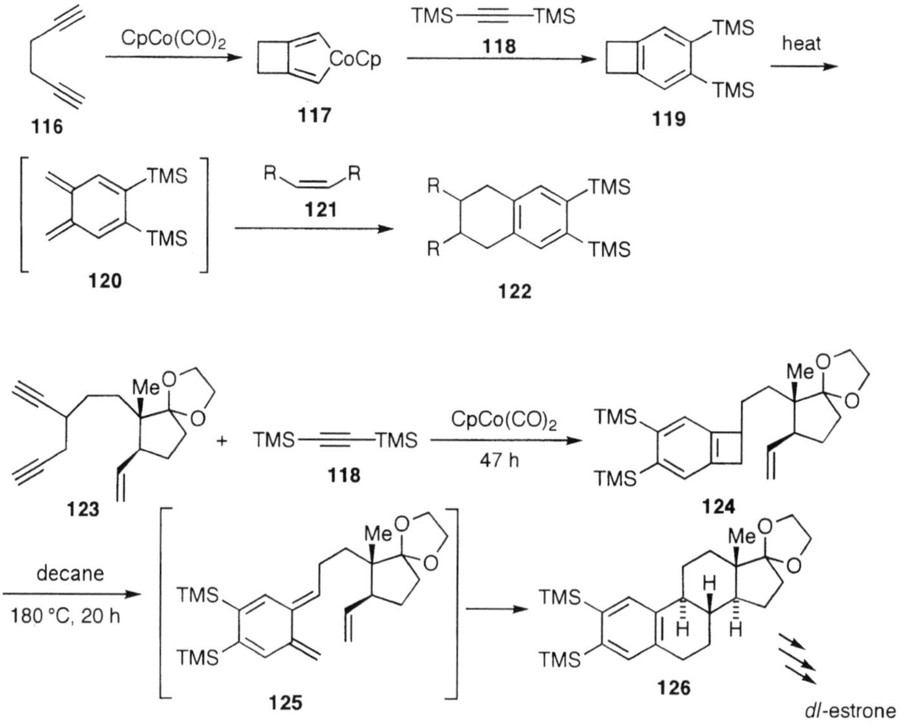

The phyllocladane skeleton **131** was constructed efficiently by stereoselective formations of six carbon–carbon bonds and four rings via a one-pot sequence of cyclizations: the ene type, [2+2+2], and [4+2] cycloadditions. In this synthesis, the Conia ene reaction of **127** takes place under mild conditions to generate **128**, and the cyclotrimerization of its diyne with **118** gives **129**. These two reactions are catalysed by CpCo(CO)$_2$. Finally, ring-opening to give **130** and intramolecular Diels–Alder reaction in the presence of DPPE produced the phyllocladane skeleton **131** in a total yield of 42% [55].

Nickel(0) complexes catalyse [2+2+2] cycloaddition. Catalytic asymmetric synthesis of isoquinoline derivative **134** is possible based on enantiotopic group-selective formation of the nickelacyclopentadiene **133** from **132** using the Ni(0) complex coordinated by a chiral ligand under acetylene atmosphere [56].

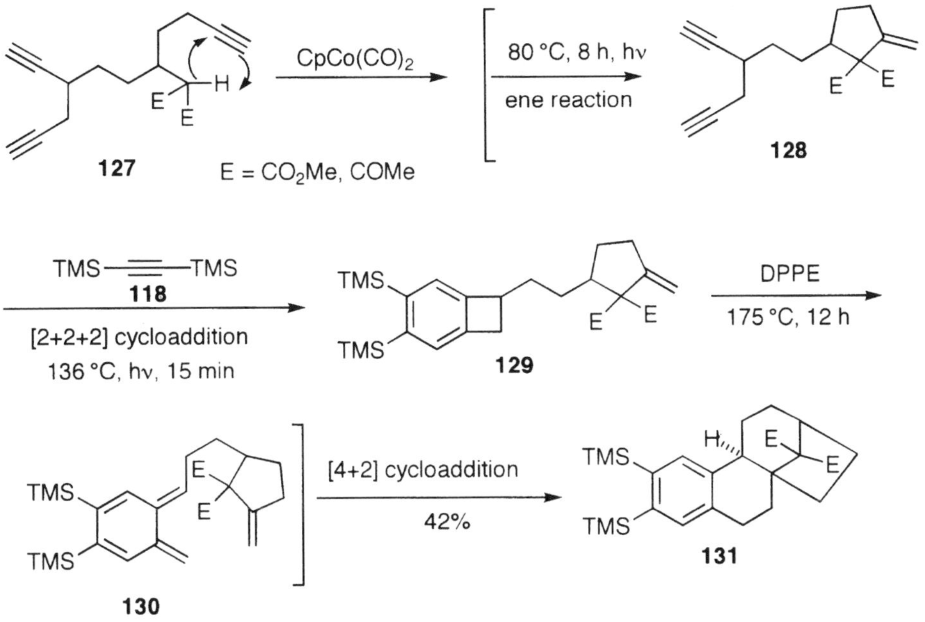

127   E = CO₂Me, COMe → CpCo(CO)₂ → [80 °C, 8 h, hv ene reaction] → **128**

TMS━━━TMS **118**

[2+2] cycloaddition
136 °C, hv, 15 min → **129** → DPPE 175 °C, 12 h

130 → [4+2] cycloaddition 42% → **131**

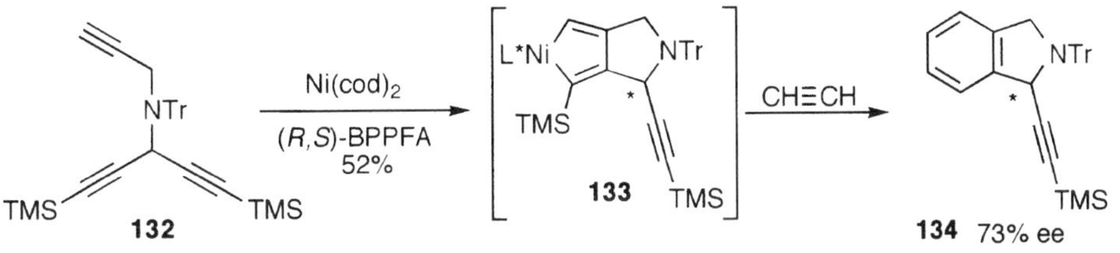

132 → Ni(cod)₂ (R,S)-BPPFA 52% → **133** → CH≡CH → **134** 73% ee

C-Arylglycoside **137** is synthesized by the RhCl(Ph₃P)₃-catalysed intermolecular reaction of **135** with **136** [57].

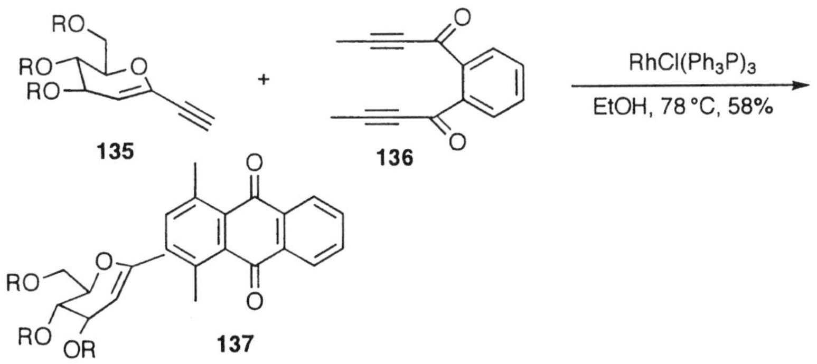

135 + **136** → RhCl(Ph₃P)₃ EtOH, 78 °C, 58% → **137**

Reppe and co-workers reported in 1948 the Ni-catalyzed cyclotetramerization of acetylene to give cyclooctatetraene (**106**) [58]. After this discovery, the reaction was expanded to monosubstituted alkynes. Monosubstituted alkyne **138** is cyclized smoothly to give tetrasubstituted cyclooctatetraene **139** [59]. Internal alkynes are difficult to cyclize to eight-membered rings, but semi-intramolecular cyclization of **140** to give **141** has been reported [60].

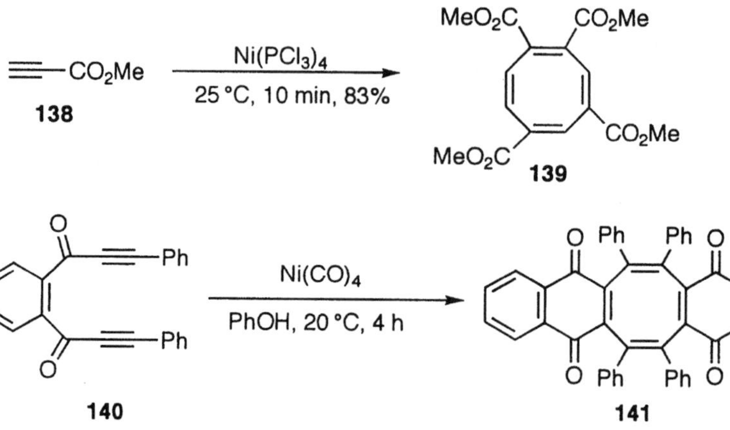

7.2.2 Formation of Five- and Six-membered Rings by the Cocyclization of Two Molecules of Alkynes with Other Unsaturated Compounds

The reactions of the metallacyclopentadiene **108** with some unsaturated bonds, such as CO, CO_2, alkenes and carbonyl groups, give rise to various five- and six-membered cyclic compounds, such as **142** and **143**.

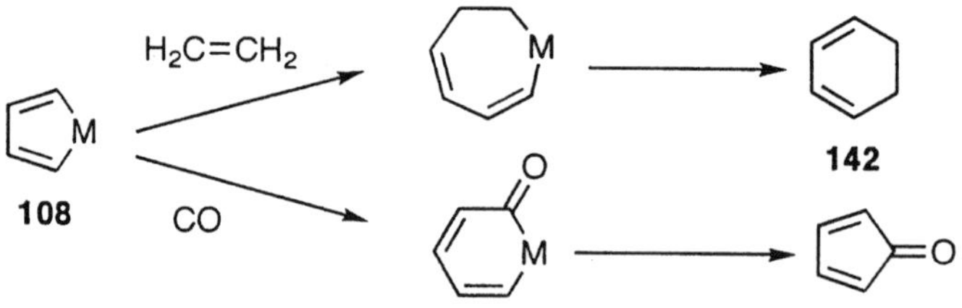

Cyclohexadiene derivatives are prepared by the insertion of alkenes. The γ-lycorane skeleton **145**, coordinated by CpCo, is constructed by [2+2+2] cycloaddition of the enamide **144** with two different triple bonds (in **144** and **118**). In this case, the Co–diene complex **145** is formed, and hence the reaction is stoichiometric. Oxidative decomplexation using $FeCl_3$ affords the lycorane skeleton [61]. The cyclohexadiene **146** is obtained by regioselective [2+2+2] cycloaddition of 1,6- or 1,7-diynes with α,β-enone using a Ni(0) catalyst generated by the treatment of $NiCl_2$ with Zn powder. Addition of $ZnCl_2$ and Et_3N assists the reaction [62].

Other transition metal complexes catalyze interesting cocyclization of alkynes and alkenes by a completely different mechanism. Efficient 6π-electrocyclization of the conjugated dienyne **147**, catalysed by an electron-deficient Ru complex, affords the angular aromatic ring system **149** in high yield [63]. The cyclization is explained by formation of the Ru dienylvinylidene complex **148** by well-known 1,2-hydride shift of the terminal alkynic hydrogen and subsequent electrocyclic reaction.

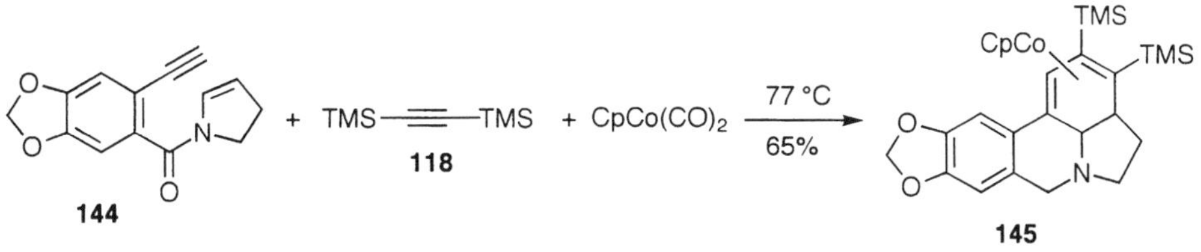

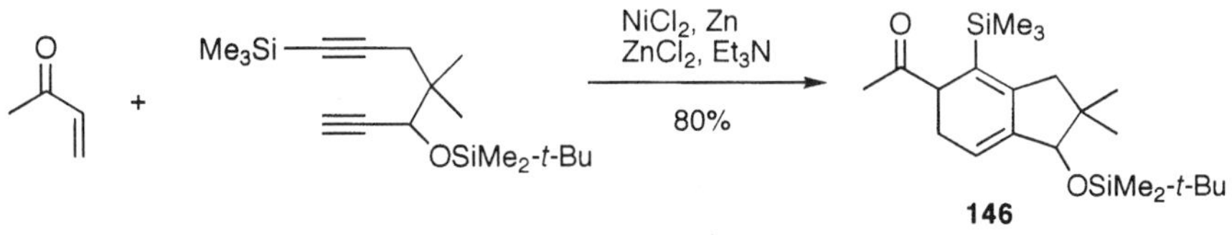

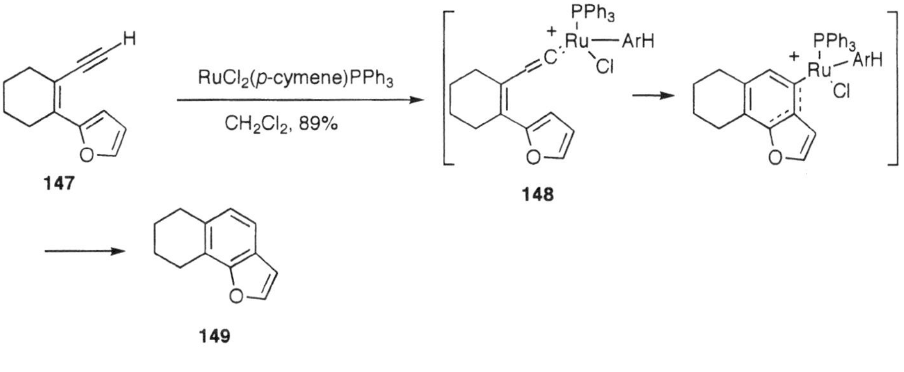

A completely new benzannulation reaction of conjugated enynes and alkynes catalysed by Pd complexes to give substituted benzenes has been developed recently. When the 1,3-enyne **150** is treated with a Pd(0) catalyst, homocyclodimerization of the enyne takes place to give the 2,6-disubstituted styrenes **151** as a single product in good yield [64]. An interesting application of this reaction is the preparation of paracyclophane **153** by the intramolecular reaction of the bis-enyne **152** [65].

Also, the [4+2] benzannulation of enyne **154** with diyne **155** proceeds chemo- and regioselectively to afford **156** [66]. A similar reaction of enyne silyl enol ether **157** with the diyne offers a good synthetic route to polysubstituted phenols **159** via **158** [67]. In this reaction enol silyl ethers having the (*E*) configuration as shown in **157** give satisfactory results.

The 1,3-diyne **160** undergoes an interesting regioselective cyclotrimerization to afford unsymmetrically 1,3,5-substituted benzene **162**. In this reaction the enetriyne **161** is formed at first by the Pd-catalysed dimerization of **160**, and the reaction

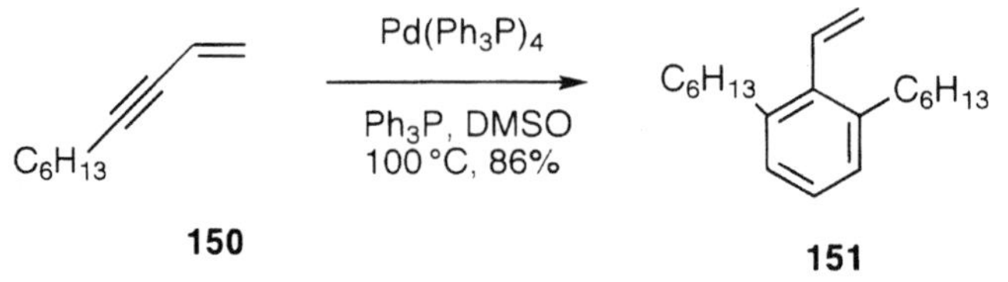

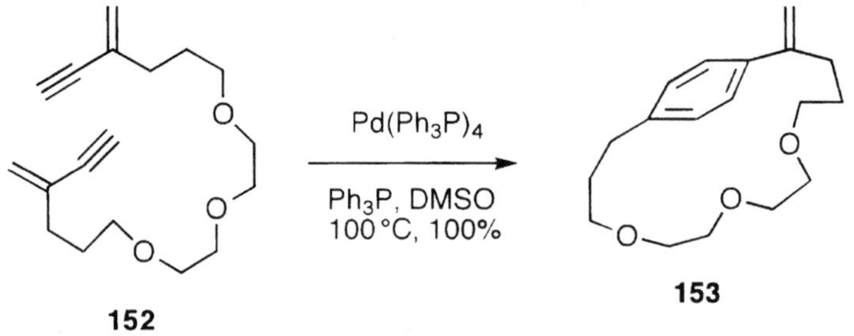

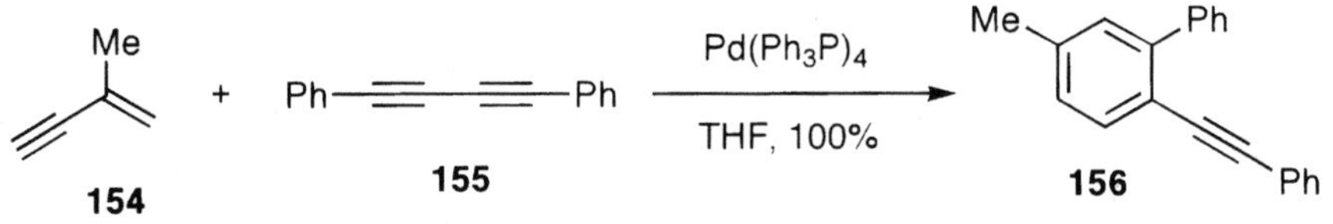

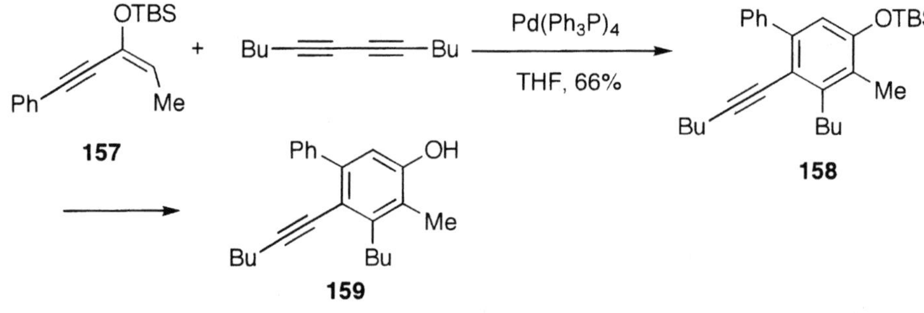

150 → **151**

Pd(Ph₃P)₄

Ph₃P, DMSO
100 °C, 86%

152 → **153**

Pd(Ph₃P)₄

Ph₃P, DMSO
100 °C, 100%

154 + **155** → **156**

Pd(Ph₃P)₄

THF, 100%

157 + Bu—≡≡—Bu → **158**

Pd(Ph₃P)₄

THF, 66%

159

of its enyne moiety with the diyne **160** gives the 1,3,5-trisubstituted benzene **162** [68].

The mechanism of these new enyne with diyne cyclizations has been studied using the deuterated enyne **163** [69]. D^1 in dideuterated **163** migrates, but D^2 does not. The deuterated (*E*)-enyne **164** cocyclizes with the diyne **165** to give **166** with migration of D. However, no migration of D was observed in the cyclized product **168** by the reaction of (*Z*)-enyne **167** with **165**.

These results are explained by the following mechanism. The reaction of **169** with the diyne starts by oxidative cyclization to generate **170**, and the 1,3-shift of D takes place to give **171**, which undergoes reductive elimination to afford **173**. As another possibility, the reductive elimination of **170** generates **172** and its 1,5-shift affords **173**.

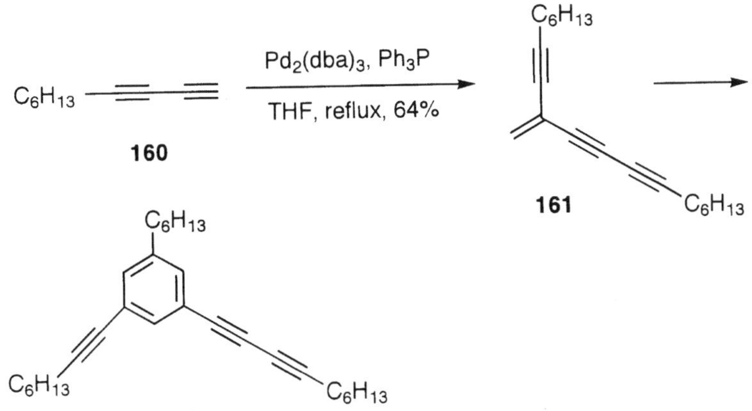

The triple bond of the nitrile group can be cotrimerized with two alkynes to produce pyridines. The cobalt-catalysed cocyclization of alkyne and nitrile in a ratio of 2:1 is a good synthetic route to pyridine derivatives [70]. Two regioisomers, **174** and **175**, are obtained by the reaction of propyne with MeCN. The reaction is carried out in a large excess of MeCN, and potentially useful for commerical production of pyridine derivatives [71]. The reaction of acetylene itself with various nitriles produces the α-substituted pyridine **176** in the presence of water under irradiation [72]. HCN cannot be used for this cocyclization. The reaction has been applied to alkaloid synthesis.

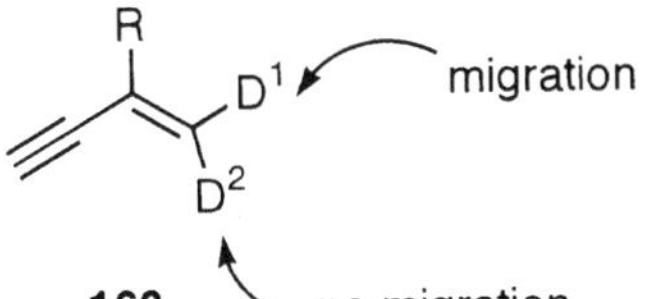

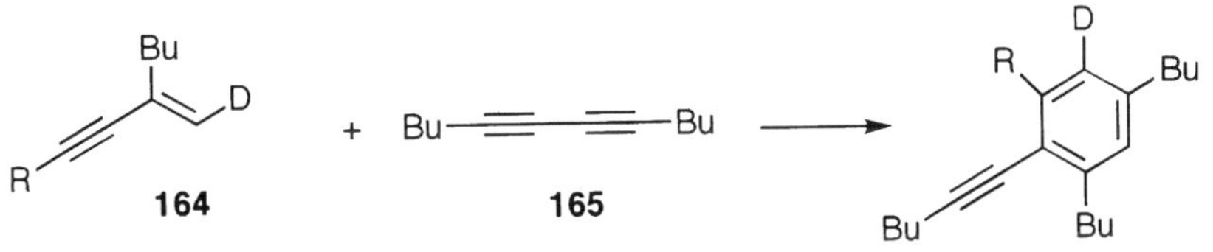

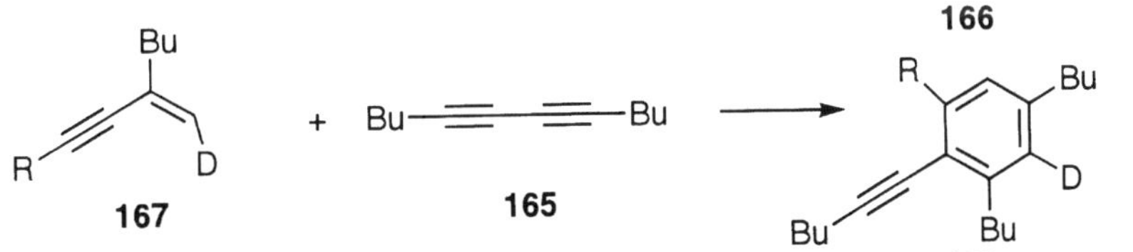

Synthesis of the lysergene and LSD skeleton **179** from **177** and **178** is an example [73].

Pyridones are prepared by insertion at the N=C bond of isocyanate using cobaltacyclopentadiene as a catalyst [74]. The reaction was applied to the synthesis of the intermediate **182** of camptothecin from the isocyanate **180** and the alkyne **181** [75].

Insertion at the C=O bond of aldehydes also occurs. The Ni(0)-catalysed reaction of 3,9-dodecadiyne (**183**) with butanal affords the α-pyran **184** [76].

Pyrones are prepared by insertion at the C=O bond of CO_2. The pyrone **188** is prepared by the Ni-catalysed reaction of 1,6-diyne **185** with CO_2 [77]. However, this reaction is explained by the carbonickelation of one of the triple bonds with CO_2 to form **186**, followed by insertion of another triple bond to give the seven-membered intermediate **187**, rather than insertion of CO_2 to nickelacyclopentadiene. In this reaction the Ni(0)-tricyclohexylphosphine complex is an active catalyst.

A labile molecule of the cyclopentadienone derivative **190**, stabilized by the complex formation with $Fe(CO)_3$, is obtained in high yield by the reaction of the 1,6-diyne **189** under moderate CO pressure [78].

The Ru carbonyl-catalysed reaction of 1,6-diynes **191**, CO and hydrosilane (6 equivalents) under rather severe conditions affords the catechol derivative **192**. This interesting reaction is explained by the following mechanism. The silyl group in the silylruthenium **193**, formed by oxidative addition, undergoes 1,3-shift from Ru to oxygen to generate the siloxycarbyne complex **194**, which reacts with CO to give the dioxyacetylene Ru complex **196** via the η^2-ketenyl complex **195**. Then [2+2+2] cycloaddition of **196** with **191** affords **192** [79].

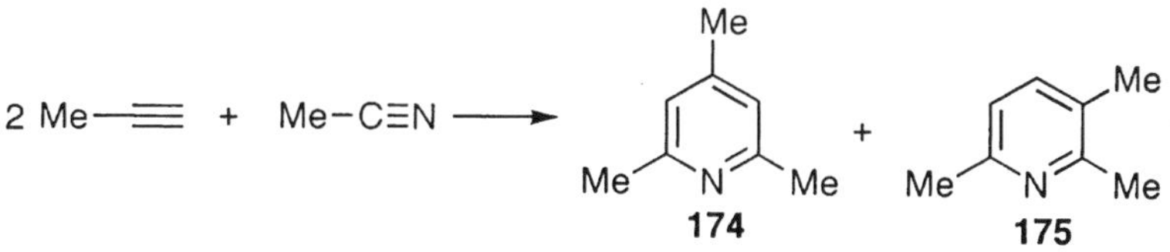

$2 \ Me\!-\!\!\equiv$ + $Me\!-\!C\!\equiv\!N$ ⟶ **174** + **175**

[CpCo(cod)], 1 atm
hv, 400 nm, 75%

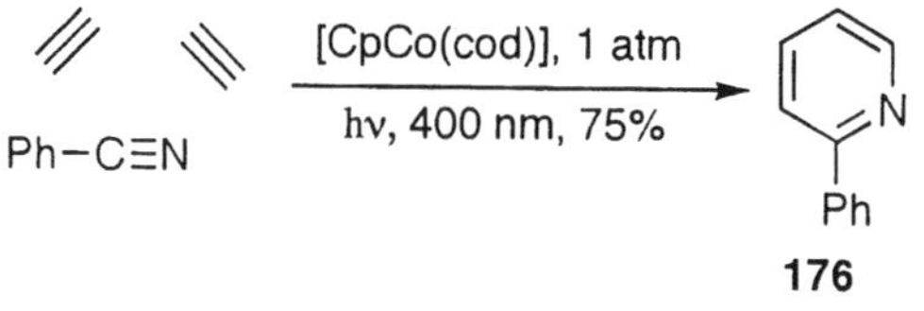

176

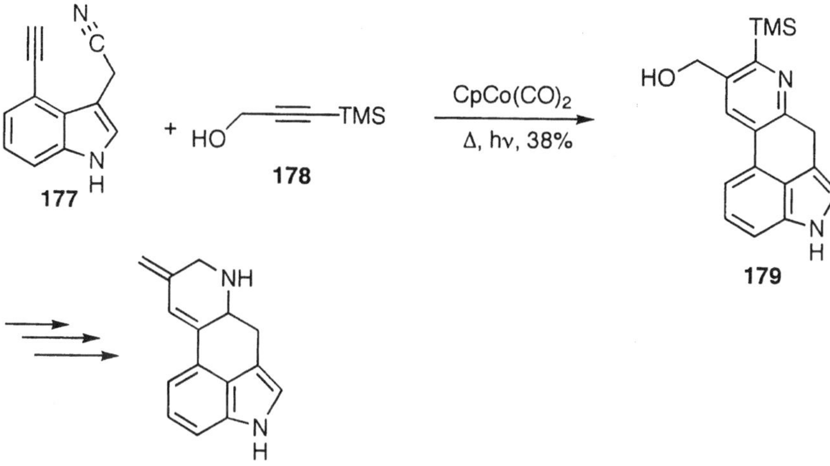

177 + **178** HO$\equiv$TMS

CpCo(CO)$_2$
Δ, hv, 38%

179

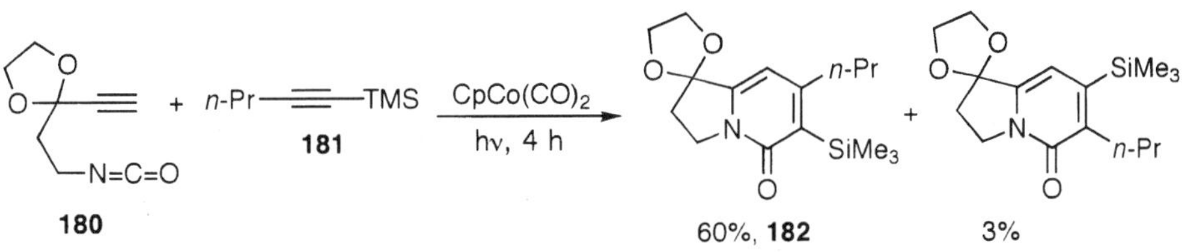

180 + **181** n-Pr$\equiv$TMS

CpCo(CO)$_2$
hv, 4 h

60%, **182** + 3%

183 + PrCHO

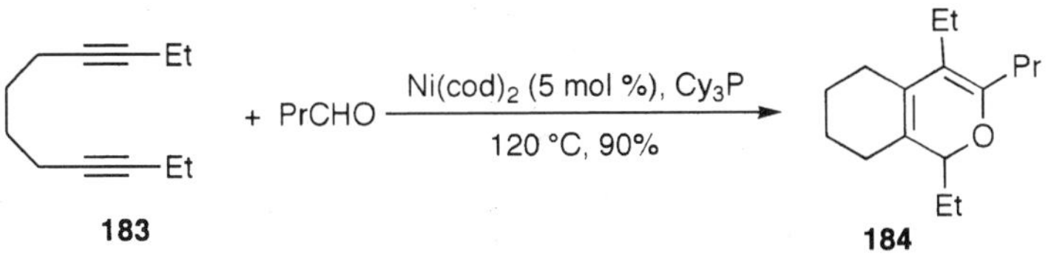

Ni(cod)$_2$ (5 mol %), Cy$_3$P
120 °C, 90%

184

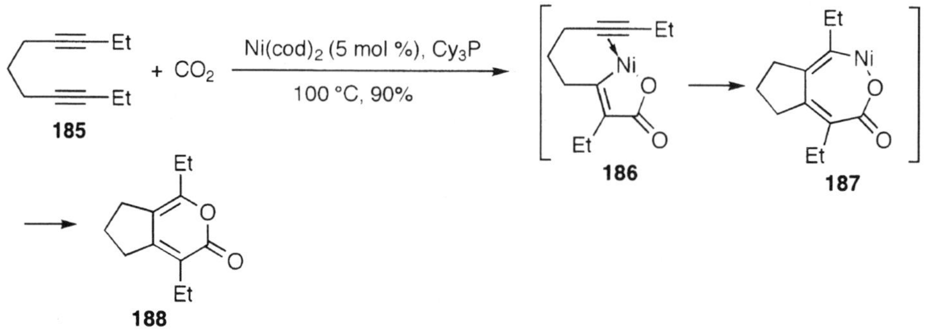
Et
+ CO_2
$Ni(cod)_2$ (5 mol %), Cy_3P
100 °C, 90%
185
Ni
186
187
188

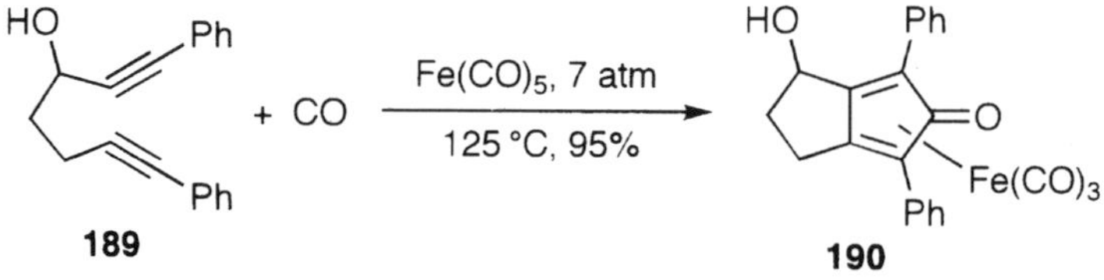
HO
Ph
+ CO
$Fe(CO)_5$, 7 atm
125 °C, 95%
189
$Fe(CO)_3$
190

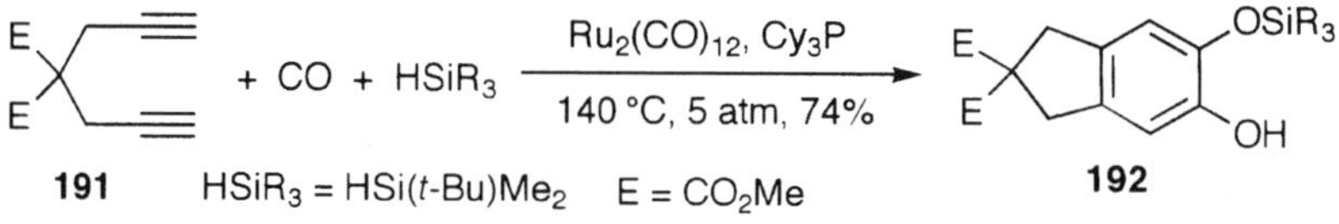
+ CO + $HSiR_3$
$Ru_2(CO)_{12}$, Cy_3P
140 °C, 5 atm, 74%
$OSiR_3$
OH
191
$HSiR_3$ = HSi(*t*-Bu)Me_2 E = CO_2Me
192

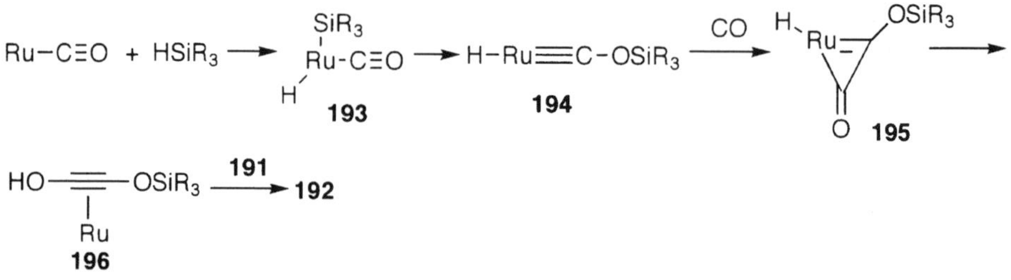
Ru−C≡O + $HSiR_3$
SiR_3
Ru-C≡O
193
H-Ru≡C-$OSiR_3$
194
CO
195
HO—≡—$OSiR_3$
Ru
196
191
192

7.2.3. Synthesis of Cyclopentenones by the Reaction of Alkyne, Alkene and Carbon Monoxide (Pauson–Khand Reaction)

Insertion of CO to the metallacyclopentenes **197** and **198** formed from enynes and metal complexes offers a useful synthetic route to the cyclopentenone derivatives **199** and **200**. This [2+2+1] cycloaddition mediated by $Co_2(CO)_8$ is called the Pauson–Khand reaction [80]. Both inter- and intramolecular versions are known.

The reaction of $Co_2(CO)_8$ with alkynes in an organic solvent affords stable complexes **201**. Two π-bonds of the alkyne bond coordinate to two Co atoms, respectively and its structure is shown by **203**. However, in this section the simplified form **204** is used for simplicity. When complex **201** is heated with alkene, the cyclopentenone **202** is formed by insertions of the alkene and CO. It is said that terminal alkynes and internal alkenes are particularly suitable for intermolecular Pauson–Khand reaction.

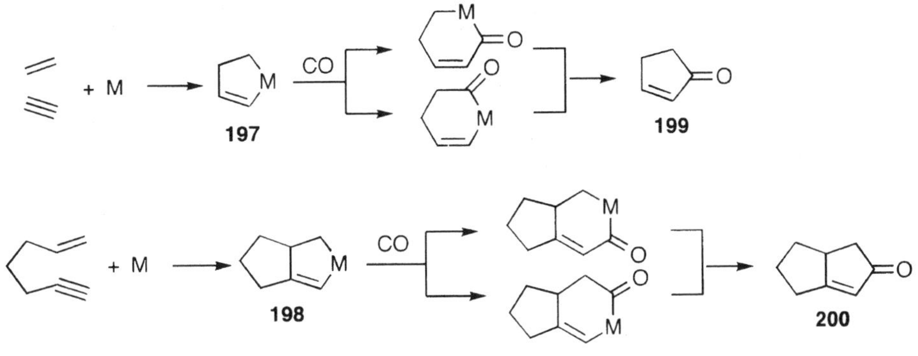

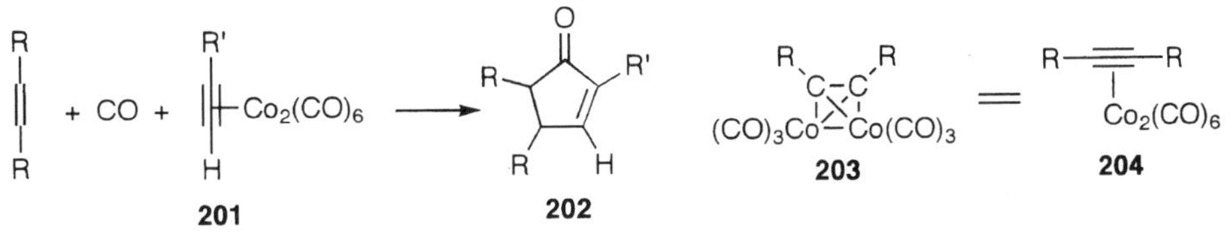

Formation of the tricyclo[3.3.0.0.]decane **209** by the reaction of [3.2.0]bicyclo-heptadiene **205** with propyne complex (**206**) is an example [81]. The Pauson–Khand reaction is explained by the following simplified mechanism. At first the oxidative cyclization of **205** and **206** generates the cobaltacyclopentene **207**, to which insertion of CO gives **208**. Finally, reductive elimination of **208** affords the cyclopentenone **209**.

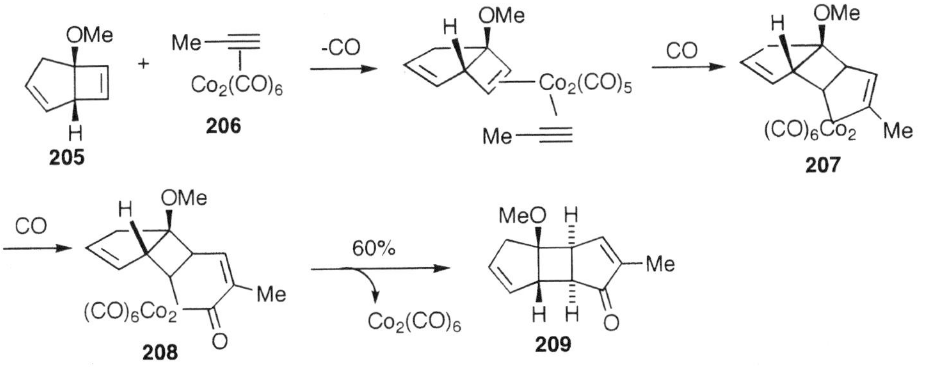

For the intermolecular version, strained alkenes such as norbornene give satisfactory results. Cyclopentenones are obtained in good yields by the reaction of alkyne–Co complex with alkene in the presence of silica. Wet alumina also gives good results [82]. Ultrasonication accelerates the reaction [83]. Reaction of ethylene with the terminal alkyne **210** proceeds under mild conditions (room temperature or 40 °C, 25 atm) in the presence of trimethylamine oxide to give the α-substituted cyclopentenone **211** in a high yield. Taylorione (**212**) has been synthesized from **211** [84].

The intramolecular version presents a very useful synthetic route to various polycyclic compounds. Even terminal alkenes give cyclopentenones in good yields, assisted by the addition of *N*-methylmorpholine *N*-oxide (NMO) [85]. The addition of trimethylamine *N*-oxide also dramatically accelerates the reaction in the presence of oxygen, and both inter- and intramolecular reactions proceed at 0 °C to room temperature [86]. The reaction was found to proceed rapidly at 25 °C by the addition of aqueous NH_4OH [87]. Numerous applications to natural product syntheses have been reported. The tri- and tetracyclic skeletons **214** for crinipellin B, from **213** [88], and the triquinacene derivative **216**, from **215**, have been constructed [89,90]. These results show that internal alkynes and terminal alkenes react smoothly in the intramolecular reactions. Domino reaction of the endiyne **217** produced the strained molecule of oxa[5.5.5.5]fenestrenedione (**219**) via **218** [91].

The Pauson–Khand reaction consumes a stoichiometric amount of $Co_2(CO)_8$. A catalytic intramolecular reaction is possible by the addition of ligands [92]. The reaction of both terminal and internal alkynes proceeds using $(PhO)_3P$ (10 mol %) and $Co_2(CO)_8$ (3 mol %) at 120 °C. Cyclization of the hept-6-en-1-yne (**220**) to give **221** is an example [93]. Also (indenyl)Co(cod) is a good catalyst [94]. Most conveniently, the catalytic cyclization of **222** proceeds smoothy using the catalyst generated *in situ* from $Co(acac)_2$ (0.046 mol %) and $NaBH_4$ (0.092 mol %) in CH_2Cl_2 [95]. Under irradiation, cyclization of **223** proceeds smoothly at 50 °C under 1 atm using $Co_2(CO)_8$ (5 mol %) of high purity [96]. A stable $Co_2(CO)_6$–alkyne complex is an active catalyst after treatment with Et_3SiH [96a].

Transition metal complexes other than $Co_2(CO)_8$ catalyse the Pauson–Khand reaction. The complex $[RhCl(CO)_2]_2$ catalyses the reaction of **224** [97], and $Ru_3(CO)_{12}$ [98] also catalyses these reactions at somewhat high temperatures. Highly enantioselective cyclization of **225** is catalysed by the chiral Ti(ebthi) complex to give **226** with 94% ee [99].

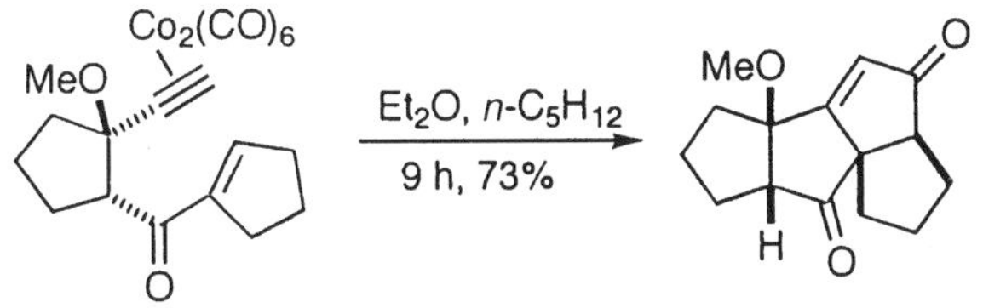
Co$_2$(CO)$_6$
MeO
Et$_2$O, n-C$_5$H$_{12}$
9 h, 73%
MeO
O
H
O
O

213 **214**

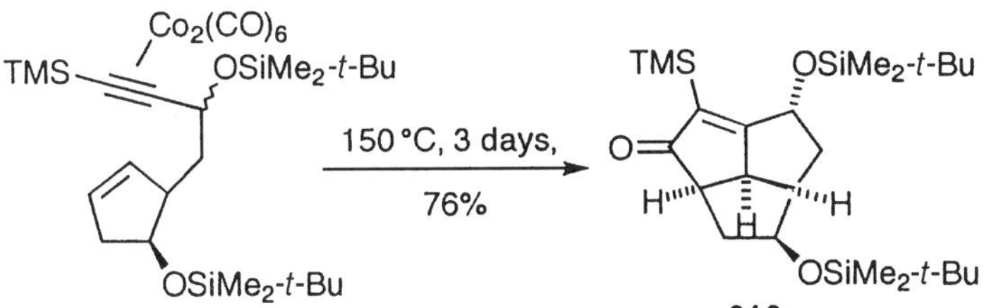
Co$_2$(CO)$_6$
TMS
OSiMe$_2$-t-Bu
OSiMe$_2$-t-Bu
150 °C, 3 days,
76%
TMS
OSiMe$_2$-t-Bu
O
H
H
H
OSiMe$_2$-t-Bu

215 **216**

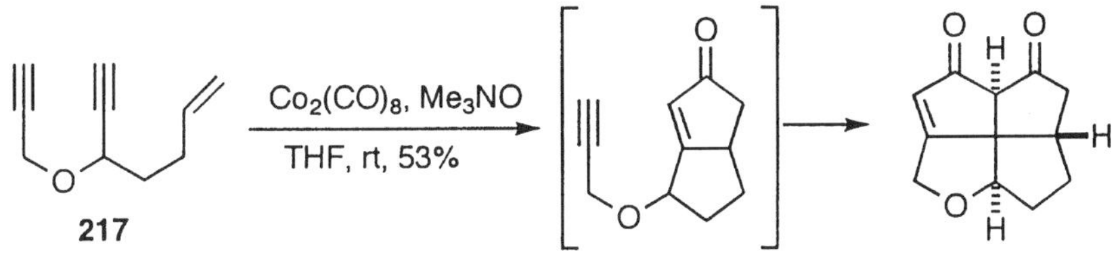
Co$_2$(CO)$_8$, Me$_3$NO
THF, rt, 53%
O
O
O
O
H
H
H
O
O

217 **218** **219**

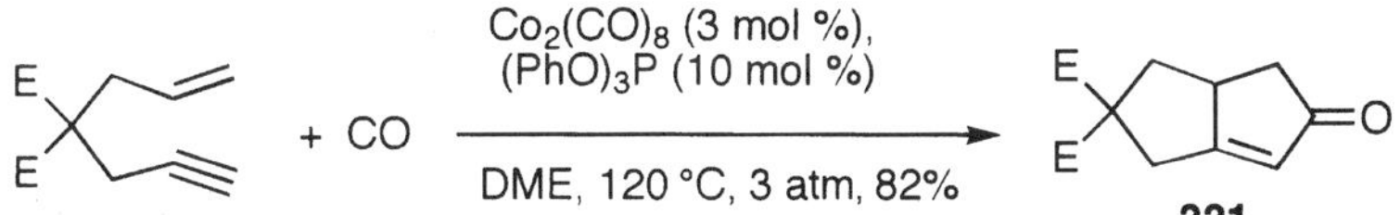
E
E
+ CO
Co$_2$(CO)$_8$ (3 mol %),
(PhO)$_3$P (10 mol %)
DME, 120 °C, 3 atm, 82%
E
E
O

220 **221**

E = CO$_2$Et

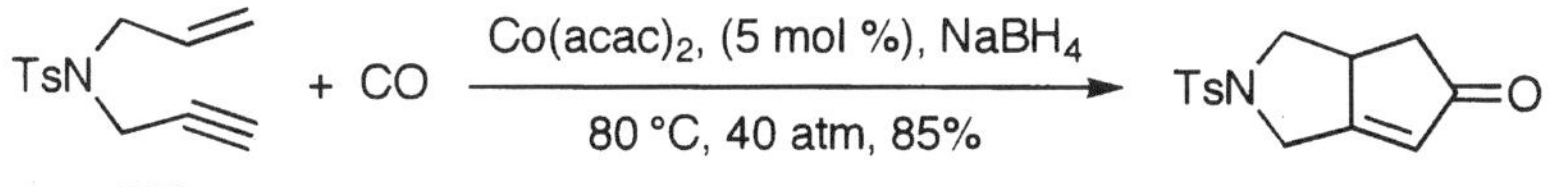
TsN
+ CO
Co(acac)$_2$, (5 mol %), NaBH$_4$
80 °C, 40 atm, 85%
TsN
O

222

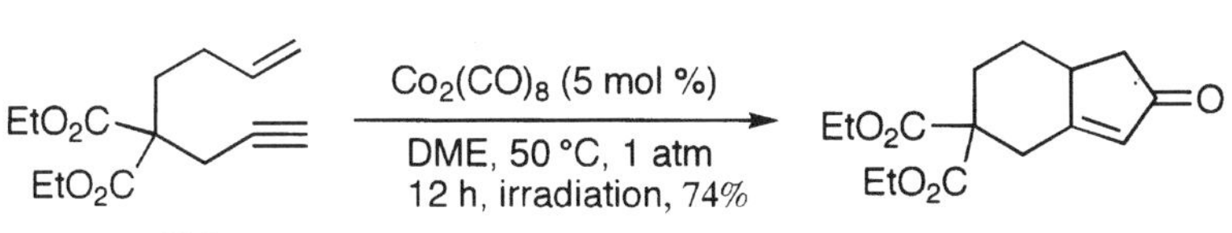
EtO$_2$C
EtO$_2$C
Co$_2$(CO)$_8$ (5 mol %)
DME, 50 °C, 1 atm
12 h, irradiation, 74%
EtO$_2$C
EtO$_2$C
O

223

Pauson–Khand-type reactions of enynes are mediated by other metal complexes, such as $Fe(CO)_5$ [100], $[W(CO)_5THF]$ and $Mo(CO)_6$ [101]. Cyclization of the allene–yne system **227** to the α-methylenecyclopentenone **228** is promoted by $Mo(CO)_6$. The products depend on the substituents of the allene, and the cyclization of **229** afforded **230** as the main product [102].

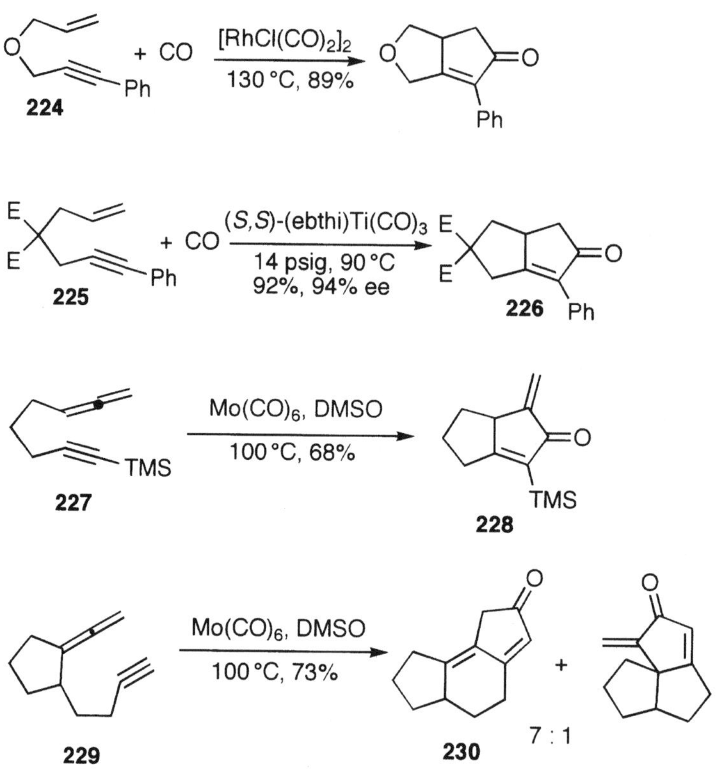

224   + CO $\dfrac{[RhCl(CO)_2]_2}{130\,°C,\ 89\%}$ Ph

225   + CO $\dfrac{(S,S)\text{-}(ebthi)Ti(CO)_3}{14\ psig,\ 90\,°C}$ **226**
92%, 94% ee   Ph

227 $\dfrac{Mo(CO)_6,\ DMSO}{100\,°C,\ 68\%}$ TMS   **228**   TMS

229 $\dfrac{Mo(CO)_6,\ DMSO}{100\,°C,\ 73\%}$ **230**   +   7 : 1

7.2.4 Reductive Cyclization of 1,6- and 1,7-Dienes, Diynes, Enynes and Arenes via Zirconacycles and Titanacycles

The 14-electron species denoted here as 'Cp₂M' (M = Ti, Zr), generated from Group 4 metallocenes, mediate useful oxidative cyclizations of dienes **231** [103,104], enynes **233** [105] and diynes **235** [105,106] via the metallacycles **232, 234** and **236**. These metallacycles are useful intermediates, which are hydrogenolysed, carbonylated, halogenated, attacked by electrophiles and converted to functionalized cyclic compounds such as **238** and **239** [107].

The complex 'Cp₂Zr', coordinated by butene (**240**) and called the Negishi reagent, is generated easily by the treatment of zirconocene dichloride with two equivalents of BuLi or Grignard reagents, and used for synthetic purposes [108]. Alkene–zirconocene complexes such as **240** and **242** having 16 valence electrons [Zr(II)] are highly reactive species, which can be regarded as the zirconacyclopropanes **241**

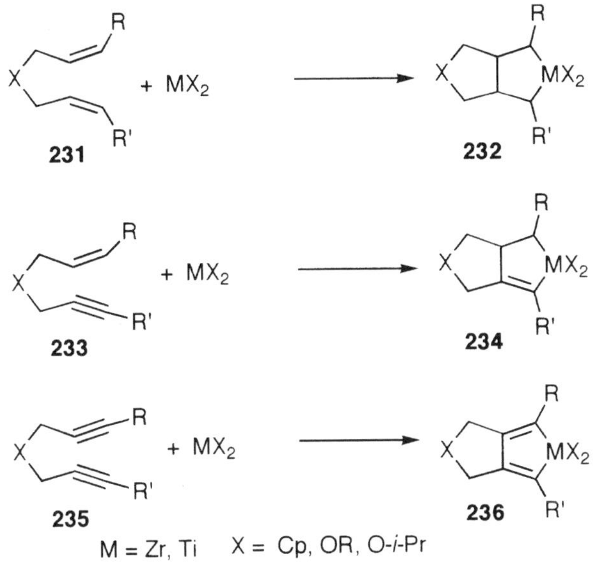

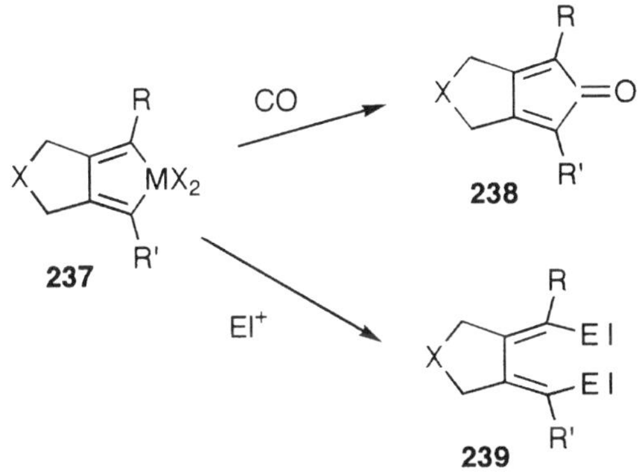

M = Zr, Ti   X = Cp, OR, O-*i*-Pr

and **243**. The zirconacyclopentanes **244** and zirconapentenes **245** [Zr(IV)] are easily prepared by the reaction of alkenes or alkynes with **242** [109,110]. The zirconacyclopentanes **244** and -pentenes **245** undergo various transformations.

Interestingly, zirconacyclopentane **246** formed by the reaction of 1,6-heptadiene with the Zr complex has the *trans* ring junction mainly [108]. It should be noted that the preparation of the *trans* ring junction in the bicyclo[3.3.0]octane system by other means is difficult. Carbonylation of **246** affords *trans*-fuzed bicyclo[3.3.0]octanone **247** [109,111]. The diacetoxy compound **248** is obtained by oxidative cleavage of **246**. Protonation affords the *trans*-dimethylcyclopentane skeleton. Similar reactions occur with 1,6-enynes, and Pauson–Khand-type cyclopentenone synthesis is possible by carbonylation.

The asymmetric synthesis of phorbol (**252**) has been carried out by the application of the Zr-mediated cyclization of 1,6-enyne **249**. Protonation of the intermediate **250** of the cyclization affords **251** in 93% yield, which was converted to **252** [112].

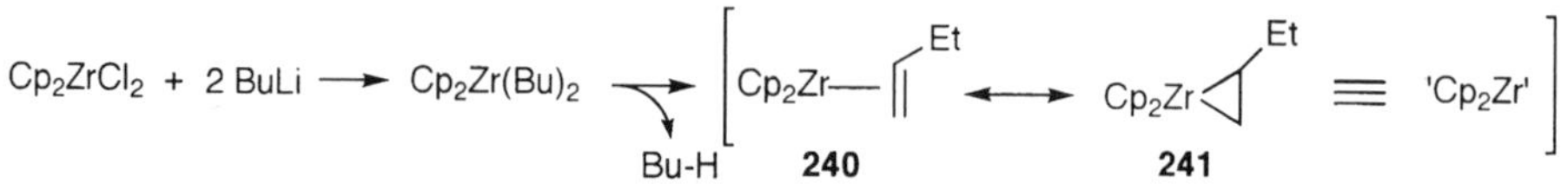
Cp2ZrCl2 + 2 BuLi
Cp2Zr(Bu)2
Bu-H
Cp2Zr
Et
'Cp2Zr'
240
241

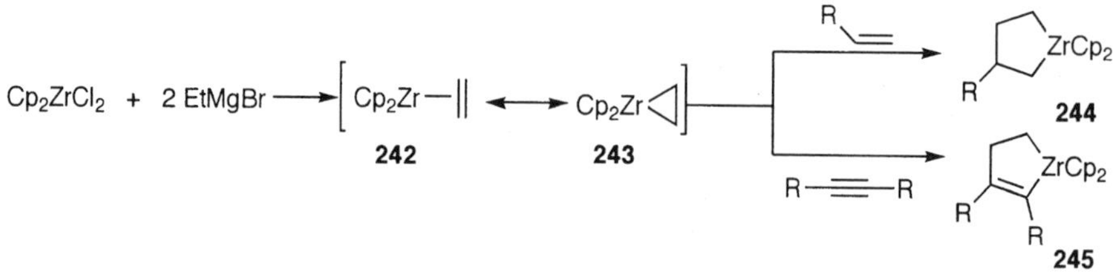
Cp2ZrCl2 + 2 EtMgBr
Cp2Zr
R
ZrCp2
R—≡—R
242
243
244
245

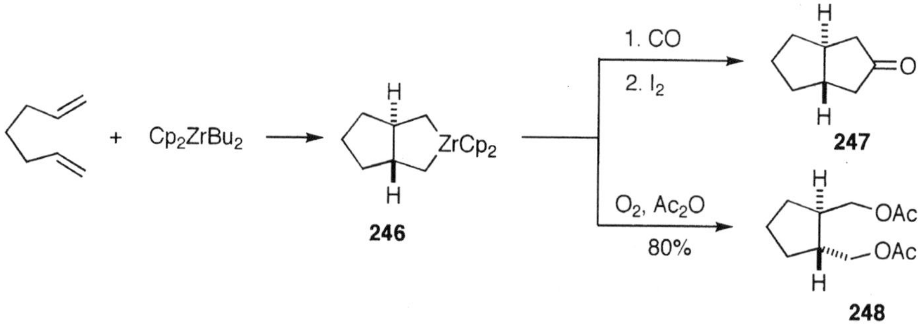
+ Cp2ZrBu2
H
ZrCp2
1. CO
2. I2
O2, Ac2O
80%
OAc
246
247
248

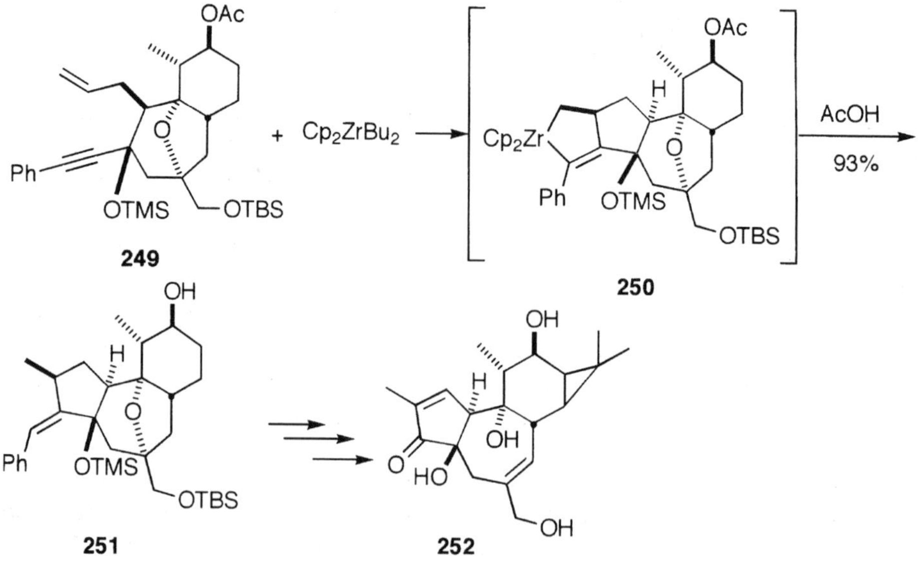
OAc
Ph
OTMS
OTBS
+ Cp2ZrBu2
Cp2Zr
AcOH
93%
249
250
OH
H
HO
251
252

The 1,6-diene **254** was prepared from (+)-carvone (**253**). Its oxidative cyclization affords the zirconacycle **255**, and its carbonylation, followed by treatment with iodine and HCl, gave the tricyclic ketone **256** [113]. The alkaloid (−)-dendrobine (**257**) was synthesized from **256**.

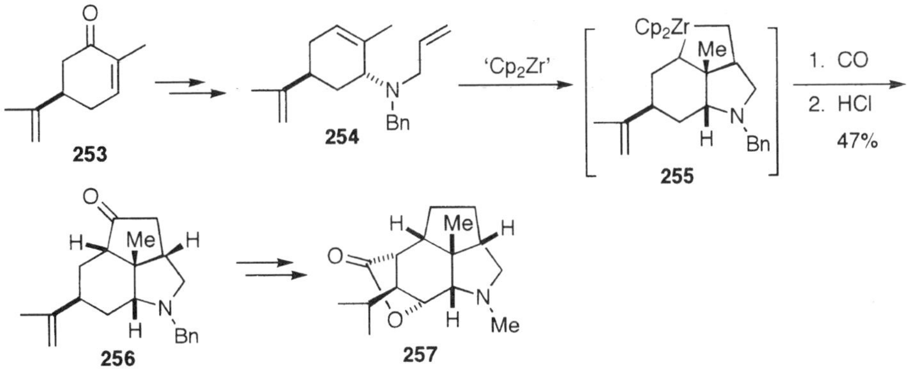

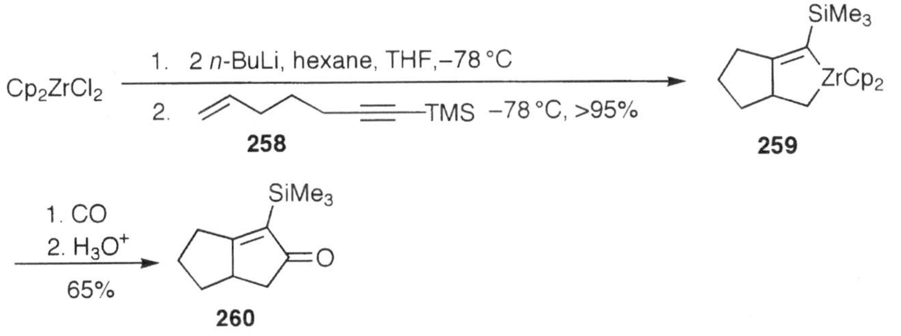

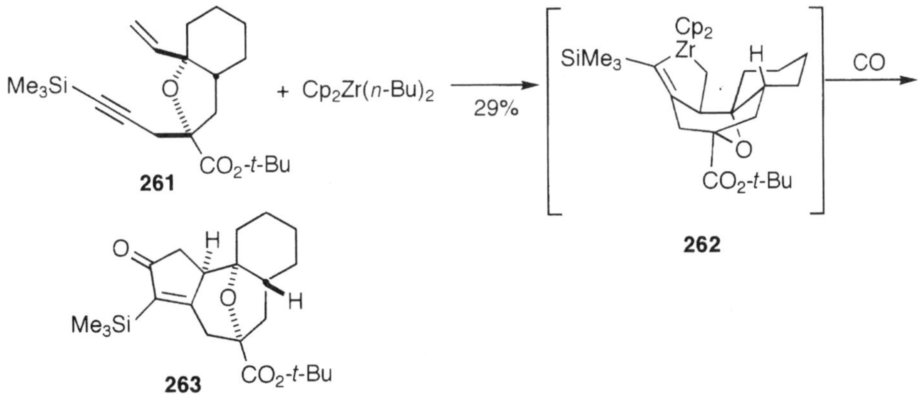

The zirconacycle **259** is prepared by the reaction of 'Cp$_2$Zr' with the 1,6-enyne **258**, and converted to the cyclopentenone **260** by the reaction of CO. The silylalkyne **258** is the appropriate substrate, because the terminal alkyne can not be used in this reaction [111]. The tricyclic skeleton of tigliane **263** has been prepared by the carbonylative cyclization of **261** via **262** [114]. The cyclopentenone **263** could not be prepared by the Co-mediated Pauson–Khand reaction.

The five-membered metallacycle **265** is formed from 1,7-diyne **264** by the reaction with the Negishi reagent. Two allyl groups are introduced by CuCl-mediated transmetallation of **265** to afford **266** [115]. The terminal alkenes in the 1,4,6,9-decatetraene **266** react again with the Negishi reagent to give the zirconabicyclo[6.3.0] system **267** [116]. Carbonylation of **267** affords ketone **269**, and protonation of the metallacycle gives **268**. Terminal alkynes can not be used for this reaction. The intermediate **271**, formed from the zirconacyclopentadiene **270** and CuCl, reacts with *o*-diiodobenzene to give the substitutd naphthalene **272** [117].

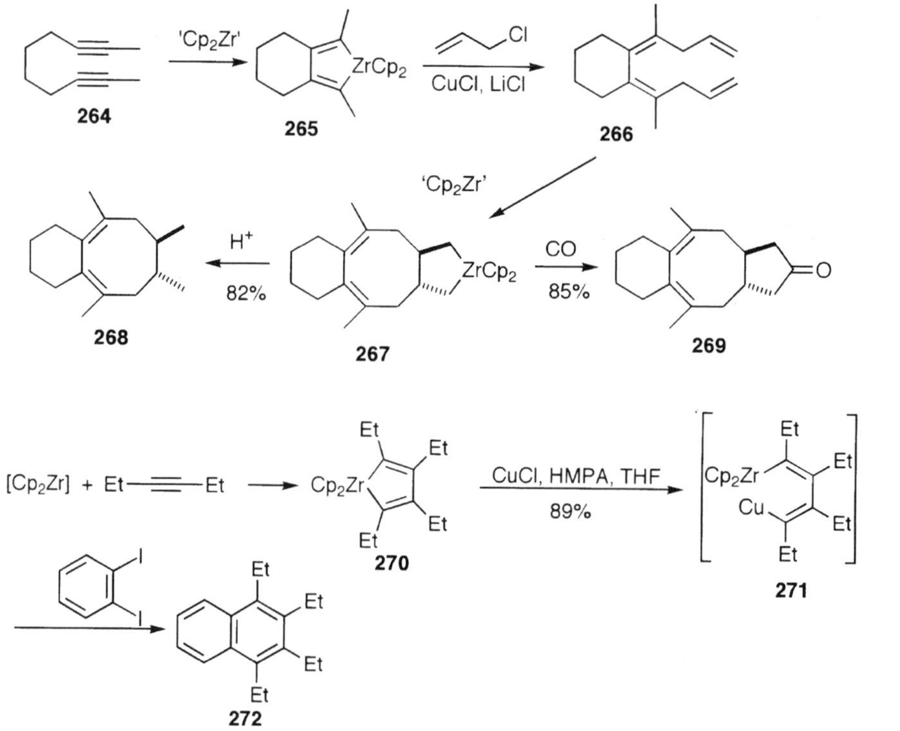

The Ti(II) reagents, related to the Negishi reagent, can be generated by the reaction of Ti(O-*i*-Pr)$_4$ with two equivalents of EtMgBr [118]. Use of *i*-PrMgBr, instead of EtMgBr and *n*-PrMgBr, is particularly important for the efficient generation of the alkene–Ti(II)(O-*i*-Pr)$_2$ reagents **273** and **274** [119]. Elimination of β-hydrogen from *i*-Pr$_2$TiX$_2$ is more facile than Et$_2$TiX$_2$ during the preparation of the Ti(II)(O-*i*-Pr)$_2$ reagent. In addition, reagent **273** is much cheaper than 'Cp$_2$Zr'.

The cyclopentenone **277** and a small amount of the cyclopentanone **278** are obtained by the carbonylation (1 atm) of titanacycle **276**, generated from 1,6-enyne **275** and **273** [120]. However, this Pauson–Khand type reaction of the 1,6-enyne proceeds with a catalytic amount of Cp$_2$Ti(CO)$_2$. Furthermore, asymmetric

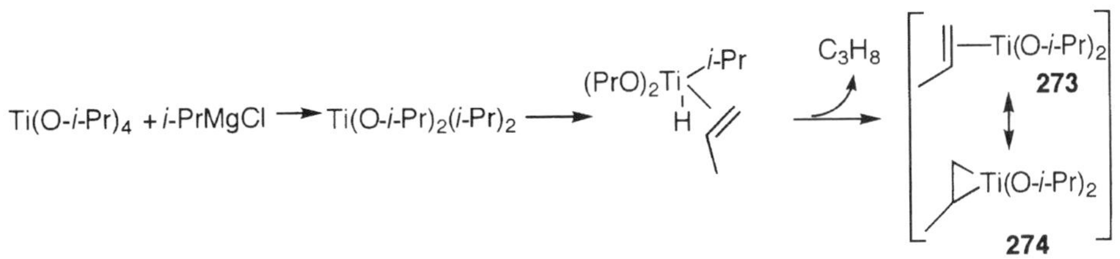

$$Ti(O\text{-}i\text{-}Pr)_4 + i\text{-}PrMgCl \longrightarrow Ti(O\text{-}i\text{-}Pr)_2(i\text{-}Pr)_2 \longrightarrow$$

carbonylation of 1,6-enyne **279** catalysed by (S,S)(ebthi)Ti(CO)$_2$ gave the cyclopentenone **281** with 94% ee in high yield via **280** [121].

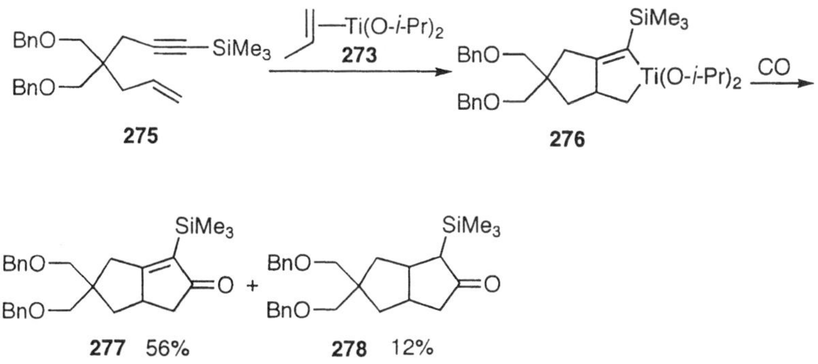

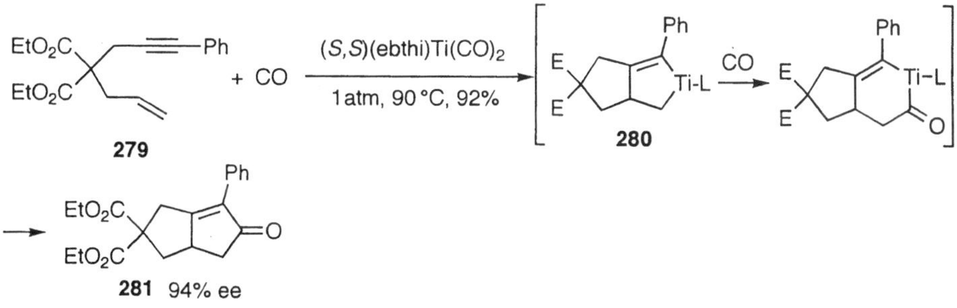

Reactions characteristic to the Ti reagent are observed with esters. Hydroxycyclo-propanation of esters catalysed this way is known as the Klinkovich reaction [118]. The titanacyclopropane **283** is generated by the reaction of EtMgBr with Ti(O-i-Pr)$_4$, and reacts with methyl acetate to give 1-methylcyclopropanol (**282**). Formation of the 1-alkylcyclopropanol **285** is explained by the generation of titanacyclopropane **283**. Then, formation of **284** by insertion of carbonyl group of the ester, and cyclopropanation of **284** by transfer of OMe group gives **286**. The reaction proceeds by the treatment of alkyl halide **287** with Mg in the presence of a catalytic amount of TiCl(O-i-Pr)$_3$ to afford the 1,2-disubstituted cyclopropanol **288** diastereoselectively. The optically active cyclopropanol **289** with 78% ee was obtained using the optically active Ti reagent **290** [122].

The 1-hydroxybicyclo[n.1.0] system is obtained by the reaction of esters bearing a terminal double bond [123]. The hydroxycyclopropanation of the unsaturated ester

$MeCO_2Me$ + 2 EtMgBr $\xrightarrow[\substack{2.\ H_2O,\ H_2SO_4 \\ 76\%}]{1.\ Ti(O\text{-}i\text{-}Pr)_4\ (10\ mol\%)}$ Me, OH

282

$Ti(O\text{-}i\text{-}Pr)_4$ $\xrightarrow{\text{EtMgBr}\quad i\text{-PrOMgBr}}$ $(R'O)_2TiEt_2$ $\xrightarrow{C_2H_6}$ $(R'O)_2Ti$ ◁

283

R OH $\xleftarrow{H_2O}$ R OMgBr ↑

285   2 EtMgBr

R' = Me or i-Pr

RCO₂Me

$(R'O)_2Ti$ ⟨ OMe / O ← R

286

$(R'O)_2Ti$ ⟨ O / O ⟩ OMe R

284

$n\text{-}PrCO_2Me$ + $EtCH_2CH_2Br$ + Mg $\xrightarrow[79\%]{(i\text{-}PrO)_3TiCl}$ HO n-Pr   H Et

287   **288**

$MeCO_2Et$ + 2 $PhCH_2CH_2MgBr$ $\xrightarrow[64\%]{Ti(OR^*)_4}$ HO Me   H Ph

289 78% ee

$Ti(OR^*)_4$ = $\left[\begin{array}{c} \text{Ti} \end{array} \right]_2$

290

291 is explained by intramolecular acylation. The titanacyclopropane **292**, formed from the unsaturated ester **291**, is converted to the ketone **293** by intramolecular acylation. The cyclopropane **294** is formed by further reaction with the ketone, and the 1-hydroxybicyclo[4.1.0]heptane **295** is obtained after hydrolysis [124].

One of the characteristic reactions of the Ti reagent is the intramolecular nucleophilic acylation of the titanacycle with esters, and 2,7- and 2,8-enyne esters afford interesting mono- and bicyclic skeletons [125]. Titanacycle **297**, formed from α,β-unsaturated ester **296**, is converted to the alkenyltitanium **300** by protonation, and the cyclopentenone **301** is formed by intramolecular acylation [126]. As **297** is a tautomeric form of the Ti enolate, it reacts with electrophiles such as aldehydes to give **298**, which cyclizes to **299**. The α,β-alkynic ester **302** generates the titanacycle **303** and is converted to the bicyclo[3.1.0]hexane system **305** via **304** [126]. The titanacycle

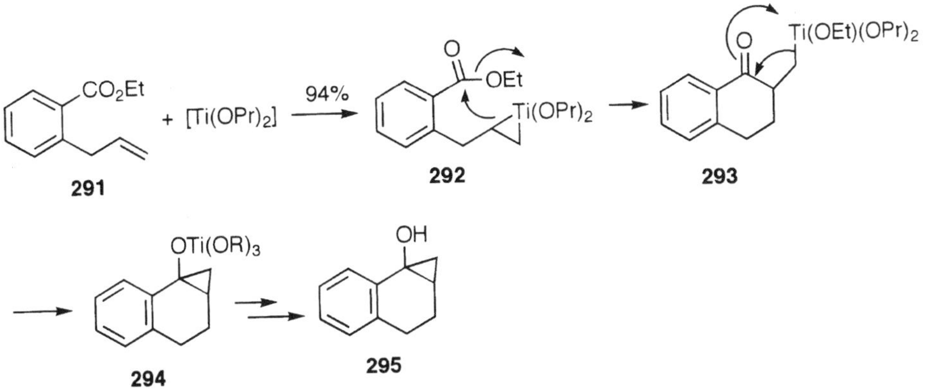

CO_2Et
+ $[Ti(OPr)_2]$
94%
OEt
$Ti(OPr)_2$
$Ti(OEt)(OPr)_2$
291
292
293
$OTi(OR)_3$
OH
294
295

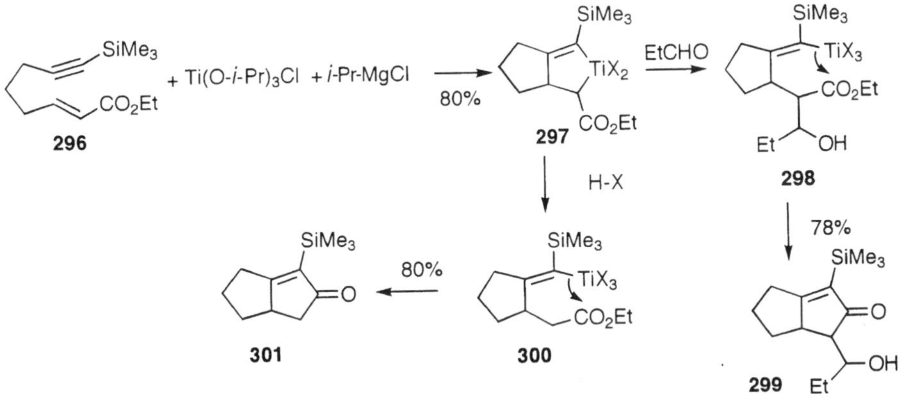

$SiMe_3$
CO_2Et
+ $Ti(O\text{-}i\text{-}Pr)_3Cl$ + i-Pr-MgCl
80%
TiX_2
EtCHO
TiX_3
Et
OH
296
297
298
H-X
78%
80%
301
300
299

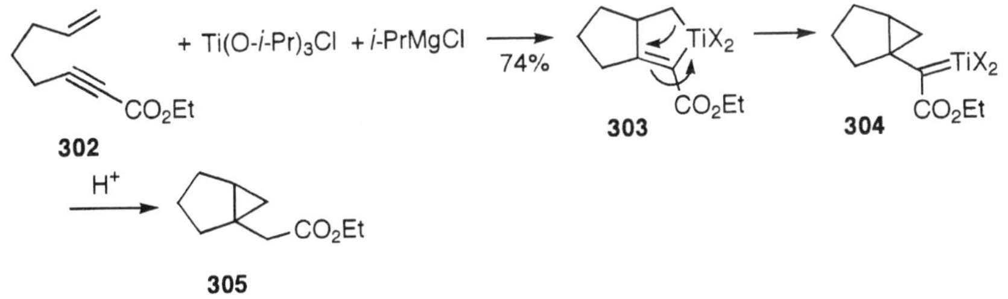

+ $Ti(O\text{-}i\text{-}Pr)_3Cl$ + i-PrMgCl
74%
TiX_2
CO_2Et
302
303
304
H^+
305

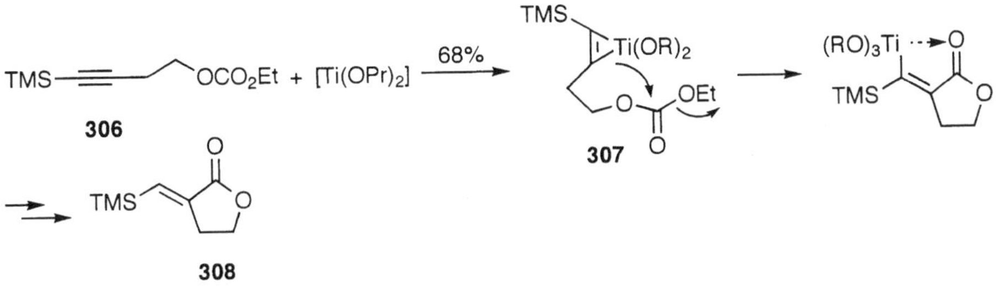

TMS
OCO_2Et + $[Ti(OPr)_2]$
68%
$Ti(OR)_2$
OEt
$(RO)_3Ti$
306
307
308

307, formed from the triple bond in **306**, is converted to the α-alkylidene-γ-lactone **308** by attacking the carbonate [127].

The reactions via zircona- and titanacycles shown above are stoichiometric. More practical Ti-catalysed syntheses of bicyclic cyclopentenones from enynes have been developed recently using either $Cp_2Ti(Me_3P)_2$ or, better, Cp_2TiCl_2-*n*-BuLi as the catalyst. The reaction of titanacycle **309** with triethylsilyl cyanide (= Et_3SiNC) affords iminocyclopentene **310** and regenerates 'Cp_2Ti' as the catalytic species. Reductive elimination and hydrolysis of **310** gives the cyclopentenone **311** [128]. The Pauson–Khand reaction of **279** catalysed by $Cp_2Ti(CO)_2$ to form **281** is the more convenient reaction [121].

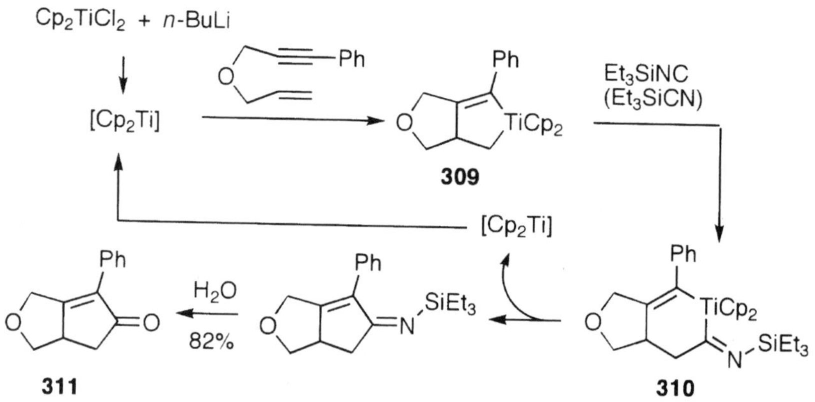

Oxametallacycles are prepared from unsaturated aldehydes or ketones. Oxidative cyclization of 6-hepten-2-one (**312**) catalysed by the Ti catalyst 'Cp_2Ti' to give cyclopentanol **315** has been developed. The key step is the cleavage of the strong Ti–O bond in the oxametallacycle **313** with oxophilic hydrosilane, and the silyl ether **314** is formed with regeneration of 'Cp_2Ti' [129,130]. Cyclization of 5-hexenal (**316**)

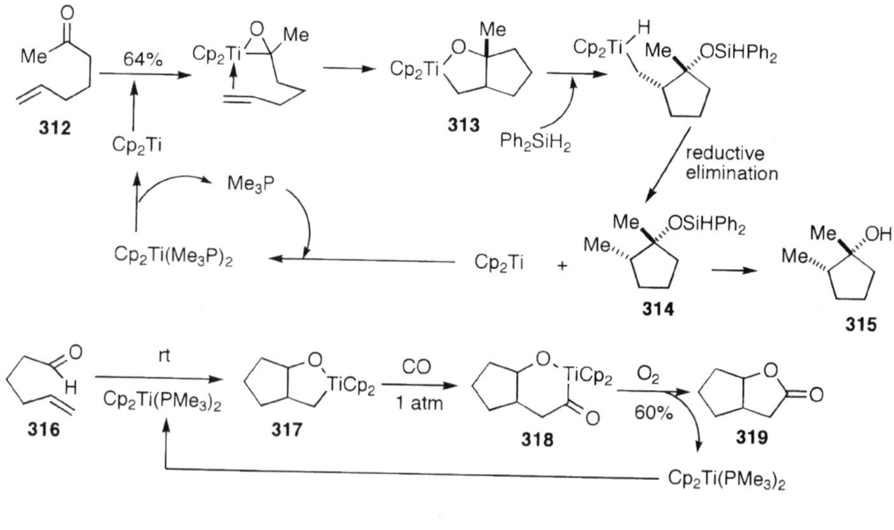

catalysed by $Cp_2Ti(PMe_3)_2$ generates **317** as an intermediate. The CO insertion to **317** gives the acyltitanium **318**, and its reductive elimination affords the γ-butyrolactone **319** under mild condition and the catalytic species is regenerated [131]. Thus this hetero-Pauson–Khand reaction proceeds with a catalytic amount of $Cp_2Ti(PMe_3P)_2$.

7.2.5 Pd-catalyzed Intramolecular Alder-ene Reaction of 1,6- and 1,7-Diynes and Enynes

Some 1,6- and 1,7-enynes undergo interesting Pd, Pt and Ru-catalysed cyclizations, which are regarded as Alder-ene reaction and metathesis. These reactions offer a useful method for the construction of polycyclic compounds [132]. These cyclizations can be understood by the following two mechanisms as shown by Scheme 7.3. As the first possibility, the oxidative cyclization of 1,6-enyne **320** generates the palladacyclopentene **321**. Elimination of two different β-hydrogens from **321** yields either **322** or **323**, which undergoes reductive elimination to produce the 1,4-diene **324** as the Alder-ene product, and the 1,3-diene **325** [133]. Of course, the 1,4-diene is the expected product of the thermal ene reaction.

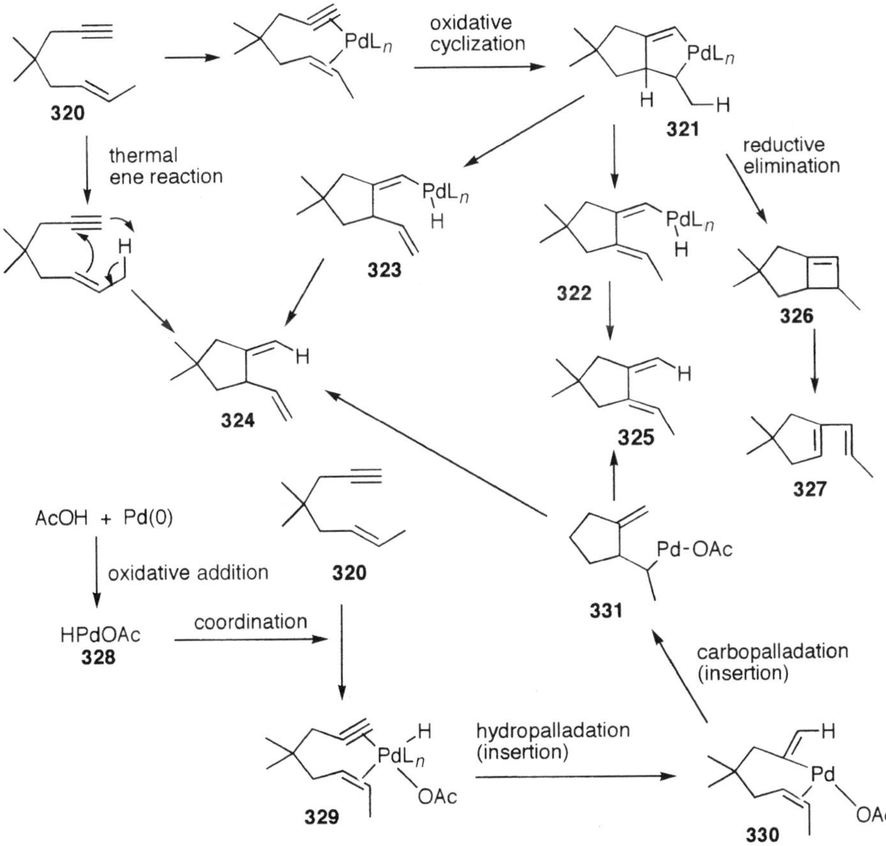

Scheme 7.3 Proposed mechanism of Pd-catalysed ene and metathesis reactions of 1,6-enynes

As another possibility, the ene-type reaction can be explained by involvement of the hydride species H−Pd−OAc **328**, generated by the oxidative addition of HOAc to Pd(0), as an initiator [134]. Preferencial insertion of the triple bond in **329** gives **330**, and subsequent intramolecular insertion of the double bond gives **331**. Finally, depending on which β-hydrogen is eliminated, either 1,4-diene **324** (ene-type), or 1,3-diene **325** is formed with regeneration of H−Pd−OAc (**328**). BBEDA [*N*,*N'*-bis(benzylidene)ethylenediamine] is used as a ligand [135,136].

Cycloisomerization or metathesis also occurs, which can be understood as the formation of cyclobutene **326** by reductive elimination of **321**. The metathesis product **327** is formed by isomerization of **326**. The metatheses involving metal–carbene complexes are discussed in Section 7.2.6. They are closely related, but somewhat different from the metathesis explained here. Balance between the ene and the metathesis reactions seems to be delicate.

The cyclization is quite general. The 1,6-enyne system of alkynyl *N*-acylenamine **332** undergoes a similar ene-type cyclization using BBEDA (5%) to give **333** [137]. The bicyclic picrotoxane skeleton **335** has been constructed by this Pd-catalysed ene-type cyclization of **334**. The cyclized product **335** was obtained in a satisfactory yield (70%) only by the combined use of the phosphine ligands **336** and **337** [138].

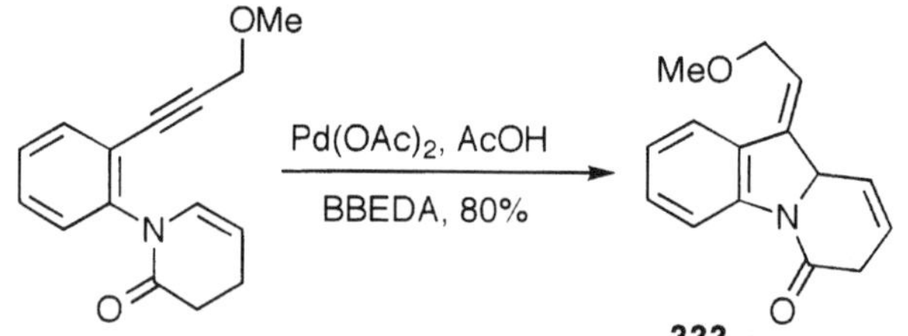

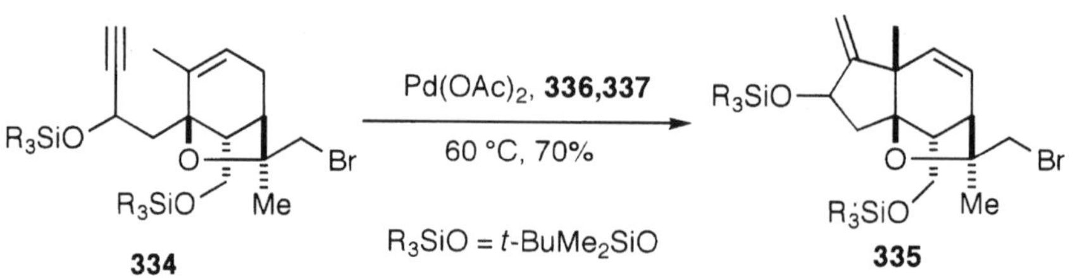

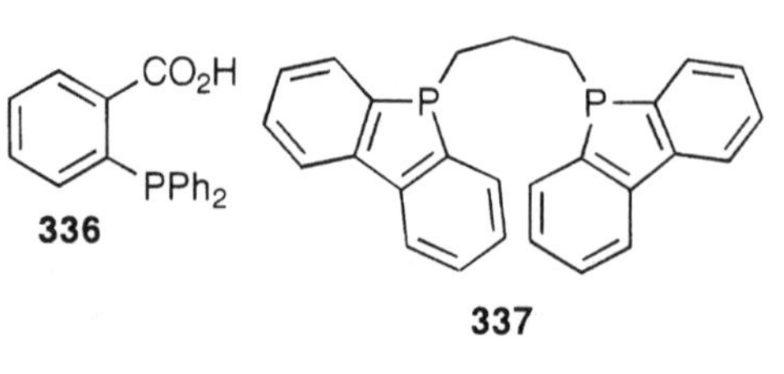

Formation of six-membered rings from 1,7-enynes is less facile than that of five-membered rings. Cyclization of the 1,7-enyne **338** bearing a hindered double bond proceeds smoothly using formic acid, a stronger acid than AcOH, and a ligandless Pd catalyst to give the six-membered ring **339** in 83% yield. Double bond migration is effected by the Pt-catalysed 1,4-addition of hydrosilane to the enone. Subsequent treatment with DDQ affords **340**, which was converted to (+)-cassiol (**341**) [139].

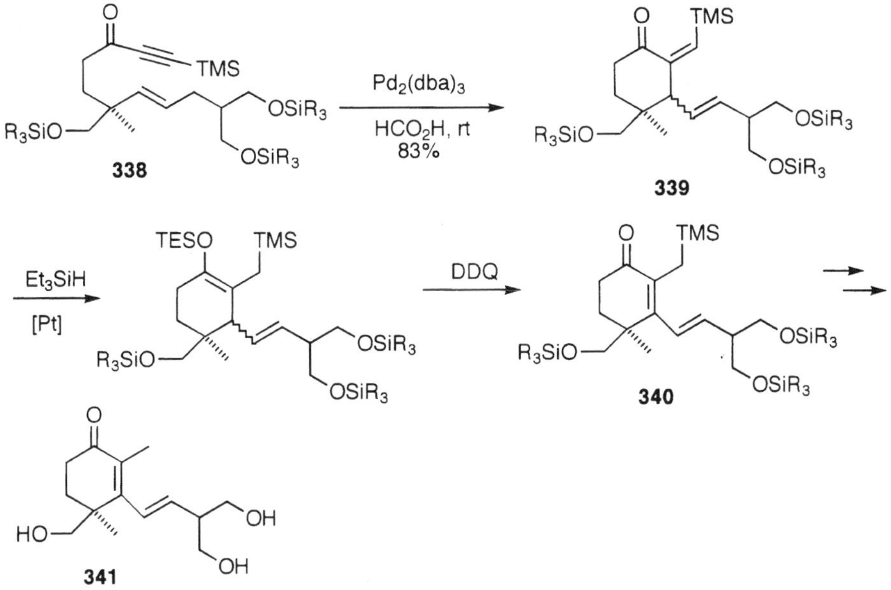

As mentioned in Scheme 7.3, although thermal ene reaction of the 1,6-enyne **320** gives only 1,4-dienes **324**, the 1,3-diene **325** is formed by the Pd-catalysed reaction. No thermal ene reaction is possible with **342**, but the smooth Pd-catalysed cyclization affords the 1,3-diene system **343**, which has a skeleton analogous to **325** in Scheme 7.3. Notably, the reaction proceeds with migration of vinylic hydrogen. The 1,3-diene system formed by the cyclization is useful for further modification, typically Diels–Alder reaction. The 1,3-diene **343** formed from **342** is converted to stereopolide (**344**) by Diels–Alder reaction [140]

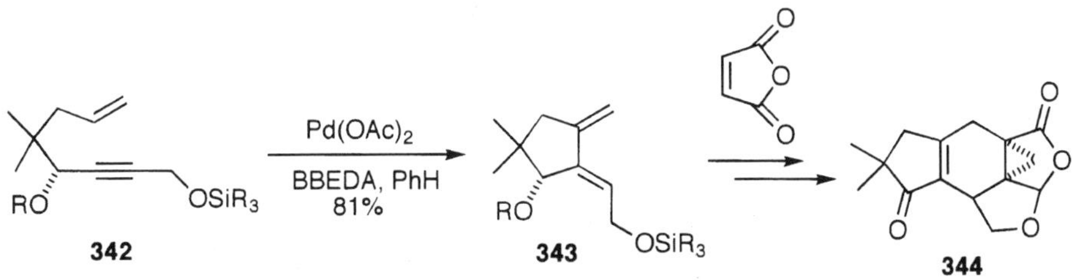

The above-mentioned cyclization starts with the insertion of alkynes to the Pd–H bond to generate alkenylpalladium species (hydropalladation). Polycyclization of polyenynes is also intiated by H–Pd–OAc species **328**. An efficient pentacyclization of the pentaenyne **345** gives **348** in 86% yield using triphenylstibine as a ligand in AcOH [141]. The reaction is called the Pd-catalysed zipper reaction. The first step is the hydropalladation of the triple bond to generate vinylpalladium, which undergoes domino insertions of alkenes to give **346** and then **347**. For this domino insertion, alkenes must be 2,2-disubstituted in order to generate the neopentylpalladiums **346** and **347** which have no β-hydrogen to be eliminated, and the insertion continues without termination. The reaction stops by the insertion of monosubstituted terminal alkene and β-elimination, giving **348**.

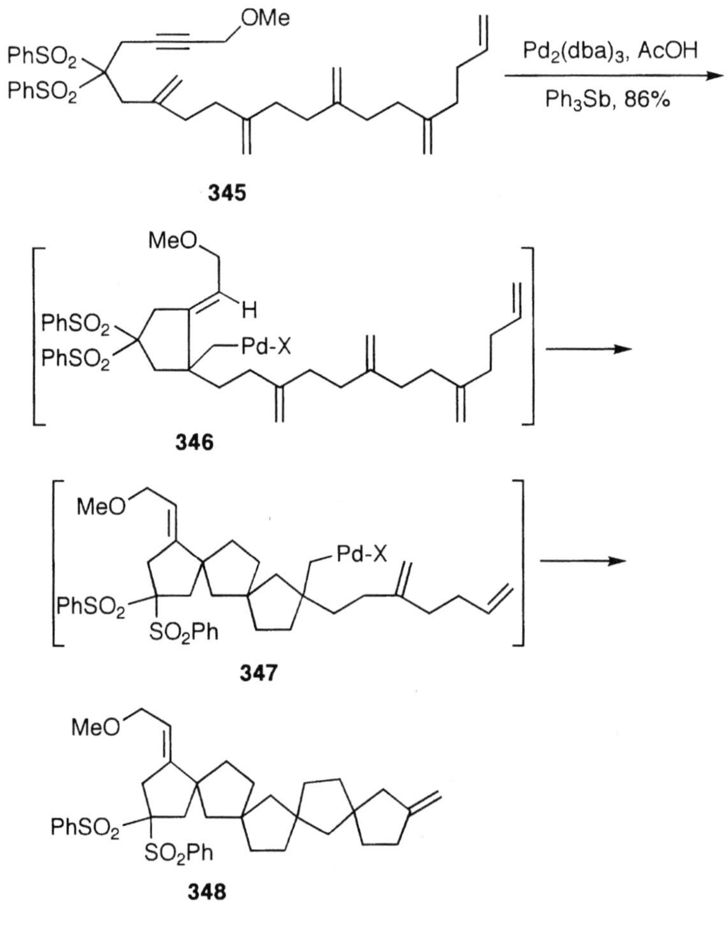

345

346

347

348

Another reaction is reductive cyclization of 1,6-diynes and 1,6-enynes using hydrosilanes as a hydrogen source in AcOH. Triethylsilane is used as the hydrogen donor for this reaction [142]. The reaction can be understood by the formation of vinylpalladium bond **350** via the insertion of alkyne **349** to the Pd–H bond in **328**. Then intramolecular alkyne insertion gives vinylpalladium **351**, which is hydrogenolysed with Si–H to give the 1,3-diene **352** by transmetallation and reductive elimination. Total synthesis of siccanin (**355**) has been achieved applying Pd-catalysed cyclization of the sterically congested 1,7-diyne **353** to give the 1,3-diene **354** using formic acid, instead of AcOH, as the key reaction [143].

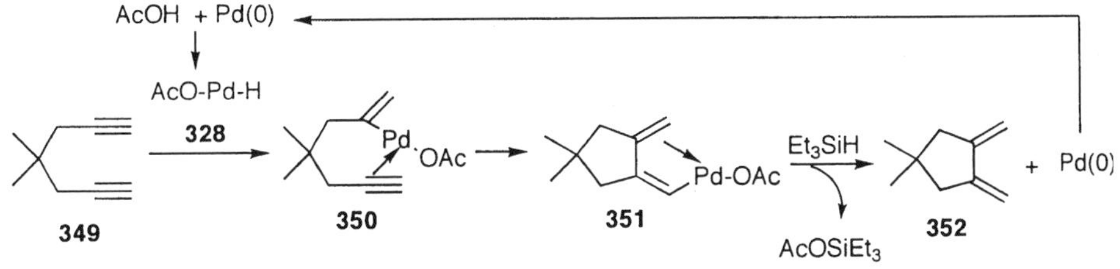

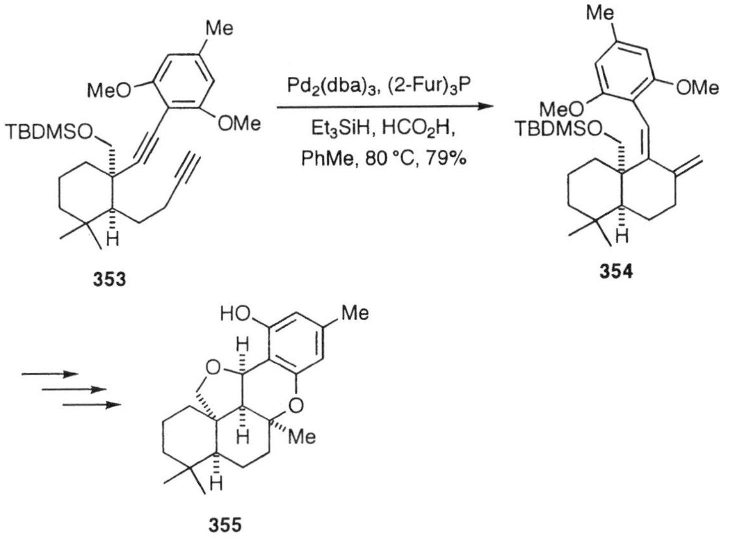

7.2.6 Skeletal Reorganization of 1,6- and 1,7-Enynes Catalysed by Pt, Ru and Pd Catalysts

Cycliation of 1,6- and 1,7-enynes **320** (Scheme 7.3) involving C−C bond cleavage and formation of the 1-vinylcyclopentene **327** is known as skeletal reorganization, cycloisomerization, cyclorearrangement or enyne metathesis [132]. The oxidative cyclization of **320** generates **321**, and its reductive elimination affords the cyclobutenes **326**. Subsequent ring cleavage to form **327** is easily understandable as the enyne metathesis as a whole. The formation of **327** from enyne **320** is called enyne metathesis because the process is explained formally by [2+2] cycloaddition of the enyne to form the cyclobutene **326**, followed by electrocyclic ring opening. However, cleavage of the cyclobutene attached to a six-membered ring is not always possible. Enyne metathesis is catalysed by Pt, Ru and Pd complexes, and their catalytic activity and selectivity are different. It is well-known that enyne metathesis proceeds via metallacyclobutene formed by cycloaddition of the Ru carbene complex (see Section 8.2.5.2). However, in the Pd-catalyzed enyne metathesis described above, the cyclobutene intermediates **326** seems to be generated without involving carbene complexes.

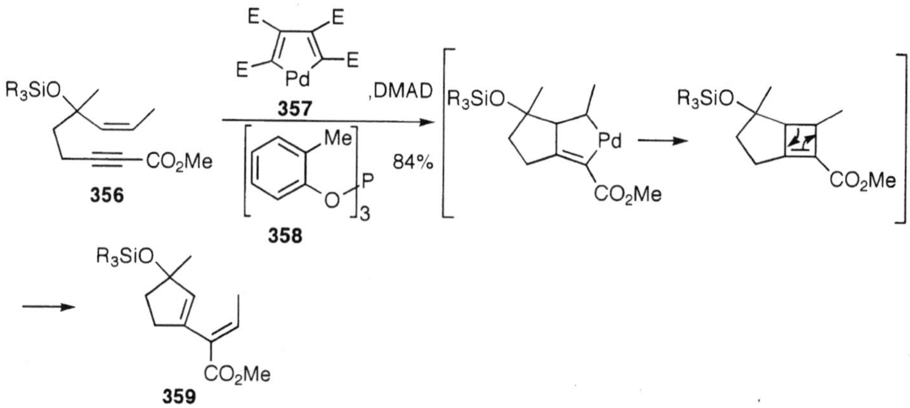

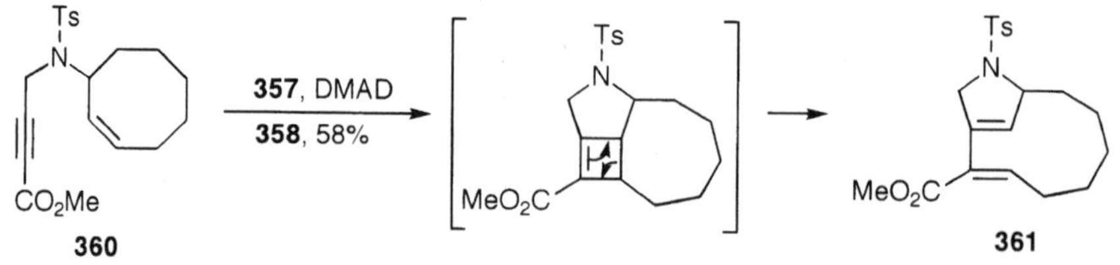

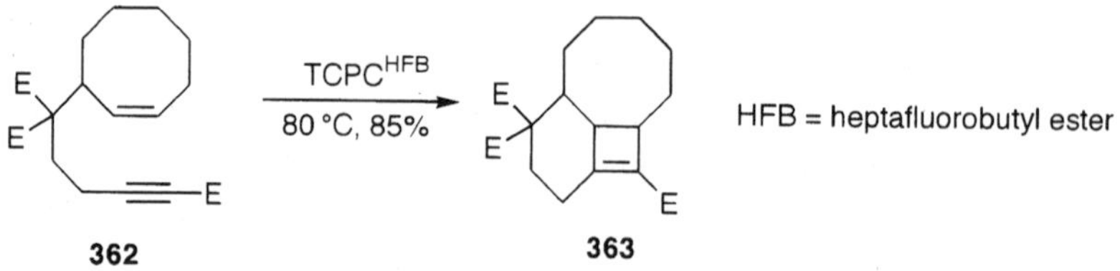

Palladium-catalysed eneyne metathesis is competitive with the intramolecular ene reaction as discussed before, and the balance between them is delicate. The presence of an ester group at the alkyne terminal and *cis*-alkene is important for the selective metathesis reaction [144,145,146]. Also, TCPC (**357**) as a catalyst and *o*-tolyl phosphite (**358**) as a ligand are used. In addition, DMAD (dimethyl acetylenedi-carboxylate) should be added. The 1,3-diene **359** is formed from **356**, and **361** is obtained from **360**. In practice, the cyclobutene attached to the six-membered ring **363** as an intermediate from **362** was isolated without ring cleavage [147]. Pd-catalysed reaction of cyclohexene derivative **364** produced the three products **367**, **368** and **369** in the ratio 2:1:1. Highly strained **367** is expected to be formed by electrocyclic opening of **366** and isomerizes to the less strained **368**. The strained molecule **369** may be formed from **366**, supporting the intermediacy of the [2+2] adduct **366** [144]. For the metathesis reaction, 2,3,4,5-tetrakis(methoxycarbonyl)palladacyclopentadiene (TCPC) (**357**), combined with *o*-tolyl phosphite (**358**), is used as the catalyst. TCPC (**357**) has been synthesized as a tetramer by the reaction of Pd$_2$(dba)$_3$ with DMAD. In particular, the esters with EWG groups such as trifluoroethyl (TCPC^{TFE}) and heptafluorobutyl (TCPC^{HFB}) esters in **357** give better results, and the metathesis becomes the main reaction path, rather than the ene-type reaction. The terminal substitutions on both alkenes and alkynes, particularly an EWG on the alkynes, is essential for smooth metathesis. The reaction is carried out in the presence of one equivalent of DMAD.

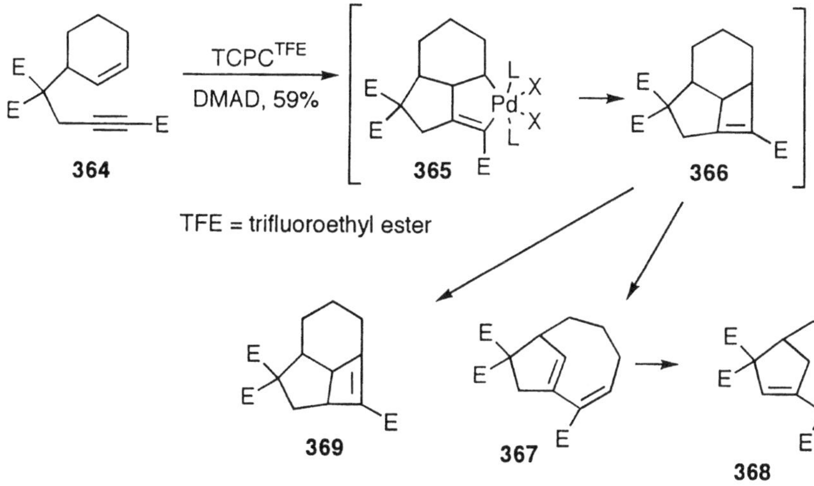

TFE = trifluoroethyl ester

367 : 368 : 369 = 2 : 1 : 1

Ru and Pt complexes also catalyse the enyne metathesis. Ru and Pt catalysts were found to be more versatile catalysts than the Pd catalysts. PtCl$_2$ is the most active and versatile catalyst without ligands [148]. With this simple catalyst, various 1,6-enynes **370** and **372** with functional groups are converted cleanly to the 1-vinylcylopentenes **371** and the bicyclic 1,3-diene **373**. Furthermore, sometimes in this reaction, products which can not be explained by formation of the cyclobutene **376** and its cleavage are obtained. For example, the unexpected product **375** was obtained selectively from **374**, and the product **377**, which is expected to be formed by cleavage of the intermediate **376**, is not formed. The mechanism for the formation of **375** is unknown [149].

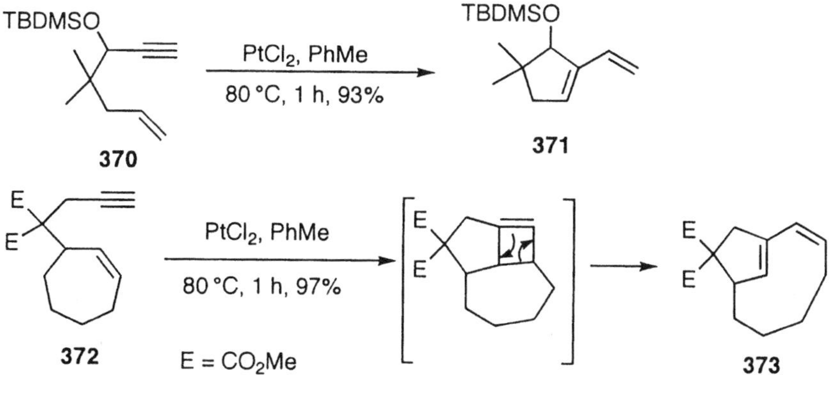

E = CO$_2$Me

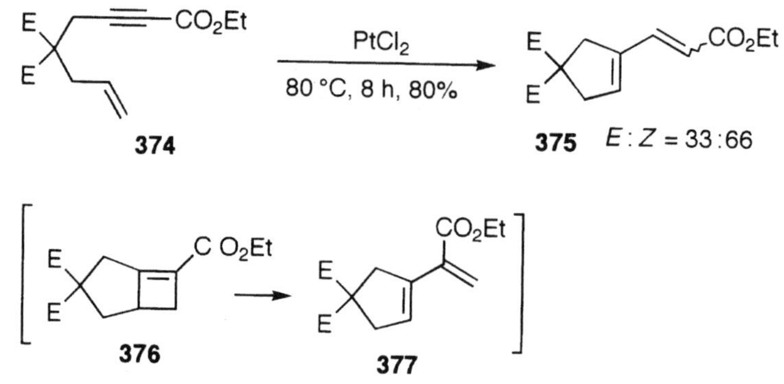

374 → **375** E : Z = 33 : 66

376 → **377**

Highly selective cyclorearrangement of 1,6-enynes is catalysed by a Ru carbonyl complex under CO atmosphere. Smooth cyclization of the 1,7-enyne **378** using [RuCl₂(CO)₃]₂ under CO atmosphere gives the 1-vinylcyclohexene **379** [150]. The 1,6-enyne ester **380** is converted to the expected 1-vinylcyclopentene **382** under similar conditions. In addition, the products **383** and **384**, unexpected from the intermediate **381**, are also obtained.

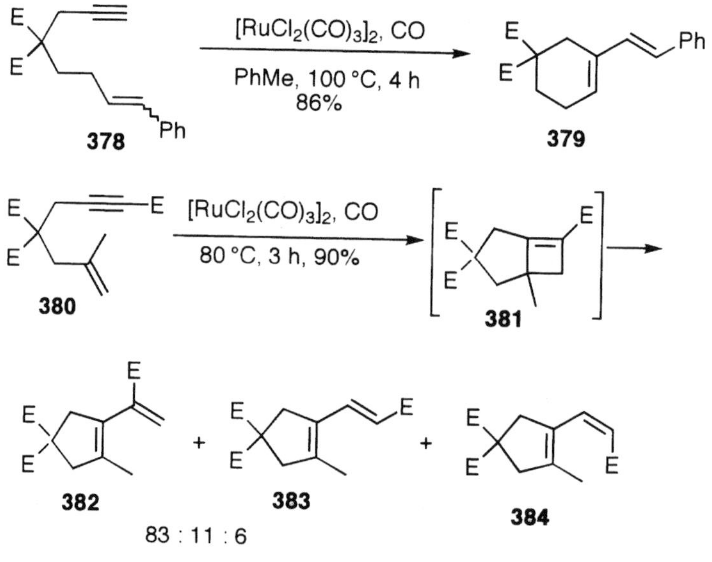

378 → **379**

380 → **381**

382 + **383** + **384**

83 : 11 : 6

The model skeleton **386** of prodiginine antibiotics such as metacycloprodigiosin **387** was constructed by the PtCl₂-catalyzed metathesis reaction of 1,6-enynes **385**, and mechanistic studies have been carried out [151].

Up to this point Pd, Pt and Ru-catalysed enyne metatheses have been explained without involvement of metal–carbenes. However, carbenoid species seem to play a key role in these metatheses (or cycloarrangements) based on the following polycyclization involving cyclopropanation. Polycyclic ring systems are constructed by the cycloisomerization of dienynes catalysed by Ru, Pt, Rh, Ir and Re complexes

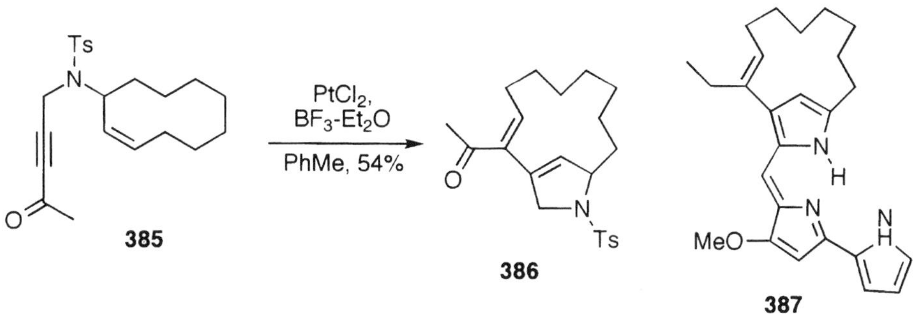

385   **386**   **387**

[152]. Treatment of the dodeca-6,11-dien-1-yne *trans* derivative **388** with [RuCl$_2$(CO)$_3$]$_2$ at 80 °C under N$_2$ atmosphere produced the tetra-cyclo[6.4.0.0.^{1,9}0^{2,4}]undecane derivative **392** in 84% yield as a single stereoisomer. No cyclization of the corresponding *cis* isomer occurred. This novel cyclization is explained by the formation of carbenoid species **390** via the polarized η^1-alkyne complex **389**, bearing a positive charge at the β-position [153]. The carbenoid **390** attacks the 6-alkene to give the cyclopropane **391** with another carbene complex, which again undergoes cyclopropanation of the terminal double bond to produce **392**.

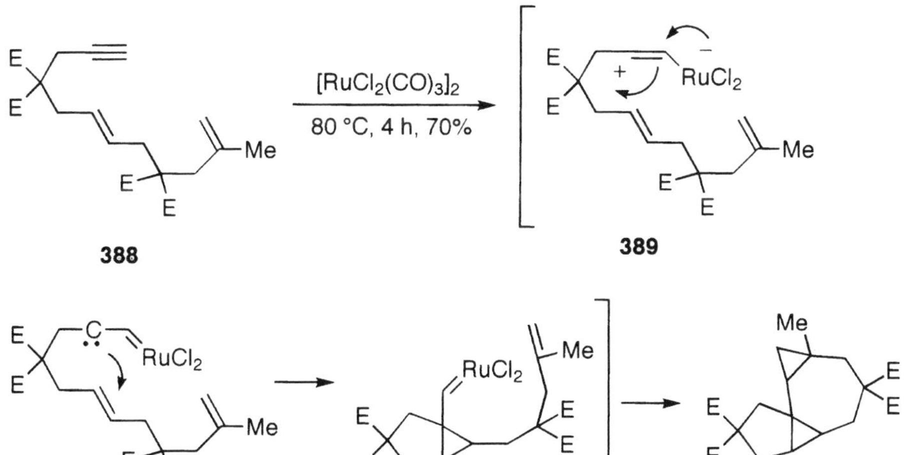

388   **389**

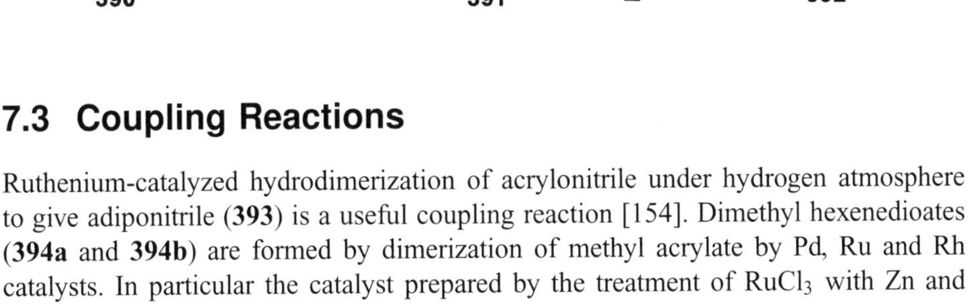

390   **391**   **392**

7.3 Coupling Reactions

Ruthenium-catalyzed hydrodimerization of acrylonitrile under hydrogen atmosphere to give adiponitrile (**393**) is a useful coupling reaction [154]. Dimethyl hexenedioates (**394a** and **394b**) are formed by dimerization of methyl acrylate by Pd, Ru and Rh catalysts. In particular the catalyst prepared by the treatment of RuCl$_3$ with Zn and coordinated by phosphite, and the pentamethylcyclopentadienyl rhodium complex, are active catalysts [155].

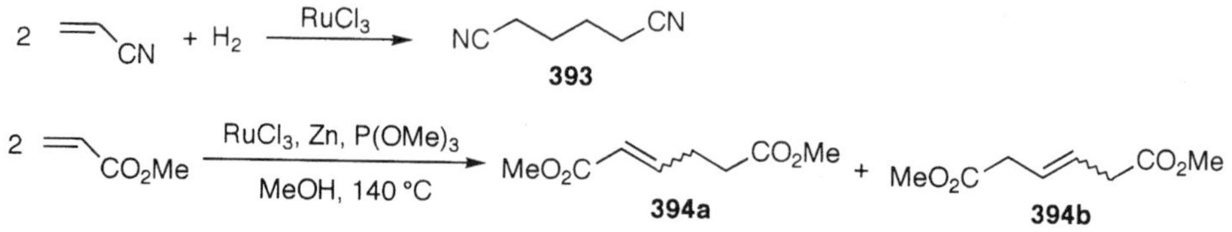

Transition metal-catalysed reactions of alkynes can be explained by the formation of either a η^2-alkyne complex (π-complex) **1**, an alkynyl complex **2** or an η^1-vinylidene complex (carbene complex) **3**. Terminal alkynes undergo a head-to-tail coupling involving transfer of hydrogen to form the conjugated enynes **395** [156]. The alkynyl complex **397** is formed by oxidative addition of the terminal alkyne **396**. Subsequent insertion of alkyne **396** generates **398**, and enyne **399** is formed by its reductive elimination [157]. However, cross-coupling of alkynyl ester **400** with terminal alkyne **401** yields **402**. The propargyl acetate **403**, derived from **402**, undergoes intramolecular Pd-catalysed reaction to give the 18-membered macrolide **404** [158]. (Section 6.7 discusses the Pd-catalysed reaction of **403**).

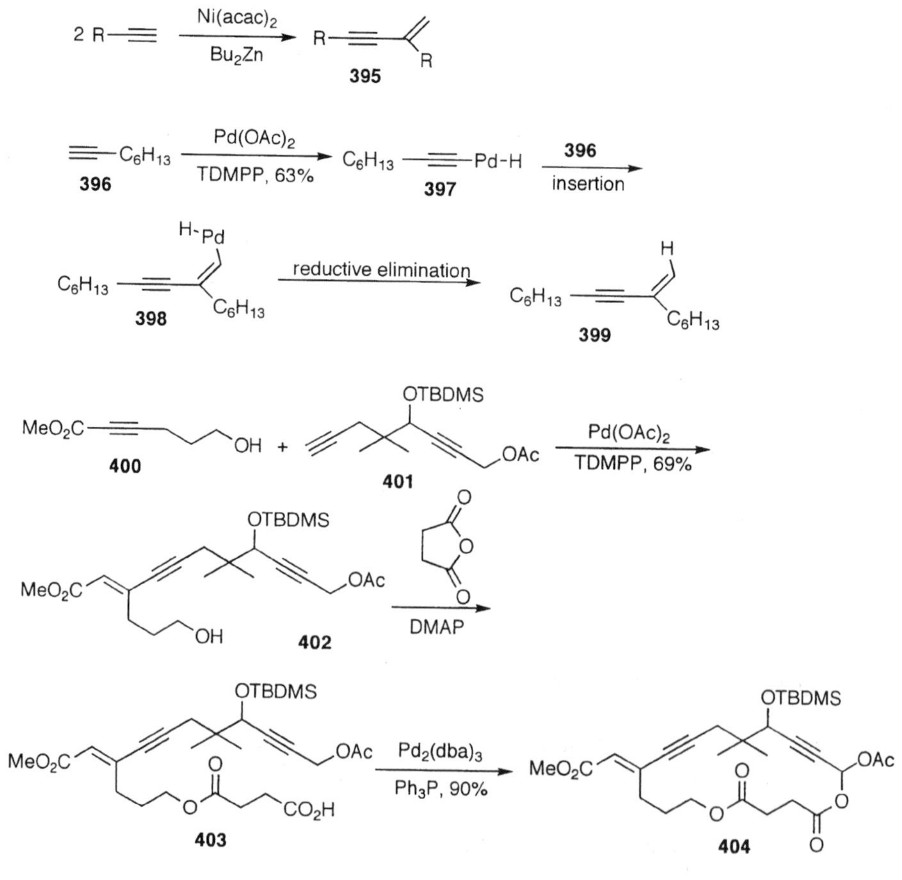

Several Ru complexes catalyse the cross-coupling of alkynes with alkenes. Ru forms six-coordinated complexes, and their catalytic activity changes drastically depending on ligands. The Ru-catalysed addition of alkyne to alkene can be understood by the oxidative cyclization to form a metallacyclopentene. The coupling of norbornene (**405**) with alkyne to form the cyclobutene **407** can be explained by the formation of ruthenacyclopentene **406** and its reductive elimination. In this case, elimination of β-hydrogen is difficult for steric reasons [159]. The conjugated diene ester **409** is obtained by the coupling of alkynes with acrylate catalyzed by Ru(cod)(cot) complex (cot = cyclooctene) [160]. The reaction may be understood by the formation of ruthenacyclopentene **408**, followed by elimination of β-hydrogen and reductive elimination.

Coupling of 1-alkyne **410** with 1-alkene **411**, catalysed by CpRu(cod)Cl in aqueous DMF, affords the diene **414** as an ene-type product in good yield. One explanation of the reaction is the formation of the (π-allyl)(η^2-alkyne)intermediate from the 1-alkene and insertion of the alkyne [161]. However, formation of the ruthenacyclopentene **412**, subsequent β-elimination to form **413**, and reductive elimination offer a more easily understandable mechanism. A formal synthesis of alternaric acid (**415**) was achieved by this reaction [162].

γ,δ-Unsaturated aldehydes **418** or ketones are formed from 1-alkynes and allylic alcohols using CpRu(cod)Cl [163, 164]. In this reaction the ruthenacyclopentene **417** is formed by insertion of allyl alcohol to the η^2 complex **416** (oxidative cyclization), and *syn*-elimination of β-hydrogen from the OH-bearing carbon in **417** gives the aldehyde **418**.

Using this reaction, the steroidal alkynyl side chain in **419** is extended with allylic alcohol to afford **420** with the Ru complex without phosphine [164]. However, formation of the β,γ-unsaturated ketone (allyl ketone) **421** and its isomerization to the α-enone are observed by using a CpRu–phosphine complex [165]. In this product, the terminal alkyne carbon becomes the ketone. These results show that Ru complexes having different ligands give different products.

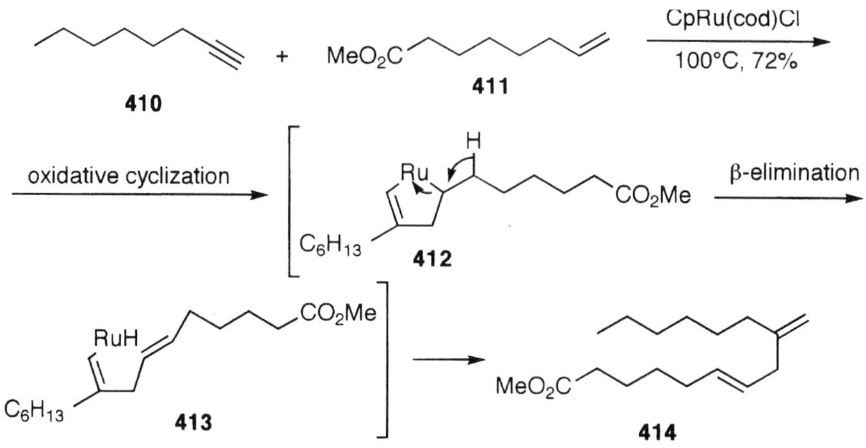

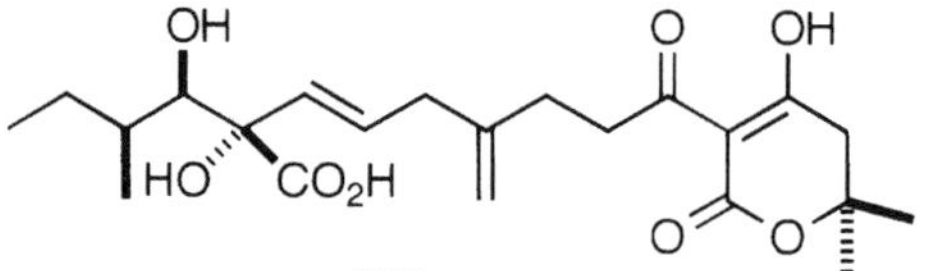

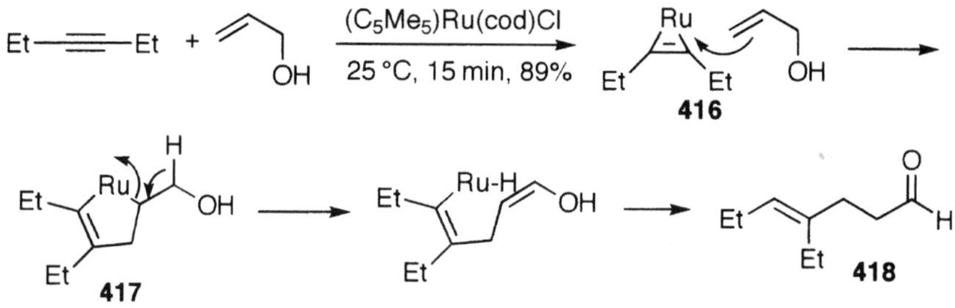

As another example of the reaction via the ruthenacyclopentene, 1,4-diketones are formed by the Ru-catalysed reaction of terminal alkynes with vinyl ketones in aqueous DMF in the presence of NH_4PF_6 and $InCl_3$. The reaction is explained by the generation of the ruthenacyclopentene **422**, followed by addition of H_2O to the double bond. Elimination of β-hydrogen and reductiuve elimination afford 1,5-diketones [165a].

The formation of the allyl ketone **421** is explained by the following mechanism. The Ru alkynyl complex **423**, formed by oxidative addition, isomerizes to the Ru vinylidene complex **424**. Nucleophilic attack of allyl alcohol to the electron-deficient *sp*-carbon of **424** generates the allyloxycarbene complex **425**, which is converted to

R = -(CH₂)₃CN, 93%

the (acyl)(π-allyl) complex **426** by C$-$O bond cleavage (metalla-Claisen rearrangement), and the allyl ketone **427** is formed by reductive elimination.

As a supporting evidence, it is well-known that the electron-rich (η^6-arene)Ru complex of terminal alkyne **428** rearranges easily by the treatment with NaPF$_6$ of the η^1-vinylidene complex **429**, which is a strongly electrophilic carbene complex. Attack of ROH on the carbene carbon generates the the alkoxycarbene complex **431** via **430** [166]. Formation of ketone **427** by attack of the allylic alcohol is understanable by this mechanism. Formation of Ru–vinylidene complex **429** from the terminal alkyne has been proposed as the intermediate **432** of the reaction of terminal alkyne, amine and CO$_2$ to form the vinyl carbamate **433** [167,168].

Even aromatic and vinylic hydrogens can be activated for coupling by a Ru hydride complex. Coupling of *o*-methylacetophenone (**434**) with the functionalized alkene **435**, catalysed by RuH$_2$(CO)(Ph$_3$P)$_3$ gave **437** [169]. This complex becomes coordinatively unsaturated after liberating hydrogen. Selective *ortho*-alkylation occurs in high yields due to a chelation (*ortho*-metallation) of the Ru complex to the carbonyl group as shown by **436**. Subsequent insertion of the alkene to the H$-$Ru bond and reductive elimination produce **437** [169]. The coupling is not limited to aromatic rings. Similarly, various cyclic enones **438** and cyclic α,β-unsaturated esters react regioselectively with the silylethylene **435** to give **439** [170]. Also the aromatic C$-$H bond of α-tetralone (**440**) adds to internal alkynes to give **441** [171]. In these reactions, oxidative addition of sp^2 C$-$H bonds to the Ru complex becomes possible by the chelation effect.

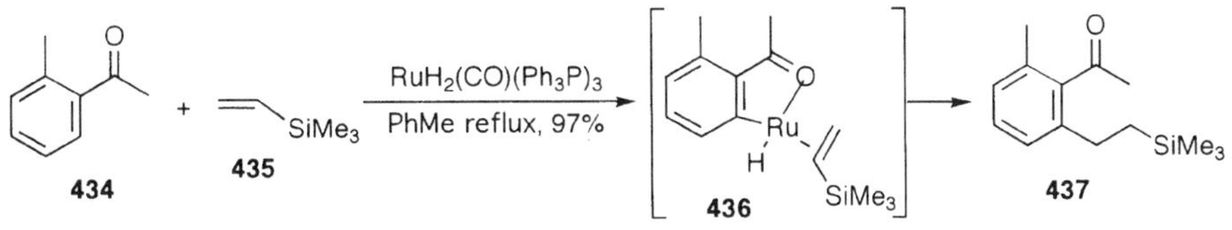

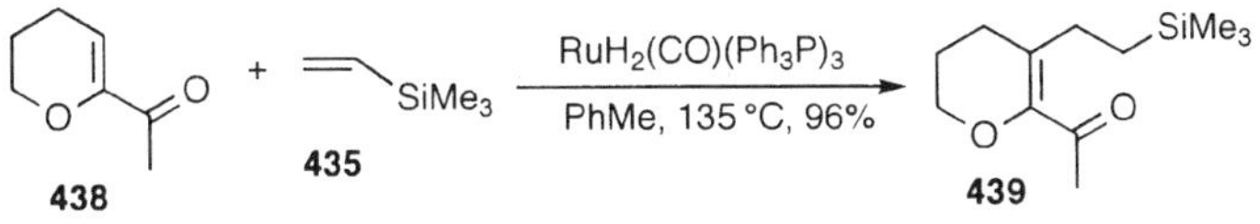

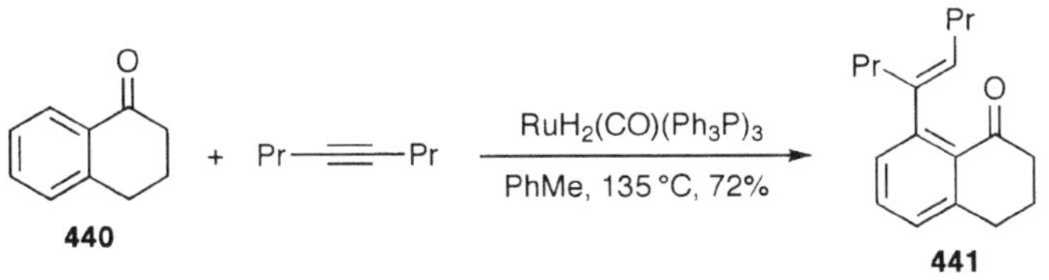

7.4 Addition of Main Group Metal Compounds

1,2-Addition reactions of three kinds of main group metal compounds, namely $R-M'X_n$ (carbometallation, when R are alkyl, alkenyl, aryl or allyl groups), $H-M'X_n$ (hydrometallation with metal hydrides) and $R-M'-M''-R$ (dimetallation with dimetal compounds) to alkenes and alkynes, are important synthetic routes to useful organometallic compounds. Some reactions proceed without a catalyst, but many are catalysed by transition metal complexes.

7.4.1 Carbometallation

1,2-Addition of organometallic compounds of main group metals $R-M'-X_n (M' = B$, Al, Zn, Mg, Sn) to alkenes and alkynes is called carbometallation. Some reactions proceed without a catalyst, but they are promoted or accelerated by transition metal catalysts.

Products of the carbometallation **442** and **443** contain very reactive metal–carbon bonds, and can not be isolated in many cases. They undergo further reactions such as protonation, transmetallation and nucleophlic attack. The preparation of tri- or tetrasubstituted alkenes **444** and **445** from alkynes via the alkenylmetals **443** is particularly useful. Only catalysed carbometallations are treated here.

cis-Carbomagnesation of internal alkynes **446** is catalysed by a Ni complex to give alkenyl Grignard reagents. The trisubstituted alkenes **447** and **448** are obtained after protonation [172]. Terminal alkynes can not be used due to the presence of an acidic hydrogen. Hydromagnesation, rather than carbomagnesation, takes place with EtMgBr.

The regioselective carbomagnesation of alkenes is catalysed by Cp_2ZrCl_2 (**449**) [173]. In this reaction, the zirconocene–ethylene complex (**450**) is generated and reacts

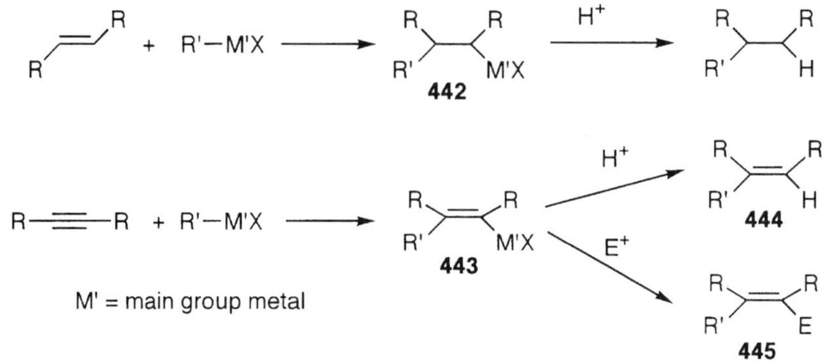

442

443

444

445

M' = main group metal

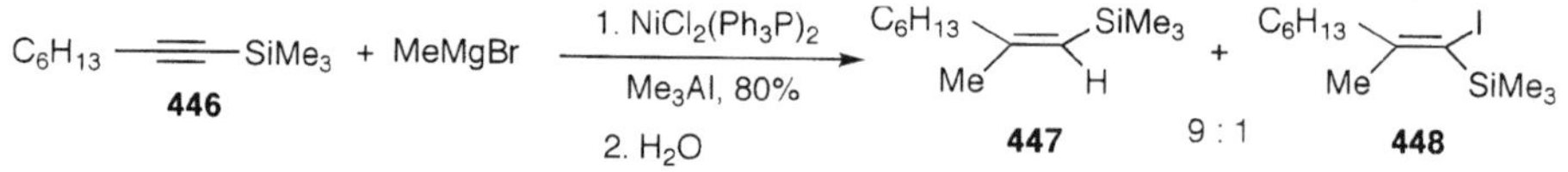

C_6H_{13} ———≡——— $SiMe_3$ + MeMgBr
446

1. $NiCl_2(Ph_3P)_2$
———————————
Me_3Al, 80%

2. H_2O

C_6H_{13} ＼＝＼ $SiMe_3$
Me　　H
447

+

C_6H_{13} ＼＝＼ I
Me　　$SiMe_3$
448

9 : 1

with alkene to form the zirconacyclopentane **451**. Transmetallation of **451** with EtMgBr generates **453** via the *ate* complex **452**. Elimination of β-hydrogen from the ethylzirconium affords the carbomagnesation product **454** with regeneration of the ethylene complex **450**.

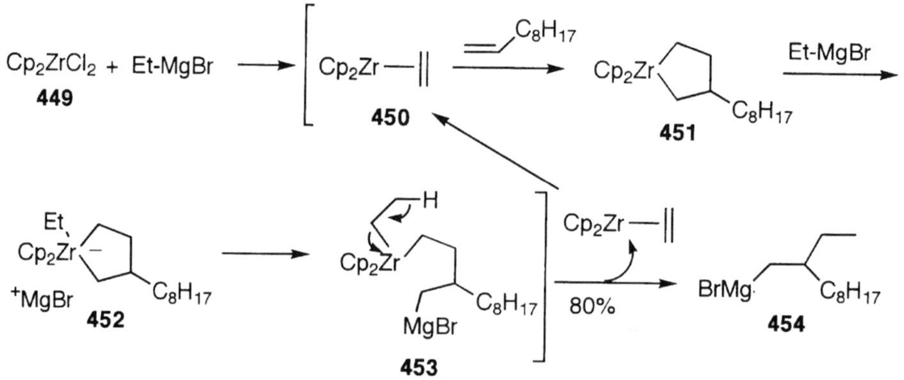

Cp_2ZrCl_2 + Et-MgBr
449

Cp_2Zr—‖
450

Cp_2Zr
451 C_8H_{17}

Et-MgBr

Et
Cp_2Zr
^+MgBr　C_8H_{17}
452

Cp_2Zr
C_8H_{17}
MgBr
453

Cp_2Zr—‖
80%

BrMg　C_8H_{17}
454

When allylic ethers **455** or allylic amines are used, the carbomagnesation product **456** undergoes β-elimination to give the displacement product **457** of the allyl ether group with EtMgBr [174]. The reaction has been extended to Zr-catalysed asymmetric carbomagnesation [175]. The Ru-catalysed metathesis of the protected allyl(5-hexenyl)amine (**458**) affords the eight-membered cyclic amine **459**. The asymmetric ethylmagnesation of **459** using (R)(ebthi)Zr-binol where ebthi = [ethylene-1,2-bis(η^5-4,5,6,7-tetrahydro-1-indenyl)] gives **461** with 98% ee via the β-elimination of **460** [176].

Carboalumination of alkynes and alkenes proceeds slowly without a catalyst. The reaction is accelerated by Cp_2ZrCl_2 [177]. Carboalumination of alkenes is catalysed by

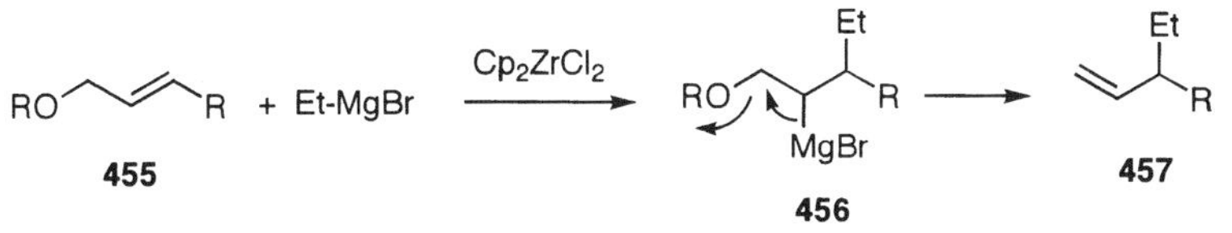

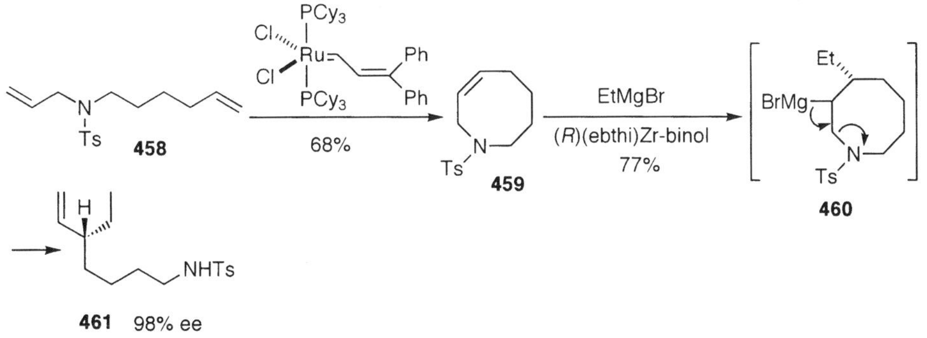

Ti complexes [178]. Domino carboaluminations of diene **462** with Et$_2$AlCl catalysed by Ti(O-*t*-Pr)$_4$ gives **464** via **463**, and oxidation of **464** produces the alcohol **465** as the main product. Also, the diol **466** is obtained by the oxidation unexpectedly as a minor product [179].

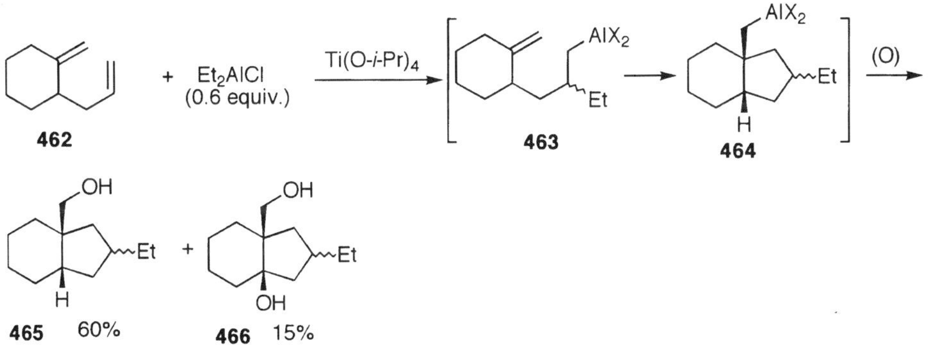

Carbozincation is catalysed by Ni complexes [180]. The 1,4-addition of organozinc compounds to α,β-unsaturated ketones (carbozincation) is catalysed by Ni complexes. Addition of Me$_2$Zn to the enone **467** is accelerated by a Ni catalyst, and the product is trapped with chlorosilane to give the silyl enol ether **468** [181]. Similarly the Ni-catalysed 1,4-addition of Me$_3$Al to enones **469** gives **470** after hydrolysis [182]. The alkenylzirconium **472**, obtained by hydrozirconation of alkynes, undergoes the Ni or Pd-catalysed conjugated addition to the enone **471** to give the prostaglandin intermediate [183]. No addition of alkylzirconium takes place.

Efficient asymmetric 1,4-addition of phenylboronic acid to cyclohexenone, followed by hydrolysis, affords 3-phenylcyclohexanone (**473**) with 97% ee using Rh-BINAP as a catalyst [183a].

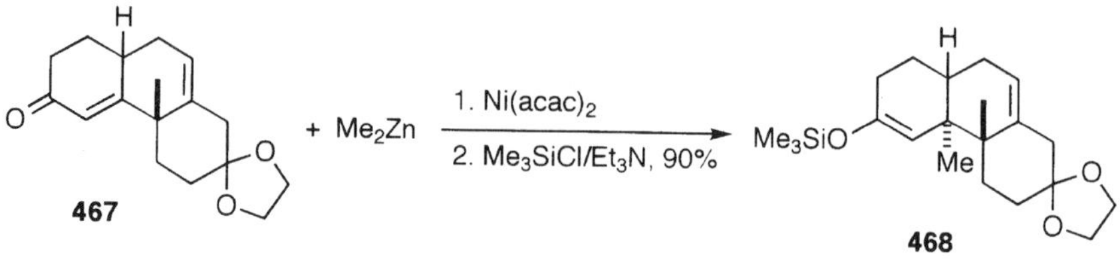

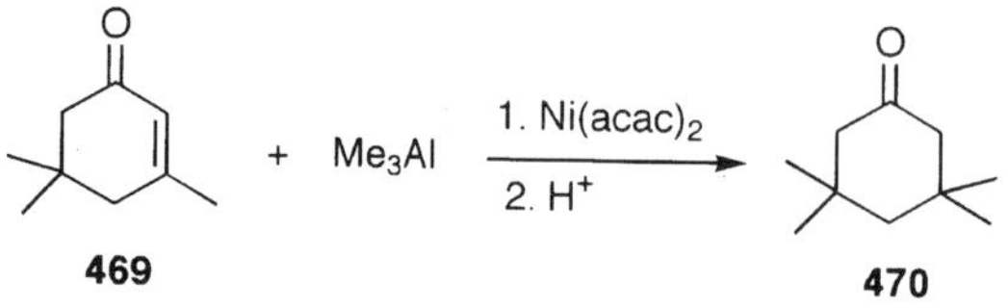

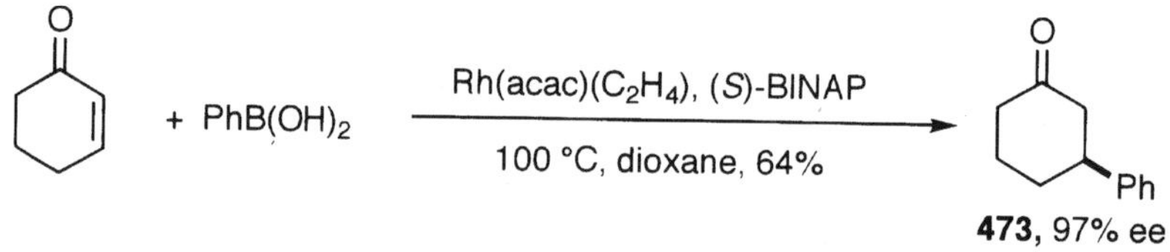

467   + Me$_2$Zn   1. Ni(acac)$_2$ / 2. Me$_3$SiCl/Et$_3$N, 90%   Me$_3$SiO   **468**

469   + Me$_3$Al   1. Ni(acac)$_2$ / 2. H$^+$ **470**

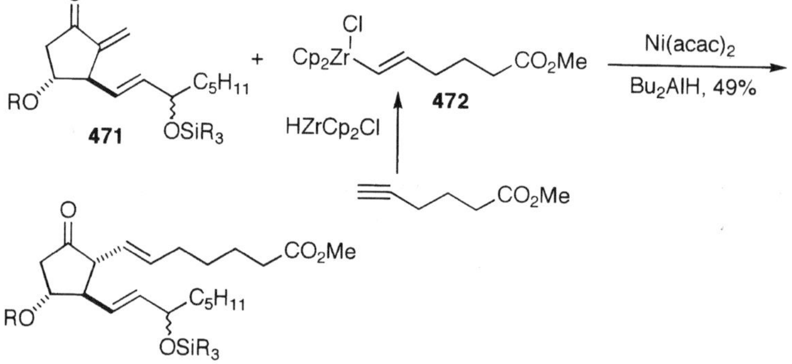

471   +   Cp$_2$Zr   CO$_2$Me   **472**   Ni(acac)$_2$ / Bu$_2$AlH, 49%

HZrCp$_2$Cl

CO$_2$Me

CO$_2$Me

RO   C$_5$H$_{11}$ ÖSiR$_3$

+ PhB(OH)$_2$ Rh(acac)(C$_2$H$_4$), (*S*)-BINAP / 100 °C, dioxane, 64%   **473**, 97% ee   Ph

Carbostannation of alkyne **474** with alkynylstannane **475**, catalysed by the Pd–iminophosphine complex **476** gives (*Z*)-stannylenyne **477** as the main product. The reaction is *syn* addition [184].

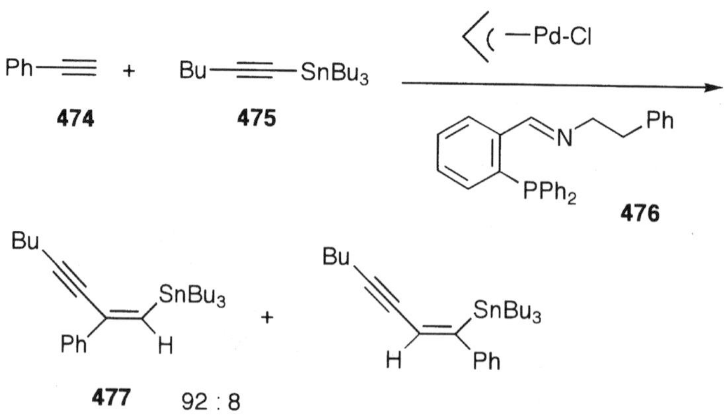

Ph   +   Bu   SnBu$_3$
474   **475**

476

Bu   SnBu$_3$ Bu   SnBu$_3$
Ph   H   H   Ph
477   92 : 8   +

7.4.2 Metalametallation (Bis-metallation)

Metalametallations of alkenes and alkynes are useful methods for the construction of 1,2-dimetala-alkanes and 1,2-dimetala-1-alkenes, which react subsequently with suitable electrophiles to form substituted alkanes and alkenes. Metalametallation is carried out usually with bimetallic reagents of the type $R_3Si-M'R_n$ or $R_3Sn-M'R_n$ in which $M' = B$, Al, Mg, Cu, Zn, Si or Sn. Some metalametallations proceed without catalysts; Cu, Ag and Pd compounds are good catalysts. The metalametallation with bimetallic compounds, such as Si−B, Si−Mg, Si−Al, Si−Zn, Si−Sn, Si−Si, Sn−Al or Sn−Sn bonds, catalysed by transition metal complexes, is explained by the oxidative addition of the bimetallic compounds to form **478**, and insertion of alkene generates **479**. Finally 1,2-dimetallic compounds **480** are formed by reductive elimination. Dimetallation of alkynes proceeds similarly to give **481**. Dimetallation is *syn* addition.

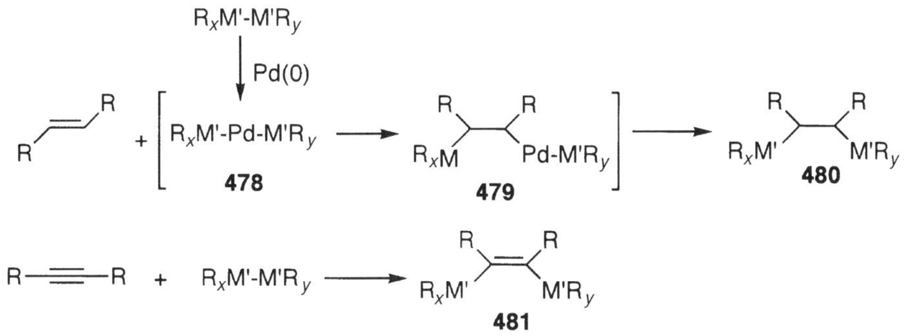

The *cis*-bis(boryl)alkenes **482** are obtained by bis-boration of terminal alkynes catalysed by a Pt complex [185]. Pd and Rh complexes are inactive. Then the (Z)-1,2-diphenylalkene **483** is prepared by Pd-catalysed Suzuki–Miyaura coupling of **482** with iodobenzene.

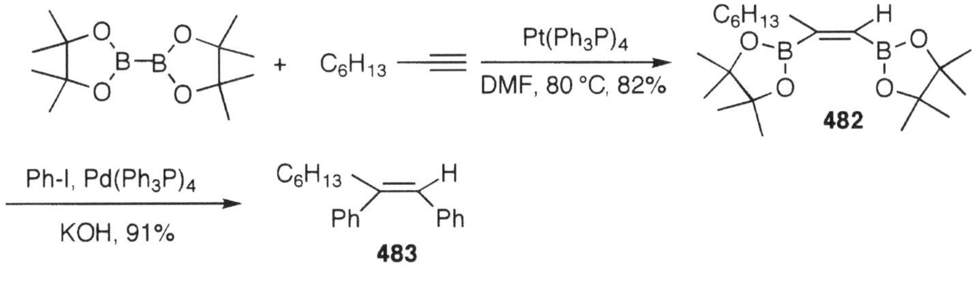

Bimetallic compounds of Mg, Zn and Al are prepared *in situ* by the reaction of R_3SiLi or R_3SnLi with RMgCl, Et_2AlCl or $ZnBr_2$, and reacted with alkynes in the presence of Pd catalysts. The reaction is not regioselective. CuI is a better catalyst, giving higher regioselectivity. In most cases, the final products are isolated after protonation of the C−M′(M′ = Mg, Al, Zn) bonds formed by the addition, and these addition reactions are used for the preparation of Si and Sn compounds. Silylmagnesation, -zincation and -alumination, as well as stannylmagnesation, -zincation and -alumination using the corresponding dimetallic compounds generated *in situ,* are catalysed mainly by Pd complexes. The Mg, Zn and Al compounds thus prepared are protonated, or treated with electrophiles to give silyl and stannyl compounds as final products [186]. For example, The silylalumination using **484** produces **485** and **486**, and they are converted to 2-silyl-1-iodoalkenes **487** and **488** by treatment with iodine. Protonation of the silylzincation product affords **489** and **490** [187].

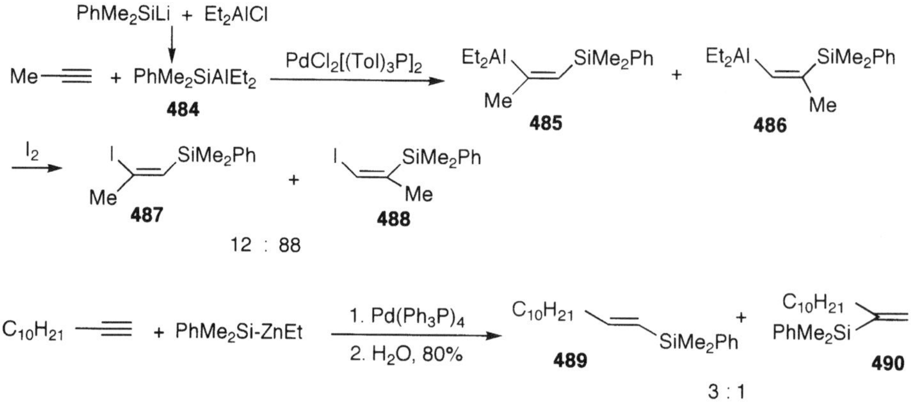

Stannation can be carried out using $R_3Sn-SnR_3$ and $R_3Sn-SiR_3$. Silylstannation of 1-alkynes with **491** gives **492** with high regio- and stereoselectivity. The reaction is *cis* addition, and SnR_3 always occupies the internal position. The addition products are converted to either vinylsilane **493** or 2-stannylalkene **494** [188]. The *trans*-1-silylalkene **497** is obtained by silylstannation of alkynes **495**, followed by treatment of **496** with hydrogen iodide [189]. Distannylalkene **498** can be prepared easily by bis-stannation of alkynes [190].

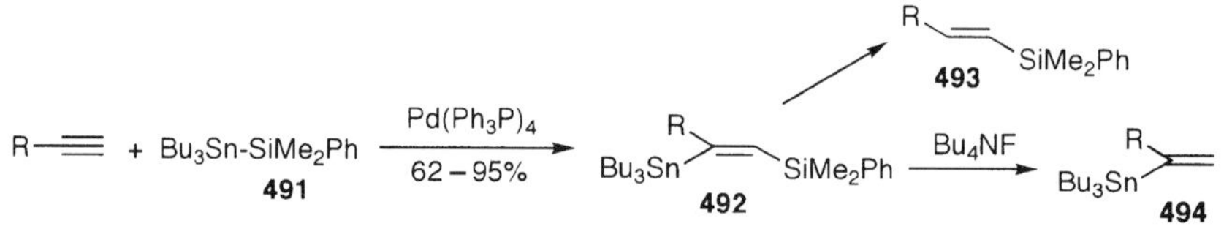

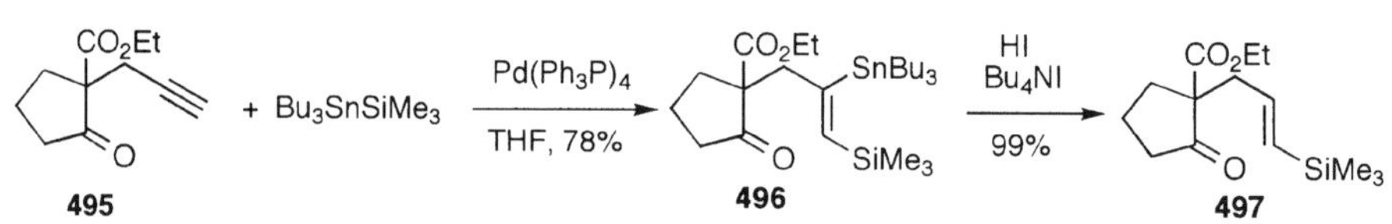

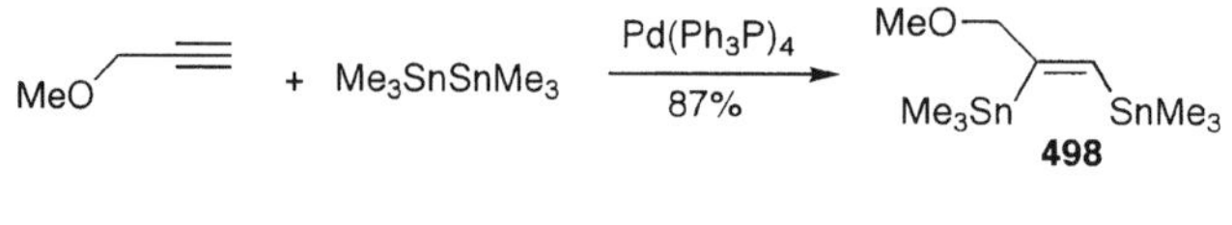

Bis-silylation of alkynes proceeds more smoothly with disilane **499** bearing electron-donating groups [191]. However, efficient bis-silylation of alkynes with hexalkyldisilane can be carried out using the sterically crowded isocyanide **501** and Pd(OAc)$_2$ [192], or Pd$_2$(dba)$_3$ and TMSPP (**XLVII**) as a ligand [193]. Intramolecular bis-silylation of the homoallyl disilyl ether **502** gives **503** with high stereoselectivity. The C–Si bond is oxidized to alcohol with retention of the stereochemistry, and the 1,2,4-triol **504** is prepared with high stereoselectivity [194]. The tertiary isonitrile **501** is a highly active ligand for the bis-silylation. Usually 10–15 equivalents of the ligand **501** to Pd are used. No bis-silylation occurs with Ph$_3$P.

Enones are also bis-silylated. Asymmetric 1,4-bis-silylation of the enone **505** catalyzed by Pd-BINAP afforded the silyl enol ether **507**, and the keto alcohol **509** with 87% ee was obtained after hydrolysis and oxidation of **508** [195]. In this case, the unsymmetric disilane **506** should be used.

7.4.3 Hydrometallation

Addition of hydride bonds of main group metals such as B−H, Mg−H, Al−H, Si−H and Sn−H to alkenes and alkynes to give **513** and **514** is called hydrometallation and is an important synthetic route to compounds of the main group metals. Further transformation of the addition product of alkenes **513** and alkynes **514** to **515, 516** and **517** is possible. Addition of B−H, Mg−H, Al−H and Sn−H bonds proceeds without catalysis, but their hydrometallations are accelerated or proceed with higher stereoselectivity in the presence of transition metal catalysts. Hydrometallation with some hydrides proceeds only in the presence of transition metal catalysts. Hydrometallation starts by the oxidative addition of metal hydride to the transition metal to generate transition metal hydrides **510**. Subsequent insertion of alkene or alkyne to the M−H bonds gives **511** or **512**. The final step is reductive elimination. Only catalysed hydrometallations are treated in this section.

7.4.3.1 Hydromagnesation and hydrozincation

Most conveniently, MgH_2 is generated *in situ* by the Ti-catalysed elimination of β-hydrogen from Grignard reagents having β-hydrogen such as isopropyl and isobutyl Grignard reagents. The new Grignard reagent **520** can be prepared from the alkene **518** based on the $TiCl_4$-catalysed reversible exchange reaction between alkenes **518** and Grignard reagents **519** bearing β-hydrogen [196]. As the Cp_2TiCl_2-catalysed exchange reaction between isopropyl Grignard reagent and styrene becomes irreversible by removing propylene, benzylic Grignard reagents can be prepared in quantitative yield by the exchange reaction [197].

These reactions can be explained by the following mechanism. At first, isopropyltitanocene (**521**) is formed by transmetallation and its β-elimination generates Cp_2Ti-H and propylene. Insertion of the alkene to Ti−H affords the alkyltitanium **522**. Then the alkyl Grignard reagent **523** is formed by transmetallation of **522** with *i*-PrMgBr. At the same time, isopropyltitanocene (**521**) is regenerated. Hydromagnesation of alkenes is also possible by an Ni catalyst [198].

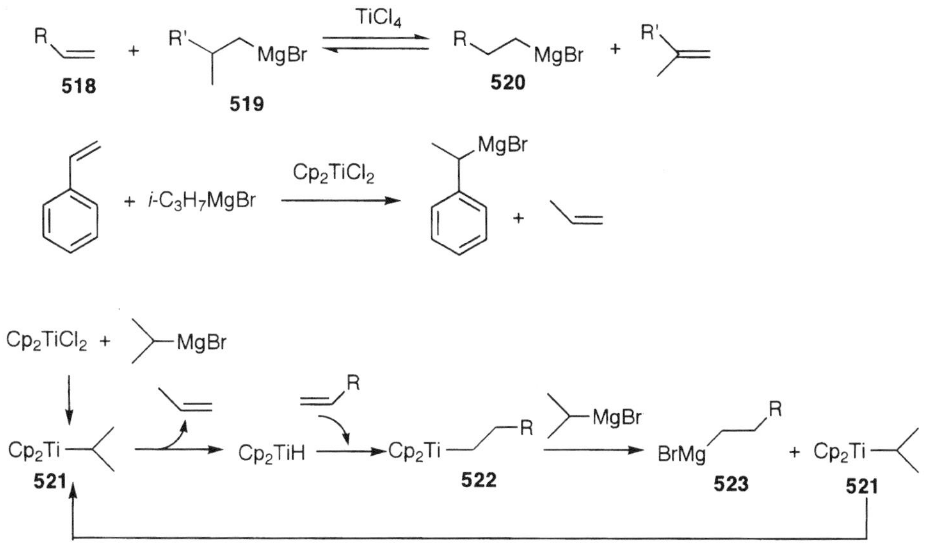

Hydrozincation of terminal alkene with Et$_2$Zn, catalysed by Ni(acac)$_2$, yields the dialkylzinc **525**, which reacts with the nucleophile **526** via transmetallation involving Cu salt to give **527** [199]. In this reaction, the Ni hydride is generated by β-elimination of EtNi(acac), and the hydronickellation product **524** is formed by alkene insertion to the hydride H−Ni(acac). Then transmetallation of **524** with Et$_2$Zn affords **525**.

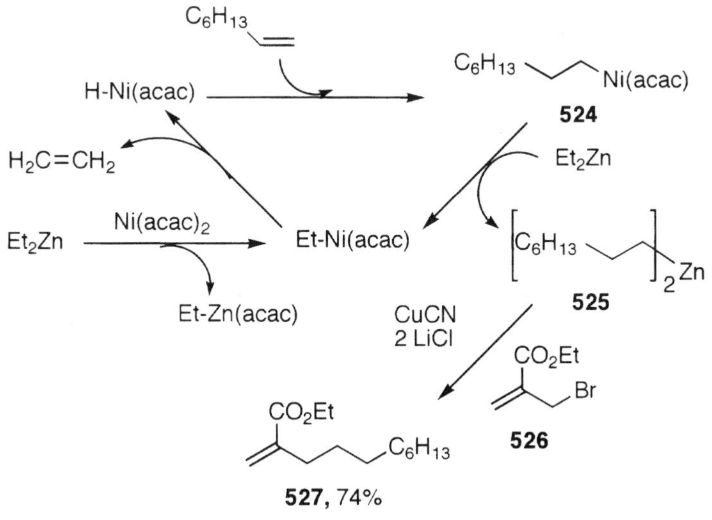

7.4.3.2   Hydroboration

Hydroboration of alkenes with BH_3 and R_2BH proceeds at room temperature without a catalyst. As an exception, the reaction of the less reactive catecholborane (CBH, **529**) proceeds at 70–100 °C. It was found that the slow reaction of CBH is accelerated by a Rh catalyst, and the catalyzed and uncatalyzed hydroborations give different products [200]. By the uncatalyzed reaction, the ketone in the enone **530** is attacked to form **531**. On the other hand, the alkene is hydroborated to give **532** by the catalysis of $RhCl(Ph_3P)_3$ [201]. Different regio- and stereoselectivities are observed in the Rh-catalysed hydroboration of cyclohexenol (**533**) with catecholborane and uncatalysed reactions with 9-BBN-H [202].

Dicarbonyltitanocene $[Cp_2Ti(CO)_2]$ is also active for the hydroboration of alkynes at room temperature. Dimethyltitanocene $[Cp_2TiMe_2]$ catalyses the hydroboration of alkene with CBH at room temperature [203].

Asymmetric hydroboration is achieved using the Rh–BINAP catalyst. Usually the hydroboration of styrene, followed by oxidation, affords 2-phenylethanol and there is no possibility of asymmetric hydroboration. Interestingly, however, hydroboration of styrene with CBH (**529**) catalysed by the Rh–BINAP complex and oxidation yields 1-phenylethanol (**534**) with 96% ee [204]. To explain this rather unusual regiochemistry, the hydroboration of 1-decene (**535**) using deuteriocatecholborane (**529D**) was carried out. It was confirmed that the recovered 1-decene (**535D**) after the reaction of 1-decene (**535**) with 0.1 equivalent of deuteriocatecholborane (**529D**) was deuterated equally at C(1) and C(2), and the 1-decanol produced was partially deuterated, as shown by **536**. From these results, it was concluded that the Rh-catalysed hydroboration is reversible [205].

529 **534** OH
96% ee

535 + **529D** → **535D** + **536**
0.1 equivalent

50%D   50%D   14%D
86%D

Pinacolborane (PBH, **537**) sluggishly hydroborates alkynes and alkenes. Hydroboration of alkyne with PBH is catalyzed by hydrozirconocene chloride (HCp$_2$ZrCl) [206], CpNi(Ph$_3$P)Cl and Rh(CO)(Ph$_3$P)$_2$Cl [207] at room temperature. Hydroboration of 4-octene with PBH at room temperature gives either terminal or internal boranes **538** or **539** regioselectively, depending on the catalyst used [207]. PBH is more stable than CBH, and easier to handle.

537   **538**   **539** B(OR)$_2$

	538	:	**539**	Yield
Rh(Ph$_3$P)$_3$Cl	100	:	0	92%
Rh(CO)(Ph$_3$P)$_2$Cl	3	:	97	94%
CpNi(Ph$_3$P)Cl	1	:	99	97%

-B⟨O⟩⟨O⟩ = -B(OR)$_2$

7.4.3.3   Hydroalumination

Whereas the hydroalumination of alkynes with H−Al(i-Bu)$_2$ gives vinylaluminiums easily without a catalyst, that of alkenes is possible only at high temperature. For example, commercial production of higher alcohols by the Ziegler method by continous insertion of ethylene to Al−H and Al−R bonds is carried out under pressure at high temperature. TiCl$_4$ and Cp$_2$TiCl$_2$ have high catalytic activity and the Ti-catalysed hydroalumination of alkenes using LiAlH$_4$ proceeds at room temperature [208]. Ethers such as THF and DME are good solvents. The reaction is explained by the following mechanism. The Ti hydride **540** is generated by transmetallation of TiCl$_n$ with LiAlH$_4$, and converted to the alkyltitanium **541** by hydrotitanation of alkene. Then its transmetallation with Al hydride **542** affords the alkylaluminium **543**, and Ti−H **540** is regenerated. The products **544** of TiCl$_4$-catalysed hydroalumination of alkenes are useful intermediates, and converted to alkanes **545** by Cu(II)-mediated oxidative coupling. The Cu-mediated carbonylation affords the symmetric ketones **546**, and saturated aldehydes **547** are obtained by the conjugate addition to α,β-unsaturated aldehydes [209].

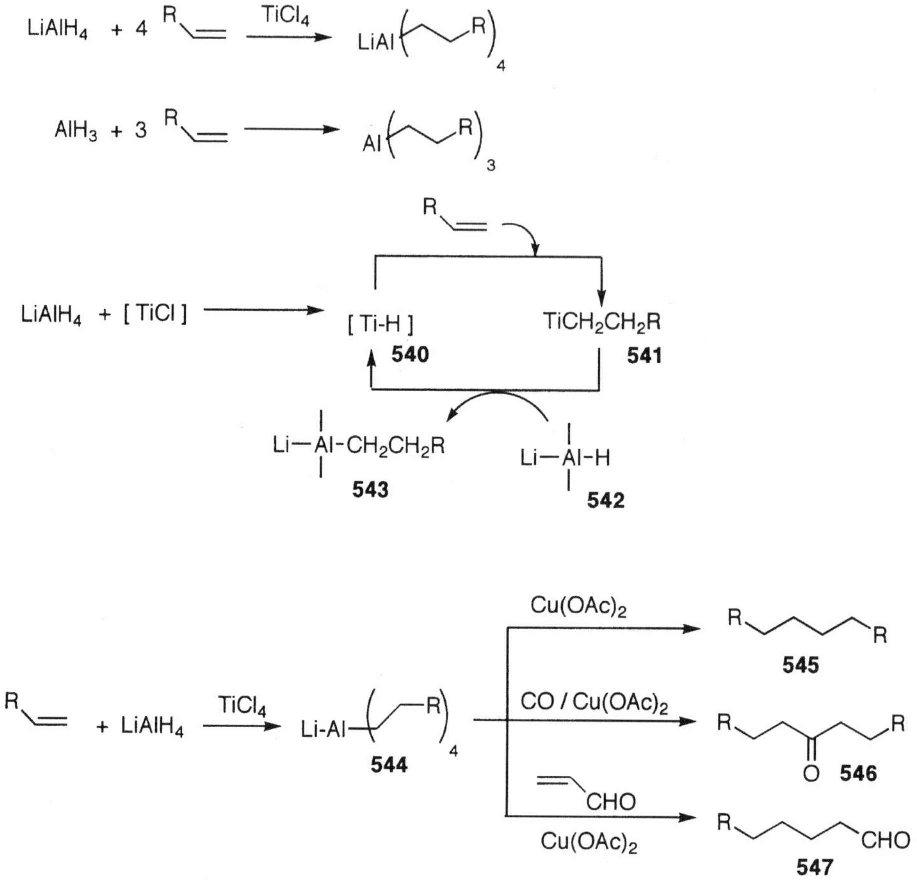

7.4.3.4 *Hydrostannation (Hydrostannylation)*

Addition of a Sn—H bond to alkenes and alkynes is called hydrostannation or hydrostannylation, and proceeds with a radical initiator. Some transition metal complexes catalyse the addition, giving products different from those of the free-radical reaction initiated by AIBN. Addition to alkynes is a good synthetic route to vinylstannanes. Whereas a *trans* adduct is obtained by the radical reaction, the *cis* adduct **548** is obtained by the Pd-catalysed reaction at room temperature. Hydrostannation of the terminal alkynes **549**, catalysed by Pd and Ni complexes, gives a mixture of the regioisomers **550** and **551** in different ratios, depending on the reaction conditions [210]. However, the reaction catalysed by RhCl(CO)(Ph$_3$P)$_2$ at 0 °C gives **552** with high regioselectivity [211]. The Pd-catalysed reaction of the 1,3-enyne **553** gives **554** selectively, whereas the terminal adduct **555** is obtained by the radical addition [212].

Chemoselective reduction of the conjugated double bond of α, β-unsaturated aldehydes such as citral (**556**) to give citronellal (**577**) is possible by Pd-catalysed hydrostannation in the presence of AcOH [213].

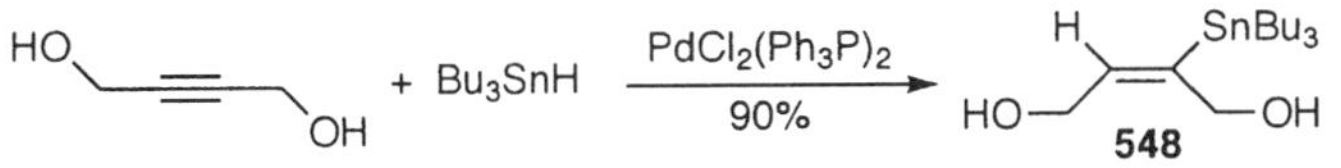

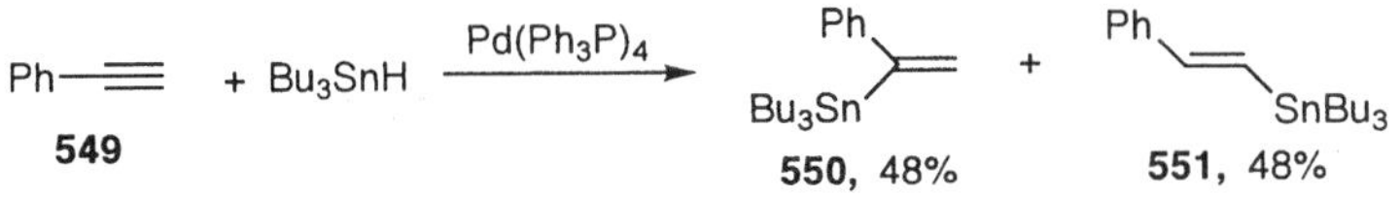

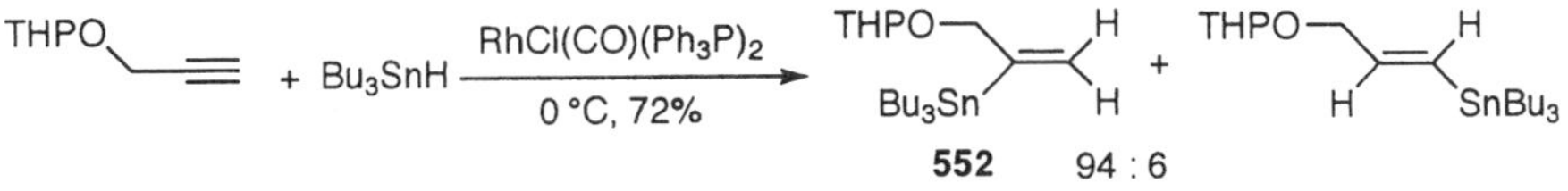

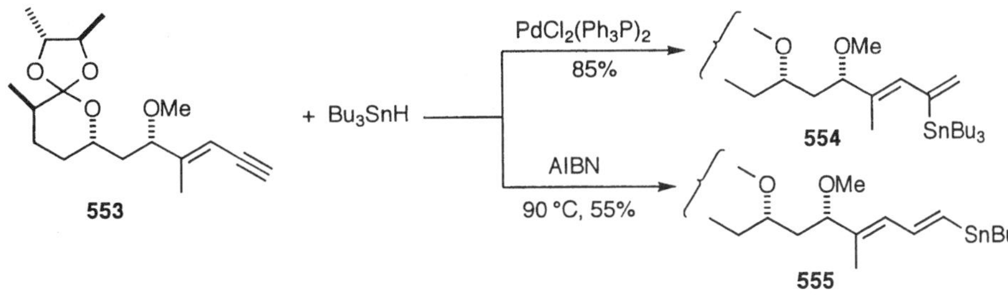

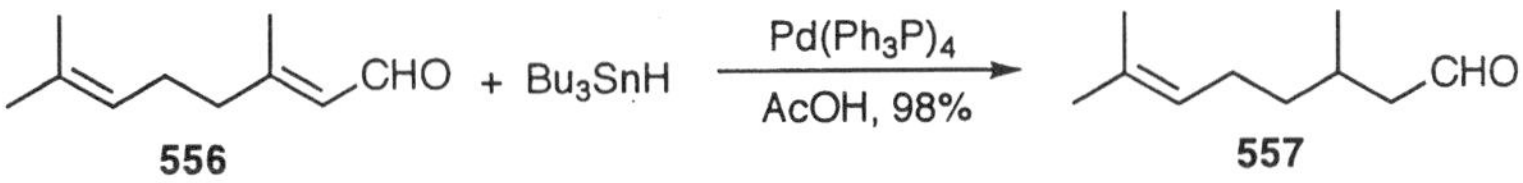

7.4.3.5 Hydrosilylation (Hydrosilation)

Addition of hydrosilane to alkenes, dienes and alkynes is called hydrosilylation, or hydrosilation, and is a commercially important process for the production of many organosilicon compounds. As related reactions, silylformylation of alkynes is treated in Section 7.1.2, and the reduction of carbonyl compounds to alcohols by hydrosilylation is treated in Section 10.2. Compared with other hydrometallations discussed so far, hydrosilylation is sluggish and proceeds satisfactorily only in the presence of catalysts [214]. Chloroplatinic acid is the most active catalyst and the hydrosilylation of alkenes catalysed by H_2PtCl_4 is operated commercially [215]. Colloidal Pt is said to be an active catalytic species. Even the internal alkenes **558** can be hydrosilylated in the presence of a Pt catalyst with concomitant isomerization of the double bond from an internal to a terminal position to give terminal silylalkanes **559**. The oxidative addition of hydrosilane to form $R_3Si-Pt-H$ **560** is the first step of the hydrosilylation, and insertion of alkenes to the Pt−H bond gives **561**, and the alkylsilane **562** is obtained by reductive elimination.

Pd, Rh, Ru, Co, and Ni complexes catalyze the hydrosilylation. But their catalytic activities are lower than that of the Pt catalyst. However, the lower catalytic activity gives higher selectivity in some cases, and the hydrosilylation catalyzed by these catalysts offers useful synthetic methods.

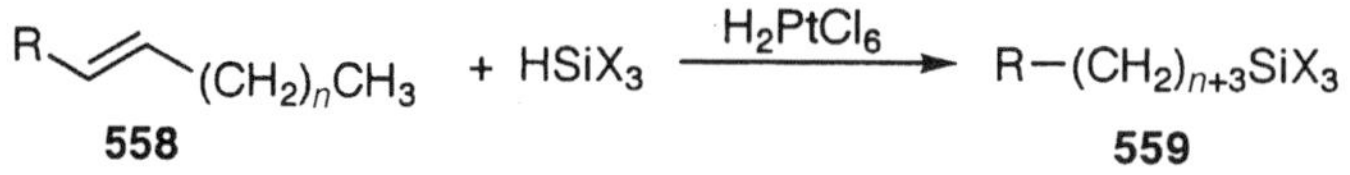

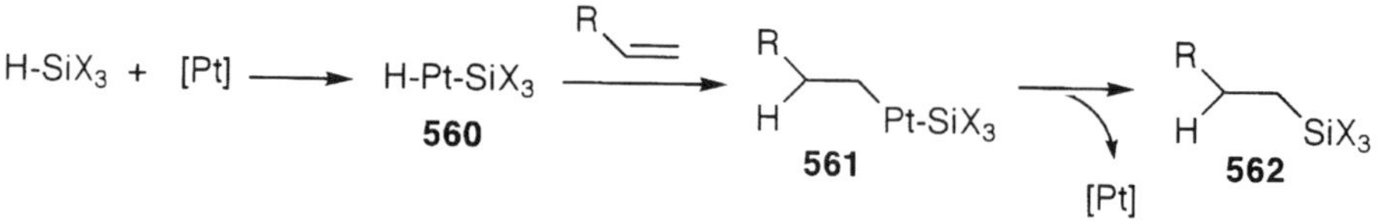

Various Pd–phosphine complexes catalyse the hydrosilylation of alkenes to give 1-silylalkanes **562** almost exclusively [216]. Mainly trichloro- or dichlorosilane is used for the reaction. The reactivity of trialkylsilanes is low. Internal alkenes are difficult to hydrosilylate with Pd catalysts. Usually, the silyl group is introduced selectively at a terminal carbon by the Pd-catalysed hydrosilylation of 1-alkenes. Surprisingly, however, the use of a Pd catalyst coordinated by optically active monodentate phosphines (MOP, **XXXVII**) as the ligand, introduces the trichlorosilyl group mainly at the internal carbon of some terminal alkenes. Moreover, this remarkable ligand accelerates the reaction, which is highly enantioselective. Thus 1-hexene is hydrosilylated to give 2-trichlorosilylhexane (**563**), and converted to 2-hexanol (**564**) with 94% ee [217].

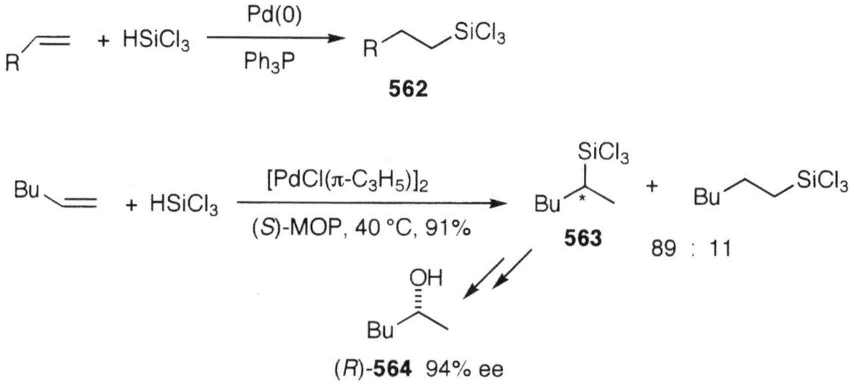

Norbornanol with 96% ee was obtained by the asymmetric hydrosilylation of norbornene. Monofunctionalization of norbornadiene (**565**) was achieved with high chemo- and enantioselectivities to give *exo*-2-trichlorosilyl-5-norbornene (**566**) with one equivalent of trichlorosilane and converted to the alcohol **567** with 95% ee. With 2.5 equivalents, the chiral disilylnorbornane **568** was obtained rather than the *meso* isomer **569** (18:1). The disilylated product **568** was converted to the diol **570** with 99% ee [218].

565, 566, 567, 568, 569, 570

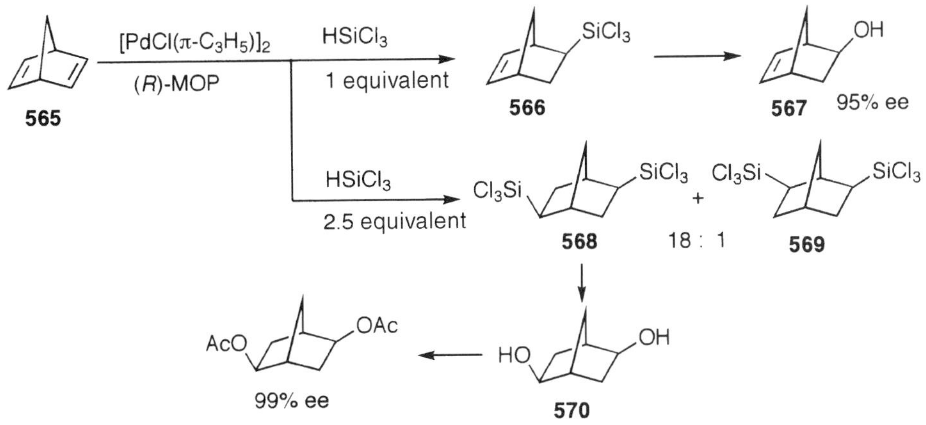

The C–Si bond formed by the hydrosilation of alkene is a stable bond. Although it is difficult to convert the C–Si bond to other functional groups, it can be converted to alcohols by oxidation with MCPBA or H_2O_2. This reaction enhances the usefulness of hydrosilylation of alkenes [219]. Combination of intramolecular hydrosilylation of allylic or homoallylic alcohols and the oxidation offers regio- and stereoselective preparation of diols [220]. Internal alkenes are difficult to hydrosilylate without isomerization to terminal alkenes. However, intramolecular hydrosilation of internal alkenes can be carried out without isomerization. Intramolecular hydrosilylation of the silyl ether **572** of the homoallylic alcohol **571** afforded **573** regio- and stereoselectively, and the Prelog–Djerassi lactone **574** was prepared by applying this method.

571, 572, 573, 574

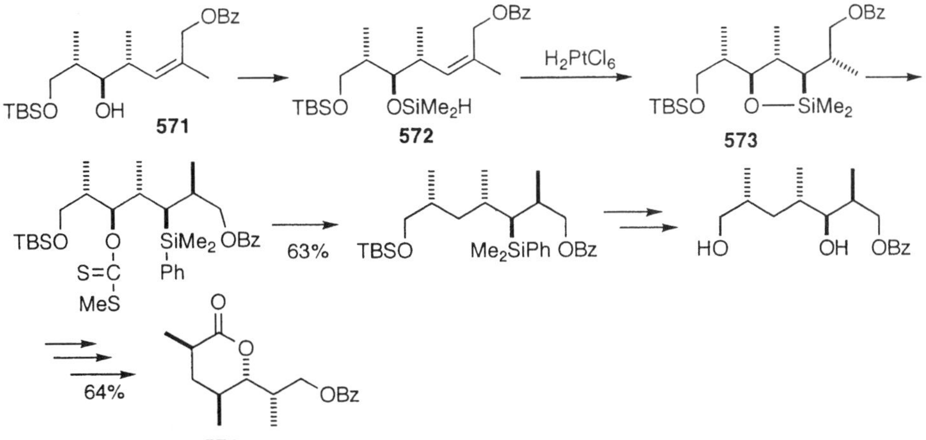

Vinylsilanes and allylsilanes are prepared by dehydrogenative silylation of alkenes catalysed by Rh and other complexes [221]. A particularly effective catalyst for alkenylsilanes is $Ru_3(CO)_{12}$ [222]. Using excess 1-hexene, 1-silyl-1-hexene **575** was obtained in 83% yield [222]. The reaction is explained by the insertion of hexene to the Ru–SiR$_3$ bond in **576** to form **577**, rather than the Ru–H bond, and β-elimination affords **575** and hydrogen. One equivalent of hexene is hydrogenated to hexane. The commercially important vinylsilane **578** can be prepared from ethylene.

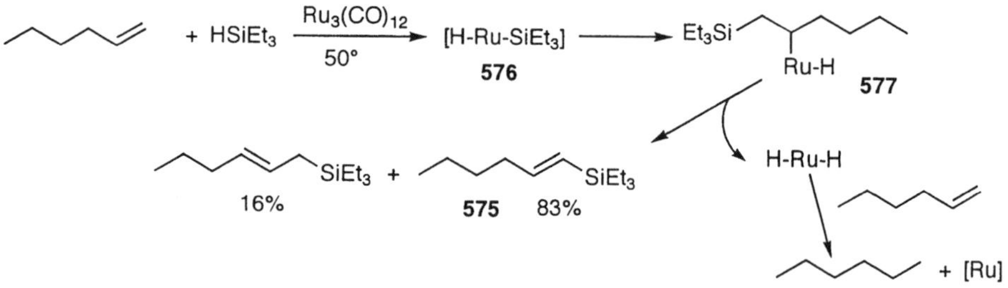

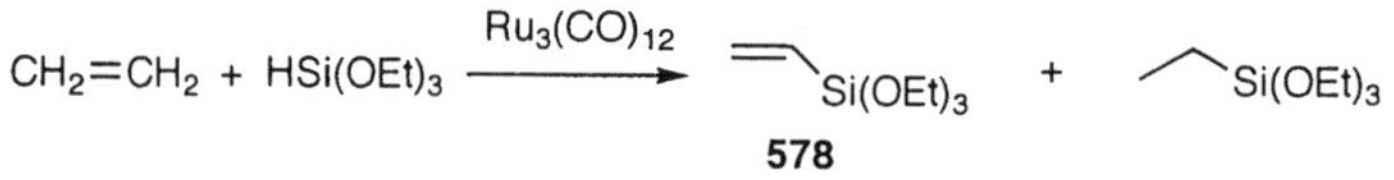

Alkynes are hydrosilylated more easily than alkenes. Hexenylsilanes **579** and **580** are prepared by the Pt-catalysed hydrosilylation of hexyne with one equivalent of hydrosilane. By the reaction of two equivalents of HSiCl$_3$, 1,6-disilylhexane **581** is obtained as a major product rather than the 1,2-adduct **582** [223].

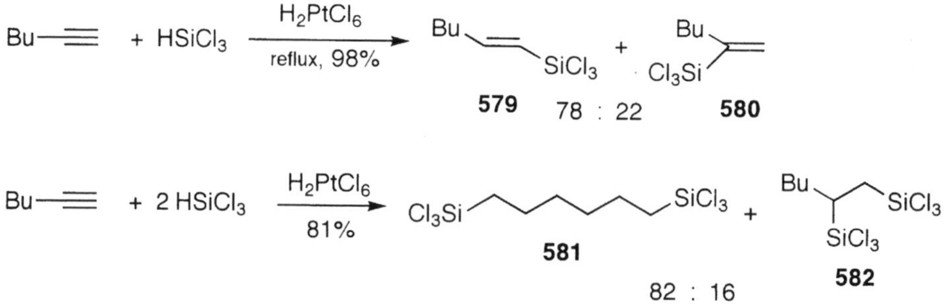

However, the Rh-catalysed hydrosilylation of terminal alkynes affords the thermodynamically unfavorable *anti* product **583** as the main product [224]. The *syn:anti* ratio changes depending on the catalysts and solvents. The *syn* adduct **584** is obtained by a cationic Rh complex in MeCN [225]. The Ru catalyst gives the *anti* adduct. Formation of the *anti* adducts is explained by the following mechanism [224, 226]. Insertion of alkyne to the R$_3$S–Rh bond generates **585** which, due to steric repulsion, isomerizes to **588** via the carbene species **586**, or the metallacyclopropene **587**, and gives the *anti* adduct **583**.

The Rh-catalysed silylcarbocyclization of the allyl propargyl ether (**589**) with silane gives **592**. The reaction starts by the insertion of the triple bond to the Rh–SiR$_3$ bond

$[Rh(cod)Cl]_2$	DMF	97%	1%	2%
$[Rh(cod)_2]BF_4$ / $2Ph_3P$	MeCN	2%	95%	3%

to give **590**, and intramolecular alkene insertion gives **591**, which is converted to **592** by reductive elimination [227].

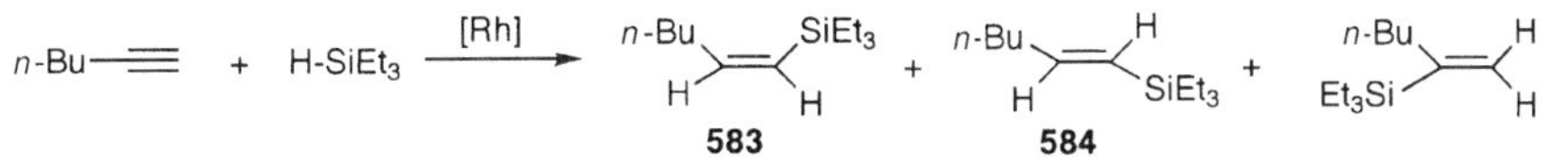

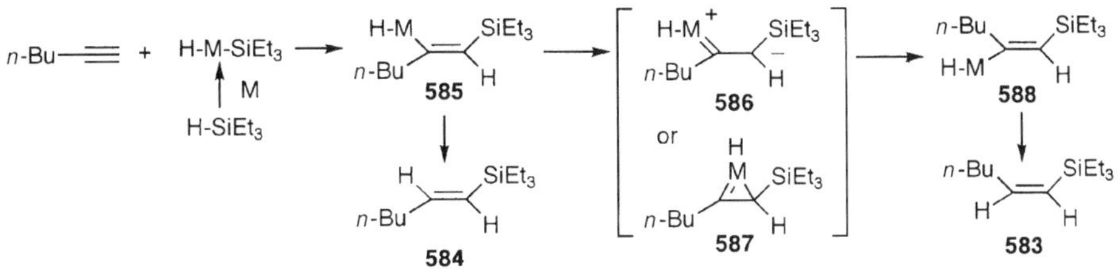

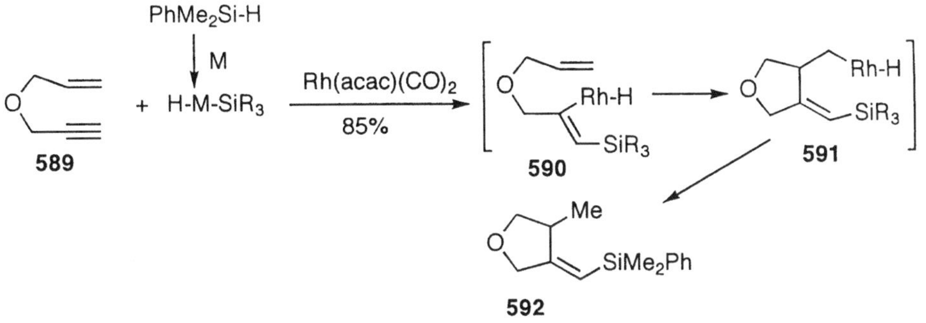

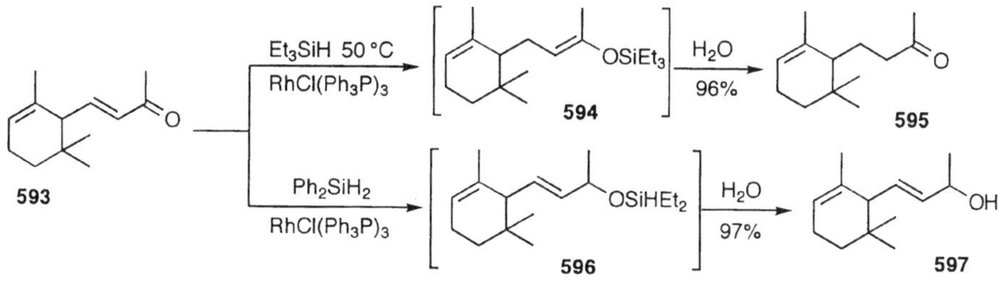

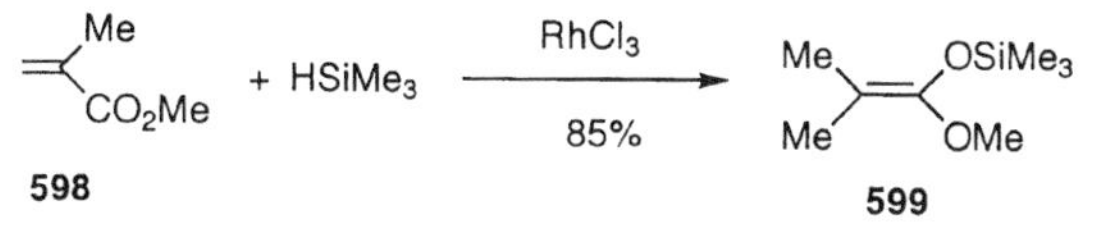

1,4-Addition of monohydrosilane to the α,β-unsaturated ketone **593** affords the silyl enol ether **594**, and its hydrolysis produces the saturated ketone **595**. However, the 1,2-adduct **596** is produced with dihydrosilane, and allylic alcohol **597** is obtained after hydrolysis [228]. The ketene silyl acetal **599** is formed by the Rh-catalysed hydrosilylation of α,β-unsaturated esters. The ketene silyl acetal **599** of high purity is obtained from methyl methacrylate (**598**) with an excess of the silane using $RhCl_3$ as the catalyst [229].

7.5 Hydroacylation of Alkenes and Alkynes

Acylmetal hydride is formed by the oxidative addition of aldehyde, and hydroacylation occurs by insertion of alkene or alkyne. The Ni-catalysed hydroacylation of internal alkyne **600** with aldehyde gave rise to the α,β-unsaturated ketone **601** [230]. The Ru-catalysed hydroacylation of cyclohexene with aldehyde **602** under CO pressure at high temperature gives the ketone **603** [231].

600 **601**

602 **603**

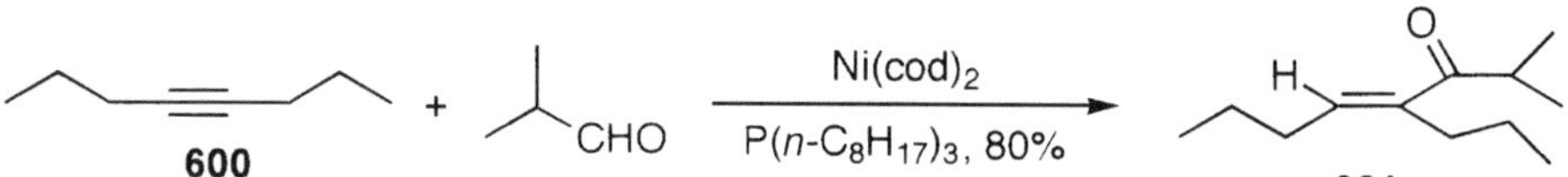

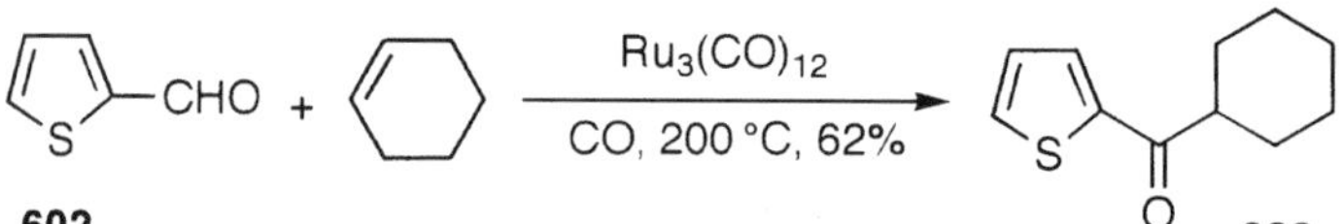

Facile decarbonylation of aldehydes with the Rh complex (Wilkinson complex) is known [43,44]. The reaction is explained by the oxidative addition of aldehyde to Rh, followed by decarbonylation and reductive elimination. However, the Rh-catalysed intramolecular reaction of some unsaturated aldehydes proceeds without the decarbonylation, and cyclic ketones are obtained. Treatment of unsaturated aldehyde

604 **605**

606 **607** **608**

609 **610**

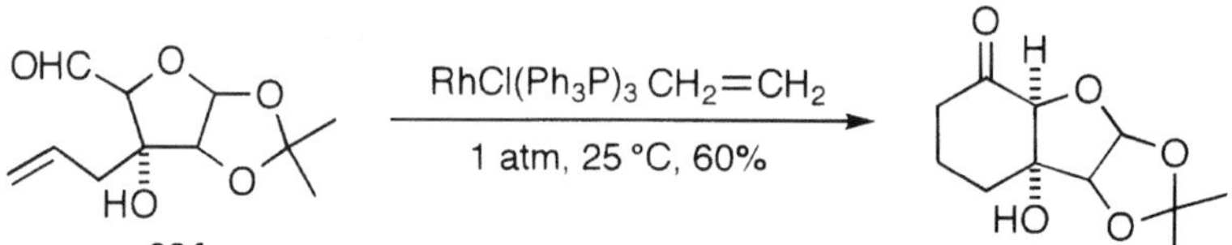

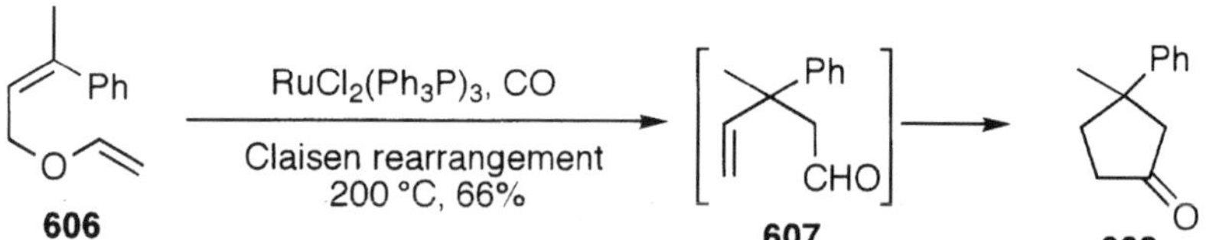

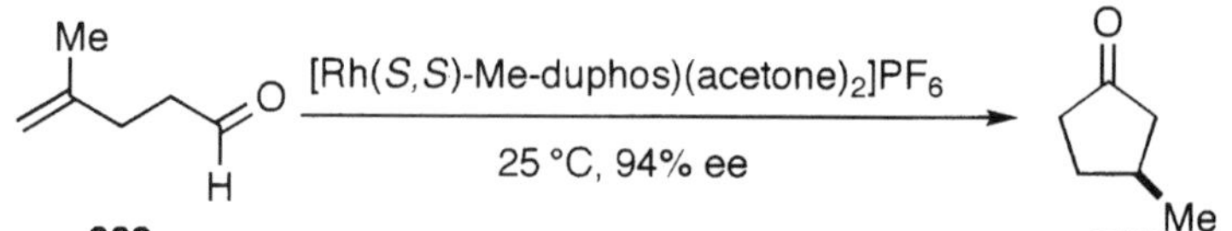

604 under ethylene atmosphere with a catalytic amount of the Rh complex afforded the cyclohexanone **605** by insertion of the double bond to the acyl Rh–H bond (hydroacylation) [232,233]. The allyl vinyl ether **606** undergoes Claisen rearrangement to give 4-enal **607**, and cyclopentanone **608** is formed by its intramolecular hydroacylation in the presence of the Ru or Rh catalyst [234]. 4-Substituted 4-enals **609**, bearing primary and secondary substituents, are converted to the corresponding cyclopentanones **610** with 93–96% ee using [Rh(S,S-Me-Duphos)(acetone)$_2$]PF$_6$ [235].

The Rh-catalyzed hydroacylation of alkynes is also possible. Reaction of salicylaldehyde (**611**) with 4-octyne using an Rh-DPPF complex gave the unsaturated ketone **612** in high yield [236].

References

1. D. J. Thompson, *Comprehensive Organic Synthesis*, vol 3, p. 1015, Pergamon Press 1991.
2. Book. H. M. Colquhoun, D. J. Thompson and M. V. Twigg, *Carbonylation*, Plenum Press, 1991.
3. J. Tsuji, M. Morikawa and J. Kiji, *Tetrahedron Lett.*, 1437 (1963).
4. K. Bittler, N. V. Kutepow, K. Neubauer and H. Reis, *Angew. Chem.*, **80**, 352 (1968).
5. Y. Sugi, K. Bando and S. Shin, *Chem. Ind.*, 397 (1975).
6. T. Hiyama, N. Wakasa and T. Kusumoto, *Synlett*, 569 (1991).
7. H. Alper, J. B. Woell, B. Despeyroux and D. J. H. Smith, *Chem. Commun.*, 1270 (1983); H. Alper and D. Leonard, *Tetrahedron Lett.*, **26**, 5639 (1985).
8. K. Inomata, S. Toda and H. Kinoshita, *Chem. Lett.*, 1567 (1990).
9. B. El Ali and H. Alper, *J. Org. Chem.*, **58**, 3595 (1993); B. El Ali, G. Vasapollo and H. Alper, *J. Org. Chem.*, **58**, 4739 (1993).
10. J. F. Knifton, *J. Org. Chem.*, **41**, 2885 (1976), *J. Am. Oil Chem. Soc.*, **55**, 496 (1978).
11. S. Brewis and P. R. Hughes, *Chem. Commun.*, 489 (1965), 71 (1967).
12. J. Tsuji, S. Hosaka, J. Kiji and T. Susuki, *Bull. Chem. Soc. Jpn.*, **39**, 141 (1966).
13. S. Brewis and P. R. Hughes, *Chem. Commun.*, 6 (1966).
14. E. Drent and P. H. M. Budzelaar, *Chem. Rev.*, **96**, 663 (1996); A. Sen, *Acc. Chem. Res.*, **26**, 303 (1993).
15. J. Tsuji and S. Hosaka, *Polymer Lett.*, **3**, 703 (1965).
16. A. Batistini, A. Consiglio and U. W. Suter, *Angew. Chem., Int. Ed. Engl.*, **104**, 306 (1992); M. Sperrle and G. Consiglio, *J. Am. Chem. Soc.*, **117**, 12130 (1995).
17. K. Nozaki, N. Sato and H. Takaya, *J. Am. Chem. Soc.*, **117**, 9911 (1995); K. Nozaki, N. Sato, Y. Tonomura, M. Yasutomi, H. Takaya, T. Hiyama, T. Matsubara and N. Koga, *J. Am. Chem. Soc.*, **119**, 12779 (1997).
18. K. Mori, T. Mizoroki and A. Ozaki, *Chem. Lett.*, 39 (1975).

19. J. F. Knifton, *J. Mol. Catal.*, **2**, 293 (1977).
20. T. Iimori and M. Shibasaki, *Tetrahedron Lett.*, **27**, 2149 (1986).
21. E. Drent, P. Arnoldy and P. H. M. Budzelaar, *J. Organometal. Chem.*, **455**, 247 (1993).
22. J. R. Norton, K. E. Shenton and J. Schwartz, *Tetrahedron Lett.*, 51 (1975); T. F. Murray and J. R. Norton, *J. Am. Chem. Soc.*, **101**, 4107 (1979); T. F. Murray, E. G. Samsel, V. Varma and J. R. Norton, *J. Am. Chem. Soc.*, **103**, 7520 (1981).
23. C. G. Chardarian, S. L. Woo, R. D. Clark and C. H. Heathcock, *Tetrahedron Lett.*, 1769 (1976)
24. T. F. Murray, V. Varma and J. R. Norton, *J. Org. Chem.*, **43**, 353 (1978); *J. Am. Chem. Soc.*, **99**, 8085 (1977).
25. Review, B. Cornils, W. A. Herrmann and M. Rasch, *Angew. Chem., Int. Ed. Engl.*, **33**, 2144 (1994).
26. E. G. Kuntz, *Chemtech*, 570 (1987); B. Cornils, *Angew. Chem., Int. Ed. Engl.*, **34**, 1575 (1995).
27. G. D. Cuny and S. L. Buchwald, *J. Am. Chem. Soc.*, **115**, 2066 (1993); *Synlett*, 519 (1995).
28. J. L. Leighton and D. N. O'Neil, *J. Am. Chem. Soc.*, **119**, 11118 (1997).
29. F. Agbossou, J. F. Carpentier and A. Mortreux, *Chem. Rev.*, **95**, 2485 (1995).
30. N. Sakai, S. Mano, K. Nozaki and H. Takaya, *J. Am. Chem. Soc.*, **115**, 7033 (1993); T. Higashizima, N. Sakai, K. Nozaki and H. Takaya, *Tetrahedron Lett.*, **35**, 2023 (1994).
31. H. Wakamatsu, J. Uda and N. Yamakami, *Chem. Commun.*, 1540 (1971); P. Pino, *J. Mol. Catal.*, **6**, 341 (1979).
32. I. Ojima, M. Tzanarioudaki and M. Eguchi, *J. Org. Chem.*, **60**, 7078 (1995).
33. G. Dyker, *Angew. Chem., Int. Ed. Engl.*, **36**, 1700 (1997); M. Beller, M. Eckert, F. Vollmüller, S. Bogdanovic and H. Geissler, *Angew. Chem., Int. Ed. Engl.*, **36**, 1494 (1997); M. Beller, W. A. Moradi, M. Eckert and H. Neumann, *Tetrahedron Lett.*, **40**, 4523 (1999).
34. K. Doyama, T. Joh, S. Takahashi and T. Shiohara, *Tetrahedron Lett.*, **27**, 4497 (1986).
35. J. R. Johnson, G. D. Cuny and S. L. Buchwald, *Angew. Chem., Int. Ed. Engl.*, **34**, 1760 (1995).
36. Y. Seki, A. Hidaka, S. Murai and N. Sonoda, *Angew. Chem., Int. Ed. Engl.*, **16**, 174 (1977); Reviews, S. Murai and N. Sonoda, *Angew. Chem., Int. Ed. Engl.*, **18**, 837 (1979); N. Chatani and S. Murai, *Synlett*, 414 (1996).
37. N. Chatani, S. Ikeda, K. Ohe and S. Murai, *J. Am. Chem. Soc.*, **114**, 9710 (1992).
38. I. Matsuda, A. Ogiso, S. Sato and Y. Izumi, *J. Am. Chem. Soc.*, **111**, 2332 (1989).
39. I. Ojima, P. Ingallina, R. J. Donovan and N. Clos, *Organometallics*, **10**, 38 (1991).
40. M. Eguchi, Q. Zeng, A. Korda and I. Ojima, *Tetrahedron Lett.*, **34**, 915 (1993).
41. I. Matsuda, J. Sakakibara and H. Nagashima, *Tetrahedron Lett.*, **32**, 7431 (1991).
42. F. Moteil, I. Matsuda and H. Alper, *J. Am. Chem. Soc.*, **117**, 4419 (1995); I. Ojima, E. Vidal, M. Tzamarioudaki and I. Matsuda, *J. Am. Chem. Soc.*, **117**, 6797 (1995).
43. J. Tsuji and K. Ohno, *Synthesis*, 157 (1969).
44. K. Ohno and J. Tsuji, *J. Am. Chem. Soc.*, **90**, 99 (1968)
45. A. M. Walborsky and L. E. Allen, *J. Am. Chem. Soc.*, **93**, 5465 (1971).
46. D. E. Iley and B. Fraser-Reid, *J. Am. Chem. Soc.*, **97**, 2563 (1975).
47. J. Tsuji and K. Ohno, *J. Am. Chem. Soc.*, **90**, 94 (1968).
48. H. E. Eschinazi and H. Pines, *J. Org. Chem.*, **24**, 1369 (1959).
49. N. E. Schore, *Chem. Rev.*, **88**, 1081 (1988); M. Lautens, W. Klute and W. Tam, *Chem. Rev.* **96**, 49 (1996); I. Ojima, M. Tzamarioudaki, Z. Li and R. J. Donovan, *Chem. Rev.*, **96**, 635 (1996); H. W. Fruhauf, *Chem. Rev.*, **97**, 523 (1997).

50. Y. Wakatsuki, T. Kuramitsu and H. Yamazaki, *Tetrahedron Lett.*, 4549 (1974).
51. A. K. Jhingan and W. F. Maier, *J. Org. Chem.*, **52**, 1161 (1987).
52. T. Takahashi, Z. Xi, A. Yamazaki, Y. Liu, K. Nakajima and M. Kotora, *J. Am. Chem. Soc.*, **120**, 1672 (1998).
53. K. P. C. Vollhardt, *Acc. Chem. Res.*, **10**, 1 (1977); *Angew. Chem., Int. Ed. Engl.*, **23**, 539 (1984).
54. R. L. Funk, C. A. Parnell and K. P. C. Vollhardt, *J. Am. Chem. Soc.*, **99**, 5483 (1977); **101**, 215 (1979); **102**, 5253 (1980).
55. P. Cruciani, R. Stammler, C. Aubert and M. Malacria, *J. Org. Chem.*, **61**, 2699 (1996).
56. Y. Sato, T. Nishimata and M. Mori, *J. Org. Chem.*, **59**, 6133 (1994).
57. F. E. McDonald, H. Y. H. Zhu and C. R. Helquist, *J. Am. Chem. Soc.*, **117**, 6605 (1995).
58. W. Reppe, O. Schlichting, K. Klager and T. Toepel, *Liebigs Ann. Chem.*, **560**, 1 (1948).
59. J. R. Leto and M. F. Leto, *J. Am. Chem. Soc.*, **83**, 2944 (1961).
60. F. Wagner and H. Meier, *Tetrahedron*, **30**, 773 (1974).
61. D. B. Grotjahn and K. P. C. Vollhardt, *Synthesis*, 579 (1993).
62. S. Ikeda, H. Watanabe and Y. Sato, *J. Org. Chem.*, **63**, 7026 (1998).
63. C. A. Merlic and M. E. Pauly, *J. Am. Chem. Soc.*, **118**, 11319 (1996).
64. S. Saito, M. M. Salter, V. Gevorgyan, N. Tsuboya, K. Tando and Y. Yamamoto, *J. Am. Chem. Soc.*, **118**, 3970 (1996); V. Gevorgyan, K. Tando, N. Uchiyama and Y. Yamamoto, *J. Org. Chem.*, **63**, 7022 (1998).
65. D. Weibel, V. Gevorgyan and Y. Yamamoto, *J. Org. Chem.*, **63**, 1217 (1998); S. Saito, N. Tsuboya and Y. Yamamoto, *J. Org. Chem.*, **62**, 5042 (1997).
66. V. Gevorgyan, N. Sadayori and Y. Yamamoto, *Tetrahedron Lett.*, **38**, 8603 (1997).
67. V. Gevorgyan, L. G. Quan and Y. Yamamoto, *J. Org. Chem.*, **63**, 1244 (1998).
68. A. Takeda, A. Ohno, I. Kadota, V. Gevorgyan and Y. Yamamoto, *J. Am. Chem. Soc.*, **119**, 4547 (1997).
69. V. Gevorgyan, A. Takeda and Y. Yamamoto, *J. Am. Chem. Soc.*, **119**, 11313 (1997).
70. Y. Wakatsuki and H. Yamazaki, *Bull. Chem. Soc. Jpn.*, **58**, 2715 (1985).
71. H. Bönnemann, *Angew. Chem., Int. Ed. Engl.*, **24**, 248 (1985).
72. B. Heller and G. Oehme, *Chem. Commun.*, 179 (1995).
73. C. Saa, D. D. Crotts, G. Hsu and K. P. C. Vollhardt, *Synlett*, 487 (1994).
74. P. Hong and H. Yamazaki, *Synthesis*, 50 (1977).
75. R. A. Earl and K. P. C. Vollhardt, *J. Org. Chem.*, **49**, 4786 (1984).
76. T. Tsuda, T. Kiyoi, T. Miyane and T. Saegusa, *J. Am. Chem. Soc.*, **110**, 8570 (1988).
77. T. Tsuda, S. Morikawa and T. Saegusa, *J. Org. Chem.*, **53**, 3140 (1988); T. Tsuda, S. Morikawa, N. Hasegawa and T. Saegusa, *J. Org. Chem.*, **55**, 2978 (1990).
78. A. J. Pearson, R. A. Dubbert and R. J. Shively, Jr., *Organometallics*, **11**, 4096 (1992).
79. N. Chatani, S. Ikeda, K. Ohe and S. Murai, *J. Am. Chem. Soc.*, **114**, 9710 (1992); N. Chatani, Y. Fukumoto, T. Ida and S. Murai, *J. Am. Chem. Soc.*, **115**, 11614 (1993).
80. N. E. Schore, *Org. React.*, **40**, 1, (1991); I. P. Khand, G. P. Knox, P. L. Pauson, W. E. Watts and M. I. Forman, *J. Chem. Soc. Perkin Trans*, I, 975 and 977 (1973); P. L. Pauson, *Tetrahedron*, **41**, 5855 (1985).
81. B. E. La Belle, M. J. Knudsen, M. M. Olmstead, H. Hope, M. D. Yanuck and N. E. Schore, *J. Org. Chem.*, **50**, 5215 (1985), V. Sampath, E. C. Lund, M. J. Knundsen, M. M. Olmstead and N. E. Schore, **52**, 3595 (1987).
82. W. A. Smit, S. L. Kireev, O. M. Nefedov and V. A. Tarasov, *Tetrahedron Lett.*, **30**, 4021 (1989); W. A. Smit, S. O. Simonyan, V. A. Tarasov, G. S. Mikaelian, A. S. Gybin, I. I. Iberagimov, R. Caple, D. Froen and A. Kreager, *Synthesis*, 472 (1989).

83. D. C. Billington, I. M. Helps, P. L. Pauson, W. Thomson and D. Willison, *J. Organometal. Chem.*, **354**, 233 (1988); P. Bladon, P. L. Pauson, H. Brunner and R. Eder, *J. Organometal. Chem.*, **355**, 449 (1988).

84. A. R. Gordon, C. Johnstone and W. J. Kerr, *Synlett*, 1083 (1995); J. G. Donkervoot, A. R. Gordon, C. Johnstone, W. J. Kerr and U. Lange, *Tetrahedron*, **52**, 7391 (1996).

85. S. Shambayati, W. E. Crowe and S. L. Schreiber, *Tetrahedron Lett.*, **31**, 5289 (1990).

86. N. Jeong, Y. K. Chung, B. Y. Lee, S. H. Lee and S. E. Yoo, *Synlett*, 204 (1991).

87. T. Sugihara, M. Yamada, H. Ban, M. Yamaguchi and C. Kaneko, *Angew. Chem., Int. Ed. Engl.*, **36**, 2801 (1997).

88. A. S. Gybin, W. A. Smit, R. Caple, A. L. Veretenov, A. S. Shashkov, L. G. Vorontsova, M. G. Kurella, V. S. Chertkov, A. A. Carapetyan, A. Y. Kosnikov, M. S. Alexanyan, S. V. Lindeman, V. N. Panov, A. V. Maleev, Y. T. Struchkov and S. M. Sharpe, *J. Am. Chem. Soc.*, **114**, 5555 (1992).

89. P. Magnus, M. J. Slater and L. M. Principe, *J. Org. Chem.*, **54**, 5148 (1989); P. Magnus, C. Exon and P. A. Robertson, *Tetrahedron*, **41**, 5861 (1985).

90. E. Carcellar, V. Centellas, A. Moyano, M. A. Pericas and F. Serratosa, *Tetrahedron Lett.*, **26**, 2475 (1985); *Tetrahedron*, **42**, 1831 (1986).

91. M. Thommen, A. L. Veretenov, R. Guidetti-Gret and R. Keese, *Helv. Chim. Acta*, **79**, 461 (1996).

92. V. Rautenstrauch, P. Megard, J. Conesa and W. Küster, *Angew. Chem., Int. Ed. Engl.*, **29**, 1413 (1990).

93. N. Jeong, S. H. Hwang, Y. Lee and Y. K. Chung, *J. Am. Chem. Soc.*, **116**, 3159 (1994).

94. B. Y. Lee, Y. K. Chung, N. Jeong, Y. Lee and S. H. Hwang, *J. Am. Chem. Soc.*, **116**, 8793 (1994).

95. N. Y. Lee and Y. K. Chung, *Tetrahedron Lett.*, **37**, 3145 (1996).

96. B. L. Pagenkopf and T. Livinghouse, *J. Am. Chem. Soc.*, **118**, 2285 (1996).

96a. D. B. Belanger and T. Livinghouse, *Tetrahedron Lett.*, **39**, 7641 (1998).

97. Y. Koga, T. Kobayashi and K. Narasaka, *Chem. Lett.*, 249 (1998).

98. T. Kondo, N. Suzuki, T. Okada and T. Mitsudo, *J. Am. Chem. Soc.*, **119**, 6187 (1997); T. Morimoto, N. Chatani, Y. Fukunoto and S. Murai, *J. Org. Chem.*, **62**, 3762 (1997).

99. F. A. Hicks and S. L. Buchwald, *J. Am. Chem. Soc.*, **118**, 11688 (1996).

100. A. J. Pearson and R. A. Dubbert, *Chem. Commun.*, 202 (1991).

101. T. R. Hoye and J. Suriano, *J. Am. Chem. Soc.*, **115**, 1154 (1993); N. Jeong, S. J. Lee, B. Y. Lee and Y. K. Chung, *Tetrahedron Lett.*, **34**, 4027 (1993).

102. J. L. Kent, H. Wan and K. M. Brummond, *Tetrahedron Lett.*, **36**, 2407 (1995); **39**, 931 (1998).

103. W. A. Nugent and D. F. Taber, *J. Am. Chem. Soc.*, **111**, 6435 (1989).

104. C. J. Rousset, D. R. Swanson, F. Lamaty and E. Negishi, *Tetrahedron Lett.*, **30**, 5105 (1989).

105. T. V. RajanBabu, W. A. Nugent, D. F. Taber and P. J. Fagan, *J. Am. Chem. Soc.*, **110**, 7128 (1988); E. Negishi, S. J. Holmes, J. M. Tour, J. A. Miller, F. E. Cederbaum, D. R. Swanson and T. Takahashi, *J. Am. Chem. Soc.*, **111**, 3336 (1989).

106. W. A. Nugent and J. C. Calabrese, *J. Am. Chem. Soc.*, **106**, 6422 (1984); W. A. Nugent, D. L. Thorn and R. L. Harlow, *J. Am. Chem. Soc.*, **109**, 2788 (1987).

107. E, Negishi and T. Takahashi, *Acc. Chem. Res.*, **27**, 124 (1994); E. Negishi, in *Comprehensive Organic Synthesis*, Vol. 5, p. 1163, Pergamon Press, 1991; S. L. Buchwald and R. B. Nielsen, *Chem. Rev.*, **88**, 1047 (1988); Special issue devoted to the chemistry of Cp_2Zr – E. Negishi, (Ed.) *Tetrahedron*, **51**, No. 15 (1995).

108. E. Negishi, D. R. Swanson, F. E. Cederbaum and T. Takahashi, *Tetrahedron Lett.*, **27**, 2829 (1986).
109. E. Negishi, F. E. Cederbaum and T. Takahashi, *Tetrahedron Lett.*, **27**, 2829 (1986); S. L. Buchwald, B. T. Watson and J. C. Huffman, *J. Am. Chem. Soc.*, **109**, 2544 (1987).
110. T. Takahashi, T. Seki, Y. Nitto, M. Saburi, C. J. Rousset and E. Negishi, *J. Am. Chem. Soc.*, **113**, 6266 (1991); P. Binger, P. Muller, R. Benn, A. Rufinska, B. Gabor, C. Kruger and P. Betz, *Chem. Ber.*, **122**, 1035 (1989).
111. E. Negishi, S. J. Holmes, J. M. Tour and J. A. Miller, *J. Am. Chem. Soc.*, **107**, 2568 (1985).
112. P. A. Wender, K. D. Rice and M. E. Schnute, *J. Am. Chem. Soc.*, **119**, 7897 (1997).
113. N. Uesaka, F. Saitoh, M. Mori, M. Shibasaki, K. Okamura and T. Date, *J. Org. Chem.*, **59**, 5633 (1994); F. Saitoh, M. Mori, K. Okamura and T. Date, *Tetrahedron*, **51**, 4439 (1995).
114. P. A. Wender and F. E. McDonald, *Tetrahedron Lett.*, **31**, 3691 (1990).
115. T. Takahashi, M. Kotora, K. Kasai and N. Suzuki, *Tetrahedron Lett.*, **35**, 5685 (1994).
116. T. Takahashi, M. Kotora, K. Kasai, N. Suzuki and K. Nakajima, *Organometallics*, **13**, 4183 (1994).
117. T. Takahashi, R. Hara, Y. Nishihara and M. Kotora, *J. Am. Chem. Soc.*, **118**, 5154 (1996).
118. O. G. Klinkovich, S. V. Sviridov and D. A. Vasilevski, *Synthesis*, 234 (1991).
119. K. Harada, H. Urabe and F. Sato, *Tetrahedron Lett.*, **36**, 3203 (1995).
120. H. Urabe, T. Hara and F. Sato, *Tetrahedron Lett.*, **36**, 4261 (1995).
121. S. L. Buchwald, F. A. Hicks and N. M. Kablaoni, *J. Am. Chem. Soc.*, **118**, 9450 (1996); F. A. Hicks and J. L. Buchwald, *J. Am. Chem. Soc.*, **118**, 11688 (1996).
122. E. J. Corey, S. A. Rao and M. C. Noe, *J. Am. Chem. Soc.*, **116**, 9345 (1994).
123. J. Lee, C. H. Kang, H. Kim and J. K. Cha, *J. Am. Chem. Soc.*, **118**, 291, 4198 (1996).
124. A. Kasatkin, K. Kobayashi, S. Okamoto and F. Sato, *Tetrahedron Lett.*, **37**, 1849 (1996).
125. H. Urabe, K. Suzuki and F. Sato, *J. Am. Chem. Soc.*, **119**, 10014 (1997).
126. K. Suzuki, H. Urabe and F. Sato, *J. Am. Chem. Soc.*, **118**, 8729 (1996).
127. S. Okamoto, A. Kasatkin, P. K. Zubaidha and F. Sato, *J. Am. Chem. Soc.*, **118**, 2208 (1996).
128. S. C. Berk, R. B. Grossman and S. L. Buchwald, *J. Am. Chem. Soc.*, **116**, 8593 (1994); F. A. Hicks, S. C. Berk and S. L. Buchwald, *J. Org. Chem.*, **61**, 2713 (1996).
129. N. M. Kablaoui and S. L. Buchwald, *J. Am. Chem. Soc.*, **117**, 6785 (1995), **118**, 3182 (1996).
130. W. E. Crowe and M. J. Rachita, *J. Am. Chem. Soc.*, **117**, 6787 (1995).
131. W. E. Crowe and A. T. Vu, *J. Am. Chem. Soc.*, **118**, 1557 (1996); N. M. Kablaoni, F. A. Hicks and S. L. Buchwald, *J. Am. Chem. Soc.*, **118**, 5818 (1996); **119**, 4424 (1997).
132. B. M. Trost and M. J. Krische, *Synlett*, 1 (1998).
133. B. M. Trost, *Acc. Chem. Res.*, **23**, 34 (1990).
134. B. M. Trost, D. C. Lee and F. Rise, *Tetrahedron Lett.*, **30**, 651 (1989); *J. Am. Chem. Soc.*, **109**, 3161 (1987).
135. B. M. Trost and M. Lautens, *J. Am. Chem. Soc.*, **107**, 1781 (1985); B. M. Trost, M. Lautens, C. Chan, D. J. Jebaratnam and T. Mueller, *J. Am. Chem. Soc.*, **113**, 636 (1991).
136. G. J. Engelbrecht and C. W. Holzapfel, *Tetrahedron Lett.*, **32**, 2161 (1991).
137. B. M. Trost and C. Pedregal, *J. Am. Chem. Soc.*, **114**, 7292 (1992).
138. B. M. Trost and M. J. Krische, *J. Am. Chem. Soc.*, **118**, 233 (1996); B. M. Trost and D. J. Jebaratnam, *Tetrahedron Lett.*, **28**, 1611 (1987).
139. B. M. Trost and Y. Li, *J. Am. Chem. Soc.*, **118**, 6625 (1996).

140. B. M. Trost, P. A. Hipskind, J. Y. L. Chung and C. Chan, *Angew. Chem., Int. Ed. Engl.*, **28**, 1502 (1989); B. M. Trost, G. J. Tanoury, M. Lautens, C. Chan and D. T. MacPherson, *J. Am. Chem. Soc.*, **116**, 4255 (1994).
141. B. M. Trost and Y. Shi, *J. Am. Chem. Soc.*, **113**, 701 (1991); **115**, 9421 (1993).
142. B. M. Trost and D. C. Lee, *J. Am. Chem. Soc.*, **110**, 7255 (1988).
143. B. M. Trost, F. J. Fleitz and W. J. Watkins, *J. Am. Chem. Soc.*, **118**, 5146 (1996).
144. B. M. Trost and M. K. Trost, *Tetrahedron Lett.*, **32**, 3647 (1991).
145. B. M. Trost and M. K. Trost, *J. Am. Chem. Soc.*, **113**, 1850 (1991).
146. B. M. Trost and G. J. Tanoury, *J. Am. Chem. Soc.*, **110**, 1636 (1988).
147. B. M. Trost, M. Yanai and K. Hoogsteen, *J. Am. Chem. Soc.*, **115**, 5294 (1993).
148. N. Chatani, N. Furukawa, H. Sakurai and S. Murai, *Organometallics*, **15**, 901 (1996).
149. N. Chatani and S. Murai, unpublished results.
150. N. Chatani, T. Morimoto, T. Muto and S. Murai, *J. Am. Chem. Soc.*, **116**, 6049 (1994).
151. A. Fürstner, H. Szillat, B. Gabor and R. Mynott, *J. Am. Chem. Soc.*, **120**, 8305 (1998).
152. N. Chatani, K. Kataoka and S. Murai, *J. Am. Chem. Soc.*, **120**, 9104 (1998).
153. D. Pilette, S. Moreau, H. Le Bozec, P. H. Dixneuf, J. F. Corrigan and A. J. Carty, *Chem. Commun.*, 409 (1994).
154. W. Strohmeier and A. Kaiser, *J. Organometal. Chem.*, **114**, 273 (1976).
155. W. A. Nugent and F. W. Hobbs, Jr., *Org. Syn.*, **66**, 52 (1987); R. J. Mckinney and M. C. Cotton, *Organometallics*, **5**, 1080 and 1752 (1986); M. Brookhart and S. Sabo-Etienne, *J. Am. Chem. Soc.*, **113**, 2777 (1991); W. A. Nugent and F. W. Hobbs, Jr., *Org. Syn.*, **66**, 52 (1987).
156. G. Giacomelli, F. Mardacci, A. M. Caporusso and L. Lardicci, *Tetrahedron Lett.*, 3217 (1979); L. Carlton and G. Read, *J. Chem. Soc., Perkin, Trans.* I, 1631 (1978).
157. B. M. Trost, C. Chan and G. Rühter, *J. Am. Chem. Soc.*, **109**, 3486 (1987); B. M. Trost, M. T. Sorum, C. Chan, A. E. Harms and G. Rühter, *J. Am. Chem. Soc.*, **119**, 698 (1997).
158. B. M. Trost, W. Brieden and K. H. Baringhaus, *Angew. Chem., Int. Ed. Engl*, **31**, 1335 (1992).
159. T. Mitsudo, K. Kokuryo, T. Shinsugi, Y. Nakagawa, Y. Watanabe and Y. Takegami, *J. Org. Chem.*, **44**, 4492 (1979); T. Mitsudo, H. Naruse, T. Kondo, Y. Ozaki and Y. Watanabe, *Angew. Chem., Int. Ed. Engl.*, **33**, 580 (1994).
160. T. Mitsudo, S. W. Zhang, M. Nagao and Y. Watanabe, *Chem. Commun.*, 598 (1991).
161. B. M. Trost and A. Indolese, *J. Am. Chem. Soc.*, **115**, 4361 (1993); B. M. Trost, A. F. Indolese, J. J. Müller and B. Treptow, *J. Am. Chem. Soc.*, **117**, 615 (1995).
162. B. M. Trost, G. D. Probst and A. Schoop, *J. Am. Chem. Soc.*, **120**, 9228 (1998).
163. S. Derien, D. Jan and P. H. Dixneuf, *Tetrahedron*, **52**, 5511 (1996).
164. B. M. Trost, J. A. Martinez, R. J. Kulawiec and A. F. Indolese, *J. Am. Chem. Soc.*, **115**, 10402 (1993).
165. B. M. Trost and R. J. Kulawiec, *J. Am. Chem. Soc.*, **114**, 5579 (1992); B. M. Trost, R. J. Kulawiec and A. Hammes, *Tetrahedron Lett.*, **34**, 587 (1993).
165a. B. M. Trost, M. Portnoy and H. Kurihara, *J. Am. Chem. Soc.*, **119**, 836 (1997).
166. M. I. Bruce, *Adv. Organometal. Chem.*, **22**, 60 (1983); *Chem. Rev.*, **91**, 197 (1991); K. Ouzzine, H. Le Bozec and P. H. Dixneuf, *J. Organometal. Chem.*, **317**, C25 (1986).
167. R. Mahe and P. H. Dixneuf, *Tetrahedron Lett.*, **27**, 6333 (1986); R. Mahe, Y. Sasaki, C. Bruneau and P. H. Dixneuf, *J. Org. Chem.*, **54**, 1518 (1989).
168. T. Mitsudo, Y. Hori, Y. Yamakawa and Y. Watanabe, *Tetrahedron Lett.*, **28**, 4417 (1987).
169. S. Murai, F. Kakiuchi, S. Sekine, Y. Tanaka, A. Kamatani, N. Sonoda and N. Chatani, *Nature*, **366**, 529 (1993); *Pure Appl. Chem.*, **66**, 1527 (1994); F. Kakiuchi, S. Sekine, Y.

Tanaka, A. Kamatani, M. Sonoda, N. Chatani and S. Murai, *Bull. Chem. Soc. Jpn.*, **68**, 62 (1995).

170. F. Kakiuchi, Y. Tanaka, T. Sato, N. Chatani and S. Murai, *Chem. Lett.*, 679 (1995).

171. F. Kakiuchi, Y. Yamamoto, N. Chatani and S. Murai, *Chem. Lett.*, 681 (1995).

172. B. B. Snider, M. Karras and R. S. E. Conn, *J. Am. Chem. Soc.*, **100**, 4624 (1978); *Tetrahedron Lett.*, 1679 (1979).

173. U. M. Dzhemilev and O. S. Vostrikova, *J. Organometal. Chem.*, **285**, 43 (1985); T. Takahashi, T. Seki, Y. Nitto, M. Saburi, C. J. Rousset and E. Negishi, *J. Am. Chem. Soc.*, **113**, 6266 (1991); K. S. Knight and R. M. Waymouth, *J. Am. Chem. Soc.*, **113**, 6268 (1991); A. H. Hoveyda and Z. Xu, *J. Am. Chem. Soc.*, **113**, 5079 (1991).

174. N. Suzuki, D. Kondakov and T. Takahashi, *J. Am. Chem. Soc.*, **115**, 8485 (1993).

175. J. P. Morken, M. T. Didiuk and A. H. Hoveyda, *J. Am. Chem. Soc.*, **115**, 6997 (1993).

176. M. S. Visser, N. M. Heron, M. T. Didiuk, J. F. Sagal and A. H. Hoveyda, *J. Am. Chem. Soc.*, **118**, 4291 (1996).

177. E. Negishi, *Pure Appl. Chem.*, 53, 2333 (1981); E. Negishi, L. F. Valente and M. Kobayashi, *J. Am. Chem. Soc.*, **102**, 3298 (1980); D. E. Van Horn and E. Negishi, *J. Am. Chem. Soc.*, **100**, 2252 (1978); R. E. Ireland and P. Wipf, *J. Org. Chem.*, **55**, 1425 (1990).

178. U. M. Dzhemilev, A. G. Ibragimov, O. S. Vostrikova, G. A. Tolsikov and L. M. Zelenova, *Izv. Akad. Nauk SSSR, Ser. Khim.*, 361 (1981).

179. E. Negishi, M. D. Jensen, D. Y. Kondakov and S. Wang, *J. Am. Chem. Soc.*, **116**, 8404 (1994); D. Y. Kondakov, S. Wang and E. Negishi, *Tetrahedron Lett.*, **37**, 3803 (1996).

180. T. Stüdemann and P. Knochel, *Angew. Chem., Int. Ed. Engl.*, **36**, 93 (1997).

181. A. E. Greene, J. P. Lansard, J. L. Luche and C. Petrier, *J. Org. Chem.*, **49**, 931 (1984); C. Petrier, J. L. Luche and C. Dupuy, *Tetrahedron Lett.*, **25**, 3463 (1984).

182. E. C. Ashby and G. Heinsohn, *J. Org. Chem.*, **39**, 3297 (1974); R. T. Hansen, D. B. Carr and J. Schwartz, *J. Am. Chem. Soc.*, **100**, 2244 (1978).

183. J. Schwartz, M. J. Loots and H. Kosugi, *J. Am. Chem. Soc.*, **102**, 1333 (1980).

183a. Y. Takaya, M. Ogasawara, T. Hayashi, M. Sakai and T. Miyaura, *J. Am. Chem. Soc.*, **120**, 5579 (1998).

184. E. Shirakawa, H. Yoshida, T. Kurahashi, Y. Nakao and T. Hiyama, *J. Am. Chem. Soc.*, **120**, 2975 (1998).

185. T. Ishiyama, N. Matsuda, N. Miyaura and A. Suzuki, *J. Am. Chem. Soc.*, **115**, 11018 (1993); T. Ishiyama, N. Matsuda, M. Murata, F. Ozawa, A. Suzuki and N. Miyaura, *Organometallics*, **15**, 713 (1996).

186. H. Hayami, M. Sato, S. Kanemoto, Y. Morizawa, K. Oshima and H. Nozaki, *J. Am. Chem. Soc.*, **105**, 4491 (1983).

187. K. Wakamatsu, T. Nonaka, Y. Okuda, W. Tuckmantel, K. Oshima, K. Utimoto and H. Nozaki, *Tetrahedron*, **42**, 4427 (1986); Y. Okuda, K. Wakamatsu, W. Tuckmantel, K. Oshima and H. Nozaki, *Tetrahedron Lett.*, **26**, 4629 (1985).

188. K. Ritter, *Synthesis*, 218 (1989).

189. M. Mori, N. Watanabe, N. Kaneta and M. Shibasaki, *Chem. Lett.*, 1615 (1991).

190. T. N. Michell, A. Amamria, H. Killing and D. Rutschow, *J. Organometal. Chem.*, **241**, C45 (1983); **304**, 257 (1986).

191. H. Sakurai, Y. Kamiyama and Y. Nakadaira, *J. Am. Chem. Soc.*, **97**, 931 (1975); H. Okinoshima, K. Yamamoto and M. Kumada, *J. Organometal. Chem.*, **86**, C27 (1975); H. Watanabe, M. Kobayashi, K. Higuchi and Y. Nagai, *J. Organometal. Chem.*, **186**, 51 (1980); H. Watanabe, M. Kobayashi, M. Saito and Y. Nagai, *J. Organometal. Chem.*, **216**, 149 (1981); **51**, 421 (1973).

192. Y. Ito, M. Suginome and M. Murakami, *J. Org. Chem.*, **56**, 1948 (1991).
193. H. Yamashita, M. Catellani and M. Tanaka, *Chem. Lett.*, 241 (1991).
194. M. Murakami, M. Suginome, K. Fujimoto, H. Nakamura, P. G. Andersson and Y. Ito, *J. Am. Chem. Soc.*, **115**, 6487 (1993).
195. T. Hayashi, Y. Matsumoto and Y. Ito, *Tetrahedron Lett.*, **50**, 335 (1994).
196. H. L. Finkbeiner and G. D. Cooper, *J. Org. Chem.*, **27**, 3395 (1962).
197. F. Sato, H. Ishikawa and M. Sato, *Tetrahedron Lett.*, 365 (1980).
198. H. Felkin and G. Swierczewski, *Tetrahedron*, **31**, 2735 (1975).
199. S. Vettel, A. Vaupel and P. Knochel, *J. Org. Chem.*, **61**, 7473 (1996).
200. K. Burgess and M. J. Ohlmeyer, *Chem. Rev.*, **91**, 1179 (1991); I. P. Beletskaya and A. Pelter, *Tetrahedron*, **53**, 4957 (1997).
201. D. Männig and H. Nöth, *Angew. Chem., Int. Ed. Engl.*, **24**, 878 (1985).
202. D. A. Evans, G. C. Fu and A. H. Hoveyda, *J. Am. Chem. Soc.*, **110**, 6917 (1988); K. Burgess and M. J. Ohlmeyer, *Tetrahedron Lett.*, 395 (1989).
203. X. He and J. F. Hartwig, *J. Am. Chem. Soc.*, **118**, 1696 (1996).
204. T. Hayashi, Y. Matsumoto and Y. Ito, *J. Am. Chem. Soc.*, **111**, 3426 (1989).
205. D. A. Evans and G. C. Fu, *J. Org. Chem.*, **55**, 2280 (1990).
206. S. Pereira and M. Srebnik, *Organometallics*, **14**, 3127 (1995); *J. Am. Chem. Soc.*, **118**, 911 (1996).
207. S. Pereira and M. Srebnik, *Tetrahedron Lett.*, **37**, 3283 (1996).
208. F. Sato, S. Sato and M. Sato, *J. Organometal. Chem.*, **131**, C26 (1977); K. Isagawa, K. Tatsumi and Y. Otsuji, *Chem. Lett.*, 1145 (1976).
209. F. Sato, T. Oikawa and M. Sato, *Chem. Lett.*, 167 (1979).
210. H. Miyake and K. Yamamura, *Chem. Lett.*, 981 (1989); H. X. Zhang, F. Guibe and G. Balavoines, *J. Org. Chem.*, **55**, 1857 (1990).
211. K. Kikukawa, F. Umekawa, G. Wada and T. Matsuda, *Chem. Lett.*, 881 (1988).
212. A. B. Smith, S. M. Condon, J. C. McCauley, J. L. Leazer, J. W. Leahy and R. E. Maleczka, *J. Am. Chem. Soc.*, **119**, 962 (1997).
213. P. Four and F. Guibe, *Tetrahedron Lett.*, **23**, 1825 (1982); E. Keinan, P. A. Gleize, *Tetrahedron Lett.*, **23**, 477 (1982).
214. T. Hiyama and T. Kusumoto, in *Comprehensive Organic Synthesis*, Vol. 8, p. 763, Pergamon Press 1991.
215. J. L. Speier, *Adv. Organomet. Chem.*, **17**, 407 (1979).
216. M. Hara, K. Ohno and J. Tsuji, *Chem. Commun.*, 247 (1971).
217. Y. Uozumi and T. Hayashi, *J. Am. Chem. Soc.*, **113**, 9887 (1991); *Pure Appl. Chem.*, **64**, 1911 (1992).
218. Y. Uozumi, S. Y. Lee and T. Hayashi, *Tetrahedron Lett.*, **33**, 7185 (1992).
219. K. Tamao, T. Kakui and M. Kumada, *J. Am. Chem. Soc.*, **100**, 2268 (1978); K. Tamao, N. Ishida, T. Tanaka and M. Kumada, *Organometal.*, **2**, 1694 (1983).
220. K. Tamao, T. Nakajima, R. Sumiya, H. Arai, N. Higuchi and Y. Ito, *J. Am. Chem. Soc.*, **108**, 6090 (1986); K. Tamao, T. Yamauchi and Y. Ito, *Chem. Lett.*, 171 (1987).
221. A. Milan, E. Towns and P. M. Maitlis, *Chem. Commun.*, 673 (1981); A. Onopchenko, E. T. Sabourin and D. L. Beach, *J. Org. Chem.*, **48**, 5101 (1983); **49**, 3389 (1984).
222. Y. Seki, K. Takeshita, K. Kawamoto, S. Murai and N. Sonoda, *J. Org. Chem.*, **51**, 3890 (1986); *Angew. Chem., Int. Ed. Engl.*, **19**, 928 (1980).
223. R. A. Benkeser, A. F. Cunico, S. Dunny, P. R. Jones and P. G. Nerlekar, *J. Org. Chem.*, **32**, 2634 (1967); I. G. lovel, Y. S. Goldberg, M. V. Shymanska and E. Lukevics, *Organometallics*, **6**, 1410 (1987).
224. I. Ojima, N. Clos, R. L. Donovan and P. Ingallina, *Organometallics*, **9**, 3127 (1990).

225. R. Takeuchi and N. Tanouchi, *J. Chem. Soc. Perkin Trans. I*, 2909 (1994).
226. C. H. Jun and R. H. Crabtree, *J. Organometal. Chem.*, **447**, 177 (1993); R. Tanke and R. H. Crabtree, *J. Am. Chem. Soc.*, **112**, 7984 (1990).
227. I. Ojima, R. J. Donovan and W. R. Shay, *J. Am. Chem. Soc.*, **114**, 6580 (1992).
228. I. Ojima, T. Kogure and Y. Nagai, *Tetrahedron Lett.*, 5035 (1972); I. Ojima and T. Kogure, *Organometallics*, **1**, 1390 (1982).
229. A. Revis and T. K. Hilty, *J. Org. Chem.*, **55**, 2972 (1990).
230. T. Tsuda, T. Kiyoi and T. Saegusa, *J. Org. Chem.*, **55**, 2554 (1990).
231. T. Kondo, M. Akazoe, Y. Tsuji and Y. Watanabe, *J. Org. Chem.*, **55**, 1286 (1990).
232. K. Sakai, J. Ide, O. Oda and N. Nakamura, *Tetrahedron Lett.*, 1287 (1972).
233. K. P. Gable and G. A. Benz, *Tetrahedron Lett.*, **32**, 3473 (1991).
234. P. Eilbrach, A. Gersmeier, D. Lennatz and T. Huber, *Synthesis*, 330 (1995).
235. R. W. Barnhart, D. A. McMorran and B. Bosnich, *Chem. Commun.*, 589 (1997).
236. K. Kokubo, K. Matsumasa, M. Miura and M. Nomura, *J. Org. Chem.*, **62**, 4564 (1997).

8

SYNTHETIC REACTIONS VIA TRANSITION METAL CARBENE COMPLEXES

8.1 Chemistry of Transition Metal Carbene Complexes

Carbenes, generated by several methods, are reactive intermediates and used for further reactions without isolation. Carbenes can also be stabilized by coordination to some transition metals and can be isolated as carbene complexes which have formal metal-to-carbon double bonds. They are classified, based on the reactivity of the carbene, as electrophilic heteroatom-stabilized carbenes (Fischer type), and nucleophilic methylene or alkylidene carbenes (Schrock type).

Fischer-type complexes such as **1** were first prepared in 1964 and their chemical properties studied [1]. Schrock-type nucleophilic complexes such as **2** were prepared later [2]. They are formed by coordination of strong donor ligands such as alkyl or cyclopentadienyl with no π-accepter ligand to metals of high oxidation states. The nucleophilic carbene complexes show Wittig's ylide-type reactivity and the structures may be considered as ylides (eq. 8.1)

$$M{=}CH_2 \longleftrightarrow M^+{-}CH_2{}^- \eqno(8.1)$$

These carbene (or alkylidene) complexes are used as either stoichiometric reagents or catalysts for various transformations which are different from those of free carbenes. Reactions involving the carbene complexes of W, Mo, Cr, Re, Ru, Rh, Pd, Ti and Zr are known. Carbene complexes undergo the following transformations: (i) alkene metathesis; (ii) alkene cyclopropanation; (iii) carbonyl alkenation; (iv) insertion to C–H, N–H and O–H bonds; (v) ylide formation; and (vi) dimerization. Their chemoselectivity depends mainly on the metal species and ligands, as discussed in the following sections.

OC, CO Ph
OC—Cr.=C
OC CO OMe

Fischer type

1

t-BuO''''Mo=C
t-BuO   t-Bu

Schrock type

2

$\delta- \quad \delta+ \diagup R$
$M=C$
$\diagdown R \ (XR)$

X = O, N

[Cr, Mo, W, Fe]

electrophilic

$M=C \diagup^R_{R}$

[Ni, Fe]

neutral

$\delta+ \quad \delta- \diagup R$
$M=C$
$\diagdown R$

[Ti, Ta, W]

nucleophilic

8.2 Catalytic Metatheses of Alkenes and Alkynes, and Their Synthetic Applications

8.2.1 Historical Background and Mechanism of Alkene Metathesis

It was known that internal alkenes could not be polymerized by Ziegler–Natta catalysts. However, in 1964 Natta and co-workers found that cyclobutene and cyclopentene could be polymerized using catalysts prepared by treating WCl_6 or $MoCl_5$ with Et_3Al [3]. This polymerization was discovered in the same year as Fischer's preparation of carbene complex **1** [1]. Strangely, presence of *cis* double bonds was observed in the polymer showing that this is not a simple vinyl polymerization of alkenes. It was subsequently found that the polymerization of cyclic alkenes involves ring-opening polymerization, as shown by the reaction of cyclopentene (**38**) to afford polymer **41** via **39** and **40**. Therefore, the presence of *cis*-alkene bonds in the resulting polymer confirmed that the W complex-catalysed polymerization was not a simple polyethylene-type polymerization, which produces saturated polymers. This reaction proved to be the first example of ring-opening metathesis polymerization (ROMP), although the Natta group was unable to determine the correct mechanism of this strange reaction. In order to explain the new polymerization, Calderon and co-workers ingeneously carried out the reaction of 2-pentene using a similar catalyst and confirmed rapid formation of an equilibrium

Me⟍=⟍Et →(WCl₆, EtAlCl₂ / rt, 3 min)→ Me⟍=⟍Me + Et⟍=⟍Et

R¹⟍=⟍R² ⇌ R¹⟍=⟍R¹ + R²⟍=⟍R²

(8.2)

mixture of 2-pentene, 3-hexene and 2-butene. Based on this result, they proposed the alkene metathesis as shown in eq. (8.2) [4,5]. Also it was known at that time that solid catalysts of W and Mo catalyse alkene metathesis in the gas phase [6].

At present, Mo, W, Re and Ru complexes are known to catalyse alkene metathesis [7]. This unique reaction, catalysed by transition metal complexes, is impossible to achieve by other means. Later, based on studies of the reactivities of Fischer-type carbene complexes, it was discovered that carbene complexes are the intermediates in alkene metatheses. WCl_6 reacts with $EtAlCl_2$ to afford the diethyltungsten complex **3** by transmetallation, and subsequent elimination of α-hydrogen generates ethane and the carbene complex **4** which is the active catalyst.

$$WCl_6 + 2\,EtAlCl_2 \longrightarrow \left[Cl_nW(CH_2CH_3)_2\right] + 2\,AlCl_3$$
$$\textbf{3}$$

$$\left[CH_3CH_2-\overset{\cdot\cdot}{W}\overset{\curvearrowleft}{CHCH_3}\right] \longrightarrow CH_3CH_2-\underset{H}{W}=CHCH_3 \longrightarrow \left[W=CHCH_3\right] + CH_3CH_3$$
$$\textbf{3} \qquad\qquad\qquad\qquad\qquad\qquad \textbf{4}$$

Metathesis of alkene **6** to give the new alkenes **11** and **15** is explained by the following mechanism. The first step is [2+2] cycloaddition between metal carbene **5** and alkene **6** to generate the metallacyclobutane **7** as an intermediate. The real catalyst **8** is generated by retrocycloaddition of the metallacyclobutane **7**. Reaction of **8** with alkene **6** generates the metallacyclobutanes **9** and **10** as intermediates. The intermediate **10** is a nonproductive intermediate, which reproduces **6** and **8**, while **9** is a productive intermediate and yields the new alkene **11** and the real catalyst **12**. Cycloaddition of **12** to alkene **6** produces the productive intermediate **14**, from which the new alkene **15** and the active catalytic species **8** are formed. The intermediate **13** is a nonproductive one.

Although not strictly correct from a mechanistic viewpoint, the course of alkene metathesis can be easily understood and the products predicted by considering the reversible formation and cleavage of the hypothetical cyclobutane **16** from alkene **6** to give the alkenes **11** and **15**. The whole process involves the cleavage of double bonds and the formation of new double bonds. Such a process cannot be realized by any other means. The cyclobutane **17** is nonproductive.

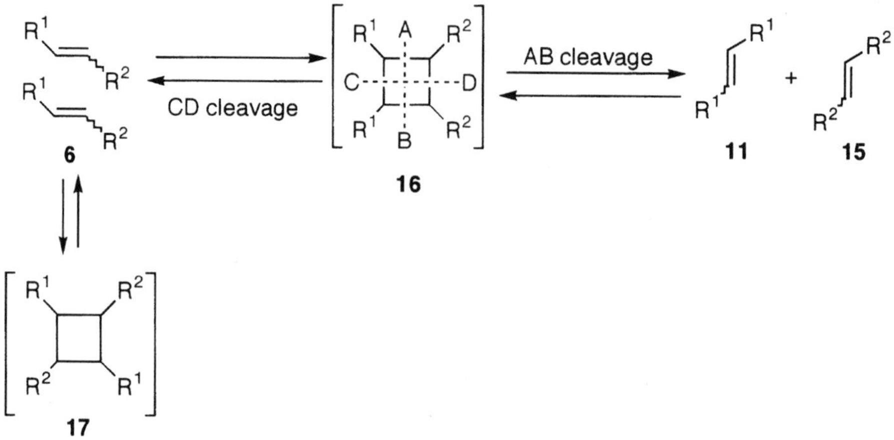

8.2.2 Development of Catalysts for Metathesis

Several catalysts have been introduced since the discovery of alkene metathesis. Catalysts generated by combinations of $MoCl_5$ and WCl_6 with alkylaluminium (R_nAlCl_{3-n}) were the first-generation catalysts, which were not suitable for functionalized alkenes.

Methyltrioxorhenium (MTO, CH_3ReO_3 **18**) is a second-generation catalyst [8]. MTO is easily prepared from Re(VII) oxide and Me_4Sn. It is stable and soluble in organic solvents. The catalyst is used as heterogeneous (MTO supported on Al_2O_3 or SiO_2) and homogeneous catalysts, and they catalyse the metathesis of functionalized alkenes such as oleate at room temperature in methylene chloride.

$$ClSiMe_3 \xrightarrow{Re_2O_7} 2\,ClReO_3 \xrightarrow{Me_4Sn} 2\,\underset{\mathbf{18}}{MeReO_3} + Me_3SnCl$$

$$Re_2O_7 + Me_4Sn \longrightarrow 2\,\underset{\mathbf{18}}{MeReO_3} + Me_3SnOReO_3$$

Further developments have been made possible by the introduction of third-generation catalysts. These are the commercially available, highly active Schrock-type Mo-based catalyst **19** and W catalyst **20**, coordinated by a bulky neopentylidene group and alkoxides [2]. They are well-defined, single-component catalysts. The W complex **20** is very active, but the Mo complex **19** is more versatile. These catalysts have some functional group tolerance, although they are sensitive to O_2 and water. These complexes show catalytic activity without activation by Lewis acids.

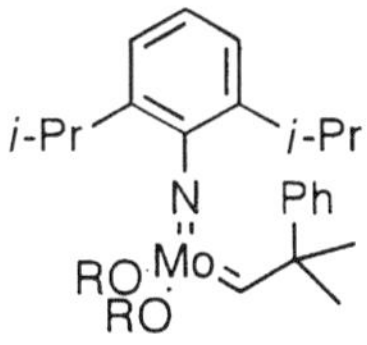

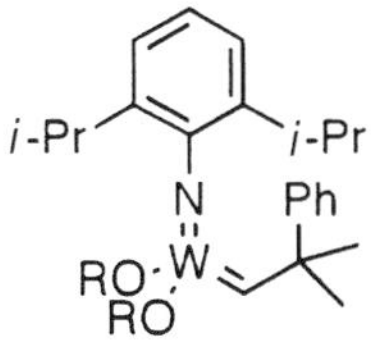

A significant breakthrough was achieved in 1993 by Grubbs and co-workers who prepared the well-defined Ru complexes **22** and **24** as the fourth generation of catalysts, which are more tolerant to functional groups [9]. Although the Ph$_3$P complex **21** formed from 3,3-diphenylcyclopropene is inactive, it becomes active by exchange with bulky tricyclohexylphosphine (Cy)$_3$P. The more active Ru catalyst **24** can be prepared by the reaction of the Ru complex with phenyldiazomethane, followed by exchange with tricyclohexylphosphine [10].

Reaction of the complex **24** with terminal alkene **25** generates styrene and the real catalytic species **27** via the ruthenacyclobutane **26**. The complex **24** is commercially available, active without rigorous exclusion of O_2 and water, and has functional group tolerance. Carbonyl alkenation is not observed with the catalysts **22** and **24**. Their introduction has enormously accelerated the synthetic applications of alkene metathesis [11].

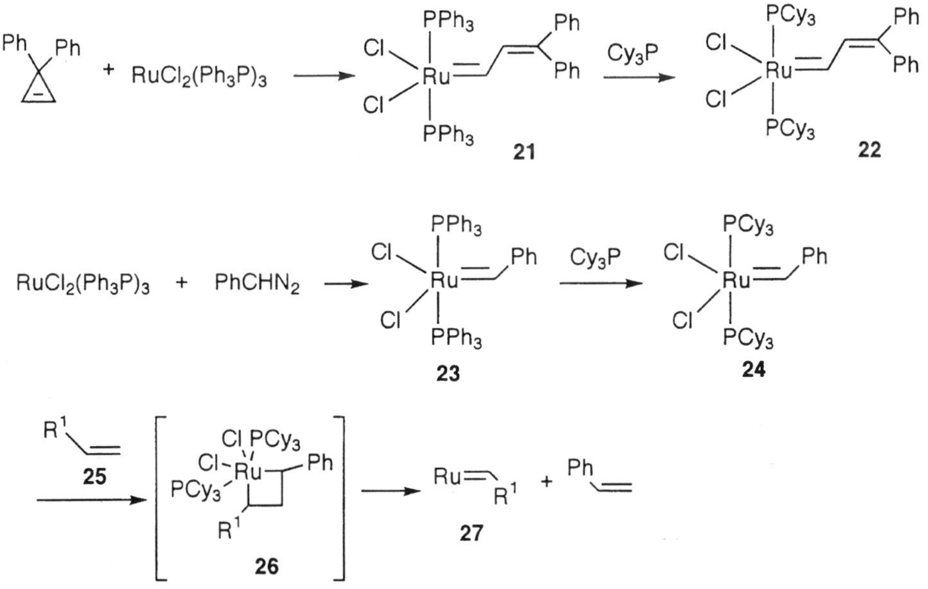

The new carbene complex **31** can be prepared more easily by reaction of Ru hydride **28** with propargyl chloride **29**. Insertion of the triple bond to Ru−H, followed by γ-chloro elimination from **30** affords the carbene complex **31** [12]. As another method, carbene complex **34** is obtained by oxidative addition of the Ru complex to benzal chloride (**32**) and α-chloro elimination [13]. It is known that $RuCl_2(Ph_3P)_2$ reacts with terminal alkyne to form alkynyl complexes, which isomerize to vinylideneruthenium complexes [14]. For example, vinylidene complex **36** is obtained from *tert*-butylacetylene via the alkynyl complex **35**. After ligand exchange of **36** with Cy_3P, the complex **37** is active for ROMP of norbornene derivatives [15].

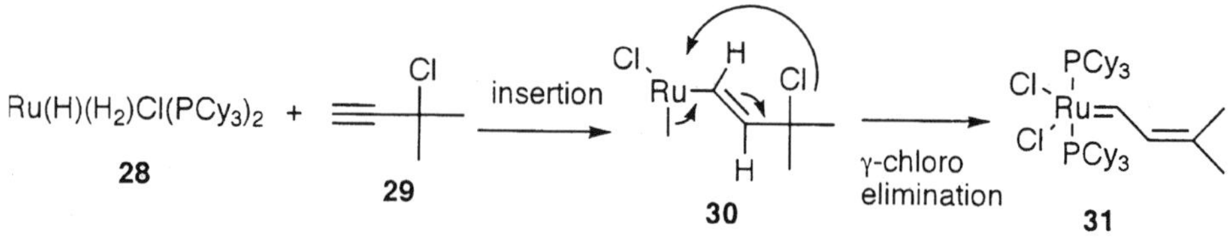

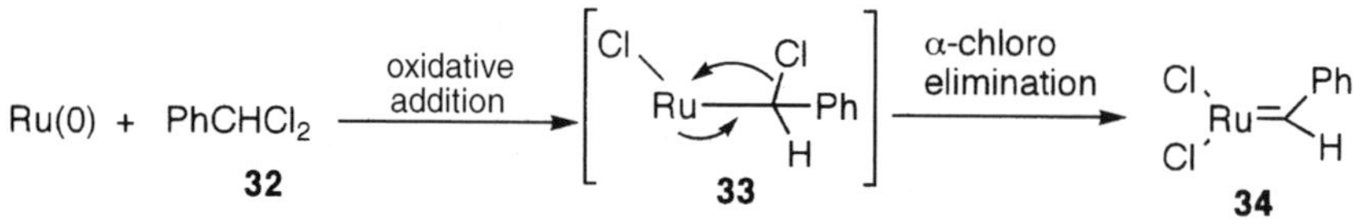

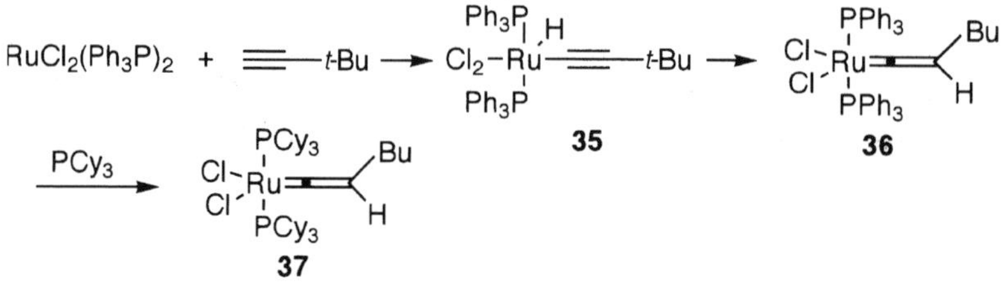

8.2.3   Classification of Alkene Metathesis

Alkene metatheses useful for organic synthesis can be classified into four types.

8.2.3.1   *Homometathesis*

The symmetric alkenes **11** and **15** are formed by homometathesis of the unsymmetric alkene **6**. Alkene metathesis is an equilibrium reaction, and the homometathesis of internal alkene **6** may be a useful one only when separation of the products **11** and **15** from the starting alkene **6** is easy, namely when R_1 and R_2 are clearly different functional groups.

Homometathesis of cyclic alkenes affords polymers that have macrocycles by ring-opening metathesis polymerization (ROMP). The polymerization of cyclopentene, carried out by the Natta group in 1964, was the first example of ROMP [3]. This would be a good synthetic method for macrocyclic compounds, if the ring-opening metathesis (ROM) of cycloalkene **38** could be stopped at an earlier stage. The 2*n*- and 3*n*-membered oligomeric rings **39** and **40** may be prepared, rather than polymer **41** from *n*-membered ring **38**.

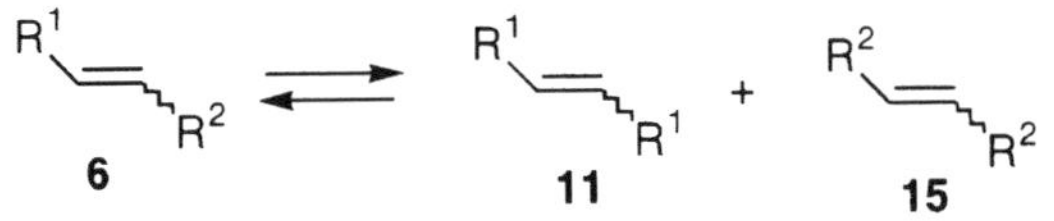

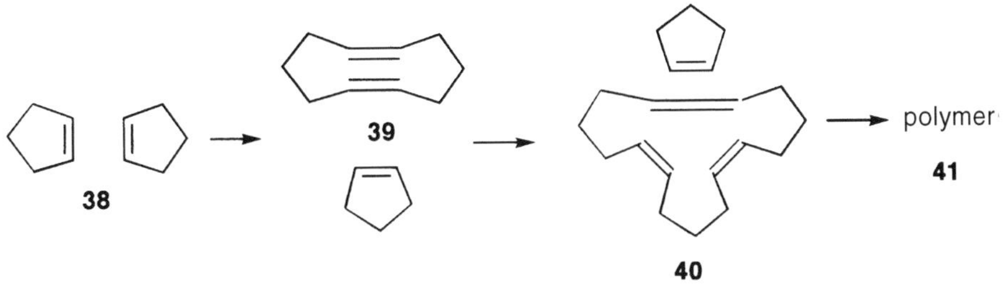

8.2.3.2   *Cross-metathesis of two alkenes and ethenolysis*

Cross-metathesis of two different alkenes **11** and **42** usually produces a mixture of products **6** and **15**. However, depending on the functional groups R^1 and R^2, the cross-product **6** is obtained with high selectivity rather than the homoproduct **15** from **11** and **42**. Some terminal alkenes, such as allylstannane [16], acrylonitrile [17,18] and allylsilane [19], undergo clean cross-metathesis to give cross-products **6** as the main product, rather than homoproducts **15**. Cross-metathesis of the cyclic alkenes **43** with terminal alkenes **42** can be used for the synthesis of dienes **44**.

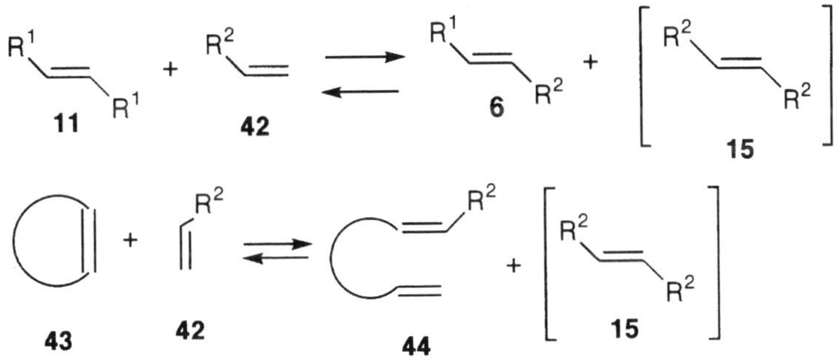

A very useful cross-metathesis is the reaction involving ethylene, which is called ethenolysis. Reaction of ethylene with internal alkenes produces the more useful terminal alkenes. Two terminal alkenes **45** and **42** are formed from the unsymmetric alkene **6** and ethylene. The symmetric alkenes **11** are converted to single terminal alkenes **45**. The terminal dienes **46** are formed by ethenolysis of the cyclic alkenes **43**.

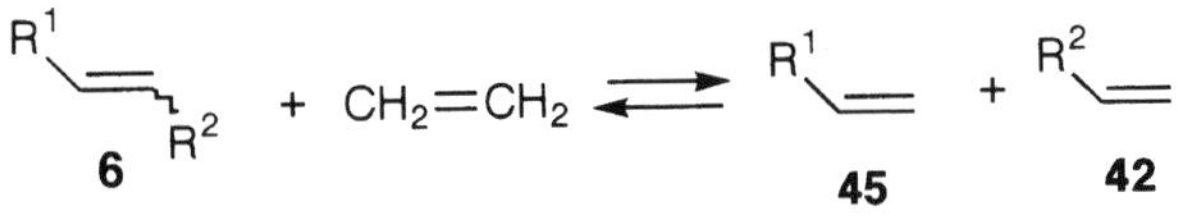

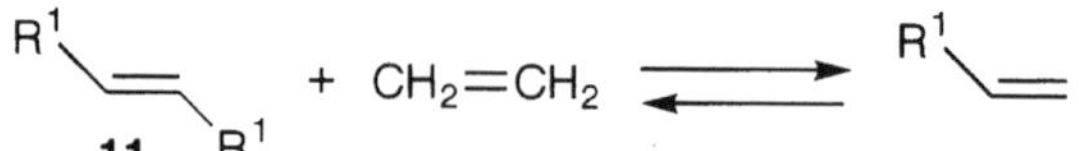

8.2.3.3   Ring-closing metathesis (RCM)

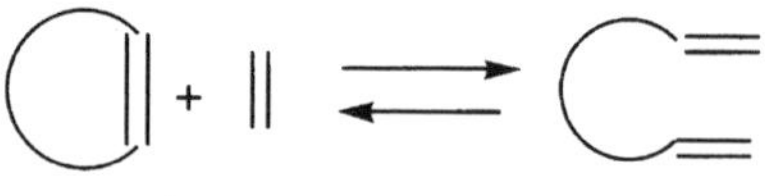

Dienes are cyclized by intramolecular metathesis. In particular, cyclic alkenes **43** and ethylene are formed by the ring-closing metathesis of the α,ω-diene **46**. This is the reverse reaction of ethenolysis. Alkene metathesis is reversible, and usually an equiliubrium mixture of alkenes is formed. However, the metathesis of α,ω-dienes **46** generates ethylene as one product, which can be removed easily from reaction mixtures to afford cyclic compounds **43** nearly quantitatively. This is a most useful reaction, because from not only five to eight membered rings, but also macrocycles can be prepared by RCM under high-dilution conditions. However, it should be noted that RCM is an intramolecular reaction and competitive with acyclic diene metathesis polymerization (ADMET), which is intermolecular to form the polymer **47**. In addition, the polymer **47** may be formed by ROMP of the cyclic compounds **43**.

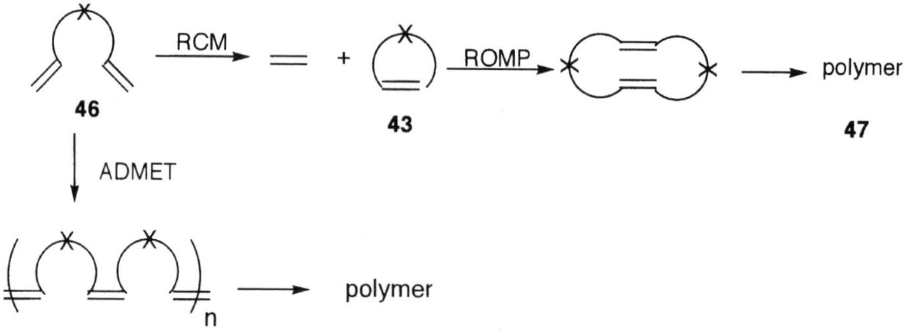

8.2.3.4   Ring-opening–closing metathesis

Intramolecular metathesis of the cyclic–acyclic diene **48** affords the rearranged products **50** by reconstruction of the ring via **49**.

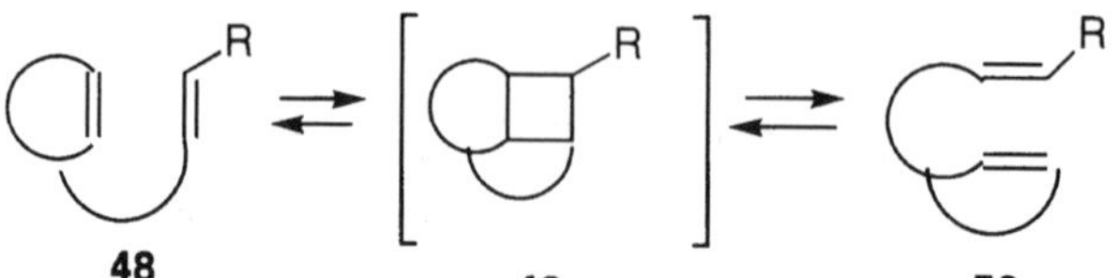

8.2.4   Synthetic Applications of Alkene Metathesis

8.2.4.1   Homometathesis

The complex $WOCl_4-Cp_2TiMe_2$ was used for the metathesis of ethyl oleate (**51**) to give diethyl 9-octadecenedioate (**52**). Civetone (**53**) was synthesized by the Dieckmann condensation of this diester, followed by decarboxylation [20]. Homometathesis of terminal alkenes is useful, because it yields symmetric internal alkenes and ethylene, which can be removed easily. Metathesis of 10-undecenoate (**54**) proceeds smoothly to give the diester **55** [20].

Homometathesis of cyclic alkenes affords larger rings. For example, cyclohexadecadiene (**57**) was prepared by liquid-phase metathesis of cyclooctene (**56**) using WCl_5/Et_3Al [21], or in the gas phase by short-time contact of cyclooctene to a supported catalyst of Re_2O_7 activated with Me_4Sn [22]. Cyclohexadecenone (**58**), a muscone-type perfume, is produced commercially from **57** by selective oxidation.

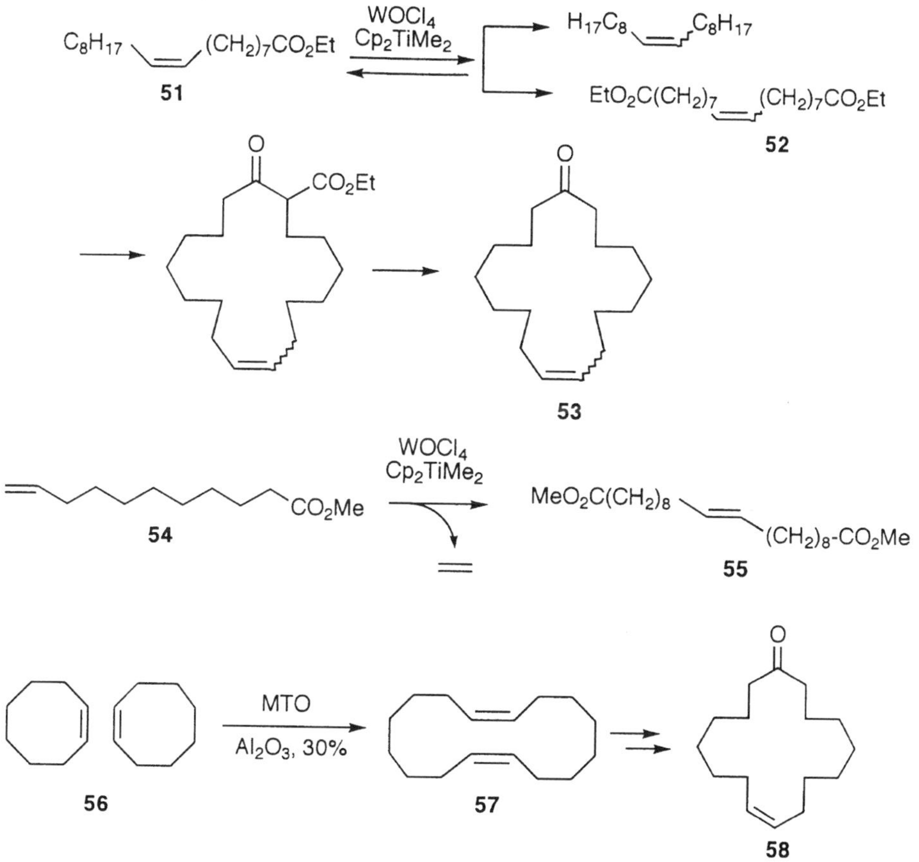

8.2.4.2   Cross-metathesis using terminal alkenes or ethylene

Formation of an equilibrium mixture of ethylene and 2-butene by the metathesis of propylene is a convenient process for controlling the relative amounts of ethylene and propylene in the petrochemical industry. The process is called the 'triolefin process' and carried out in gas phase using a supported catalyst of molybdenum oxide [6].

$$\text{=\!\!\!/CH}_3 \quad \underset{\text{Mo}}{\rightleftharpoons} \quad = \quad + \quad \text{CH}_3\text{=\!\!\!\diagdown CH}_3$$

The Mo-catalysed cross-metathesis of acrylonitrile (**59**) [17,18] and allylsilane (**60**) [19] with alkenes **61** and **62** produced cross-products **63** and **64** with high selectivity. Reaction of 1-octene with 2 equivalents of styrene (**65**) afforded **66** in 89% yield. Only small amounts of stilbene (**68**) and **67** as the homoproducts were formed [23].

$$\text{=\!\!\!\diagdown CN} \quad + \quad \overset{\text{61}}{\underset{\text{CO}_2\text{Me}}{\bigg|}} \quad \xrightarrow[\text{92\%}]{\text{Mo 19}} \quad \overset{\text{63}}{\underset{\text{CO}_2\text{Me}}{\bigg|}}\text{CN} \quad + \quad =$$

59

61

63

$$\diagup\!\!\!\diagdown\text{TMS} \quad + \quad \diagup\!\!\!\diagdown\text{CO}_2\text{Me} \quad \xrightarrow[\text{87\%}]{\text{Mo}} \quad \text{MeO}_2\text{C}\diagup\!\!\!\diagdown\!\!\!\diagup\text{TMS} \quad + \quad =$$

60   **62**   **64**

$$\text{Ph}\diagdown\!\!= \quad + \quad \text{C}_6\text{H}_{13}\diagup\!\!= \quad \underset{}{\overset{\text{Mo 19}}{\rightleftharpoons}} \quad \left[\begin{array}{c}\text{C}_6\text{H}_{13}\\\square\!\!-\!\text{Mo}\\\text{Ph}\end{array}\right] \quad \longrightarrow$$

65

$$\text{Ph}\diagdown\!\!=\!\!\diagup\text{C}_6\text{H}_{13} \quad + \quad \text{C}_6\text{H}_{13}\diagup\!\!=\!\!\diagdown\text{C}_6\text{H}_{13} \quad + \quad \text{Ph}\diagdown\!\!=\!\!\diagup\text{Ph}$$

66 89%   **67** 2%   **68** trace

Ethenolysis is synthetically very useful. The reaction of stilbene (**68**) with ethylene is attracting attention as a potential commercial process for styrene (**65**). The α,ω-dienes **46** are formed from cyclic alkenes **43** and ethylene. Ethenolysis of the bicyclo[2.2.0]hexene **71**, formed from **69** via **70**, afforded the 1,5-diene **72**, which underwent Cope rearrangement to give the cyclooctadiene **73** [24].

$$\text{Ph}\diagup\!\!=\!\!\diagup^{\text{Ph}} \quad + \quad = \quad \longrightarrow \quad 2 \; \text{Ph}\diagup\!\!=$$

68   **65**

8.2.4.3   Preparation of cyclic compounds by RCM of dienes

Intramolecular metathesis of dienes leads to ring-closing metathesis (RCM), offering a useful synthetic route to cyclic compounds. The Ru and Mo catalysts are extremely useful for RCM of dienes as a synthetic method for cycloalkenes, cylic ethers and cyclic amines, from small to large rings. Carboxylic acids, aldehydes and amine hydrochlorides do not destroy the catalytic activity. However, the reaction is competitive with the intermolecular metathesis to form polymeric products by acyclic diene metathesis polymerization (ADMET), and further complicated by possible ring-opening metathesis polymerization (ROMP). Thus high dilution condition is necessary for selective RCM.

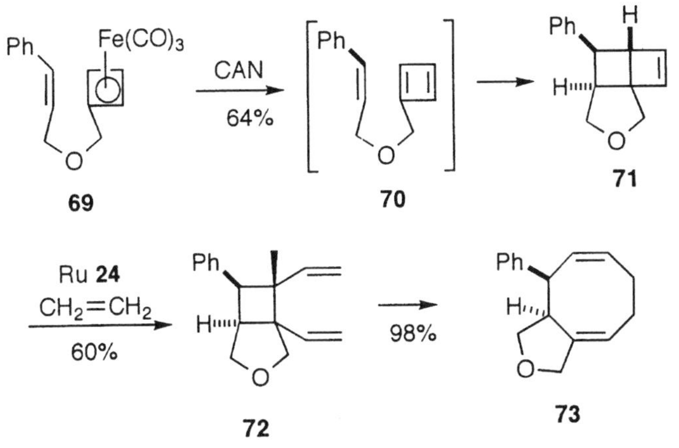

Preparation of five to eight-membered rings   A number of common ring amines and ethers have been prepared [25]. The five-membered amine is obtained from the diallylamine **74** [26,27]. (+)-Citronelene (**75**), 0.75M in toluene, was converted to 3-methylcyclopentene quantitatively in 30 min with retention of stereochemistry using

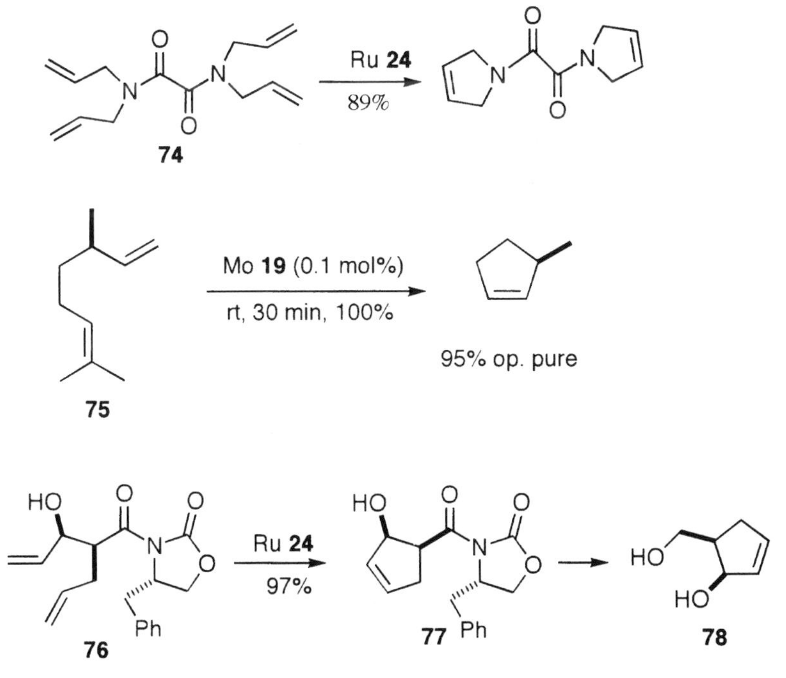

0.1 mol% of Mo catalyst [28]. However, racemization occurred at higher concentrations. Ru-catalysed RCM of the (*S*)-4-benzyl-2-oxazolidinone **76** afforded the cyclopentene **77**, which was converted to 5-(hydroxymethyl)-2-cyclopenten-1-ol (**78**) with 99.5% ee [29].

Enantioselective RCM is achieved using the chiral Mo complex **79** [30]. Kinetic resolution occurred in the reaction of the racemic diene **80** catalysed by **79**, and the cyclized product **81** with 93% ee was obtained, and the unreacted diene **80** (19%) of 99% ee was recovered. Also the optically active dihydrofuran **83** with 93% ee was obtained in 85% yield by enantioselective desymmetrization through RCM of triene **82** using the Mo complex **79** [30a].

79

Ar = 2,5-(*i*-Pr)$_2$C$_6$H$_3$

80
unreacted
19%, 99% ee

81
43%, 93% ee

82

83
93% ee

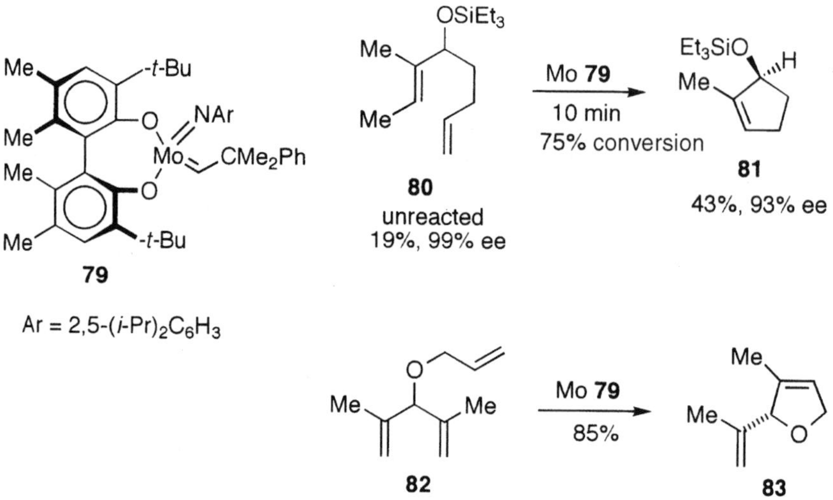

The six-membered ring **85** is obtained from the allylamine **84** [31]. The sulfur-containing ring **87** was obtained from **86** using the Mo catalyst. The Ru catalyst is not active for this reaction [32]. The (*S, R*)-chromene derivative **89** was obtained in 97% yield by the Mo-catalysed intramolecular metathesis of (*S,R*)-cycloheptenyl styrenyl ether **88** under an atmosphere of ethylene. In the absence of ethylene, **89** and its dimer were obtained. The enantioselective total synthesis of (*S,R,R,R*)-nebivolol (**90**) has been carried out from **89** [33]. No cyclization of the cyclopentene **91** was observed, because the highly strained cyclobutane intermediate **92** is difficult to form.

84 **85**

86 **87**

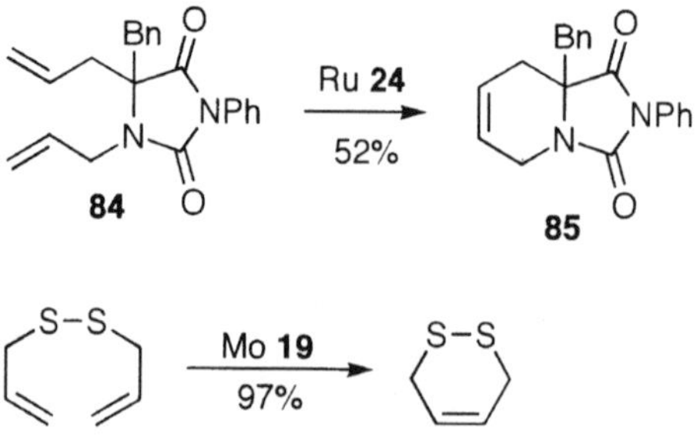

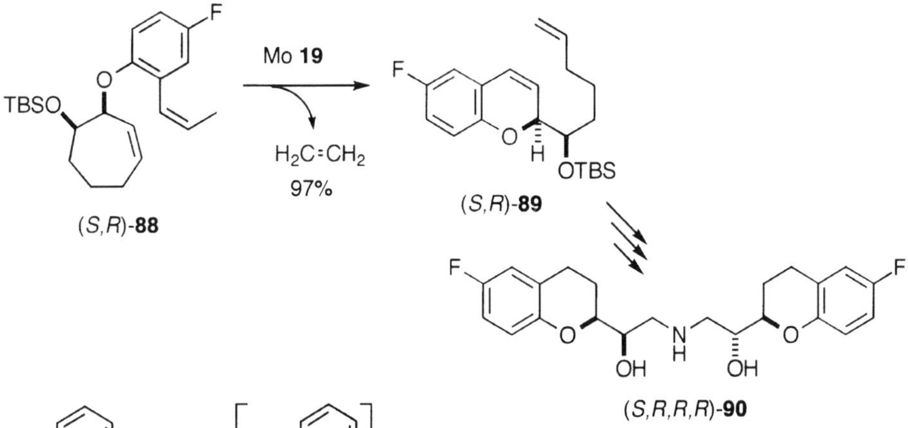

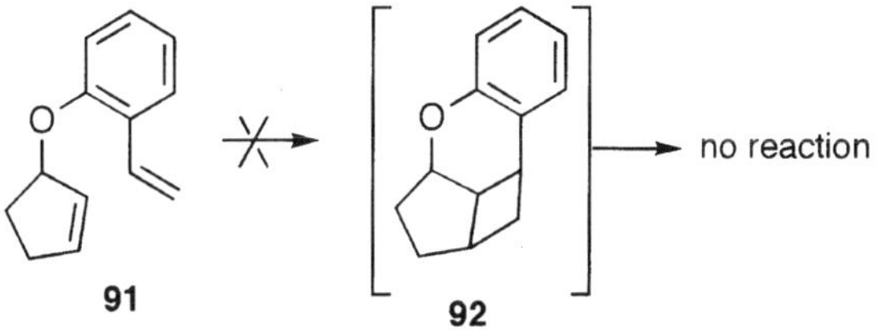

The dihomoallyl ether (**93**) is converted to the seven-membered ether **94** [25]. The hydroazulene **96** is obtained by the metathesis of the 1,8-diene **95** using MTO as a catalyst in solution [34]. Similarly, the seven-membered lactam **98** is prepared from **97** [35].

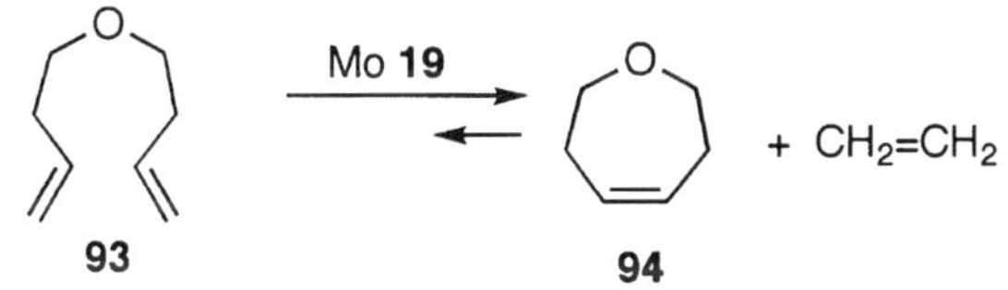

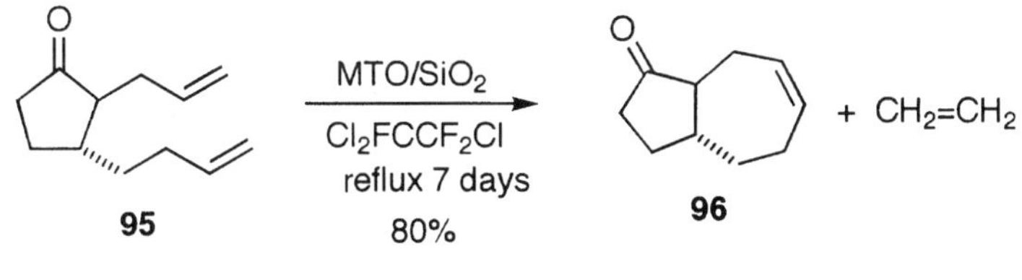

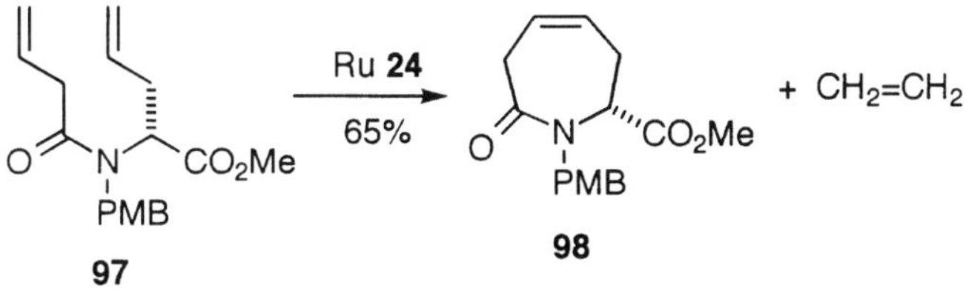

Preparation of eight-membered rings is expected to be rather difficult by conventional methods. However, the eight-membered precursor **100** of the antitumer agent FR900482 is prepared in a good yield by the metathesis of terminal diene **99** catalysed by the Mo complex [36]. The eight-membered lactam **102** was prepared in 63% yield from terminal diene **101** using the Mo-based catalyst **19** in the synthesis of Manzamine A (**103**) [37].

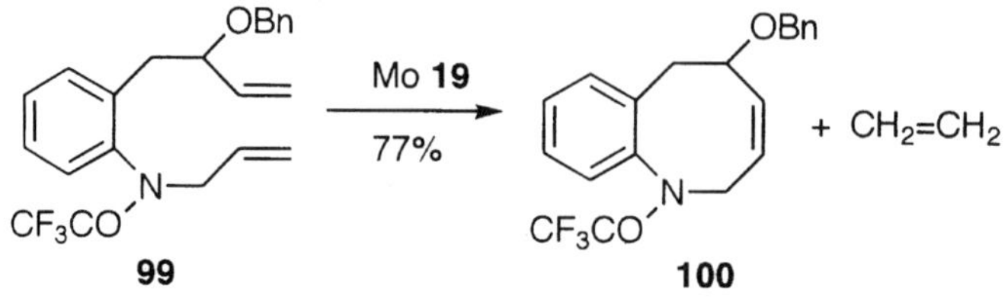

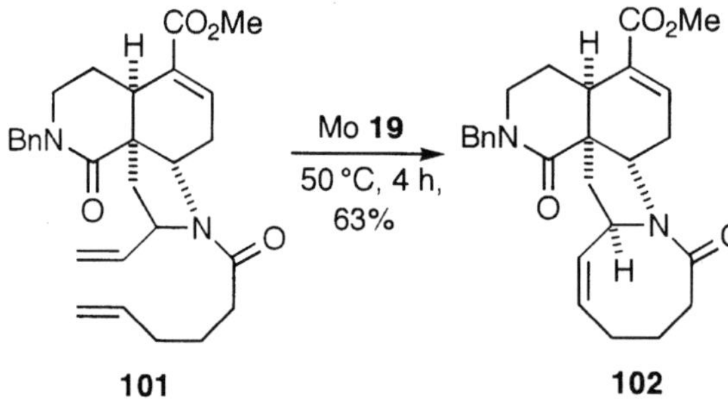

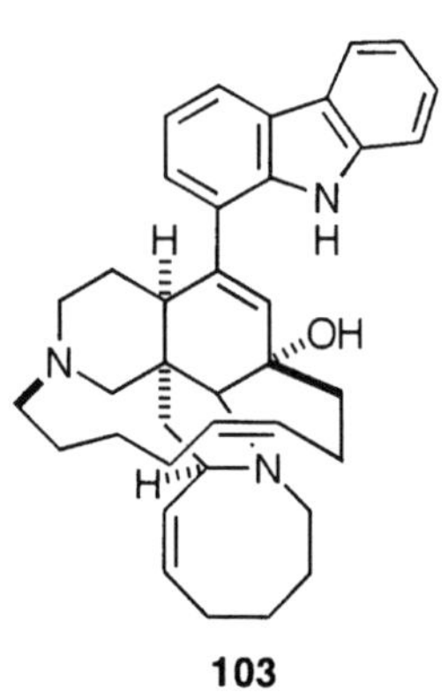

Preparation of macrocyclic compounds Development of efficient RCM, catalysed by Ru and Mo complexes, has revolutionized the synthesis of macrocyclic compounds. As an early example of RCM, intramolecular metathesis of oleyl oleate (**104**) using $WCl_6 - Cp_2TiMe_2$ afforded the 19-membered lactone **105**, although yield was not satisfactory [20].

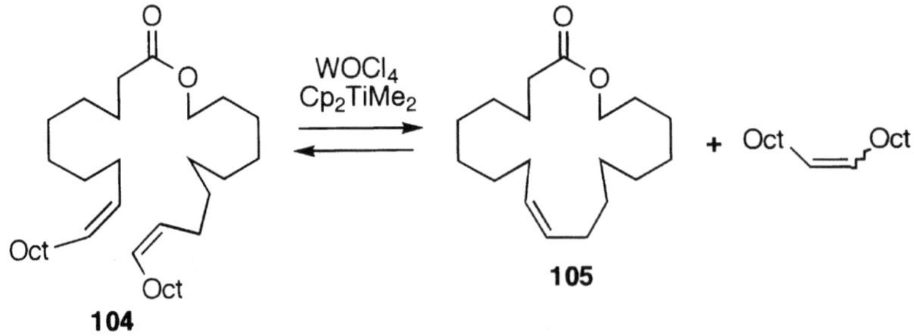

The 12-membered lasiodiplodine ring **107** was prepared in 94% yield by slow addition of terminal diene **106** to a solution of the Ru catalyst. The equilibrium was shifted to completion by bubbling with argon to remove the ethylene [38].

The disaccharide fragment of tricolorin A **110**, which is a 19-membered macrolactone, was synthesized by efficient RCM of terminal diene **108** to give **109** and its hydrogenation in 77% yield [39]. The presence of sugar groups as a polar 'relay' substituent, its proper distance to the alkene groups, and low steric hindrance close to the double bonds are decisive parameters for the efficient RCM of **108**.

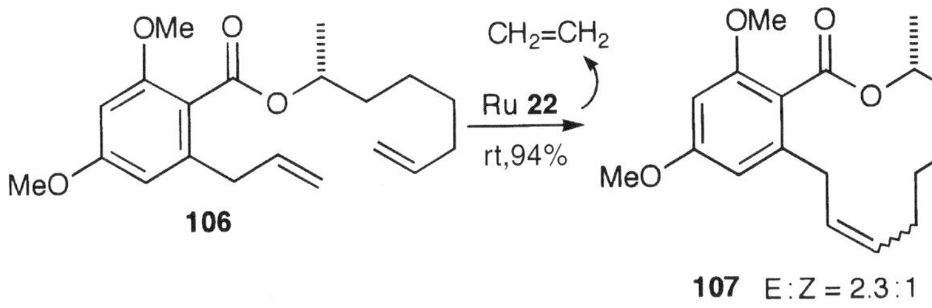

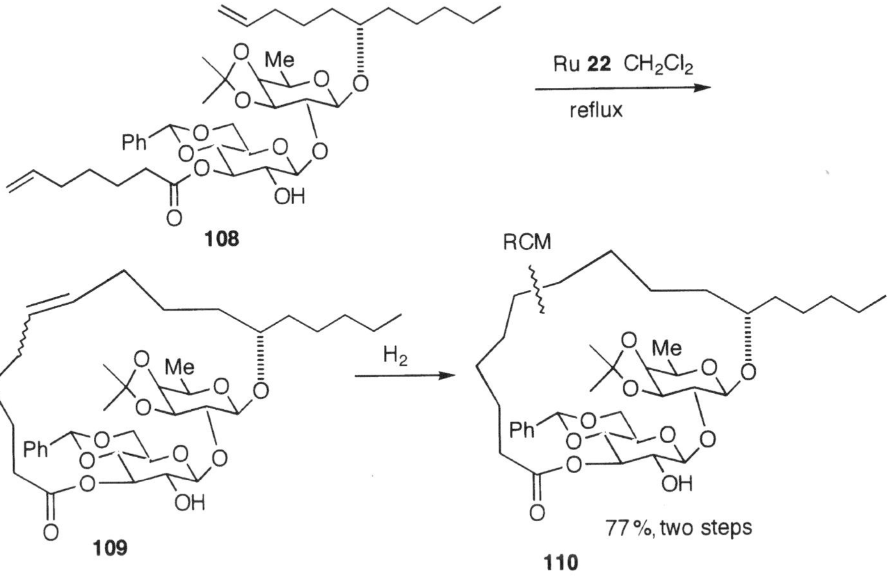

RCM of the (*S,S,S*)-tetrapeptide **111**, catalysed by Ru complex **24**, gives the 14-membered (*S,S,S*)-macrocycle **112** in 60% yield. The covalently stabilized β-turn structure in the tetrapeptide **112** serves as a model for **113** [40]. The correct configuration of peptide **111** is crucial for efficient RCM, and the allyl groups in the tetrapetpide, having only the (*S,S,S*) structure **111**, gave **112** in high yield. Hydrogen bonding facilitates the cyclization. As another example, the 14-membered lactam ring **115** which is the aglycon of Sch 38516 and fluvirucin B, was obtained in 90% yield from diene amide **114** using Mo catalyst **19** [41].

The 16-membered polyfunctionalized antitumor macrolide, epothilone A (**118**), was synthesized by the Ru-catalysed RCM of **116** to give **117**, followed by epoxidation [42–44]. The (*E*):(*Z*) ratio in **117** may be changed by protecting groups.

Substrates that are devoid of any conformational constraint can be efficiently cyclized by RCM to macrocyclic compounds. For example, the 21-membered lactone **120** was prepared in 71% yield by slow addition of 10-undecenyl 10-undecenoate (**119**) to a solution of Ru complex **24**. 20-Icosanolide was obtained by hydrogenation of **120** [45].

Efficient synthesis of 7–20-membered crown ethers **122** having a *trans*-alkene bond was achieved by the Ru-catalysed metathesis of diallyl ethers **121** [46]. A remarkable

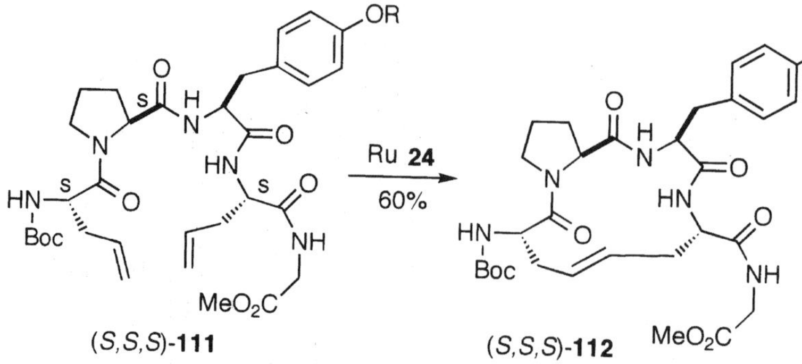

(S,S,S)-**111** $\xrightarrow[\text{60%}]{\text{Ru }\textbf{24}}$ (S,S,S)-**112**

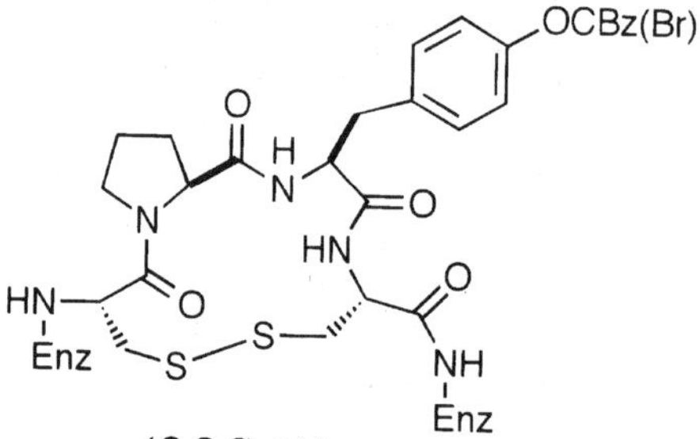

(S,S,S)-**113**

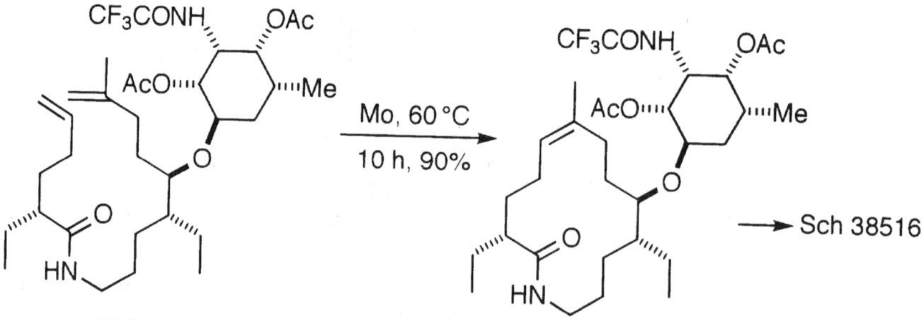

114 $\xrightarrow[\text{10 h, 90%}]{\text{Mo, 60 °C}}$ **115** $\longrightarrow$ Sch 38516

116 $\xrightarrow[\text{94%}]{\text{Ru }\textbf{24}}$ **117**   E:Z = 1:1

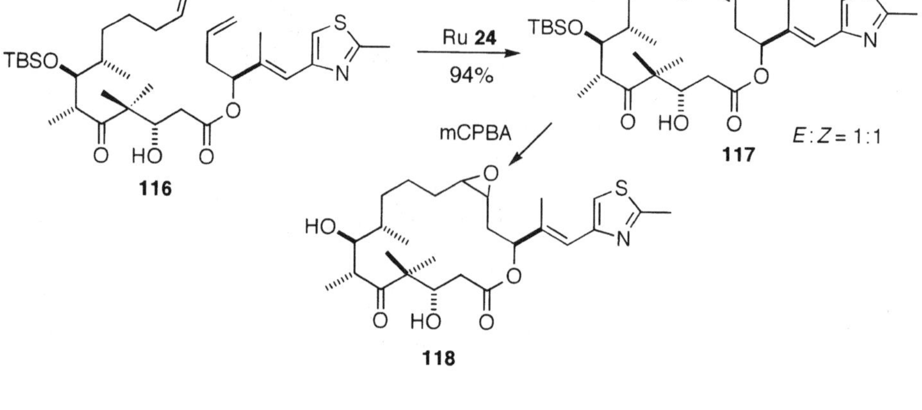

mCPBA

118

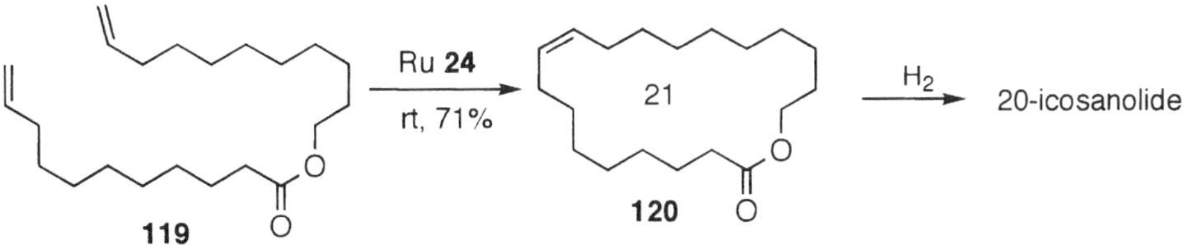

template effect was observed in the formation of the crown ethers [47]. Crown ether **124** with a *cis* bond was obtained in 95% yield by RCM of **123** in the presence of 5 equivalents of LiClO$_4$. In the absence of the Li salt, the yield was 39% and a *cis–trans* mixture was formed. RCM of neat **123** afforded the polymer **125** by ADMET. Ru-catalysed depolymerization of **125** in 0.02 M solution produced **124** in the presence of the Li salt.

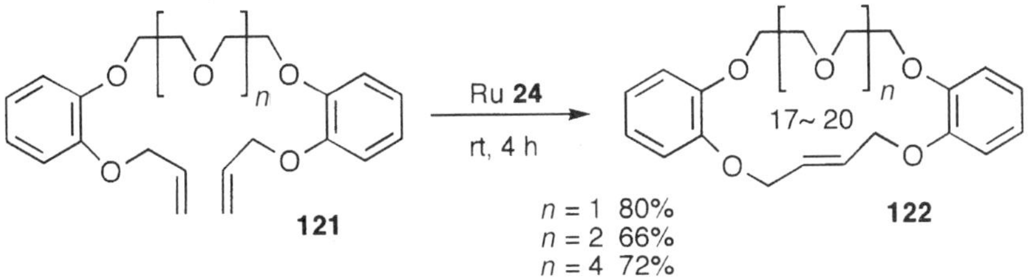

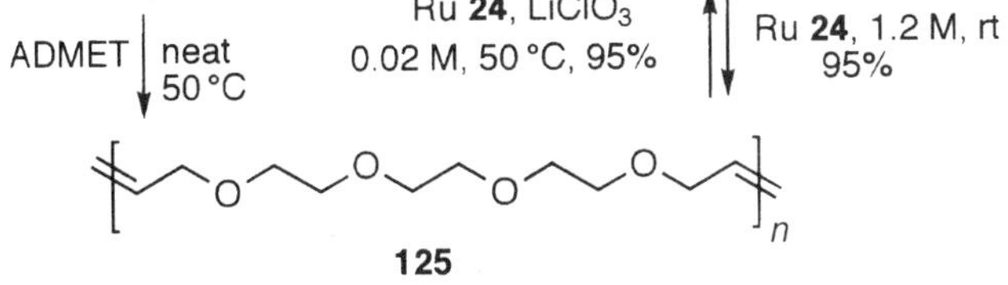

124	*cis:trans*
LiClO$_4$ 95%	100:0
NaClO$_4$ 42%	62:38
KClO$_4$ 36%	36:64
none 39%	38:62

8.2.4.4 *Ring-opening–closing metathesis*

Ru-catalysed domino ring-opening–closing metathesis of strained cyclic alkenes with terminal alkenes gives the less-strained rings. Thus cyclobutene **126** is converted to two five-membered ether rings **127**. Although no metathesis of cyclohexene usually occurs, under high-dilution conditions diallyl ether **128** undergoes the smooth ring-opening–closing metathesis to afford **129** in 73% yield with evolution of ethylene. Also the *N,N*-diallyl diamide of the 1,4-dihydronaphthalenedicarboxylic acid **130** was transformed to **135** in 95% yield as explained by steps **131–134** [48].

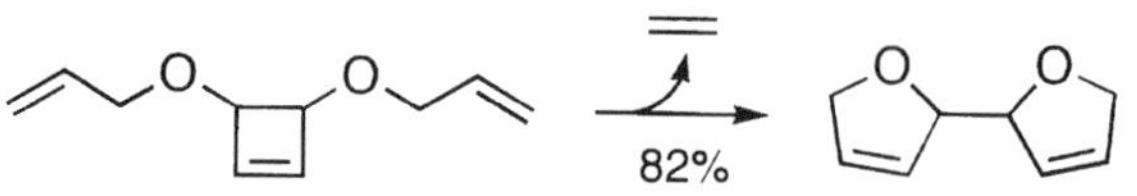

126   82%   **127**

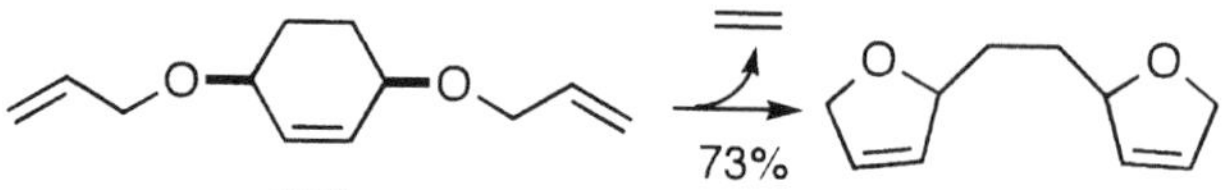

128   73%   **129**

130   **131**   **132**

133   **134**   **135**   95%

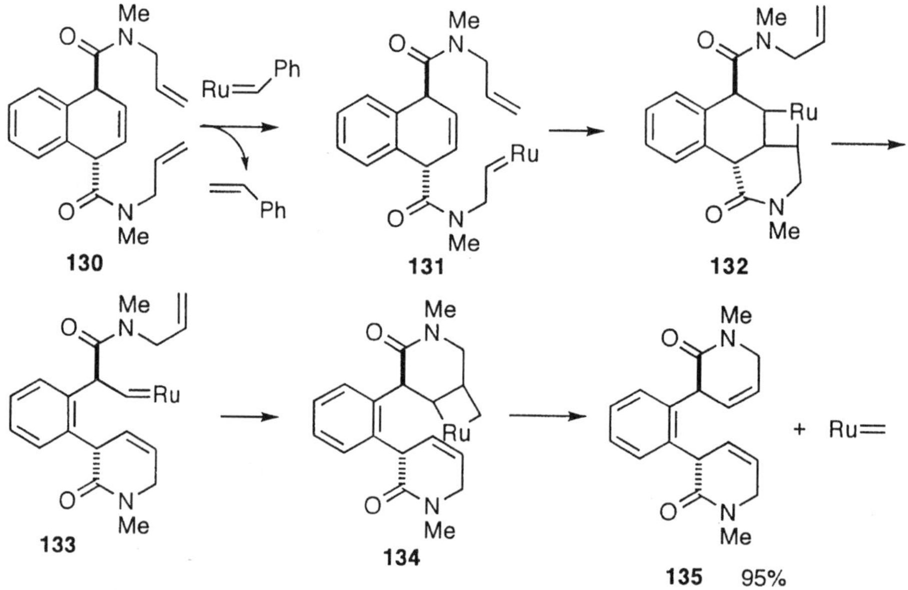

8.2.5   Metathesis of Akynes and Enynes

8.2.5.1   *Metathesis of alkynes*

Although studies are still limited, alkynes undergo metathesis using W and Mo catalysts. Addition of phenol in an equivalent amount to the alkynes gives better results [49]. The metathesis of unsymmetric alkyne **136** can be explained by the formation of carbyne complex **137**, and proceeds via reversible formation of metallacyclobutadienes **138a** and **138b** as intermediates to afford the symmetric alkyne **140** and the carbyne complex **139**. The metathesis of alkynes **141** bearing an ester group was carried out with $Mo(CO)_6$ as a catalyst in the presence of *p*-chlorophenol to give two symmetric internal alkynes [50]. Cross-metathesis of diphenylacetylene (**142**) in a large excess of the symmetric alkyne **143** gave unsymmetric alkyne **144** in 74% yield [51]. RCM of the internal diyne ester **145** catalysed by tungsten alkylidene complex **146** under high dilution conditions afforded lactam **147** in 69% yield [51a]. Based on this reaction, the lactam with a *cis* double bond can be prepared after hydrogenation.

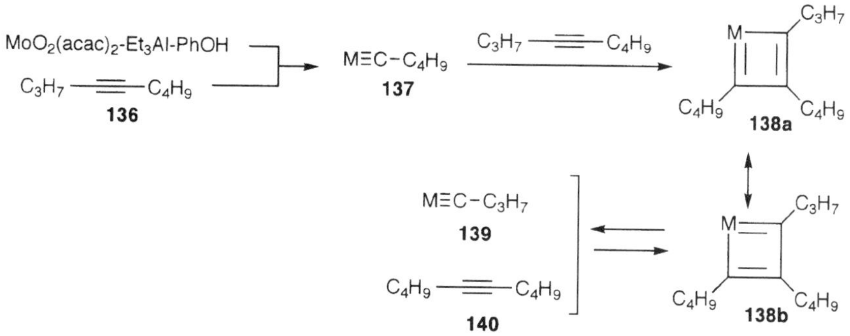

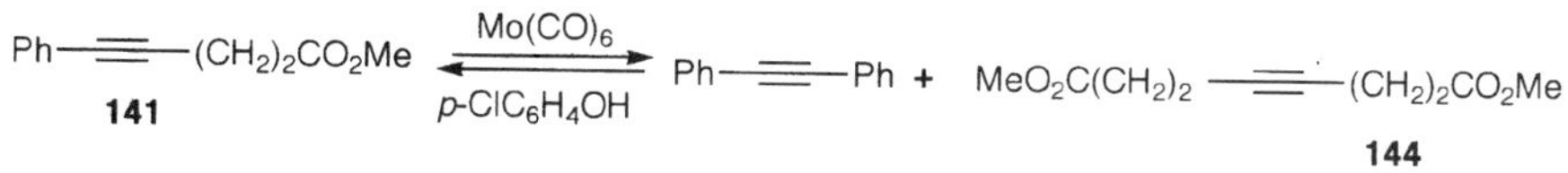

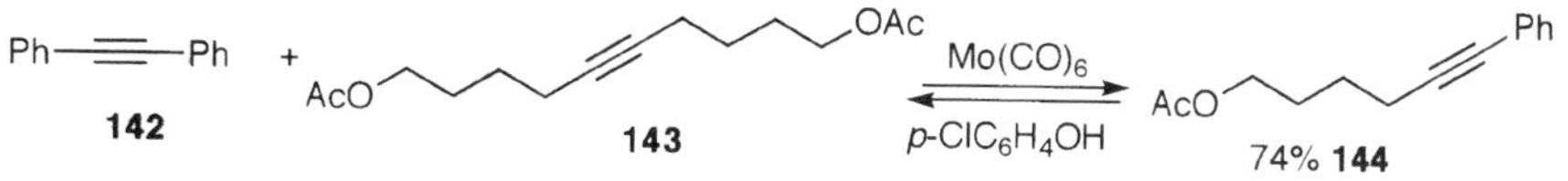

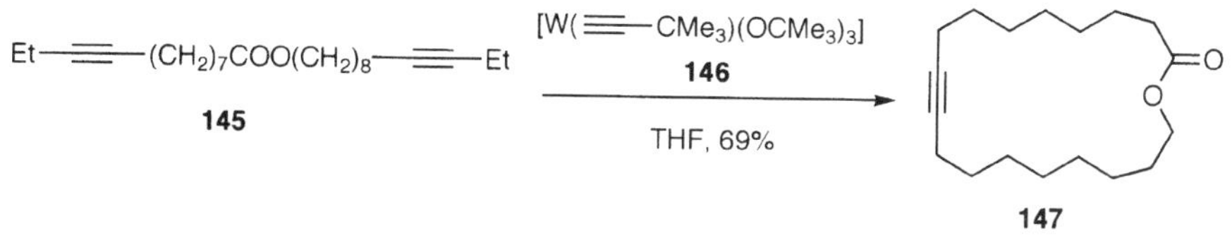

8.2.5.2 *Domino metathesis of enynes*

Reaction of the carbene complex **148** with alkyne affords vinylcarbene **150** via metallacyclobutene **149**. In the intramolecular reaction of enyne **152**, catalysed by carbene complex **151**, the triple bond is converted to vinylcarbene **153** which then reacts with the double bond to give the conjugated diene **154**. Generation of **154** is expected by the formation and cleavage of cyclobutene **155** as a hypothetical intermediate. Based on this reaction, Ru-catalysed intramolecular metathesis of enyne **156** gave the N-containing cyclic diene **157**, from which (−)-stemoamide (**158**) was synthesiszed. The reaction can be understood by assuming the formation of the hypothetical cyclobutene **159** from **156** [52].

Selective conversion of acyclic dienynes to fused bicyclic rings containing five-, six- and seven-membered rings is efficiently catalysed by Ru complex **22**. The reaction of dienyne **160** having two terminal alkene chains gave the two products **162** and **164**

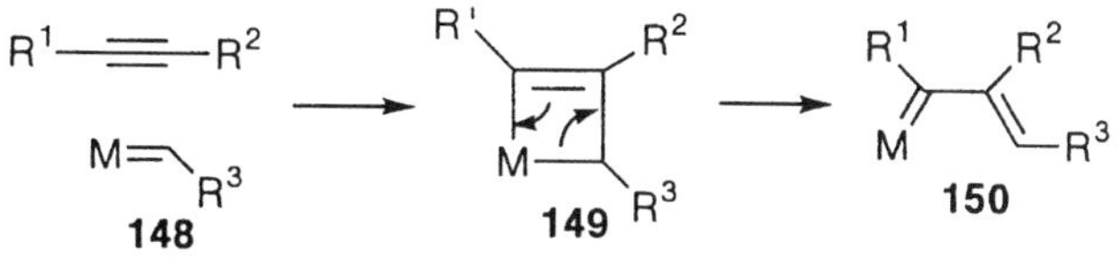

162 : 164 = 1 : 1

in 1:1 ratio in 86% yield via **161** and **163**. However, metathesis of dieneyne **165**, with the terminal and internal double bonds, starts from the terminal alkene, and leads selectively to the bicyclic structure **166** [53]. Also, the tricyclic compound **171** was obtained from **167** in 70% yield. The reaction can be understood by the following sequence starting from **167** via **168, 169** and **170**. This sequence of the domino reactions is a good example of useful metathesis chemistry [54].

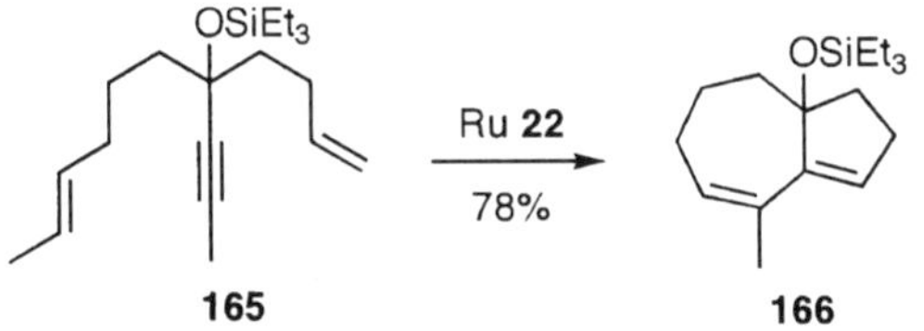

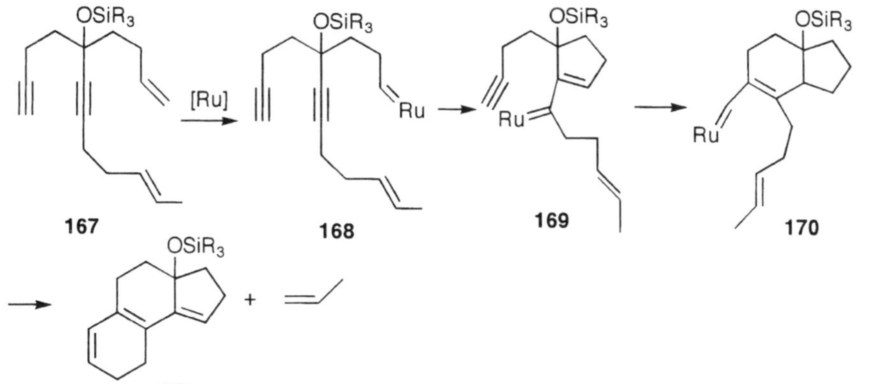

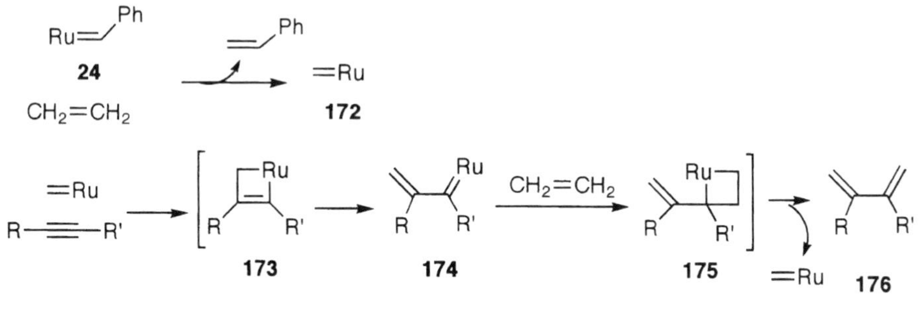

Unlike the efficient intramolecular domino reaction of enynes, intermolecular enyne metathesis is of limited application because a complex mixture is obtained. Selective intermolecular enyne metathesis has been carried out using ethylene as an alkene, offering a useful synthetic method for the conjugated dienes **176**. At first the methylene complex **172** is generated by the reaction of ethylene with Ru complex **24**, and the ruthenacyclobutene **173** is formed, which is cleaved to give **174**. Again ruthenacyclobutane **175** is formed by the reaction of ethylene, from which the conjugated diene **176** is produced. As shown by **177**, diene **178** is prepared by introduction of two methylene groups from ethylene to the alkyne bond. As an example, the Ru-catalyzed reaction of the diacetate of butynediol (**179**) under an ethylene atmosphere affords the conjugated diene **180** in high yield [55].

8.3 Carbonyl Alkenation Reactions via Carbene Complexes

Wittig-type alkenation of the carbonyl group is possible with Ti carbene compounds [56]. The reaction is explained by the formation of nucleophilic carbene complexes of Ti, although they are not isolated. In the carbonyl alkenation, the oxametallacyclobutane intermediate **182** is formed by [2+2] cycloaddition of the carbene complex **181** with the carbonyl group. This intermediate is converted to the new alkene **183** and the Ti(IV) oxo species **184**, which is a stable compound, and hence the carbonyl alkenation requires a stoichiometric amount of the Ti complex. Also, ester **185** is converted to the enol ether **187** via **186**.

The Ti carbene complex **188** is prepared easily *in situ* by the reaction of CH_2Br_2 with a low-valent Ti species generated by treatment of $TiCl_4$ with Zn. Without isolation, the complex **188** is used for the alkenation. The presence of a small amount of Pb in Zn was found to be crucial [57]. It is synthetically equivalent to '$Cl_2Ti=CH_2$'. The reaction is explained by the formation of the four-membered metallacycle **189** as an intermediate. However, in this case, the reaction may proceed probably via the formation of the dimetal compound **190**, rather than the carbene complex **188** to give **192** via **191**. The reagent reacts with ketones, but not with esters. Usefulness of this reaction is shown by application to synthesis of the prostaglandin derivative **194** without attacking other labile functional groups in **193** [58].

The Wittig reaction is carried out under strongly basic conditions, and is not possible with the five-membered keto ester **195**, because an enolate of the ketone is formed. The Tebbe reagent, as described later, reacts with both esters and ketones to give a mixture of products. Selective reaction of the ketone in **195** without attacking the ester group to give **196** is possible only with the $CH_2Br_2-TiCl_4-Zn$ reagent **190** [59].

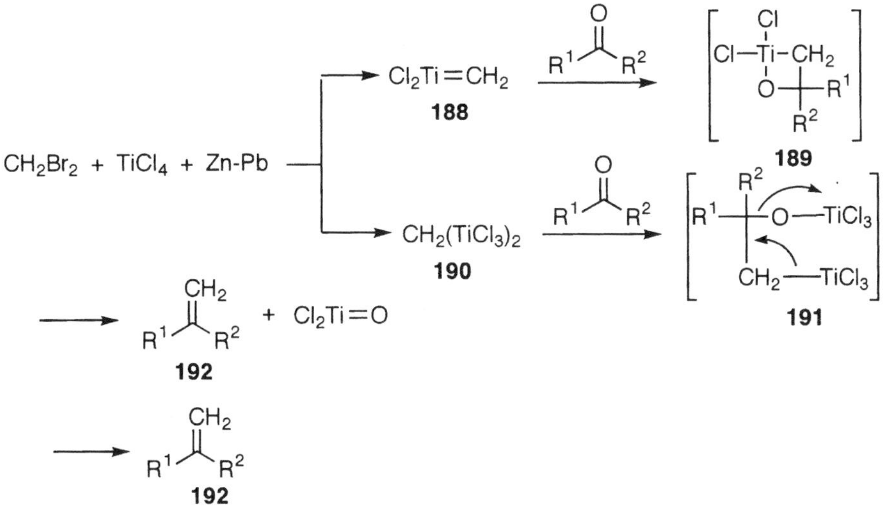

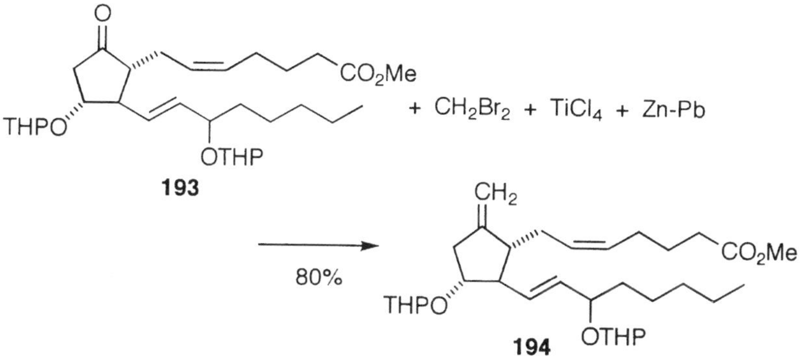

$$CH_2Br_2 + TiCl_4 + Zn\text{-}Pb$$

$$Cl_2Ti=CH_2 \quad \textbf{188}$$

$$\begin{bmatrix} Cl \\ Cl-Ti-CH_2 \\ O\!\!+\!\!R^1 \\ R^2 \end{bmatrix} \quad \textbf{189}$$

$$CH_2(TiCl_3)_2 \quad \textbf{190}$$

$$\begin{bmatrix} R^2 \\ R^1\!\!+\!\!O-TiCl_3 \\ CH_2-TiCl_3 \end{bmatrix} \quad \textbf{191}$$

$$\overset{CH_2}{\underset{R^1 \quad R^2}{\|}} + Cl_2Ti=O \quad \textbf{192}$$

$$\overset{CH_2}{\underset{R^1 \quad R^2}{\|}} \quad \textbf{192}$$

193 + CH_2Br_2 + $TiCl_4$ + Zn-Pb

$\xrightarrow{80\%}$ **194**

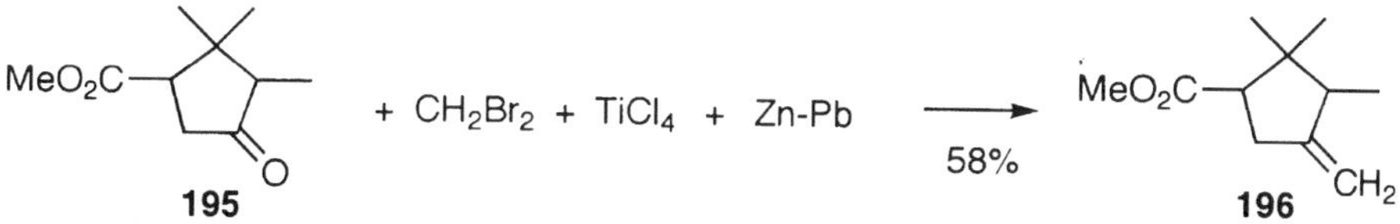

195 + CH_2Br_2 + $TiCl_4$ + Zn-Pb $\xrightarrow{58\%}$ **196**

The yields from aldehyde alkylidenation is somewhat lower due to the reductive dimerization of aldehydes with low-valent Ti. Alkylidenation of esters is possible by the reaction of 1,1-dibromoalkane, $TiCl_4$ and Zn in the presence of TMEDA to give (Z) vinyl ethers [60]. Cyclic vinyl ethers are prepared from unsaturated esters in two steps. The first step is formation of the acyclic enol ethers using a stoichiometric amount of the Ti reagent, and the second step is ring-closing alkene metathesis catalysed by Mo complex **19**. Thus the benzofuran moiety of sophora compound I (**199**, R = H) was synthesized by the carbonyl alkenation of ester in **197** with the Ti reagent prepared *in situ,* and the subsequent catalytic RCM of the resulting enol ether **198** catalysed by **19** [61].

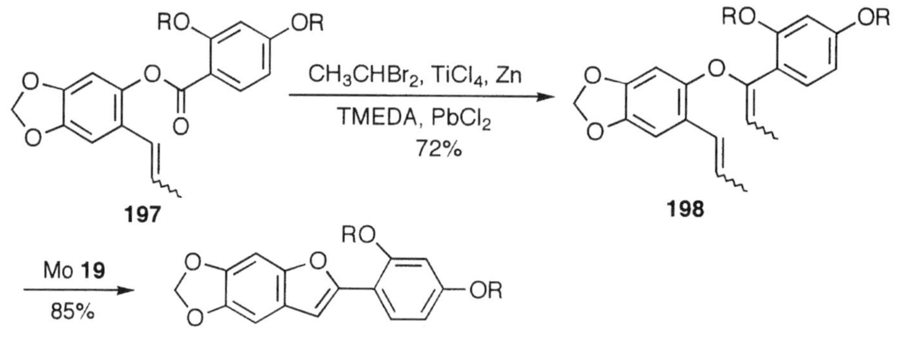

197   **198**

199

The Ti–methylene complex **200**, stabilized by forming a bridged structure with a Lewis acid, is isolated by the reaction of Cp_2TiCl_2 with Me_3Al, and called the Tebbe reagent [56,62,63]. This is synthetically equivalent to $Cp_2Ti=CH_2$ **201**. Due to high oxophilicity of Ti, it reacts smoothly with ketones, esters and lactones to form oxometallacycles **202**. It was shown that the Tebbe reagent is superior to the Wittig reagent, because methylenation of ketones takes place smoothly under nonbasic conditions, and reaction of hindered ketones proceeds without racemization of chiral ketones. This reagent reacts with esters **203** to give vinyl ethers **204** [64].

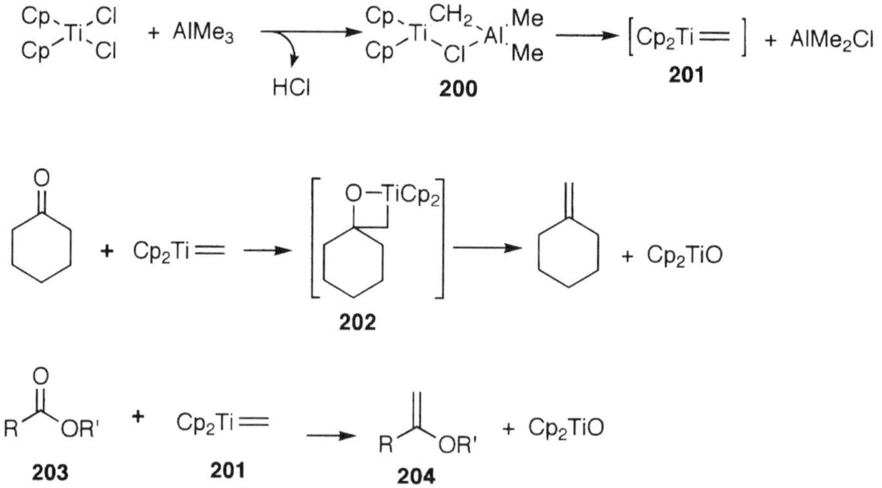

200   **201**

202

203   **201**   **204**

The Tebbe reagent reacts with some alkenes. The tricyclo[5.3.0.0] ring **207** was obtained nearly quantitatively by domino alkene metathesis and carbonyl alkenation of the norbornene-type ester **205** with the Tebbe reagent. This interesting reaction to give the intermediate **206** can be explained by the kinetic preference of the Tebbe reagent for the strained double bond over the ester. Alkenation of the ester in **206** produces **207**. Capnellene (**208**) has been synthesized by applying this reaction as a key reaction [65].

Direct conversion of the ester **209** to the cyclic enol ether **211** via enol ether **210** by ester alkenation and subsequent RCM has been achieved by using 3–6 equivalents of the Tebbe reagent or Cp_2TiMe_2 (**213**), and applied to the synthesis of polycyclic polyether systems **212** [66]

Dimethyltitanocene (**213**), called the Petasis reagent, can be used for alkenation of carbonyls (aldehydes, ketones, esters, thioesters and lactones). This reagent is prepared more easily than the Tebbe reagent by the reaction of titanocene dichloride with MeLi. However, this reagent may not be a carbene complex and its reaction may be explained as a nucleophilic attack of the methyl group at the carbonyl [67]. Alkenylsilanes are prepared from carbonyl compounds. Tri(trimethylsilyl)titanacyclobutene (**216**), as a precursor of the carbene complex **215**, is formed by the reaction of bis(trimethylsi-lylmethyl)titanocene (**214**) with bis(trimethylsilyl)acetylene under mild conditions. The lactone **217** is converted to the alkenylsilane **218** by reaction with **216** [68].

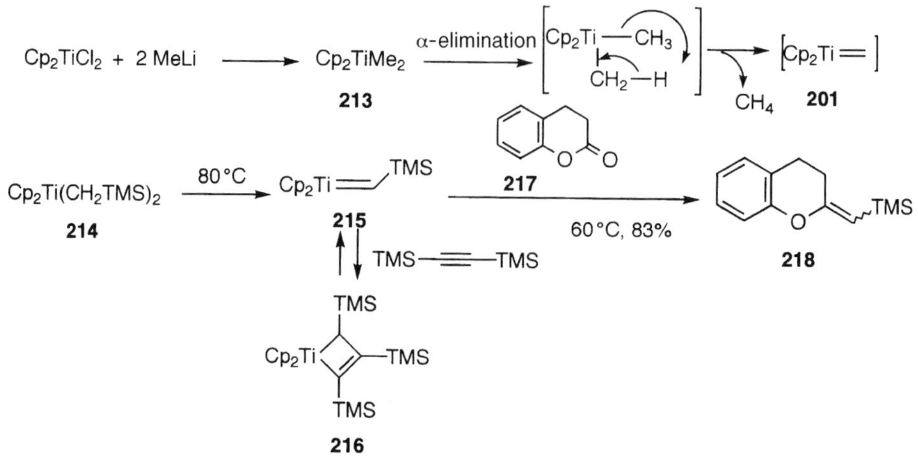

Instead of titanocene dichloride, zirconocene dichloride can be used for the preparation of the carbene complex **219**. The Lewis acidity of Zr is lower than that of Ti, and the Zr reagent **219** reacts smoothly with aldehydes and ketones, but not with esters [69].

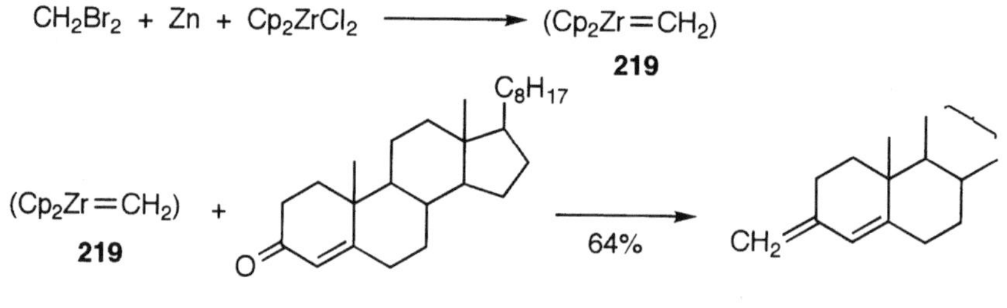

Methylenation of aldehydes and ketones can be carried out more easily by the W or Mo reagent **221**, generated *in situ* by the reaction of $MoCl_5$ or WCl_5 with MeLi via **220**. The reagent does not react with esters [70].

Formation of the Mo carbene complex **223** from **222** and the Mo carbene complex **19**, followed by the alkenation of the ketone in **223** gives seven-membered cycloalkene **224** [71]. The ester in **225** undergoes carbonyl alkenation with the more reactive W complex **20** to afford the cyclic enol ethers **226** [72].

Ethylidenation is not possible with the Tebbe-type reagents. The dimetal compound **228** can be generated from Cr(II) salts and diiodoalkanes **227**. Alkylidenation of ketones and aldehydes can be carried out with this reagent to afford **229**. Alkylidenation with other diiodoalkanes proceeds smoothly in the presence of DMF to afford the (*E*)-alkene selectively. Reaction of aldehydes with the reagent, generated from (dibromomethyl)trimethylsilane (**230**) and $CrCl_2$, offers a good synthetic route to (*E*)-alkenylsilane **231** [73]. The (*E*)-alkenyl iodides **232** can be prepared by the reaction of aldehyde, triiodomethane and $CrCl_2$.

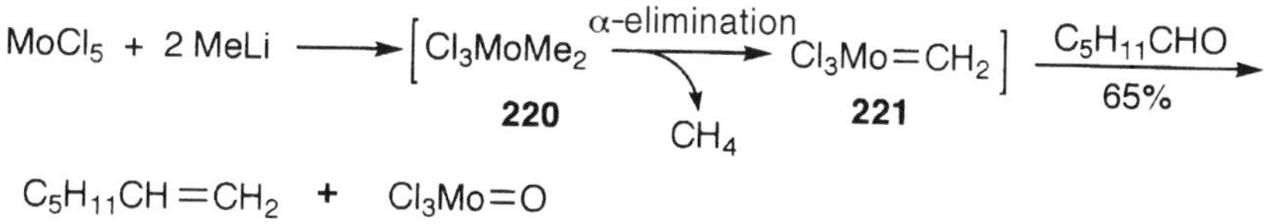

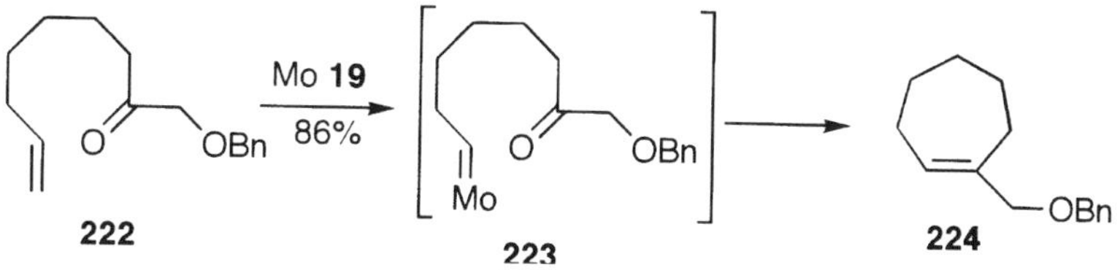

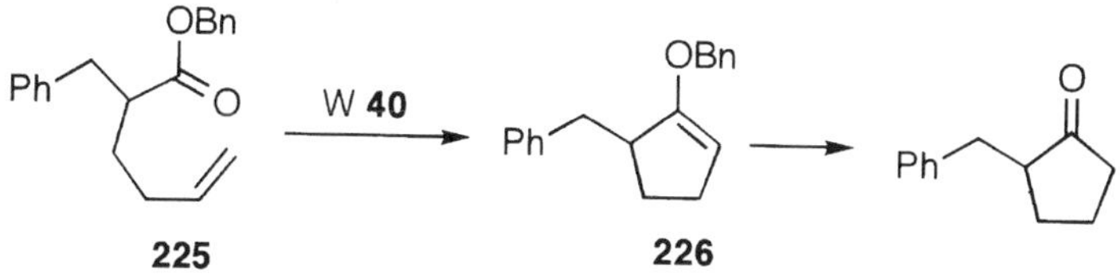

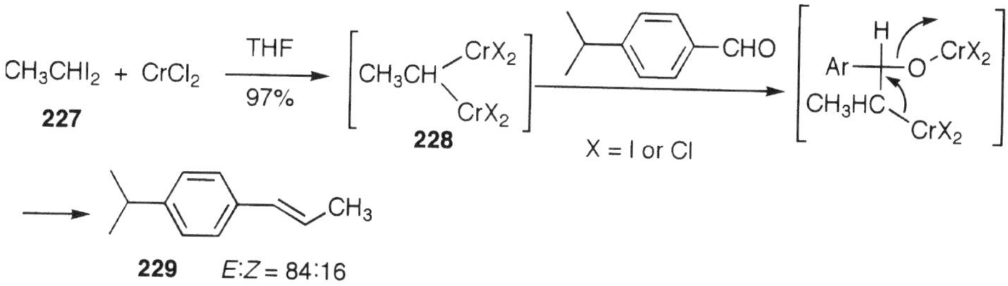

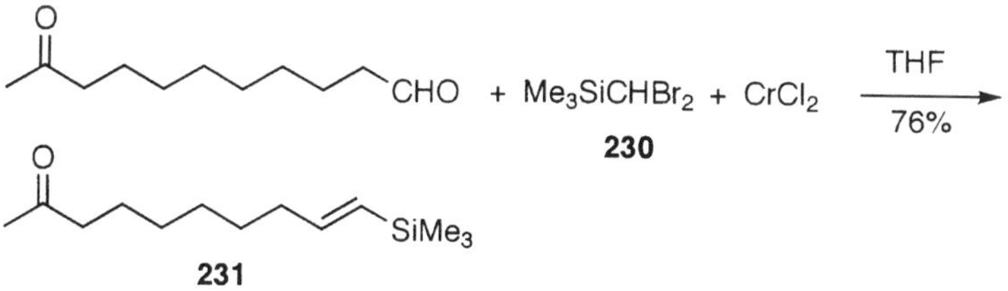

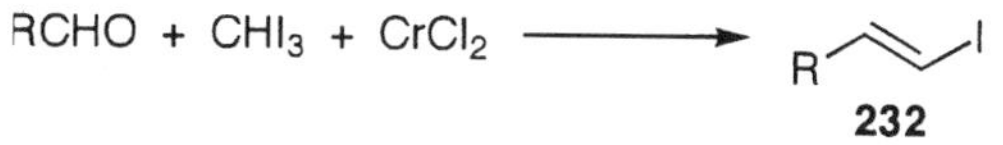

$$MoCl_5 + 2\ MeLi \longrightarrow [Cl_3MoMe_2 \xrightarrow{\alpha\text{-elimination}} Cl_3Mo{=}CH_2] \xrightarrow[65\%]{C_5H_{11}CHO}$$

220 ↓ CH₄ **221**

$$C_5H_{11}CH{=}CH_2 \ + \ Cl_3Mo{=}O$$

222 → Mo 19, 86% → **223** (Mo) → **224** —OBn

225 → W 40 → **226** (OBn) → Ph cyclopentanone

$$CH_3CHI_2 + CrCl_2 \xrightarrow[97\%]{THF} \left[CH_3CH{\begin{smallmatrix}CrX_2\\CrX_2\end{smallmatrix}} \right] \xrightarrow{\text{iPr-C}_6H_4\text{-CHO}} \left[Ar{-}\overset{H}{\underset{CH_3HC}{|}}{-}O{\begin{smallmatrix}CrX_2\\CrX_2\end{smallmatrix}} \right]$$

227 **228** X = I or Cl

$$\longrightarrow \text{iPr-C}_6H_4\text{-CH=CH-CH}_3$$

229   *E:Z* = 84:16

$$\cdots CHO \ + \ Me_3SiCHBr_2 \ + \ CrCl_2 \xrightarrow[76\%]{THF}$$

230

231 ···SiMe₃

$$RCHO + CHI_3 + CrCl_2 \longrightarrow R{\sim}{\sim}I$$

232

8.4 Synthetic Reactions Using Carbene Complexes of Metal Carbonyls as Stoichiometric Reagents

Stable electrophilic carbene complexes **234** are prepared from $Cr(CO)_6$, $Mo(CO)_6$ and $W(CO)_6$. They are electrophilic carbenes due to the strong electron-withdrawing effect of CO. The complexes can be isolated and used as useful synthetic reagents offering unique synthetic methods.

8.4.1 Reactions of Electrophilic Carbene Complexes of Cr, Mo, W, Fe and Co

Electrophilic carbene complexes are prepared from $M(CO)_6$ (M = Cr, Mo, W) in the following way. An organolithium compound attacks one of the six coordinated CO in $M(CO)_6$ (M = Cr, Mo, W) to give the anionic lithium acyl 'ate' complex **233** due to anion-stabilizing and delocalizing effects of the remaining five π-accepting electron-withdrawing CO groups. Acidification of the product, followed by treatment with diazomethane, affords the methoxycarbene complex **234** [1]. The reactivity of the carbene complexes **235** can be understood in the following way. The carbon atom of the carbene is electron defficient due to the electron-attracting CO groups, and the electron-donating OR group stabilizes the carbene. They are strongly electrophilic carbenes and attacked by nucleohiles. In addition, the H on the α-carbon is acidic and a carbanion is generated by deprotonation with a base, and hence electrophiles attack the α-carbon. The alkoxycarbene complex **236** resembles an reactive organic ester, and in practice the ester is prepared by the oxidation of **236** with Ce(IV).

Based on these reactivities various derivatives of carbenes, such as the aminocarbene **238**, are prepared by displacement of the OR group in **237** with amine via addition–elimination, analogous to transesterification [74,75]. As an example the carbanion **240**, generated by deprotonation of **239**, attacks ethylene oxide to give the lactone equivalent **241**, which is further alkylated by chloromethyl methyl ether, again at the α-position. Finally the α-methylene-γ-lactone **242** is obtained by oxidative demetallation with a Ce(IV) salt [76].

Due to the strongly electron-withdrawing character of the $Cr(CO)_5$ unit, the alkyne in **243** becomes a good dienophile and undergoes a smooth Diels–Alder reaction at 50 °C to give **244**, and subsequent annulation with alkyne and CO affords the dihydronaphthol complex **245**. The reaction can be carried out as a one-pot reaction [77].

Various cyclic compounds are prepared by the reaction of these carbene complexes with various unsaturated compounds [78–80]. The metallacyclobutane **246** is generated by [2+2] cycloaddition with electron-rich alkenes, and its reductive elimination affords the cyclopropanes **247**.

No [4+2] cycloaddition of the vinylcarbene complex **249** with the optically active diene **248** takes place. Instead, the cyclopropane **250** is obtained, and the Cope

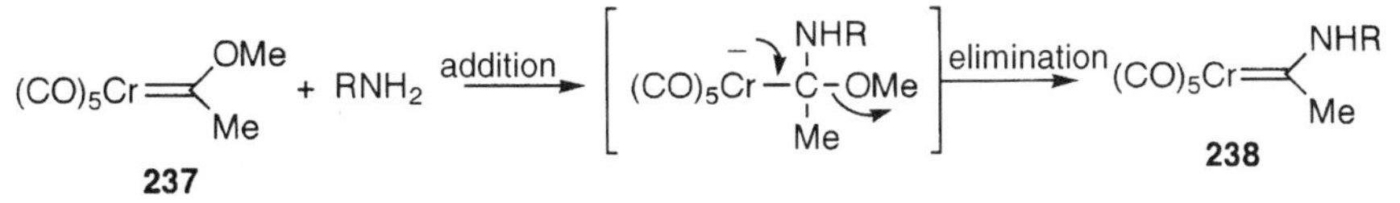

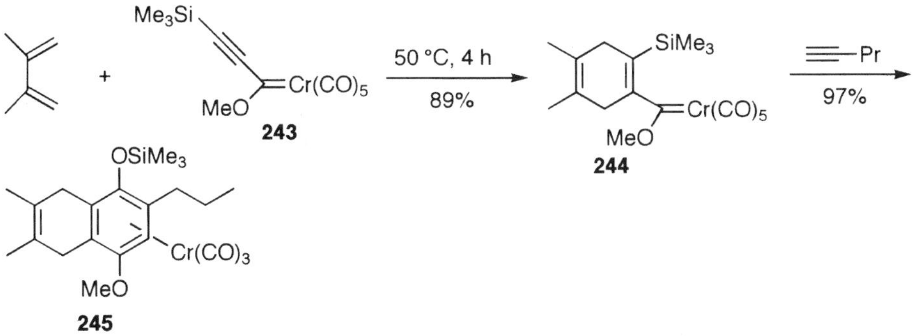

$(CO)_5Cr$ =, OMe, Me **237** + RNH_2 — addition → [$(CO)_5Cr$ –C–OMe, NHR, Me] — elimination → $(CO)_5Cr$ =, NHR, Me **238**

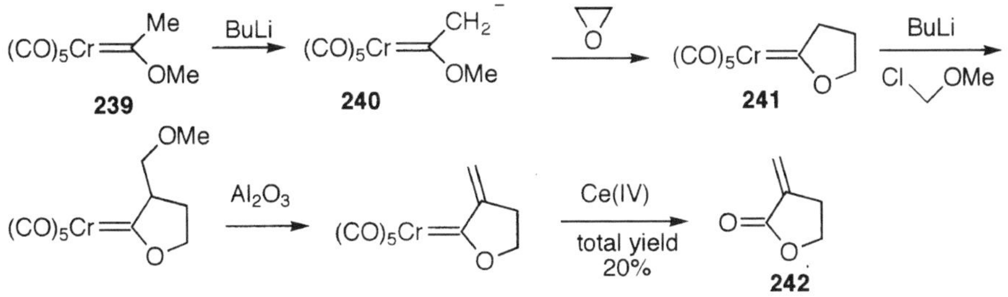

$(CO)_5Cr$ =, Me, OMe **239** — BuLi → $(CO)_5Cr$ =, CH_2^-, OMe **240** → $(CO)_5Cr$ =, O **241** — BuLi → Cl, OMe

$(CO)_5Cr$ =, OMe, O — Al_2O_3 → $(CO)_5Cr$ =, O — Ce(IV), total yield 20% → O=, O **242**

Me_3Si, MeO =Cr(CO)_5 **243** + — 50 °C, 4 h, 89% → SiMe_3, MeO, =Cr(CO)_5 **244** — ≡–Pr, 97% →

OSiMe_3, Cr(CO)_3, MeO **245**

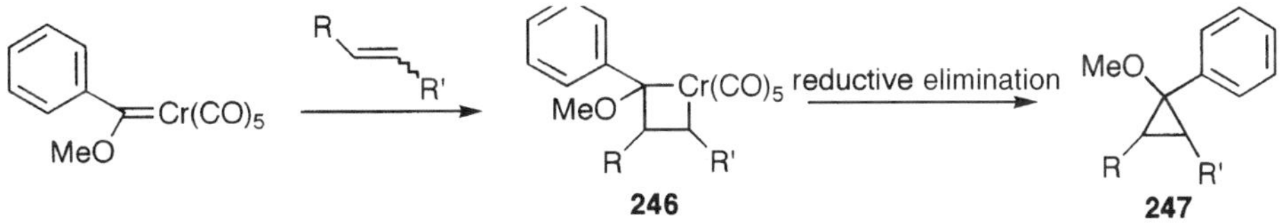

$(CO)_5Cr$ =, MeO — R, R' → Cr(CO)_5, MeO, R, R' **246** — reductive elimination → MeO, R, R' **247**

rearrangement of the divinylcyclopropane and hydrolysis afford the cycloheptanedione **252** with 86% ee [81].

The carbene complex **253** reacts with alkyne to give vinylcarbene complex **255** via the metallacyclobutene **254**. The triple bond in allylpropargylamine **256** reacts at first to form vinylcarbene **257**, and cyclopropanation of the double bond gives **258** [82].

The reaction of carbene complexes with alkynes offers useful synthetic methods. The formation of various cyclic compounds by the reaction of alkynes with the alkoxycarbene complex **259** can be summarized by the following scheme, which is simplified for easy understanding, although the explanation is not exactly correct mechanistically.

At first, cycloaddition of alkyne to the carbene complex **259** gives the chromacyclobutene **260**, which is cleaved to form the vinylcarbene complex **261**. It is claimed that vinylcarbenes **255** and **261** are formed directly without forming chromacyclobutenes **254** and **260** (M = Cr) [83]. The 6π-electrocyclization of **261** involving the benzene ring generates the chromacyclohexadiene **262** and its reductive elimination affords indene **263**. Hydrolysis of **263** yields indanone **264**. The vinyl ketene **265** is formed by insertion of CO to the vinyl carbene **261** and the 6π-electrocyclization of **265** gives rise to the cyclohexadienone **266**, which isomerizes to the 1,4-dihydroxynaphthalene monoether **270**. Cyclization of the vinyl ketene **265** generates **267**, which is converted to the furan **268** by demethoxylation. Another route is the formation of cyclobutenone **269**. Selectivity among these various possible reactions is controlled by the reaction conditions, particularly by solvents and irradiation.

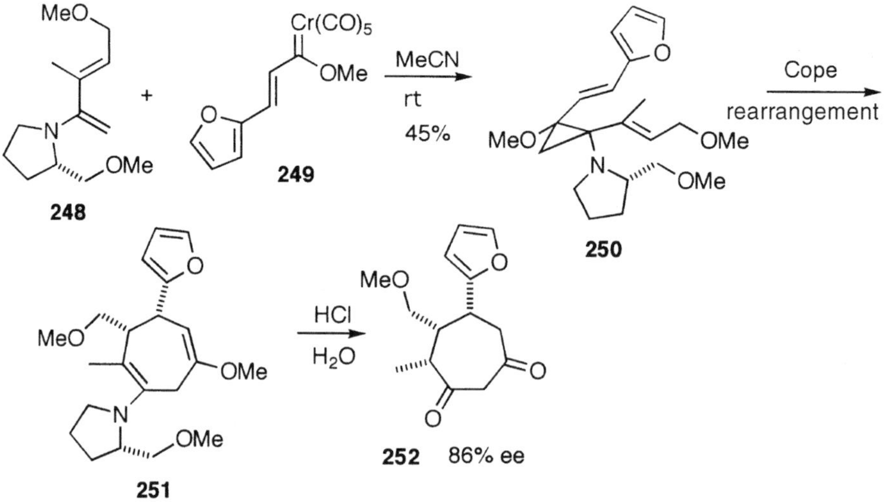

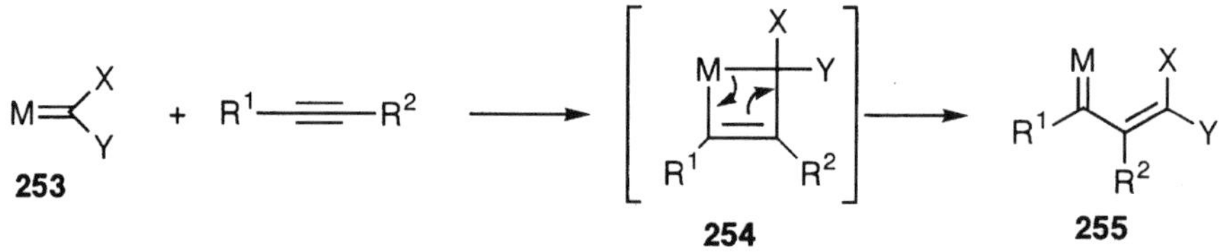

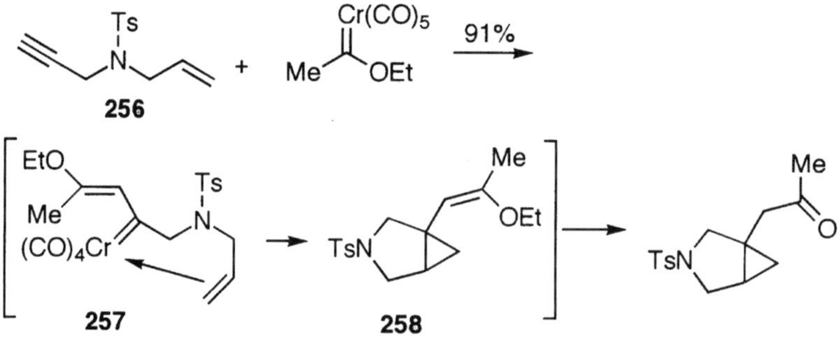

Well-established is the formation of hydroquinone and phenol derivatives **273** from alkynes. This reaction is called the Dötz reaction [78,79]. The reaction of carbene complex **271** to give **273** can be expressed by the general scheme **272**.

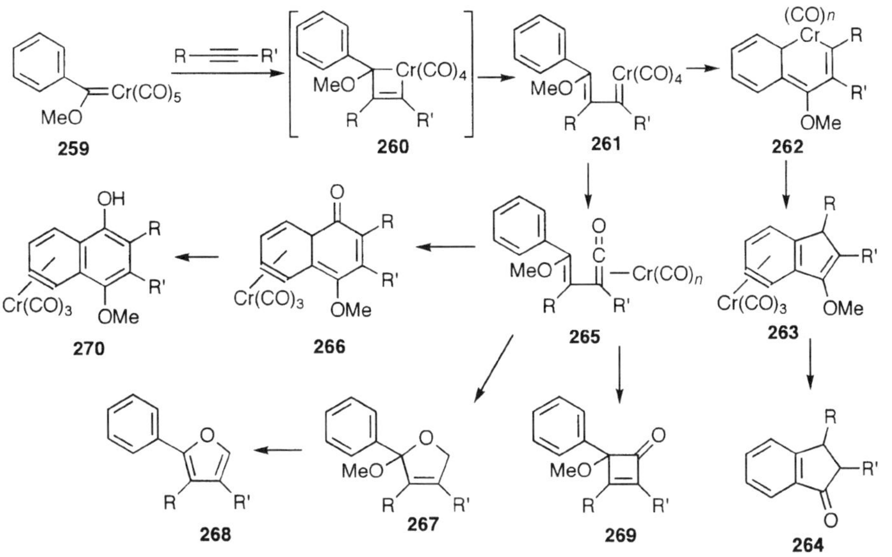

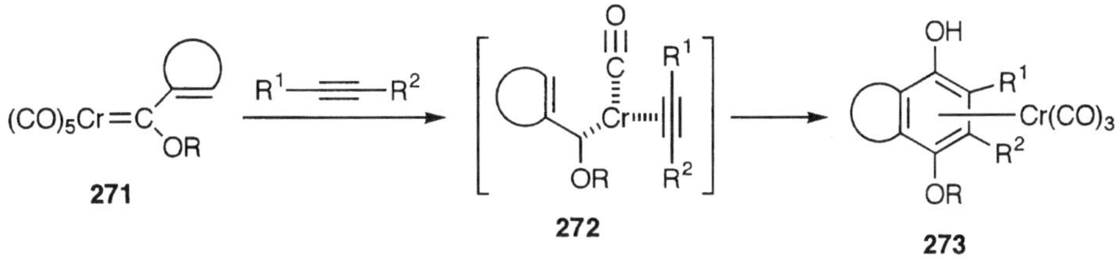

The Dötz reaction of the carbene complex **274** with phenylacetylene proceeds smoothly to give metal-free phenol derivatives **275** in high yields by photo-irradiation employing a xenon lamp without the oxidative workup [84]. Hydroquinone **277** is formed from the vinylmethoxycarbene complex **276**, and the reaction is applied to the synthesis of the tetracycline **278** [85]. The triple bond activated by the carbene in the triyne chain of the W alkynylcarbene complex **280** undergoes Diels–Alder reaction with the diene **279** selectively to give **281**, and the intramolecular reaction with the remaining two triple bonds generates the vinylcarbene complex **282** and then **283**, which is converted to the ketene **284** by CO insertion. Cyclization of the ketene **284** affords the steroid skeleton **285** in 63% yield [86]. CO is isoelectronic with isonitrile. The indolocarbazole **287** is constructed by aminobenzannulation of carbene complex **286** when isonitrile is used instead of CO [87].

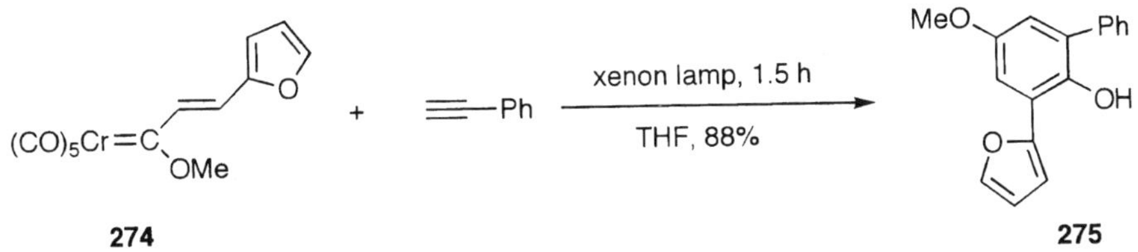

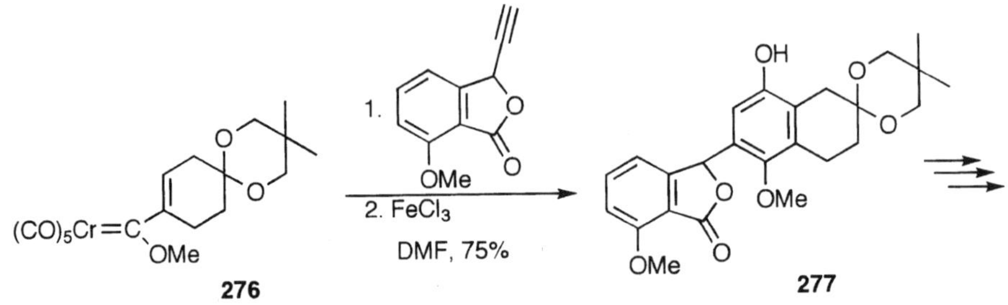
1.
OMe
2. FeCl3
DMF, 75%
(CO)5Cr=C
OMe
OH
OMe
276
277

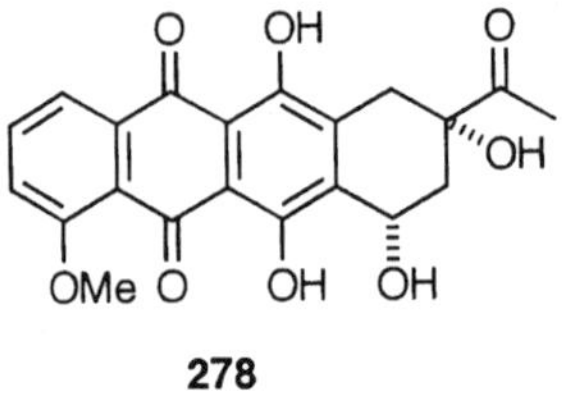
O
OH
OH
OMe O
OH
OH
278

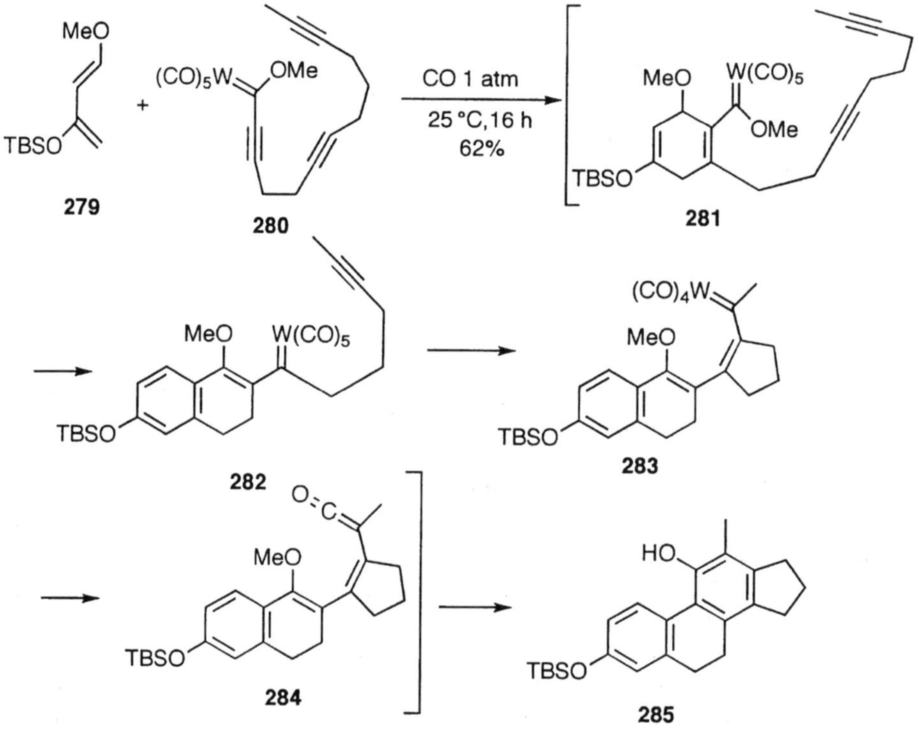
MeO
TBSO
+
(CO)5W
OMe
CO 1 atm
25 °C,16 h
62%
279
280
281
282
283
284
285
W(CO)5
(CO)4W
HO

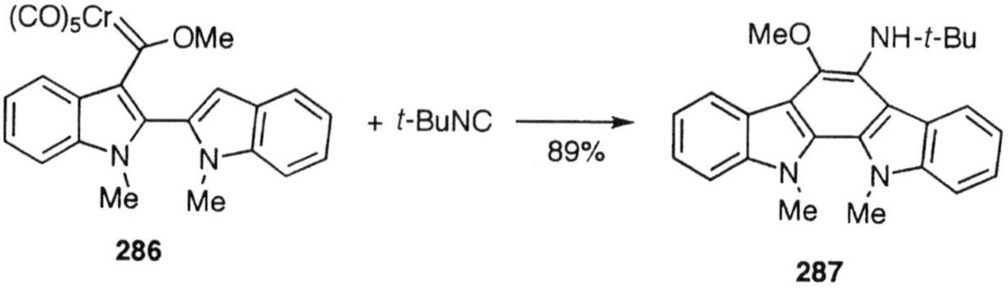
(CO)5Cr
OMe
+ t-BuNC
89%
MeO
NH-t-Bu
Me
286
287

Facile addition of either alcohols or amines to carbene complex **288** produces the β-amino-α,β-unsaturated carbene complexes **289**, which are useful for the preparation of cyclopentenone derivatives. Insertion of alkyne to **289** gives the 1,3-dienylcarbene complex **290**, and its formal [3+2] cycloaddition gives cyclopentadiene **291**. Under different conditions, [2+2+1] cycloaddition of vinylketene **292** produces the 5-methylene-2-cyclopentenone derivative **293** [88]. The cyclopentenone **295**, isomeric to **293**, was obtained by the reaction of complex **294** with trimethylsilylacetylene, and oudenone (**296**) was synthesized by its hydrolysis [89].

The methoxyketene **297**, coordinated to Cr carbonyl, is formed from methoxy-carbene easily by insertion of CO under irradiation [90]. An ester is formed by the reaction of ketene with alcohol. The aminocarbene complex **298** was prepared from benzamide and converted to phenylalanine ester **300** under irradiation of sunlight in alcohol via ketene **299** [91]. The eight-membered lactone **304** was prepared in high yield by the reaction of the alkyne **301** having the OH group in a tether with Cr carbene without irradiation. The vinylcarbene **302** is formed at first and converted to the vinylketene intermediate **303** as expected. The keto lactone **304** is formed from **303** by intramolecular reaction with the OH group and hydrolysis [92].

$$R\underset{MeO}{\overset{CO}{=}}Cr(CO)_4 \quad \underset{hv}{\rightleftarrows} \quad \left[R\underset{MeO}{\overset{O}{=}}Cr(CO)_4 \right] = \left[R\underset{MeO}{\overset{Cr(CO)_4}{=}}C=O \right]$$
297

$$Na_2Cr(CO)_5 + PhCONH_2 \xrightarrow{TMSCl} (CO)_5Cr=\overset{NH_2}{\underset{Ph}{<}} \quad \xrightarrow[CO]{hv} \quad \left[(CO)_4Cr-\overset{Ph}{\underset{C=O}{<}}NH_2 \right]$$
298 · · · **299**

$$\xrightarrow{MeOH} \quad Ph-\overset{NH_2}{\underset{|}{CH}}CO_2Me$$
300

$$\mathbf{301} \quad + \quad (CO)_5Cr=\overset{OEt}{\underset{Me}{<}} \quad \xrightarrow[0.5\ h,\ 81\%]{MeCN\ reflux} \quad \left[\ldots \right] \rightarrow$$
E = CO₂tBu → $E = CO_2tBu$

302 **303** **304**

Irradiation of carbene complexes in CO atmosphere generates the ketene **305** and its [2+2] cycloaddition to alkene gives the cyclobutanone **306** [93]. Total synthesis of (+)-cerulenin (**310**) has been carried out by the formation of cyclobutanone **309** by cycloaddition of **307** to the double bond of **308** as the key reaction without attacking the triple bond. Then cyclobutanone **309** was converted to (+)-cerulenin (**310**) via regioselective Bayer–Villiger reaction of **309**, and side-chain elongation using π-methallylnickel bromide, epoxidation and hydrolysis [94].

Cycloaddition of a ketene complex with unsaturated bonds other than alkenes and alkynes is also possible. The ketene **312**, formed from **311**, adds to imine **313** to give the β-lactam **314** under sunlight photolysis. The optically active β-lactam **314** was prepared from the optically active carbene complex **311** with 99% ee, and converted to **315** [95]. Irradiation of carbene complex **316** generates ketene **317**, which cyclizes to the o-hydroquinone derivative **318** [96].

Carbene complexes of Fe and Co carbonyls are also prepared. Unlike the Cr carbene complexes, no cyclopropanation of alkenes occurs with these carbene complexes. Furans are formed by the reaction of alkynes involving rearrangement of methoxy group. The 2-aminofuran **323** is formed by the reaction of the dimethylaminocarbene complex **319** of Fe carbonyl, via rearrangement of the amino group. Under CO pressure, pyrone **324** is the main product [97]. In these reactions, the ketene complex **321** is formed by CO insertion to the metallacyclobutene **320**. The rearranged amino or methoxy group attacks the ketene **321** to give amide or ester with the regeneration of the Fe carbene complex **322**. Its cyclization affords the furan **323**, and the pyrone **324** is formed by CO insertion and cyclization.

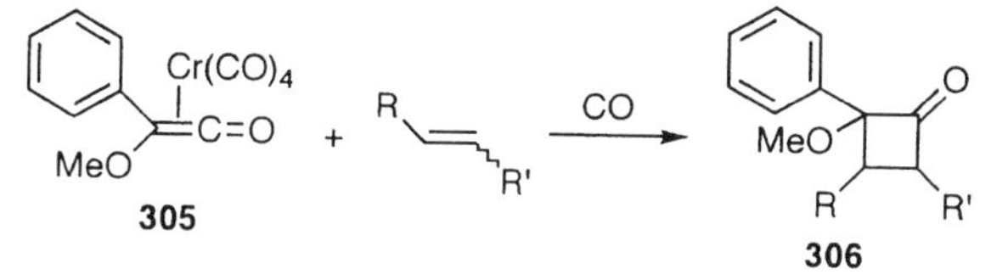

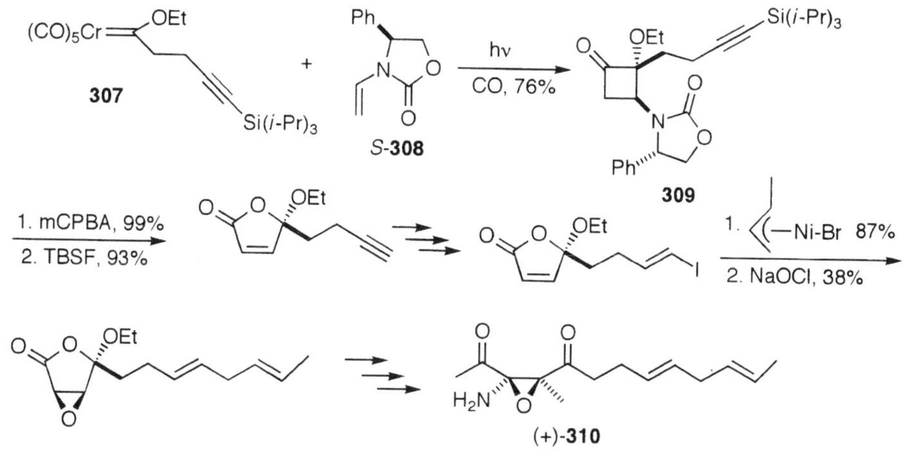

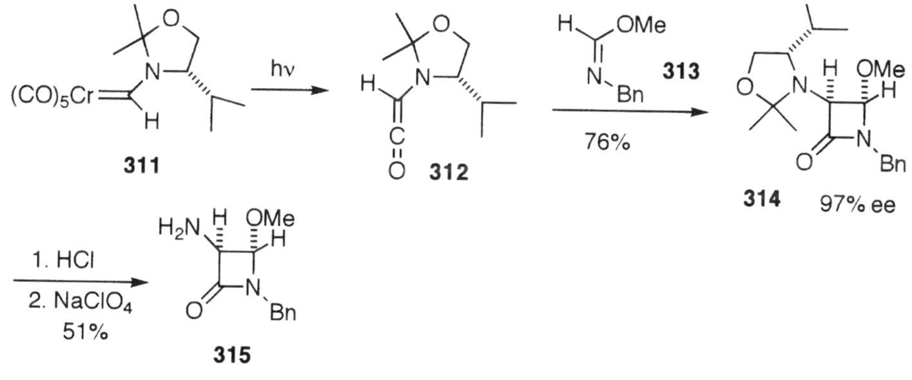

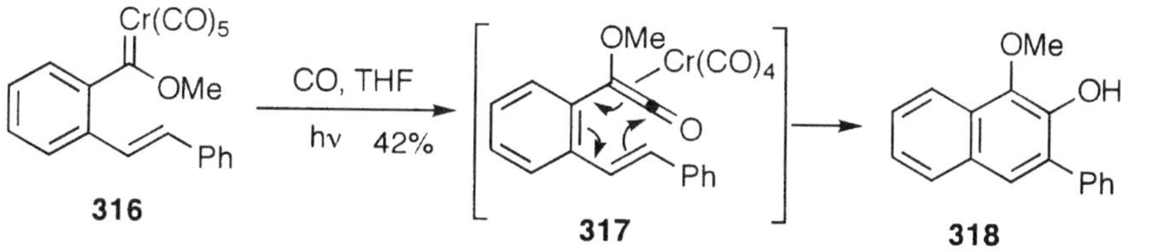

8.5 Rh and Pd-catalysed Reactions of Diazo Compounds via Electrophilic Carbene Complexes

Transition metal Lewis acids of Cu(I), Pd(II) and Rh(II) catalyze decomposition of diazo compounds **325** via unstabilized electrophilic metal carbene **326**, offering useful synthetic methods [98]. In particular, relatively stable α-diazocarbonyl compounds are used extensively. Rh(II) carboxylates are by far the most useful catalysts in synthesis [99]. Rh(II) is a d^7 metal and $Rh_2(OAc)_4$ is a binuclear compound with four bridging acetate ligands (**328**) and possesses one vacant axial coordination site per metal atom. The Rh-catalysed decomposition of α-diazo carbonyl compounds has wide synthetic application [98]. Rh carbene **326** can be regarded as an ylide of inverted polarity or inverse ylide, as shown in **327**, which may be stabilized by electron donation from Rh, but destabilized by the adjacent electron-withdrawing carbonyl group. This type of inverse ylide has never been isolated and does not induce alkene metathesis. The purity of $Rh_2(OAc)_4$ seems to be crucial for high catalytic activity, particularly for intramolecular reactions. The reactions include: (i) cyclopropanation of alkenes (ii) insertion to C–H, N–H and O–H bonds (iii) generation of ylides and (iv) β-hydride elimination. The insertion and generation of ylides are particularly useful, because they are not possible with other metal–carbene complexes [98].

Well-known is the cyclopropanation of various alkenes. As shown by **329**, cyclopropanation starts by electrophilic attack to the alkene. Electron-rich alkenes have higher reactivity. Numerous applications of intramolecular cyclopropanation to syntheses of natural products have been reported. Optically active cyclopropanes are prepared by enantioselective cyclopropanation [100]. As the first successful example, asymmetric synthesis of chrysanthemic acid (**331**) was carried out by cyclopropanation of 2,5-dimethyl-2,4-hexadiene (**330**) with diazoacetate, catalysed by the chiral copper imine complex **332**, prepared from salicylaldehyde and the optically active amino alcohol as the key step [101]. Later, highly enantioselective cyclopropanation was achieved using copper complexes of bidentate nitrogen ligands such as semicorrin [102], 5-aza-semicorrin [103] and bisoxazoline [104,105].

A = electron-withdrawing groups

The enantioselective intramolecular cyclopropanation of alkenyl diazoacetates to afford fused bicyclic products catalysed by chiral Rh catalysts is particularly useful. Highly efficient asymmetric cyclization of the allyl or homoallyl diazo esters **333** to bicyclic γ- or δ-lactones is achieved using Rh complexes of chiral carboxamide ligands such as methyl 2-pyrrolidone-5-carboxylate [(5*S*)-MEPY] (**335**) and 4-alkyloxazoli-dinones [106,107]. Although the Rh carboxamides are less reactive toward diazo compounds than $Rh_2(OAc)_4$, they provide higher selectivities. The bicyclolactone **334** of 98.6% ee was obtained from the allylic ester of diazoacetic acid **333** catalysed by the Rh complex **335** [106]. Intramolecular cyclopropanation of farnesyl diazoacetate (**336**,) catalysed by $Rh_2[(5S)\text{-MEPY}]_4$ occurred at the neighbouring allylic double bond in 96% yield to give the optically active bicyclic lactone containing quarternary carbon **338** with 94% ee, and efficient asymmetric synthesis of presqualene diphosphate has been achieved [107]. Interestingly, the 13-membered macrocyclic lactone **337** was obtained in 63% yield by the regioselective cyclopropanation at the remote double bond when $Rh_2(OAc)_4$ was used [107].

1,4-Cycloheptadiene (**340**) is obtained by the Cope rearrangement of *cis*-divinylcyclopropane (**339**.) Based on this reaction, highly diastereoselective and enantioselective construction of the 1,4-cycloheptadiene **343** (98% ee) was achieved by domino asymmetric cyclopropanation to generate *cis*-divinylcyclopropane

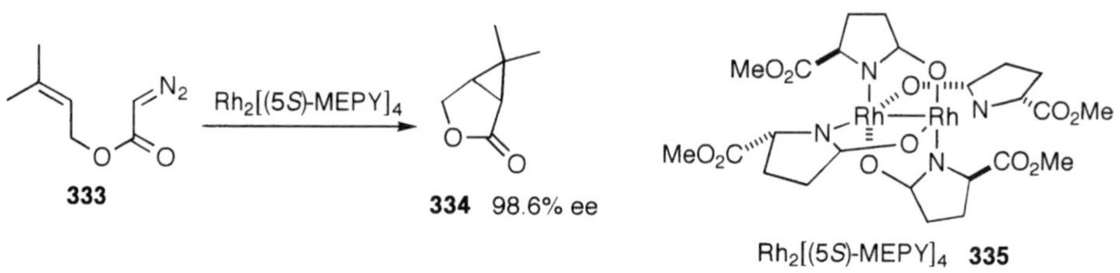

333   **334** 98.6% ee   Rh$_2$[(5S)-MEPY]$_4$ **335**

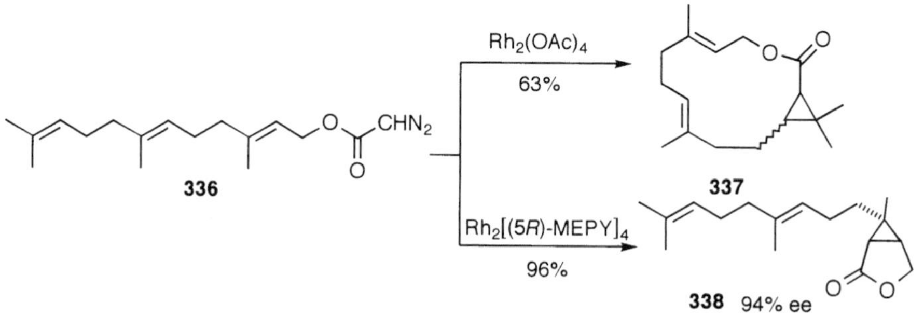

336   **337**   **338** 94% ee

from diene **341** and vinyldiazoacetate **342** using Rh$_2$[(S)-DOSP]$_4$, DOSP = [(N-dodecylbenzenesulfonyl)prolinate] as a catalyst, and subsequent Cope rearrangement of the divinylcyclopropane [108].

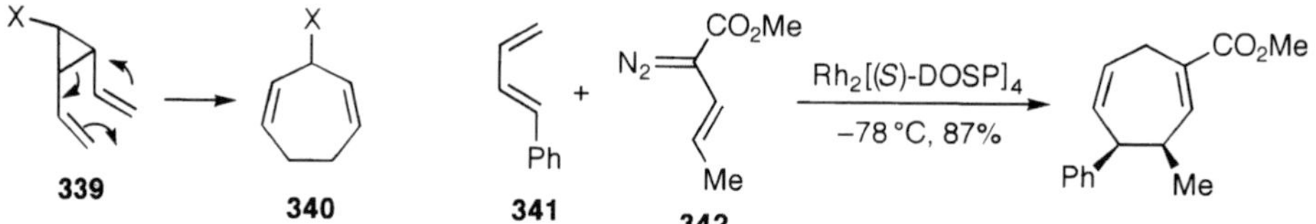

339   **340**   **341**   **342**   **343** 98% ee

Pd(OAc)$_2$ catalyses cyclopropanation of terminal alkene or α,β-unsaturated enones [109]. Selective cyclopropanation of the terminal alkene in the macrolactam **344** gives **345** in high yield [110].

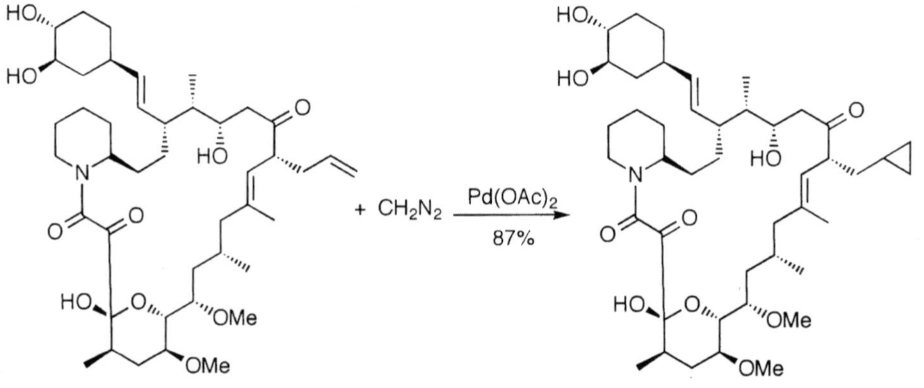

344   **345**

The intramolecular selective insertion of carbenes to C−H is synthetically important, and has evolved with the use of Rh carboxylates. Easily available α-diazo derivatives of β-keto esters and β-diketones are used extensively [111]. Formation of five-membered rings is the favoured process, and cyclopentanone derivatives are easily prepared [112]. The reactivity of C−H bonds is in the order of tertiary > secondary > primary. Reaction of **346** with TPA (triphenylacetic acid) salt gives the five-membered ring **347** with high selectivity, whereas the four-membered ring **348** is obtained with Rh₂(OAc)₄ [113]. The construction of six-membered rings is also possible. The synthesis of pentalenolactone (**350**) from **349** is an example [114]. Four-membered ring formation from the β-keto amide **351** gives β-lactam **352** as a *trans* isomer in a nearly quantitative yield [115]. Efficient asymmetric insertion based on the differentiation of enantiotopic aliphatic C−H bonds in a molecule with prochiral quarternary carbons is possible, and the highly optically pure bicyclic lactone **353** was obtained by this method [116].

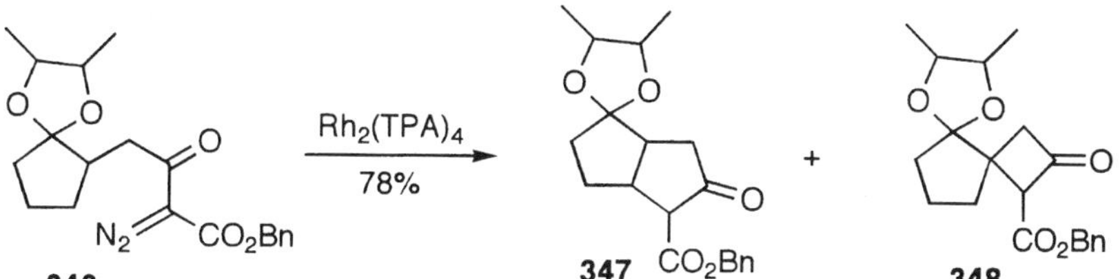

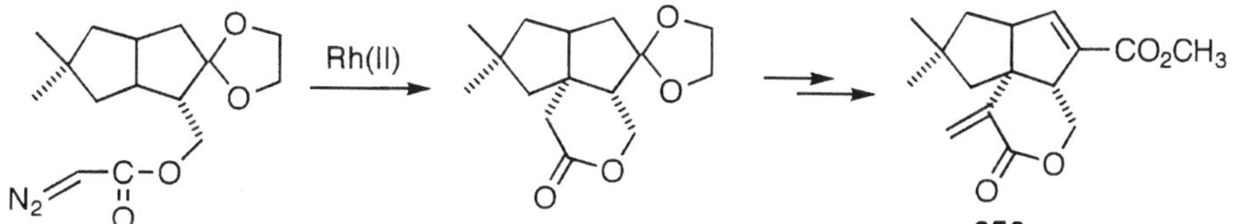

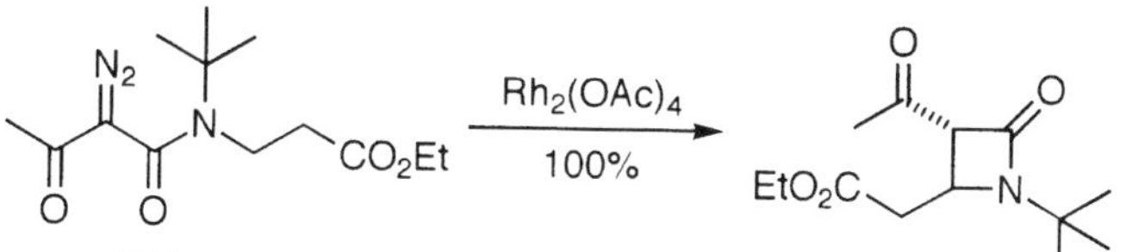

Intermolecular insertion to aryl C−H bonds is possible. The asymmetric intramolecular reaction of the α-diazo compound **354** catalysed by Rh₂[(S)-PTTL]₄, Rh₂[(S)−PTTL]₄ = dirhodium tetrakis[N-phthaloyl(S)−t−leucinate], afforded indane

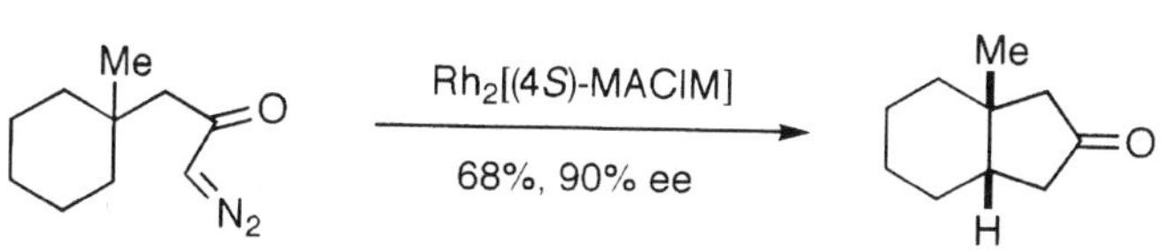

355 with 93% ee and the reaction was applied to the asymmetric total synthesis of FR115427, an NMDA receptor antagonist **356** [117]. Ring enlargement of benzene derivatives by the reaction of α-keto carbenes generated from the diazo compound **357** (Buchner reaction) takes place to generate norcaradiene **358** as an intermediate, which rearranges to 3,8a-dihydroazulene-1(2H)-one (**359**). The reaction has been applied to the synthesis of the sesquiterpene confertin (**360**) [118].

The insertion of keto carbene to the N–H bond in **361** to form the carbapenem ring system **362** is a commercially established synthesis [119].

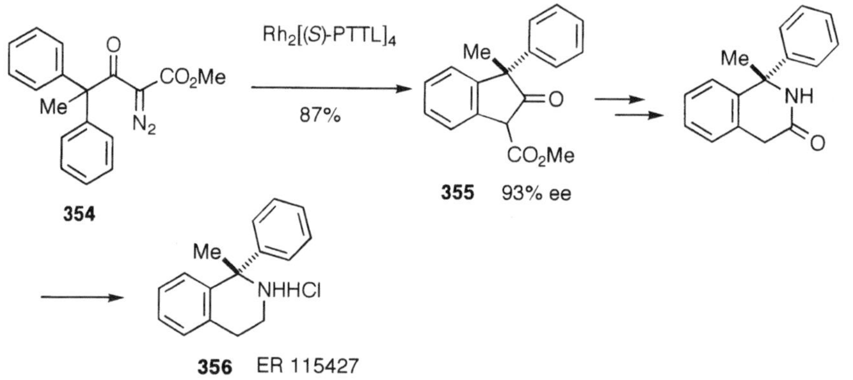

Rh2[(S)-PTTL]4
87%
354
355 93% ee
356 ER 115427

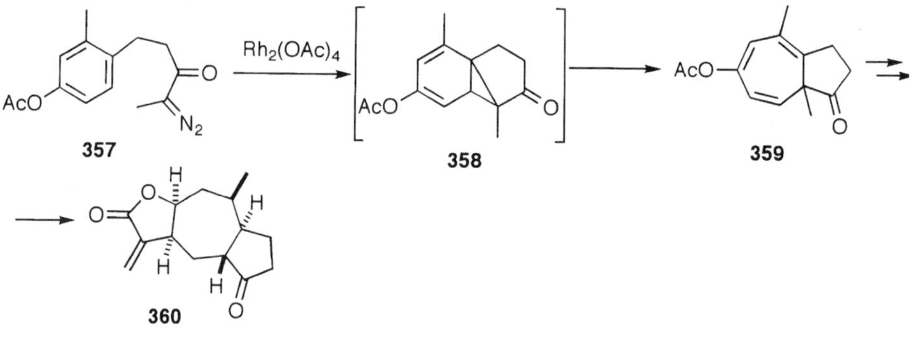

Rh2(OAc)4
357
358
359
360

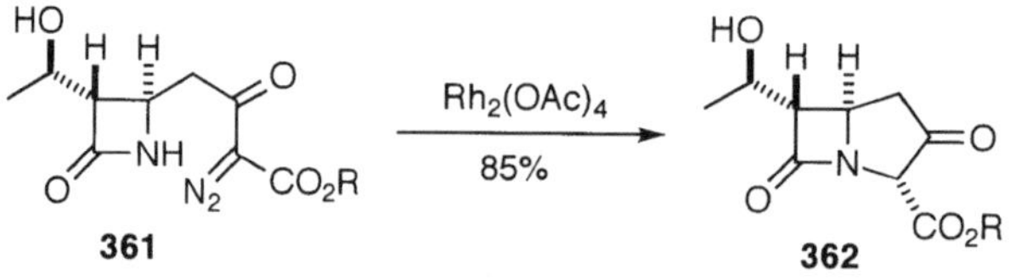

Rh2(OAc)4
85%
361
362

Intermolecular insertion to Si–H bonds offers a synthetic route to allylsilanes. The Rh-catalyzed reaction of the α-vinyldiazoacetate **363** with hydrosilane gives the allylsilane **364** [120].

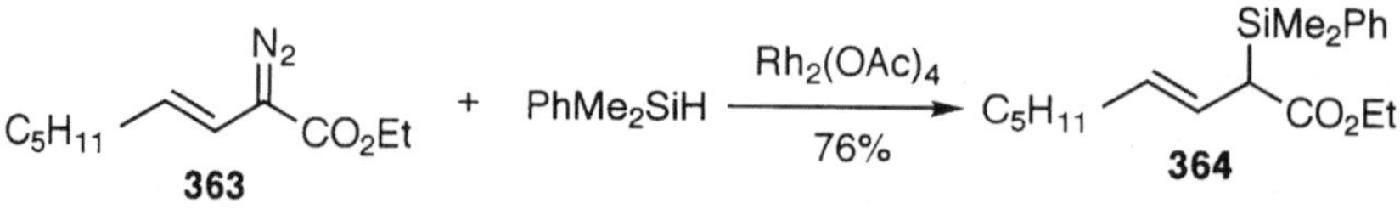

PhMe2SiH
Rh2(OAc)4
76%
363
364

The cyclic ylide intermediate **366**, as a 1,3-dipole, is generated by intramolecular reaction of Rh–carbene with the ketone in **365**, and undergoes cycloaddition with π-bonds to give the adduct **367** [121]. When α-diazocarbonyls have additional unsaturation, domino cyclizations occur to produce polycyclic compounds. The Rh–carbene method offers a powerful tool for the construction of complex polycyclic molecules in short steps, and has been applied to elegant syntheses of a number of complex natural products.

The decomposition of **368** catalysed by Rh perfluorobutyrate and subsequent intramolecular cycloaddition give **370** in high yield (93%) as the key step in the total synthesis of lysergic acid (**371**), and is believed to involve the intramolecular reaction of the ylide intermediate **369** at the alkene. No C−H insertion takes place [122]. Another elegant example is the efficient construction of the aspidosperma alkaloid skeleton **374**. The Rh-catalysed domino cyclization–cycloaddition of diazo imide **372** afforded cycloadduct **374** in 95% yield as a single diastereomer via the dipole **373**, and desacetoxy-4-oxo-6,7-dihydrovindorosine (**375**) has been synthesized from **374** [123].

Intermolecular cycloaddition also proceeds smoothly. The 2,8-dioxabicyclo-[3.2.1]octane core system **379** of zaragozic acid **380** was constructed by the intramolecular carbonyl ylide formation from **376** catalysed by $Rh_2(OAc)_4$, followed by intermolecular 1,3-dipolar cycloaddition of the electron-deficient dipolarophile **377** as shown by **378** as a single diastereomer out of four possible diastereomers [124].

The Rh-catalysed intramolecular reaction of the diazocarbonyl with the alkyne in **381** generates vinyl ketocarbenes **382**, which undergo further transformations. The Rh(II)-catalysed domino cyclization–cycloaddition sequence, in particular the intramolecular reaction of α-diazocarbonyl with alkynes, is useful for efficient construction of polycyclic compounds. *o*-Alkynyl-α-diazopropiophenone **383** containing tethered carbonyl and alkyne groups underwent Rh(II)octanoate-catalysed alkyne–carbene metathesis as shown by **384** and **385** to give the vinyl carbene **386**, which cyclized onto the neighbouring carbonyl group to give the resonance-stabilized ylide dipole **387**. Finally, intramolecular dipolar cycloaddition to the alkene afforded the cycloadduct **388** in a remarkably high yield of 97% [125].

The oxonium ylide **390** is generated by the interaction of carbene with the unshared electron pair of the oxygen atom of ether **389**, and subsequent sigmatropic rearrangement affords **391** [126]. The reaction was applied to the diastereoselective construction of 2,8-dioxabicyclo[3.2.1]octane, the core system **394** of zaragozic acid. The Rh-catalysed reaction of diazo ester **392** generates the bicyclic oxonium ylide **393** from the acetal, and its exocyclic 2,3-shift affords **394** [127].

368

369

370 **371**

372 **373**

374 **375**

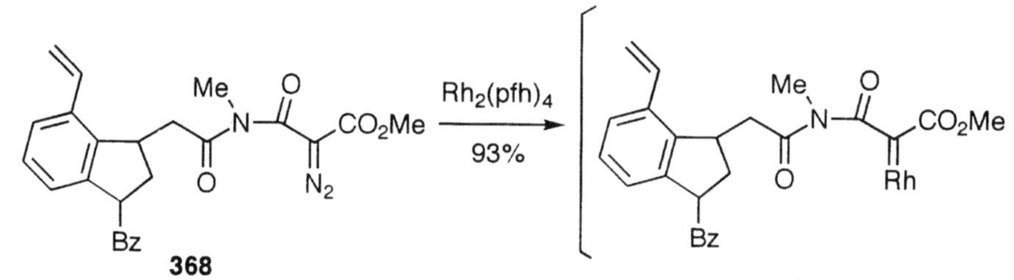

α-Diazo esters and ketones undergo carbene–alkene rearrangement to afford (*Z*)-α, β-unsaturated esters and ketones [128]. Elimination of N_2 from **395** proceeded selectively at −78°C to give the α, β-unsaturated ester **396** with high (*Z*) selectivity, particularly when Rh trifluoroacetate is used [128].

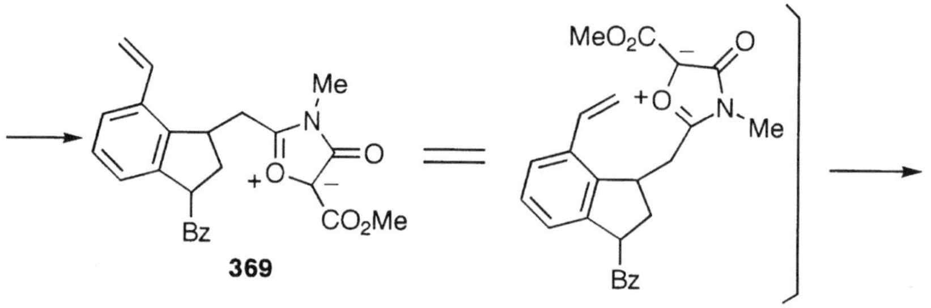

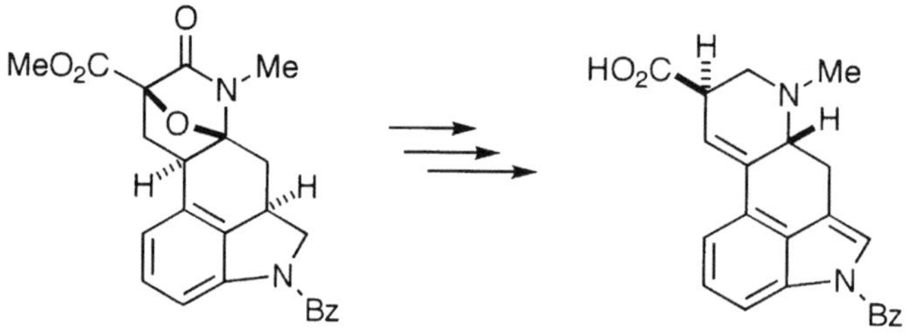

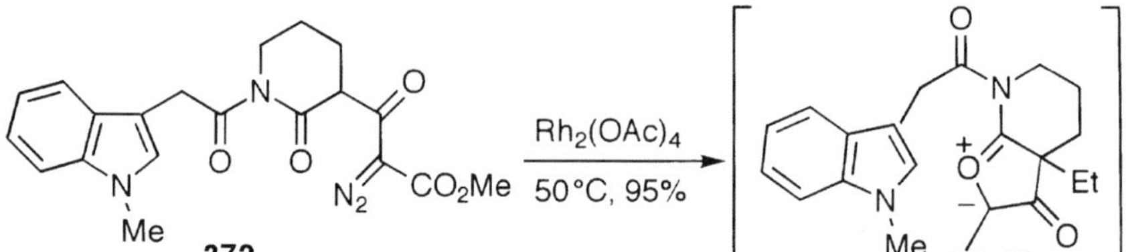

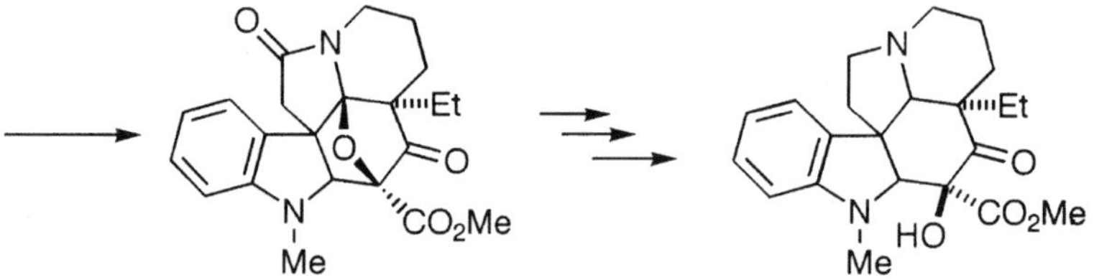

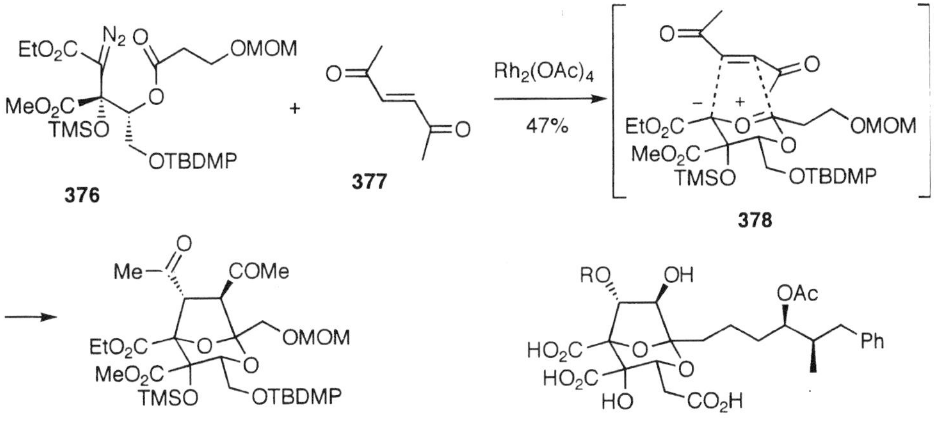

376

377

378

379

380

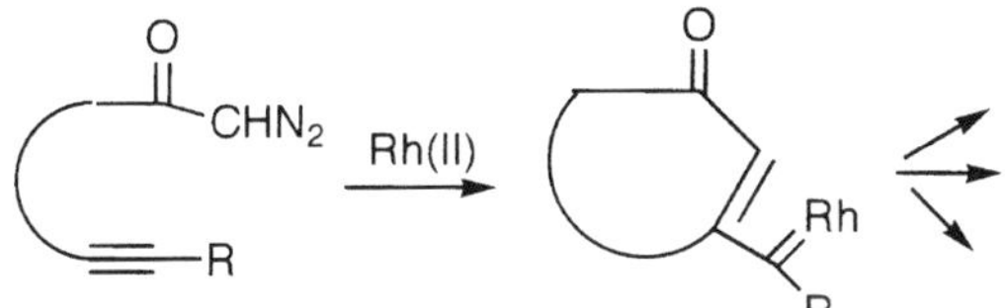

381

382

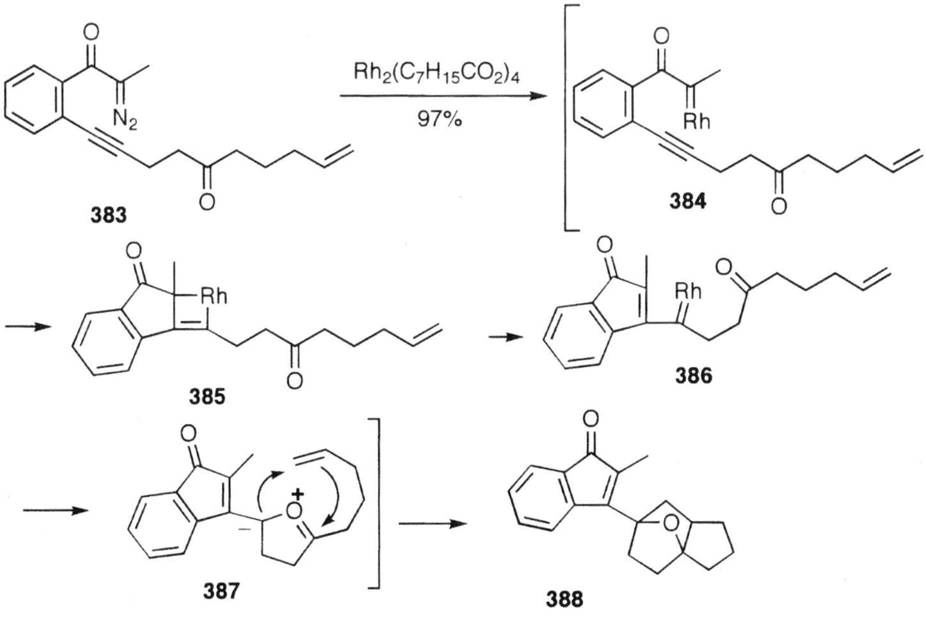

383

384

385

386

387

388

8.6 Other Reactions

Some reactions proceed by the formation of carbene complexes and their cycloaddition. Carbodiimide **398** is prepared in good yields by heating *o*-tolylisocyanate (**397**) with a catalytic amount of $Fe(CO)_5$ or $Mo(CO)_6$, with evolution of CO_2 [129]. The reaction can be understood by the following mechanism. Bonding between transition metals and CO and isonitrile can be expressed by the resonance forms **399** and **400**. These carbene-type complexes, possessing cumulative double bonds, undergo the cycloaddition with isocyante. The first step is the cycloaddition of metal carbonyl **401** with isocyanate **402** to form four-membered metallacycles **403**. Decarboxylation of the cycloadduct **403** generates the Fe–isonitrile (or carbene) complex **404**, which then reacts with the isocyanate to form the four-membered metallacycle **405**. Retrocycloaddition of the four-membered ring **405** affords carbodiimide **406**. At the same time, $Fe(CO)_5$ (**401**) is regenerated and the reaction proceeds catalytically.

Treatment of diphenylketene (**407**) with a catalytic amount of $Co_2(CO)_8$ produces tetraphenylethylene (**410**,) involving a carbene complex as an intermediate. In this reaction carbene complex **408** is formed from **407** and $Co_2(CO)_8$, the cobaltacyclobutanone **409** is generated by cycloaddition of **407** and **408**, and cleaved to give **410** [130].

It is known that vinylidene complexes **412** are formed by isomerization of the terminal alkynyl complexes **411** [131]; their reactions are treated in Section 3.5.2.2.

Complexes of Pd, Pt and Ru catalyse enyne metathesis, giving similar products to those obtained by the Ru–carbene complex **22** as described in this chapter. These enyne metatheses are discussed in Section 7.2.6. Other mechanisms, without involving carbene complexes as intermediates, have been proposed.

$$2 \ \text{Ar}-\text{N}=\text{C}=\text{O} \xrightarrow[\substack{180-250\,°C,\ 35\ min \\ 85\%}]{Fe(CO)_5,\ 1\ wt\ \%} \text{Ar}-\text{N}=\text{C}=\text{N}-\text{Ar} \ + \ CO_2$$

397   **398**

Ar = *o*-tolyl

$$M=C=O \ \longleftrightarrow \ \overset{+}{M}-C\equiv O\overset{-}{:}$$

399

$$M=C=\overset{\cdot\cdot}{\underset{R}{N}} \ \longleftrightarrow \ \overset{-}{M}-C\equiv\overset{+}{N}-R$$

400

$$(CO)_4Fe=C=O$$
401

$+ \ R-N=C=O$
402

$\longrightarrow$

$$\begin{array}{c} (CO)_4Fe\!-\!C\!=\!O \\ | \quad\quad | \\ R\!-\!N\!-\!C\!-\!O \end{array}$$
403

$\longrightarrow \ (CO)_4Fe=C\equiv N-R \ + \ CO_2$
404

$$(CO)_4Fe=C=N-R$$
404

$+ \ O=C=N-R$
402

$\longrightarrow$

$$\begin{array}{c} (CO)_4Fe\!-\!C\!=\!N\!-\!R \\ | \quad\quad | \\ O\!=\!C\!-\!N\!-\!R \end{array}$$
405

$\longrightarrow \ (CO)_4Fe=C=O \ + \ R-N=C=N-R$
         401   **406**

$$\underset{Ph}{\overset{Ph}{>}}C=C=O \xrightarrow{Co_2(CO)_8} \underset{Ph}{\overset{Ph}{>}}C=Co(CO)_n \xrightarrow{\underset{Ph}{\overset{Ph}{>}}C=C=O \ \textbf{407}}$$
    407   **408**

$$\begin{array}{c} Ph \\ | \\ Ph\!-\!C\!-\!Co(CO)_n \\ | \\ Ph\!-\!C \\ | \quad \searrow O \\ Ph \end{array} \longrightarrow \underset{Ph}{\overset{Ph}{>}}C=C\underset{Ph}{\overset{Ph}{<}} \ + \ Co(CO)_{n+1}$$
   409   **410**

$$R-\!\!\equiv\!\!-M-H \longrightarrow \underset{H}{\overset{R}{>}}C=\bullet=M$$
   411   **412**

References

1. E. O. Fischer and A. Maasbol, *Angew. Chem., Int Ed., Engl.* **3**, 580 (1964); E. O. Fischer, *Angew. Chem.*, **86**, 651 (1974).
2. R. R. Schrock, *Acc. Chem. Res.*, **23**, 158 (1990); R. R. Schrock, J. S. Murdzek, G. C. Bazan, J. Robbins, M. DiMare and M. O'Regan, *J. Am. Chem. Soc.*, **112**, 3875 (1990).
3. G. Natta, G. Dall'Asta and G. Mazzanti, *Angew. Chem.*, **76**, 765 (1964); *Angew. Chem., Int. Ed. Engl.*, **3**, 723 (1964).
4. N. Calderon, N. Y. Chen and K. W. Scott, *Tetrahedron Lett.*, 3327 (1967).
5. N. Calderon, *Acc. Chem. Res.*, **5**, 1279 (1979); N. Calderon, J. P. Lawrence and E. A. Ofstead, *Adv. Organometal. Chem.*, **17**, 449 (1979); J. C. Mol, *J. Mol. Catal.*, **15**, 35 (1982); *ChemTech*, 250 (1983); R. L. Banks, *ChemTech*, 494 (1979); J. M. Basset and M. Leconte, *ChemTech*, 762 (1980).
6. R. L. Banks and G. C. Bailey, *Ind. Eng. Chem. Prod. Res. Develop.*, **3**, 170 (1964).
7. R. H. Grubbs, *Comprehensive Organometallic Chemistry*, Vol 8, 499, Pergamon Press 1982; R. H. Grubbs and S. H. Pine, *Comprehensive Organic Synthesis*, Vol. 5, Chapter 9, Pergamon Press, (1991); K. J. Ivin, *Olefin Metathesis*, Academic Press, 1983.

8. W. A. Herrmann, W. Wagner, U. N. Flessner, U. Volkhardt and H. Komber, *Angew. Chem. Int. Ed. Engl.*, **30**, 1636 (1991); W. A. Herrmann and F. E. Kühn, *Acc. Chem. Res.*, **30**, 169 (1997).

9. S. T. Nguyen, R. H. Grubbs and J. W. Ziller, *J. Am. Chem. Soc.*, **115**, 9858 (1993).

10. P. Schwab, M. B. France, J. W. Ziller and R. H. Grubbs, *Angew. Chem., Int. Ed. Engl.*, **34**, 2039 (1995).

11. R. H. Grubbs, S. J. Miller and G. C. Fu, *Acc. Chem. Res.*, **28**, 446 (1995); S. K. Armstrong, *J. Chem. Soc., Perkin Trans. I.* 371 (1998); R. H. Grubbs and S. Chang, *Tetrahedron*, **54**, 4413 (1998).

12. T. E. Wilhelm, T. R. Belderrain, S. N. Brown and R. H. Grubbs, *Organometallics*, **16**, 3867 (1997).

13. T. R. Belderrain and R. H. Grubbs, *Organometallics*, **16**, 4001 (1997).

14. M. I. Bruce, *Chem. Res.*, **91**, 197 (1991).

15. H. Kitayama and F. Ozawa, *Chem. Lett.*, 67 (1998).

16. J. Feng, M. Schuster and S. Blechert, *Synlett*, 129 (1997)

17. O. Brümmer, A. Rückert and S. Blechert, *Chem. Eur. J.*, **3**, 441 (1997).

18. W. E. Crowe and D. R. Goldberg, *J. Am. Chem. Soc.*, **117**, 5162 (1995).

19. W. E. Crowe, D. R. Goldberg and Z. J. Zhang, *Tetrahedron Lett.*, **37**, 2117 (1996).

20. J. Tsuji and S. Hashiguchi, *Tetrahedron Lett.*, **21**, 2955 (1980); *J. Organometal. Chem.*, **218**, 69 (1981).

21. L. G. Wildman, *J. Org. Chem.*, **33**, 4541 (1968).

22. S. Warwel and H. Katker, *Synthesis*, 935 (1987).

23. W. E. Crowe and Z. J. Zhang, *J. Am. Chem. Soc.*, **115**, 10998 (1993).

24. M. L. Snapper, J. A. Tallarico and M. L. Randall, *J. Am. Chem. Soc.*, **119**, 1478 (1997).

25. G. C. Fu and R. H. Grubbs, *J. Am. Chem. Soc.*, **114**, 5426, 7324 (1992), **115**, 3800 (1993).

26. C. M. Huwe and S. Blechert, *Synthesis*, 61 (1997).

27. G. C. Fu, S. T. Nguyen and R. H. Grubbs, *J. Am. Chem. Soc.*, **115**, 9856 (1993); G. C. Fu and R. H. Grubbs, *J. Am. Chem. Soc.*, **114**, 7324 (1992).

28. L. R. Sita, *Macromol.*, **28**, 656 (1995).

29. M. T. Crimmins and B. W. King, *J. Org. Chem.*, **61**, 4192 (1996).

30. J. B. Alexander, D. S. La, D. R. Cefalo, A. H. Hoveyda and R. R. Schrock, *J. Am. Chem. Soc.*, **120**, 4041 (1998).

30a. D. S. La, J. B. Alexander, D. R. Cefalo, D. D. Graf, A. H. Hoveyda and R. R. Schrock, *J. Am. Chem. Soc.*, **120**, 9720 (1998).

31. A. B. Dyatkin, *Tetrahedron Lett.*, **38**, 2065 (1997).

32. Y. S. Shon and T. R. Lee, *Tetrahedron Lett.*, **38**, 1283 (1997).

33. J. P. Harrity, M. S. Visser, J. D. Gleason and A. H. Hoveyda, *J. Am. Chem. Soc.*, **119**, 1488 (1997); C. W. Johannes, M. S. Visser, G. S. Weatherhead and A. H. Hoveyda, *J. Am. Chem. Soc.*, **120**, 8340 (1998).

34. H. Junga and S. Blechert, *Tetrahedron Lett.*, **34**, 3731 (1993).

35. F. P. J. T. Rutjes and H. E. Schoemaker, *Tetrahedron Lett.*, **38**, 677 (1997).

36. S. F. Martin, Y. Liao, H. J. Chen, M. Pützel and M. N. Ramser, *Tetrahedron Lett.*, **35**, 6005 (1994); S. F. Martin and A. S. Wagman, *Tetrahedron Lett.*, **36**, 1169 (1995).

37. S. F. Martin, Y. Liao and T. Rein, *Tetrahedron Lett.*, **35**, 691 (1994); S. F. Martin, H. J. Chen, A. K. Courtery, Y. Liao, M. Pätzel, M. N. Ramser and A. S. Wagman, *Tetrahedron*, **52**, 7251 (1996).

38. S. Fürstner and N. Kindler, *Tetrahedron Lett.*, **37**, 7005 (1996).

39. A. Fürstner and T. Müller, *J. Org. Chem.*, **63**, 424 (1998).

40. S. J. Miller, H. E. Blackwell and R. H. Grubbs, *J. Am. Chem. Soc.*, **118**, 9606 (1996).

41. A. F. Houri, Z. Xu, D. A. Cogan and A. H. Hoveyda, *J. Am. Chem. Soc.*, **117**, 2943 (1995); Z. Xu, C. W. Johannes, S. S. Salman and A. H. Hoveyda, *J. Am. Chem. Soc.*, **118**, 10926 (1996).

42. K. C. Nicolaou, Y. He, D. Vourloumis, H. Vallberg and Z. Yang, *Angew. Chem., Int. Ed. Engl.*, **36**, 166 (1997); K. C. Nicolaou, Y. He, F. Roschanger, N. P. King, D. Vourloumis and T. Li, *Angew. Chem., Int. Ed. Engl.*, **37**, 84 (1998).

43. D. Meng, D. S. Su, A. Balog, P. Bernato, E. J. Sorensen, S. J. Danishefsky, Y. H. Zheng, T. C. Chou, L. He and S. B. Horwitz, *J. Am. Chem. Soc.*, **119**, 2733, 10073 (1997).

44. D. Schinzer, A. Limberg, A. Bauer, O. M. Böhm and M. Cordes, *Angew. Chem., Int. Ed. Engl.*, **36**, 523 (1997).

45. A. Fürstner and K. Langemann, *J. Org. Chem.*, **61**, 3942 (1996).

46. B. König and C. Horn, *Synlett*, 1013 (1996).

47. M. J. Marsella, H. D. Maynard and R. H. Grubbs, *Angew. Chem., Int. Ed. Engl.*, **36**, 1101 (1997).

48. W. J. Zuercher, M. Hashimoto and R. H. Grubbs, *J. Am. Chem. Soc.*, **118**, 6634 (1996).

49. A. Bencheick, M. Petit, A. Mortreux and F. Petit, *J. Mol. Catal.*, **15**, 93 (1982); M. Petit, A. Mortreux and F. Petit, *Chem. Commun.*, 1385 (1982).

50. D. Villemin and P. Cadiot, *Tetrahedron Lett.*, **23**, 5139 (1982).

51. N. Kaneta, K. Hikichi, S. Asaka, M. Uemura and M. Mori, *Chem. Lett.*, 1055 (1995).

51a. A. Fürstner and G. Seidel, *Angew. Chem., Int. Ed. Engl.*, **37**, 1734 (1998).

52. A. Kinoshita and M. Mori, *Synlett*, 1020, 1994; *J. Org. Chem.*, **61**, 8356 (1996); S. Watanuki, N. Ochifuji and M. Mori, *Organometallics*, **13**, 4129 (1994).

53. S. H. Kim, W. J. Zuercher, N. B. Bowden and R. H. Grubbs, *J. Org. Chem.*, **61**, 1073 (1996).

54. W. J. Zuercher, M. Scholl and R. H. Grubbs, *J. Org. Chem.*, **63**, 4291 (1998).

55. A. Kinoshita, N. Sakakibara and M. Mori, *J. Am. Chem. Soc.*, **119**, 12388 (1997).

56. S. H. Pine, *Org. React.*, **43**, 1 (1993). John Wiley & Sons, Inc., New York.

57. K. Takai, Y. Hotta, K. Oshima and H. Nozaki, *Tetrahedron Lett.*, 2417 (1978); *Bull. Chem. Soc. Jpn.*, **53**, 1698 (1980).

58. M. Shibasaki, Y. Torisawa and S. Ikegami, *Tetrahedron Lett.*, **24**, 3493 (1983); Y. Ogawa and M. Shibasaki, *Tetrahedron Lett.*, **25**, 1067 (1984).

59. R. T. Jacobs, G. I. Feutrill and J. Meinwald, *J. Org. Chem.*, **55**, 4051 (1990).

60. T. Okazoe, K. Takai, K. Oshima and K. Utimoto, *J. Org. Chem.*, **52**, 4410 (1987); K. Takai, Y. Kataoka, T. Okazoe and K. Utimoto, *Tetrahedron Lett.*, **29**, 1065 (1988).

61. O. Fujimura, G. C. Fu and R. H. Grubbs, *J. Org. Chem.*, **59**, 4029 (1994).

62. F. N. Tebbe, G. W. Parshall and G. S. Reddy, *J. Am. Chem. Soc.*, **100**, 3611 (1978).

63. F. N. Tebbe, G. W. Parshall and D. W. Overall, *J. Am. Chem. Soc.*, **101**, 5074 (1979).

64. S. H. Pine, R. Zahler, D. A. Evans and R. H. Grubbs, *J. Am. Chem. Soc.*, **102**, 3270 (1980).

65. J. R. Stille and R. H. Grubbs, *J. Am. Chem. Soc.*, **108**, 855 (1986); J. R. Stille, B. D. Santarsiero and R. H. Grubbs, *J. Org. Chem.*, **55**, 843 (1990).

66. K. C. Nicolaou, M. H. D. Postema and C. F. Claiborne, *J. Am. Chem. Soc.*, **118**, 1565 (1996).

67. N. A. Petasis and E. I. Bzowej, *J. Am. Chem. Soc.*, **112**, 6392 (1990); *J. Org. Chem.*, **57**, 1327 (1992).

68. N. A. Petasis, J. P. Staszewski and D. K. Fu, *Tetrahedron Lett.*, **36**, 3619 (1995).

69. J. M. Tour, P. V. Bedworth and R. Wu, *Tetrahedron Lett.*, **30**, 3927 (1989).

70. T. Kauffmann, B. Ennen, J. Sander and R. Wieschollek, *Angew. Chem., Int. Ed. Engl.*, **22**, 244 (1983).

71. G. C. Fu and R. H. Grubbs, *J. Am. Chem. Soc.*, **115**, 3800 (1993).
72. R. R. Schrock, R. T. DePue, J. Feldman, K. B. Yap, D. C. Yang, W. M. Davis, L. Park, M. DiMare, M. Schofield, J. Anhaus, E. Walborsky, E. Evitt, C. Kruger and P. Betz, *Organometallics*, **9**, 2262 (1990).
73. T. Okazoe, K. Takai and K. Utimoto, *J. Am. Chem. Soc.*, **109**, 951 (1987); K. Takai, Y. Kataoka, T. Okazoe and K. Utimoto, *Tetrahedron Lett.*, **28**, 1443 (1987).
74. K. Weiss and E. O. Fischer, *Chem. Ber.*, **106**, 1277 (1973).
75. C. P. Casey and T. J. Burkhardt, *J. Am. Chem. Soc.*, **95**, 5833 (1973); C. P. Cassey, R. A. Boggs and R. L. Anderson, *J. Am. Chem. Soc.*, **94**, 8947 (1972).
76. C. P. Casey and W. R. Brunsvold, *J. Organometal. Chem.*, **102**, 175 (1975).
77. W. D. Wulff and D. C. Young, *J. Am. Chem. Soc.*, **106**, 7565 (1984).
78. K. H. Dötz, *Angew. Chem., Int. Ed. Engl.*, **14**, 644 (1975); **23**, 587 (1984).
79. W. D. Wulff, P. C. Tang, K. S. Chan, J. S. McCallum, D. C. Yang, S. R. Gilbertson, *Tetrahedron*, **41**, 5813 (1985); W. D. Wulff, *Comprehensive Organic Synthesis*, Vol. 5, p. 1065, Pergamon Press 1991.
80. D. F. Harvey and D. M. Sigano. *Chem. Rev.*, **96**, 271 (1996).
81. J. Barluenga, F. Aznar, A. Martin and J. T. Vazques, *J. Am. Chem. Soc.*, **117**, 9419 (1995).
82. M. Mori and S. Watanuki, *Chem. Commun.*, 1082 (1992).
83. P. Hofmann and Mämmerle, *Angew. Chem., Int. Ed. Engl.*, **28**, 908 (1989).
84. Y. H. Choi, K. S. Rhee, K. S. Kim, G. C. Shin and S. C. Shin, *Tetrahedron Lett.*, **36**, 1871 (1995).
85. W. D. Wulff and P. C. Tang, *J. Am. Chem. Soc.*, **106**, 434 (1984).
86. J. Bao, V. Dragisich, S. Wenglowaky and W. D. Wulff, *J. Am. Chem. Soc.*, **113**, 9873 (1991).
87. C. A. Merlic, D. M. McInnes and Y. You, *Tetrahedron Lett.*, **38**, 6787 (1997).
88. B. L. Flynn, F. J. Funke, C. C. Silveria and A. de Meijere, *Synlett*, 1007 (1995); M. Deuetsch, S. Vidoni, F. Stein, F. Funke, M. Noltemeyer and A de Meijere, *Chem. Commun.*, 1679 (1994).
89. B. L. Flynn, C. C. Silveira and A. de Meijere, *Synlett*, 812 (1995).
90. L. S. Hegedus, *Tetrahedron*, **53**, 4105 (1997).
91. L. S. Hegedus, M. A. Schwindt, S. DeLombaert and R. Imwinkelried, *J. Am. Chem. Soc.*, **112**, 2264 (1990).
92. M. Mori, T. Norizuki and T. Ishibashi, *Heterocycles*, **47**, 651 (1998).
93. M. A. Sierra and L. S. Hegedus, *J. Am. Chem. Soc.*, **111**, 2335 (1989).
94. T. E. Kedar, M. W. Miller and L. S. Hegedus, *J. Org. Chem.*, **61**, 6121 (1996).
95. L. S. Hegedus, G. de Weck and S. D'Andrea, *J. Am. Chem. Soc.*, **110**, 2122 (1988); L. S. Hegedus, R. Imwinkelried, M. A. Sargent, D. Dvorak and Y. Satoh, *J. Am. Chem. Soc.*, **112**, 1109 (1990).
96. C. A. Merlic and D. Xu, *J. Am. Chem. Soc.*, **113**, 7418 (1991).
97. M. F. Semmelhack and J. Park, *Organometallics*, **5**, 2550 (1986); M. F. Semmelhack, R. Tamura, W. Schnatter and J. Springer, *J. Am. Chem. Soc.*, **106**, 5363 (1984).
98. M. P. Doyle, *Comprehensive Organometalic Chemistry*, Vol. 2, Chapter 5.; M. P. Doyle, *Acc. Chem. Res.*, **19**, 348 (1986); *Chem. Rev.*, **86**, 919 (1986); A. Padwa and S. F. Hornbuckle, *Chem. Rev.*, **91**, 263 (1991); A. Padwa and M. D. Weingarten, *Chem. Rev.*, **96**, 223 (1996); T. Ye and M. A. Mckervey, *Chem. Rev.*, **94**, 1091 (1994); A. Padwa and K. E. Krumpe, *Tetrahedron*, **48**, 5385 (1992); S. Hashimoto, N. Watanabe, M. Anada and S. Ikegami, *Tetrahedron*, **54**, 988 (1996); G. A. Sulikowski, K. L. Cha and M. M. Sulikowski, *Tetrahedron, Asymmetry*, **9**, 3145 (1998).

99. R. Paulissen, H. Reimlinger, E. Hayez, A. J. Hubert and Ph. Teyssie, *Tetrahedron Lett.*, 2233 (1973); R. Paulissen, E. Hayes, A. J. Hubert and Ph. Teyssie, *Tetrahedron Lett.*, 607 (1974).

100. M. P. Doyle, in *Catalytic Asymmetric Synthesis*, (Ed.) I. Ojima, Chapter 3, VCH, 1993.

101. T. Aratani, *Pure Appl. Chem.*, 1839 (1985); T. Aratani, Y. Yoneyoshi and T. Nagase, *Tetrahedron Lett.*, 1707 (1975).

102. H. Fritschi, U. Leutenegger and A. Pfaltz, *Angew. Chem., Int. Ed. Engl.*, **25**, 1005 (1986); D. Muller, G. Umbricht, B. Weber and A. Pfaltz, *Helv. Chim. Acta*, **74**, 232 (1991).

103. U. Leutenegger, G. Umbricht, C. Fahrni, P. von Matt and A. Pfaltz, *Tetrahedron*, **48**, 2143 (1992).

104. R. E. Lowenthal, A. Abiko and S. Masamune, *Tetrahedron Lett.*, **31**, 6005 (1990).

105. D. A. Evans, K. A. Woerpel and M. M. Hinman, *J. Am. Chem. Soc.*, **113**, 726 (1991).

106. M. P. Doyle, R. J. Pieters, S. F. Martin, R. E. Austin, C. J. Oalmann and P. Miller, *J. Am. Chem. Soc.*, **113**, 1423 (1991); S. F. Martin, C. J. Oalmann and S. Liras, *Tetrahedron Lett.*, **33**, 6727 (1992).

107. D. H. Rogers, E. C. Yi and C. D. Poulter, *J. Org. Chem.*, **60**, 941 (1995); M. P. Doyle, M. N. Protopopova, C. D. Poulter and D. H. Rogers, *J. Am. Chem. Soc.*, **117**, 7281 (1995).

108. H. M. L. Davies, D. G. Stafford, B. D. Doan and J. H. Houser, *J. Am. Chem. Soc.*, **120**, 3326 (1998); *J. Org. Chem.*, **63**, 657 (1998).

109. A. J. Anciaux, A. J. Hubert, A. F. Noels, N. Petiniot and Ph. Teyssie, *J. Org. Chem.*, **45**, 695 (1980); R. Paulissen, A. J. Hubert and Ph. Teyssie, *Tetrahedron Lett.*, 1465 (1972); M. Suda, *Synthesis*, 714 (1981); M. W. Majchrzak, A. Kotelko and J. B. Lambert, *Tetrahedron Lett.*, **24** 469 (1983); U. Mende, B. Radüchel, W. Skuballa and H. Vorbüggen, *Tetrahedron Lett.*, 629 (1975).

110. A. J. F. Edmunds, K. Baumann, M. Grassberger and G. Schluz, *Tetrahedron Lett.*, **32**, 7039 (1991).

111. H. L. Davies, *Comprehensive Organic Synthesis*, Vol. 4, p. 1031, Pergamon Press, 1991.

112. D. F. Taber and E. H. Petty, *J. Org. Chem.*, **47**, 4808 (1982); M. P. Doyle, A. V. Kalinin and D. G. Ene, *J. Am. Chem. Soc.*, **118**, 8837 (1996).

113. S. Hashimoto, N. Watanabe and S. Ikegami, *Tetrahedron Lett.*, **33**, 2709 (1992).

114. D. E. Cane and P. J. Thomas, *J. Am. Chem. Soc.*, **106**, 5295 (1984).

115. M. P. Doyle, J. Taunton and H. Q. Pho, *Tetrahedron Lett.*, **30**, 5397 (1989); M. P. Doyle, M. S. Shanklin, S. M. Oon, H. Q. Pho, F. R. Van der Heide and W. R. Veal, *J. Org. Chem.*, **53**, 3384 (1988).

116. M. P. Doyle, Q. L. Zhou, C. E. Raab and G. H. Roos, *Tetrahedron Lett.*, **36**, 4745 (1995).

117. N. Watanabe, Y. Ohtake, S. Hashimoto, M. Shiro and S. Ikegami, *Tetrahedron Lett.*, **36**, 1491 (1995); N. Watanabe, T. Ogawa, Y. Ohtake, S. Ikegami and S. Hashimoto, *Synlett*, 85 (1996).

118. M. Kennedy and M. A. McKervey, *Chem. Commun.*, 1028 (1988); *J. Chem. Soc. Perkin Trans I*, 2565 (1991).

119. T. N. Salzmann, R. W. Ratcliff, B. G. Christensen and F. A. Bouffard, *J. Am. Chem. Soc.*, **102**, 6161 (1980); D. H. Sih, F. Baku, L. Cama and B. G. Christiansen, *Heterocycles*, **21**, 29 (1984).

120. P. Bulugahapitiya, Y. Landais, L. Parra-Rarpado, D. Planchenault and V. Weber, *J. Org. Chem.*, **62**, 1630 (1997).

121. A. Padwa, A. T. Price and L. Zhi, *J. Org. Chem.*, **61**, 2283 (1996).

122. J. P. Marino, Jr. M. H. Osterhout and A. Padwa, *J. Org. Chem.*, **60**, 2704 (1995).

123. A. Padwa and A. T. Price, *J. Org. Chem.*, **60**, 6258 (1995), **63**, 556 (1998).

124. O. Kataoka, S. Kitagaki, N. Watanabe, J. Kobayashi, S. Nakamura, M. Shiro and S. Hashimoto, *Tetrahedron Lett.*, **39**, 2371 (1998).
125. A. Padwa, J. M. Kassir, M. A. Semones and M. D. Weingarten, *Tetrahedron Lett.*, **34**, 7853 (1993).
126. A. Padwa and M. D. Weingarten, *Chem. Rev.*, **96**, 223 (1996).
127. J. B. Brogan and C. K. Zercher, *Tetrahedron Lett.*, **39**, 1691 (1998).
128. D. F. Taber, R. J. Herr, S. K. Pack and J. M. Geremia, *J. Org. Chem.*, **61**, 2908 (1996).
129. H. Ulrich, B. Tucker and A. A. R. Sayigh, *Tetrahedron Lett.*, 1731 (1967).
130. K. Hong, K. Sonogashira and N. Hagihara, *J. Chem. Spc. Jpn.* (Japanese), **89**, 74 (1968).
131. M. I. Bruce, *Chem. Rev.*, **91**, 197 (1991).

9

PROTECTION AND ACTIVATION BY COORDINATION

Some transition metal carbonyls form stable complexes with alkenes, conjugated dienes, alkynes and aromatic rings. The electron density of these unsaturated compounds is decreased by the coordination of strongly electron-withdrawing metal carbonyls, and their reactivity is modified. Complex formation can be used in two ways, as protection or as activation. Reactive unsaturated bonds are masked by the coordination and protected for some reagents, such as reducing agent and electrophiles. As it is difficult to protect alkenes, conjugated dienes and alkynes by conventional methods, protection by coordination offers a useful method. Aromatic rings are activated toward nucleophiles by coordination. The transition metals in these complexes have low valences, and they lose the ability to coordinate by mild oxidation to higher valence states, and facile oxidative deprotection is possible, liberating the unsaturated compounds.

9.1 Protection and Activation of Alkenes by the Coordination of Iron Carbonyls

The carbonyls $Fe(CO)_5$ and $[CpFe(CO)_2]^+$ (2) form stable cationic complexes with alkenes, which are used for both protection and activation of alkenes [1]. $[CpFe(CO)_2]^+$ (2; abbreviated as Fp^+) is prepared by the reaction of cyclopentadienyl anion (1) with $Fe(CO)_5$, followed by oxidative cleavage with bromine, and used for the protection of alkenes. The electron density of the double bond is decreased by the coordination of $[CpFe(CO)_2]^+$ and hence this bond is activated to nucleophilic attacks. Introduction of nucleophiles, such as the carbon nucleophile of malonate, to cyclopentene becomes possible via the formation of the complex 3, and the stable *trans*-σ-alkyliron complex 4 of cyclopentane is prepared. The vinyl ether complex 6 is obtained easily from the α-bromoacetal 5, and reacts with an enolate of ketone 7 as an electrophile to give **8**, which is converted to the alkene complex **9** by the treatment with HBF$_4$, and deprotected to give **10** [2]. In this way, the vinyl ether complex **6** can be regarded as a vinyl cation equivalent, which reacts with enolates as the nucleophile.

Alkenes are also protected by coordination against electrophilic attacks. The less-hindered bond of the two double bonds in vinylcyclohexene (**11**) is protected preferentially by the formation of the complex **13** by exchange reaction with the isobutylene complex of FpX **12**, and the unprotected alkene in **13** can be hydrogenated selectively. After the hydrogenation, the monoene **14** is deprotected by the treatment with NaI [3]. The double bond in the enyne **16** is protected selectively with **12** as shown by **17**. The triple bond of **17** is hydrogenated and alkene **18** is obtained by deprotection. Monoprotection of norbonadiene, followed by selective bromination of the remaining double bond, gives **19** without forming cyclopropane ring **20** [3]. Bromination of the double bond in eugenol (**21**) with bromine gives **22**, whereas selective bromination of the aromatic ring is achieved by protection of the double bond as complex **23**. After the bromination to give **24**, facile deprotection with NaI generates the free alkene **25** [4].

Cyclobutadiene (**26**) is antiaromatic and its isolation is not possible. However, it can be stabilized by η^2-coordination of Fp$^+$ to one of the double bonds to give **27**, and the uncomplexed double bond in **27** undergoes Diels–Alder reaction with cyclopentadiene to give **28** [4]. As described in Section 9.2, cyclobutadiene (**26**) can be stabilized as a diene by the η^4-coordination of Fe(CO)$_3$.

9.2 Protection and Activation of 1,3-Dienes by the Coordination of Iron Carbonyls

Fe(CO)$_3$ forms stable η^4 complexes of conjugated dienes, and acts as a useful protecting group for dienes, preventing reactions normally associated with carbon–

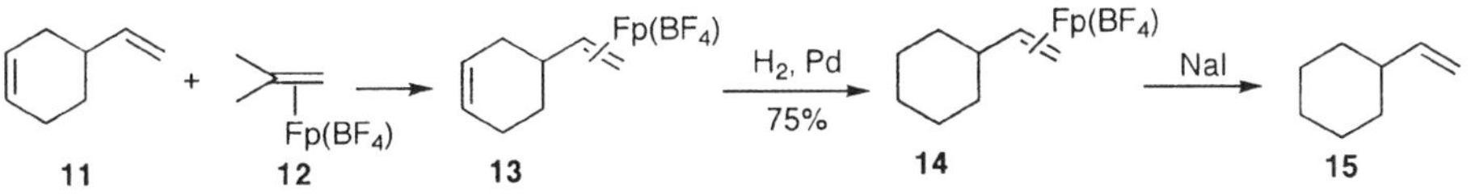

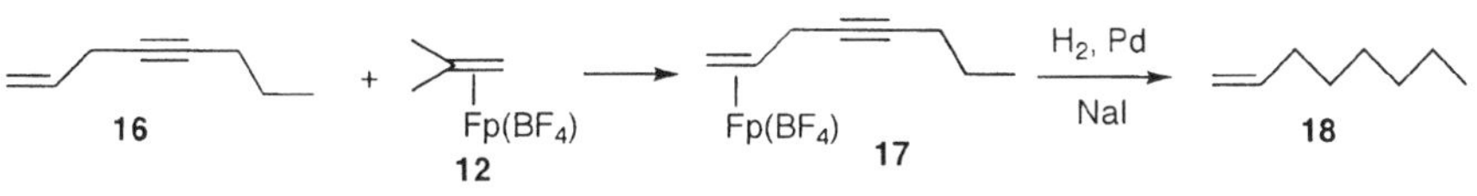

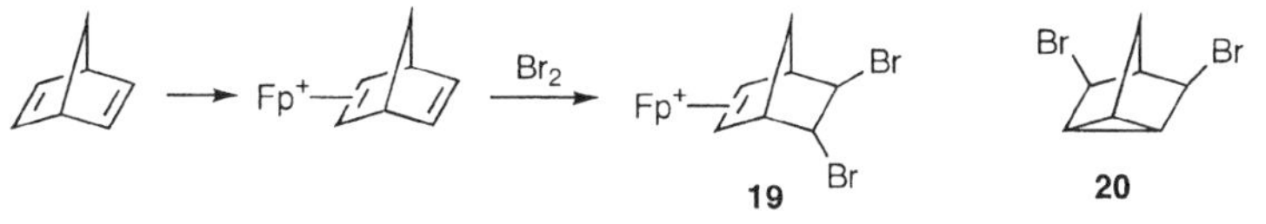

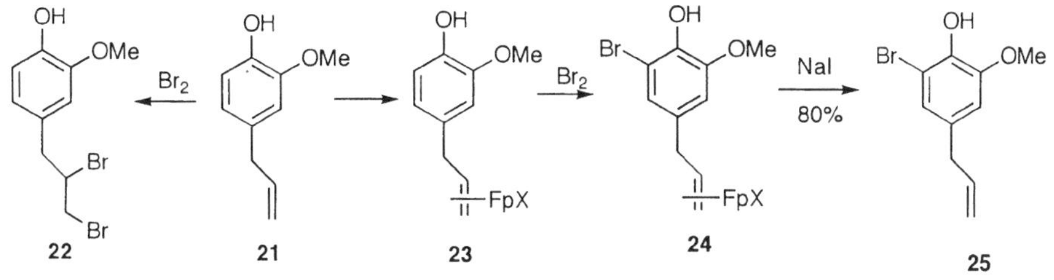

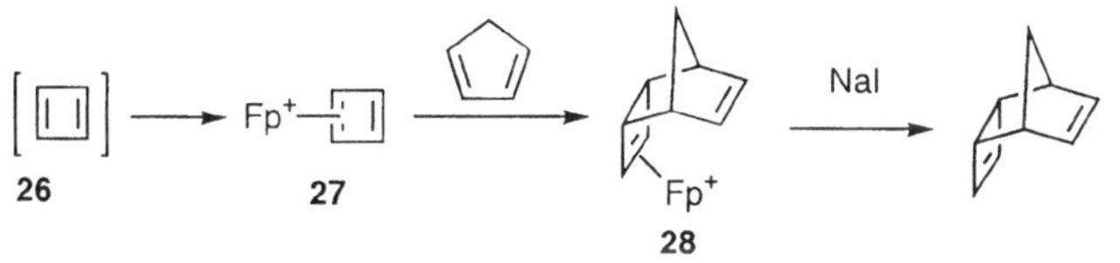

carbon double bonds (hydrogenation, hydroboration, hydride reduction, osmylation), and with 1,3-dienes (Diels–Alder reaction) [5,6]. These complexes withstand the reaction conditions of aldol and Wittig reactions. In addition, the complexation stabilizes the dienes, moderating their reactivity toward electrophiles, and enables the addition of nucleophiles that are not possible under normal conditions.

Very stable diene complexes can be prepared by the reaction of conjugated dienes with $Fe(CO)_5$, $Fe_2(CO)_9$ or $Fe_3(CO)_{12}$ by heating or under irradiation. For synthetic purposes, the complexes of acyclic 1,3-dienes, 1,3-cyclohexadienes and 1,3-cycloheptadienes are useful.

Unconjugated dienes form the 1,3-diene complexes after isomerization to conjugated dienes. Formation of the stable conjugated diene complexe is the driving force of the isomerization. For example, the 1,4-diene in the synthetic intermediate **29** of prostaglandin A can be protected as the diene complex **30** after isomerization to the conjugated diene when it is treated with $Fe_2(CO)_9$. This method was applied to the synthesis of prostaglandin C (**13**). The diene complex **30** is stable for the oxidation of the lactol and introduction of the α-chain [7].

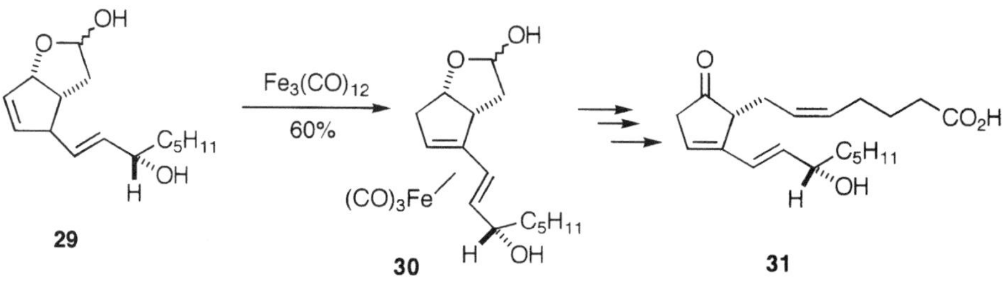

Free 1,3-dienes can be regenerated from the complexes by oxidative decomplexation using Ce(IV), H_2O_2, peracids or O_3. However, osmylation (OsO_4, t-BuO_2H) and periodate cleavage of glycols can be carried out without decomplexation [8]. The 1,3-diene of ergosterol is protected as complex **32** for osmylation, and only the double bond in the side chain is oxidized selectively to **33**. The alcohol is oxidized to ketone **34** and its reduction with a bulky reducing agent occurs from the normally hindered β-face to afford *epi*-ergosterol (**35**), because the α-face is blocked by Fe(CO)$_3$ [9].

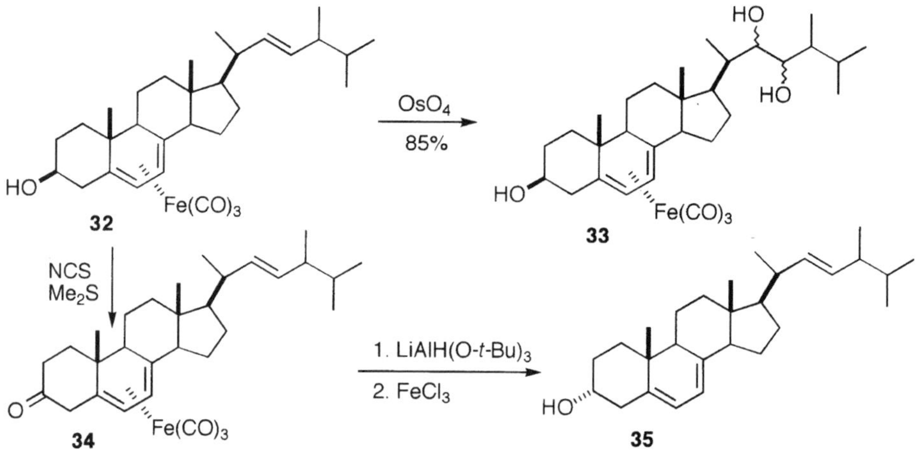

Isolation of cyclobutadiene (**26**) is not possible. It can be prepared by dechlorination of 3,4-dichlorocyclobutene (**36**) with $Fe_2(CO)_9$, and isolated as the stable cyclobutadieneiron tricarbonyl (**37**) [10,11]. The coordinated cyclobutadiene (**37**) shows aromaticity and undergoes Friedel–Crafts reaction to give **39**. Formation of the stable cationic π-allyliron **38** as an intermediate is the driving force [11]. Oxidative decomplexation of the complex **37** with Ce(IV) salt generates free cyclobutadiene (**26**), which undergoes immediate cycloaddition with unsaturated bonds. Using this method, the highly strained Dewar benzene **40** and cubane (**41**) were synthesized [12]. Also, intramolecular [2 + 2] cycloaddition of **42** produced the bicyclo[2.2.0]hexane ring **43**, which isomerized to the 1,3-cyclohexadiene **44** upon heating [13].

1,4-Cyclohexadienes are avaiable by the Birch reduction of aromatic compounds, and converted to 1,3-diene complexes by heating with Fe(CO)$_5$. 1-Methoxy-1,4-

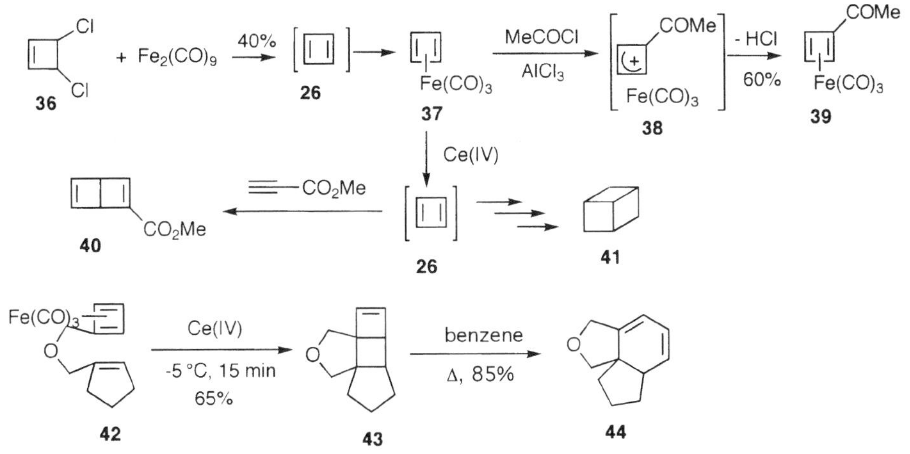

cyclohexadiene (**45**) is isomerized to the two separable conjugated diene complexes **46** and **47**. Hydride abstraction from the complexes **46** and **47** affords cationic complexes **48** and **48a**. Unstable 2,4-cyclohexadien-1-one, a keto form of phenol, can be prepared as complex **49** by hydrolysis of the methoxy-substituted cyclohexadienyl–Fe(CO)$_3$ complex **48** [14]. The complex reacts with Reformatsky reagent to give **50** and then **51**. Also, complex **49** can be used for the Wittig reaction, showing strong protection of the diene by Fe(CO)$_3$.

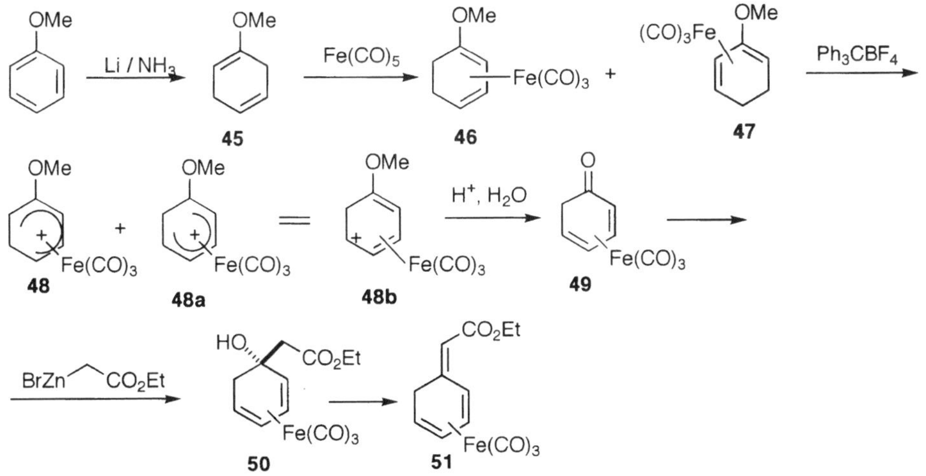

Complex **49** was converted to the oxime complex **52**. Reaction of **52** with organocuprate, followed by treatment with Ac$_2$O and CO, afforded the interesting but rather unstable [(1,2,3,4-η)-1-(*N*-acetoxy-*N*-methoxyamino)-5-*endo*-acyl-1,3-cyclohexadiene] complex **55** via **53** and **54** [15].

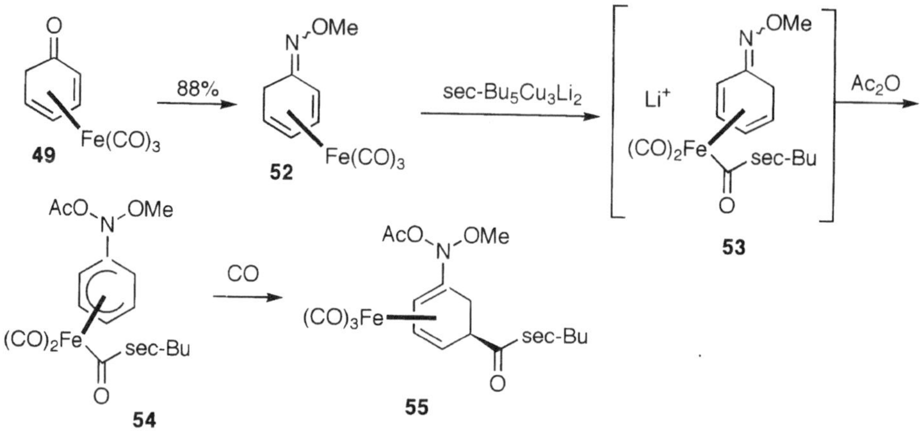

In addition to protection, a change of diene reactivity is effected by coordination to carbonyl. Butadiene forms the very stable complex **56** and its reactions are different from those of free butadiene. Electrophiles attack C(1) or C(4) of the complexed dienes, and reactions that are impossible with uncomplexed dienes now become possible.

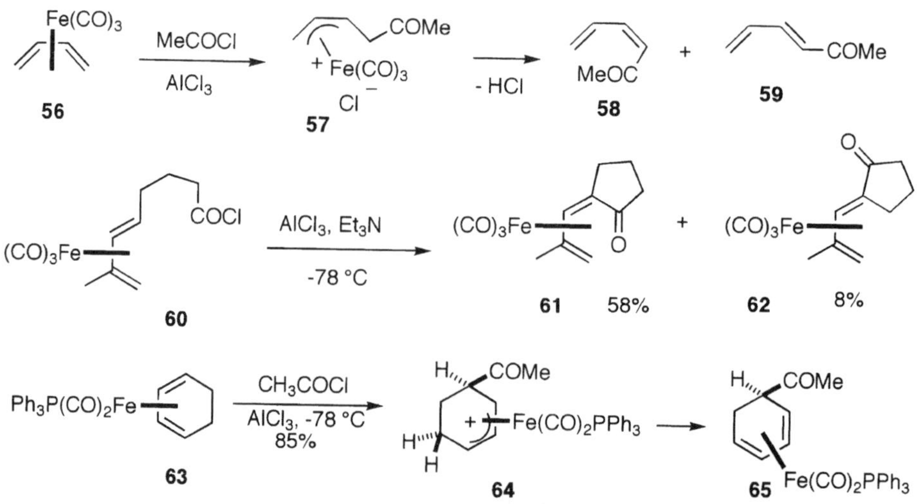

Friedel–Crafts acetylation of butadiene complex **56** proceeds smoothly to give a mixture of 1-acetyldienes **58** and **59** via the cationic π-allyl complex **57** [16]. Intramolecular Friedel–Crafts acylation with the acid chloride of the diene complex **60**, promoted by deactivated AlCl$_3$ at 0 °C, gave the cyclopentanones. The (Z)-dienone complex **61** was the major product and the (E)-dienone **62** the minor one [17]. Acetylation of the 1,3-cyclohexadiene phosphine complex **63** proceeded easily at −78 °C to give the rearranged complex **65** in 85% yield. Without phosphine coordination, poor results were obtained [18,19]. In this reaction, the acetyl group at first coordinates to Fe, and attacks at the terminal carbon of the diene from the same side of Fe to generate the cationic π-allyliron **64**. Then deprotonation occurs from the Fe side to produce **65** selectively. The acidic proton attached to the acetyl-bearing carbon in **64** is not eliminated because it is opposite to the Fe. As a result, rearrangement of the diene system occurs.

Nucleophilic attack occurs at C(2) of the diene. The 1,3-cyclohexadiene complex **66** is converted to the homoallyl anionic complex **67** by nucleophilic attack, and the 3-alkyl-1-cyclohexene **68** is obtained by protonation. Insertion of CO to **67** generates the acyl complex **69**, and its protonation and reductive elimination afford the aldehyde **70** [20]. Reaction of the butadiene complex **56** with an anion derived from ester **71** under CO atmosphere generates the homoallyl complex **72** and then the acyl complex **73** by CO insertion. The cyclopentanone complex **74** is formed by intramolecular insertion of alkene, and the 3-substituted cyclopentanone **75** is obtained by reductive elimination. The intramolecular version, when applied to the 1,3-cyclohexadiene complex **76** bearing an ester chain at C(5), offers a good synthetic route to the bicyclo[3.3.1]nonane system **78** via intermediate **77** [21].

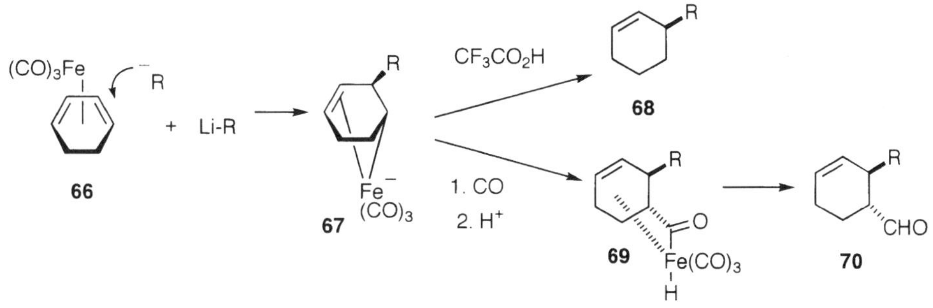

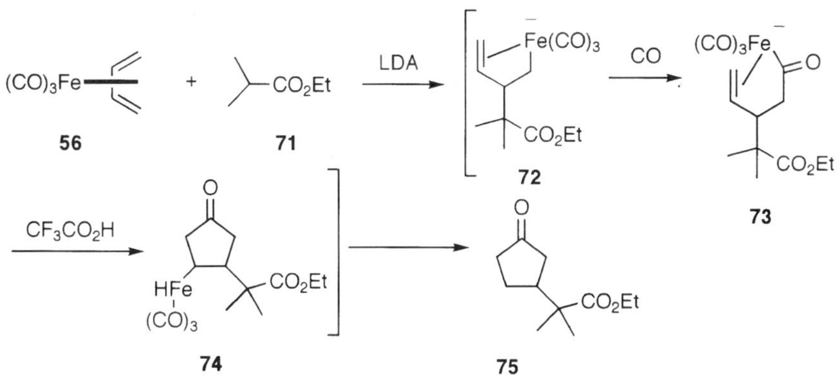

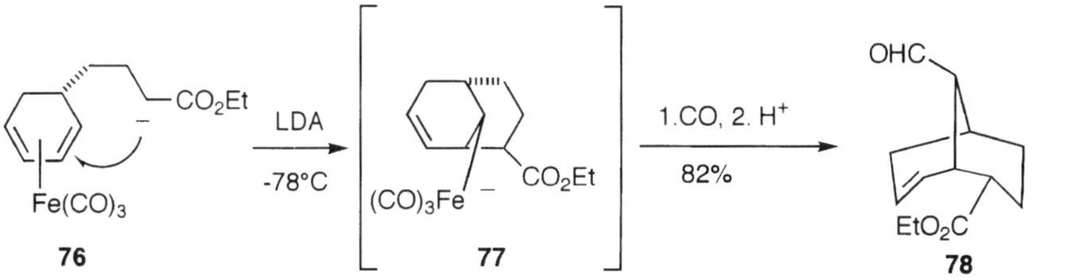

The 1,3-cyclohexadiene complex **66** was expanded to cycloheptadienone (**80**) by an interesting reaction of CO mediated by $AlCl_3$ via **79**. The bicyclo[3.2.1]octenedione **81** was prepared by the twofold carbonylation of **66** under high pressure of CO via the intermediate **79** [22,23]. This interesting transformation has been applied to the stereoselective construction of the dicyclopenta[*a,d*]cyclooctene core **83** of ceroplastin terpene from **82** under 5 atm of CO [24].

Complexes of unsymmetrically substituted conjugated dienes are chiral. Racemic planar chiral complexes are separated into their enantiomers **84** and **85** by chiral HPLC on commercially available β-cyclodextrin columns and used for enantioseletive synthesis [25]. Kinetic resolution was observed during the reaction of the *meso*-type complex **86** with the optically pure allylboronate **87** [26]. The (2*R*) isomer reacted much faster with **87** to give the diastereomer **88** with 98% ee. The complex **88** was converted to **89** by the reaction of meldrum acid. Stereoselective Michael addition of vinylmagnesium bromide to **89** from the opposite side of the coordinated Fe afforded **90**, which was converted to **91** by acetylation of the 8-OH group and displacement with Et_3Al. Finally, asymmetric synthesis of the partial structure **92** of ikarugamycin was achieved [27].

Addition to unsaturated centres (C=O, C=N, C=C) adjacent to the diene can occur in a diastereoselective fashion, and asymmetric synthesis can be carried out if the diene complex is optically active. As $Fe(CO)_3$ coordinates from one face of the unsymmetrically substituted conjugated dienes, the complexes are chiral and can be resolved to the optically active forms **93** and **94**, which are used for asymmetric synthesis. The optically active acetyldiene complex **95**, obtained by the acetylation of the optically active diene complex **94**, reacts diastereoselectively with PhLi to give **96**. The optically active tertiary alcohol **97** is obtained by its decomplexation. The enantiomer **100** can be synthesized by the opposite operation; namely the benzoylation of **94** to give **98**, and subsequent reaction of MeLi gives **99**. The enantiomer **100** is obtained by decomplexation [16].

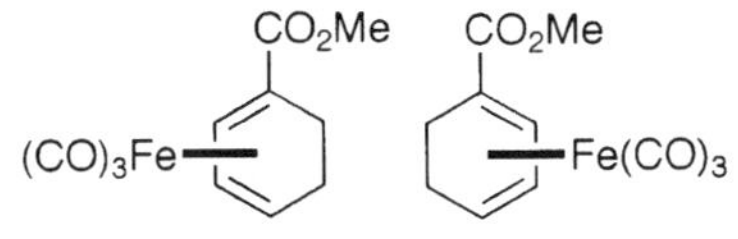
CO_2Me
$(CO)_3Fe$
CO_2Me
$Fe(CO)_3$

84 **85**

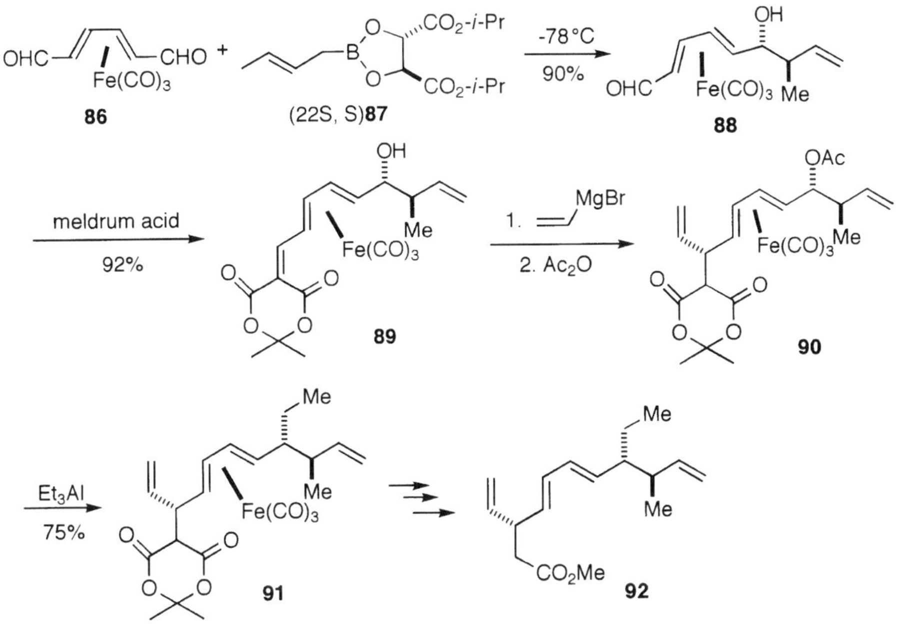
OHC
CHO
$Fe(CO)_3$
86
CO_2-*i*-Pr
CO_2-*i*-Pr
(22S, S)87
-78 °C
90%
OH
OHC
$Fe(CO)_3$ Me
88
meldrum acid
92%
89
1. MgBr
2. Ac_2O
OAc
90
Et_3Al
75%
91
CO_2Me
92

$(CO)_3Fe$
$(CO)_3Fe$

94 **93**

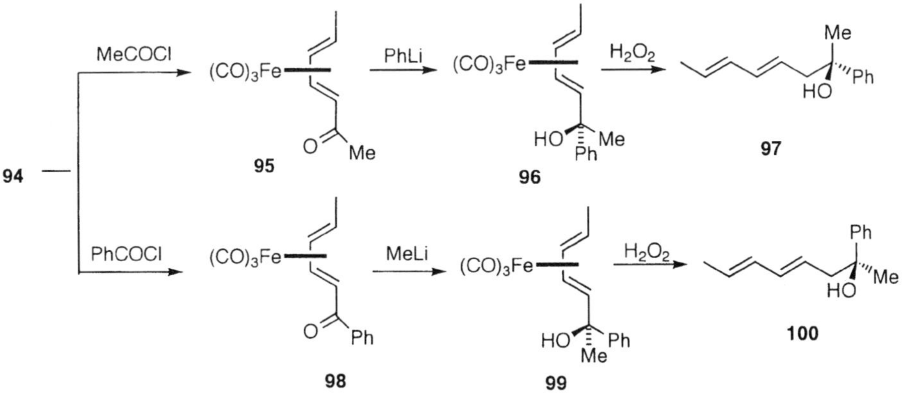
94
MeCOCl
95
PhLi
96
H_2O_2
97
PhCOCl
98
MeLi
99
H_2O_2
100

The usefulness of 1,3-cyclohexadiene complexes is enhanced by their conversion to stable cationic complexes. The η^5-cationic complex **102** is prepared as a stable salt by the hydride abstraction from the neutral complex **66** via **101**, and its highly regio- and stereoselective reaction with nucleophiles is used for synthetic purposes. Complex **102** reacts with nucleophiles such as amines, active methylenes, alkyl copper or alkoxides at C(1) or C(5) from the uncomplexed *exo* side. In other words, the nucleophilic attack occurs regioselectively at a dienyl terminus, and stereoselectively *anti* to Fe(CO)$_3$ to give **103**. Hydride abstraction from **103** affords **104**, which reacts with a nucleophile to form **105**. Decomplexation of **105** produces the 5,6-disubstituted-1,3-cyclohexadiene **106**.

66—Fe(CO)$_3$ + Ph$_3$C$^+$BF$_4^-$ $\longrightarrow$ 101 [+]—Fe(CO)$_3$ $\longrightarrow$ 102 Fe(CO)$_3$ BF$_4^-$ + Ph$_3$CH

102 Fe(CO)$_3$ BF$_4^-$ + Nu^1 $\longrightarrow$ 103 Nu^1—Fe(CO)$_3$ $\xrightarrow{\text{Ph}_3\text{C}^+\text{BF}_4^-}$ 104 Fe(CO)$_3$ BF$_4^-$ Nu^1 $\xrightarrow{\text{Nu}^2}$

105 Nu^2, Nu^1—Fe(CO)$_3$ $\longrightarrow$ 106 Nu^2, Nu^1

The cationic diene complex **107** prepared from *p*-methylanisole reacts with nucleophiles regioselectively at the methyl-substituted terminal carbon of the dienyl group and stereoselectively from the opposite side of the coordinated iron to give **108**. The following three transformations are possible, depending on subsequent decomplexation and/or dehydrogenation. As deprotection and hydrolysis of **108** afford the 4,4-disubstituted cyclohexenone **109**, umpolung of the enone γ-carbon is possible as shown by **110**, and the nucleophile is introduced at this carbon. Similarly reaction of the nucleophile at the terminal carbon of cationic complex **111** affords **112**. 1,5-Disubstituted-1,3-cyclohexadiene **113** is obtained by decomplexation. This transformation corresponds to the introduction of the nucleophile to the cyclohexadienyl cation **114**. Also, reaction of **115** with the nucleophile gives **116**, and *p*-substituted benzene **117** is obtained by decomplexation and dehydrogenation. In this way, introduction of the nucleophile to the *para* position of monosubstituted benzene is possible as shown by **118**. The products **109** and **117** represent the umpolung. They are synthetic equivalents **110** and **118**. Many synthetic applications of these species have been reported.

The following application of the synthetic equivalent **110** was carried out. Complex **107** was utilized for the total syntheses of trichothecene (**122**), trichodiene and trichodermol, applying the reaction of cationic complex **107** with a β-keto ester or tin enolate [28]. The tin enolate of cyclopentanone **119** reacted with the complex **107** with high diastereoselectivity to give the diene complex **120** in high yield. After

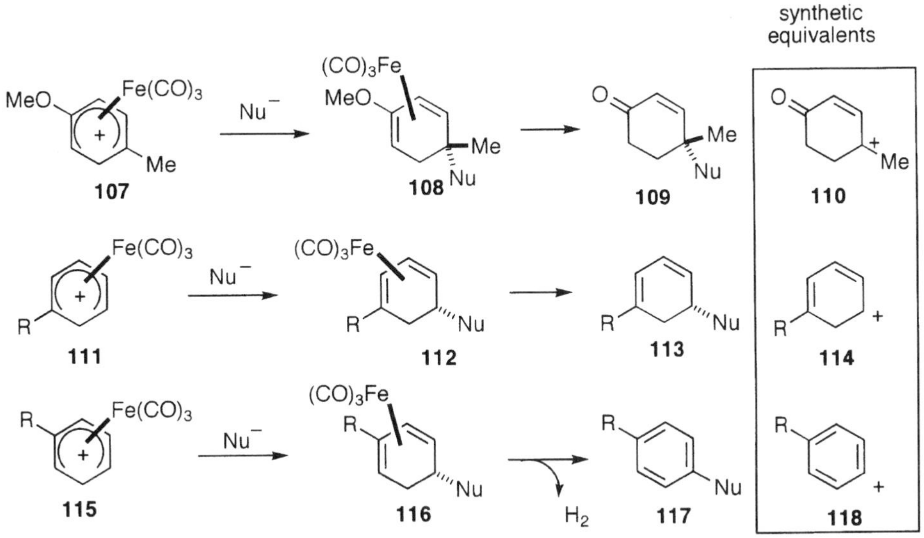

synthetic
equivalents

introduction of an oxygen function at the α-carbon of the ketone, oxidative decomplexation with $CuCl_2$ afforded the enone **121**, from which trichodermol (**122**) has been synthesized after several steps.

Nucleophilic attack of the electron-rich aromatic ring **124** to the cationic complex **123**, and intramolecular amination afforded the intermediate **125** for the synthesis of discorhabdin and prianosin alkaloids [29].

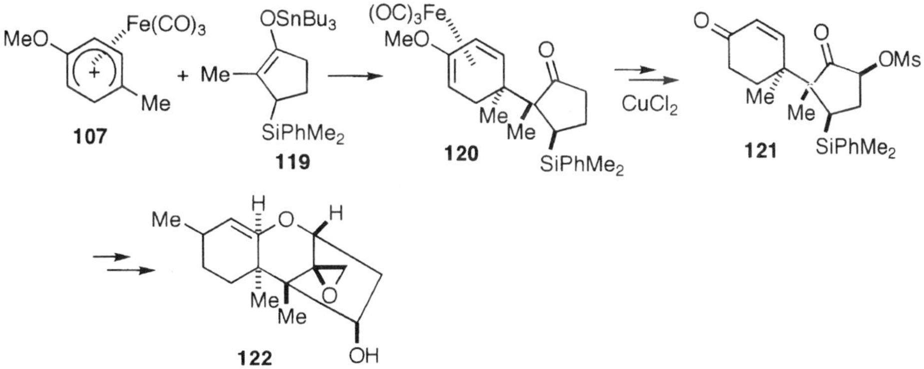

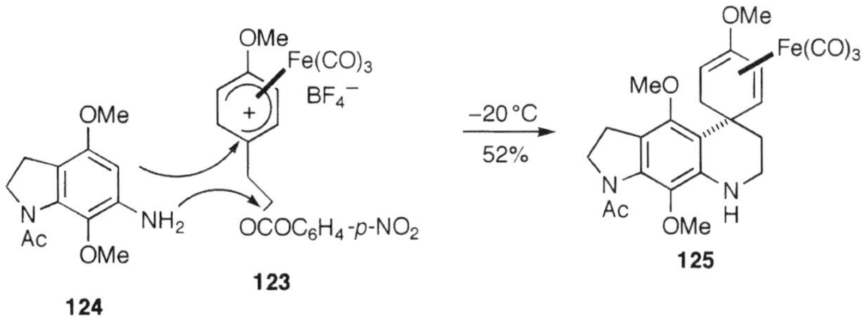

As an example of conversion of complex **111** to **113**, optically pure complex **126** was used for the enantioselective total synthesis of shikimic acid (**129**) [30]. The hydroxy-substituted diene complex **127** was prepared from **126**. Silylation and decomplexation of **127** gave **128**. Stereoselective dihydroxylation of the more reactive double bond of the decomplexed silyl ether derivative **128**, followed by desilylation afforded (−)-methyl shikimate (**129**).

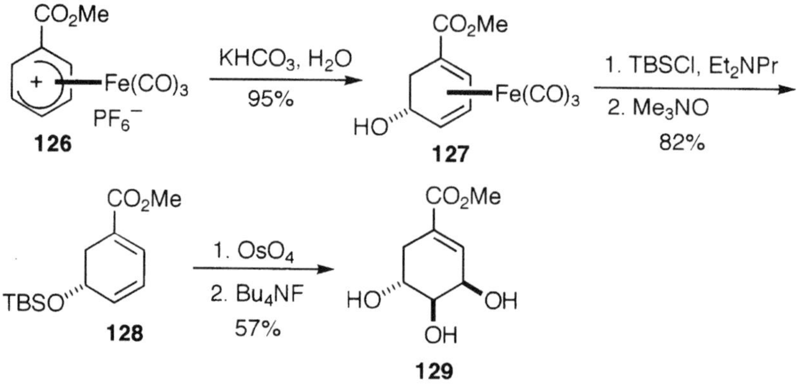

9.3 Protection and Activation of Alkynes by the Coordination of Cobalt Carbonyl

The carbonyl $Co_2(CO)_6$ forms stable π-complexes of alkynes (η^2 complexes). Four effects on alkyne reactivity are expected from this coordination: (i) protection of the triple bond; (ii) stabilization of the carbonium ion on the α-carbon (or propargylic cations; (iii) syntheses of common and medium-size cycloalkynes; and (iv) steric effects.

As a result of complex formation, the normally linear digonally hybridized triple bond bends to approximately 145°. Two π-bonds in the triple bond coordinate to two Co atoms, respectively, as shown by **130**. In the following discussion, the simplified form **131** is used instead of **130** in most cases for simplicity.

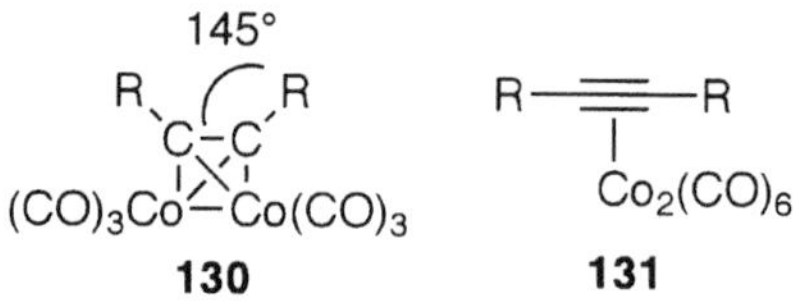

The 1,2-shift of a cyclopropyl group is well-known as shown by **132**. Similarly, propargyl-type alkyne Co complexes which have the cobaltacyclopropyl structure **130** undergo a facile 1,2-shift. The Co complex **134** of the alkynylchlorohydrin **133** undergoes facile 1,2-shift promoted by Me_3Al at low temperature to give **135**, and the α-alkynyl ketone **136** is obtained by decomplexation [31]. Rearrangement of the free alkyne **133** to **136** is not possible.

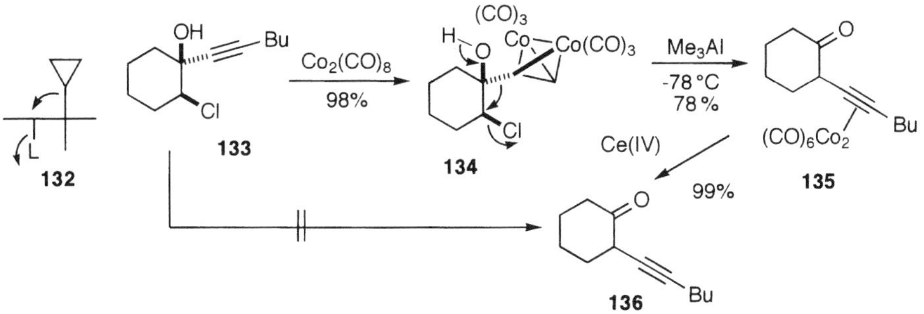

As regards the protecting effect, the complex is stable to Lewis acids. Also, no addition of BH_3 occurs. As $Co_2(CO)_6$ can not coordinate to alkene bonds, selective protection of the triple bond in enyne **137** is possible, and hydroboration or diimide reduction of the double bond can be carried out without attacking the protected alkyne bond to give **138** and **139** [32]. Although diphenylacetylene cannot be subjected to smooth Friedel–Crafts reaction on benzene rings, facile *p*-acylation of the protected diphenylacetylene **140** can be carried out to give **141** [33]. The deprotection can be effected easily by oxidation of coordinated low-valent Co to Co(III), which has no ability to coordinate to alkynes, with CAN, Fe(III) salts, amine *N*-oxide or iodine.

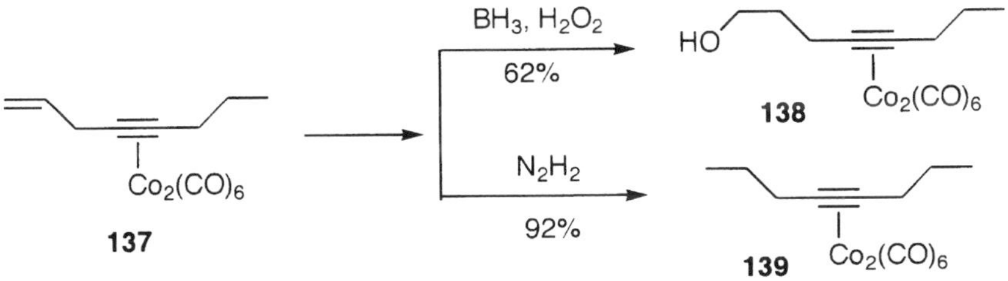

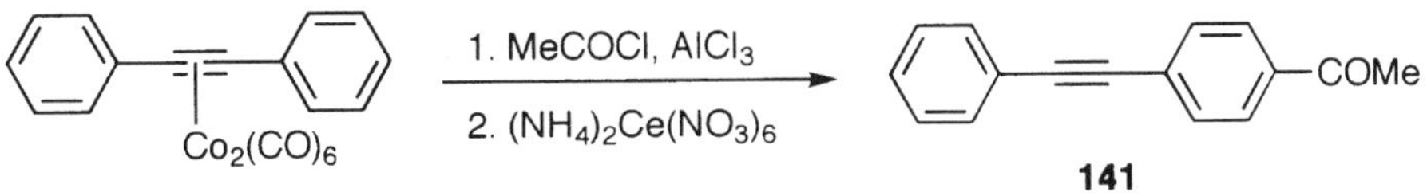

The coordination stablizes the α-carbonium ion. This stabilization is due to delocalization of the positive charge of the propargyl cation by the Co cluster as shown by **142** and **143**. Metal carbonyls are strong electron-withdrawing groups, and generally decrease the electron density of alkynes, alkenes and arenes by coordination. Therefore, stabilization of the propargyl anion is expected by the coordination of $Co_2(CO)_6$. In practice, however, the propargyl cation, rather than anion, is stabilized by the contribution of **142** and **143**. Thus propargylic alcohols undergo facile dehydration as a result of coordination and react easily with many nucleophiles. The reaction at the Co-stablized propargylic cation with nucleophiles is called the Nicholas reaction [34,35]. Friedel–Crafts reaction of propargylic alcohol **144** with activated aromatic compounds such as anisole gives **145** by this coordination [36].

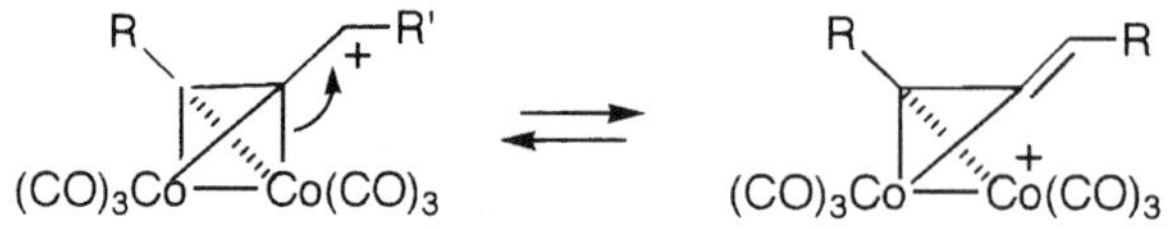

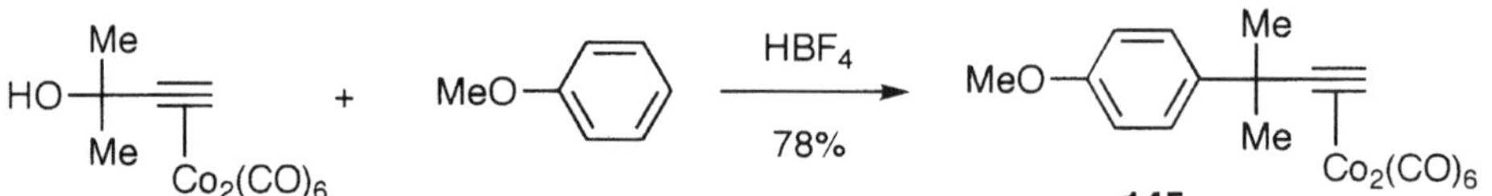

According to the Baldwin rule, the *exo* cyclization mode is favored in intramolecular reactions of alkynyloxiranes with alcohols to afford cyclic ethers. However, the unfavorable *endo* cyclization mode is observed by the complexation. Thus exclusive *endo* cyclization of epoxide complex **146** takes place regioselectively to give **147**, without forming the five-membered ether **148** by *exo* mode reaction [37].

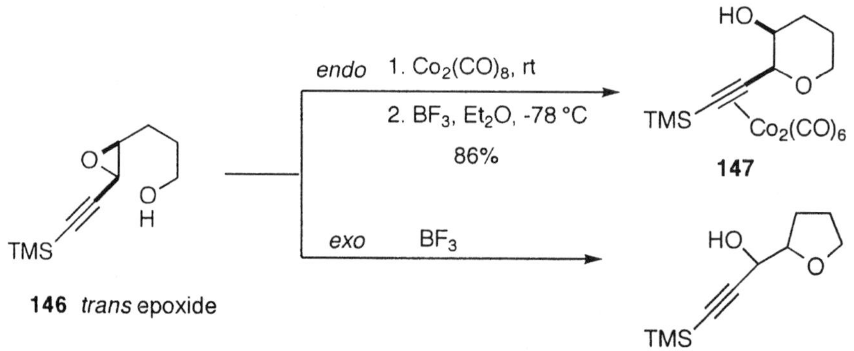

Coordination to alkynes distorts the triple bond character nearer to that of a double bond, decreasing the linearity. Utilization of the coordination effect makes it feasible to prepare cyclic alkynes whose synthesis is difficult to achieve. Highly strained cyclooctyne can be prepared by coordination. As an example, starting from (R)-pulegone, three of the four rings of the epoxydictymene skeleton **152** were constructed by the consecutive Lewis acid-promoted Nicholas reaction of allylic silane **149** to form **150**, and the intramolecular Pauson–Khand reaction of **151**. The total synthesis of (+)-epoxydictymene **153** from **152** has been achieved [38].

Intramolecular Lewis acid-promoted reaction of coordinated propargylic ether with the silyl enol ether in **158** has been applied successfully to the construction of the highly strained 10-membered cyclic enediyne system **159**, present in esperamycin and calicheamycin [39,40]. The enediyne system **157** was prepared by the Pd-catalysed Sonogashira coupling of (Z)-1,2-dichloroethylene (**154**) with two different terminal alkynes **155** and **156**.

The seven-membered cyclic ether **161** containing the Co-stablized triple bond was prepared by the Nicholas reaction of **160**. The decomplexation and reduction of the triple bond to the double bond to produce **162** were achieved by Rh-catalysed hydrogenation. The method was used for the preparation of the A/B fragment of ciguatoxin [41].

The highly strained molecule of 3,4-diphenylcycloocta-1,5-diyne (**165**), stabilized by the coordination of $Co_2(CO)_6$, was prepared from the bis-propargylic alcohol **163**. Cyclization of the propargylic dication complex **164**, induced by reduction with Zn, afforded the cyclocta-1,5-diyne ring **165** in 48% yield. The complex of more strained 7,8-diphenylcyclooct-3-ene-1,5-diyne was prepared similarly [42].

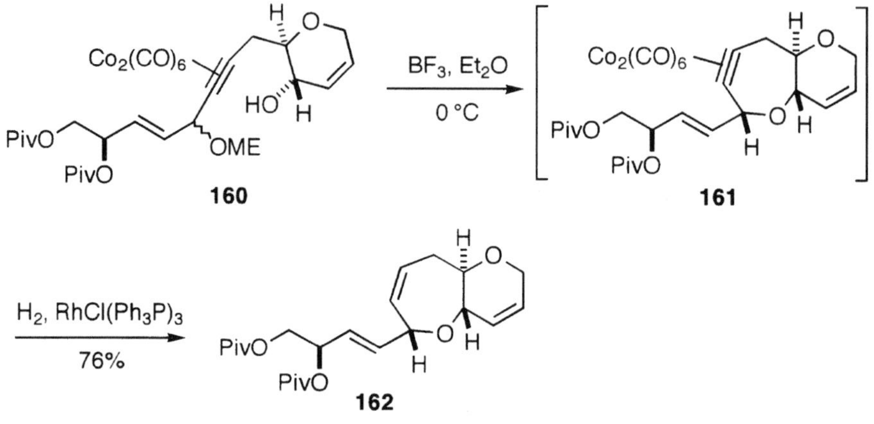

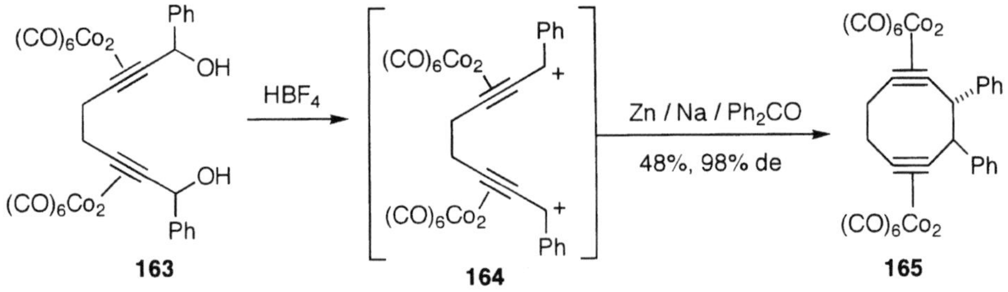

The α-oriented alkynyl sugar **166**, formed originally, is transformed to the β-isomer **169** by forming the Co complex **167**. Due to the stable propargylic cation of the Co complex, facile opening and closing of the dihydropyran ring in **167** yield the thermodynamically stable β-isomer **168** [43].

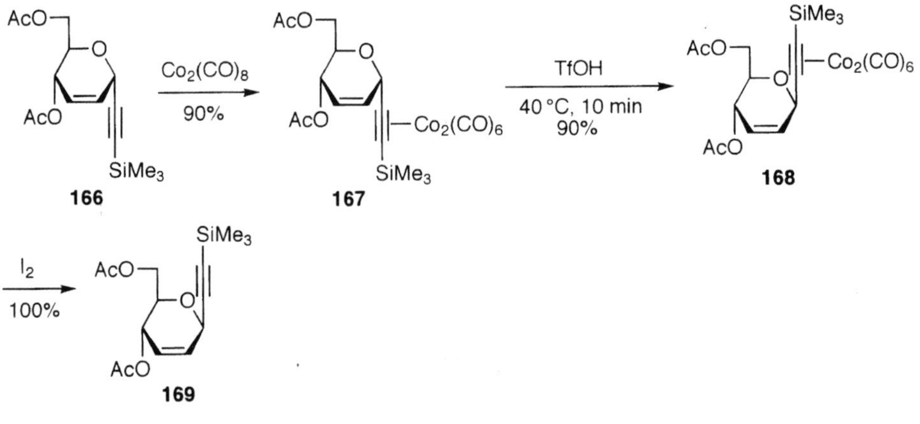

9.4 Activation of Arenes and Cycloheptatrienes by Coordination of Chromium Carbonyl and Other Metal Complexes

Complexes of Cr, W, Mo, Fe, Ru, V, Mn and Rh form stable, isolable arene η^6-complexes. Among them, arene complexes of $Cr(CO)_3$ have high synthetic uses. When benzene is refluxed with $Cr(CO)_6$ in a mixture of dibutyl ether and THF, three coordinated CO molecules are displaced with six-π-electrons of benzene to form the stable η^6-benzene chromium tricarbonyl complex (**170**) which satisfies the 18-electron rule (6 from benzene + 6 from Cr(0) + 6 from 3 CO = 18). Complex formation is facilitated by electron-donating groups on benzene, and no complex of nitrobenzene is formed. Complex formation has a profound effect on reactivity of arenes, and the resulting complexes are used in synthetic reactions. The metal-free reaction products can be isolated easily after decomplexation by mild oxidation using low-valent Cr. Cycloheptatriene also forms a stable complex with $Cr(CO)_3$ and its synthetic applications are discussed below.

Through this coordination, the electron density of the benzene ring is considerably reduced. The electron-withdrawing effect of coordinated $Cr(CO)_3$ is similar to that of a nitro group, as is apparent from the following pK_a values: pK_a of $PhCO_2H = 5.68$; $p\text{-}NO_2C_6H_4CO_2H = 4.48$; and $PhCO_2HCr(CO)_3 = 4.77$. The basicity of aniline is decreased from 11.70 (pK_b) to 13.31 by complex formation. In addition to the electron-withdrawing effect, coordinated $Cr(CO)_3$ has a large steric influence by blocking one face of the arenes. These electronic and steric effects of the coordination bring about the following changes to aromatic rings:

1. Aromatic nucleophlic substitution with stabilized carbanions is possible, and nucleophilic substitution products and cyclohexadiene derivatives are preparable.
2. The displacement of aromatic halogens is facilitated, and its order is F > Cl > Br > I.
3. The acidity of aromatic hydrogens is increased and lithiation of the aromatic ring is facilitated.

4. The acidity of the benzylic hydrogen is increased, and hence the benzylic carbanion is stablized and substitution at this carbon with electrophiles is accelearated. In addition, the benzylic cation is stabilized.
5. Due to the coordination of bulky Cr(CO)$_3$, nucleophiles react from the opposite side of the ring, (i.e. the *exo* side) with high stereoselectivity.
6. Another important feature is the planar chirality of complexes of *ortho* and *meta* disubstituted arenes, and asymmetric synthesis is possible using optically active complexes.

9.4.1 Reactions of Carbanions

Arenes usually undergo electrophilic substitution, and are inert to nucleophilic attack. However, nucleophilc attack on arenes occurs by complex formation. Fast nucleophilic substitution with carbanions with pK_a values > 22 has been extensively studied [44]. The nucleophiles attack the coordinated benzene ring from the *exo* side, and the intermediate η^2-cyclohexadienyl anion complex **171** is generated. Three further transformations of this intermediate are possible. When Cr(0) is oxidized with iodine, decomplexation of **171** and elimination of hydride occur to give the substituted benzene **172**. Protonation with strong acids, such as trifluoroacetic acid, followed by oxidation of Cr(0) gives rise to the substituted 1,3-cyclohexadiene **173**. The 5,6-*trans*-disubstituted 1,3-cyclohexadiene **174** is formed by the reaction of an electrophile.

The carbon nucleophiles listed in eq. (9.1) are known to react. But no reaction occurs with some nucleophiles such as malonate and Grignard reagents.

$$\text{LiCH}_2\text{CO}_2\text{R, LiCH}_2\text{CN, KCH}_2\text{CO-}t\text{-Bu,LiCH(CN)(OR), LiCH}_2\text{SPh}$$
$$\text{2-Li-1,3-dithiane, LiCH=CH}_2, \text{LiPh, LiCH}_2\text{CH=CH}_2, \text{LiCMe}_3, \quad \text{Li} \equiv\!\!\equiv\!\!-\text{R} \qquad (9.1)$$

First, the preparation of the substituted benzene **172** is explained. In the reaction of substituted benzene complex **175** with carbanions, the *meta* orientation to give **176** is observed even in the presence of *ortho*- and *para*-orienting electron-donating groups, such as methoxy and amino groups [45]. Using this property, the nucleophilic substitution reaction, complementary to ordinary electrophilic substitution reaction, is made possible by complex formation. As shown by the following example, substitution at C(1) in the A ring of an aromatic steroid is difficult. However, by the coordination of Cr(CO)$_3$, the 1,3-dithiane anion **178** was introduced at C(1), *meta* to the methoxy group in **177** at −78 °C in THF. Oxidative decomplexation of the product **179** with iodine and hydrogenolysis with Raney Ni produced the 1-methyl steroid **180** [46].

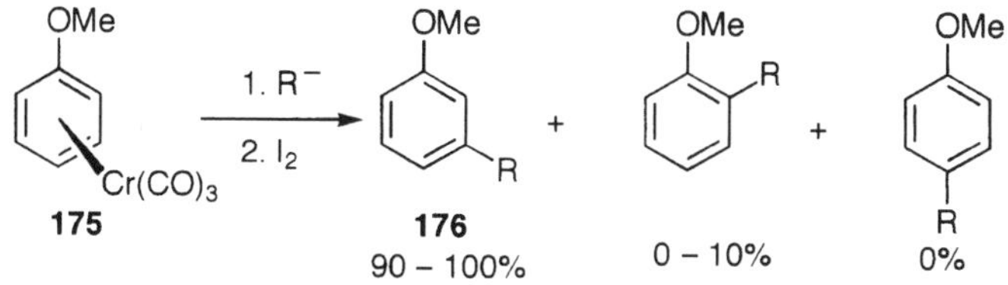

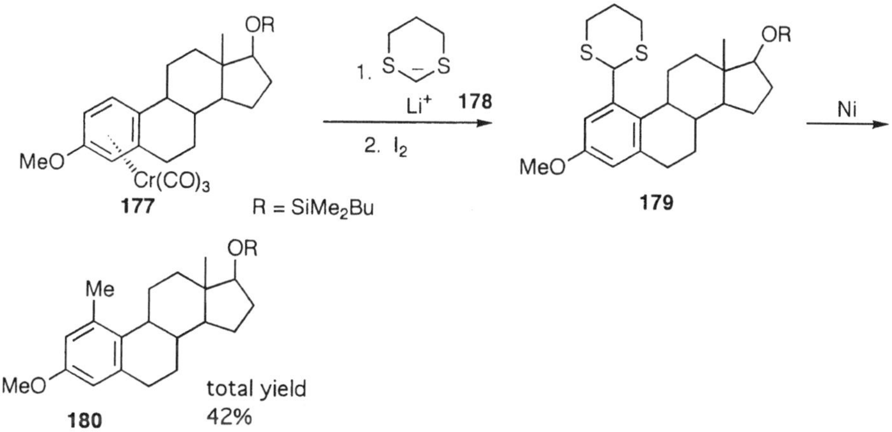

Cr(CO)$_3$ coordinates to the benzene ring of indole (**181**) selectively and the reaction occurs mainly at the normally inaccessible C(4) carbon to give **182**, and substitution at C(7) to give **183** is a minor reaction [47]. With the uncomplexed indole, substitution at C(2) is common.

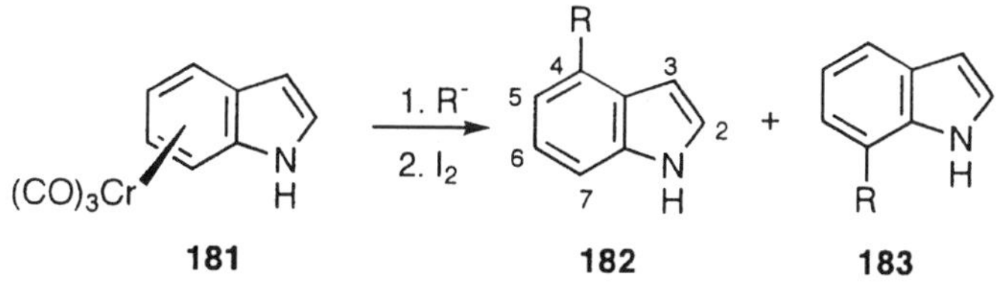

	182		**183**	
R = LiC(CH$_3$)$_2$CO$_2$-*t*-Bu	99	:	1	92% yield
= LiCH$_2$CN	75	:	25	43% yield

Formation of cyclohexadiene **173** by protonation is a useful reaction. Cyclohexenone can be prepared from anisole by this reaction. *Meta*-substitution of the complexed anisole (**175**) with a nucleophile, followed by protonation and hydrolysis of the anionic intermediate **184**, gives rise to the two cyclohexenone derivatives **185** and **186**; their ratio depends on the conditions [44].

Synthesis of acrorenone is an interesting application of the modified reactivity of *o*-methylanisole (**187**) by coordination of $Cr(CO)_3$ [48]. After the coordination, the carbanion of cyanohydrin **188** was introduced at the *meta* position to the methoxy group to produce **189** after hydrolysis, and modification of the side chain was carried out. The recoordination of $Cr(CO)_3$ gave rise to two diastereomers of **190** due to coordination from both sides of the ring. Intramolecular nucleophilic alkylation of one of the diastereomers **190** took place again at the *meta* position to generate the anionic cyclohexadiene complex with a spiro ring. Protonation of the complex gave cyclohexadiene **191** and its hydrolysis produced cyclohexenone **192**, which was converted to acrorenone (**193**).

5,6-Disubstituted-1,3-cycolohexadiene can be obtained by the irreversible reaction of a reactive nucleophile, followed by the attack of electrophiles other than a proton. For example, MeLi attacked from the *exo* side of **194** to give **195**, which was converted to the Cr–acyl complex **196** by ethylation and CO insertion. When **196** was warmed, acylation of the ring from the same side as $Cr(CO)_3$ by reductive elimination occurred to give the *trans*-5,6-disubstituted cyclohexadiene **197**, stereo- and regioselectively [49]. α-Ethylation of the ketone **197** afforded **198**. Reaction of an electrophile on the dienylanion complex **195** usually regenerates the original $Cr(CO)_3$–arene complex. However, if the reaction of nucleophile is irreversible, the disubstituted cyclohexadiene is formed.

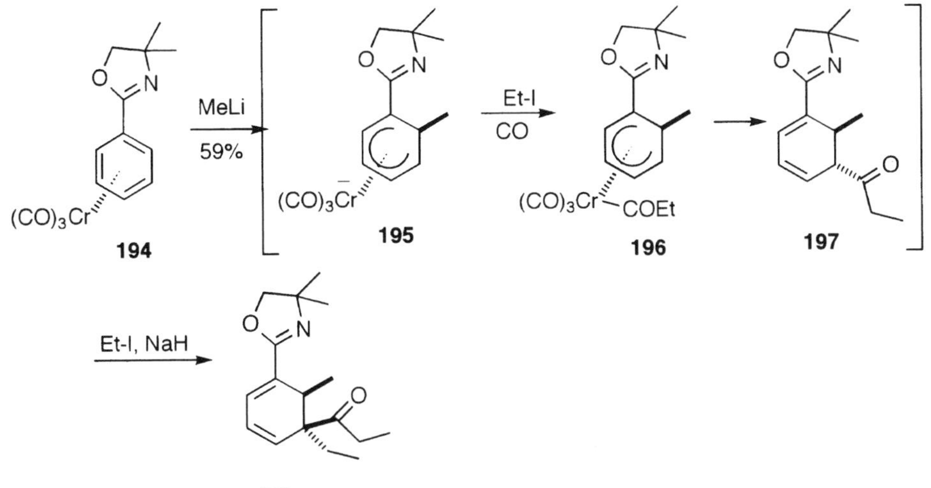

Asymmetric addition of PhLi, coordinated by the chiral ligand **200**, to the prochiral imine complex **199** generates **201**. Discrimination between enantiotopic sites at C(2) and C(6) occurs. Then the 5,6-*trans*-disubstituted 1,3-cyclohexadienal **202** was prepared with 93% ee by electrophilic attack of propargyl bromide [50].

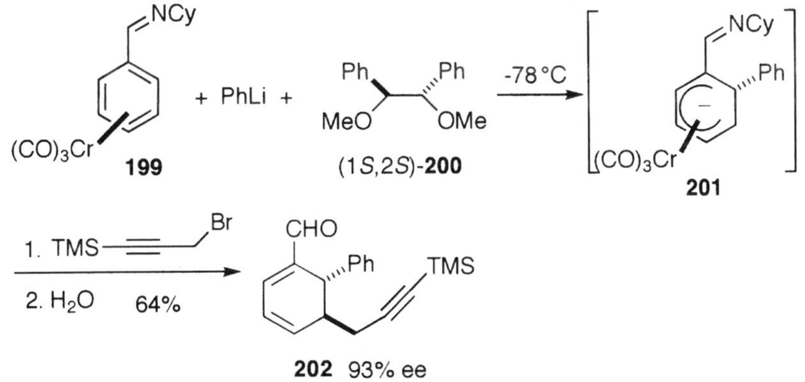

The complex of *o*-substituted anisole **203** is planar chiral, and can be used for diastereoselective generation of two new stereogenic centres in the products. Propargylation and allylation of **203** gave **204** regio- and stereoselectively. Hydrolysis of **204** afforded the cyclohexenone **205**, and its intramolecular Pauson–Khand reaction gave **206** diastereoselectively. The two reactions were completely diastereoselective, and the planar chirality in **203** was efficiently transferred to the three new stereogenic centers in **206** [51].

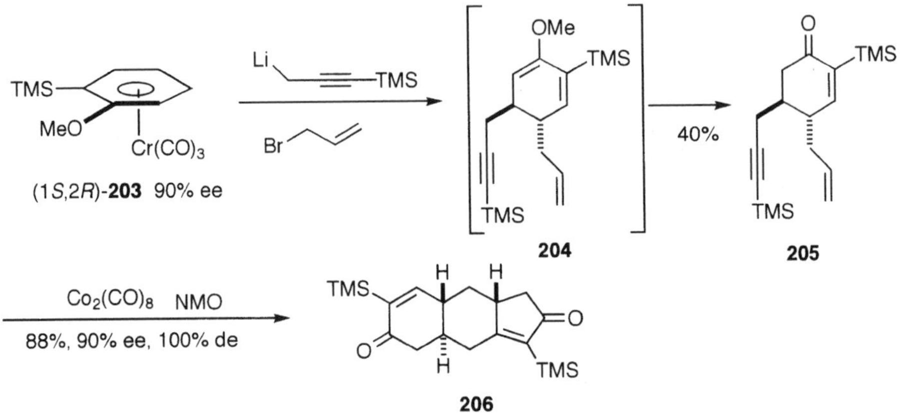

9.4.2 Nucleophilic Substitution of Aromatic Chlorides

Although chlorobenzene is rather inactive in usual reactions, its activity is enhanced by complex formation, and two products are formed by the reaction of stabilized carbanions on the complexed chlorobenzene **207**, depending on the conditions [44]. The anion of α-methylpropionitrile reacts at the *meta* position at −78 °C, and the *meta*-substituted product **208** is obtained by oxidation with I₂. However, equilibration (rearrangement) of the carbanion occurs at 25 °C, because the attack of the carbanion is reversible, and the substitution product **209** of the chlorine is obtained. The fluorobenene **210**, coordinated by Cr(CO)₃, is very reactive. Reaction of γ-butyrolactone to the *o*-lithiated fluorobenzene **211** gives rise to the alkoxide **212**, which displaces the fluoride intramolecularly to give the cyclic ether **213** [52]. In other words, the complex **211** can be regarded as the 1,2-dipolar synthon **214**. However, Cr(CO)₃-complexed aromatic bromide and iodide can not be used for the nucleophilic substitution.

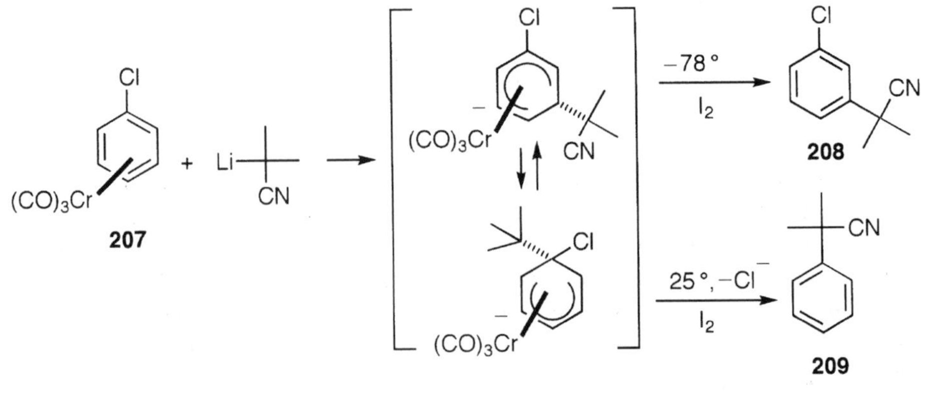

Although as described in Section 3.1.1.1, oxidative addition of chlorobenzene to Pd(0) is difficult, chlorobenzene **207**, coordinated by Cr(CO)₃, undergoes the facile oxidative addition to Pd(0). Then alkene insertion (Heck reaction) [53] and coupling with terminal alkynes occurs to afford aryl alkyne **215** [54]. In the Suzuki–Miyaura reaction of chloroanisole (**216**) with *p*-bromophenylboronate (**217**), the chloride in **216**, although further deactivated by the methoxy group, reacts with the boronate **217** to give **218**. No reaction of the bromides in **217** and **218** occurs [55]. Carbonylation of chlorobenzene to form benzoate is made possible by coordination [56].

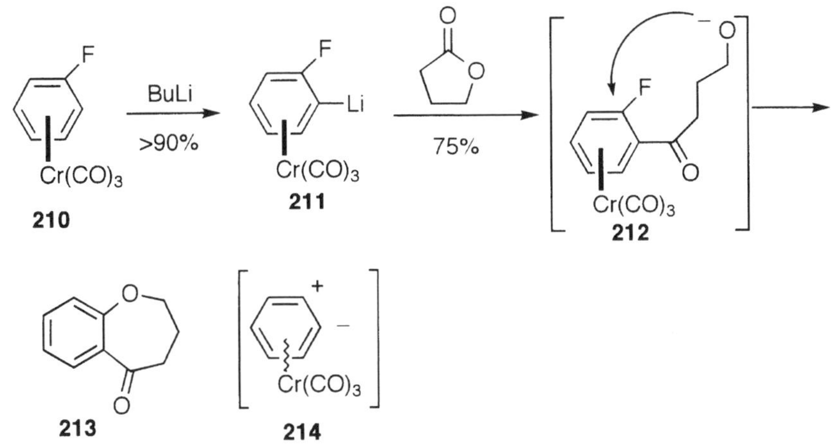

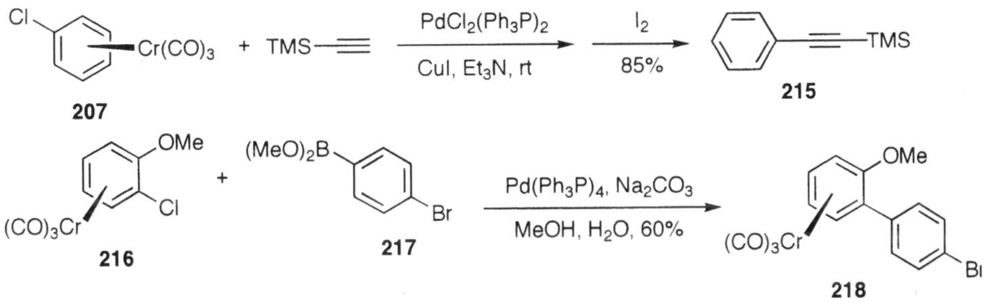

The coordination of Cr(CO)$_3$ does not activate aryl chloride sufficiently for Williamson diaryl ether formation to occur. Smooth formation of aryl ether **222** proceeds by reacting the easily prepared arene–Ru complex **220** of the highly functionalized aryl chloride with phenol **219**. Decomplexation of **221** by irradiation gives **222**, and the product is used for the synthesis of the BCF rings of ristocetin A [57].

9.4.3 Lithiation of Aromatic Rings

Another effect of coordination is facilitated lithiation with *n*-BuLi. Lithiation at the *ortho* position takes place with anisole, chlorobenzene and fluorobenzene **223**, and reaction of the lithiation products **224** with electrophiles gives the *ortho* product **225** after decomplexation [58]. This selective lithiation was applied to the synthesis of the

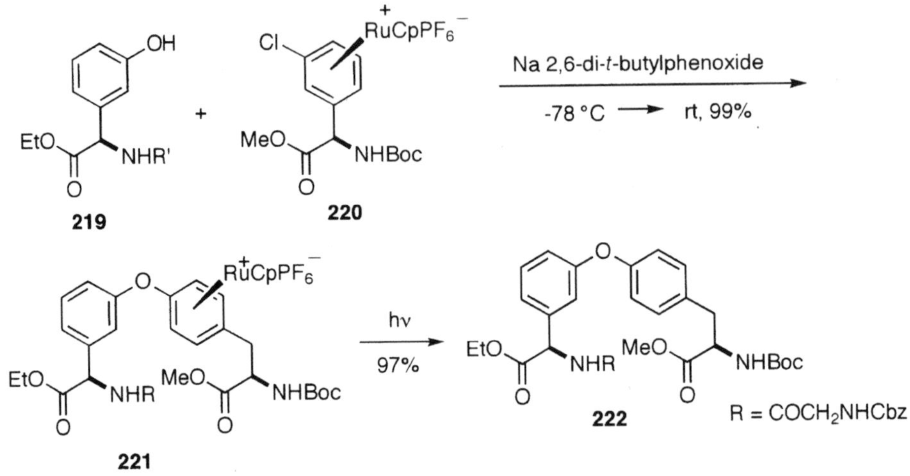

OH
EtO
NHR'
O
219
+
Cl
RuCpPF6
MeO
NHBoc
O
220
Na 2,6-di-t-butylphenoxide
-78 °C → rt, 99%
RuCpPF6
O
EtO
NHR
MeO
NHBoc
O
221
hν
97%
O
EtO
NHR
MeO
NHBoc
222
R = COCH2NHCbz

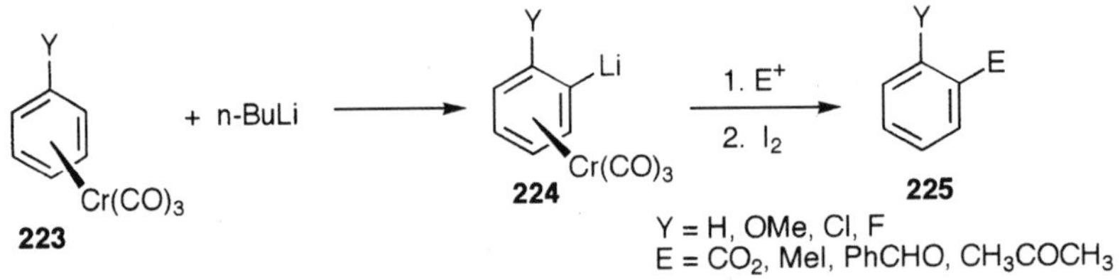

Y
Cr(CO)3
223
+ n-BuLi
Li
Cr(CO)3
224
1. E+
2. I2
E
225
Y = H, OMe, Cl, F
E = CO2, MeI, PhCHO, CH3COCH3

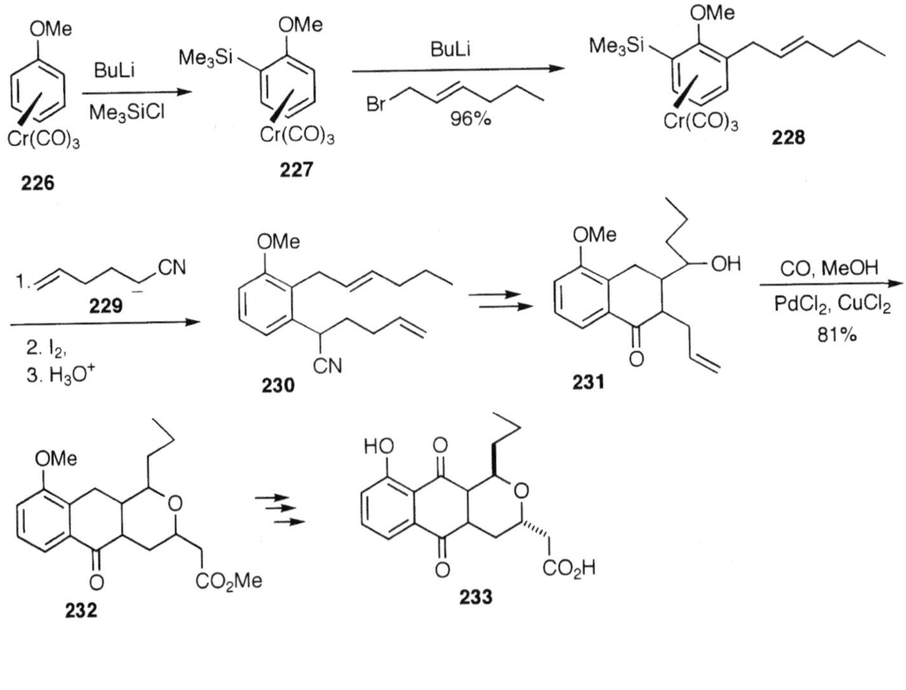

OMe
Cr(CO)3
226
BuLi
Me3SiCl
Me3Si
227
BuLi
Br
96%
228
1.
CN
229
2. I2,
3. H3O+
230
OH
231
CO, MeOH
PdCl2, CuCl2
81%
232
CO2Me
HO
CO2H
233

frenolicin intermediate **233**. One side of the methoxy group in the anisole complex (**226**) was blocked with a TMS group to give **227**, and the alkylated complex **228** was prepared after *ortho*-lithiation. Then introduction of the carbanion **229** to the less-hindered site, *meta* to the methoxy group, followed by oxidation with I$_2$, and desilylation, afforded **230**, which was converted to the unsaturated alcohol **231**. The final step is the oxidative intramolecular oxycarbonylation of the unsaturated alcohol **231** using PdCl$_2$ and CuCl$_2$ to yield **232**. Its hydrolysis afforded frenolicin (**233**) [58].

9.4.4 Activation of Benzylic Carbons by Coordination

The negative charge on the benzylic position of the side chain is stabilized by the electron-withdrawing effect of coordinated Cr(CO)$_3$. In complex **234**, attack of a carbanion occurs at the β-carbon of the double bond to generate the stabilized anion **235** at the benzylic carbon, which is trapped by an electrophile such as MeI. Thus two substituents are introduced from the opposite side of Cr(CO)$_3$ to give the *cis* product **236** [59].

Utilizing the activation of benzylic carbon, the model compound **245** of helioporing E has been synthesized stereo- and regioselectively [60]. At first the complex **238** was prepared from the optically active complex **237**. Hydrogenation of **238** proceeded stereoselectively from the opposite side of Cr(CO)$_3$. The lithiation occurred on the aromatic ring at 1 and 4 positions, which were trapped by silylation to give **239**. Then the less hindered benzylic carbon was regioselectively alkylated as shown by **240**, and Michael addition to **241** at the remaining benzylic carbon afforded **242**. The ketone **243** was obtained by intramolecular acylation, and converted to the methyl group **244** by methylation with Me$_3$Al of the acetoxy group from the opposite side of Cr(CO)$_3$ by the Uemura's method [61]. Further methylation afforded **245**.

When there is a substituent on the benzene ring, a benzylic methylene proton at the *meta* position is more activated than the *para* position. Reaction of formaldehyde on the complexed estradiol derivative **246** occurred regioselectively and stereoselectively from the opposite side of Cr(CO)$_3$ to give **247** [62].

Another effect of the coordination is that the benzylic cation is also stabilized [63]. This stabilization is explained by delocalization of the positive charge due to the interaction of the d-orbital of Cr with the π-orbital of the benzylic carbon, caused by the coordination of Cr(CO)$_3$. Facile stereospecific Friedel–Crafts-type cyclization of the complex of the optically active benzyl alcohol **248** gave the tetrahydrobenzazepine **249** with retention of the stereochemistry, and the free amine **250** with 98% ee was prepared by decomplexation [64]. Although the cyclized product **250** was obtained without coordination of Cr(CO)$_3$, complete racemization occurred and the yield was low.

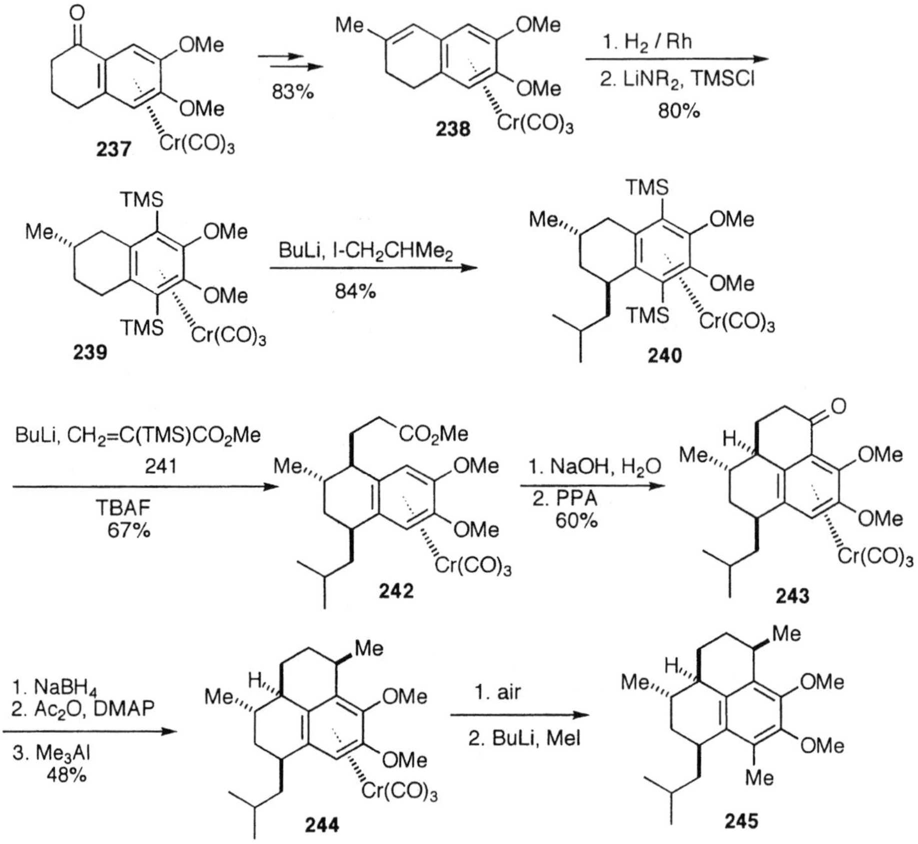

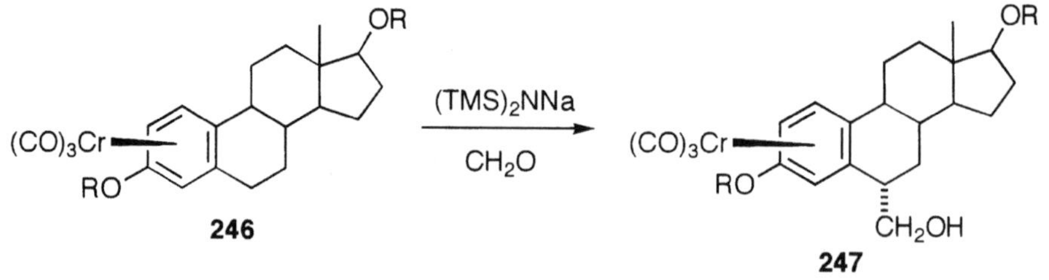

As an example of the activation of the side-chain α-carbon by the coordination, the alkylation of **251** with 1,3-dibromopropane produced **252**. No such an alkylation is possible using NaH without complex formation [65].

Attack of a carbanion at the β-carbon of styrene **253** becomes possible by complex formation, and the carbanion **254**, generated at the α-carbon, can be trapped by an electrophile (MeI) to give **255** [59].

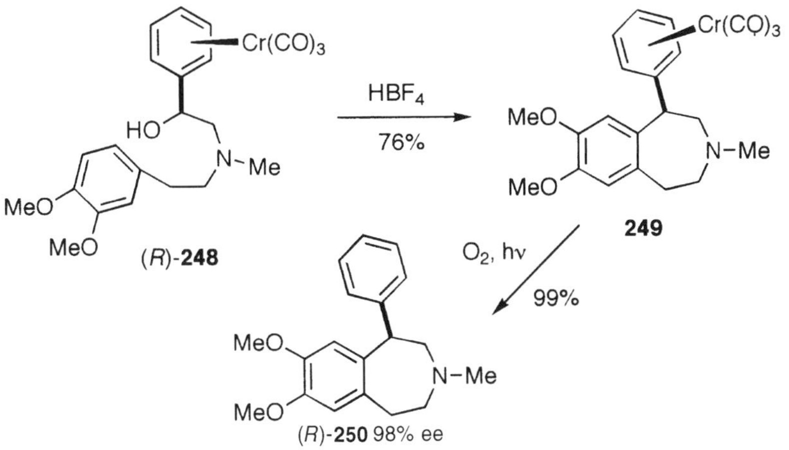

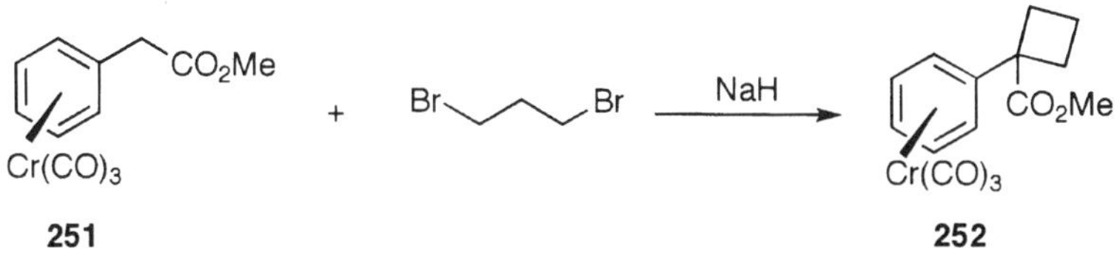

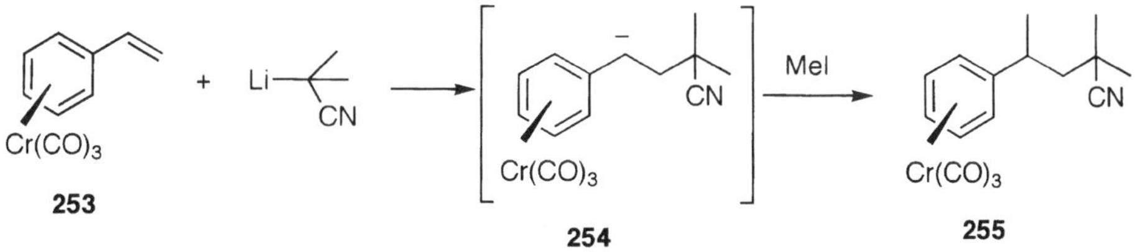

(R)-248 → HBF₄ 76% → 249

O₂, hv 99% → (R)-250 98% ee

251 + Br⌒⌒Br → NaH → 252

253 + Li—⟨CN → [254]⁻ → MeI → 255

9.4.5 Steric Effect of Coordination

Stereic hindrance due to the coordination affects the orientation of substitutions. As described before, the substitution occurs from the *exo* (opposite) side of the coordination. Utilizing this effect, two methylated products **258** and **260** of opposite stereochemistry were prepared from the α-tetralone complex **256** [61]. The ketone **256** was reduced and acetylated to give **257**. The methylated product **258** was prepared by displacement of the acetoxy with the methyl anion of Me₃Al from the back side of the coordination. However, MeLi attacked **256** from the opposite side of the coordinated Cr(CO)₃ to give **259**, and the benzyl cation generated from **259** was attacked by hydride from the back to afford **260**. Similarly, complex **261** was converted to **264** by the displacement of the acetoxy group of **262** with allylsilane **263** from the *exo* side, and the ketone in **264** was converted to a methyl group giving **265**. As described before, the *meta* substitution of **265** occurs to give **266**. Hydroboration of the double bond in the isobutenyl group in the *meta*-substituted product **266** with 9-BBN-H, and Pd-catalysed Suzuki–Miyaura coupling with alkenyl bromide **267** afforded **268**, from which dihydroxyserrulatic acid (**269**) [66] was synthesised [63].

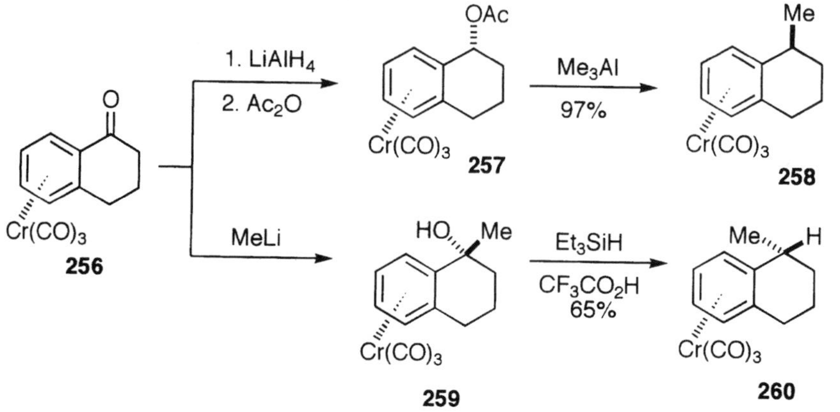

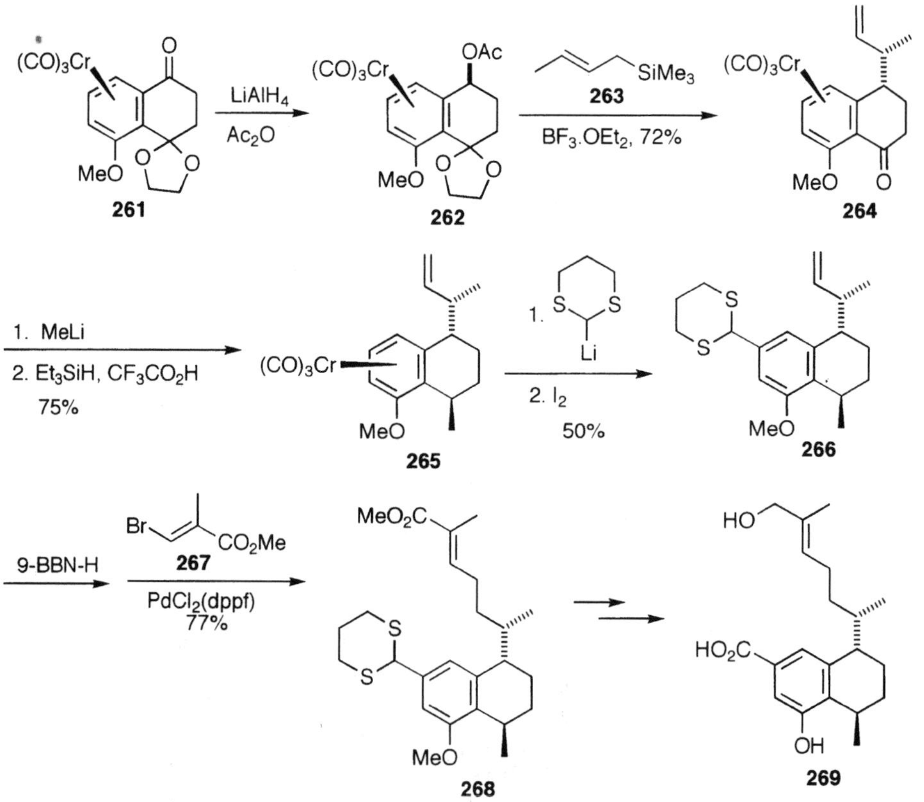

Stereoselectivity in the reaction of acyclic ketone **270** is different from that of the cyclic ketone **256**. The acetate in **271**, prepared by reduction of the ketone **270** to alcohol with LiAlH$_4$ and acetylation, was displaced with Me$_3$Al from the *exo* side to give **272** with retention of the stereochemistry. No racemization of benzyl cation was observed. However, reaction of **270** with MeLi gave **274**. The OH group of **274** was removed with hydride from the less hindered side as shown by **275** to give **276** with retention of the stereochemistry. In this way, the stereoisomers **273** and **277** were prepared selectively.

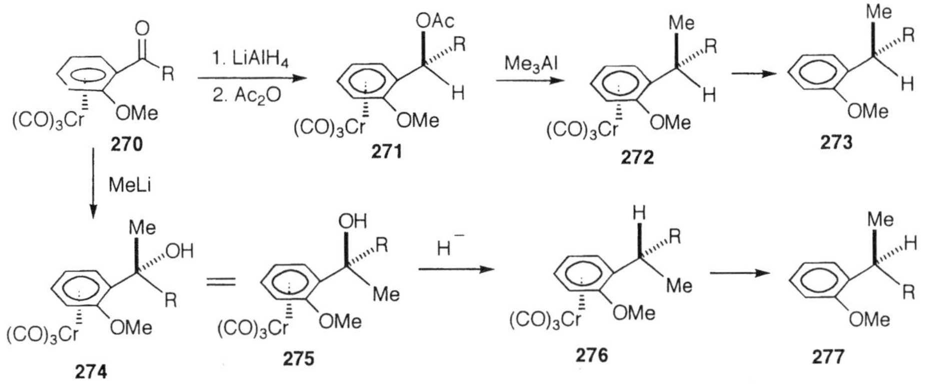

Complexation of **278** with naphthalene-Cr(CO)$_3$ gave one isomer **279** with high diastereoselectivity (89:11) due to coordination effect of OH group. After hydrolysis of the acetal, **279** was converted to **281** via **280**. The acetate **281** was converted to the *syn* product **282** by displacement with Me$_3$Al in a similar way (**271–272**) as described before. On the other hand, **283** was obtained from **280**, and the *anti* product **284** was prepared stereoselectively by selective displacement with hydride [61,67]

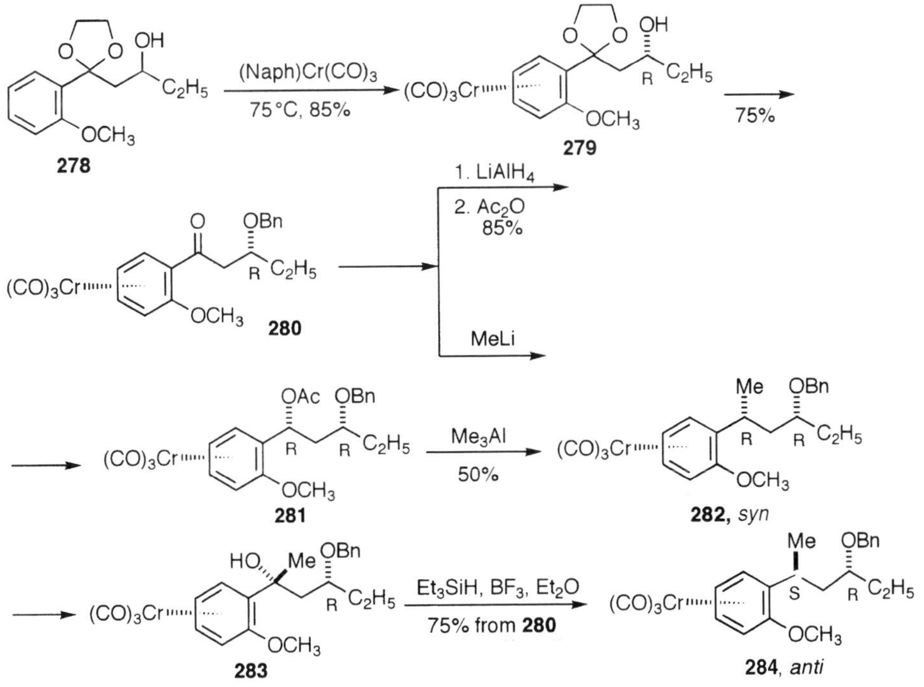

9.4.6 *Asymmetric Synthesis using Chiral* Cr(CO)$_3$–*arene Complexes*

Cr(CO)$_3$ coordinates from either the top or bottom side of aromatic rings, bearing two different substituents in *ortho* or *meta* position, so that the enantiomers **285** and **286** are obtained. Optical resolution of the enantiomers is carried out by recrystallization, or column chromatography. The racemic complex of benzyl alcohol derivative **287** was separated to **288** and **289** by lipase-catalysed acetylation [68]. Enzymes recognize Cr(CO)$_3$ as a bulky group. Chiral Cr(CO)$_3$–arene complexes are used for asymmetric synthesis [68a].

As reactions occur from the face opposite to the metal, highly enantioselective

reactions of the chiral complexes are possible, and are used for asymmetric synthesis. The imine, obtained from the racemic complex of *o*-methoxybenzaldehyde **290** and chiral L-valinol, is a mixture of diastereomers, and can be separated to **291** and **292** by column chromatography. Their hydrolysis gave **293** and **294** in optically pure forms [66]. Complex formation by ligand exchange of the optically active alcohol of tetralin **295** with the naphthaline complex **296** gave the optically active complex **297** diastereoselectively from the same side of the hydroxy group [69].

Although the optical yield was unsatisfactory, stereoselective monocoupling of the *o*-dichlorobenzene complex **298** with phenylboronic acid catalysed by an optically active Pd catalyst gave rise to the optically active complex **299** [70].

9.4.7 *Reactions of Cycloheptatriene Complexes*

In addition to benzene rings, cycloheptatriene is activated or protected by forming the stable η^6 complex **300**. An example of the strong stabilization effected by coordination is shown by isolation of the optically active 1,3,5-cycloheptatrien-3-ols **301**, **304** and **305** as their enol forms. 1,3,5-Cycloheptatrien-3-ol was isolated as complex **301** by hydrolysis of silyl enol ether **300**. The triene system is stabilized by coordination, without forming the keto form [71]. The complex **301** was converted to a mixture of diastereomers **302** and **303** by esterification with chiral α-methoxyphenylacetic acid, and the mixture was separated into the components. Hydrolysis of **302** and **303** afforded the optically active trienol complexes **304** and **305**.

The cycloheptatriene complex undergoes cycloadditions with various unsaturated bonds, such as monoenes, dienes and alkynes, under mild conditions to afford interesting polycyclic compounds. In particular, the intramolecular version offers efficient methods for the construction of strained polycyclic skeletons, which are difficult to synthesize by other means [72].

Thus [6+2] cycloaddition of alkene with complex **306**, bearing an optically active side chain, under irradiation at room temperature afforded the bicyclic compound **307** in 98% de [73]. According to the Woodward–Hoffmann rule, the [6+2] cycloaddition proceeds by irradiation, and is thermally forbidden. However, the cycloheptatriene complex **308** underwent 1,5-hydride shift, followed by [6+2] cycloaddition by heating, to give the tricyclic compound **309** in 90% yield [74]. The cycloaddition was applied to the synthesis of β-cedrene [75].

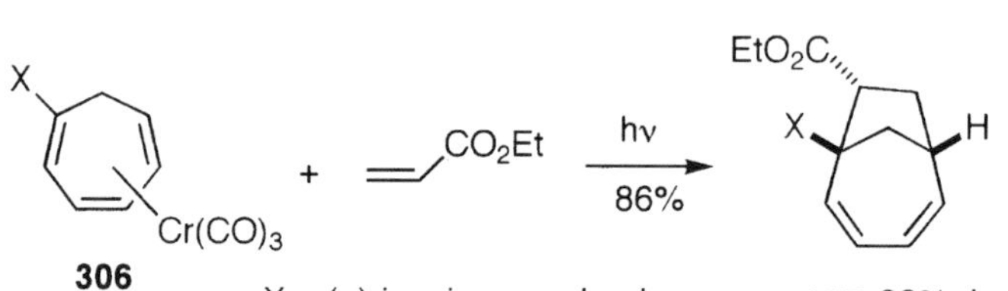

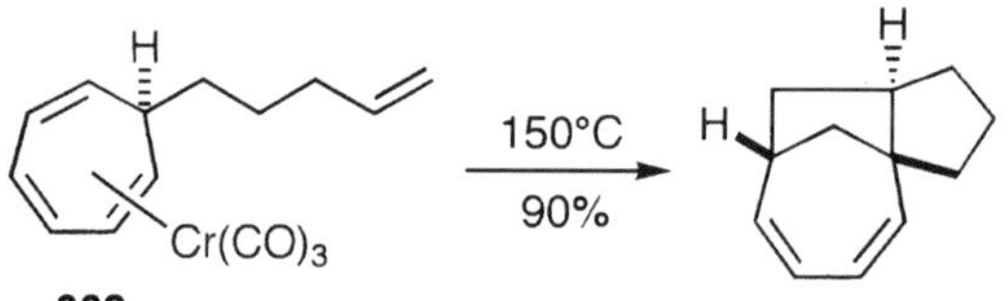

The bicyclo[4.4.1]undecane derivative **311** was obtained by [6+4] cycloaddition of **310** with a conjugated diene [72]. The [6+4] cycloaddition is a thermally allowed reaction, and free cycloheptatriene undergoes the [6+4] cycloaddition by heating. However, the [6+4] cycloaddition of cycloheptatriene coordinated by Cr(CO)₃ proceeds at 0 °C under irradiation. These results show the profound effect of coordination of Cr(CO)₃ on reactivity.

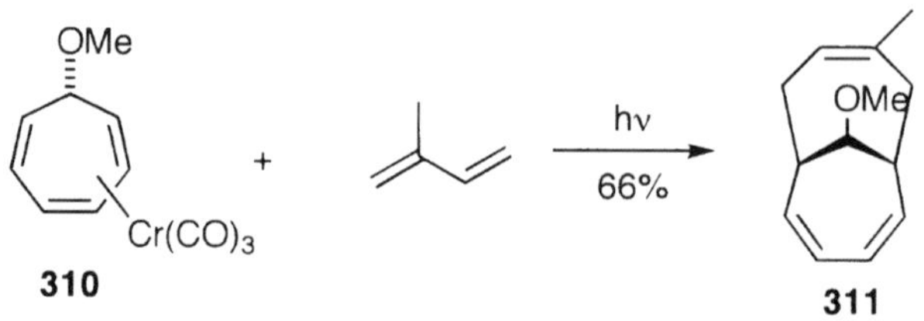

The [6+4] cycloaddition reaction was applied to the construction of a taxane skeleton [76]. Adduct **314** was obtained by the reaction of complex **312** with diene

313, and converted to **315**. The ABC ring system of the taxane model (**316**) was prepared by the rearrangement of **315**, promoted by (*i*-PrO)₃Al. Similar [6+4] cycloaddition of cycloheptatriene proceeds with a catalytic amount of cyclohepta-triene–Cr(CO)₃ in the presence of Mg powder as a reducing agent [77]. Intramolecular version of the [6+4] cycloaddition was applied to the synthesis of a tricyclic model skeleton **321** of ingenol (**317**), which has highly strained *trans*-intrabridgehead stereochemical relationship. The cycloheptatriene derivative **318** underwent 1,5-H shift in refluxing dioxane to generate **319**, and its cyclization proceeded under irradiation to give the tricyclic compound **320** after demetallation as a single diastereomer, which was converted to **321** [78]

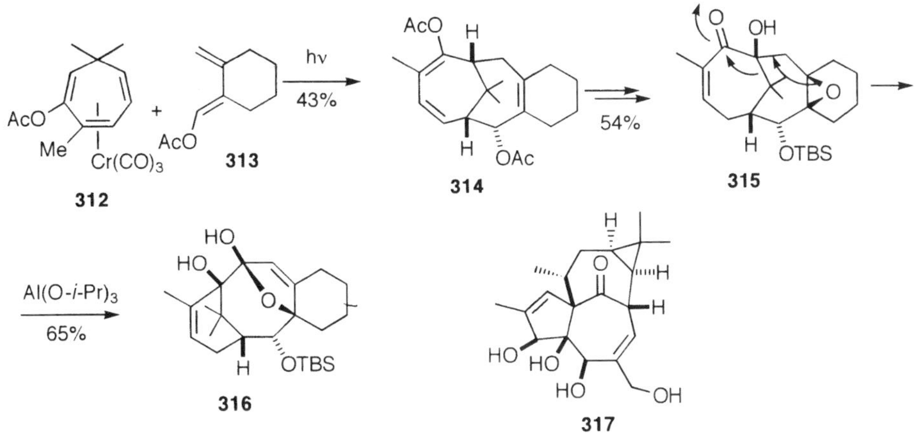

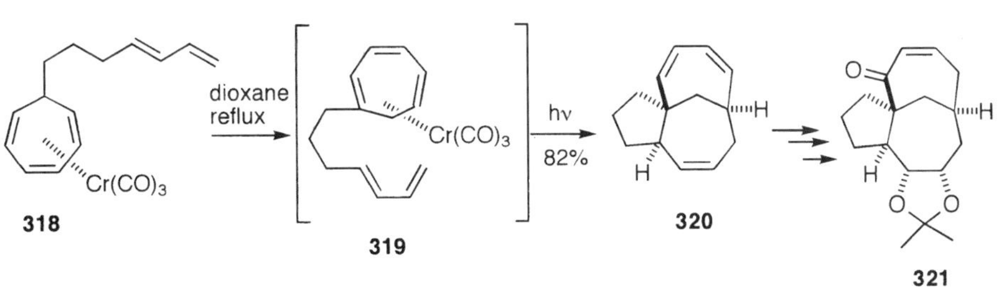

The bicyclo[4.2.1]nonatriene **323** was prepared by the [6+2] cycloaddition of internal alkyne with the complex **322** under irradiation [79]. Ligand exchange of **323** with toluene liberated **324**. The complex **325** underwent the [6+2] cycloaddition with two moles of terminal alkyne to give the tetracyclic compound **327** via **326**. The [6+2] cycloaddition of the complex **322** and 1,7-octadiyne (**328**) afforded **329** as a primary product, which was converted further to **330** in 56% yield by further intramolecular [6+2] cycloaddition [80]. The tropone complex **331** underwent intramolecular [6+2] cycloaddition under irradiation to give the strained tricyclic compound **332** in moderate yield [81].

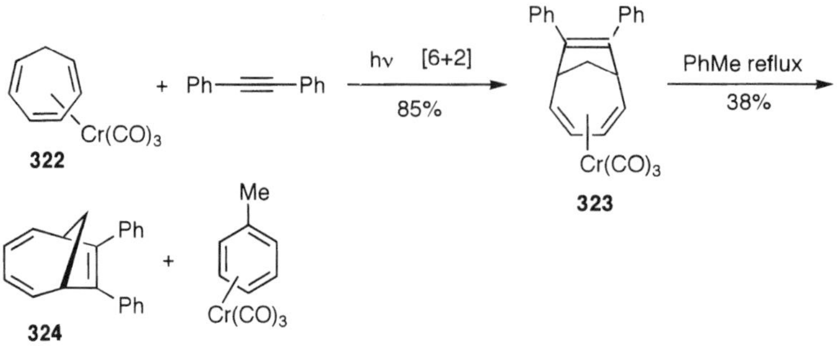

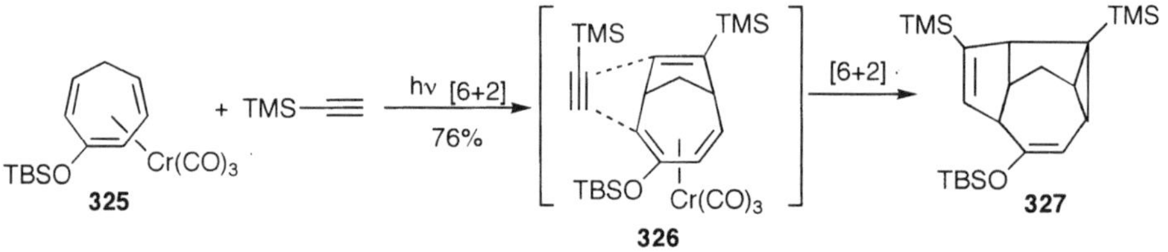

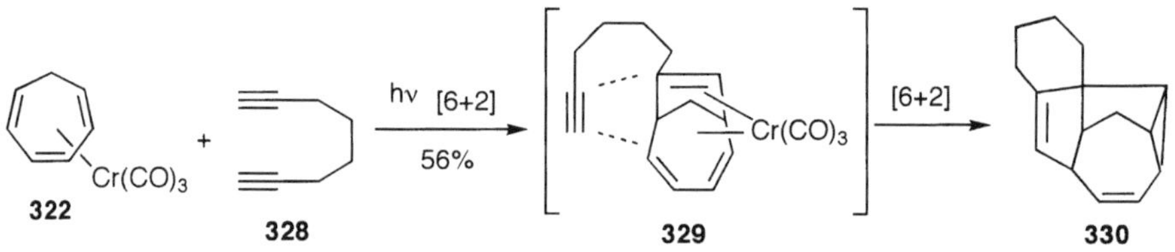

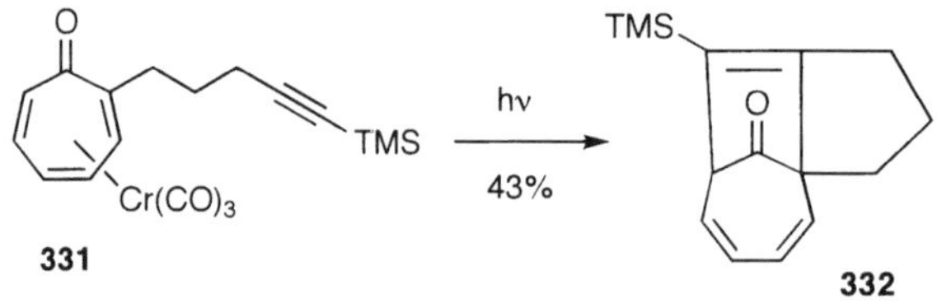

322 + Ph——≡——Ph → hv [6+2] 85% → **323** Ph Ph Cr(CO)₃ → PhMe reflux 38%

324 + Me Cr(CO)₃

325 TBSO Cr(CO)₃ + TMS——≡ → hv [6+2] 76% → **326** → [6+2] → **327** TMS TMS TBSO

322 Cr(CO)₃ + **328** → hv [6+2] 56% → **329** Cr(CO)₃ → [6+2] → **330**

331 Cr(CO)₃ + TMS → hv 43% → **332** TMS

9.4.8   Activation of Arenes by the Coordination of an Osmium Complex

A peculiar complex is formed by η^2 coordination of Os(II) ammine complex to one of the double bonds of benzene rings, rather than η^6 coordination, and the coordinated benzene rings show interesting reactivity [82]. For example, Os(II) coordinates regioselectively to the 2,3-double bond of anisole to form the complex **333**, and hence localization of the remaining π-electrons occurs. As a result, at 20 °C an electrophile attacks easily at C(4) due to electron-donation of the methoxy group. The 4H-cationic intermediate **334** is stabilized by backdonation from the metal, and the monosubstitution product **334** is formed without deprotonation. The *para*-substituted anisole **335** is obtained by deprotonation with amine and decomplexation. These results show that the η^2 coordination of Os(II) is strong enough to localize π-electrons in aromatic rings.

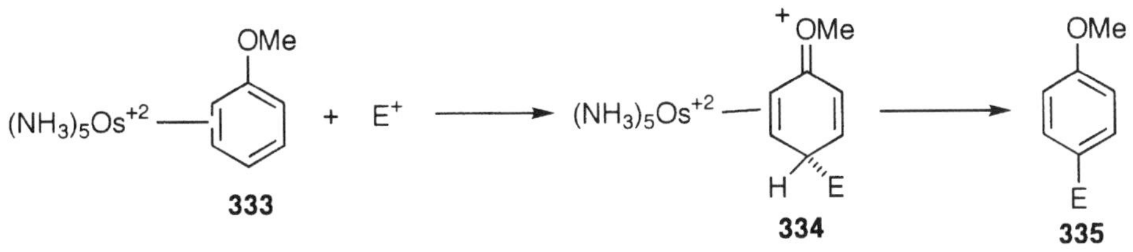

The η^2 complex **333** can be prepared in 98% yield by the reaction of anisole with Os(NH$_3$)$_5$(OTf)$_3$ in the presence of Mg. Michael addition of methyl vinyl ketone to the complex at 20 °C using TfOH afforded **336**, which was converted to **337** by deprotonation with tertiary amine [83]. The diketone **340** was obtained by the Michael addition of methyl vinyl ketone to C-4 of the 4-methylanisole complex **338** to generate **339**, followed by intramolecular nucleophilic attack of the keto enolate in **339**.

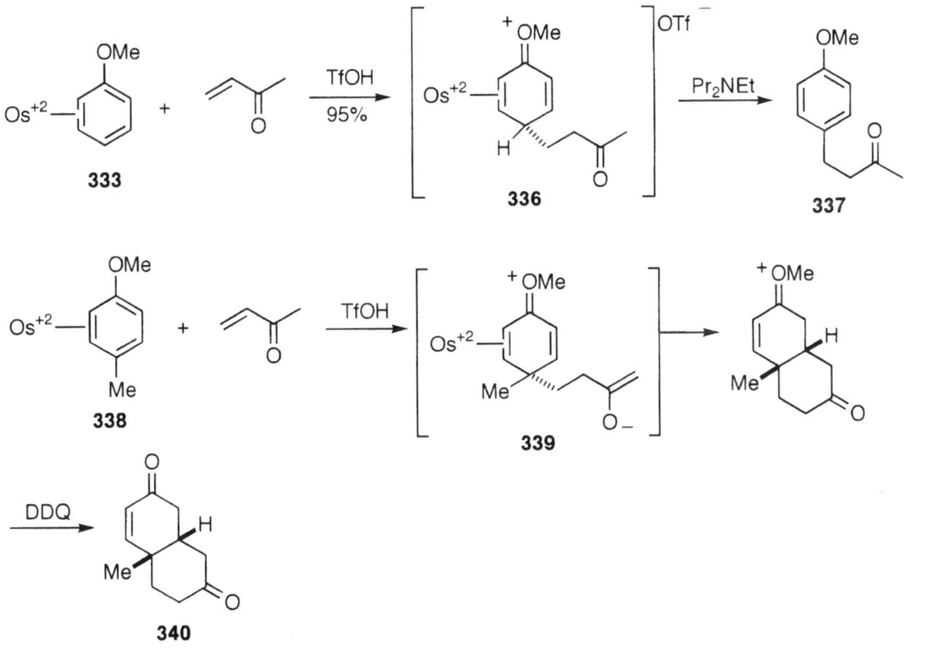

Reaction of the pyrrole complex **341** with acrylate gives **343** via **342**, and indole ring **344** is formed by oxidation with DDQ [84]. Similarly, the aniline complex **345** reacts with acrylate to afford **347** via **346** [85].

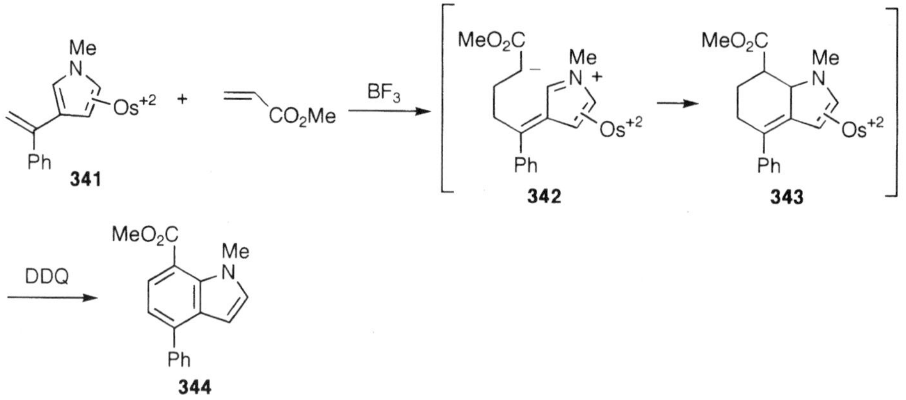

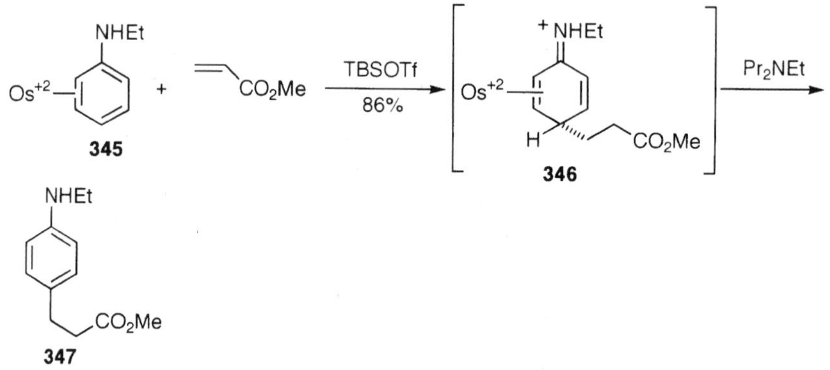

References

1. A. J. Pearson, *Comprehensive Organometallic Chemistry*, Vol. 8, p. 939, Pergamon Press, 1982; A. J. Pearson, *Iron Compounds in Organic Synthesis*, Academic Press, 1994; M. Rosenblum, *Acc. Chem. Res.*, **7**, 122 (1974).
2. T. C. T. Chang, M. Rosenblum and S. B. Samuels, *J. Am. Chem. Soc.*, **102**, 5930 (1980); *J. Org. Chem.*, **46**, 4103 (1981).
3. K. M. Nicholas, *J. Am. Chem. Soc.*, **97**, 3254 (1975).
4. A. Sanders and W. P. Giering, *J. Am. Chem. Soc.*, **97**, 919 (1975).
5. A. J. Pearson, *Iron Compounds in Organic Synthesis*, Academic Press, 1994.
6. R. Gree, *Synthesis*, 341 (1989).
7. E. J. Corey and G. Moinet, *J. Am. Chem. Soc.*, **95**, 7185 (1973).
8. W. A. Donaldson and L. Shang, *Tetrahedron Lett.*, **37**, 423 (1996).
9. D. H. R. Barton, A. A. L. Gunatilaka, T. Nakanishi, H. Patin, D. A. Widdowson and B. R. Worth, *J. Chem. Soc., Perkin Trans. I*, 821 (1987).
10. G. F. Emerson, L. Watts and R. Pettit, *J. Am. Chem. Soc.*, **87**, 131 (1965).
11. R. Pettit, *J. Organometal. Chem.*, **100**, 205 (1975).
12. L. Watts, J. D. Fritzpatrick and R. Pettit, *J. Am. Chem. Soc.*, **87**, 3253, 3254 (1965); J. C. Barborak, L. Watts and R. Pettit, *J. Am. Chem. Soc.*, **88**, 1328 (1966).
13. J. A. Tallarico, M. L. Randall and M. L. Snapper, *J. Am. Chem. Soc.*, **118**, 9196 (1996).

14. A. J. Birch, K. B. Chamberlain, M. A. Haas and D. J. Thompson, *J. Chem. Soc. Perkin Trans. I*, 1882 (1973); 1006 (1981); *Org. Synth. Coll.* **6**, 996 (1988).

15. S. H. Ban, Y. Hayashi and K. Narasaka, *Chem. Lett.*, 393 (1998).

16. M. Franck-Neumann, P. Chemla and D. Martina, *Synlett*, 641 (1990).

17. M. Franck-Neumann, L. Miesch-Gross and O. Nass, *Tetrahedron Lett.*, **37**, 8763 (1996).

18. A. J. Pearson, *Acc. Chem. Res.*, **13**, 463 (1980).

19. B. F. G. Johnson, J. Lewis and D. G. Parker, *J. Organometal. Chem.*, **141**, 319 (1977).

20. M. F. Semmelhack and J. W, Herndon, *Organometallics*, **2**, 363 (1983); *J. Am. Chem. Soc.*, **105**, 2497 (1983).

21. M. C. P. Yeh, B. A. Sheu, H. W. Fu, S. I. Tau and L. W. Cuang, *J. Am. Chem. Soc.*, **115**, 5941 (1993).

22. B. F. Johnson, K. D. Karlin and J. Lewis, *J. Organometal. Chem.*, **145**, C23 (1978).

23. P. Eilbracht, R. Jelitte and P. Trabold, *Chem. Ber.*, **119**, 169 (1986).

24. L. A. Paquette, S. Liang and H. L. Wang, *J. Org. Chem.*, **61**, 3268 (1996).

25. H. J. Knölker, P. Gonser and T. Koegler, *Tetrahedron Lett.*, **37**, 2405 (1996).

26. W. R. Roush and J. C. Park, *Tetrahedron Lett.*, **31**, 4707 (1990).

27. W. R. Roush and C. K. Wada, *J. Am. Chem. Soc.*, **116**, 2151 (1994).

28. A. J. Pearson and M. K. O'Brien, *J. Org. Chem.*, **54**, 4663 (1989).

29. H. J. Knölker, *Synlett*, 371 (1992).

30. A. J. Birch, L. F. Kelly and D. V. Weerasuria, *J. Org. Chem.*, **53**, 278 (1988).

31. T. Nagasawa, K. Taya, M. Kitamura and K. Suzuki, *J. Am. Chem. Soc.*, **118**, 8949 (1996).

32. K. M. Nicholas and R. Pettit, *Tetrahedron Lett.*, 3475 (1971).

33. D. Seyferth and A. T. Wehman, *J. Am. Chem. Soc.*, **92**, 5520 (1970).

34. K. M. Nicholas, *Acc. Chem. Res.*, **20**, 207 (1987); V. Varghese, M. Saha and K. M. Nicholas, *Org. Synth.*, **67**, 141 (1988).

35. W. A. Smit, R. Caple and I. P. Smoliakova, *Chem. Rev.*, **94**, 2359 (1994).

36. R. Lockwood and K. M. Nicholas, *Tetrahedron Lett.*, 4163 (1977).

37. C. Mukai, Y. Ikeda, Y. Sugimoto and M. Hanaoka, *Tetrahedron Lett.*, **35**, 2183 (1994); C. Mukai and M. Hanaoka, *Synlett*, 11 (1996).

38. T. F. Jamison, S. Shambayati, W. E. Crowe and S. L. Schreiber, *J. Am. Chem. Soc.*, **119**, 4353 (1997).

39. P. Magnus, R. T. Lewis and J. C. Huffman, *J. Am. Chem. Soc.*, **110**, 6921 (1988); P. Magnus, H. Annoura and J. Harling, *J. Org. Chem.*, **55**, 1709 (1990); P. Magnus, *Tetrahedron*, **50**, 1397 (1994).

40. K. Tomioka, H. Fujita and K. Koga, *Tetrahedron Lett.*, **30**, 851 (1989).

41. S. Hosokawa and M. Isobe, *Synlett*, 1179 (1995).

42. G. G. Melikyan, M. A. Khan and K. M. Nicholas, *Organometallics*, **14**, 2170 (1995).

43. S. Tanaka, T. Tsukiyama and M. Isobe, *Tetrahedron Lett.*, **34**, 5757 (1993).

44. M. F. Semmelhack, G. R. Clark, J. L. Harrison, Y. Thebtaranonth, W. Wulff and A. Yamashita, *Tetrahedron*, **37**, 3957 (1981).

45. M. F. Semmelhack and G. Clark, *J. Am. Chem. Soc.*, **99**, 1675 (1977).

46. H. Kunzer and M. Thiel, *Tetrahedron Lett.*, **29**, 3223 (1988).

47. M. F. Semmelhack, W. Wulff and J. L. Garcia, *J. Organometal. Chem.*, **240**, C5 (1982).

48. M. F. Semmelhack and A. Yamashita, *J. Am. Chem. Soc.*, **102**, 5924 (1980).

49. E. P. Kündig, A. Ripa, R. Liu and G. Bernardinelli, *J. Org. Chem.*, **59**, 4773 (1994).

50. D. Amurrio, K. Khan and E. P. Kündig, *J. Org. Chem.*, **61**, 2258 (1996).

51. A. Quattropani, G. Anderson, G. Bernardinelli and E. P. Kündig, *J. Am. Chem. Soc.*, **119**, 4773 (1997).

52. M. Ghavshou and D. A. Widdowson, *J. Chem. Soc., Perkin I*, 3065 (1983).

53. W. J. Scott, *Chem. Commun.*, 1755 (1987).
54. D. Villemin and S. Endo, *J. Organometal. Chem.*, **293**, C10 (1985). M. E. Wright, *J. Organometal. Chem.*, **376**, 353 (1989).
55. M. Uemura, H. Nishimura, K. Kamikawa, K. Nakayama and Y. Hayashi, *Tetrahedron Lett.*, **35**, 1909 (1994).
56. R. Mutin, C. Luca, J. Thivolle-Cazat, V. Dufaud, F. Dang and J. M. Basset, *Chem. Commun.*, 896 (1988).
57. A. J. Pearson and J. G. Park, *J. Org. Chem.*, **57**, 1745 (1992), A. J. Pearson and K. Lee, *J. Org. Chem.*, **60**, 7153 (1995).
58. M. F. Semmelhack and A. Zask, *J. Am. Chem. Soc.*, **105**, 2034 (1983).
59. M. F. Semmelhack, W. Suefert and L. Keller, *J. Am. Chem. Soc.*, **102**, 6584 (1980).
60. H. G. Schmalz, A. Schwarz and G. Durner, *Tetrahedron Lett.*, **35**, 6861 (1994).
61. M. Uemura, K. Isobe and Y. Hayashi, *Tetrahedron Lett.*, **26**, 767 (1985).
62. G. Jaouen, S. Top, A. Laconi, D. Couturier and J. Brocard, *J. Am. Chem. Soc.*, **106**, 2207 (1984).
63. M. Uemura, M. Nishimura, T. Minami and Y. Hayashi, *Tetrahedron Lett.*, **31**, 2319 (1990); *J. Am. Chem. Soc.*, **113**, 5402 (1991).
64. S. J. Coote, S. G. Davies, D. Middlemiss and A. Naylor, *Tetrahedron Lett.*, **30**, 3581 (1989).
65. G. Jaouen, A. Meyer and G. Simmonneaux, *Chem. Commun.*, 813 (1975).
66. S. G. Davies and C. L. Goodfellow, *Synlett*, 59 (1989).
67. M. Uemura, T. Minami and Y. Hayashi, *J. Am. Chem. Soc.*, **109**, 5277 (1987).
68. K. Nakamura, K. Ishihara, A. Ohno, M. Uemura, H. Nishimura and Y. Hayashi, *Tetrahedron Lett.*, **31**, 3603 (1990).
68a. A. Solladie-Cavallo, in *Advances in Metal-Organic Chemistry*, Ed. L. Liebeskind, Vol 1, p. 99, JAI Press, 1989.
69. A. Alexakis, P. Mangeney, I. Marek, F. Rose-Munch, E. Rose, A. Semra and F. Robert, *J. Am. Chem. Soc.*, **114**, 8288 (1992).
70. M. Uemura, H. Nishimura and T. Hayashi, *J. Organometal. Chem.*, **473**, 129 (1994).
71. J. H. Rigby, N. M. Niyaz and P. Sugathapala, *J. Am. Chem. Soc.*, **118**, 8178 (1996).
72. J. H. Rigby, *Acc. Chem. Res.*, **26**, 579 (1993); J. H. Rigby and H. S. Ateeq, *J. Am. Chem. Soc.*, **112**, 6442 (1990).
73. J. H. Rigby, H. S. Ateeq, N. R. Charles, J. A. Henshilwood, K. M. Short and P. M. Sugathapala, *Tetrahedron*, **49**, 5495 (1993).
74. J. H. Rigby, S. D. Rege, V. P. Sandanayaka and M. Kirova, *J. Org. Chem.*, **61**, 842 (1996); K. M. Short, H. S. Ateeq and J. A. Henshilwood, *J. Org. Chem.*, **57**, 5290 (1992).
75. J. H. Rigby and M. Kirova-Snover, *Tetrahedron Lett.*, **38**, 8153 (1997).
76. J. H. Rigby, N. M. Niyaz, K. Short and M. J. Heeg, *J. Org. Chem.*, **60**, 7720 (1995).
77. J. H. Rigby and C Fiedler, *J. Org. Chem.*, **62**, 6106 (1997).
78. J. H. Rigby, J. Hu and M. J. Heeg, *Tetrahedron Lett.*, **39**, 2265 (1998).
79. K. Chaffee, P. Huo, J. B. Sheridan, A. Barbieri, A. Aistars, R. A. Lalancette, R. L. Ostrander and A. L. Rheingold, *J. Am. Chem. Soc.*, **117**, 1900 (1995).
80. J. H. Rigby, N. C. Warshakoon and M. J. Heeg, *J. Am. Chem. Soc.*, **118**, 6094 (1996).
81. J. H. Rigby, M. Kirova, N. Niyaz and F. Mohammadi, *Synlett*, 805 (1997).
82. W. D. Harman, *Chem. Rev.*, **97**, 1953 (1997).
83. S. P. Kolis, M. E. Kopach, R. Liu and W. D. Harman, *J. Org. Chem.*, **62**, 130 (1997).
84. L. M. Hodges, J. Gonzalez, J. I. Koontz, W. H. Myers and W. D. Harman, *J. Org. Chem.*, **60**, 2125 (1995).
85. S. P. Kolis, J. Gonzales, L. M. Bright and W. D. Harmon, *Organometallics*, **15**, 245 (1996).

10

CATALYTIC HYDROGENATION, TRANSFER HYDROGENATION AND HYDROSILYLATION

The hydrogenation of unsaturated bonds has been carried out for many years using supported or non-supported metals such as Ni, Pd and Pt as solid catalysts under heterogeneous conditions. Transition metal complexes were found to be active catalysts for hydrogenation under homogeneous conditions in 1960s. Active studies on homogeneous hydrogenation were initiated by introduction of the Wilkinson complex $RhCl(Ph_3P)_3$ [1,2]. This complex is soluble in organic solvents, and homogeneous hydrogenation of alkenes and alkynes can be carried out under mild conditions. Since then, remarkable progress has been made in transition metal complex-catalysed homogeneous hydrogenation. Particularly noteworthy are the advances in asymmetric hydrogenation using complexes coordinated by chiral phosphines, and a number of optically active compounds with nearly 100% ee have been synthesized. Asymmetric hydrogenation of carbonyl and imine bonds has been achieved using Ru and Rh catalysts, and now it is possible to reduce some ketones and imines by hydrogenation, without applying stoichiometric reduction with $NaBH_4$ and other metal hydrides.

10.1 Homogeneous Hydrogenation of Alkenes

The homogeneous hydrogenation of alkenes is explained by two mechanisms. The first is the dihydride mechanism, in which the dihydride **1** is formed by oxidative addition of H_2, and the hydrogenation proceeds by the insertion of alkene to the metal hydride bond, followed by reductive elimination (Scheme 10.1). The other hydrogenation is explained by the formation of the monohydride **2** (Scheme 10.2). Insertion of alkenes

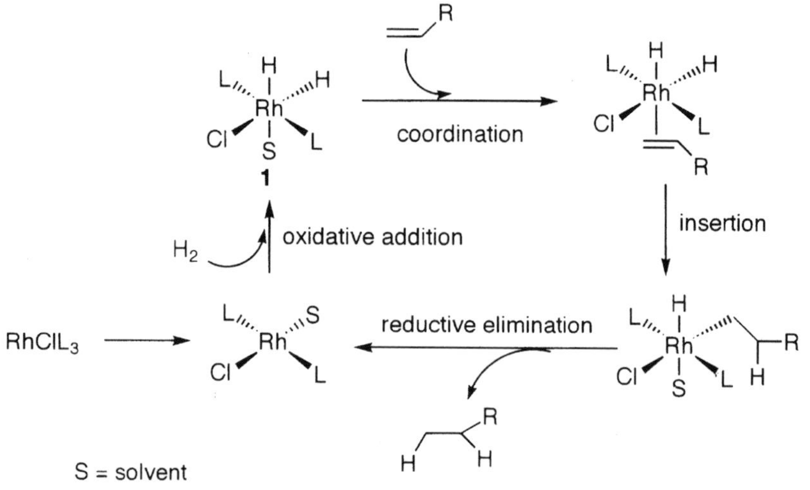

Scheme 10.1   Monohydride hydrogenation mechanism

to the metal hydride is followed by attack of H_2 to give the hydrogenation product **3** with regeneration of the monohydride **2**. Hydrogenation proceeds by one of these mechanisms, depending on catalyst species.

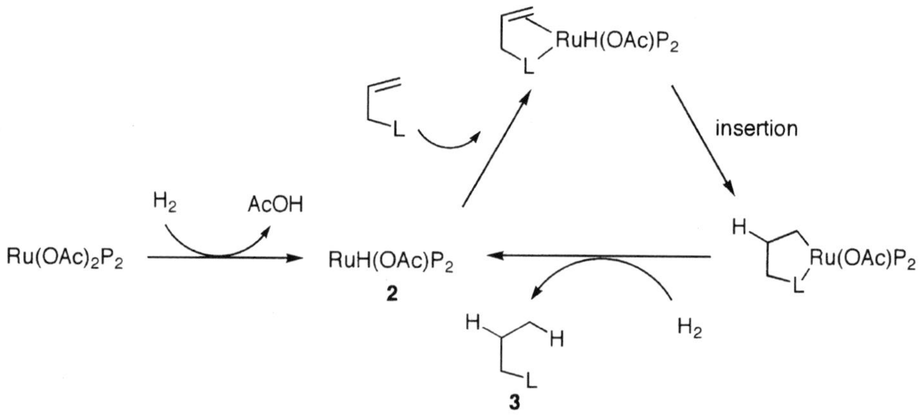

Scheme 10.2   Dihydride hydrogenation mechanism

As steric hindrance has a strong effect on the alkene insertion step, Rh complex-catalysed selective hydrogenation is possible. For example, only the isopropenyl group in carvone (**4**) is hydrogenated selectively at room temperature under 1 atm of H_2 to give **5** [3]. Unsaturated bonds of carbonyl, nitrile and nitro groups are not reduced with the Wilkinson complex, and β-nitrostyrene (**6**) is reduced to 1-phenyl-2-nitroethane [4]. $RuCl_2(Ph_3P)_3$ is a good catalyst of homogeneous hydrogenation of alkenes, and selective hydrogenation of the terminal alkene without attacking the internal alkene in ethyl 2,7-octadienylacetoacetate (**7**) at room temperature under pressure is possible [5].

Remarkable advances have been achieved in the homogeneous asymmetric hydrogenation of prochiral alkenes by using chiral complexes of Rh and Ru, which

(chemical scheme with structures **4** and **5**)

RhCl(Ph₃P)₃ / PhH, 94%

+ H₂

(chemical scheme with structure **6**)

RhCl(Ph₃P)₃

+ H₂ → Ph‿‿NO₂

(chemical scheme with structure **7**)

+ H₂

RuCl₂(Ph₃P)₃ / Et₃N, 30 atm / 93%

are prepared by replacing Ph₃P in their parent complexes with various chiral phosphines [6,7]. A number of effective chiral phosphines so far reported are listed on the inside back cover of this book. Simple alkenes are not good substrates. High enantiomeric excesses (ee) have been obtained in the hydrogenation of alkenes bearing chelating groups. Initially, high ee values were obtained as a breakthrough in the hydrogenation of α-acetylaminocinnamic acid **8**, or its ester to give phenylalanine **9**, or its ester using (R,R)-DIOP (**V**) [8] and (R,R)-DIPAMP (**II**) [9], and the reaction is applied to commercial production of L-DOPA (2,4-dihdroxyphenylalanine). At present the acid **8** is regarded as a standard substrate with which to evaluate and compare effectiveness of chiral ligands. The high ee values, obtained in the Rh-catalysed hydrogenation of **8** with some representative chiral bidentate ligands are given.

(chemical scheme with structures **8** and (R)-**9**)

Rh-L*

+ H₂

(R, R)-DIOP	85% ee(R)
(R,R)-DIPAMP	96% ee(S)
(S,S)-NORPHOS	95% ee(R)
(S)-BINAP	100% ee(R)
BisP	99.9% ee
BICP	96.8% ee

For asymmetric hydrogenation, the monodentate tertiary phosphine, *o*-anisylcyclo-hexylmethylphosphine (**I**) which has a chiral center on P atom was used [10,11], followed by the bidentate phosphine DIPAMP (**II**), which has the chiral center on P [9]. The bidentate ligand DIOP (**V**) was prepared by Kagan from cheeply available optically active tartaric acid, and found to be an excellent chiral ligand for the hydrogenation of α-acylaminocinnamate [8]. Successive asymmetric hydrogenation using DIOP as a ligand suggested that bidentate phosphines, which have a chelating effect and *C₂* symmetry, are particularly suitable. An industrial process for the production of optically active L-DOPA has been developed using DIPAMP. As another example, (S)-N-acetylphenylalanine was obtained in 99% ee by using 1/10 000 equivalent of (S,S)-PYRPHOS (**XX**) [12].

A number of bidentate phosphines with C_2 symmetry, typically BINAP (**XXXI**) [13] and DuPHOS (**XI, XII**) [14] have been introduced. However, although powerful, these ligands are not totally effective for every prochiral substrate, and improvements in the ee can be achieved by some modification of the parent phosphines. Reports on new chiral phosphines are still appearing. For example, bisphosphines, (**III, IV**) were recently reported as easily preparable P-chiral bidentate ligands [15]. These ligands with a bulky (adamantyl or *t*-Bu) and the smallest alkyl (methyl) group are highly basic, and hence active for oxidative addition. BICP (**IX**), a new type of bisphosphine containing a cyclic backbone with four stereogenic carbon centers is an interesting ligand [16]. A chiral phosphinite ligand with a rigid spiro-nonane backbone, called spirOP (**X**), has been synthesized [17].

It should be emphasized that in asymmetric hydrogenation, selection of optimum reaction conditions – solvents, hydrogen pressure, temperature – are critically important. The preparative method and purity of catalyst precursors are also crucial for achieving high ee values, and the reaction must be carried out carefully in order to reproduce good results. The mechanism of asymmetric hydrogenation using bidentate ligands has been studied [18].

Asymmetric hydrogenation of the ester **10** using the Rh complex **11**, coordinated by bidentate chiral phosphine, is explained by Scheme 10.3. The more stable intermediate **12** is formed by the coordination of **10** to the complex **11**. But the formation of hydrogenation product **14** from **12** is considered to be slow. However, the less-stable intermediate **13** is formed as the diastereomer of **12** in a smaller amount, although its rate of hydrogenation is faster. The intermediates **12** and **13** are in rapid equilibrium. As a result, the hydrogenation product **15** is formed with high ee as the main product; the enantiomer **14** is the minor product.

In the following sections, efficient asymmetric hydrogenation of various alkenes using Rh and Ru catalysts are illustrated by selecting only a few representative examples for each type of the hydrogenation.

Esters of *β*-branched amino acids are obtained by highly enantioselective hydrogenation of *α*-acylaminoacrylates bearing different substituents at the *β*-carbon using Rh complexes of Me-DuPHOS (**XI**) and Me-BPE (**XV**). The D-*allo*-isoleucine derivative **17** with 98.2% ee is obtained from the (*Z*)-enamide **16**, and the D-isoleucine

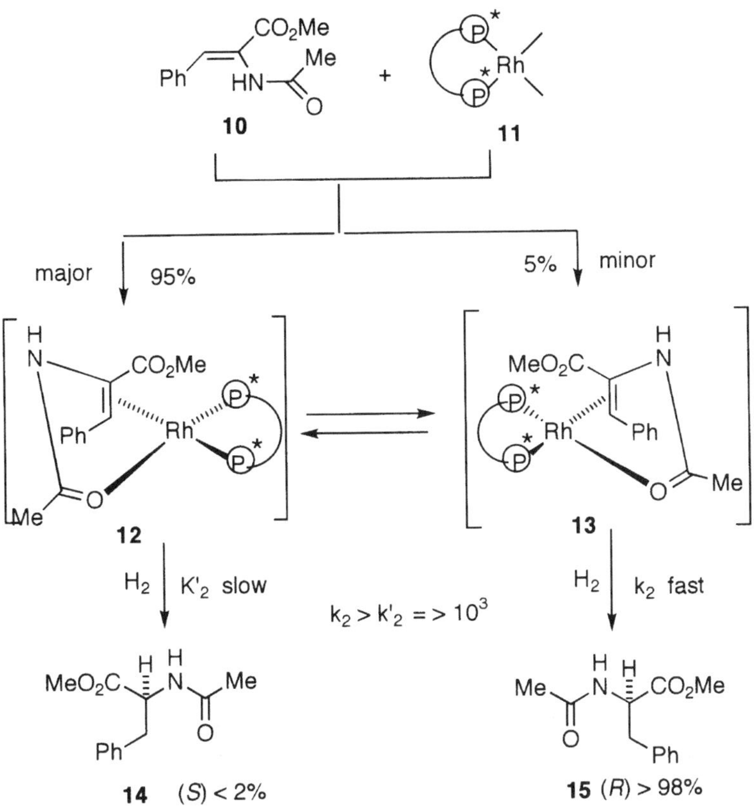

Scheme 10.3 Asymmetric hydrogenation mechanism

derivative with 98.3% ee is obtained from (*E*)-enamide [19]. Similar highly enantioselective hydrogenation is possible with Rh-TRAP complexes. TRAP (**XXVIII**) is a rare bidentate ligand, which is *trans*-chelating. The (2*S*,3*R*) ester **19** is obtained from the (*Z*)-enamide **18**, and the (2*S*,3*S*)-ester is formed from the (*E*)-enamide [20].

Asymmetric hydrogenation offers a useful synthetic route to chiral amines. Although the mechanism is unknown, only the (*R*)-*N*-acetyl-1-arylalkylamine **21** with 95% ee was obtained by the hydrogenation of a mixture of (*E*)- and (*Z*)-enamides **20a** and **20b** using Rh–Me-DuPHOS (**XI**). The *N*-acetyl enamines **20a,b** are prepared by the reduction of oximes with Fe powder in acetic anhydride [21]. Also the acetamide **23** was obtained from **22** [22].

The amine **25** was prepared with excellent enantioselectivity by the hydrogenation of **24** with Rh–H$_8$-BDPAB (**XXXV**) [23]. Beside these ligands, few suitable ligands are known for the hydrogenation of these substrates.

Highly enantioselective hydrogenation of cyclic enol acetates was achieved using the complex Rh–PennPhos (**XIII**), offering a good synthetic method for optically pure alcohols [23a].

In addition to Rh complexes, Ru complexes are also good catalysts for asymmetric hydrogenation. BINAP (**XXXI**) is a particularly good chiral ligand [6,24]. Several prochiral alkenes are hydrogenated with high ee using Ru dicarboxylate complexes **27** and **28**, coordinated by (*R*)- and (*S*)-BINAP [6, 25]. The precursor **26** of the Ru–BINAP catalysts is prepared from the Ru–COD complex as shown [13,26,27], and used for the hydrogenation of β-keto esters to β-hydroxy esters. By contrast, the carboxylate complexes **27** and **28**, prepared from the chloride **26**, are good catalysts for the hydrogenation of alkenes.

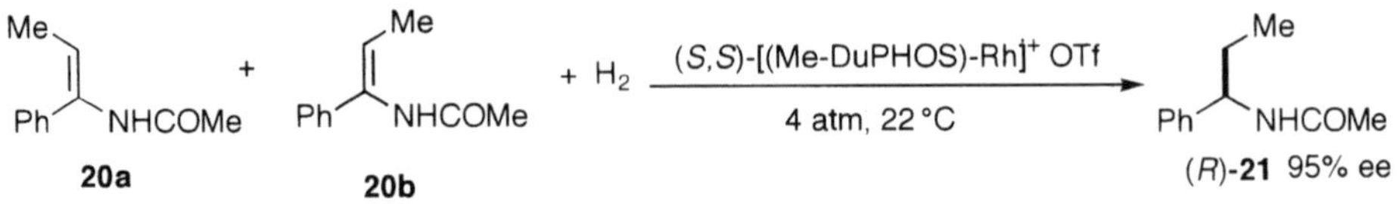

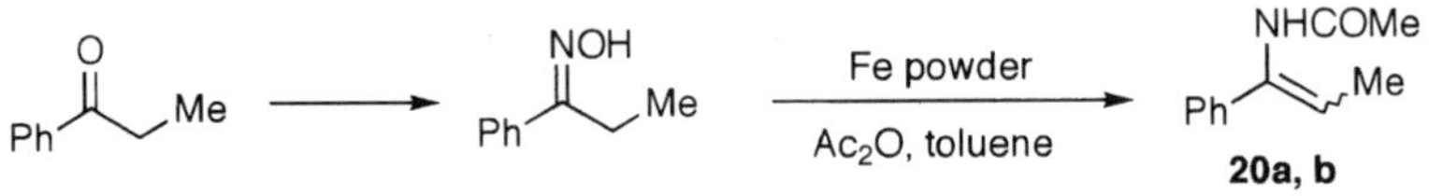

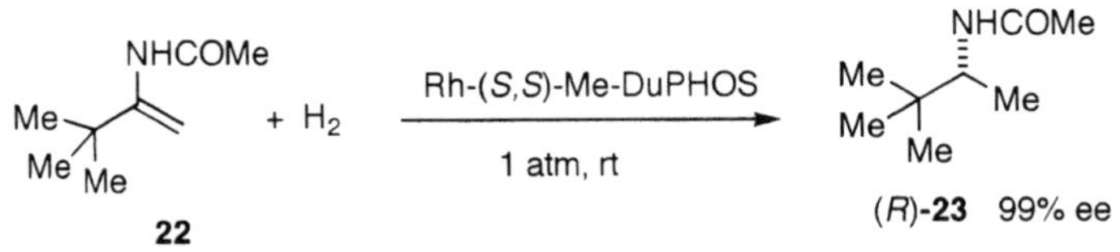

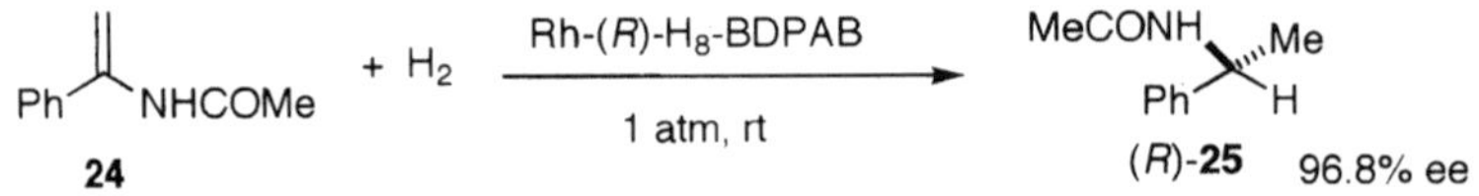

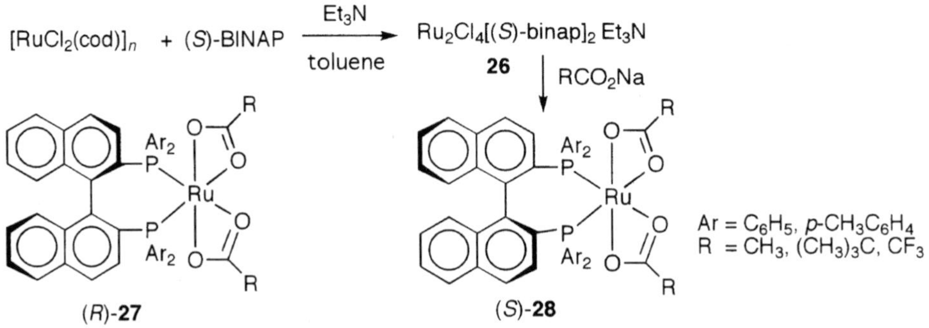

The optically active isoquinoline derivative **30** was prepared by asymmetric hydrogenation of the 1-benzylidene-1,2,3,4-tetrahydroisoquinoline **29** catalysed by **27**, and optically pure tetrahydropapaverine (**31**) is synthesized by this method [28].

Asymmetric hydrogenation of α,β-unsaturated acids is carried out successfully using Ru–BINAP and Ru–H$_8$–BINAP (**XXXIII**). (*S*)-Ibuprofen (**33**) with 97% ee is obtained by the hydrogenation of the simple derivative of acrylic acid **32**. Higher H_2 pressure gives the higher ee in this hydrogenation [29].

(S)-Methylsuccinic acid (**35**) with above 90% ee was obtained by hydrogenation of itaconic acid (**34**) using the Rh complex coordinated by (2*S*,4*S*)-BCPM (**VIII**) [30] or (*R*)-BICHEP (**XL**) [31]. Reduction of itaconic acid (**34**) to methylsuccinic acid (**35**) with 97% ee was achieved by asymmetric hydrogen transfer from formic acid using the Rh-(2*S*,4*S*)-BPPM (**VII**) catalyst in the presence of (*S*)-1-phenethylamine [32]. Hydrogenation of the itaconic acid derivative **36** proceeds using MOD-DIOP (3,5-dimethylanisole derivative of DIOP) as a ligand to give **37** with 93% ee, and optically active deoxypodophyllotoxin (**38**) was synthesized [33].

Double bonds in the allylic alcohol moieties in geraniol (**39**) and nerol (**40**) are hydrogenated regioselectively with Ru–BINAP without attacking the simple double bond in the same molecules under 30–100 atm of H_2 to give (*R*)- or (*S*)-citronellol (**41** and **42**) with 99% ee [34]. The stereochemical relationships between substrates, chiralities of the ligands and products are shown. Asymmetric hydrogenation of the allylic alcohol moiety in **43** was achieved with 99.8% ee using (*R*)-Tol-BINAP (**XXXII**) and applied to the synthesis of 1*β*-methylcarbapenem **44** [35].

Kinetic resolution is observed in the hydrogenation of a racemate of 3-methyl-2-cyclohexen-1-ol (**45**). After 46% conversion, *trans*-(1*R*,3*R*)-3-methylcyclohexanol (**47**) with 95% ee was obtained using Ru–(*R*)-BINAP. The unreacted (*S*)-allylic alcohol **46** (54%) had 80% ee. When the conversion was 54%, the (*S*)-cyclohexenol **46** with 99% ee was obtained. Also, hydrogenation of the (*S*)-allyl alcohol **46** (80% ee) using (*S*)-BINAP afforded the (1*S*,3*S*)-alcohol **48** with 99% ee after 68% conversion [36].

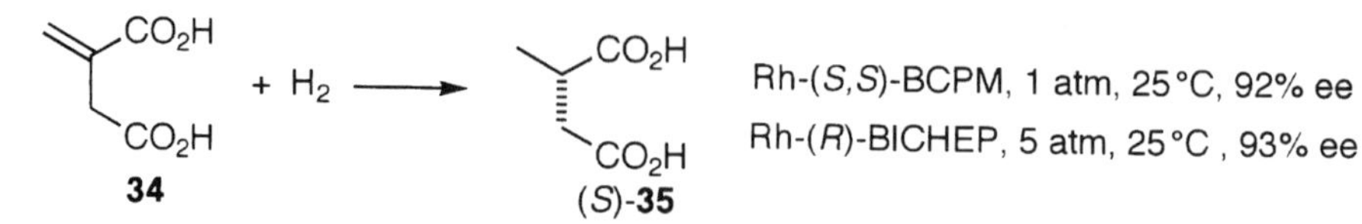
34
+ H2
(S)-35
Rh-(S,S)-BCPM, 1 atm, 25 °C, 92% ee
Rh-(R)-BICHEP, 5 atm, 25 °C, 93% ee

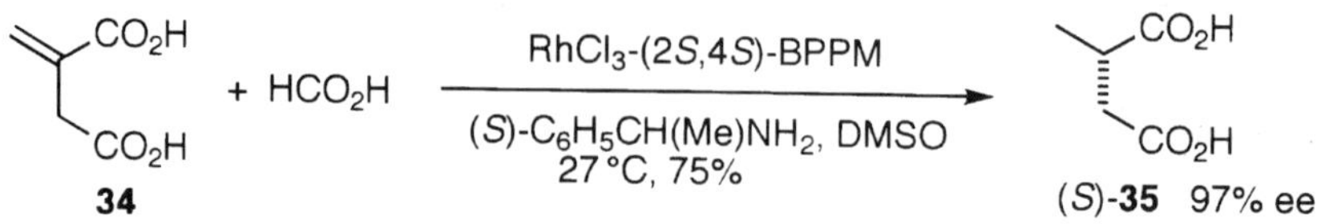
34
+ HCO2H
RhCl3-(2S,4S)-BPPM
(S)-C6H5CH(Me)NH2, DMSO
27 °C, 75%
(S)-35 97% ee

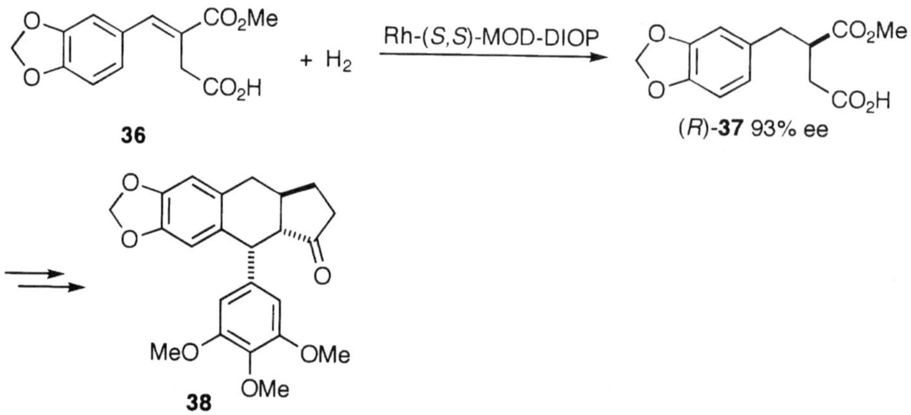
CO2Me
CO2H
36
+ H2
Rh-(S,S)-MOD-DIOP
(R)-37 93% ee
MeO
OMe
OMe
38

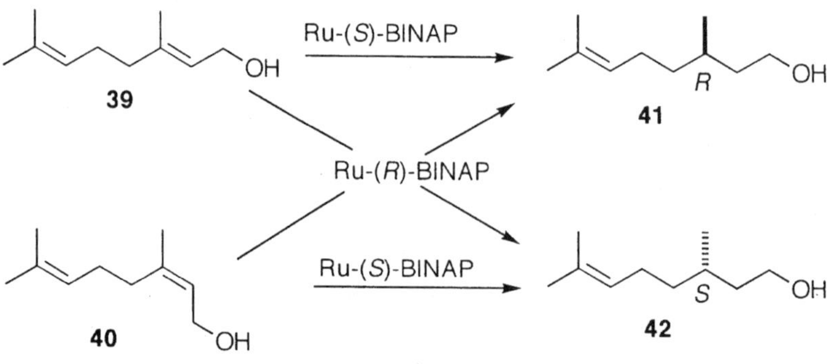
OH
39
Ru-(S)-BINAP
R
41
Ru-(R)-BINAP
40
Ru-(S)-BINAP
S
42

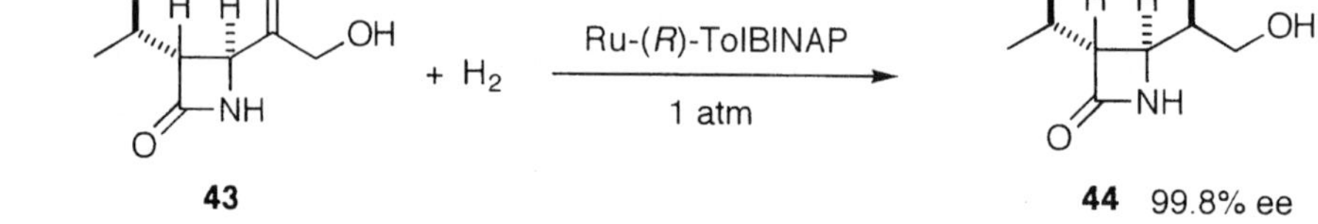
TBDMSO
NH
43
+ H2
Ru-(R)-TolBINAP
1 atm
44 99.8% ee

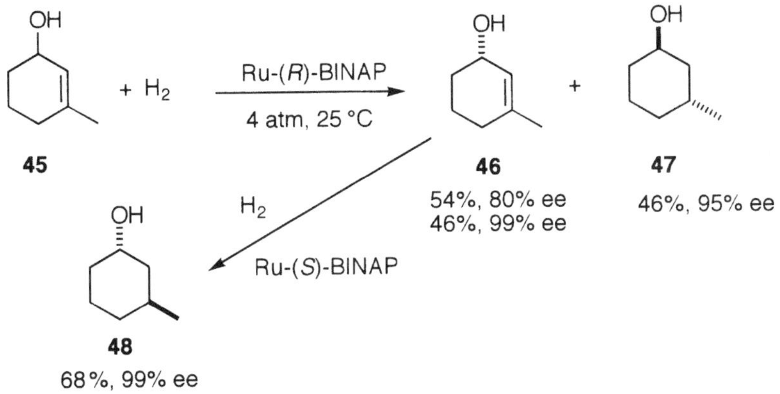

The interesting phenomena of chiral poisoning and kinetic resolution of racemic 2-cyclohexenol (**49** and **50**) are observed in the Ru-catalysed hydrogenation of the racemate using racemic BINAP by preferential deactivation of one enantiomer of the catalyst with an enantiomerically pure chiral poison. Poisoning of the racemic Ru catalyst with (1*R*,2*S*)-ephedrine (**52**) provided (*R*)-2-cyclohexenol (**49**) with 95% ee after 77% conversion. The result shows that only **50** is selectively hydrogenated by the unpoisoned chiral catalyst possibly arising from selective poisoning of the other enantiomeric one [37].

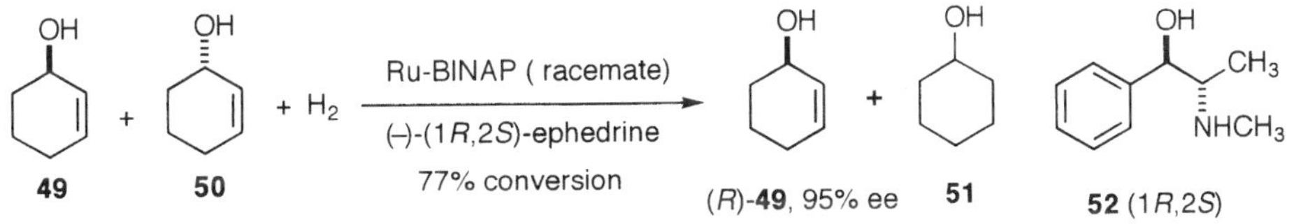

Enol-type double bonds are hydrogenated. β-Methylbutyrolactone (**54**) with 92% ee was prepared by asymmetric hydrogenation of diketene (**53**) with Ru BINAP [38]. Ring-opening polymerization of the hydrogenated lactone **54**, catalysed by distannoxane, produces poly(3-hydroxybutylate) (**55**) of high molecular weights, which is an important biodegradable polyester [39]

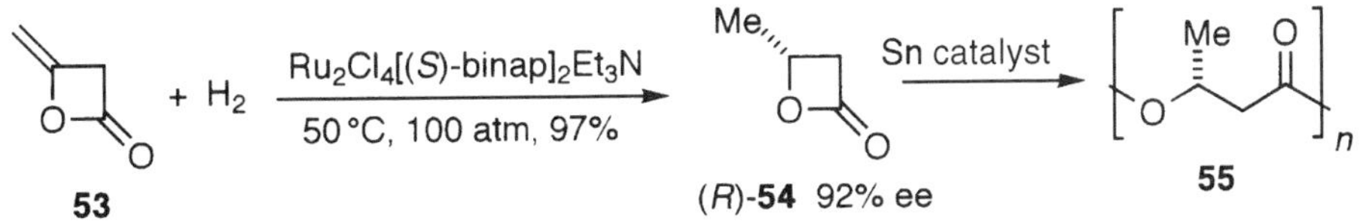

The chiral complex EBTHI−Ti is an excellent chiral catalyst [40]. This complex is a derivative of titanocene and used as the Kaminsky catalyst, which has brought epoch-making progress in polypropylene production. The chiral bridged titanocene complex is used for the production of optically active polypropylene arising from the helical structure of the polymer chain. The chiral complex also behaves as an excellent chiral catalyst for efficient asymmetric syntheses, such as asymmetric Ti-catalysed Pauson–Khand reaction described in Section 7.2.4 [41,42].

Chiral Ti complexes are prepared by the following method [43]. Coordination of ethylenebis(tetrahydroindene) (**56**; abbreviated as EBTHI) to TiCl$_4$ produces the racemic complex **57** and the *meso* complex **58** depending on which side (the same or opposite side) is occupied by the cyclohexane ring. Optical resolution of the racemate **57** is carried out in the following way. First, 0.6 equivalents of the dilithium salt of (*R*)-2,2′-dihydroxy-1,1′-binaphthyl (**59** abbreviated as (*R*)-binol) and 1.0 equivalents of *p*-aminobenzoic acid are added to the racemate **57**, and the mixture is heated in toluene with Et$_3$N. When cooled, (*S*)-ebthi-Ti(ArCO$_2$)$_2$ complex **62** precipitates, and the optically pure (*S*)-ebthi-TiCl$_2$ (**63**) is obtained in 33% yield by treatment with HCl. The pure (*R,R*)-ebthi-Ti binolate (**60**) is obtained by concentration of the filtrate and recrystallization, from which the chiral Ti hydride **61** is prepared as a precursor of the active catalyst.

Asymmetric hydrogenation of trisubstituted alkenes without a chelating effect of functional groups is not easy with Ru catalysts. Hydrogenation of 1,2-diphenylpro-

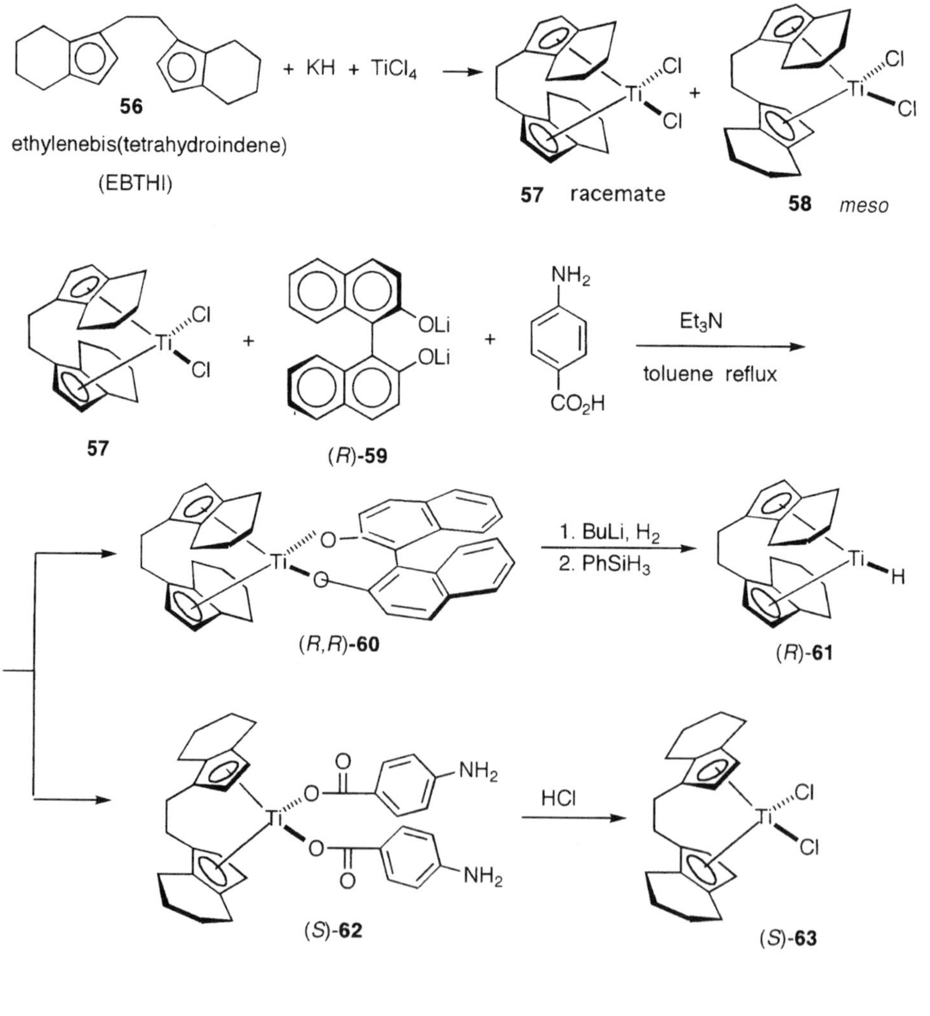

56

ethylenebis(tetrahydroindene)
(EBTHI)

+ KH + TiCl$_4$ →

57 racemate

58 *meso*

57   +   (*R*)-**59**   +   [NH$_2$ ... CO$_2$H]

Et$_3$N

toluene reflux

(*R,R*)-**60**

1. BuLi, H$_2$
2. PhSiH$_3$

(*R*)-**61**

(*S*)-**62**

HCl

(*S*)-**63**

pylene (**64**), bearing no functional group, has been achieved to give 1,2-diphenylpropane (**65**) with high ee using the chiral EBTHI–Ti catalyst **61** [44]. The α-methylbenzylamine derivative **67** with 92% ee was obtained by asymmetric hydrogenation of the enamine **66** with the same catalyst [45].

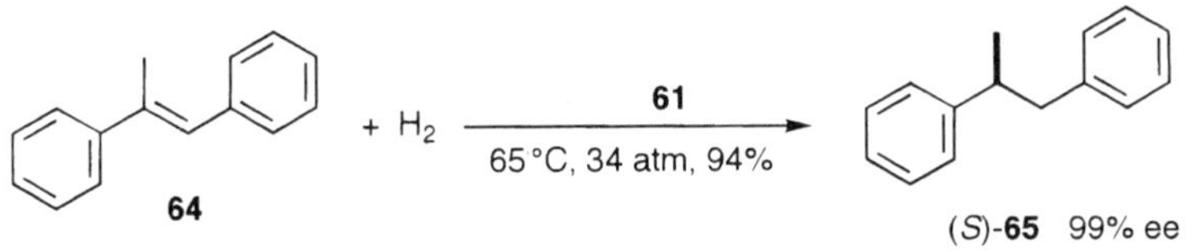

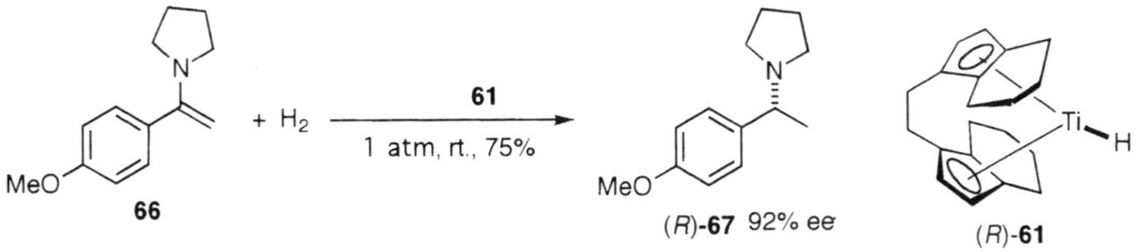

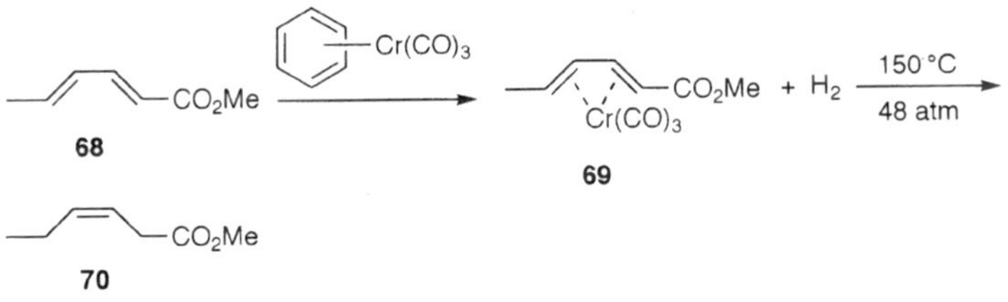

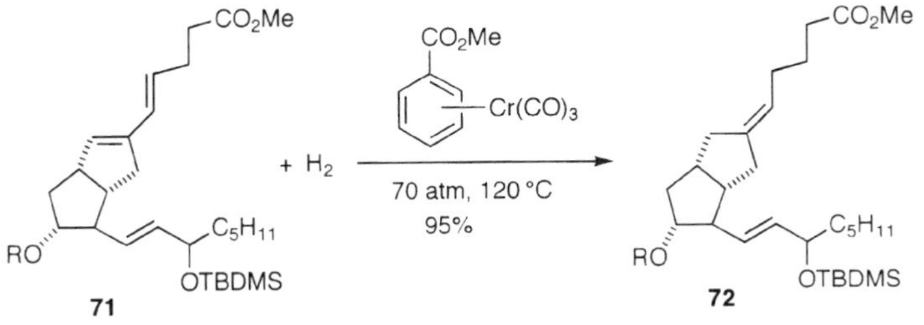

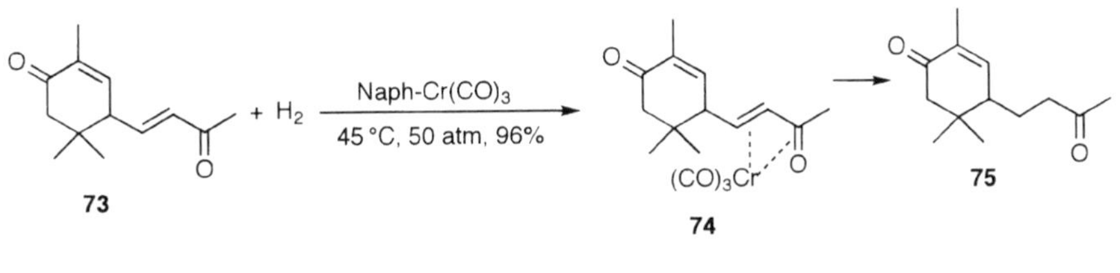

Selective 1,4-hydrogenation of 1,3-dienes to (*Z*)-alkenes is possible with benzenechromium tricarbonyl as a catalyst precursor. Only those conjugated dienes that can adopt a *cisoid* conformation, and hence chelate to the metal, are hydrogenated. Methyl sorbate (**68**) is reduced to methyl (*Z*)-3-hexenoate (**70**) [46,47]. The reaction proceeds by the coordination of coordinatively unsaturated $Cr(CO)_3$ to the diene as shown by **69**. Selective (*Z*)-alkene formation is applied to the preparation of (*E*)-trisubstituted *exo*-alkene in the carbacycline **72** from the 1,3-diene **71**. This reaction proceeds at 70 atm of H_2 at 120 °C using methyl benzoate-chromium tricarbonyl [48]. Naphthalene-chromium complex as the precursor has higher catalytic activity and the hydrogenation proceeds in THF at room temperature under 1 atm of H_2. α,β-Unsaturated carbonyl compounds can be hydrogenated when they can take a (*S*)-*cis* conformation. For example, one of the two enones in **73** is selectively reduced to **75** via the coordination as shown by **74** [49].

10.2 Asymmetric Reduction of Carbonyl and Imino Groups by Homogeneous Hydrogenation, Transfer Hydrogenation and Hydrosilylation

Although catalytic asymmetric homogeneous hydrogenation of carbonyl groups has been regarded as difficult to achieve, Ru–BINAP complexes have been found to be active for asymmetric hydrogenation of some types of ketones, and high ee values are observed in a number of cases [50, 51]. First, the ketone group of ethyl acetoacetate is reduced selectively under 86 atm of H_2 to give 3-hydroxybutanoate with 99% ee using 1/1000 equivalent of Ru–BINAP as the catalyst [26,50,52]. The Ru catalyst, active under 3 atm, is prepared from $[RuCl_2(cod)]_n$ and BINAP, and the keto ester **76** was reduced to **77** with 98% ee [53]. Hydrogenation of acetoacetate proceeds in 97% ee under atmospheric pressure of hydrogen using the Ru catalyst prepared from $(cod)Ru(2\text{-methallyl})_2$ and BINAP [54].

Acetylacetone (**78**) is hydrogenated with Ru–(*R*)-BINAP exclusively to give (*R, R*)-2,4-pentanediol (**79**) with 100% ee. Formation of the *meso* isomer **80** was only 1% [50,52]. Various important chiral pharmaceutical compounds are produced by Ru-catalysed asymmetric hydrogenation of carbonyl groups. α-Amino alcohols with high ee are obtained by the asymmetric hydrogenation of α-amino ketones using Rh complex of (*R,S*)-BPPFOH (**XXIV**) [55] and Ru–BINAP catalyst [52]. For example, (*S*)-propranolol (**82**) with 90.8% ee was synthesized by the hydrogenation of 1-naphthoxymethyl-*N*-isopropylaminomethyl ketone (**81**) using the Rh complex of (*2S,4S*)-MCCPM (**VI**) [56].

(*R*)-SEGPHOS (**XLI**) has axial chirality because free rotation is not possible and is an optically stable ligand. This ligand gives better results than BINAP for some prochiral compounds [57]. The hydrogenation of α-hydroxyacetone (**83**) with this ligand affords (*2R*)-1,2-propanediol (**84**) of 98.5% ee with a substrate-to-catalyst ratio up to 10 000, while BINAP gives **84** with 89% ee. The diol **84** is a chiral building block for levofloxacin (**85**).

The chlorohydrin **87** was obtained with 97% ee by the Ru-catalyzed hydrogenation of γ-chloroacetoacetate (**86**) at 100°C. The % ee was lower at lower temperature. The chlorohydrin **87** was converted to (*R*)-carnitine (**88**) [58]. Asymmetric hydrogenation of the α-bromo-β-keto phosphonate **89** with Ru(*S*)-BINAP affords (1*R*,2*S*)-α-bromo-β-hydroxyphosphonate **90** with 98% ee which is converted to fosfomycin (**91**) [59].

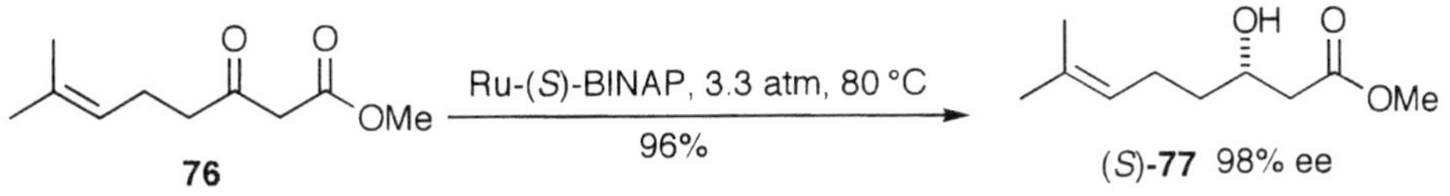

76 → Ru-(*S*)-BINAP, 3.3 atm, 80 °C / 96% → (*S*)-**77** 98% ee

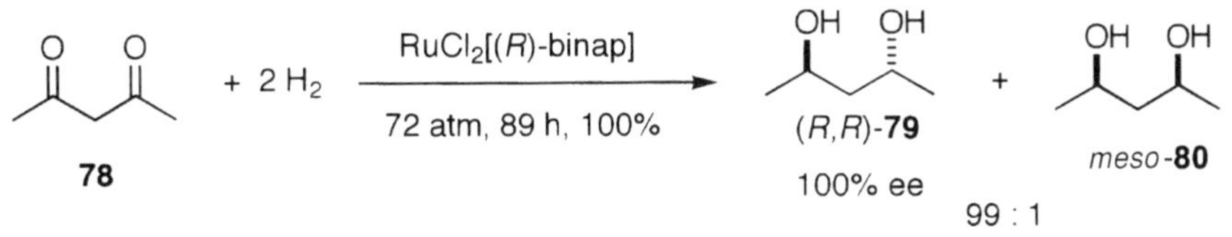

78 + 2 H$_2$ → RuCl$_2$[(*R*)-binap] / 72 atm, 89 h, 100% → (*R,R*)-**79** 100% ee + *meso*-**80**   99 : 1

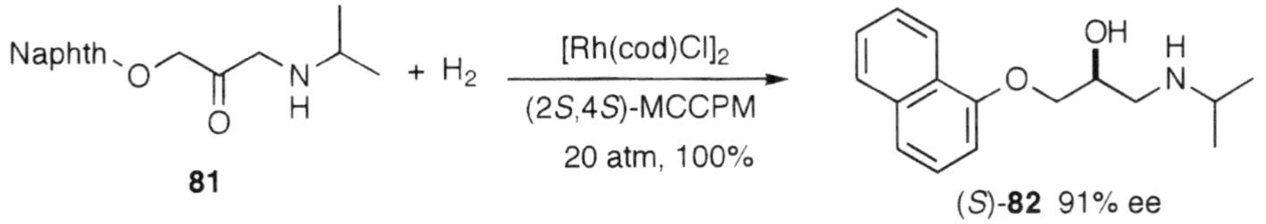

81 + H$_2$ → [Rh(cod)Cl]$_2$ / (2*S*,4*S*)-MCCPM / 20 atm, 100% → (*S*)-**82** 91% ee

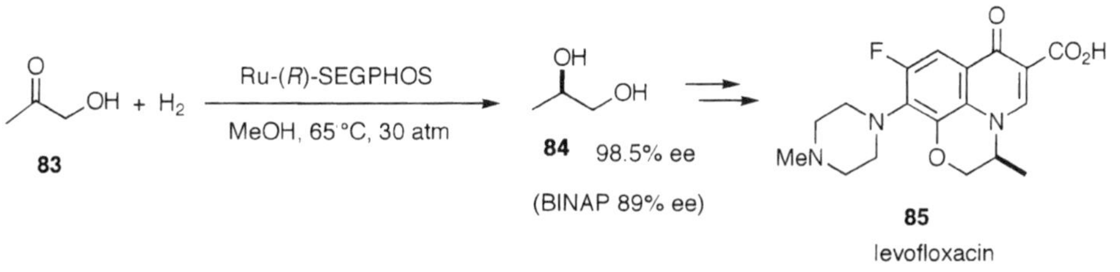

83 + H$_2$ → Ru-(*R*)-SEGPHOS / MeOH, 65 °C, 30 atm → **84** 98.5% ee (BINAP 89% ee) → **85** levofloxacin

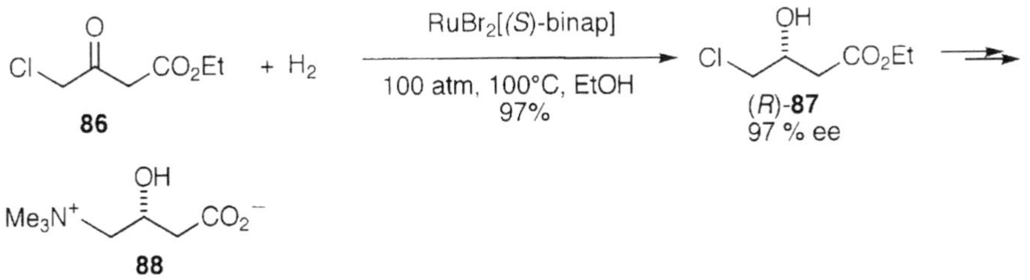

86 + H$_2$ → RuBr$_2$[(*S*)-binap] / 100 atm, 100°C, EtOH / 97% → (*R*)-**87** 97 % ee →

88

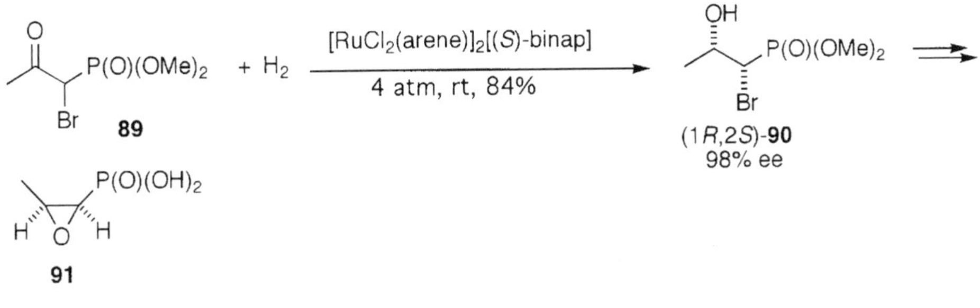

89 + H$_2$ → [RuCl$_2$(arene)]$_2$[(*S*)-binap] / 4 atm, rt, 84% → (1*R*,2*S*)-**90** 98% ee →

91

Although formation of a mixture of diastereomers is expected by the hydrogenation of α-substituted β-keto esters, in practice the hydrogenation of the keto group in α-substituted β-keto esters catalysed by **95** proceeds with epimerization of the α-carbon under the reaction conditions, and only one diastereomer is obtained in high selectivity. Thus methyl β-*N*-benzoylaminomethylacetoacetate (**92**) is reduced selectively to give **93** with 98% ee in 88% de and used for the commercial production of the carbapenem precursor **94** [60]. The high diastereoselectivity is achieved in (7:1) CH$_2$Cl$_2$:MeOH using the Ru complex coordinated by the BINAP derivative **95**.

Aromatic ketones are hydrogenated with Rh, Ir and Ru complexes. Addition of an amine as a ligand and KOH is crucial. Thus it is possible to reduce these ketones without using a stoichiometric amount of metal hydrides reagents. The activity of RuCl$_2$(Ph$_3$P)$_3$ for the hydrogenation of aromatic ketones is considerably enhanced by the addition of one equivalent of ethylenediamine and 2.8 mM solution of KOH in 2-propanol. Highly efficient enantioselective hydrogenation of aromatic ketones such as **96** is carried out using RuCl$_2$[(*S*)-binap](dmf)$_n$ in 2-propanol in the presence of (*S*,*S*)-1,2-diphenylethylenediamine (**98**) and KOH (1:1:2) under mild conditions (room temperature, 4 atm) to afford the alcohol **97** with 94% ee [61]. In the presence of the amine and KOH, only the ketone and aldehyde are hydrogenated without attacking alkene and alkynes. This is due to the affinity of the Ru—amine complex with carbonyl group, and the carbonyl group is hydrogenated 370 000 times faster by the Ru—amine complex than the Ru complex without amine. Preparation of the active Ru precatalysts has been reported [62].

The Rh complex of PennPhos (**XIII**) is a good catalyst for highly enantioselective hydrogenation of ketones, and acetophenone (**99**) was hydrognated to *sec*-phenethyl alcohol (**100**) with 95% ee in 97% yield in the presence of 2,6-lutidine as an important additive [63].

Usually, a stoichiometric amount of metal hydride, such as a selectride reagent, is required for diastereoselective reduction of aliphatic ketones to secondary alcohols. Now the same purpose can be achieved by the Ru-catalysed diastereoselective hydrogenation [64]. Using $RuCl_2(Ph_3P)_3$, $NH_2(CH_2)_2NH_2$, and KOH (1:1:2) as a catalyst precursor, 2-methylcyclohexanone (**101**) is hydrogenated to give *cis*-2-methylcyclohexanol (**102**) in 98% stereoselectivity. As the two methyl groups in 2,6-dimethylcyclohexanone (**103**) are in rapid *cis* and *trans* equilibrium in alkali-containing 2-propanol, and the hydrogenation of the *cis* isomer proceeds much faster than the *trans* isomer, *cis*,*cis*-1,6-dimethylcyclohexanol (**104**) is obtained in 98.7% stereoselectivity.

Moreover, when $(1R,4S)$-$(-)$-menthone (**105**), equilibrating with isomenthone, the 4*R* isomer, is hydrogenated with (R)-BINAP and (S,S)-1,2-diphenylethylenediamine (abbreviated as DPEN), $(+)$-neomenthol (**106**) is formed exclusively.

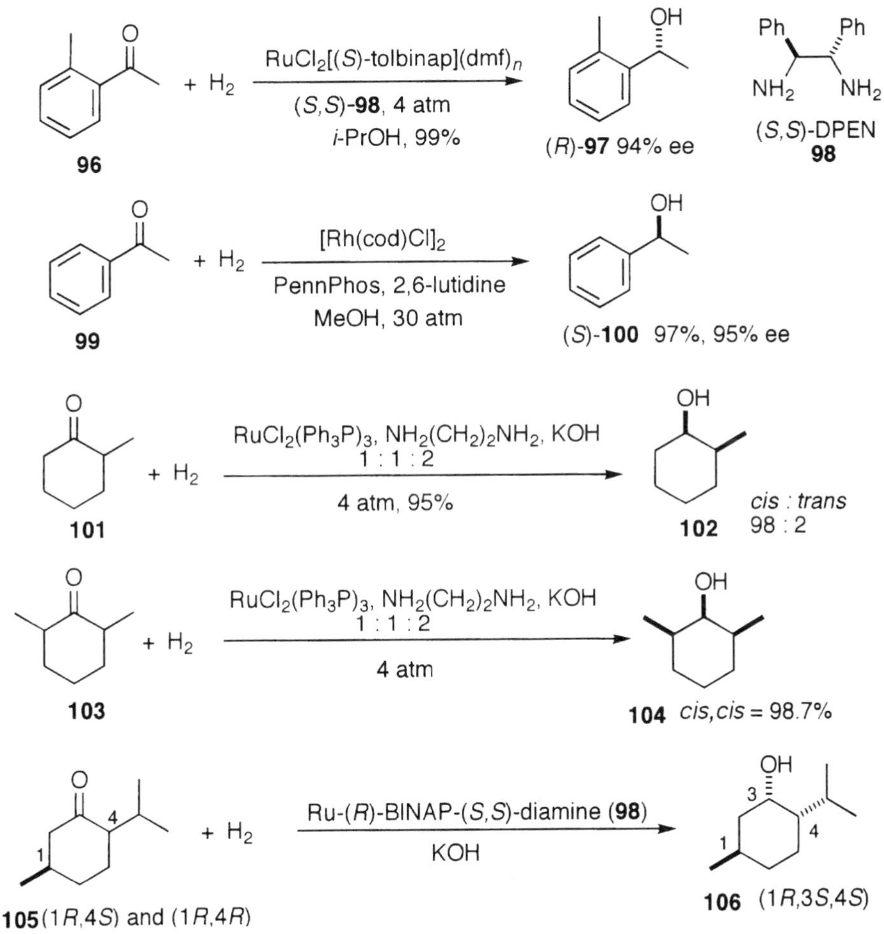

96 + H_2 → $RuCl_2[(S)$-tolbinap](dmf)$_n$ / (S,S)-**98**, 4 atm / *i*-PrOH, 99% → (R)-**97** 94% ee ; (S,S)-DPEN **98**

99 + H_2 → $[Rh(cod)Cl]_2$ / PennPhos, 2,6-lutidine / MeOH, 30 atm → (S)-**100** 97%, 95% ee

101 + H_2 → $RuCl_2(Ph_3P)_3$, $NH_2(CH_2)_2NH_2$, KOH 1:1:2 / 4 atm, 95% → **102** *cis* : *trans* 98 : 2

103 + H_2 → $RuCl_2(Ph_3P)_3$, $NH_2(CH_2)_2NH_2$, KOH 1:1:2 / 4 atm → **104** *cis*,*cis* = 98.7%

105 $(1R,4S)$ and $(1R,4R)$ + H_2 → Ru-(R)-BINAP-(S,S)-diamine (**98**) / KOH → **106** $(1R,3S,4S)$

An interesting asymmetric activation of Ru complexes for enantioselective hydrogenation of ketones has been observed [65]. Hydrogenation of 2,4,4-trimethyl-2-cyclohexenone using racemic $RuCl_2$(tolbinap)(dmf)$_n$, (S,S)-DPEN and KOH (1:1:2) afforded the alcohol **107** with 95% ee in 100% yield. When (R)-TolBINAP and (S,S)-DPEN were used, the alcohol **107** with 96% ee was obtained. On the other hand, only

26% ee was obtained by the combination of (*R*)-TolBINAP and (*R,R*)-DPEN and the reaction was slow. These results clearly show that (*R*)-TolBINAP is activated more efficiently by (*S,S*)-DPEN than (*R,R*)-DPEN.

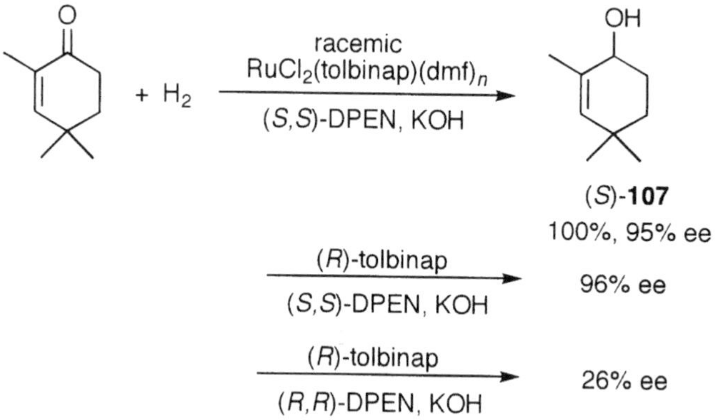

(*S*)-**107**
100%, 95% ee

96% ee

26% ee

Furthermore, efficient enantioselective hydrogenation was achieved using a conformationally flexible, and racemic bis(diarylphosphino)biphenyl (BIPHEP) −RuCl$_2$, coordinated by the enantiopure diamine (**98**). Thus hydrogenation of the ketone **108** using a mixture of racemic and proatropisomeric DM-BIPHEP-RuCl$_2$, enantiopure (*S,S*)-DPEN (**98**), and KOH (1:1:2) in 2-propanol afforded the (*R*)-alcohol with 92% ee in 99% yield. In this case, (*S*)-DM-BIPHEP-RuCl$_2$/(*S,S*)-**98** can isomerize gradually to (*R*)-DM-BIPHEP RuCl$_2$/(*S,S*)-**98**, and the efficient enantio-selective hydrogenation proceeded by the enriched (*R*)-DM-BIPHEP RuCl$_2$/(*S,S*)-**98** catalyst. These results show a general strategy for the use of not only racemic but also conformationally flexible ligands for efficient enantioselective reactions [65a].

Ketones are reduced by asymmetric hydrogen transfer from either HCO$_2$H or 2-propanol as hydrogen sources, catalysed by chiral Ru complexes [66]. HCO$_2$H is used

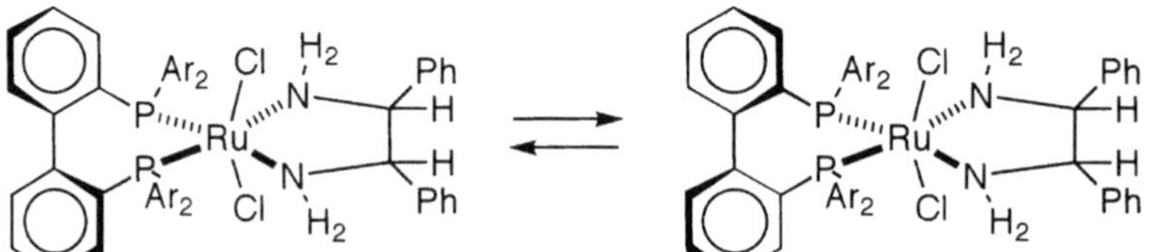

(*S*)-DM-BIPHEP / (*S,S*)-**98**

Ar = 3,5-dimethyl

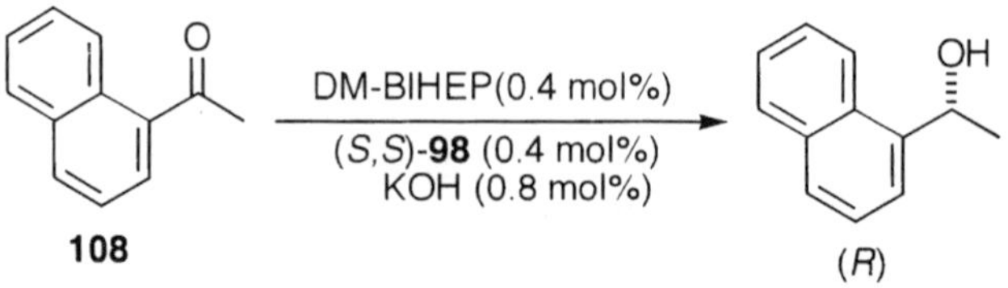

(*R*)-DM-BIPHEP / (*S,S*)-**98**

108

DM-BIHEP(0.4 mol%)

(*S,S*)-**98** (0.4 mol%)
KOH (0.8 mol%)

(*R*)

92% ee, 99%

successfully for Rh-catalysed asymmetric transfer hydrogenation of itanonic acid (**34**) in the presence of a chiral amine as described before [32]. In principle, transfer hydrogenation is reversible, but the reduction with HCO_2H proceeds irreversibly. Similarly, the efficient irreversible asymmetric transfer hydrogenation of acetophenone (**99**) to give (*S*)-1-phenylethanol (**100**) is catalysed by chiral Ru complexes, prepared from $[(\eta^6\text{-arene})RuCl_2]_2$ and the chiral 1,2-diamine (1*S*,2*S*)-*N*-*p*-toluenesulfonyl-1,2-diphenylethylenediamine (TsDPEN) **109**. An azeotropic mixture of HCO_2H–Et_3N (5:2) is used as the hydrogen source [67].

Asymmetric hydrogen transfer from 2-propanol to aromatic ketones such as acetophenone (**99**) has been achieved by using the same chiral Ru complex in 2-propanol containing KOH at room temperature, and (*S*)-1-phenylethanol (**100**) with 98% ee was obtained [68,69]. Similarly, efficient Ru-catalysed transfer hydrogenation of aromatic ketones using the cyclic amino alcohol [(1*S*,3*R*,4*R*)-2-azanorbornylmethanol] (**110**) [70] and bis(oxazolinylmethyl) amine (**111**) [71] was reported.

The catalyst precursors **112** and **114**, the true catalysts **113** and **115**, and the reactive intermediate **116** in the transfer hydrogenation were isolated and the mechanism of the transfer hydrogenation has been clearly established [69]. The catalyst precursor **114**, the 18-electron complex, was prepared by reacting $[RuCl_2(\eta^6\text{-}p\text{-cymene})]_2$, (*S*,*S*)-TsDPEN and KOH (1:1:1) as orange crystals. Elimination of HCl from **114** by treatment with one equivalent of KOH produces the true catalyst **115** as the 16-electron, neutral Ru(II) complex. The complex **115** shows distinct dehydrogenative activity for 2-propanol. Rapid formation of acetone occurs to produce the Ru-hydride species **116** as yellow needles when **115** is treated with 2-propanol at room temperature in the absence of base. In turn, treatment of **116** with excess acetone leads instantaneously to the true catalyst **115** and 2-propanol. It was confirmed that the complex **115** indeed catalyses the asymmetric reduction of acetonephenone (**108**) in 2-propanol without KOH to afford (*S*)-1-phenylethanol (**109**) with 97% ee. Similar 16- and 18-electron complexes of Cp–Rh and Cp–Ir were prepared, and they are efficient catalysts for the transfer hydrogenation of ketones [72].

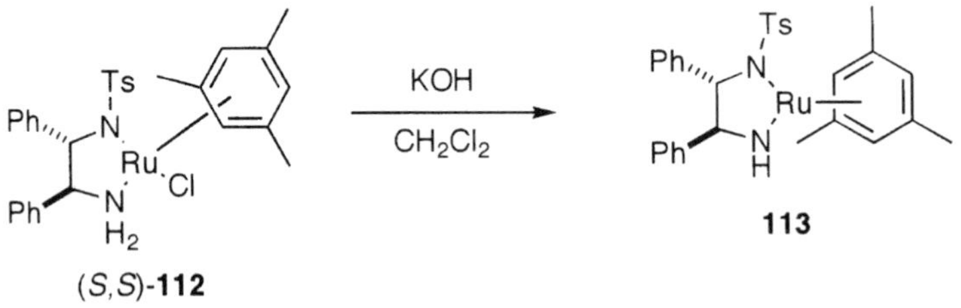

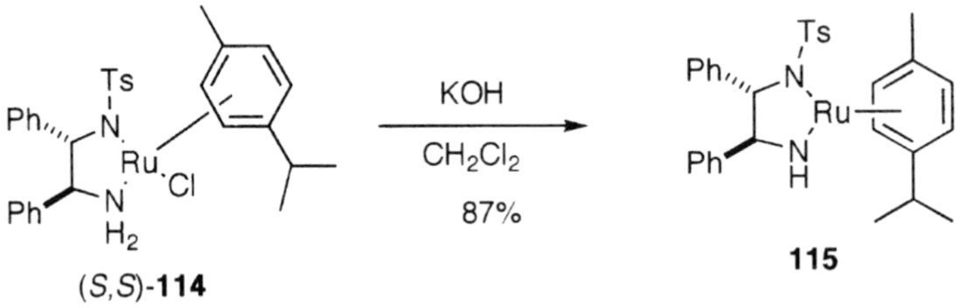

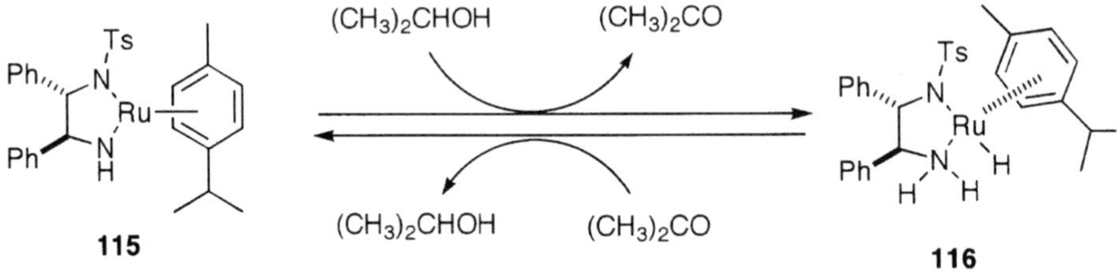

As the hydrogen transfer from 2-propanol is reversible, the efficiency is highly dependent on the redox properties of the alcohols formed. Therefore, highly enantioselective transfer hydrogenation of ketones with an electron-donating group on the aromatic ring is not possible. However, this tendency provides an opportunity for kinetic resolution of secondary alcohols in acetone [73]. Thus when the racemic benzyl alcohol with an electron-donating group **117** in the presence of the true Ru catalyst **115** was left in acetone for 22 h, the (*R*)-alcohol **118** with 92% ee was recovered in 47% yield in addition to the acetophenone **119**. The use of **115**, which has a unique 16-electron configuration, is the key for the successful kinetic resolution under neutral conditions.

In addition, based on the reversibility of the transfer hydrogenation, very unique desymmetrization of *meso* substrates is possible using cheaply available acetone as the

117 (R)-118 92% ee, 47% 119

120 *meso* (R)-121 96% ee, 70%

hydrogen accepter. When the *meso* diol **120** was treated with the Ru catalyst **113** in acetone, the (*R*)-keto alcohol **121** with 96% ee was obtained in 70% yield.

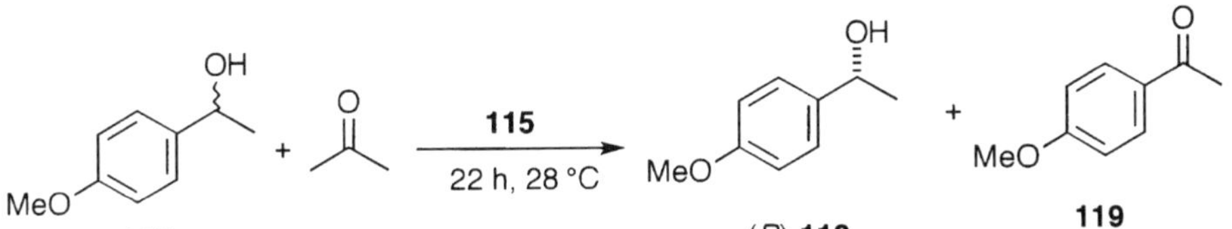

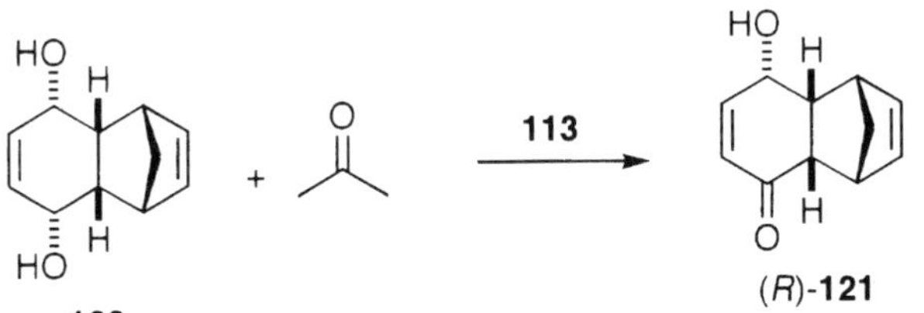

The successful Ru-catalysed transfer hydrogenation of α,β-alkynyl ketones, without reducing the triple bond in 2-propanol, offers a good synthetic route to chiral propargyl alcohols [74]. Transfer hydrogenation of the chiral (*S*)-ketone **122** using the (*R,R*)-Ru catalyst corresponding to (*S,S*)-**115** proceeds in 2-propanol at room temperature to afford (3*S*,4*S*)-amino alcohol **123** with 99% ee in 97% yield, whereas the reduction of **122** with the enantiomeric (*S,S*)-Ru catalyst **115** gives the (3*R*,4*S*)-amino alcohol **124** with 99% ee in 97% yield, showing that the carbonyl diastereo faces in **122** are efficiently differentiated by the chirality of the Ru template; the adjacent N-substituted stereogenic center does not play a significant role.

(S)-122 (3S,4S)-123 99% ee, 97% (3R,4S)-124 99% ee, 97%

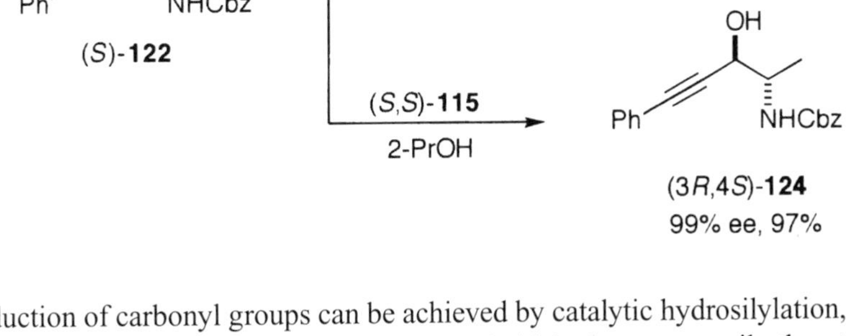

Reduction of carbonyl groups can be achieved by catalytic hydrosilylation, followed by hydrolysis. Hydrosilanes add to ketones and aldehydes more easily than to alkenes using Rh and Ru catalysts. Oxophilic Si adds to oxygen of the carbonyl groups to afford silyl ethers, which can be hydrolysed easily to alcohols. In the reduction of camphor (125) catalysed by RhCl(Ph$_3$P)$_3$, ratios of borneol (126) and isoborneol (127) depend on a kind of silanes used, and the more stable isomer 127 is obtained with the bulky silane [75].

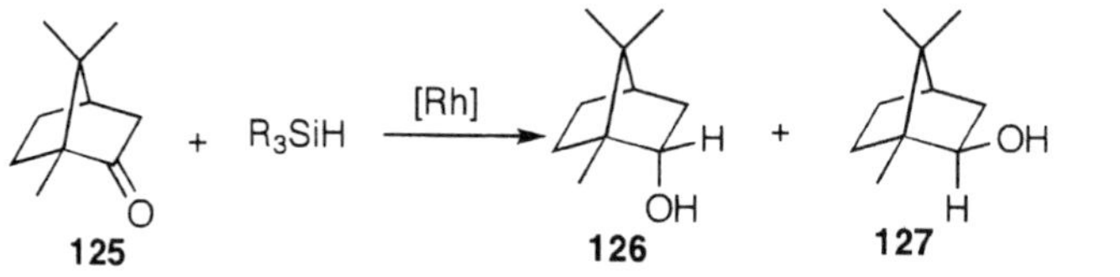

R$_3$SiH	126 : 127
PhSiH$_3$	90 : 10
n-Pr$_3$SiH	30 : 70

Chiral alcohols are prepared by asymmetric hydrosilylation. The nitrogen-containing polydentate ligands 128, 129 and 130 are better than chiral phosphine ligands for Rh-catalysed asymmetric hydrosilyation of ketones [76]. Highly efficient (≈99% ee) asymmetric hydrosilylation of 1-tetralone (131) with Ph$_2$SiH$_2$ was achieved to give 132 with 99% ee using optically active bis(oxazolinyl)pyridine (pybox) (128) as a ligand [77]. Pythia (129) [78] and pymox (130) [79] are also good ligands.

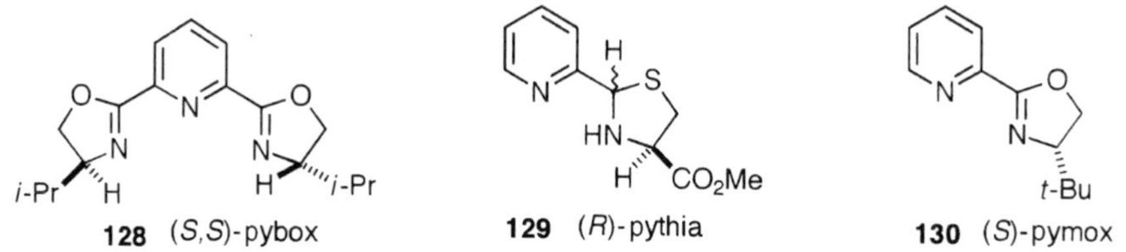

128 (S,S)-pybox   129 (R)-pythia   130 (S)-pymox

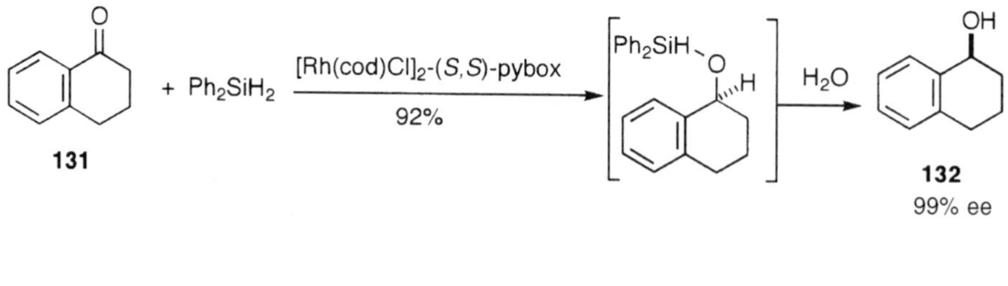

131   132
99% ee

The chiral titanocene complex 61 is an excellent catalyst for the enantioselective hydrosilylation of the ketone 131 with PMHS (133) to afford the alcohol 132 with 91% ee [80]. Efficient asymmetric hydrosilylation of symmerical diketones is catalysed by the Rh complex coordinated by EtTRAP. Biacetyl (134) was converted to (2S,3S)-2,3-butanediol (135) with 95% ee in 69% yield [81].

Chiral amines can be prepared by asymmetric hydrogenation, transfer hydrogenation and hydrosilylation of imines. The piperidine 137 with 98% ee was obtained by highly efficient asymmetric hydrogenation of the cyclic imines 136 catalysed by the Ti catalyst 61 [82]. Pyrrolidine 139 with 99% ee was obtained in 34% after 50%

131 **133** **132** 91% ee

134 **135** 95% ee

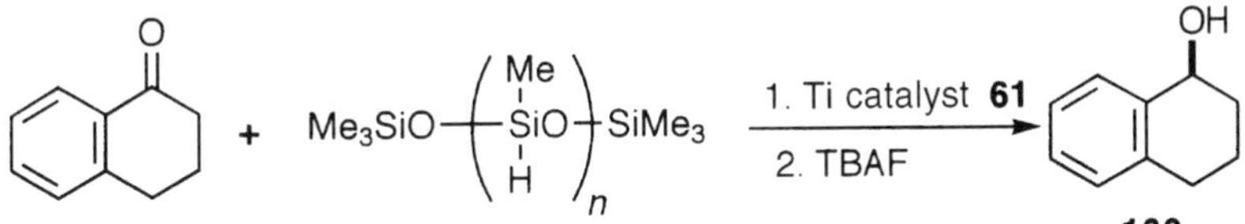

conversion of the asymmetric hydrogenation of the racemic disubstituted 1-pyrroline **138** using by the titanocene catalyst **61**. The kinetic resolution occurs during the hydrogenation [83]. Imine **140** with 99% ee was recovered in 37%. Amine **142** with 80% ee was obtained by the asymmetric hydrogenation of **141** using 1/1 000 000 equivalent of Ir–josiphos (**XXVI**). The herbicide (*S*)-metolachlor (**143**) is produced commercially from **142** [84].

136 **137** 98% ee

138 **139** 34%, 99% ee **140** 37%, 99% ee

141 **142** 80% ee **143**

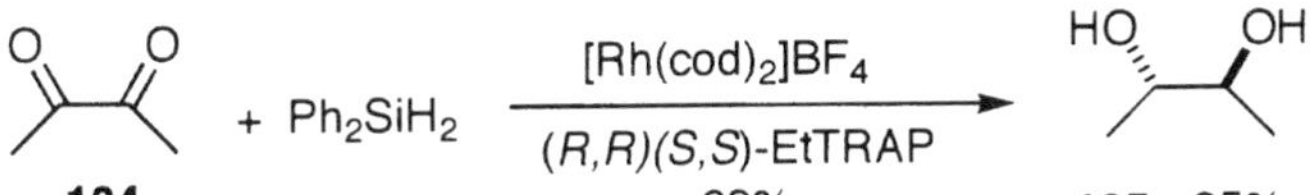

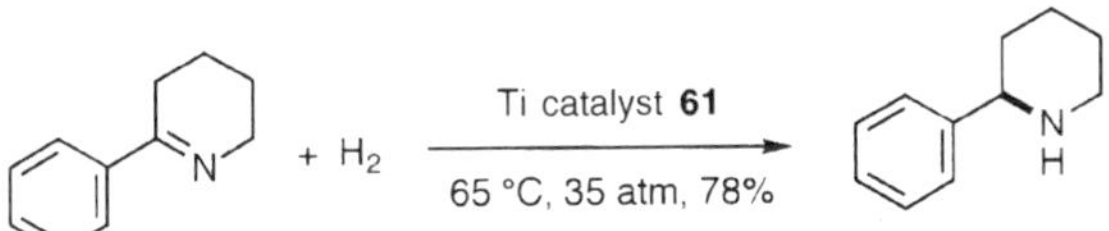

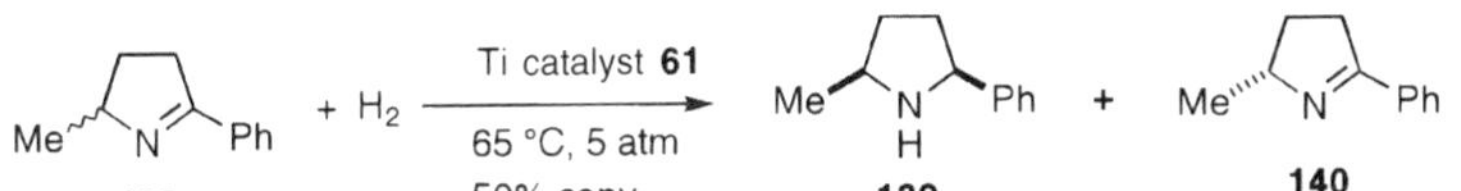

Highly enantioselective hydrosilylation of the imine **144** was achieved using (*S*,*S*)-(ebthi)titanium difluoride as a catalyst precursor to give the optically active benzylamine **145** with 99% ee in 95% after acidic workup [85].

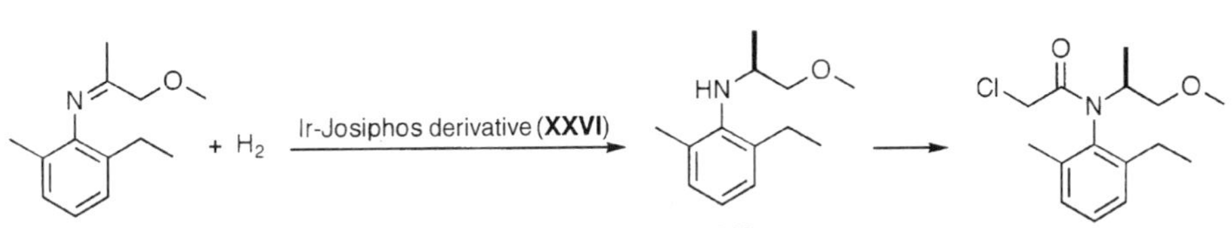

Transfer hydrogenation of the imine bond in **146** from HCO_2H, catalysed by the Ru chiral amine complex **114**, afforded salsolidine (**147**) [86].

$$Cp_2TiF_2 \ + \ PhSiH_3 \longrightarrow \left[Cp_2TiH \right]$$

144   + PhSiH₃

1. TiF₂-(*S,S*)-(ebthi), rt

2. H⁺, H₂O   95%

145   99% ee

146   + HCO₂H

[RuCl₂(*p*-cymene)]₂ (*S,S*)-TsDPEN

Et₃N, MeCN, 99%

(*R*)-**147**   95% ee

References

1. J. A. Osborn, F. H. Jardine, J. F. Young and G. Wilkinson, *J. Chem. Soc. A*, 1711 (1966); 1574 (1967).
2. A. J. Birch and D. H. Williamson, *Org. React.*, **24**, 1 (1976).
3. R. E. Ireland and P. Bey, *Org. Synth., Coll.* **6**, 459 (1988).
4. R. E. Harmon, J. L. Parsons, D. W. Cooke, S. K. Gupta and J. Schoolenberg, *J. Org. Chem.*, **34**, 3684 (1969).
5. J. Tsuji and H. Suzuki, *Chem. Lett.*, 1085 (1977).
6. I. Ojima, N. Clos and C. Bastos, *Tetrahedron*, **45**, 6901 (1989); R. Noyori and H. Takaya, *Acc. Chem. Res.*, **23**, 345 (1990).
7. R. Noyori, *Asymmetric Catalysis in Organic Synthesis*, p. 16, John Wiley & Sons, 1994; H. Takaya, T. Ohta and R. Noyori, in *Catalytic Asymmetric Synthesis*, (Ed.) I. Ojima, Chapter 3, VCH, 1993.
8. T. P. Dang and H. B. Kagan, *Chem. Commun.*, 481 (1971); *J. Am. Chem. Soc.*, **94**, 6429 (1972); T. P. Dang, J. C. Poulin and H. B. Kagan, *J. Organometal. Chem.*, **91**, 105 (1975).
9. W. S. Knowles, M. J. Sabacky, B. D. Vineyard and D. J. Weinkauff, *J. Am. Chem. Soc.*, **97**, 2567 (1975); W. S. Knowles, *Acc. Chem. Res.*, **16**, 106 (1983).
10. W. S. Knowles and M. J. Sabsacky, *Chem. Commun.*, 1445 (1968); W. S. Knowles, M. J. Sabacky and B. D. Vineyard, *Chem. Commun.* 10 (1972).
11. L. Horner, H. Siegel and H. Büthe, *Angew. Chem., Int. Ed. Engl.*, **7**, 942 (1968).
12. U. Nagel, E. Kinzel, J. Andrade and G. Prescher, *Chem. Ber.*, **119**, 3326 (1986); U. Nagel, *Angew. Chem., Int. Ed. Engl.*, **23**, 435 (1984).
13. A. Miyashita, H. Takaya, T. Souchi and R. Noyori, *Tetrahedron*, **40**, 1245 (1984); T. Ikariya, Y. Ishii, H. Kawano, T. Arai, M. Saburi, S. Yoshikawa and S. Akutagawa, *Chem. Commun.*, 922 (1985).
14. M. J. Burk, J. E. Feaster and R. L. Harlow, *Organometallics*, **9**, 2653 (1990); M. J. Burk, *J. Am. Chem. Soc.*, **113**, 8518 (1991).
15. T. Imamoto, J. Watanabe, Y. Wada, H. Masuda, H. Yamada, H. Tsuruta, S. Matsukawa and K. Yamaguchi, *J. Am. Chem. Soc.*, **120**, 1635 (1998).
16. G. Zhu, P. Cao, Q. Jiang and X. Zhang, *J. Am. Chem. Soc.*, **119**, 1799 (1997).

17. A. S. C. Chan, W. Hu, C. C. Pai, C. P. Lau, Y. Jiang, A. Mi, M. Yan, J. Sun, R. Lou and J. Deng, *J. Am. Chem. Soc.*, **119**, 9570 (1997).

18. C. R. Landis and J. Halpern, *J. Am. Chem. Soc.*, **109**, 1746 (1987).

19. M. J. Burk, M. F. Gross and J. P. Martinez, *J. Am. Chem. Soc.*, **117**, 9375 (1995).

20. M. Sawamura, R. Kuwano and Y. Ito, *J. Am. Chem. Soc.*, **117**, 9602 (1995).

21. M. J. Burk, G. Casy and N. B. Johnson, *J. Org. Chem.*, **63**, 6084 (1998).

22. M. J. Burk, Y. M. Wang and J. R. Lee, *J. Am. Chem. Soc.*, **118**, 5142 (1996).

23. F. Y. Zhang, C. C. Pai and A. S. C. Chan, *J. Am. Chem. Soc.*, **120**, 5808 (1998).

23a. Q. Jiang, D. Xiao, Z. Zhang, P. Cao and X. Zhang, *Angew. Chem., Int. Ed. Engl.*, **38**, 516 (1999).

24. H. Kumobayashi, *Rec. Trav. Chim. Pays-Bas*, **115**, 201 (1996); S. Akutagawa, *Appl. Catalysis. A*, **128**, 171 (1995).

25. T. Ohta, H. Takaya and R. Noyori, *Inorg. Chem.*, **27**, 566 (1988).

26. M. Kitamura, M. Tokunaga, T. Ohkuma and R. Noyori, *Org. Syn.*, **71**, 1 (1993).

27. H. Kawano, T. Ikariya, Y. Ishii, M. Saburi, S. Yoshikawa, Y. Uchida and H. Kumobayashi, *J. Chem. Soc., Perkin I*, 1571 (1989).

28. R. Noyori, M. Ohta, Y. Hsiao, M. Kitamura, T. Ohta and H. Takaya, *J. Am. Chem. Soc.*, **108**, 7117 (1986); M. Kitamura, Y. Hsiao, M. Ohta, M. Tsukamoto, T. Ohta, H. Takaya and R. Noyori, *J. Org. Chem.*, **59**, 297 (1994).

29. T. Ohta, H. Takaya, M. Kitamura, K. Nagai and R. Noyori, *J. Org. Chem.*, **52**, 3174 (1987); T. Uemura, X. Zhang, K. Matsumura, N. Sayo, H. Kumobayashi, T. Ohta, K. Nozaki and H. Takaya, *J. Org. Chem.*, **61**, 5510 (1996).

30. H. Takahashi and K. Achiwa, *Chem. Lett.*, 1921 (1987).

31. T. Chiba, A. Miyashita, H. Nohira and H. Takaya, *Tetrahedron Lett.*, **32**, 4745 (1991).

32. H. Brunner, E. Graf, W. Leitner and K. Wutz, *Synthesis*, 743 (1989).

33. T. Morimoto, W. Chiba and K. Achiwa, *Tetrahedron Lett.*, **31**, 261 (1990); **30**, 735 (1989).

34. H. Takaya, T. Ohta, N. Sayo, H. Kumobayashi, S. Akutagawa, S. Inoue, I. Kasahara and R. Noyori, *J. Am. Chem. Soc.*, **109**, 1596, 4129 (1987).

35. M. Kitamura, K. Nagai, Y. Hsiao and R. Noyori, *Tetrahedron Lett.*, **31**, 549 (1990).

36. M. Kitamura, I. Kasahara, K. Manabe, R. Noyori and H. Takaya, *J. Org. Chem.*, **53**, 708 (1988).

37. J. W. Faller and M. Tokunaga, *Tetrahedron Lett.*, **34**, 7359 (1993).

38. T. Ohta, T. Miyake and H. Takaya, *Chem. Commun.*, 1725 (1992); T. Ohta, T. Miyake, N. Seido, H. Kumobayashi and H. Takaya, *J. Org. Chem.*, **60**, 357 (1995).

39. Y. Hori, M. Suzuki, A. Yamaguchi and T. Nishishita, *Macromol.*, **26**, 5533 (1993).

40. A. H. Hoveyda and J. P. Morken, *Angew. Chem., Int. Ed. Engl.*, **35**, 1262 (1996).

41. F. A. Hicks, N. M. Kablaoui and S. L. Buchwald, *J. Am. Chem. Soc.*, **118**, 9450 (1996).

42. F. A. Hicks and S. L. Buchwald, *J. Am. Chem. Soc.*, **118**, 11688 (1996).

43. B. Chin and S. L. Buchwald, *J. Org. Chem.*, **61**, 5650 (1996).

44. R. D. Broene and S. L. Buchwald, *J. Am. Chem. Soc.*, **115**, 12569 (1993).

45. N. E. Lee and S. L. Buchwald, *J. Am. Chem. Soc.*, **116**, 5985 (1994).

46. E. N. Frankel, E. Selke and C. A. Glass, *J. Am. Chem. Soc.*, **90**, 2446 (1968); M. Cais, E. N. Frankel and A. Rejoan, *Tetrahedron Lett.*, 1919 (1968).

47. M. Sodeoka and M. Shibasaki, *Synthesis*, 643 (1993).

48. M. Shibasaki, M. Sodeoka and Y. Ogawa, *J. Org. Chem.*, **49**, 4096 (1984).

49. M. Sodeoka and M. Shibasaki, *J. Org. Chem.*, **50**, 1147 (1985).

50. R. Noyori, T. Ohkuma, M. Kitamura, H. Takaya, N. Sayo, H. Kumobayashi and S. Akutagawa, *J. Am. Chem. Soc.*, **109**, 5856 (1987).

51. D. J. Ager and S. A. Laneman, *Tetrahedron: Asymm.*, **8**, 3327 (1997).

52. M. Kitamura, T. Ohkuma, S. Inoue, N. Sayo, H. Kumobayashi, S. Akutagawa, T. Ohta, H. Takaya and R. Noyori, *J. Am. Chem. Soc.*, **110**, 629 (1988).
53. D. F. Taber and L. J. Silverberg, *Tetrahedron Lett.*, **32**, 4227 (1991).
54. J. P. Genet, V. Ratovelomanana-Vidal, M. C. Cano de Andrade, X. Pfister, P. Guerreiro and J. Y. Lenoir, *Tetrahedron Lett.*, **36**, 4801 (1995).
55. T. Hayashi, A. Katsumura, M. Konishi and M. Kumada, *Tetrahedron Lett.*, 425 (1979).
56. H. Takahashi, S. Sakuraba, H. Takeda and K. Achiwa, *J. Am. Chem. Soc.*, **112**, 5876 (1990).
57. T. Saito, T. Yokozawa, T. Ishizaki, X. Zhang and N. Sayo, unpublished results.
58. M. Kitamura, T. Ohkuma, H. Takaya and R. Noyori, *Tetrahedron Lett.*, **29**, 1555 (1988).
59. M. Kitamura, M. Tokunaga and R. Noyori, *J. Am. Chem. Soc.*, **117**, 2931 (1995).
60. R. Noyori, T. Ikeda, T. Ohkuma, M. Widholm, M. Kitamura, H. Takaya, S. Akutagawa, N. Sayo, T. Saito, T. Taketomi and H. Kumobayashi, *J. Am. Chem. Soc.*, **111**, 9134 (1989); K. Mashima, Y. Matsumura, K. Kusano, H. Kumobayashi, N. Sayo, Y. Hori, T. Ishizaki, S. Akutagawa and H. Takaya, *Chem. Commun.*, 609 (1991).
61. T. Ohkuma, H. Ooka, S. Hashiguchi, T. Ikariya and R. Noyori, *J. Am. Chem. Soc.*, **117**, 2675 (1995); T. Ohkuma, H. Ooka, T. Ikariya and R. Noyori, *J. Am. Chem. Soc.*, **117**, 10417 (1995).
62. H. Doucet, T. Ohkuma, K. Murata, T. Yokozawa, M. Kozawa, E. Katayama, A. F. England, T. Ikariya and R. Noyori, *Angew. Chem., Int. Ed. Engl.*, **37**, 1703 (1998); T. Ohkuma, M. Koizumi, H. Doucet, T. Pham, M. Kozawa, K. Murata, K. Katayama, T. Yokozawa, T. Ikariya and R. Noyori, *J. Am. Chem. Soc.*, **120**, 13529 (1998).
63. Q. Jiang, Y. Jiang, D. Xiao, P. Cao and X. Zhang, *Angew. Chem., Int. Ed. Engl.*, **37**, 1100 (1998).
64. T. Ohkuma, H. Ooka, M. Yamakawa, T. Ikariya and R. Noyori, *J. Org. Chem.*, **61**, 4872 (1996).
65. T. Ohkuma, H. Doucet, T. Pham, K. Mikami, T. Korenaga, M. Terada and R. Noyori, *J. Am. Chem. Soc.*, **120**, 1086 (1998).
65a. K. Mikami, T. Korenaga, M. Terada, T. Ohkuma, T. Pharm and R. Noyori, *Angew. Chem., Int. Ed. Engl.*, **38**, 495 (1999).
66. R. Noyori and S. Hashiguchi, *Acc. Chem. Res.*, **30**, 97 (1997).
67. A. Fujii, S. Hashiguchi, N. Uematsu, T. Ikariya and R. Noyori, *J. Am. Chem. Soc.*, **118**, 2521 (1996).
68. S. Hashiguchi, A. Fujii, J. Takehara, T. Ikariya and R. Noyori, *J. Am. Chem. Soc.*, **117**, 7562 (1995).
69. K. J. Haack, S. Hashiguchi, A. Fujii, T. Ikariya and R. Noyori, *Angew. Chem., Int. Ed. Engl*, **36**, 285 (1997).
70. D. A. Alonso, D. Guijarro, P. Pinho, O. Temme and P. G. Andersson, *J. Org. Chem.*, **63**, 2749 (1998).
71. Y. Jiang, Q. Jiang and X. Zhang, *J. Am. Chem. Soc.*, **120**, 3817 (1998).
72. K. Mashima, T. Abe and K. Tani, *Chem. Lett.*, 1199, 1201 (1998).
73. S. Hashiguchi, A. Fujii, F. J. Haack, K. Matsumura, T. Ikariya and R. Noyori, *Angew. Chem., Int. Ed. Engl.*, **36**, 288 (1997).
74. K. Matsumura, S. Hashiguchi, T. Ikariya and R. Noyori, *J. Am. Chem. Soc.*, **119**, 8738 (1997).
75. I. Ojima, M. Nihonyanagi and Y. Nagai, *Chem. Commun.*, 938 (1972), *Bull. Chem. Soc. Jpn.*, **45**, 3722 (1972).
76. H. Brunner, H. Nishiyama and K. Itoh, in *Catalytic Asymmetric Synthesis*, (Ed.) I. Ojima, Chapter 6, VCH, 1993.

77. H. Nishiyama, H. Sakaguchi, T. Nakamura, M. Horihata, M. Kondo and K. Itoh *Organometallics*, **8**, 846 (1989); **10**, 500 (1991); H. Nishiyama, S. Yamaguchi, M. Kondo and K, Itoh, *J. Org. Chem.*, **57**, 4306 (1992).

78. H. Brunner, G. Riepl and H. Weitzer, *Angew. Chem., Int. Ed. Engl.*, **22**, 331 (1983).

79. H. Brunner and U. Obermann, *Chem. Ber.*, **122**, 499 (1989).

80. M. B. Carter, B. Schiott, A. Gutierrez and S. L. Buchwald, *J. Am. Chem. Soc.*, **116**, 11667 (1994).

81. R. Kuwano, M. Sawamura, J. Shirai, M. Takahashi and Y. Ito, *Tetrahedron Lett.*, **36**, 5239 (1995).

82. C. A. Willoughby and S. L. Buchwald, *J. Am. Chem. Soc.*, **114**, 7562 (1992): **116**, 8952 and 11703 (1994): *J. Org. Chem.*, **58**, 7627 (1993).

83. A. Viso, N. E. Lee and S. L. Buchwald, *J. Am. Chem. Soc.*, **116**, 9373 (1994).

84. F. Spindler, B. Pugin and H. U. Blaser, *Angew. Chem., Int. Ed. Engl.*, **29**, 558 (1990); A. Togni, *Angew. Chem., Int. Ed. Engl.*, **35**, 1475 (1996).

85. X. Verdaguer, U. E. W. Lange, M. T. Reding and S. L. Buchwald, *J. Am. Chem. Soc.*, **118**, 6784 (1996).

86. N. Uematsu, A. Fujii, S. Hashiguchi, T. Ikariya and R. Noyori, *J. Am. Chem. Soc.*, **118**, 4916 (1996).

11

REACTIONS PROMOTED AND CATALYSED BY Pd(II) COMPOUNDS

Palladium is the most widely used transition metal for organic synthesis. Its compounds are used in two ways. One is catalysis by Pd(0) complexes, which starts from oxidative addition to substrates, and Pd(0) is regenerated after the reaction. Thus Pd(0) complexes are used as catalysts as described in Chapters 3–7. The other is catalysis by Pd(II) compounds, involving the oxidation of substrates. Many unique oxidation reactions (or dehydrogenations) specific to Pd(II) salts are treated in this chapter. In the oxidation of substrates, Pd(II) is reduced to Pd(0). If a stoichiometric amount of expensive Pd(II) compounds is consumed, the reaction cannot be regarded as a truly useful synthetic method. Interestingly, *in situ* reoxidation of Pd(0) to Pd(II) is possible using $CuCl_2$ and some other inorganic compounds such as $Cu(OAc)_2$, $Cu(NO_3)_2$, $FeCl_3$, dichromate, HNO_3 and MnO_2 under certain conditions. Also, organic oxidants such as benzoquinone (BQ) and organic peroxides are used for the *in situ* oxidation of Pd(0). Alkyl nitrites are unique oxidants and used in some industrial processes [1,1a]. Efficient reoxidation of Pd(0) with O_2 alone without other reoxidants is said to be possible in DMSO in some cases [2,3].

The *in situ* regeneration of Pd(II) from Pd(0) should not be counted as being an easy process, and the appropriate solvents, reaction conditions, and oxidants should be selected to carry out smooth catalytic reactions. In many cases, an efficient catalytic cycle is not easy to achieve, and stoichiometric reactions are tolerable only for the synthesis of rather expensive organic compounds in limited quantities. This is a serious limitation of synthetic applications of oxidation reactions involving Pd(II). However it should be pointed out that some Pd(II)-promoted reactions have been developed as commercial processes, in which supported Pd catalysts are used. For example, vinyl acetate, allyl acetate and 1,4-diacetoxy-2-butene are commercially produced by oxidative acetoxylation of ethylene, propylene and butadiene in gas or liquid phases using Pd supported on silica. It is likely that $Pd(OAc)_2$ is generated on the surface of the catalyst by the oxidation of Pd with AcOH and O_2, and reacts with alkenes.

11.1 Oxidative Reactions of Alkenes

Palladium (II) compounds coordinate to alkenes to form π-complexes. Roughly speaking, the decrease of alkene electron density caused by coordination to an electrophilic Pd(II) compound enables an attack by nucleophiles on the coordinated alkenes. The attack of a nucleophile with concomitant formation of a carbon–palladium σ-bond **1** is called the palladation of alkenes. This reaction is similar to the mercuration reaction. Unlike the products of mercuration which are stable and isolable, palladation product **1** is usually unstable and undergoes rapid decomposition. Palladation is followed by two reactions. The elimination of H-Pd-X from **1** to form the vinyl compounds **2** is one path, resulting in nucleophilic substitution of the alkene. Displacement of the Pd in **1** by an other nucleophile effects nucleophilic addition of the alkene to give **3**. Depending on the reactants and conditions, either nucleophilic substitution of the alkene or the nucleophilic addition to the alkene takes place [4,5].

AH, BH = nucleophiles, H_2O, ROH, RCO_2H, RNH_2, CH_2E_2

The oxidative reactions of alkenes can be classified further based on the attacking species.

11.1.1 Reactions with Water

Formation of acetaldehyde and precipitation of metallic Pd by passing ethylene into an aqueous solution of $PdCl_2$ was reported by Phillips in early 1894 [6] and used for quantitative analysis of Pd(II) [7]. The reaction has been highlighted by the development of an industrial process for acetaldehyde production from ethylene based on this reaction. The process is called the Wacker process [8–10]. The Wacker process is the first example of oxidation of organic compounds with Pd(II) in a homogeneous phase. The great success of the Wacker process relies on its ingeneous catalytic cycle, in which the reduced Pd(0) is reoxidized *in situ* to Pd(II) by a base metal salt, such as $CuCl_2$ which is reduced to CuCl. In turn, CuCl is easily reoxidized to $CuCl_2$ with oxygen. As a result, ethylene is oxidized indirectly with oxygen without consuming $PdCl_2$ and $CuCl_2$ by the combination of these redox reactions. Palladium is called a noble metal because it is not easily oxidized, whereas Cu is a base metal because it is easily oxidized. Therefore, the oxidation of Pd(0) with a base metal salt seems unexpected [11]. The very small equilibrium constant calculated for the oxidation of metallic Pd with free Cu(II) ion suggests the difficulty of oxidizing Pd(0) with Cu(II) salts. The reaction becomes somewhat easier in the presence of chloride ion, which stabilizes Pd(II) and Cu(I) states by complex formation [12]. In the Wacker process, the oxidation is actually carried out in aqueous HCl at high chloride ion concentration and elevated temperature. A high concentration of $CuCl_2$ and a low concentration of $PdCl_2$ are factors that shift the equilibrium further in the right direction.

Extensive studies on the Wacker process have been carried out in industrial laboratories, and many mechanistic and kinetic studies have been published [13–16]. Several interesting observations have been made on the oxidation of ethylene. Most importantly, it has been established that no incorporation of deuterium occurs by the reaction carried out in D_2O, and the four hydrogens in ethylene are retained in acetaldehyde, indicating that a hydride shift occurs. Therefore, free vinyl alcohol (4) is not an intermediate [8,9]. One possible explanation is oxypalladation of ethylene to generate 5, followed by the hydride shift as shown by 6 to afford acetaldehyde, rather than β-elimination of 5 to give vinyl alcohol (7).

The attack of OH^- anion obeys the Markovnikov rule. The oxidation of propylene affords acetone. Propionaldehyde is not formed. Higher alkenes are oxidized to ketones. This means that the oxidation of terminal alkenes affords methyl ketones 8, which are useful synthetic intermediates. Based on this reaction, terminal alkenes can be regarded as masked methyl ketones which are stable to acids, bases and nucleophiles [17,18]. The oxidation of higher alkenes is carried out in organic solvents which can mix with both alkenes and water. DMF is widely used for this purpose [19]. Although the oxidation proceeds faster in alcohols, extensive double bond isomerization also occurs [20]. The oxidation of ethylene proceeds even in aqueous ammonia to afford the two pyridine derivatives **9** and **10** selectively via acetaldehyde [21].

In addition to $CuCl_2$, several other oxidizing agents of Pd(0) are used in reactions of alkenes. Sometimes chlorination of carbonyl compounds occurs with $CuCl_2$. Use of CuCl, after facile preoxidation to Cu(II) with oxygen, is recommendable because no chlorination of ketones takes place with CuCl [18,22]. A typical oxidation procedure of 1-decene to give 2-decanone in 68–73% yield was reported [23]. Some organic compounds are used as stoichiometric oxidants of Pd(0). Benzoquinone is widely used [24]. The oxidation can be carried out using a catalytic amount of benzoquinone by combination of electrochemical oxidation [25], or with iron phthalocyanin [26]. Peroxides such as H_2O_2 [27,28] and *t*-butyl hydroperoxide [29,30] are good oxidants.

The oxidation of terminal alkenes to methyl ketones has been extensively applied to syntheses of many natural products [31]. Several 1,4-dicarbonyl compounds are prepared based on this oxidation. Typically, the 1,4-diketone **11** or the 1,4-keto aldehyde **13** can be prepared by the allylation of ketone [18] or aldehyde [32,33], followed by the oxidation. The reaction provides a good annulation method to prepare cyclopentenones **12** and **14**.

7-Acetoxy-1,11-dodecadien-3-one (**17**) is a synthetic equivalent of 1-dodecene-3,7,11-trione (**18**), and prepared from 1,7-octadien-3-one (**16**), which is obtained from the butadiene telomer **15** (see Section 5.2). The acetoxy group and the double bond in **17** are precursors of two carbonyl groups. The reagent **17** is used for steroid synthesis.

After Michael addition of **17** to 2-methyl-1,3-cyclohexanedione (**19**) and the first aldol condensation, the double bond in **20** is reduced to give **21**. The acetoxy group in **21** is unmasked by hydrolysis and oxidation, and subsequent aldol condensation gives **22**. Again, the terminal double bond in **22** is oxidized with PdCl$_2$ to methyl ketone, and subsequent hydrogenation of the internal double bond affords the trione **23**, from which the steroid A-ring **24** is formed by aldol condensation [34].

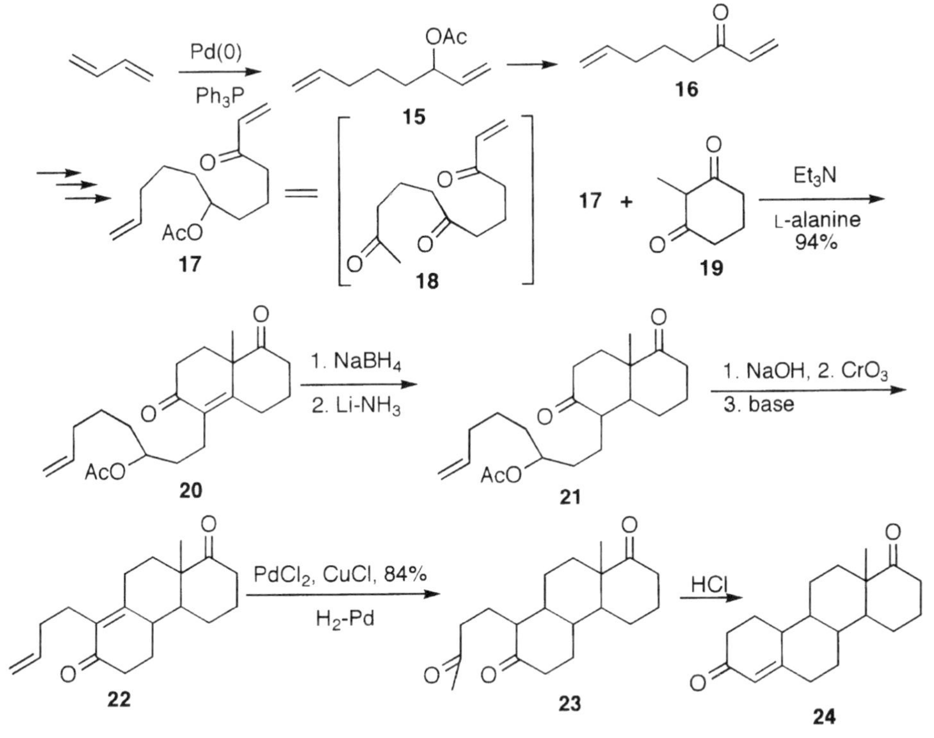

The 1,5-diketone **26** is prepared by 3-butenylation of ketone, followed by the Pd-catalyzed oxidation of **25** and annulated to form cyclohexenone **27** [18]. In this method, the 3-butenyl group is the masked methyl vinyl ketone. The 3-butenyl group attached to the B ring of the baccatin skeleton **28** was oxidized to the methyl ketone **29** in 98% yield, whereas the corresponding allyl group was oxidized to aldehyde [35].

The oxidation of simple internal alkenes is very slow. Clean selective oxidation of the terminal double bond in **30** to the methyl ketone **31** in the presence of the internal double bond is possible under normal conditions [36]. The oxidation of cyclic alkenes is difficult. Addition of strong mineral acids such as HClO$_4$, H$_2$SO$_4$ or HBF$_4$ accelerates the oxidation of cyclohexene and cyclopentene [25,37]. PdCl$_2$–CuCl$_2$ in EtOH is used for the oxidation of cyclopentene and cyclohexene [38].

11.1.2 Reactions with Oxygen Nucleophiles

Oxidation of ethylene in alcohol with PdCl$_2$ in the presence of a base gives the acetal of acetaldehyde and vinyl ether [39,40]. The reaction of alkenes with alcohols mediated by PdCl$_2$ affords the acetals **32** as major products and the vinyl ethers **33** as minor products. No deuterium incorporation was observed in the acetal **34** formed from ethylene and MeOD, indicating that a hydride shift takes place as shown by **34**. The acetal is not formed by the addition of methanol to methyl vinyl ether [41].

The syntheses of brevicomin (**36**) [42,43] and frontalin [44,45] have been achieved as an elegant application of the intramolecular acetal formation from the double bond with the diol in **35** in dry DME.

The oxidation of terminal alkenes with an EWG in alcohols or ethylene glycol affords acetals of aldehydes chemoselectively. Acrylonitrile is converted to 3,3-dimethoxypropionitrile (**38**), which is produced commercially as an intermediate of vitamin B$_1$ in MeOH using methyl nitrite (**37**) as a unique reoxidant of Pd(0). Methyl nitrite (**37**) is regenerated by oxidation of NO with oxygen in MeOH. Methyl nitrite is a gas, which can be separated easily from water formed by the oxidation [1a].

Oxypalladation of vinyl ether, followed by alkene insertion, is an interesting synthetic route to functionalized cyclic ethers. In prostaglandin synthesis, the oxypalladation of ethyl vinyl ether (**40**) with the protected cyclopentenediol **39** generates **41** and its intramolecular alkene insertion generates **42**. The intermolecular insertion of the alkene **43**, and β-elimination of **44** occurred as one-pot reaction at room temperature, giving the final product **45** in 72% yield [46]. The stereochemistry of the product shows that the alkene insertion (carbopalladation of **41**) is *syn*. It should be noted that the elimination of β-hydrogen from the intermediate **42** is not possible, because there is no β-hydrogen *syn* coplanar to the Pd and, instead, the insertion of alkene **43** occurs.

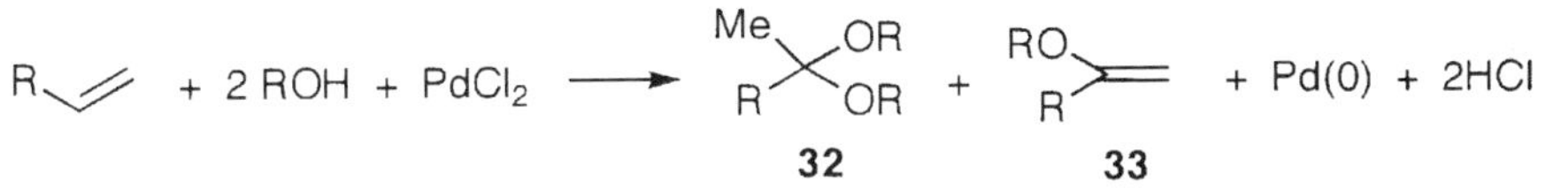

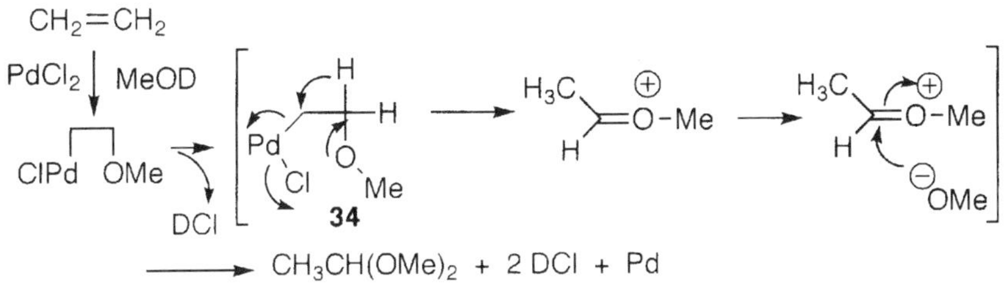

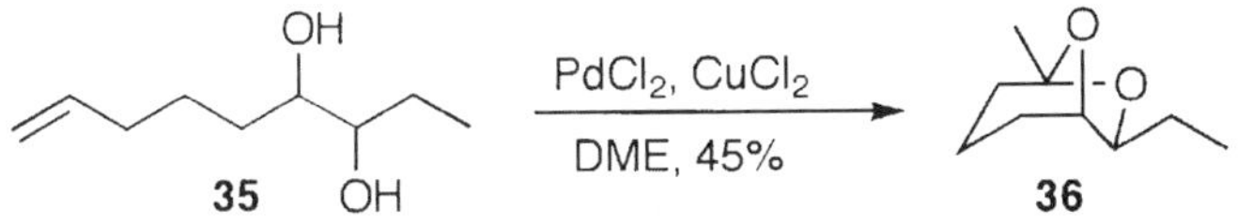

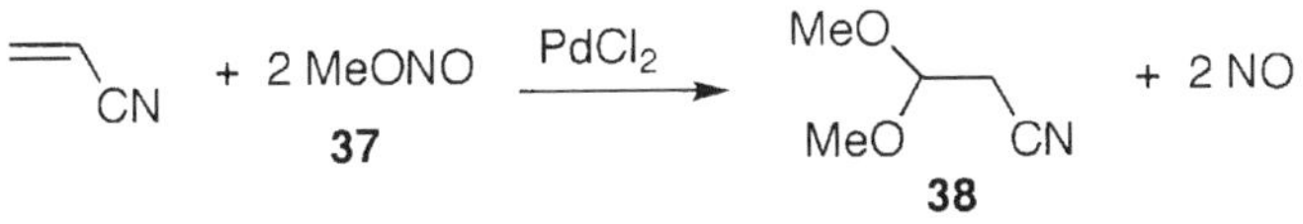

R⚬ + 2 ROH + PdCl₂ ⟶ (structure **32**) + (structure **33**) + Pd(0) + 2HCl

CH₂=CH₂

PdCl₂ | MeOD

[(structure **34**) ⟶ (cation structure) ⟶ (cation structure)]

CH₃CH(OMe)₂ + 2 DCl + Pd

(structure **35**) $\xrightarrow[\text{DME, 45\%}]{\text{PdCl}_2,\ \text{CuCl}_2}$ (structure **36**)

(structure) + 2 MeONO $\xrightarrow{\text{PdCl}_2}$ (structure **38**) + 2 NO

37

2 NO + 2 MeOH + 1/2 O₂ ⟶ 2 MeONO + H₂O

37

(structure **39**) + (structure **40**) $\xrightarrow[\text{NaI, rt, 72\%}]{\text{Pd(OAc)}_2,\ \text{AcONa}}$ [(structure **41**) ⟶ (structure **42**)]

(structure **43**) ⟶ (structure **44**) ⟶ (structure **45**)

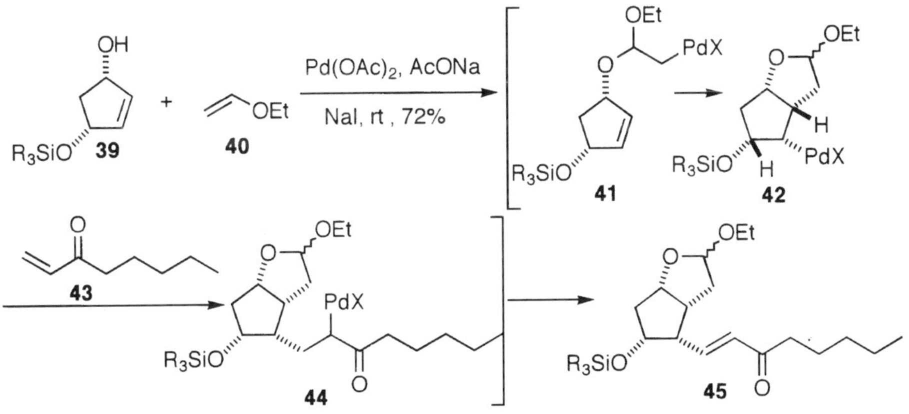

The direction of elimination of the β-hydrogen to give either enol ethers or allylic ethers from alkenes can be controlled by using DMSO as a solvent. Formation of the allylic ether **47** was utilized in the synthesis of the tetronomycin precursor **47** from **46** [47]. The oxidation of optically active 3-alkene-1,2-diol **48** afforded the 2,5-dihydrofuran **50** with high ee. It should be noted that β-OH in **49** is eliminated to

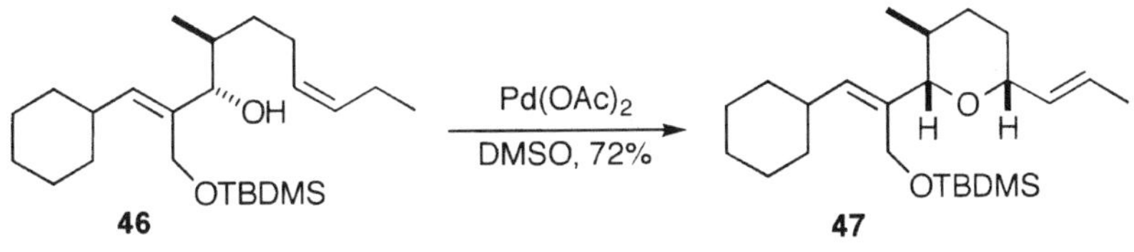

(structure **46**) $\xrightarrow[\text{DMSO, 72\%}]{\text{Pd(OAc)}_2}$ (structure **47**)

generate Pd(II), rather than β-H, at the end of the reaction [48]. Similarly, the dehydrative cyclization of *cis*-4-alkyl-2-cyclohexenol **51** gives the furan **53** and the tricyclic spiroacetal **54** as shown by **52** with generation of Pd(II) in one-pot using a catalytic amount of PdCl$_2$ without a reoxidant [49].

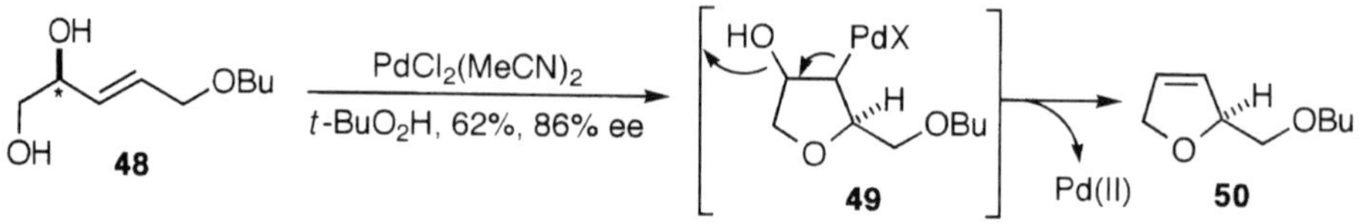

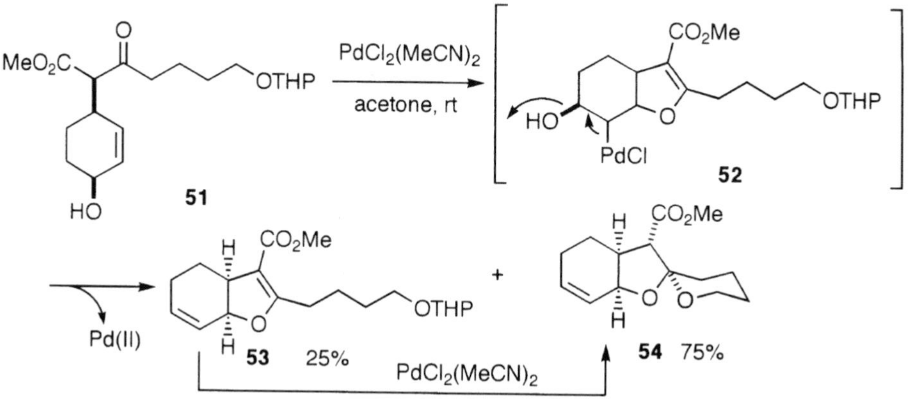

A phenolic oxygen participates in facile oxypalladation. 2-Allylphenol (**55**) undergoes clean cyclization to 2H-1-benzopyran (**56**) in DMSO under air with a catalytic amount of Pd(OAc)$_2$ without a reoxidant. 2-Methylbenzofuran (**57**) is obtained when PdCl$_2$ is used [50]. But different chemoselectivity with the Pd(II) salts was also reported [51]. Catalytic asymmetric cyclization of the tetrasubstituted 2-allylic phenol **58** using the binaphthyl-type chiral ligand **60**, called (*S,S*)-ip-borax, afforded the furan **59** with 96% ee. Use of Pd(CF$_3$CO$_2$)$_2$ as a catalyst is essential in the presence of benzoquinone [52]. Formation of the benzofuran **62** from **61** has been utilized in the synthesis of aklavinione [53]. The intramolecular reaction of 2-hydroxychalcone (**63**) produces the flavone **64** [54].

Soon after the invention of the Wacker process, the formation of vinyl acetate (**65**) by the reaction of ethylene with PdCl$_2$ in AcOH in the presence of sodium acetate was reported [39,40]. No reaction takes place in the absence of base. Also, the reaction of

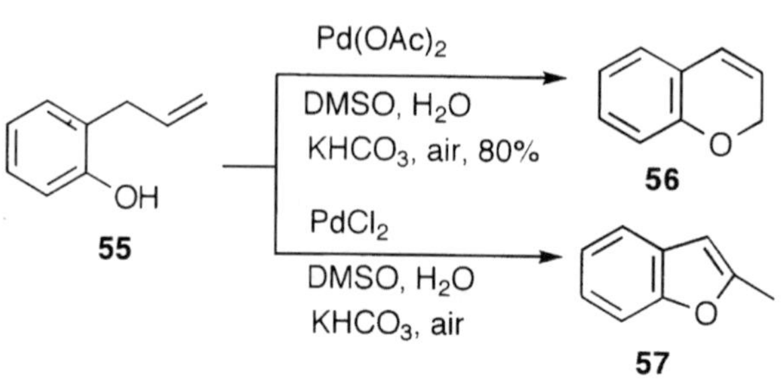

58 **59** 96% ee

60

61 **62**

63 **64**

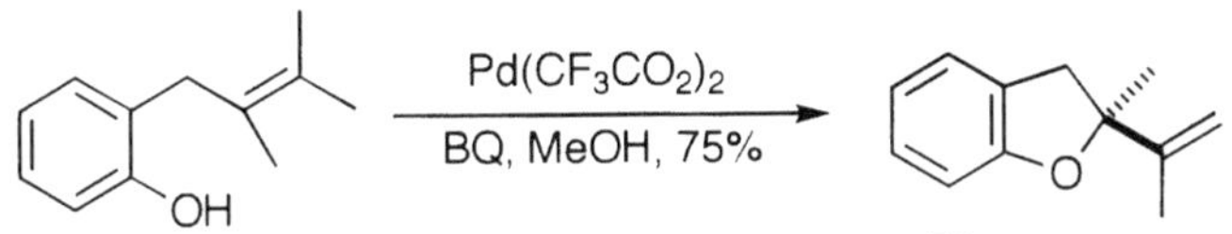

$Pd(OAc)_2$ with ethylene forms vinyl acetate. Industrial production of vinyl acetate from ethylene and AcOH has been developed by Imperial Chemical Industries, initially in the liquid phase [55]. However, due to operational problems caused mainly by corrosion, the liquid-phase process was abandoned. Then a gas-phase process using a supported Pd catalyst was developed [56]. Vinyl acetate is now produced commercially, based on this reaction in the gas phase, using Pd supported on alumina or silica as a catalyst in the absence of any Cu(II) salt [57]. It seems likely that $Pd(OAc)_2$ is generated from the supported Pd by the reaction with AcOH and O_2 at high temperature.

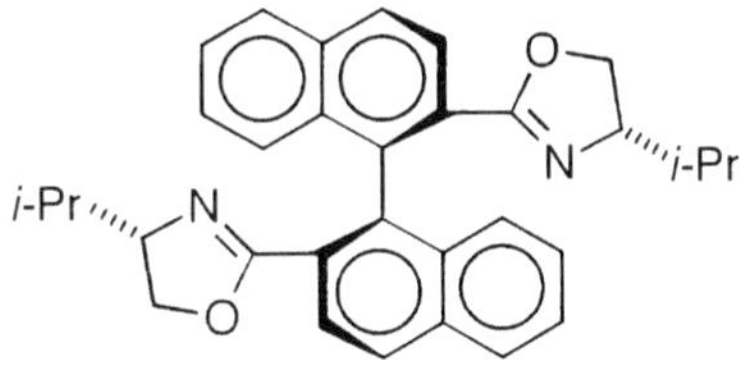

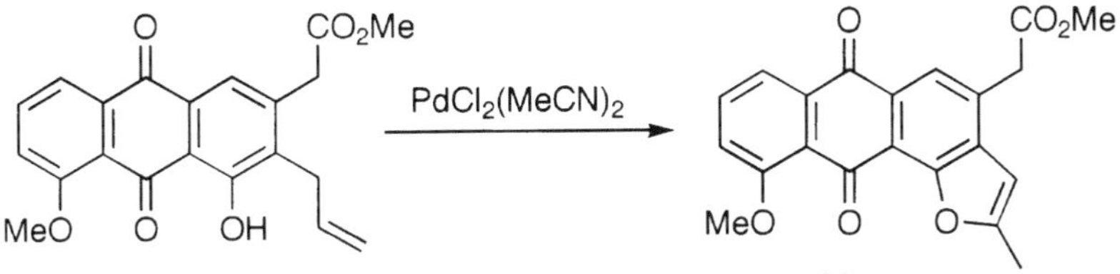

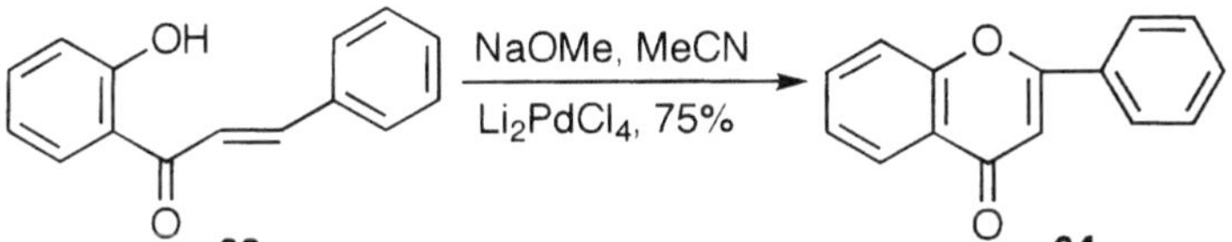

$$CH_2{=}CH_2 + Pd(OAc)_2 \longrightarrow \underset{\mathbf{65}}{CH_2{=}CHOAc} + Pd(0) + AcOH$$

$$CH_2{=}CH_2 + AcOH + O_2 \xrightarrow[\text{gas phase}]{Pd\text{-}SiO_2 \text{ (or } Al_2O_3)} \underset{\mathbf{65}}{CH_2{=}CHOAc} + H_2O$$

With higher alkenes, alkenyl acetates, allylic acetates and dioxygenated products are obtained [58]. The reaction of propylene gives two propenyl acetates, **66** and **67**, and allyl acetate (**68**) by the acetoxypalladation to form two intermediates, followed by elimination of β-hydrogens. The chemoselective formation of **68** by a gas-phase reaction using a supported Pd catalyst has been developed as a commercial process [59].

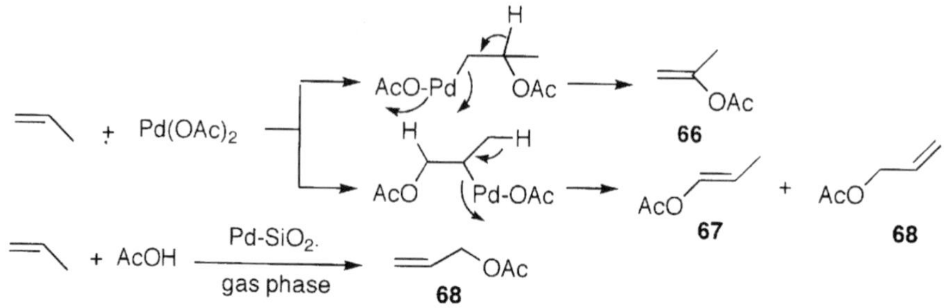

The oxidation of cycloalkenes to cyclic ketones with PdCl$_2$ is difficult, but its allylic acetoxylation with Pd(OAc)$_2$ proceeds smoothly [60–63]. In the reaction of cycloheptene with Pd(OAc)$_2$, MnO$_2$ and benzoquinone (BQ), *syn* acetoxypalladation gives **69**, and subsequent *syn* elimination of the β-hydrogen in **69** yields the allylic acetate 3-acetoxycycloheptene (**70**). 1-Acetoxycycloheptene (**71**) is not formed, because no β-hydrogen *syn* to Pd is available on the acetoxy-bearing carbon [64]. Reaction of cyclohexene gives 3-acetoxycyclohexene [65].

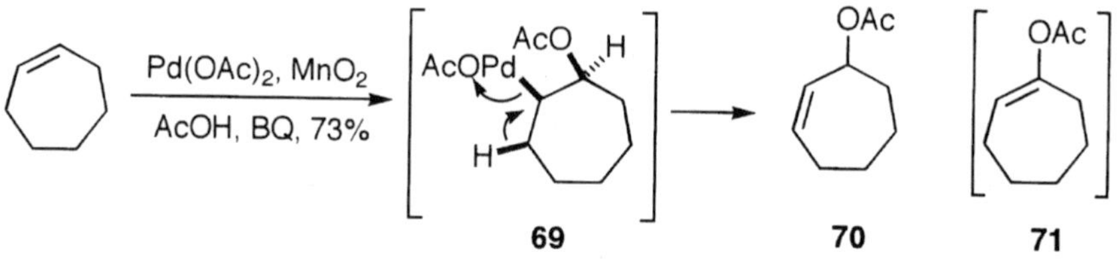

1,2-Dioxygenation by nucleophilic addition to alkenes is the main path in the presence of nitrate anion. The reaction of ethylene with Pd(OAc)$_2$ in the presence of LiNO$_3$ affords ethylene glycol monoacetate (**73**). The primary product may be glycol ester **72** of AcOH and HNO$_3$, which is hydrolysed easily to give **73** [66]. Propylene glycol monoacetates are formed from propylene in the presence of LiNO$_3$. These reactions attract attention as being potentially useful for the commercial production of ethylene glycol (**74**) and propylene glycol.

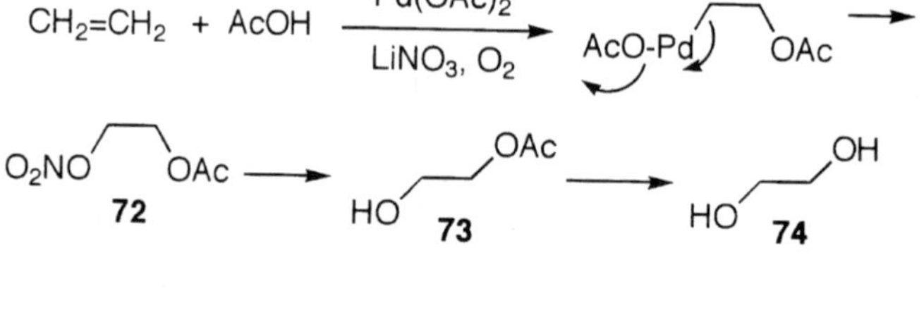

Acetoxylchlorination of norbornene (**75**), followed by skeletal rearrangement of **76** in an excess of CuCl$_2$, gives *exo*-2-chloro-*syn*-7-acetoxynorbornane (**77**). This is a good synthetic method for *syn*-7-norbornenol (**78**) [67]. Similarly, the brendane derivative tricyclo[4.2.1.0^{3,7}]nonane **80** was prepared in one step by the oxidative acetoxychlorination of the commercially available vinylnorbornene (**79**) using PdCl$_2$–CuCl$_2$ as the catalyst [68].

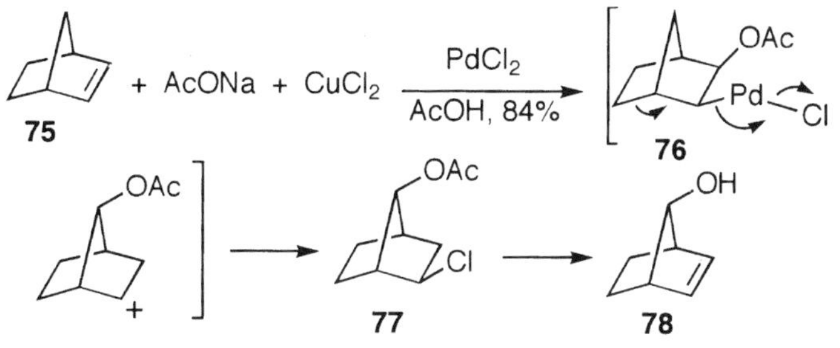

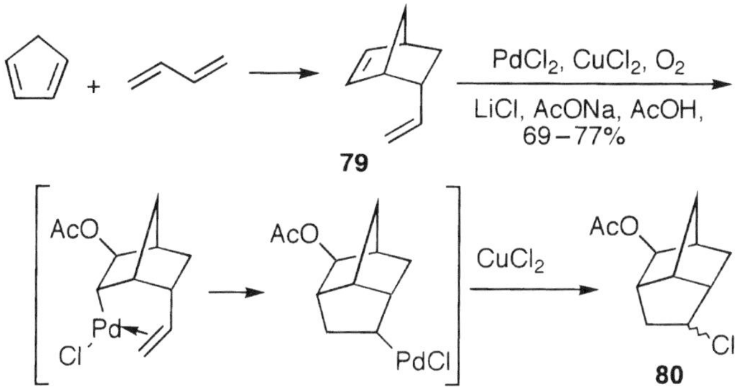

The intramolecular reaction of alkenes with various O and N functional groups offers useful synthetic routes to heterocycles [69–71]. The isocoumarin **83** is prepared by the intramolecular reaction of 2-(2-propenyl)benzoic acid (**81**) with one equivalent of PdCl$_2$(MeCN)$_2$. However, the (*Z*)-phthalide **82** is obtained from the same acid with a catalytic amount of Pd(OAc)$_2$ under 1 atm of O$_2$ in DMSO. O$_2$ alone is remarkably efficient in reoxidizing Pd(0) in DMSO. The isocoumarin **83** is obtained by the reaction of 2-(1-propenyl)benzoic acid (**84**) under the same conditions [2]. 2-Vinylbenzoic acid was converted to the isocoumarin, but not to the five-membered lactone [72,73].

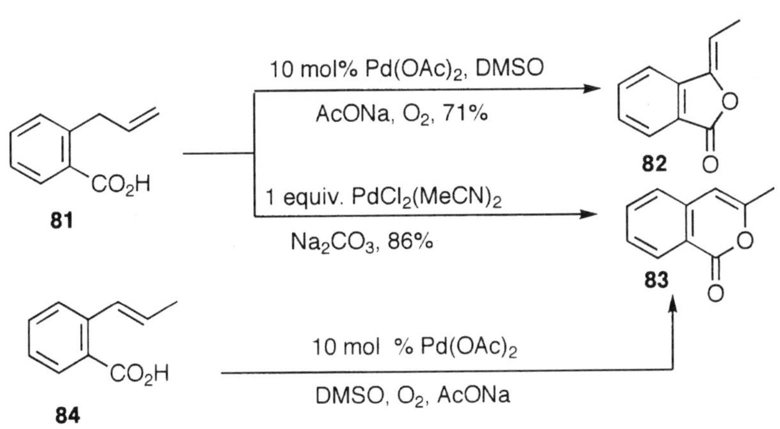

11.1.3   Reactions with Amines

Enamine formation is expected from the reaction of alkenes with amines using Pd(II) salts via aminopalladation, but enamine synthesis is not easy. Aliphatic amines are rather strong ligands to electrophilic Pd(II), and the aminopalladation is possible only in a special case. Amide nitrogen reacts more easily than the amines to form the enamide **85**, because amidation of amines reduces the complexing ability of the amines. Unlike the intermolecular reaction, intramolecular aminopalladation proceeds more easily [69–71]. Methylindole (**87**) is obtained by intramolecular *exo*-amination of 2-allylaniline (**86**). If there is another double bond in the same molecule, the aminopalladation product **88** undergoes intramolecular domino alkene insertion to give the tricyclic compound **89** [74]. Oxidative amination proceeds smoothly with aromatic amines which are less basic than aliphatic amines, whereas it is difficult to use aliphatic amines for cyclization under similar conditions. Successful oxidative amination is possible with the tosylated aliphatic amine **90** to give **91**. The tosylation of amines also reduces the strong complexing ability of aliphatic amines, which is why amides or tosylamides are used for smooth reaction with alkenes [75].

The usefulness of Pd-catalysed reactions is demonstrated amply in the total synthesis of clavicipitic acid [76]. The first step is intramolecular aminopalladation of the 2-vinyltosylamide **92** with Pd(II) to give the indole **93**. Then stepwise Heck reactions of the iodide and bromide of **94** with two different alkenes **95** and **96** in the absence and presence of a phosphine ligand give **97**. In the last step of the synthesis, the intramolecular aminopalladation of **97** with a catalytic amount of Pd(II) gives the cyclized product **99**. It should be noted that the aminopalladation is a stoichiometric reaction in nature. However, instead of the elimination of the β-hydrogen, the Pd(II) species HO−PdCl is generated as the elimination product, as shown by **98**, and reoxidation of Pd(0) to Pd(II) is not necessary; hence the reaction is catalytic without a reoxidant.

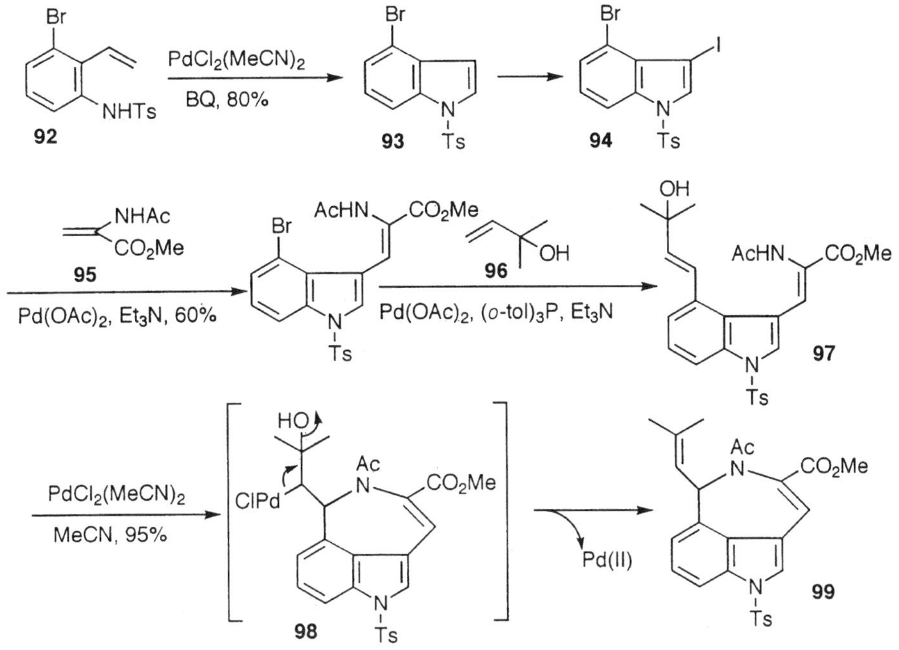

11.1.4 Reactions with Carbon Nucleophiles

Carbopalladation occurs with soft carbon nucleophiles. The $PdCl_2$ complex of COD (**100**) is difficult to dissolve in organic solvents. However, when a heterogeneous mixture of the complex, malonate and Na_2CO_3 in ether is stirred at room temperature, the new complex **101** is formed. This reaction is the first example of C−C bond formation and carbopalladation in the history of organopalladium chemistry. The double bond becomes electron deficient by the coordination of Pd(II), and attack of the carbon nucleophile becomes possible. The Pd−carbon σ-bond in complex **101** is stabilized by coordination of the remaining alkene. The carbanion is generated by treatment of **101** with a base, and the cyclopropane **102** is formed by intramolecular nucleophilic attack. Overall, the cyclopropanation occurs by attack of the carbanion twice on the alkenic bond activated by Pd(II). The bicyclo[3.3.0]octane **103** was obtained by intermolecular attack of malonate on the complex **101** [77].

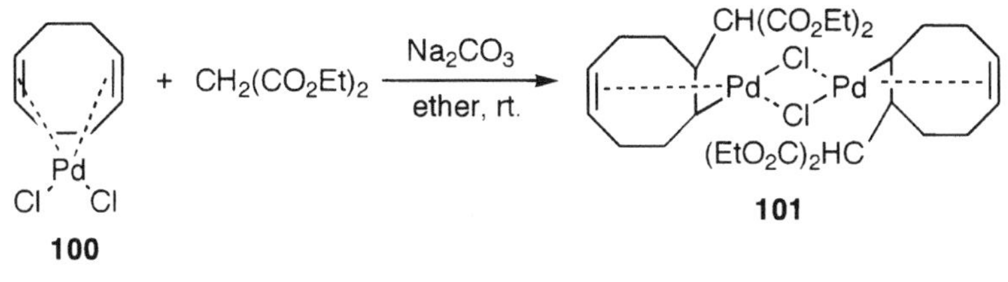

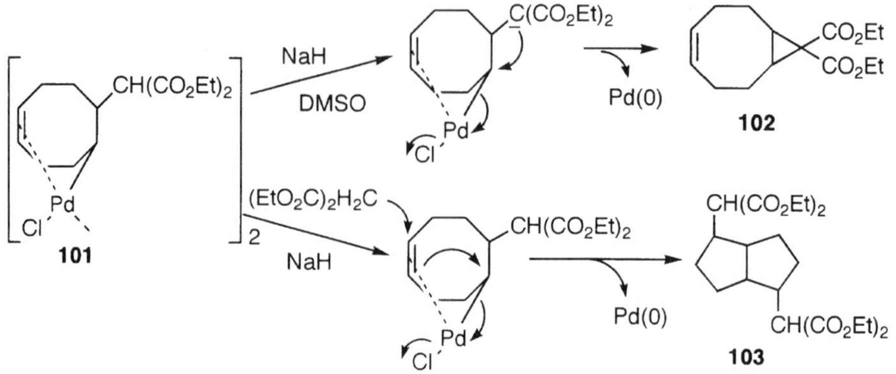

Another type of carbopalladation is observed in the oxidation of the 2,2-disubstituted alkene attached to the cyclobutanol **104**. The shift of the carbon–carbon bond as shown by **105** and ring expansion generate **106**. Intramolecular alkene insertion gives the bicyclo[4.3.0]nonane system **107**. Finally, **108** is obtained [78].

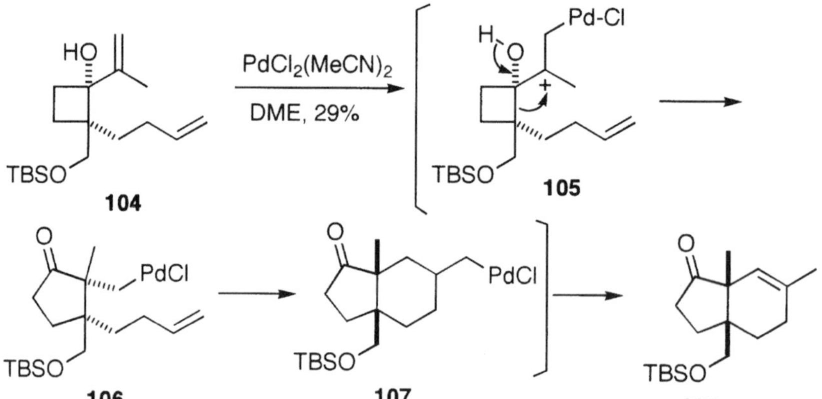

11.1.5 Oxidative Carbonylation

The unique oxidative carbonylation of alkenes is possible with Pd(II) salts, which is mechanistically different from the hydrocarboxylation of alkenes catalysed by Pd(0) (Section 7.1.1). The first example of oxidative carbonylation is the reaction of alkenes with CO in benzene in the presence of PdCl$_2$, affording β-chloroacyl chloride **109** [79]. The oxidative carbonylation of alkene in alcohol gives the α,β-unsaturated ester **110** and β-alkoxy ester **111** by monocarbonylation, and succinate **112** by dicarbonylation, depending on the reaction conditions [80–82]. Oxidative carbonylation in alcohol can be understood in the following way. The reaction starts by the formation of alkoxycarbonylpalladium **113**. Carbopalladation of the alkene (alkene insertion) with **113** gives **114**. Then elimination of β-hydrogen from the intermediate **114** yields the α,β-unsaturated ester **110**. Further CO insertion to **114** gives the acylpalladium intermediate **115** and its alcoholysis yields the succinate derivative **112**.

The β-alkoxy ester **111** is formed by nucleophilic substitution of **114** with alkoxide. Formation of **109**, the esters **111**, and **112** can be regarded as the nucleophilic addition to alkenes promoted by Pd(II).

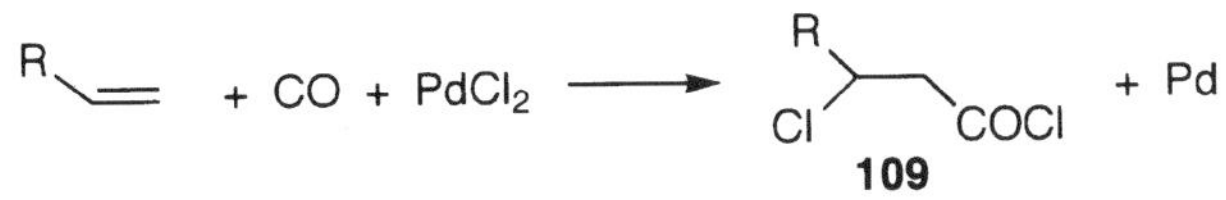

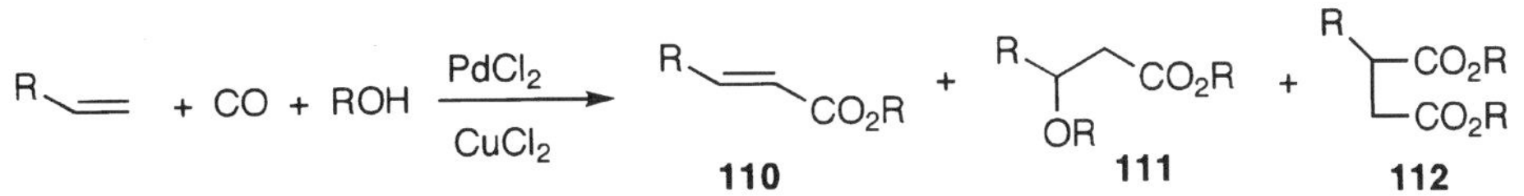

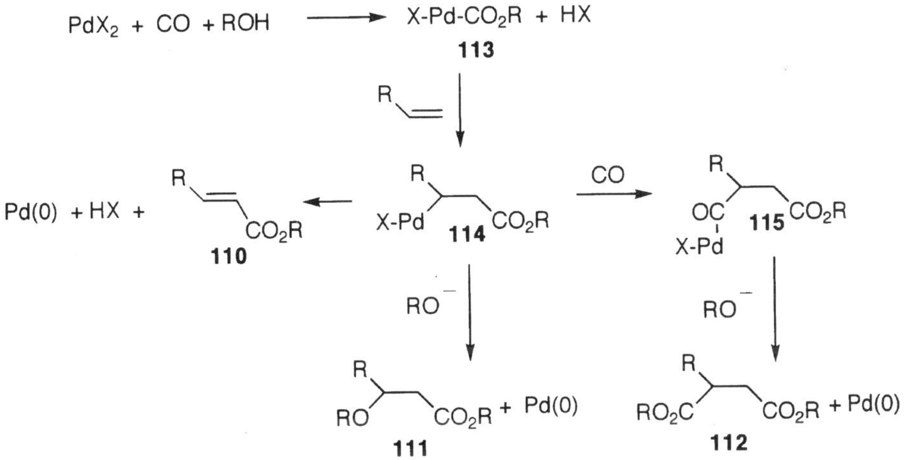

Scope of oxidative carbonylation has been studied [83]. The synthesis of acrylic acid or its ester (**116**) from ethylene has been investigated in AcOH from the standpoint of its commercial production [84]. The carbonylation of styrene is a promising commercial process for cinnamate (**117**) [80,85,86]. Succinate formation occurs at room temperature and 1 atm of CO using Pd on carbon as a catalyst in the presence of an excess of $CuCl_2$, although the reaction is slow (100% conversion after 9 days) [87].

Asymmetric carbonylation of styrene with $Pd(acac)_2$ and benzoquinone in the presence of TsOH using 2,2′-dimethoxy-6,6′-bis(diphenylphosphino)-biphenyl (**119**) as a chiral ligand gave dimethyl phenylsuccinate (**118**) with 93% ee, although yield was not satisfactory, showing that phosphine coordination influences the stereochemical course of the oxidative carbonylation with Pd(II) salts [88].

The dicarboxylation of cyclic alkenes is a useful reaction. Only *exo*-methyl-7-oxabicyclo[2.2.1]heptane-2,3,5,6-tetracarboxylate (**121**) was prepared from the cyclic alkene **120** using Pd on carbon and $CuCl_2$ in MeOH at room temperature with high diastereoselectivity [89]. Methyl nitrite **37** is used for efficient oxidative carbonylation of alkenes to produce the succinate derivatives **122** [1a,90]. It was claimed that the

$$CH_2{=}CH_2 \ + \ CO \ + \ ROH \quad + \ O_2 \quad \xrightarrow[\text{CuCl}_2]{\text{PdCl}_2} \quad \text{\\/=\\\\CO}_2R$$
116

$$Ph\diagdown{=} \ + \ CO \ + \ MeOH \quad \xrightarrow[\text{LiCl}]{\text{PdCl}_2, \ \text{CuCl}_2} \quad Ph\diagdown{=}\diagdown CO_2Me$$
117

$$Ph\diagdown{=} \ + \ 2\,CO \ + \ MeOH \quad \xrightarrow{\text{Pd(acac)}_2, \ \text{TsOH, BQ}} \quad$$

Ph—CO$_2$Me
—CO$_2$Me
118

MeO   PPh$_2$
MeO   PPh$_2$

48% conversion
93% ee

119

oxidative carbonylation using alkyl nitrites is different mechanistically from the carbonylation catalysed by PdCl$_2$/CuCl$_2$ [91].

The carbonylation of alkene in AcOH–acetic anhydride in the presence of NaCl affords the β-acetoxy carboxylic anhydride **123** in good yields and the method offers a good synthetic route to the β-hydroxycarboxylic acid **124** [92].

$$\text{120} \ + \ CO \ + \ MeOH \quad \xrightarrow[\text{AcONa, rt, 71\%}]{\text{Pd/C, CuCl}_2} \quad \text{121}$$

MeO$_2$C   CO$_2$Me
MeO$_2$C   CO$_2$Me

$$\diagdown{=} \ + \ 2\,CO \ + \ 2\,MeONO \quad \xrightarrow{\text{PdCl}_2} \quad$$ —CO$_2$Me / —CO$_2$Me $\ + \ 2\,NO$

37   **122**

$$\diagdown\diagdown{=} \ + \ CO \ + \ Ac_2O \ + \ 1/2\,O_2 \quad \xrightarrow[\text{AcOH, 80°, 1 atm., 84\%}]{\text{PdCl}_2, \ \text{CuCl}_2, \ \text{NaCl}}$$

CO$_2$Ac
OAc
123

$\xrightarrow{\text{H}_2\text{O}}$

CO$_2$H
OH
124

Intramolecular oxidative carbonylation has wide synthetic application. The γ-lactone **128** is prepared by intramolecular oxycarbonylation of the alkenediol **125** with a stoichiometric amount of Pd(OAc)$_2$ at atmospheric pressure [93]. The intermediate **126** is formed by oxypalladation, and subsequent CO insertion gives the acylpalladium **127**. The oxycarbonylation of alkenol and alkenediol can be carried out with a catalytic amount of $PdCl_2$ and a stoichiometric amount of $CuCl_2$, and has been applied to the synthesis of frenolioin [94] and frendicin B [95].

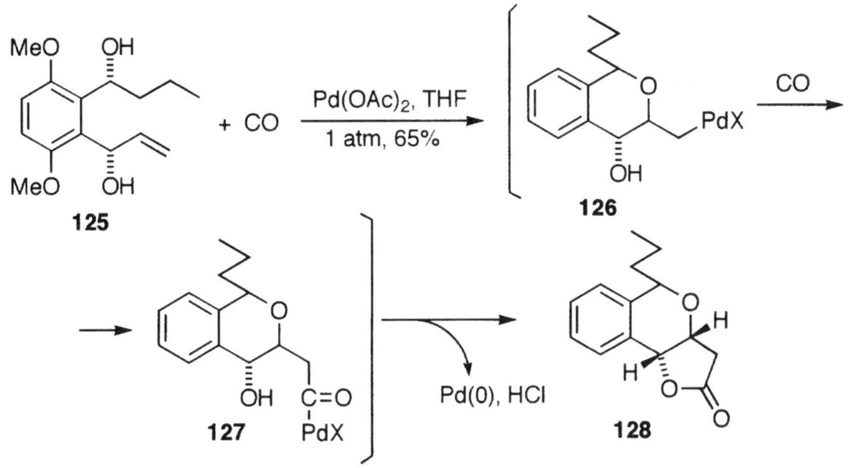

The aminopalladation and subsequent carbonylation proceed smoothly. The aminopalladation of one of the double bonds of allene **129** generates **130**, and CO insertion affords the ester **131** [96]. The carbopalladation of the optically active *N*-vinyl carbamate **132** with malonate afforded **133**. Then CO insertion and trapping with vinylstannane produced the ketone **134** with 95% de, which was converted to **135** [97].

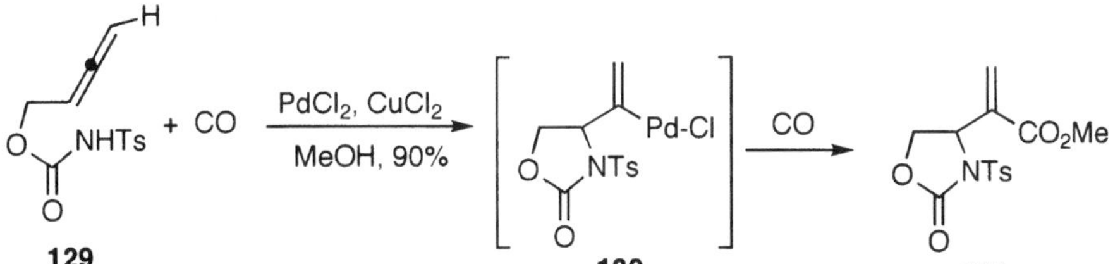

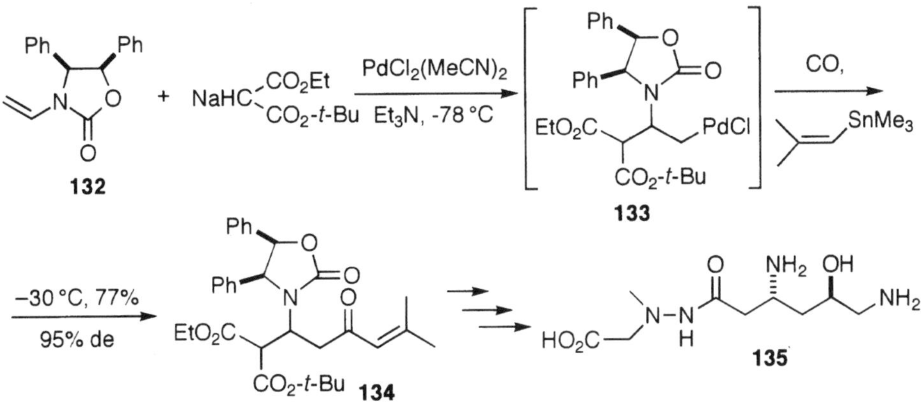

11.2 Difunctionalization of Conjugated Dienes

Palladium(II)-promoted oxidative 1,4-difunctionalization of conjugated dienes with various nucleophiles is a useful reaction [98]. The reaction is stoichiometric with respect to Pd(II) salts, but it can be made catalytic by use of Pd(0) reoxidants. 1,4-Difunctionalization with the same or different nucleophiles has wide synthetic application. The oxidative diacetoxylation of butadiene with Pd(OAc)$_2$ proceeds by acetoxypalladation to generate the π-allylpalladium **136**, which is attacked by acetoxy anion as the nucleophile, and (E)-1,4-diacetoxy-2-butene (**137**) is formed with 3,4-diacetoxy-1-butene (**138**) as the minor product. The commercial process for 1,4-diacetoxy-2-butene (**137**) by the reaction of butadiene, AcOH and O$_2$ has been developed using a supported Pd catalyst containing Te. 1,4-Butanediol (**139**) and THF are produced commercially from 1,4-diacetoxy-2-butene (**137**) [99].

trans-3,6-Diacetoxycyclohexene (**141**) is obtained by the reaction of 1,3-cyclohexadiene (**140**) with Pd(OAc)$_2$ (catalytic amount) and benzoquinone (BQ, stoichiometric amount) in the presence of LiOAc. When LiCl (0.2 equivalent) is added, *cis*-3,6-diacetoxycyclohexene (**142**) is obtained. *cis*-3-Acetoxy-6-chlorocyclohexene (**143**) is formed selectively by the addition of 2 equivalents of LiCl [100]. In these reactions, acetoxypalladation occurs in *trans* fashion to afford π-allylpalladium intermediate **144**. As the acetoxy anion migrates from Pd, the *trans* product **141** is formed. With a small amount of LiCl, displacement of chloride ion with acetoxy ion takes place to generate **145**, and the *cis*-product **142** is obtained by external attack of acetoxy ion. Chloride ion attacks the intermediate **145** from the opposite side of Pd, and *cis*-3-acetoxy-6-chlorocyclohexene (**143**) is obtained at high LiCl concentration. The same type of stereoselectivity is observed with cyclic as well as linear dienes.

The stereoselective *cis* diacetoxylation of 5-carbomethoxy-1,3-cyclohexadiene (**146**) has been applied to the synthesis of *dl*-shikimic acid (**147**).

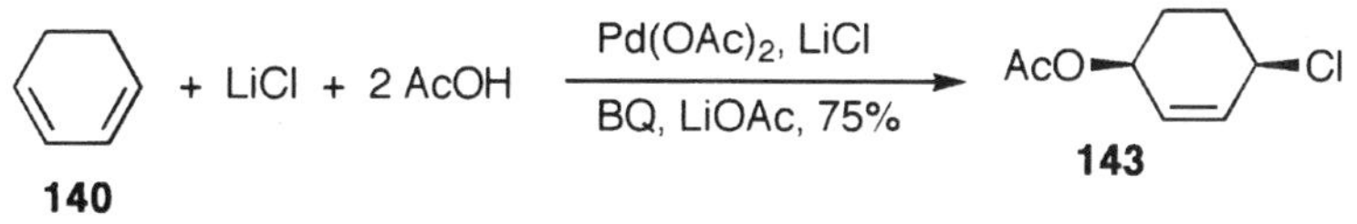

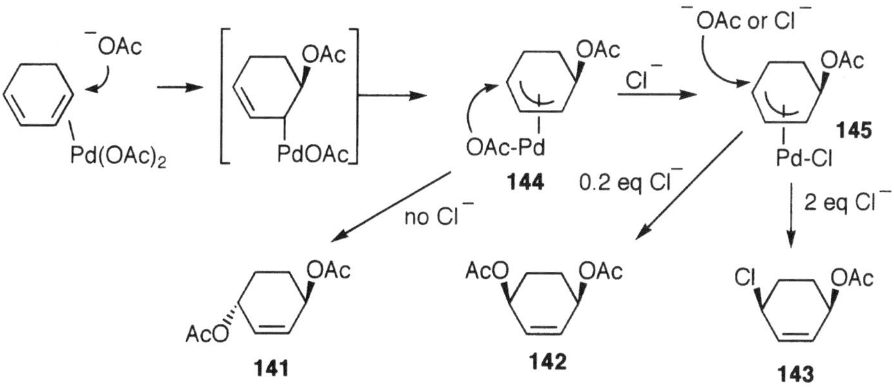

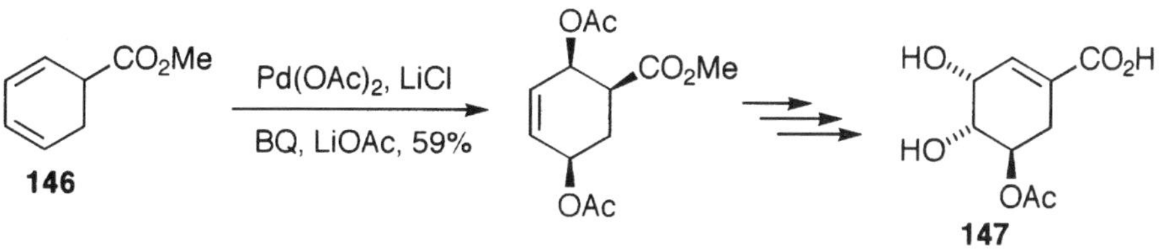

The following Pd-catalysed stereoselective transformations of **142** and **143** are possible. The Pd-catalysed reaction of the *cis* product **143** with malonate gives the *cis* product **148** with retention of the stereochemistry. However, reaction of **143** without the Pd catalyst affords the *trans* product **149**. The *cis* product **142** is a *meso* form and can be converted to the chiral half ester **150** by enzyme-catalysed partial hydrolysis.

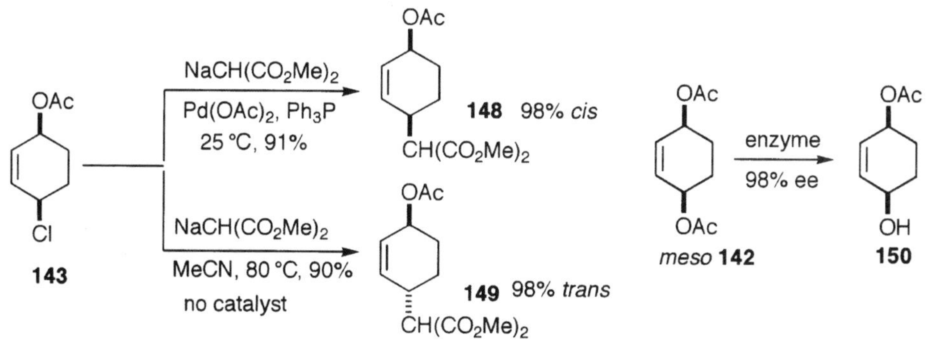

By changing the relative reactivity of the allylic leaving groups, namely acetate and the more reactive carbonate, either enantiomer of 4-substituted cyclohexenyl acetate is accessible by choice. The Pd-catalysed reaction of the allylic acetate moiety of **150** with malonate affords **151**. Acetylation of 151 and Pd-catalysed 1,4-elimination of

AcOH, followed by hydrolysis, gives the (*R*)-1,3-dienecarboxylic acid **152**. Then the acetoxylactone **153** can be prepared by the Pd-catalysed intramolecular *trans*-1,4-functionalization of the 1,3-cyclohexadiene **152**. On the other hand, the acetate **155** is obtained by the Pd-catalysed chemoselective displacement of the allylic carbonate moiety in **154** with malonate under neutral conditions. The (*S*)-1,3-dienecarboxylic acid **156** is obtained by Pd-catalysed 1,4-elimination of **155**. The Pd-catalysed 1,4-functionalization of the 1,3-cyclohexadiene **156** and acetylation, afford **157**, which is an enantiomer of **153** [101].

150 **151** **153** **154** **155** **157**

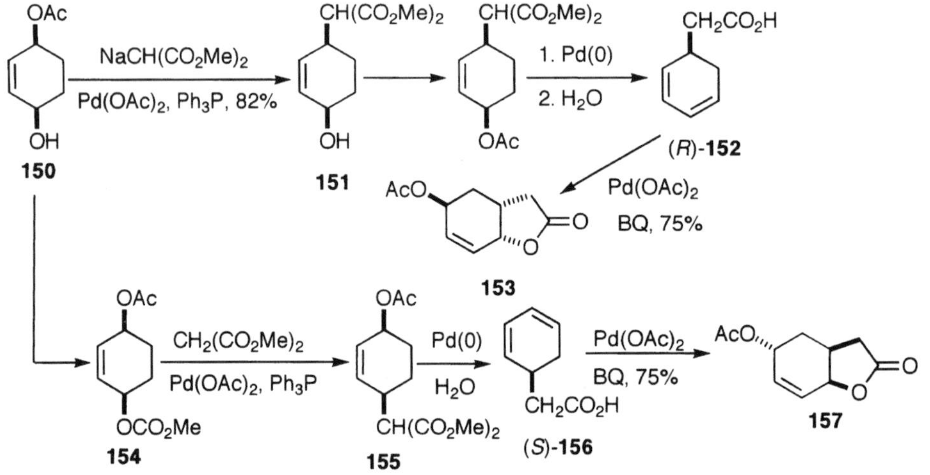

The intramolecular version offers useful synthetic methods for heterocycles. The total syntheses of α- and γ-lycoranes, **160a** and **160b**, have been carried out by applying the intramolecular aminochlorination of the carbamate of 5-(2-aminoethyl)-1,3-cyclohexadiene **158** to give **159** as the key reaction [102]. Interestingly, the 4,6-diene amide **161** undergoes intramolecular amination twice via the π-allylpalladium **162** to form the skeleton of pyrrolizidinone **163**, showing that the amide group is reactive [103].

158 **159** **160a** **160b**

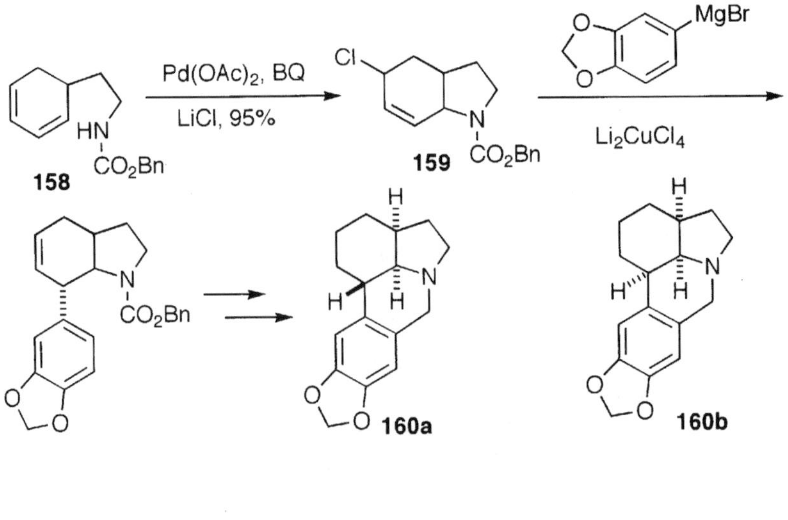

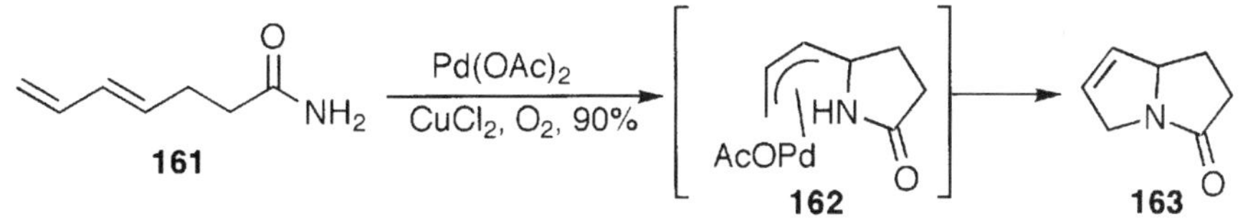

11.3 Reactions of Aromatic Compounds

The unstable phenylpalladium intermediate **164** is formed by palladation of benzene with Pd(OAc)$_2$, and then the following three reactions occur.

1. homocoupling to form biphenyl (**165**);
2. formation of phenyl acetate (**166**) by displacement (reductive elimination);
3. formation of the styrene derivative **167** by the coupling with alkene (insertion and β-elimination).

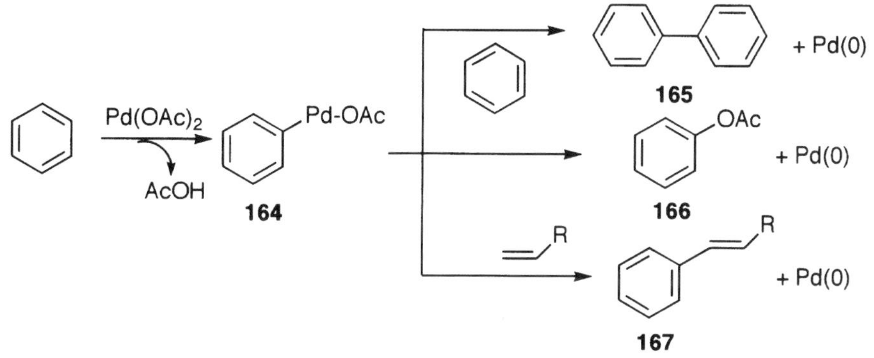

Oxidative homocoupling of aromatic and heteroaromatic rings proceeds with Pd(OAc)$_2$ in AcOH. Biphenyl (**165**) is prepared by the oxidative coupling of benzene [104,105]. The reaction is accelerated by the addition of perchloric acid. Biphenyl-tetracarboxylic acid (**169**), used for polyimide synthesis, is produced from dimethyl phthalate (**168**) commercially [106]. Intramolecular coupling of the indole rings **170** is useful for the synthesis of staurosporine aglycone **171** [107].

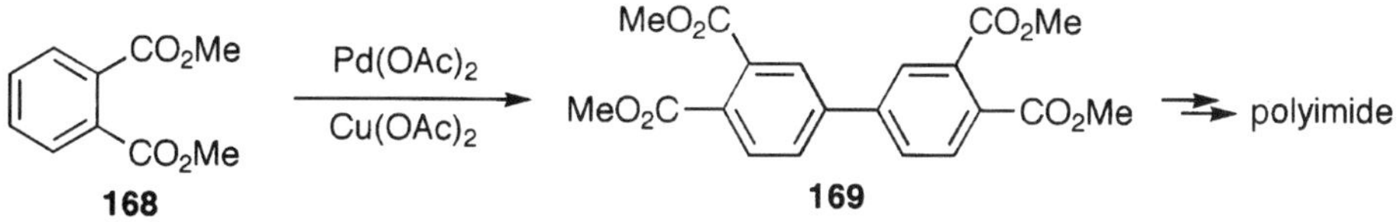

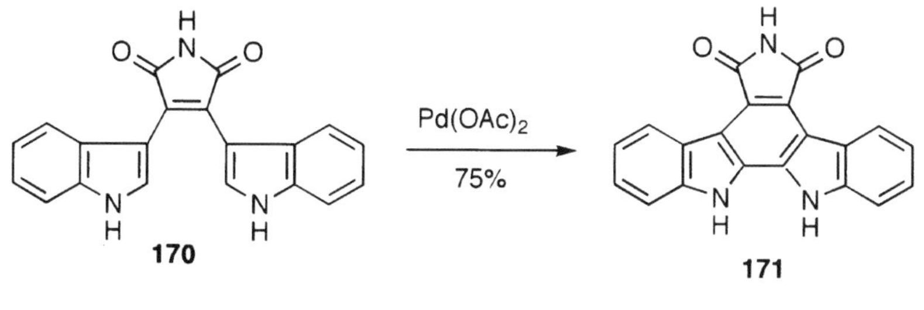

Biaryls and 1,3-dienes can be synthesized by the Pd-promoted oxidative homocoupling of different substrates. The homocoupling of the arylstannane **172** and the alkenylstannane **175** gives the biaryl **173** and 1,3-diene **176** using a catalytic amount of Pd(OAc)$_2$ and O$_2$ in the presence of the iminophosphine ligand **174**, or in the absence of ligand [107a].

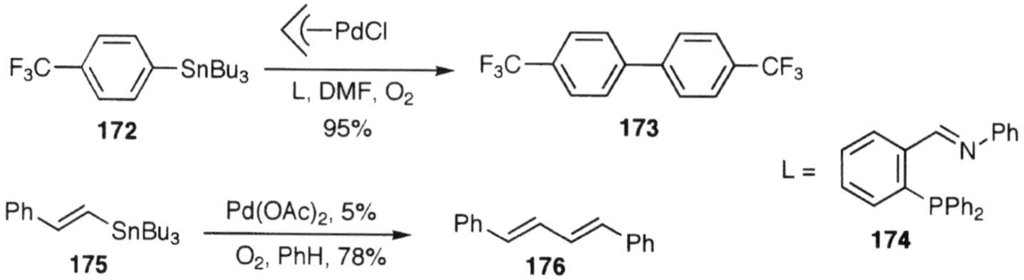

The formation of phenyl acetate (**166**) from benzene is attracting attention as a promising commercial process for phenol. Phenol is also obtained by the oxidation of benzene with O$_2$ under CO atmosphere [108].

Oxidative cross-coupling with alkenes is possible with Pd(OAc)$_2$ [109]. The reaction proceeds by the palladation of benzene to form phenylpalladium acetate (**164**), followed by alkene insertion and elimination of β-hydrogen. Heteroaromatics such as furan and thiophene react more easily than benzene [109]. Stilbene (**177**) is formed by the reaction of benzene and styrene. The complex skeleton of paraberquamide **179** was obtained in 80% yield by the Pd(II)-promoted coupling of the indole ring with the double bond in **178**, followed by reduction of the intermediate with NaBH$_4$ [110].

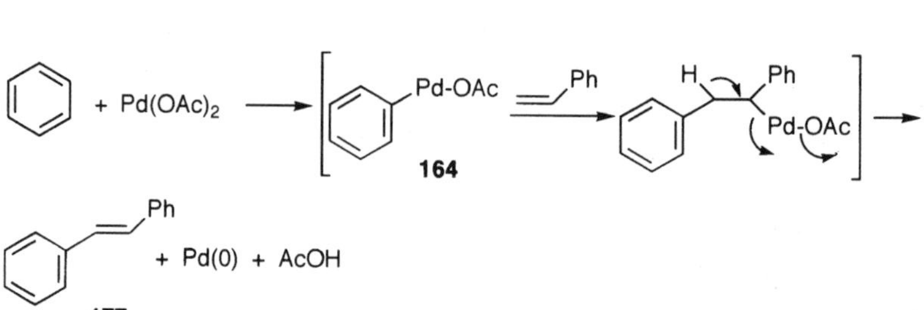

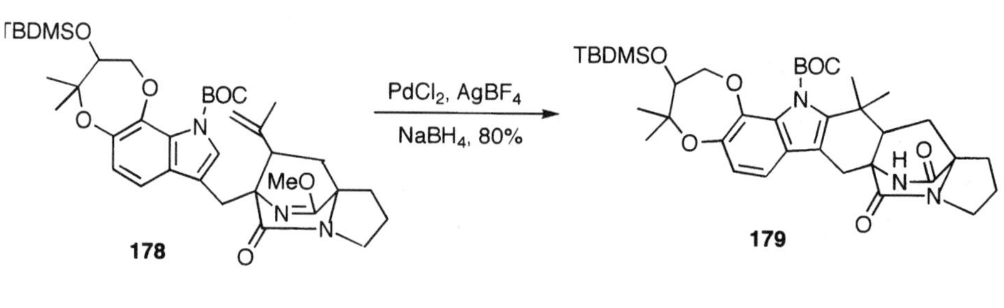

11.4 Synthetic Reactions Based on the Chelation of Heteroatoms

11.4.1 ortho-Palladation of Aromatic Compounds

Aromatic compounds with substituents containing heteroatoms, such as N, P, S or O, at positions suitable for forming mainly five-membered or sometimes six-membered chelating rings, undergo the cyclopalladation at an *ortho* position to form a σ-arylpalladium bond by virtue of the stabilization due to the chelation of these heteroatoms. The *ortho*-palladation products are stable and can be isolated [111,112]. After the first report of the preparation of the PdCl$_2$ complexes of *N,N*-dimethylbenzylamine and azobenzene complexes **180** and **181** [113], numerous *ortho*-palladation complexes have been prepared. The amide carbonyl of acetanilide is able to form the six-membered chelate ring **182** [114]. In addition to sp^2 aromatic carbons, even the sp^3 carbon of the methyl group in 8-methylquinoline (**183**) [115,116] and *ortho-N,N*-dimethylaminotoluene (**184**) [117] form σ-bonds to Pd.

The σ-arylpalladium bonds in these complexes are reactive and undergo insertion and substitution reactions. The reactions offer useful methods for regiospecific functionalization of aromatic rings, although the reactions are difficult to make catalytic in most cases. Insertion of styrene to *N,N*-dimethylbenzylamine complex (**180**) to form the stilbene derivative **185** occurs smoothly at room temperature in AcOH [5]. The reaction has been extended to the functionalization of the dopamine analogue (*N,N*-dimethyl-2-arylethylamine) **187** via the six-membered *ortho*-palladated complex **186** [118].

Facile insertion of CO takes place. The 2-aryl-3-indazolone **189** is obtained in high yields from azobenzene complex **188** in alcohol or water [119]. For unsymmetrically substituted 4-methyl, 4-chloro- or 4-methoxyazobenzenes, σ-bond formation with Pd was found to take place mainly with substituted benzene rings, because the cobalt-catalysed carbonylation of **189** gives **190**, and subsequent hydrolysis affords 5-methylanthranilic acid (**191**) and aniline. The results show that σ-bond formation is an electrophilic substitution of PdCl$_2$ on the benzene ring [119,120].

Chlorination of the azobenzene complex **181** with chlorine produces monochlor-oazobenzene (**192**) with regeneration of PdCl$_2$. The complex of the chlorinated azobenzene **193** is formed again. By this sequence, finally the tetrachloroazobenzene (**194**) is obtained using a catalytic amount of PdCl$_2$. The reaction, carried out by passing chlorine gas into an aqueous dioxane solution of azobenzene and PdCl$_2$ for 16 h, gives a mixture of polychlorinated azobenzenes [121].

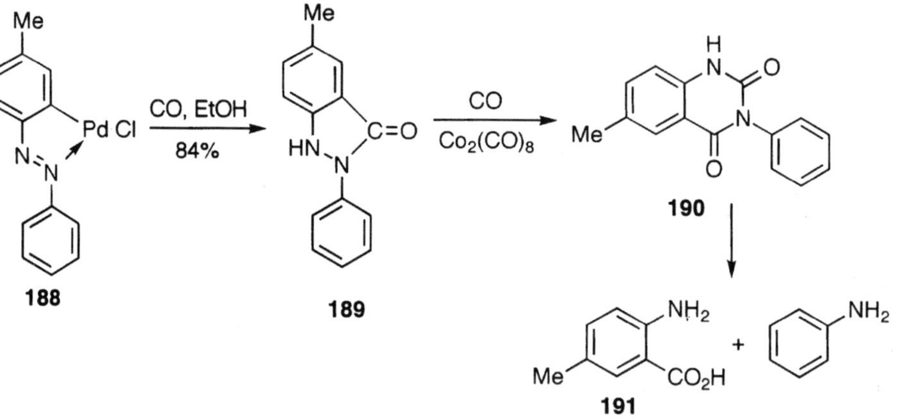

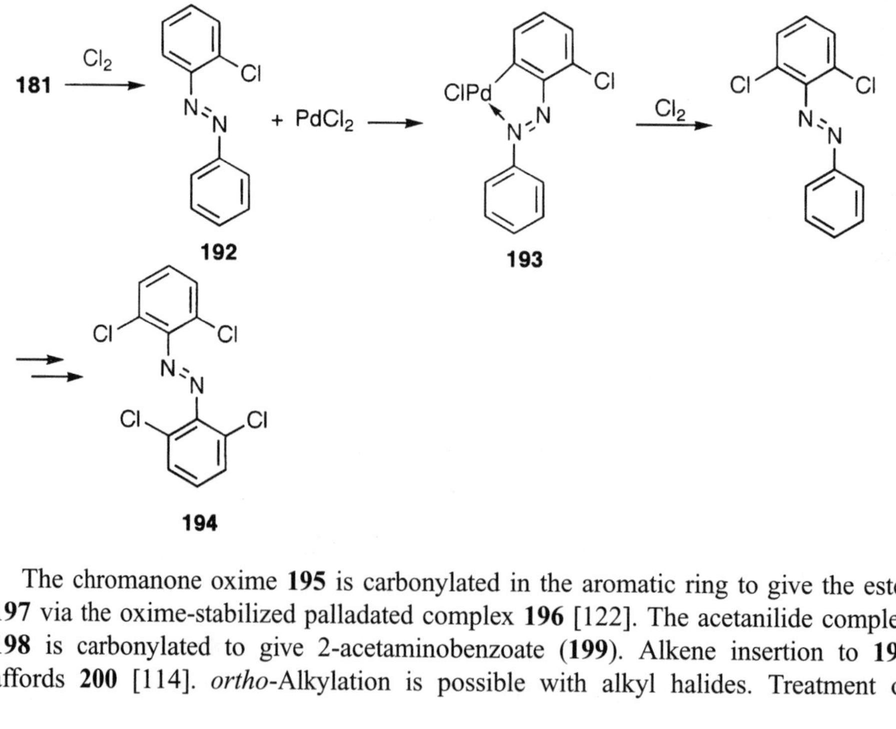

The chromanone oxime **195** is carbonylated in the aromatic ring to give the ester **197** via the oxime-stabilized palladated complex **196** [122]. The acetanilide complex **198** is carbonylated to give 2-acetaminobenzoate (**199**). Alkene insertion to **198** affords **200** [114]. *ortho*-Alkylation is possible with alkyl halides. Treatment of complex **198** with three equivalents of Pd(OAc)$_2$ and excess MeI affords 2,6-dimethylacetanilide (**202**) by stepwise *ortho*-palladation and methylation via **201** [123].

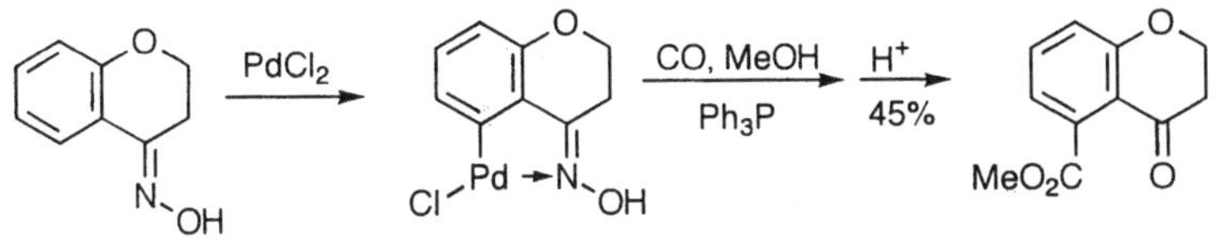

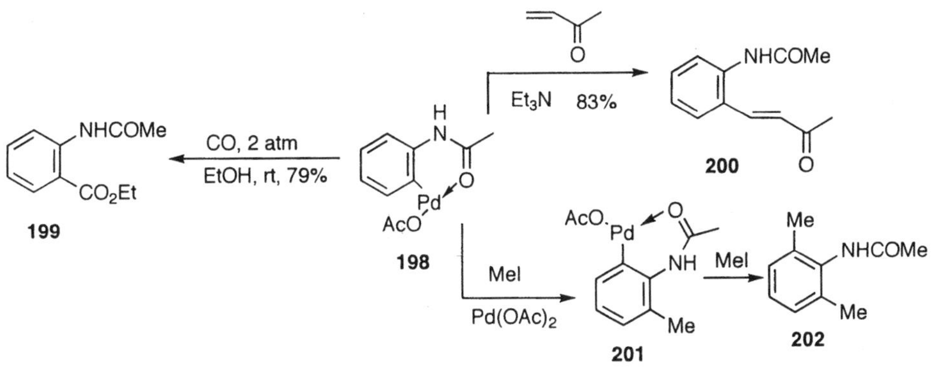

11.4.2 Reactions of Allylic Amines

The facile cyclopalladation of allylamine proceeds due to the chelating effect of the nitrogen. Carbopalladation of allylamine with malonate affords the chelating complex **203**, which undergoes insertion of methyl vinyl ketone to form the amino enone **204** [124]. Homoallylamines and allyl sulfides behave similarly [125].

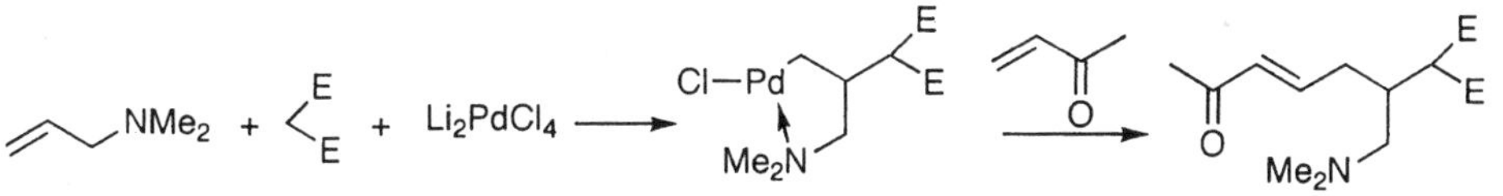

E = CO$_2$Me

An ingenious application of the facile palladation of allylic amines is the synthesis of a prostaglandin derivative starting from 3-(dimethylamino)cyclopentene (**205**) [126]. The key steps in this synthesis are the facile and stereoselective introductions of a carbanion and an oxy anion into the cyclopentene ring by virtue of the chelating effect of the amino group, and the alkene insertion to the Pd–carbon σ-bond. The first step is the stereodefined carbopalladation with malonate to generate **206**. The subsequent elimination of β-hydrogen gives the 3-substituted 4-aminocyclopentene **207** in 92% yield. The attack of the malonate occurs *anti* to the amino group. Further treatment of the amino ester **207** with Li$_2$PdCl$_4$, 2-chloroethanol and diisopropylethylamine in DMSO gives rise to the oxypalladation product **208**, which is immediately treated with pentyl vinyl ketone. The insertion of this alkene affords the desired enone **209** in 50% yield. This enone **209** was converted to the intermediate **210** for prostaglandin synthesis.

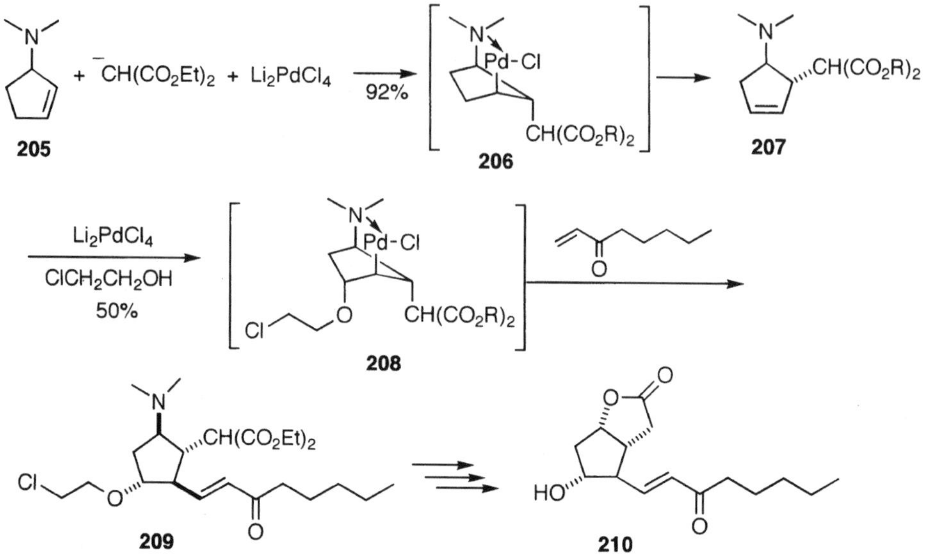

11.5 Reactions of Alkynes

Alkynes undergo stoichiometric oxidative reactions with Pd(II). A useful reaction is oxidative carbonylation. Two types of oxidative carbonylation of alkynes are known. The first one is the synthesis of alkynic carboxylates **211** by the oxidative carbonylation of terminal alkynes using PdCl$_2$ and CuCl$_2$ in the presence of a base [127]. Dropwise addition of the alkyne is recommended as a preparative-scale procedure of this reaction, in order to minimize the oxidative dimerization of alkynes as a competitive reaction [128]. The reaction has been applied to the synthesis of the intermediate **213** of the carbapenem **214** from the terminal alkyne **212** [129].

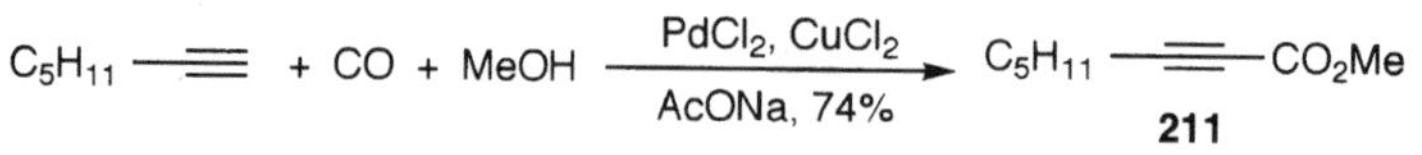

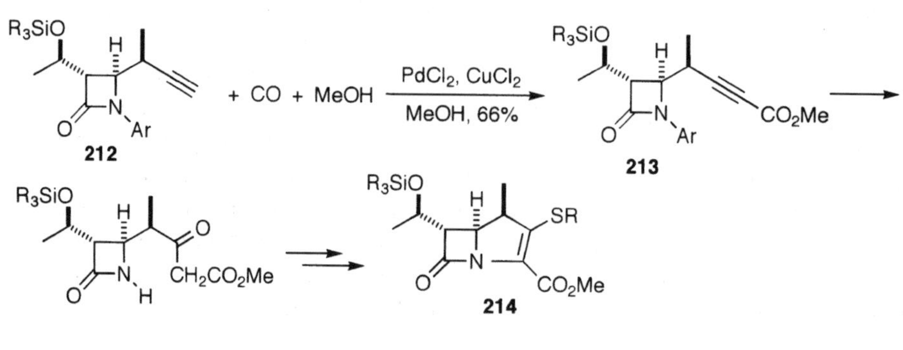

The second type is oxidative dicarbonylation of acetylene with $PdCl_2$ in benzene, which produces chlorides of maleic, fumaric and muconic acids [130]. Methyl maleate (**215**), fumarate (**216**) and muconate (**217**) are obtained in MeOH containing thiourea by passing acetylene and oxygen in the presence of a catalytic amount of $PdCl_2$ [131].

Aconitate is obtained as a minor product in the carbonylation of propargyl alcohol [132]. However, in the two-step synthesis of methyl aconitate (**220**) from propargyl alcohol (**218**) in 70% overall yield, the first step is the oxidative carbonylation under CO and air using PdI_2 and KI to give dimethyl hydroxymethylbutenedioate (**219**), which is carbonylated further to give trimethyl aconitate (**220**) by using $[Pd(tu)_4]I_2 (tu = thiourea)$ as a catalyst [133]. The oxidative cyclocarbonylation of dipropargylamine (**221**) affords the muconate derivative **222**, which is isomerized to 3,4-bis(methoxycarboylmethyl)pyrrole (**223**) [134].

$$HC{\equiv}CH + CO + MeOH \xrightarrow{PdCl_2}$$

$$\underset{\textbf{215}}{MeO_2C\diagup{=}\diagdown CO_2Me} \;+\; \underset{\textbf{216}}{MeO_2C\diagup{=}\diagup^{CO_2Me}} \;+\; \underset{\textbf{217}}{MeO_2C\diagup{=}\diagup{=}\diagup^{CO_2Me}}$$

$$\underset{\textbf{218}}{{\equiv}\diagdown_{OH}} + 2\,CO + 2\,MeOH \xrightarrow[20\,°C,\,90\%]{PdI_2,\,KI,\,air} \underset{\textbf{219}}{MeO_2C\diagup{=}\diagdown^{OH}_{CO_2Me}} \xrightarrow[tu\,=\,thiourea]{\underset{[Pd(tu)_4]I_2,\,73\%}{CO,\,MeOH}}$$

$$\underset{\textbf{220}}{\overset{MeO_2C}{}\diagdown{=}\diagup^{CO_2Me}_{CO_2Me}}$$

$$\underset{\textbf{221}}{AcN{<}^{\equiv}_{\equiv}} + CO + MeOH \xrightarrow[DMSO,\,Et_3N,\,82\%]{Pd/C,\,O_2,\,KI} \left[\underset{\textbf{222}}{AcN{<}\diagup^{CO_2Me}_{CO_2Me}} \right] \longrightarrow$$

$$\underset{\textbf{223}}{AcN{<}\diagup^{CO_2Me}_{CO_2Me}}$$

The propargyl esters **224** undergo an interesting oxidative rearrangement in the presence of a catalytic amount of $PdBr_2$ under an oxygen atmosphere to form the α-acyloxy-α,β-unsaturated aldehydes **227**. Oxypalladation of **224** generates **225**, and rearrangement occurs by the attack of hydroxy ion to give **227** via **226**. Interestingly, reoxidation of Pd(0) takes place smoothly under oxygen without using other reoxidants [135]. The reaction was applied to the propargyl ester **229**, prepared from the steroidal ketone **228**, and corticosteroid (**231**) was prepared via **230** [136].

11.6 Oxidative Carbonylation Reactions

Oxidative carbonylation of MeOH with PdCl$_2$ affords dimethyl carbonate (**233**) and dimethyl oxalate (**232**) [137,138]. Selectivity of the mono- and dicarbonylation depends on CO pressure and reaction conditions.

An industrial process for oxalate (**232**) from CO, alcohol and oxygen, catalysed by Pd, has been developed by Ube Industries [1a,139]. An ingeneous point in this process is use of alkyl nitrite **37** as a reoxidant of Pd(0). Although the mechanism of the reaction is not completely clear, formally Pd(0) is oxidized to Pd(II) easily with alkyl nitrite and then oxidative carbonylation proceeds to give oxalate and Pd(0). NO gas is generated from alkyl nitrite. In turn, NO is reconverted to the alkyl nitrite **37** by the reaction of oxygen and alcohol. Most importantly from the standpoint of the commercial process, the water, formed by the oxidation, can be separated easily from the alkyl nitrite. Otherwise, water reacts with CO to give CO$_2$. The industrial production of dimethyl carbonate (**233**) by oxidative carbonylation was developed by Ube Industries using alkyl nitrite as the reoxidant. Another promising reaction is the preparation of commercially important diphenyl carbonate (**234**) by oxidative carbonylation of phenol [140].

Carbamates are produced by the oxidative carbonylation of amines in alcohol, and active research for commercial production of carbamates **235** as precursors of isoyanates **236** based on this reaction has been carried out. As one example, ethyl phenylcarbamate is produced in a high yield (95%) with a selectivity higher than 97% by the reaction of aniline with CO in EtOH at 150 °C and 50 atm. Pd on carbon is the catalyst and KI is added as a promoter [141]. The reaction proceeds even at room temperature and 1 atm of CO [142].

$$2\ MeONO\ +\ 2\ CO \xrightarrow{PdCl_2} \begin{array}{c} CO_2Me \\ | \\ CO_2Me \end{array} +\ 2\ NO$$

37   **232**

$$2\ MeONO\ +\ CO \xrightarrow{[Pd]} O{=}{<}^{OMe}_{OMe} +\ 2\ NO$$

37   **233**

$$2\ NO\ +\ 2\ MeOH\ +1/2\ O_2 \longrightarrow 2\ MeONO\ +\ H_2O$$

37

$$2\ PhOH\ +\ CO\ +\ PdCl_2 \longrightarrow O{=}{<}^{OPh}_{OPh} +\ Pd(0)\ +\ 2\ HCl$$

234

$$Ph\text{-}NH_2\ +\ CO\ +\ MeOH \xrightarrow{PdCl_2} PhNHCO_2Me \longrightarrow PhN{=}C{=}O$$

235   **236**

As an extension of the oxidative carbonylation using alkyl nitrite, malonate **238** can be prepared by the oxidative carbonylation of ketene (**237**) [1a,143]. Also, acetonedicarboxylate (**240**) is prepared by the Pd-catalysed, alkyl nitrite-mediated oxidative carbonylation of diketene (**239**) [1a,144].

$$CH_2{=}C{=}O\ +\ CO\ +\ 2\ MeONO \xrightarrow{PdCl_2} \left[\begin{array}{c} Pd\text{-}Cl \\ | \\ CH_2\text{-}CO_2Me \end{array}\right] \longrightarrow H_2C{<}^{CO_2Me}_{CO_2Me} +\ 2\ NO$$

237   **238**

$$239 + CO + 2\ MeONO \xrightarrow{PdCl_2} MeO_2C{\diagdown}\overset{O}{\underset{}{\diagup}}{\diagup}CO_2Me\ +\ 2\ NO$$

239   **240**

Although turnover of the catalyst is low, even unreactive cyclohexane and its derivatives are oxidatively carbonylated to cyclohexanecarboxylic acid using $K_2S_2O_8$ as a reoxidant in 565% yield based on Pd(II) [145]. Methane and propane are similarly converted to acetic acid and butyric acid in 1520% and 5500% yields respectively [146,147].

$$CH_4\ +\ CO \xrightarrow[K_2S_2O_8,\ CF_3CO_2H]{Pd(OAc)_2,\ Cu(OAc)_2} CH_3CO_2H$$

$$CH_3CH_3\ +\ CO \xrightarrow[CF_3CO_2H]{Pd(OAc)_2,\ K_2S_2O_8} CH_3CH_2CO_2H$$

11.7 Reactions via Pd(II) Enolates

Transmetallation of silyl enol ethers of ketones and aldehydes with Pd(II) generates Pd(II) enolates, which are usefull intermediates. Pd(II) enolates undergo alkene insertion and β-elimination. The silyl enol ether of 5-hexen-2-one (**241**) was converted to the Pd enolate **242** by transmetallation with Pd(OAc)$_2$, and 3-methyl-2-cyclopentenone (**243**) was obtained by intramolecular insertion of the double bond and β-elimination [148]. Formally this reaction can be regarded as carbopalladation of alkene with carbanion. Preparation of the stemodin intermediate **246** by the reaction of the silyl enol ether **245**, obtained from **244**, is one of the many applications [149]. Transmetallation and alkene insertion of the silyl enol ether **249**, obtained from cyclopentadiene monoxide (**247**) via **248**, afforded **250**, which was converted to the prostaglandin intermediate **251** by further alkene insertion. In this case *syn* elimination from **250** is not possible [150]. However, there is a report that the reaction proceeds by oxypalladation of alkene, rather than transmetallation of silyl enol ether with Pd(OAc)$_2$ [151].

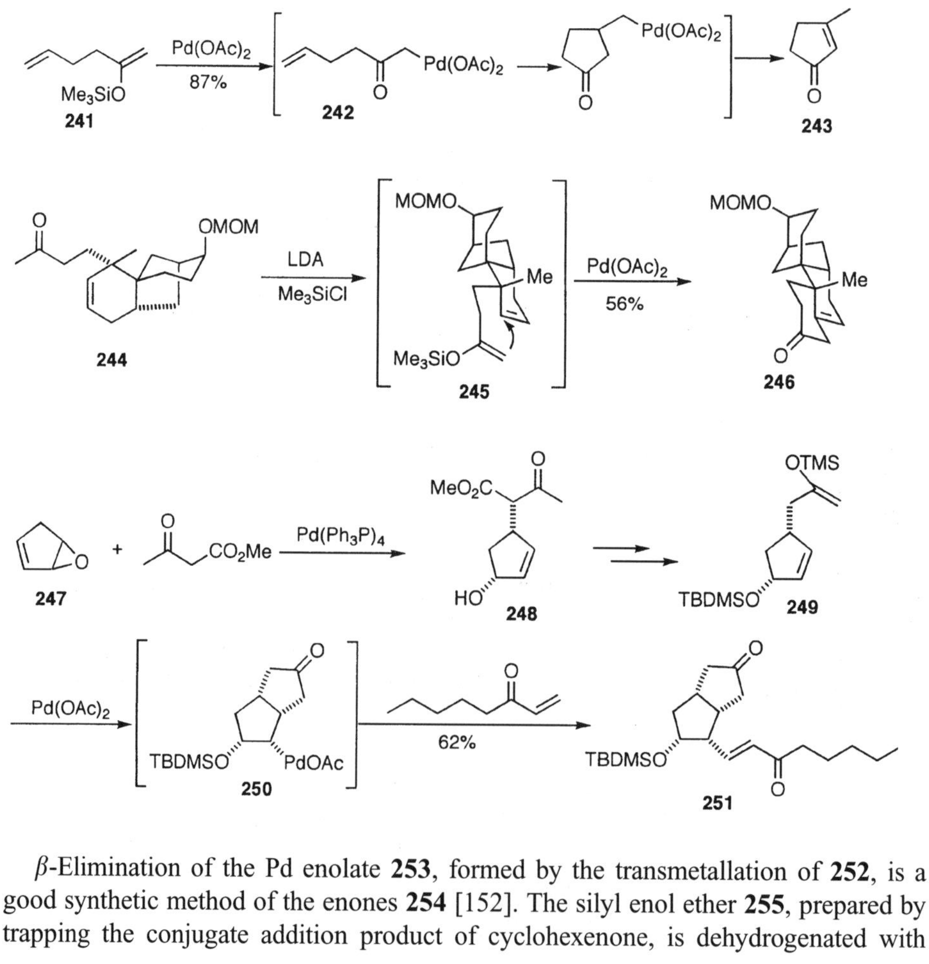

β-Elimination of the Pd enolate **253**, formed by the transmetallation of **252**, is a good synthetic method of the enones **254** [152]. The silyl enol ether **255**, prepared by trapping the conjugate addition product of cyclohexenone, is dehydrogenated with

Pd(OAc)₂ to give the cyclohexenone **256**. In clavulone synthesis, only the silyl enol ether in **257** reacts with Pd(OAc)₂ to give the enone **258** [153]. The dehydrogenation can be carried out with a catalytic amount of Pd(OAc)₂ using benzoquinone as the reoxidant. Cyclopentenone (**260**) is prepared from cyclopentanone (**259**) by using a supported Pd catalyst under O₂ atmosphere [154]. The enone **261** is converted to the dienone **263** via the dienol silyl ether **262** [155].

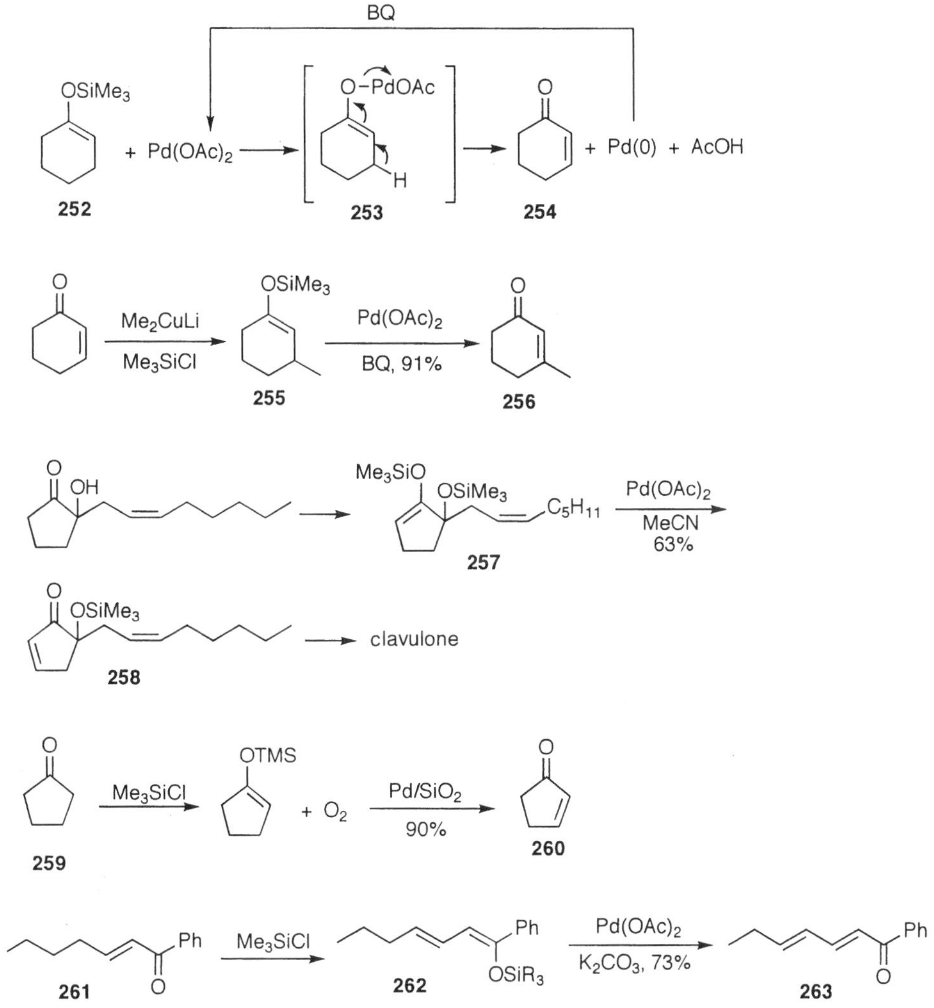

An interesting application of Pd(II) enolates is catalytic asymmetric aldol condensation and enantioselective addition to imines. Reaction of the silyl enol ether of benzophenone **264** with benzaldehyde gave aldols **265** and **266** in high optical yields using a catalyst [Pd(R)-binap)(H₂O)]²⁺(BF₄⁻)₂ **267**, generated from PdCl₂[(R)-binap)] and AgOTf [156]. The presence of water is essential. In this reaction, the Pd(II) enolate **268** reacts with aldehyde to generate the Pd alkoxide **269**, which undergoes transmetallation with the silyl enol ether **264** to give the aldol product **265**. Addition of the silyl enol ether **264** to the imine **270**, catalysed by the chiral complex [Pd(R)-tol-binap)(H$_2$O)]$^{2+}$(BF$_4^-$)$_2$, gave **271** in 95% yield and 90% ee [157].

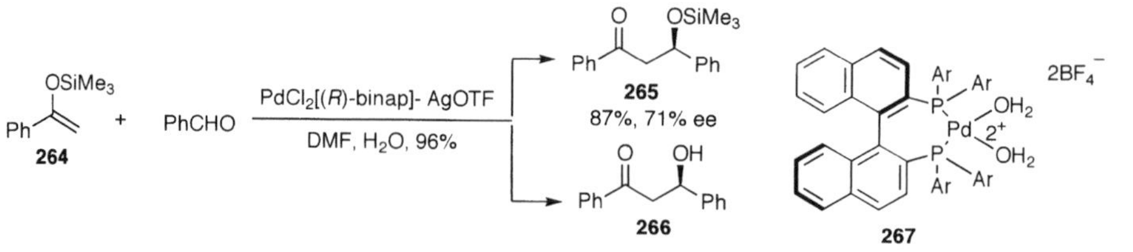

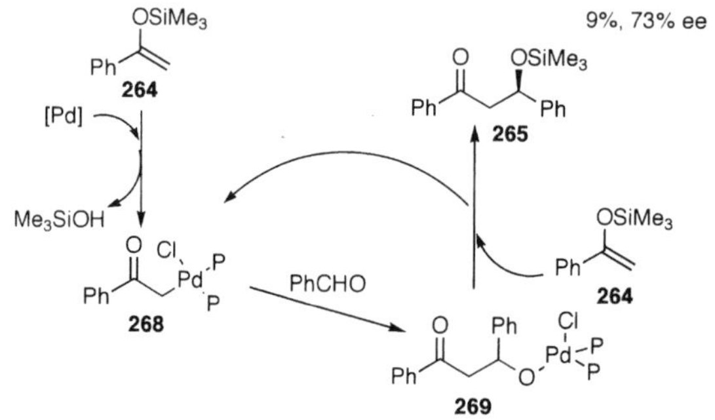

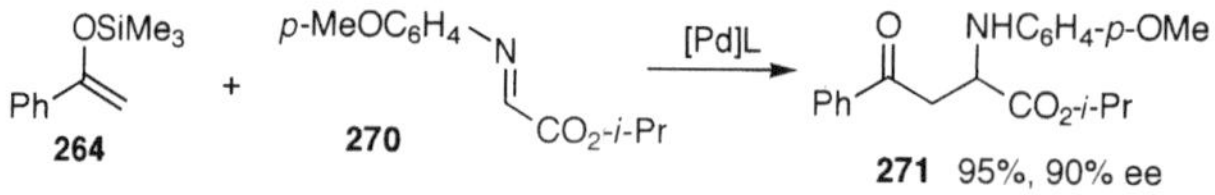

References

1. K. Matsui, S. Uchiumi, A. Iwayama and T. Umezu (Ube Industries Ltd) *Eur. Pat. Appl.*, EP. 55108; *Chem. Abst.*, **97**, 162364 (1982).

1a. J. Tsuji (Ed), *Perspectives In Organopalladium Chemistry for the XXI Century*, pp. 279–289, Elsevier 1999.

2. R. C. Larock and T. R. Hightower, *J. Org. Chem.*, **58**, 5298 (1993); **61**, 3584 (1996).

3. R. A. T. M. van Benthem and W. N. Speckamp, *J. Org. Chem.*, **57**, 6083 (1992); R. A. T. M. van Benthem, H. Hiemstra, J. J. Michels and W. N. Speckamp, *Chem. Commun.*, 357 (1994); R. A. T. M. van Benthem, H. Hiemstra, G. R. Longarela and W. N. Speckamp, *Tetrahedron Lett.*, **35**, 9281 (1994).

4. P. M. Henry, *Palladium Catalyzed Oxidation of Hydrocarbons*, Reidel, Dordrecht, 1980.

5. J. Tsuji, *Acc. Chem. Res.*, **2**, 144 (1969); *Palladium Reagents and Catalysts*, Chapter III, John Wiley, 1995.

6. F. C. Philips, *Am. Chem. J.*, **16**, 255 (1894).

7. S. C. Ogburn, Jr. and W. C. Brastow, *J. Am. Chem. Soc.*, **55**, 1307 (1933).

8. J. Smidt, W. Hafner, R. Jira, J. Sedlmeier, R. Sieber, R. Ruttinger and H. Kojer, *Angew. Chem.*, **71**, 176 (1959);. J. Smidt, W. Hafner, R. Jira, R. Sieber, J. Sedlmeier and J. Sabel, *Angew. Chem., Int. Ed. Engl.*, **1**, 80 (1962); J. Smidt, *Chem. Ind.*, 54 (1962).

9. W. Hafner, R. Jira, J. Sedlmeier and J. Smidt, *Chem. Ber.*, **95**, 1575 (1962).

10. R. Jira, J. Sedlmeier and J. Smidt, *Liebigs Ann. Chem.*, **693**, 99 (1966).

11. C. Chapman, *Analyst*, **29**, 348 (1904).

12. R. Jira and W. Freiesleben, *Organometal. React.*, **3**, 5 (1972).

13. P. M. Henry, *J. Org. Chem.*, **38**, 2415 (1973); **32**, 2575 (1967); *J. Am. Chem. Soc.*, **86**, 324 (1964); **94**, 4437 (1972); **88**, 1595 (1966).

14. I. I. Moiseev, O. G. Levanda and M. N. Vargaftik, *J. Am. Chem. Soc.*, **96**, 1003 (1974).

15. J. E. Bäckvall, B. Akermark and S. O. Ljunggren, *J. Am. Chem. Soc.*, **101**, 2411 (1979).

16. J. S. Coe and J. B. J. Unsworth, *J. Chem. Soc., Dalton Trans*, 645 (1975).

17. J. Tsuji, *Synthesis*, 369 (1984); *Comprehensive Organic Synthesis*, vol 7, p. 449, Pergamon Press, 1991.

18. J. Tsuji, I. Shimizu and K. Yamamoto, *Tetrahedron Lett.*, 2975 (1976).

19. W. H. Clement and C. M. Selwitz, *J. Org. Chem.*, **29**, 241 (1964).

20. D. G. Lloyd and B. J. Luberoff, *J. Org. Chem.*, **34**, 3949 (1969).

21. Y. Kusunoki and H. Okazaki, *Hydrocarbon Processing*, 129 (1974).

22. F. J. McQuillin and D. G. Parker, *J. Chem. Soc. Perkin Trans. I*, 809 (1974).

23. J. Tsuji, H. Nagashima and H. Nemoto, *Org. Synth.*, **62**, 9 (1984).

24. I. I. Moiseev, M. N. Vargaftik and Y. K. Syrkin, *Dokl. Akad. Nauk SSSR*, **130**, 820 (1960); *Chem. Abst.*, **54**, 24350 (1960).

25. J. Tsuji and M. Minato, *Tetrahedron Lett.*, **28**, 3683 (1987).

26. J. E. Bäckvall and R. B. Hopkins, *Tetrahedron Lett.*, **29**, 2885 (1988).

27. I. I. Moiseev, M. N. Vargaktik and Y. K. Syrkin, *Dokl. Akad. Nauk SSSR*, **130**, 820 (1960); *Chem. Abst.*, **54**, 24350 (1960).

28. M. Roussel and H. Mimoun, *J. Org. Chem.*, **45**, 5387 (1980).

29. J. Tsuji, H. Nagashima and K. Hori, *Chem. Lett.*, 257 (1980).

30. H. Mimoun, *J. Mol. Catal.*, **7**, 1 (1980); M. Roussel and H. Mimoun, *J. Org. Chem.*, **45**, 5387 (1980); *J. Am. Chem. Soc.*, **102**, 1047 (1980).

31. J. Tsuji, *Pure Appl. Chem.*, **53**, 2371 (1981); *Top. Current Chem.*, **91**, 30 (1980).

32. P. D. Magnus and M. S. Nobbs, *Synth. Commun.*, **10**, 273 (1980).

33. D. Pauley, F. Anderson and T. Hudlicky, *Org. Synth.*, **67**, 121 (1988).

34. J. Tsuji, Y. Kobayashi and T. Takahashi, *Tetrahedron Lett.*, 483 (1980).

35. I. Shiina, H. Iwadare, H. Sakoh, M. Hasegawa, Y. Tani and T. Mukaiyama, *Chem. Lett.*, 1 (1998).

36. J. Tsuji and T. Mandai, *Tetrahedron Lett.*, 1817 (1978); T. Takahashi, K. Kasuga and J. Tsuji, *Tetrahedron Lett.*, 4917 (1978).

37. D. G. Miller and D. D. M. Wayner, *J. Org. Chem.*, **55**, 2924 (1990).

38. K. Takehira, H. Orita, I. H. Ho, C. R. Leobardo, G. C. Martinez, M. Shimadzu, T. Hayakawa and T. Ishikawa, *J. Mol. Catal.*, **42**, 247 (1987); K. Takehira, T. Hayakawa and H. Orita, *Chem. Lett.*, 1853 (1985); K. Takehira, I. H. Ho, V. C. Martinez, R. S. Chavira, T. Hayakawa, H. Orita, M. Shimadzu and T. Ishikawa, *J. Mol. Catal.*, **42**, 237 (1987).

39. I. I. Moiseev, M. N. Vargaftik and Ya. K. Syrkin, *Dokl. Akad. Nauk SSSR*, **133**, 377 (1960).

40. E. W. Stern and M. L. Spector, *Proc. Chem. Soc.*, 370 (1961).

41. I. I. Moiseev and M. N. Vargaftik, *Izv. Akad. Nauk SSSR*, 759 (1965).

42. N. T. Byrom, R. Grigg and B. Kongkathip, *Chem. Commun.*, 216 (1976); *J. Chem. Soc. Perkin Trans.*, 1643 (1984).

43. K. Mori and Y. B. Seu, *Tetrahedron*, **41**, 3429 (1985).

44. B. Kongkathip, R. Sookkho and N. Kongkathip, *Chem. Lett.*, 1849 (1985).

45. T. Hosokawa, Y. Makabe, T. Shinohara and S. Murahashi, *Chem. Lett.*, 1529 (1985).

46. R. C. Larock and N. H. Lee, *J. Am. Chem. Soc.*, **113**, 7815 (1991).

47. M. F. Semmelhack, C. R. Kim, W. Dobler and M. Meier, *Tetrahedron Lett.*, **30**, 4925 (1989).
48. S. Saito, T. Hata, N. Takahashi, M. Hirai and T. Moriwake, *Synlett*, 237 (1992).
49. A. Tanaglia and F. Kammerer, *Synlett*, 576 (1996).
50. R. C. Larock, L. Wei and T. R. Hightower, *Synlett*, 522 (1998).
51. T. Hosokawa, K. Maeda, K. Koga and I. Moritani, *Tetrahedron Lett.*, 739 (1973); T. Hosokawa, S. Yamashita, S. Murahashi and A. Sonoda, *Bull. Chem. Soc. Jpn.*, **49**, 3662 (1976); T. Hosokawa, H. Ohkata and I. Moritani, *Bull. Chem. Soc. Jpn.*, **48**, 1533 (1975); T. Hosokawa, S. Miyagi, S. Murahashi and A. Sonoda, *J. Org. Chem.*, **43**, 2752 (1978).
52. Y. Uozumi, K. Kato and T. Hayashi, *J. Am. Chem. Soc.*, **119**, 5063 (1997); *J. Org. Chem.*, **63**, 5071 (1998).
53. B. A. Pearlman, J. M. McNamara, I. Hasan, S. Hatakeyama, H. Sekizaki and Y. Kishi, *J. Am. Chem. Soc.*, **103**, 4248 (1981).
54. A. Kasahara, T. Izumi and M. Ooshima, *Bull. Chem. Soc. Jpn.*, **47**, 2526 (1974).
55. ICI, *UK Patent* 964,001 (1964).
56. Bayer, *German Patent*, 1,185,604 (1965); National Distillers, *US Patent*, 3,190,912.
57. W. Schwerdtel, *Chem. Ind.*, 1559 (1968); *Hydrocarbon Process.*, **47**, 187 (1968).
58. W. Kitching, Z. Rappoport, S. Winstein and W. C. Young, *J. Am. Chem. Soc.*, **88**, 2054 (1966).
59. N. Nagato, K. Maki, K. Uematsu and R. Ishioka (Showa Denko), *Jpn. Pat. Kokai*, 60-32747 (1985).
60. C. B. Anderson and S. Winstein, *J. Org. Chem.*, **28**, 605 (1963).
61. N. Green, R. N. Haszeldine and J. Lindley, *J. Organometal. Chem.*, **6**, 107 (1966).
62. S. Wolfe and P. G. C. Campbell, *J. Am. Chem. Soc.*, **93**, 1497 and 1499 (1971).
63. P. M. Henry, *J. Am. Chem. Soc.*, **94**, 7305 (1972); P. M. Henry and G. Ward, *J. Am. Chem. Soc.*, **93**, 1494 (1971); S. Hansson, A. Heumann, T. Rein and B. Akermark, *J. Org. Chem.*, **55**, 975 (1990); *Angew. Chem., Int. Ed. Engl.*, **23**, 453 (1984); S. E. Bystrom, E. M. Lasson and B. Akermark, *J. Org. Chem.*, **55**, 5674 (1990).
64. A. Heumann, B. Akermark, S. Hansson and T. Rein, *Org. Synth.*, **68**, 109 (1990).
65. E. M. Larsson and B. Akermark, *Tetrahedron Lett.*, **34**, 2523 (1993).
66. Kuraray, French Pat., 1,509,372 (1967), M. Tamura and A. Yasui, *Chem. Commun.*, 1209 (1968); *J. Ind. Chem. (in Japanese)*, **72**, 575 and 578 (1969); M. Tamura and A. Yasui, *J. Ind. Chem.* (in Japanese), **72**, 581 (1969).
67. W. C. Baird, Jr., *J. Org. Chem.*, **31**, 2411 (1966).
68. A. Heumann, S. Kaldy and A. Tenaglia, *Tetrahedron*, **50**, 539 (1994).
69. L. S. Hegedus, *Comprehensive Organic Synthesis*, Vol 4, pp. 551 and 571, Pergamon Press, 1991; *J. Mol. Catal.*, **19**, 201 (1983); *Angew. Chem., Int. Ed. Engl.*, **27**, 1113 (1988).
70. T. Hosokawa and S. Murahashi, *Heterocycles*, **33**, 1079 (1992); *Acc. Chem. Res.*, **23**, 49 (1990).
71. G. Cardillo and M. Orena, *Tetrahedron*, **46**, 3321 (1990).
72. A. Kasahara, T. Izumi, K. Sato, K. Maemura and T. Hayasaka, *Bull. Chem. Soc. Jpn.*, **50**, 1899 (1977).
73. D. E. Korte, L. S. Hegedus and R. K. Wirth, *J. Org. Chem.*, **42**, 1329 (1977).
74. L. S. Hegedus, G. F. Allen and E. L. Waterman, *J. Am. Chem. Soc.*, **98**, 2674 (1976), L. S. Hegedus, G. F. Allen, J. J. Bozell and E. L. Waterman, *J. Am. Chem. Soc.*, **100**, 5800 (1978); L. S. Hegedus, G. F. Allen and D. J. Olsen, *J. Am. Chem. Soc.*, **102**, 3583 (1980).
75. L. S. Hegedus and J. M. McKearin, *J. Am. Chem. Soc.*, **104**, 2444 (1982).

76. P. J. Harrington, L. S. Hegedus and K. F. McDaniel, *J. Am. Chem. Soc.*, **109**, 4335 (1987).

77. J. Tsuji and H. Takahashi, *J. Am. Chem. Soc.*, **87**, 3275 (1965); **90**, 2387 (1968).

78. H. Nemoto, J. Miyata, M. Yoshida, N. Raku and K. Fukumoto, *J. Org. Chem.*, **62**, 6450 and 7850 (1997); *Tetrahedron*, **50**, 10391 (1994).

79. J. Tsuji, M. Morikawa and J. Kiji, *Tetrahedron Lett.*, 1061 (1963); *J. Am. Chem. Soc.*, **86**, 4851 (1964).

80. T. Yukawa and S. Tsutsumi, *J. Org. Chem.*, **34**, 738 (1969).

81. D. M. Fenton and P. J. Steinwand, *J. Org. Chem.*, **37**, 2034 (1972).

82. R. F. Heck, *J. Am. Chem. Soc.*, **94**, 2712 (1972).

83. J. K. Stille and R. Divakarumi, *J. Org. Chem.*, **44**, 3474 (1979).

84. K. L. Olivier, D. M. Fenton and J. Biale, *Hydrocarbon Processing*, No. 11, 95 (1972).

85. G. Cometti and G. P. Chiusoli, *J. Organometal. Chem.*, **181**, C14 (1979).

86. K. Wada and Y. Hara, (Mitsubishi Kasei), *Jpn. Pat., Tokukaisho*, 56-71039 (1981); 57-21342 (1982).

87. K. Inomata, S. Toda and H. Kinoshita, *Chem. Lett.*, 1567 (1990).

88. S. C. A. Nefkens, M. Sperrle and G. Consiglio, *Angew. Chem., Int. Ed. Engl.*, **32**, 1719 (1993).

89. P. R. Ashton, G. R. Brown, N. S. Issacs, D. Giuffrida, F. H. Kohnke, J. P. Mathias, A. M. Z. Slawin, D. R. Smith, J. F. Stoddart and D. J. Williams, *J. Am. Chem. Soc.*, **114**, 6330 (1992).

90. Ube Industries, *US Patent* 4138580 (1979); 4234740 (1980); *German Patent*, 2853178; *Chem. Abstr.*, **91**, 123415 (1979).

91. P. Brechot, Y. Chauvin, D. Commereuc and L. Saussine, *Organometallics*, **9**, 26 (1990).

92. H. Urata, A. Fujita and T. Fuchikami, *Tetrahedron Lett.*, **29**, 4435 (1988).

93. M. F. Semmelhack, C. Bodurow and M. Baum, *Tetrahedron Lett.*, **25**, 3171 (1984); *J. Am. Chem. Soc.*, **106**, 1496 (1984).

94. M. F. Semmelhack and A. Zask, *J. Am. Chem. Soc.*, **105**, 2034 (1983); M. F. Semmelhack, J. J. Bozell, T. Sato, W. Wulff, E. Spiess and A. Zask, *J. Am. Chem. Soc.*, **104**, 5850 (1982).

95. G. Kraus, J. Li, M. S. Gordon and J. H. Jensen, *J. Am. Chem. Soc.*, **115**, 5859 (1993).

96. M. Kimura, N. Saeki, S. Uchida, H. Harayama, S. Tanaka, K. Fugami and Y. Tamaru, *Tetrahedron Lett.*, **34**, 7611 (1993).

97. J. J. Masters and L. S. Hegedus, *J. Org. Chem.*, **58**, 4547 (1993).

98. J. E. Bäckvall, *Acc. Chem. Res.*, **16**, 335 (1983); *Pure Appl. Chem.*, **55**, 1669 (1983); *New J. Chem.*, **14**, 447 (1990); *Adv. Metal-Organic Chem.*, **1**, 135 (1989).

99. Mitsuibishi Chem. Corp., *ChemTech*, 759 (1988).

100. J. E. Bäckvall and R. E. Nordberg, *J. Am. Chem. Soc.*, **103**, 4959 (1981); J. E. Bäckvall, S. E. Bystrom and R. E. Nordberg, *J. Org. Chem.*, **49**, 4619 (1984); J. E. Bäckvall, J. Vagberg, and R. E. Nordberg, *Tetrahedron Lett.*, **25**, 2717 (1984); J. E. Bäckvall and J. O. Vagberg, *Org. Synth.*, **69**, 38 (1990).

101. J. E. Bäckvall, R. Gatti and H. E. Schink, *Synthesis*, 343 (1993); J. E. Bäckvall, K. L. Granberg, P. G. Andersson, R. Gatti and A. Gogoll, *J. Org. Chem.*, **58**, 5445 (1993).

102. J. E. Bäckvall, *Pure Appl. Chem.*, **64**, 429 (1992); J. E. Bäckvall, P. G. Andersson, G. B. Stone and A. Gogoll, *J. Org. Chem.*, **56**, 2988 (1991); J. E. Bäckvall and P. G. Andersson, **112**, 3683 (1990).

103. P. G. Andersson and J. E. Bäckvall, *J. Am. Chem. Soc.*, **114**, 8696 (1992).

104. R. Van Helden and G. Verberg, *Rec. Trav. Chim.*, **84**, 1263 (1965).

105. J. M. Davidson and C. Triggs, *J. Chem. Soc.*, (A), 1324 (1968).

106. A. Shiotani, H. Itatani and T. Inagaki, *J. Mol. Catal.*, **34**, 57 (1986); H. Itatani and T. Sakakibara, *Synthesis*, 607 (1978).

107. W. Harris, C. H. Hill, E. Keech and P. Malsher, *Tetrahedron Lett.*, **34**, 8361 (1993).

107a. E. Shirakawa, Y. Murata, Y. Nakao and T. Hiyama, *Synlett*, 1143 (1997); L. Alcaraz and R. J. K. Taylor, *Synlett*, 791 (1997).

108. T. Jintoku, K. Nishimura, K. Takai and Y. Fujiwara, *Chem. Lett.*, 1687 (1990).

109. I. Moritani and Y. Fujiwara, *Synthesis*, 524 (1973); I. Moritani and Y. Fujiwara, *Tetrahedron Lett.*, 1119 (1967); Y. Fujiwara, I. Moritani and M. Matsuda, *Tetrahedron*, **24**, 4819 (1968).

110. T. D. Cushing, J. F. Sanz-Cerrera and R. M. Williams, *J. Am. Chem. Soc.*, **115**, 9323 (1993).

111. M. I. Bruce, *Angew. Chem., Int. Ed. Engl.*, **16**, 73 (1977); I. Omae, *Chem. Rev.*, **79**, 287 (1987); G. R. Newkome, W. E. Puckatt, V. K. Gupta and G. E. Kiefer, *Chem. Rev.*, **86**, 451 (1986); A. D. Ryabov, *Synthesis*, 233 (1985).

112. I. Omae, *Organometallic Intramolecular Coordination Compounds*, Elsevier, 1986.

113. A. C. Cope and R. D. W. Siekman, *J. Am. Chem. Soc.*, **87**, 3272 (1965); A. C. Cope and E. C. Friedrich, *J. Am. Chem. Soc.*, **90**, 909 (1968).

114. H. Horino and N. Inoue, *J. Org. Chem.*, **46**, 4416 (1981).

115. G. E. Hartwell, R. V. Lawrence and M. J. Smas, *Chem. Commun.*, 912 (1970).

116. A. J. Deeming and I. P. Rothwell, *Chem. Commun.*, 344 (1978); *J. Organometal. Chem.*, **205**, 117 (1981).

117. M. Pfeffer, E. Wehman and G. van Koten, *J. Organometal. Chem.*, **282**, 127 (1985).

118. C. D. Liang, *Tetrahedron Lett.*, **27**, 1971 (1986).

119. H. Takahashi and J. Tsuji, *J. Organometal. Chem.*, **10**, 511 (1967).

120. M. I. Bruce, B. L. Goodall and F. G. A. Stone, *Chem. Commun.*, 558 (1973).

121. D. R. Fahey, *Chem. Commun.*, 417 (1970); *J. Organometal. Chem.*, **27**, 283 (1971).

122. T. Izumi, T. Katou, A. Kasahara and K. Hanaya, *Bull. Chem. Soc. Jpn.*, **51**, 3407 (1978).

123. S. J. Tremont and H. U. Rahman, *J. Am. Chem. Soc.*, **106**, 5759 (1984).

124. R. A. Holton and R. A. Kjonaas, *J. Am. Chem. Soc.*, **99**, 4177 (1977).

125. R. A. Holton and R. A. Kjonaas, *J. Organometal. Chem.*, **142**, C15 (1977).

126. R. A. Holton, *J. Am. Chem. Soc.*, **99**, 8083 (1977).

127. J. Tsuji, M. Takahashi and T. Takahashi, *Tetrahedron Lett.*, **21**, 849 (1980).

128. S. F. Vasilevsky, B. A. Trofimov, A. G. Mal'kina and L. Brandsma, *Synth. Commun.*, **24**, 85 (1994).

129. J. S. Prasad and L. S. Liebeskind, *Tetrahedron Lett.*, **28**, 1859 (1987).

130. J. Tsuji, N. Iwamoto and M. Morikawa, *J. Am. Chem. Soc.*, **86**, 2095 (1964).

131. G. P. Chiusoli, C. Venturello and S. Merzoni, *Chem. Ind.*, 977 (1968).

132. T. Nogi and J. Tsuji, *Tetrahedron*, **25**, 4099 (1969).

133. B. Gabriele, M. Costa, G. Salerno. and G. P. Chiusoli, *Chem. Commun.*, 1007 (1992).

134. G. P. Chiusoli, M. Costa and S. Reverberi, *Synthesis*, 262 (1989).

135. H. Kataoka, K. Watanabe and K. Goto, *Tetrahedron Lett.*, **31**, 4181 (1990).

136. H. Kataoka, K. Watanabe, K. Miyazaki, S. Tahara, K. Ogu, R. Matsuoka and K. Goto, *Chem. Lett.*, 1705 (1990).

137. D. M. Fenton and P. J. Steinwand, *J. Org. Chem.*, **39**, 701 (1974).

138. M. Graziani, P. Uguagliati and G. Carturan, *J. Organometal. Chem.*, **27**, 275 (1971).

139. Ube Industries, Ltd., *Belgium Patent*, 870,268 (1979).

140. J. E. Hallgren and R. O. Mattews, *J. Organometal. Chem.*, **175**, 135 (1979); J. E. Hallgren, G. M. Lucas and R. O. Mathews, *J. Organometal. Chem.*, **204**, 135 (1981); J. E. Hallgren and G. M. Lucas, **212**, 135 (1981).

141. S. Fukuoka, M. Chono and M. Kohno, *J. Org. Chem.*, **49**, 1458 (1984); *ChemTech*, 670 (1984).
142. H. Alper, G. Vasapollo, F. W. Hartstock, M. Mlekuz, D. J. H. Smith and G. E. Morris, *Organometallics*, **6**, 2391 (1987); H. Alper and F. W. Hartstock, *Chem. Commun.*, 1141 (1985).
143. K. Nishimura, S. Furusaki, Y. Shiomi, K. Fujii, K. Nishihara and M. Yamashita (Ube Industries Ltd.) *Eur. Pat. Appl.* EP 77542; *Chem. Abstr.*, **99**, 87654 (1983).
144. S. Uchiumi, K. Ataka, K. Sataka, *Eur. Pat. Appl.* EP, 1088332; *Chem. Abstr.*, **101**, 90431 (1984).
145. Y. Fujiwara, K. Takaki, J. Watanabe, Y. Uchida and H. Tamiguchi, *Chem. Lett.*, 1687 (1989).
146. K. Satoh, J. Watanabe, K. Takaki and Y. Fujiwara, *Chem. Lett.*, 1433 (1991); K. Nakata, J. Watanabe, K. Takaki and J. Fujiwara, *Chem. Lett.*, 1437 (1991).
147. Y. Fujiwara, K. Takaki and Y. Taniguchi, *Synlett*, 591 (1996).
148. Y. Ito, H. Aoyama and T. Saegusa, *J. Am. Chem. Soc.*, **102**, 4519 (1980); Y. Ito, H. Aoyama, T. Hirao, A. Mochizuki and T. Saegusa, *J. Am. Chem. Soc.*, **101**, 494 (1979).
149. M. Toyota, T. Seishi and K. Fukumoto, *Tetrahedron Lett.*, **34**, 5947 (1993).
150. R. C. Larock and N. H. Lee, *Tetrahedron Lett.*, **32**, 5911 (1991).
151. A. S. Kende, B. Roth, P. J. Santilippo and T. J. Blacklock, *J. Am. Chem. Soc.*, **104**, 5808 (1982).
152. Y. Ito, T. Hirao and T. Saegusa, *J. Org. Chem.*, **43**, 1011 (1978).
153. M. Shibasaki and Y. Ogawa, *Tetrahedron Lett.*, **26**, 3841 (1985).
154. T. Baba, K. Nakano, S. Nishiyama and S. Tsuruya, M. Masai, *Chem. Commun.*, 1697 (1989), *J. Chem. Soc. Perkin II*, 1113 (1990).
155. M. Ihara, S. Suzuki, T. Taniguchi and K. Fukumoto, *Synlett*, 435 (1993).
156. M. Sodeoka, K. Ohrai and M. Shibasaki, *J. Org. Chem.*, **60**, 2648 (1995).
157. E. Hagiwara, A. Fujii and M. Sodeoka, *J. Am. Chem. Soc.*, **120**, 2474 (1998).

INDEX

Note: Index entries referring to Figures are indicated by *italic page numbers*